AF361875

Fatigue and Fracture of Non-metallic Materials and Structures

Fatigue and Fracture of Non-metallic Materials and Structures

Special Issue Editor

Andrea Spagnoli

MDPI • Basel • Beijing • Wuhan • Barcelona • Belgrade • Manchester • Tokyo • Cluj • Tianjin

Special Issue Editor
Andrea Spagnoli
University of Parma
Italy

Editorial Office
MDPI
St. Alban-Anlage 66
4052 Basel, Switzerland

This is a reprint of articles from the Special Issue published online in the open access journal *Applied Sciences* (ISSN 2076-3417) (available at: https://www.mdpi.com/journal/applsci/special_issues/Non-metallic_Materials_and_Structures).

For citation purposes, cite each article independently as indicated on the article page online and as indicated below:

LastName, A.A.; LastName, B.B.; LastName, C.C. Article Title. *Journal Name* **Year**, *Article Number*, Page Range.

ISBN 978-3-03928-778-9 (Hbk)
ISBN 978-3-03928-779-6 (PDF)

Contents

About the Special Issue Editor

Andrea Spagnoli is an Associate Professor in solid and structural mechanics. Born in Bologna on 24 March, 1967, Andrea received a M.S. (Laurea) in civil engineering, University of Parma (1992) and Ph.D. in structural engineering, Imperial College, University of London, U.K. (1997). Andrea was an Assistant Professor of Structural Mechanics, University of Parma (2000–2005). From 2005 to the present, Andrea as been a Associate Professor of Structural Mechanics, University of Parma and anuthor of more than 200 scientific papers in refereed journals and proceedings of international conferences. According to Scopus, he has 130 publications (1996–2019), 2172 citations and an h-index of 29. Andrea has been an academic visitor at the Department of Civil Engineering, Imperial College, London (1997, 1999); Department of Mechanical Science and Engineering, Kyushu University (2000–2001); and Department of Civil and Environmental Engineering, University of Michigan (2007). He has been a Guest Professor in the Division of Solid Mechanics, Lund University, Sweden, since 2018. He has participated in several domestic (funded by the Italian Ministry of Education and by private companies) and international research projects (funded by Europen Community) since 1992. He has been the coordinator of a project financed by the Italian Ministry of Education (PRIN2009) and co-editor of the book Biaxial/Multiaxial Fatigue and Fracture, Elsevier, 2003. He has been co-editor of Special Issues of the journal Engineering Fracture Mechanics in 2008, 2010, 2011, 2014, and 2017. He is a member of the Editorial Board of the journal Fatigue and Fracture of Engineering Materials and Structures (2004–present), and of the journal Structural Engineering and Mechanics, (2013–present). Andrea is the Secretary of the Technical Committee Fatigue of Engineering Materials and Structures, ESIS (2000–present) and is the corresponding member of ECCS Technical Working Group 8.4 'Stability of steel shells' (2004–present), as well as a member of the Executive Committee of the Italian Group of Fracture (2011–present).

Editorial

Special Issue on Fatigue and Fracture of Non-Metallic Materials and Structures

Andrea Spagnoli

Department of Engineering and Architecture, University of Parma, 43124 Parma, Italy; andrea.spagnoli@unipr.it

Received: 17 February 2020; Accepted: 25 February 2020; Published: 7 March 2020

Abstract: This Special Issue covers the broad topic of structural integrity of non-metallic materials, and it is concerned with the modelling, assessment and reliability of structural elements of any scale. In particular, the articles being contained in this issue concentrate on the mechanics of fracture and fatigue in relation to applications to a variety of non-metallic materials, including concrete and cementitious composites, rocks, glass, ceramics, bituminous mixtures, composites, polymers, rubber and soft matters, bones and biological materials, advanced and multifunctional materials.

Keywords: Fatigue; Fracture mechanics; Structural integrity; Polymers; Composites; Ceramics; Concrete; Rock; Soft matter; Advanced materials.

---

1. Introduction

The mechanics of fracture and fatigue have produced a huge body of research work in relation to applications to metal materials and structures. However, a variety of non-metallic materials (e.g., concrete and cementitious composites, rocks, glass, ceramics, bituminous mixtures, composites, polymers, rubber and soft matters, bones and biological materials, advanced and multifunctional materials) have received comparatively less attention, despite their attractiveness for a large spectrum of applications related to the components and structures of diverse engineering branches, applied sciences and architecture, and to the load-carrying systems of biological organisms.

This special issue covers the broad topic of structural integrity of non-metallic materials and is is concerned with the modeling, assessment, and reliability of structural elements of any scale. Original contributions from engineers, mechanical materials scientists, computer scientists, physicists, chemists, and mathematicians are presented, following both experimental and theoretical approaches.

2. Fracture

A number of papers in this special issue are specifically devoted to fracture mechanics problems. Different approaches have been used, including experimental investigations, theoretical models, and numerical simulations. Papers [1–5] are related to concrete material, from papers [6–12] to rocks. Other contributions investigate the fracture behavior of functionally graded materials [13], glass [14], laminate composites [15], refractories [16], and soft matter [17]. Concrete material is also investigated in relation to impact resistance [18] and self-healing properties [19]. Impact behavior is studied with reference to laminate composites [20,21]. Other papers devoted to rocks concern their cutting resistance [22] and their multiaxial response under dry and saturated conditions [23]. Finally, the mechanical behavior of a monomer structural component is studied in [24].

3. Fatigue

Fatigue is investigated from both a experimental and theoretical point-of-view in different papers. In [25–30] in particular, concrete behavior under fatigue loading is described, with an emphasis on bridge applications [25–27] and concrete reinforced with fibers and rebars [25,29]. Fatigue of asphalt and of rocks is investigated in [31,32], respectively. Finally, an account on fatigue life assessment of wind turbine blades is presented in [33].

Funding: This research received no external funding.

Acknowledgments: This special issue would not have been made possible without the irreplaceable contributions of valuable authors coming from many different countries, hardworking and professional reviewers, and the dedicated editorial team of Applied Sciences, an international, peer-reviewed journal that is free for readers embracing all aspects of applied sciences.

Conflicts of Interest: The author declares no conflict of interest.

References

1. Huang, C.; Wu, C.; Lin, S.; Yen, T. Effect of Slag Particle Size on Fracture Toughness of Concrete. *Appl. Sci.* **2019**, *9*, 805. [CrossRef]
2. Saleem, M.; Qureshi, H.; Amin, M.; Khan, K.; Khurshid, H. Cracking Behavior of RC Beams Strengthened with Different Amounts and Layouts of CFRP. *Appl. Sci.* **2019**, *9*, 1017. [CrossRef]
3. Gao, X.; Liu, C.; Tan, Y.; Yang, N.; Qiao, Y.; Hu, Y.; Li, Q.; Koval, G.; Chazallon, C. Determination of Fracture Properties of Concrete Using Size and Boundary Effect Models. *Appl. Sci.* **2019**, *9*, 1337. [CrossRef]
4. Wu, C.; Huang, C.; Kan, Y.; Yen, T. Effects of Fineness and Dosage of Fly Ash on the Fracture Properties and Strength of Concrete. *Appl. Sci.* **2019**, *9*, 2266. [CrossRef]
5. Xiong, X.; Xiao, Q. Meso-Scale Simulation of Concrete Based on Fracture and Interaction Behavior. *Appl. Sci.* **2019**, *9*, 2986. [CrossRef]
6. Yang, G.; Leung, A.; Xu, N.; Zhang, K.; Gao, K. Three-Dimensional Physical and Numerical Modelling of Fracturing and Deformation Behaviour of Mining-Induced Rock Slopes. *Appl. Sci.* **2019**, *9*, 1360. [CrossRef]
7. Li, Y.; Cai, W.; Li, X.; Zhu, W.; Zhang, Q.; Wang, S. Experimental and DEM Analysis on Secondary Crack Types of Rock-Like Material Containing Multiple Flaws Under Uniaxial Compression. *Appl. Sci.* **2019**, *9*, 1749. [CrossRef]
8. Shu, J.; Jiang, L.; Kong, P.; Wang, Q. Numerical Analysis of the Mechanical Behaviors of Various Jointed Rocks under Uniaxial Tension Loading. *Appl. Sci.* **2019**, *9*, 1824. [CrossRef]
9. Shu, J.; Jiang, L.; Kong, P.; Wang, P.; Zhang, P. Numerical Modeling Approach on Mining-Induced Strata Structural Behavior by Considering the Fracture-Weakening Effect on Rock Mass. *Appl. Sci.* **2019**, *9*, 1832. [CrossRef]
10. Masłowski, M.; Kasza, P.; Czupski, M.; Wilk, K.; Moska, R. Studies of Fracture Damage Caused by the Proppant Embedment Phenomenon in Shale Rock. *Appl. Sci.* **2019**, *9*, 2190. [CrossRef]
11. Jin, L.; Sui, W.; Xiong, J. Experimental Investigation on Chemical Grouting in a Permeated Fracture Replica with Different Roughness. *Appl. Sci.* **2019**, *9*, 2762. [CrossRef]
12. Han, Z.; Zhang, L.; Zhou, J. Numerical Investigation of Mineral Grain Shape Effects on Strength and Fracture Behaviors of Rock Material. *Appl. Sci.* **2019**, *9*, 2855. [CrossRef]
13. Cho, J. A Numerical Evaluation of SIFs of 2-D Functionally Graded Materials by Enriched Natural Element Method. *Appl. Sci.* **2019**, *9*, 3581. [CrossRef]
14. Yang, G.; Shao, Y.; Yao, K. Understanding the Fracture Behaviors of Metallic Glasses—An Overview. *Appl. Sci.* **2019**, *9*, 4277. [CrossRef]
15. Bennati, S.; Fisicaro, P.; Taglialegne, L.; Valvo, P. An Elastic Interface Model for the Delamination of Bending-Extension Coupled Laminates. *Appl. Sci.* **2019**, *9*, 3560. [CrossRef]
16. Stückelschweiger, M.; Gruber, D.; Jin, S.; Harmuth, H. Wedge-Splitting Test on Carbon-Containing Refractories at High Temperatures. *Appl. Sci.* **2019**, *9*, 3249. [CrossRef]
17. Spagnoli, A.; Terzano, M.; Brighenti, R.; Artoni, F.; Carpinteri, A. How Soft Polymers Cope with Cracks and Notches. *Appl. Sci.* **2019**, *9*, 1086. [CrossRef]
18. Soleimani, S.; Sayyar Roudsari, S. Analytical Study of Reinforced Concrete Beams Tested under Quasi-Static and Impact Loadings. *Appl. Sci.* **2019**, *9*, 2838. [CrossRef]
19. Kang, C.; Kim, T. Curable Area Substantiation of Self-Healing in Concrete Using Neutral Axis. *Appl. Sci.* **2019**, *9*, 1537. [CrossRef]

20. Jia, S.; Wang, F.; Huang, W.; Xu, B. Research on the Blow-Off Impulse Effect of a Composite Reinforced Panel Subjected to Lightning Strike. *Appl. Sci.* **2019**, *9*, 1168. [CrossRef]
21. Sellitto, A.; Saputo, S.; Di Caprio, F.; Riccio, A.; Russo, A.; Acanfora, V. Numerical–Experimental Correlation of Impact-Induced Damages in CFRP Laminates. *Appl. Sci.* **2019**, *9*, 2372. [CrossRef]
22. Sun, Y.; Li, X.; Guo, H. Failure Probability Prediction of Thermally Stable Diamond Composite Tipped Picks in the Cutting Cycle of Underground Roadway Development. *Appl. Sci.* **2019**, *9*, 3294. [CrossRef]
23. Li, D.; Sun, Z.; Zhu, Q.; Peng, K. Triaxial Loading and Unloading Tests on Dry and Saturated Sandstone Specimens. *Appl. Sci.* **2019**, *9*, 1689. [CrossRef]
24. Kim, Y.; Hwang, E.; Jeon, E. Optimization of Shape Design of Grommet through Analysis of Physical Properties of EPDM Materials. *Appl. Sci.* **2019**, *9*, 133. [CrossRef]
25. Fathalla, E.; Tanaka, Y.; Maekawa, K. Fatigue Life of RC Bridge Decks Affected by Non-Uniformly Dispersed Stagnant Water. *Appl. Sci.* **2019**, *9*, 607. [CrossRef]
26. Fathalla, E.; Tanaka, Y.; Maekawa, K. Effect of Crack Orientation on Fatigue Life of Reinforced Concrete Bridge Decks. *Appl. Sci.* **2019**, *9*, 1644. [CrossRef]
27. Lantsoght, E.; Koekkoek, R.; van der Veen, C.; Sliedrecht, H. Fatigue Assessment of Prestressed Concrete Slab-Between-Girder Bridges. *Appl. Sci.* **2019**, *9*, 2312. [CrossRef]
28. Shan, Z.; Yu, Z.; Li, X.; Lv, X.; Liao, Z. A Damage Model for Concrete under Fatigue Loading. *Appl. Sci.* **2019**, *9*, 2768. [CrossRef]
29. Mínguez, J.; Gutiérrez, L.; González, D.; Vicente, M. Plain and Fiber-Reinforced Concrete Subjected to Cyclic Compressive Loading: Study of the Mechanical Response and Correlations with Microstructure Using CT Scanning. *Appl. Sci.* **2019**, *9*, 3030. [CrossRef]
30. Soleimani, S.; Boyd, A.; Komar, A.; Roudsari, S. Fatigue in Concrete under Low-Cycle Tensile Loading Using a Pressure-Tension Apparatus. *Appl. Sci.* **2019**, *9*, 3217. [CrossRef]
31. Li, K.; Huang, M.; Zhong, H.; Li, B. Comprehensive Evaluation of Fatigue Performance of Modified Asphalt Mixtures in Different Fatigue Tests. *Appl. Sci.* **2019**, *9*, 1850. [CrossRef]
32. Fu, H.; Wang, S.; Pei, X.; Chen, W. Indices to Determine the Reliability of Rocks under Fatigue Load Based on Strain Energy Method. *Appl. Sci.* **2019**, *9*, 360. [CrossRef]
33. Loza, B.; Pacheco-Chérrez, J.; Cárdenas, D.; Minchala, L.; Probst, O. Comparative Fatigue Life Assessment of Wind Turbine Blades Operating with Different Regulation Schemes. *Appl. Sci.* **2019**, *9*, 4632. [CrossRef]

Article

Effect of Slag Particle Size on Fracture Toughness of Concrete

Chung-Ho Huang [1], Chung-Hao Wu [2,*], Shu-Ken Lin [3] and Tsong Yen [3]

[1]   Department of Civil Engineering, National Taipei University of Technology, Taipei City 106, Taiwan; cdewsx.hch@gmail.com
[2]   Department of Civil Engineering, Chienkuo Technology University, Changhua City 500, Taiwan
[3]   Department of Civil Engineering, National Chung-Hsing University, No.145, Xingda Road, Taichung 402, Taiwan; sklin@nchu.edu.tw (S.-K.L.); tyen@dragon.nchu.edu.tw (T.Y.)
*   Correspondence: chw@ctu.edu.tw; Tel.: +886-4-7111111 (ext. 3428)

Received: 1 February 2019; Accepted: 20 February 2019; Published: 25 February 2019

Abstract: The effects of particle size of ground granulated blast furnace slag (GGBS) on the fracture energy, critical stress intensity, and strength of concrete are experimentally studied. Three fineness levels of GGBS of 4000, 5000, 6000 cm^2/g were used. In addition to the control mixture without slag, two slag replacement levels of 20% and 40% by weight of the cementitious material were selected for preparing the concrete mixtures. The control mixture was designed to have a target compressive strength at 28 days of 62 MPa, while the water to cementitious material ratio was selected as 0.35 for all mixtures. Test results indicate that using finer slag in concrete may improve the filling effect and the reactivity of slag, resulting in a larger strength enhancement. The compressive strength of slag concrete was found to increase in conjunction with the fineness level of the slag presented in the mixture. Use of finer slag presents a beneficial effect on the fracture energy (G_F) of concrete, even at an early age, and attains a higher increment of G_F at later age (56 days). This implicates that the finer slag can have a unique effect on the enhancement of the fracture resistance of concrete. The test results of the critical stress intensity factor (K^S_{IC}) of the slag concretes have a similar tendency as that of the fracture energy, indicating that the finer slag may present an increase in the fracture toughness of concrete.

Keywords: fracture toughness; blast furnace slag; particle size; compressive strength; concrete

1. Introduction

Ground granulated blast furnace slag (GGBS) is a byproduct of iron making, which is produced by water quenching of molten blast furnace slag that turns out to be a glassy material [1]. It is ground to improve the reactivity during cement hydration. GGBS is one such supplementary material which can be used as a cementitious ingredient in either cement or concrete composites. Research to date suggests that the supplementary cementitious materials improve many of the performance characteristics of concrete, such as strength, workability, and corrosion resistance [2,3]. Some of the effective parameters like chemical composition, fineness, and hydraulic reactivity have been carefully examined by many earlier studies [4]. Among these, the reactive glass content and fineness of GGBS alone will influence the cementitious/pozzolanic efficiency or its reactivity in concrete composite significantly.

It is generally agreed that the use of fine GGBS improves the properties of concrete. K. Tan and X. Pu [5] investigated the effect of finely ground GGBS on the compressive strength of concrete. Test results showed that incorporating 20% finely ground GGBS can significantly increase the strength after 3 days. Mantel [6] reported that the slag activity depends on the particle size distribution (fineness) of slag and the cement used and showed that this ranges from 62%–115% at 28 days. In addition, the investigation of Yüksel et al. [7] reported that the increase in fineness of GGBS improves compressive strength due to the pozzolanic reaction causing a reduction in permeability, signifying that finer slag can provide higher resistance against deteriorations from chemical or physical attacks.

The traditional properties, such as compressive strength, have been assessed to be in close relation to the pore characteristic and interfacial transition zone (ITZ) of concrete. The test results of Duan, Gao, and Tan et al. [5,8,9] show that mineral admixtures have a positive impact on pore refinement and ITZ enhancement, especially at later curing stages. The development of compressive strength is significantly related to the evolution of ITZ and the pore structure. In fact, the cementitious and pozzolanic behavior of GGBS is essentially similar to that of high-calcium fly ash. Previous reports [5,9–11] have primarily indicated that increasing the fineness of fly ash and slag may increase the strength and durability of concrete attributed to the microcracking resistance of cement paste, particularly in ITZ.

The interfacial transition zone in concrete that is often modeled as a three-phase material is represented as the third phase [12]. Existing as a thin shell around larger aggregate, the ITZ is generally weaker than either of the two main components, the hydrated cement paste, and aggregate, of concrete. The ITZ has less crack resistance, and accordingly, the factor occurs preferentially. A major factor responsible for the poor strength of the ITZ is the presence of microcracks. Much of the physical nature of the response of concrete under loading can be described in terms of microcracks that can be observed at relatively low magnification. Cracking in concrete is mostly due to the tensile stress that occurs under load or environmental changes. As such, the failure of concrete in tension is governed by the microcracking, associated particularly with the ITZ [13,14].

The tensile strength of concrete is a very essential property. Not only the tensile strength but also the behavior at the tensile fracture is of importance, particularly the toughness. When concrete fails in tension, its behavior is characterized by both the peak stress and the energy required to fully generate a crack. Fracture mechanics could in principle be a suitable basis for analyzing the tensile toughness of concrete [15]. The defects and the stress concentration are the main factors causing failure. Concrete that exhibits defects or is subjected to stress concentrations easily cracks and results in a reduction of strength. It is important to identify the fracture behavior and toughness for widely used high-performance concrete that normally incorporates slag or fly ash [16,17].

Because of the non-homogeneous characteristics of concrete, its fracture behavior is quite complicated. It was proved that the methods developed in conventional fracture mechanics are unsuitable for the analysis of the influence of fracture toughness on the behavior of concrete structures [15]. Many theoretical models have been developed to make such an analysis possible. Among them, three well-known models are the fictitious crack model, the crack band model, and the two-parameter fracture model. By means of numerical techniques, for example, in the finite element method, it is possible using these models to make a theoretical analysis for the development of the damage zone and the complete behavior of the structure. As for the fracture mechanics of concrete, RILEM has put forward several methods to determine the fracture properties and parameters of concrete [18]. Hillerborg [15] determine the fracture energy according to the fictitious crack model. Jeng and Shah [19] used the two-parameter fracture model to determine the critical stress intensity factor.

A number of past studies on the concrete containing GGBS have been conducted, mainly dealing with the influence of the fineness of slag on the strength and durability of concrete. More rarely, investigations were conducted on the fracture behaviors of concrete incorporating finer GGBS. Consequently, there is a need for exploring the effects of the particle size of GGBS on the fracture properties of concrete. This study experimented using the three-point bend test to analyze the fracture energy and the critical stress intensity factor for evaluating the fracture mechanics or fracture toughness of concrete containing finer GGBS.

2. Research Significance

Laboratory investigations have shown that when the slag particle size is reduced, its mechanical performance in concrete is improved. The finer slag with the larger surface area has an intensive reaction which may lead to higher enhancement of the strength, the fracture energy, and the critical stress intensity factor of concrete. Other than the research by Jensen and Hansen [20], who observed a dependence of the fracture energy on the aggregate type and independence from the compressive strength of concrete, this study found that an increase in concrete compressive strength of 10% resulted in an increase in fracture energy of around 18%.

3. Experimental Details

3.1. Materials

The materials used included a Type I Portland cement, river sand, crushed coarse aggregate with a maximum size of 10mm, a naphthalene-sulfonate-based superplasticizer, and GGBS of three fineness levels. The chemical composition of cement and GGBS are presented in Table 1. Three different fineness levels of GGBS of 4000, 5000, and 6000 cm^2/g with a specific gravity of 2.85 were selected in this study.

Table 1. Chemical analysis of cement and ground granulated blast furnace slag (GGBS).

Chemical Analysis (%)	SiO_2	Al_2O_3	Fe_2O_3	CaO	MgO	SO_3	Na_2O	K_2O	LOI
Cement	20.9	5.62	3.21	63.6	2.52	2.16	0.27	0.52	0.92
GGBS	33.1	15.6	0.33	40.7	7.7	0.1	-	-	0.12

3.2. Mixture Proportions

Seven mixtures were prepared for testing in this research: A reference mixture (R0) without GGBS and six slag mixtures, designated as S4R2, S4R4, S5R2, S5R4, S6R2, and S6R4. S4, S5, and S6 refer to slag fineness of 4000, 5000 and 6000 cm^2/g, respectively. R2 and R4 refer to the replacement ratios of 20% and 40%, respectively, by weight of the cementitious material. The water to cementitious material ratio (w/cm) was kept at 0.35 for all mixtures. All concretes were designed to have a target compressive strength level at 28 days of 62 MPa. The dosage of superplasticizer was adjusted to produce a designed slump of 250 $\pm$ 20 mm and a slump flow of 600 $\pm$ 50 mm for all mixtures. The mixture proportions are shown in Table 2.

Table 2. Mixture proportions and properties of fresh concrete.

Mixture No.	w/cm	Water	Cement	Slag	Sand	Aggregate	SP.	Slump	Slump Flow
					kg			mm	mm
R0			520	0	850	783	8.1	230	625
S4R2			416	104	850	803	7.5	230	625
S4R4			312	208	870	773	6.8	265	580
S5R2	0.35	182	416	104	850	803	7.8	250	620
S5R4			312	208	860	783	6.8	255	610
S6R2			416	104	850	803	7.5	255	585
S6R4			312	208	840	793	7.0	235	580

3.3. Specimen Fabrication

Cylinder specimens (ψ100 $\times$ 200 mm) were cast from each mixture for compression testing; three cylinders each for testing at 4 ages (7, 14, 28, 56 days). Prism specimens with diminution of 50 $\times$ 50 $\times$ 650 mm and 80 $\times$ 150 $\times$ 700 mm were cast for the determination of fracture energy (G_F) and the critical stress intensity factor (K^S_{IC}), respectively. Four specimens were fabricated for each mixture that tested the two specimens at 2 ages (14 and 56 days). The specimens and concrete mixtures were prepared in accordance with ASTM C192. After casting, test specimens were covered with plastic sheets and left in the casting room for 24 h. They were then demolded and put into a 100% RH moist-curing room at about 23 °C until time for testing. One day before testing, the fracture energy specimen and the critical stress intensity factor specimen, as shown in Figures 1 and 2, were prepared by cutting a notches 25 mm deep by 2 mm width and 50 mm deep by 2 mm width, respectively, at the middle of the beam specimens.

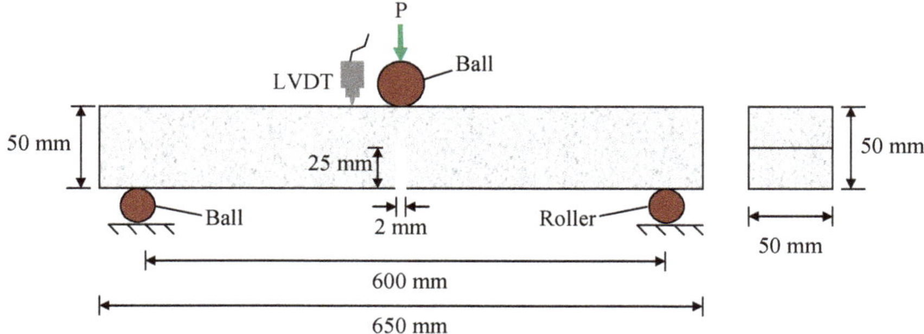

Figure 1. The test set up for G_F–determination.

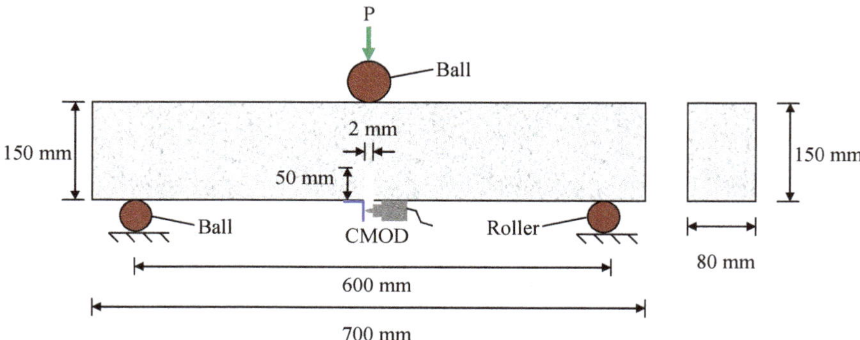

Figure 2. The test set up for K^S_{IC}-determination.

3.4. Testing Procedure

The compression test was carried out in accordance with ASTM C39. The three-point bending test was performed on the notched beam as follows to determine the G_F and K^S_{IC}:

3.4.1. The Test for Determining G_F

The fracture energy test followed the guidelines established by RILEM [21] using a closed-loop testing machine. The testing arrangement is shown in Figure 1. A linear variable differential transformer (LVDT) was installed at mid-span of the beam to measure the deflection. The loading velocity was chosen so that the maximum load was reached 30 s after the loading started. The loading rate selected was 0.25 mm/min. A load-deflection (F-δ) curve was then plotted, with the energy "W_0" representing the area under the curve. The G_F (N-m) can be calculated using the following expression [15,19]:

$$G_F = W_0 + mg\frac{\delta_0}{A} \tag{1}$$

where W_0 = area under the load-deflection curve (N-m);

m = mass of the beam between the supports (kg);

g = acceleration due to gravity;

δ_0 = final deflection of the beam (m);

A = cross-sectional area of the beam above the notch (m^2).

3.4.2. Test for Determining K^S_{IC}

The test for the determination of K^S_{IC} followed the test method suggested by Jenq and Shah [19] using a closed-loop testing machine under crack mouth opening displacement (CMOD) control. The test setup is shown in Figure 2. A clip gauge was used to measure the CMOD. The CMOD and the applied load were recorded continuously during the test. The test procedure included two steps of loading and unloading. The first step was started from loading until the peak load was reached; the applied load was manually reduced (also termed unloading) when the load passed the maximum load and was about 95% of the peak load. When the applied load was reduced to zero, reloading was applied. Each loading and unloading cycle was finished in about 1 min. Only one cycle of loading-unloading was required for the test. The K^S_{IC} is calculated using the following equation:

$$K^S_{IC} = \frac{3P_{max}S}{2bd^2}\sqrt{(\pi\alpha_c)}F(\alpha) \tag{2}$$

in which,

$$F(\alpha) = 1.99 - \frac{\alpha(1-\alpha)(2.15 - 3.93\alpha + 2.7\alpha^2)}{\sqrt{\pi}(1+2\alpha)(1-\alpha)^{3/2}} \tag{3}$$

where P_{max} = the measured maximum load (N);
S = the span of the beam; b = beam width; d = beam depth;
α_c = critical effective crack length;
$\alpha = \alpha_c/d$.

4. Results and Discussion

4.1. Compressive Strength

Compressive strength was determined at the ages of 7, 14, 28 and 56 days on concrete stored under moist-curing condition. The results are shown in Table 3. It is seen that all mixtures were proportioned to have equivalent workability and target strengths of 62 MPa at the age of 28 days. This was generally achieved except for in mixtures S4R2 and S4R4. In this case, the strength equivalence was achieved at the age of 56 days, which was also exceeded by that of the reference mixture (R0).

The development of compressive strength for each concrete mixture is shown in Figure 3. At early ages (7 days), as expected, the reference mixture without slag exhibited higher strength than the other slag mixtures. The early strength gain of the reference mixture was superior to that of the slag mixtures, indicating that replacing any amount of cement with GGBS of various fineness will reduce the strength of concrete. Moreover, the concrete incorporating more slag (40%) displayed lower strength at an early age than that with less slag content (20%). However, the curve slope of the slag mixtures in Figure 3 tends to be steeper than that of the reference mixture at later ages. In other words, the strength increment versus age for the slag mixtures is obviously larger than that of the reference mixture. Consequently, the strength of each slag mixture exceeds that of the reference mixture at the age of 56 days.

On the other hand, it can be found from Table 3 that under the same replacement ratio (20% and 40%) of slag, the mixtures S6R2 and S6R4 presented the highest strengths at an early age (7 days), followed by S5R2 and S5R4, and the S4R2 and S4R4 mixtures are the lowest. This is due to the fact that incorporating finer slag into concrete may fill the micro-voids much better, exhibit higher reactivity and producing higher packing density, resulting in a larger strength enhancement. In addition, a beneficial effect of the fineness of slag on concrete strength can be seen in Figure 4; the compressive strength of the slag concrete increases as the fineness level of the slag incorporated into the mixture increases for each age and various slag replacement ratios. It is also found in Table 3 that there are significant strength gains from 7–28 days and again from 28–56 days for the mixture containing finer slag. For example, the strength increment of the mixtures S6R2 and S6R4 that contain finer slag displayed an increase rate from 100% at 7 days to 121.8% and 122.7% at 28 days, and 125.5% and 128.9% at 56 days, respectively. These are obviously larger than the corresponding mixtures of S4R2 and S4R4 that increased from 100% at 7 days to 114.9% and 115.1% at 28 days, and 124.2% and 126.2% at 56 days, respectively. At 28 days, the strength of the finer-slag concrete is shown consistently to exceed that of the reference concrete, except in the mixtures of S4R2 and S4R4. This is particularly significant considering that the cement replacement of finer slag in each case is up to 40%. In addition, the highest strength achieved at 56 days was 70.5 MPa for the mixture S6R4 with 40% slag replacement, while the strength obtained for the corresponding reference mixture was a relatively lower value of 65.5 MPa.

Table 3. The compressive strength of concrete cylinders.

Mixture No.	w/cm	Compressive Strength (MPa)			
		7 Days	14 Days	28 Days	56 Days
S0		56.2 (100%)	60.7 (108.0%)	63.2 (112.5%)	65.5 (116.5%)
S4R2		54.5 (100%)	59.2 (108.6%)	62.6 (114.9%)	67.7 (124.2%)
S4R4	0.35	53.0 (100%)	58.7 (110.8%)	61.0 (115.1%)	66.9 (126.2%)
S5R2		55.0 (100%)	60.1 (109.3%)	65.5 (119.1%)	69.1 (125.6%)
S5R4		54.2 (100%)	59.5 (109.8%)	64.8 (119.6%)	68.4 (126.2%)
S6R2		56.0 (100%)	62.3 (111.3%)	68.2 (121.8%)	70.3 (125.5%)
S6R4		54.7 (100%)	61.0 (111.5%)	67.1 (122.7%)	70.5 (128.9%)

* Average value of three specimens

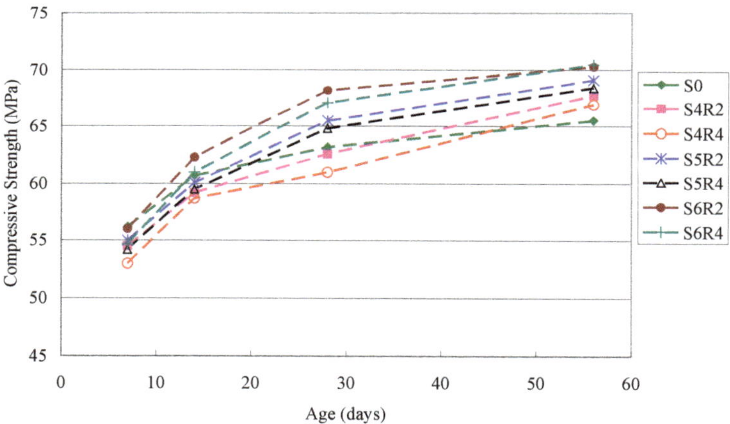

Figure 3. Compressive strength development for cylindrical specimens.

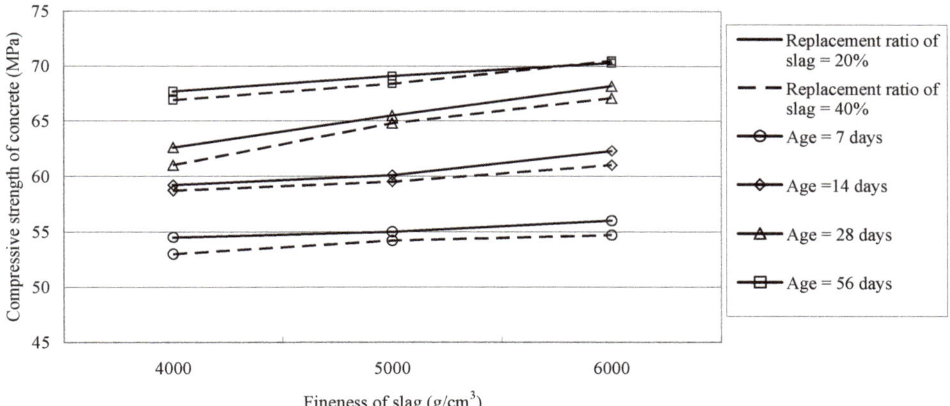

Figure 4. The average compressive strength of slag concrete versus the fineness of ground granulated blast furnace slag (GGBS).

4.2. Fracture Energy

In this research, fracture energy (G_F) was determined using a notched beam in a three-point bending test. Load-deflection curves were plotted from the tests. The energy absorbed in the test to failure is represented by the area below the load-deflection curve for the specimen. The area represents the fracture energy per unit area of the fracture surface. The G_F was calculated using Equation (1). Concrete specimens were tested at 14 and 56 days. Results are presented in Table 4. At earlier ages (14 days), the G_F values of each slag mixture are mostly less than that of the reference mixture, except slag mixture S6R2. In addition, the concrete containing more slag (40%) presented smaller G_F than that with less slag content (20%) for various ages. Nevertheless, the presence of the finer slag has a beneficial effect on the G_F of concrete. As shown in Figure 5, increasing the fineness level of slag leads to an increase of the G_F value of concrete. Even at an earlier age of 14 days, the G_F value of the mixture with finer slag (S6R4) is in turn higher than that of the mixtures S5R4 and S4R4 for a similar slag replacement ratio of 40%. This implies that although the pozzolanic reaction of slag does not yet fully occur at early ages, the superior filling effect and more active hydration reaction of the finer slag can increase the density of concrete, resulting in an increased fracture resistance. After 56 days, the G_F value of the slag mixtures almost exceeds that of the reference mixture. In addition, it is also found in Table 4 that the increment of fracture energy for each slag concrete from 14 days to 56 days is attained by 18–24%, which is much higher than that of reference concrete (10.1%). This signifies that, during the period, the pozzolanic reaction of slag activates, associating with the filling effect to enhance the fracture toughness of the slag concrete.

Table 4. Fracture energy of the concrete.

Mixture No.	Compressive Strength (MPa)		Fracture Energy (N/m)		
	14 Days	56 Days	14 Days	56 Days	Increment N/m (%)
S0	60.7	65.5	74.3	81.8	7.5 (10.1)
S4R2	59.2	67.7	71.2	84.5	13.3 (18.7)
S4R4	58.7	66.9	65.7	78.4	12.7 (19.3)
S5R2	60.1	69.1	72.1	89.1	17.0 (23.5)
S5R4	59.5	68.4	69.8	85.3	15.5 (22.2)
S6R2	62.3	70.3	76.5	94.6	18.1 (23.7)
S6R4	61.0	70.5	73.9	92.1	18.2 (24.6)

* Average value of two specimens

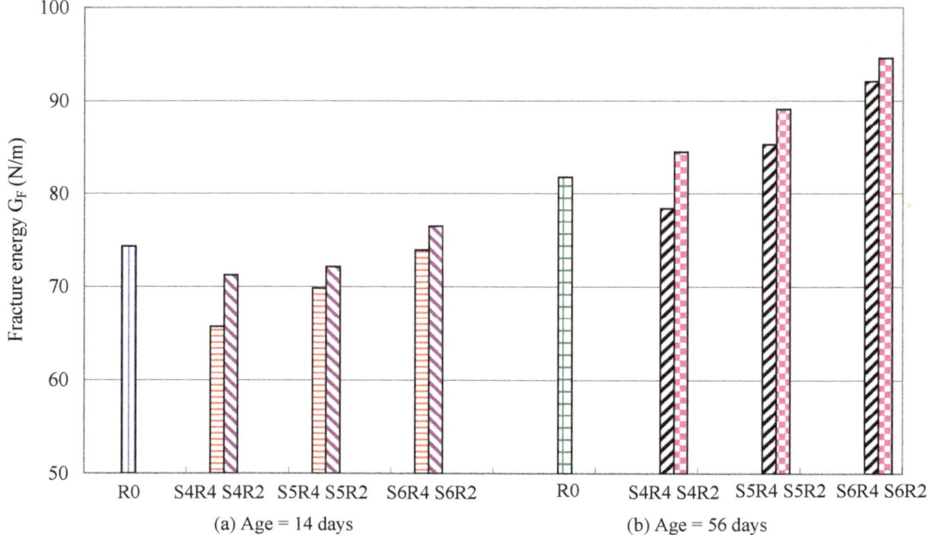

Figure 5. Fracture energy for concretes with various fineness levels and replacement ratios of slag at ages of 14 days and 56 days.

Figure 6 compares fracture energy with the compressive strength of all slag concretes in the study. The G_F is found to increase in conjunction with the compressive strength. This compares the trend favorably with the studies by Giaccio [22], Gettu [23], and Xie [24]. Nevertheless, the other studies observed smaller increases in G_F with the increases in the compressive strength of high-strength concrete. Giaccio, Rocco, and Zerbino [22] found that G_F increased as compressive strength increased, but only at a fraction of the rate. Getta, Bažant, and Kerr [23] reported that an increase in compressive strength of 160% resulted in an increase in G_F of only 12%. Xie, Elwi, and MacGregor [24] found that increases in compressive strength of 29% and 35% resulted in a G_F increase of 11% and 13%, respectively. In this study, an increase in compressive strength of 10% resulted in a larger increase in G_F of around 18%. This implicates the unique effect of using finer slag on the enhancement of the fracture resistance of concrete.

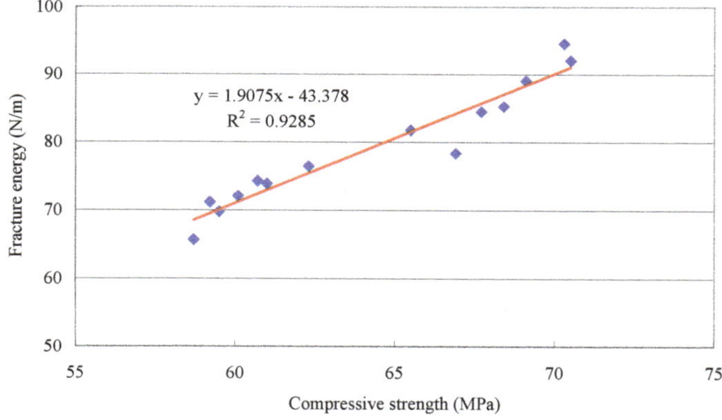

Figure 6. Fracture energy versus compressive strength for slag concretes.

4.3. Critical Stress Intensity Factor

The critical stress intensity factor (K^S_{IC}) was experimentally determined with a notch beam in a three-point bend test using a closed-loop testing machine under the crack mouth opening displacement control. Load-CMOD curves were plotted from the tests. The K^S_{IC} was calculated using Equation (2). Concrete specimens were tested at 14 and 56 days. Table 5 summarizes the results. It shows that the related tendency of the K^S_{IC} of concrete is similar to that of the G_F. At an early age (14 days), the K^S_{IC} values of slag concrete are less that of the reference concrete (R0), but exceed that of R0 after 56 days. This implies that the additional pozzolanic reaction of slag with $Ca(OH)_2$ in concrete becomes active during the period between 14 days and 56 days, which leads to the enhancement of K^S_{IC} of the slag concretes. Moreover, it is seen that the concrete incorporating more slag (40%) presented a lower KSIC value than that with less slag content (20%) for various ages. On the other hand, the concrete containing finer slag can exhibit larger K^S_{IC} values. It can be found from Figure 7 that the increase of the fineness level of the slag leads to an increase in the K^S_{IC} value of concrete for various ages. In other words, the K^S_{IC} value of the mixture with finer slag (S6R2) is in turn higher than those of S5R2 and S4R2 for the similar slag replacement ratio of 20%. Also, the K^S_{IC} of S6R4 > S5R4 > S4R4 for the similar slag replacement ratio of 40%. The reason for this result is that the finer slag particle has a larger surface area, presenting a more active pozzolanic reaction, thus resulting in an increase of strength and fracture toughness.

Table 5. Critical stress intensity factor of the concrete.

Mixture No.	Compressive Strength (MPa)		Critical Stress Intensity Factor (MPa $\times$ m$^{0.5}$)	
	14 Days	56 Days	14 Days	56 Days
S0	60.7	65.5	0.248	0.261
S4R2	59.2	67.7	0.246	0.276
S4R4	58.7	66.9	0.239	0.265
S5R2	60.1	69.1	0.247	0.282
S5R4	59.5	68.4	0.241	0.278
S6R2	62.3	70.3	0.252	0.286
S6R4	61.0	70.5	0.244	0.280

* Average value of two specimens.

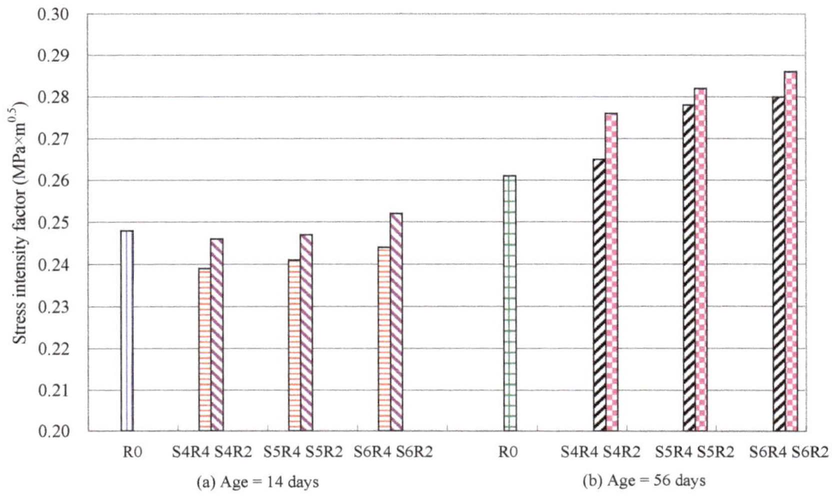

Figure 7. The stress intensity factor for concretes with various fineness levels and replacement ratios of slag at the ages of 14 days and 56 days.

Figure 8 compares the critical stress intensity factor with compressive strength for all slag-concrete specimens in the study. As shown in the figure, the two values of stress are nearly linearly related. The relationships shown in Figure 8 are significant because, based on the close relationship between the particle size of slag and compressive strength (Figure 4), the critical stress intensity factor will increase with the increase of the fineness level of the slag.

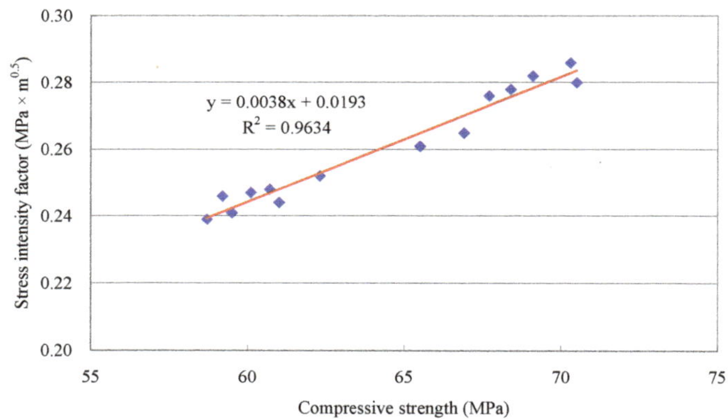

Figure 8. Critical stress intensity factor versus compressive strength for slag concretes.

5. Conclusions

The following conclusions were made based on the experimental results and the findings of the study:

1. The filling effect and the reactivity of slag can be improved by reducing its particle size. Incorporating finer slag into concrete may lead to larger early strength gains and larger strength increments of the concrete at later ages. The compressive strength of slag concrete was found to increase in conjunction with the fineness level of the slag incorporated into the mixture.

2. The use of finer slag presents a beneficial effect on the fracture energy (G_F) of concrete, even at early ages (14 days), due to superior filling effect. Increasing the fineness level of the incorporated slag leads to an increase of the G_F value of concrete or an enhanced fracture toughness.

3. The increment of the fracture energy of all the slag concretes measured in this study from 14–56 days was attained by 18–24%, which is found much higher than that of reference concrete (10.1%), and accordingly, the G_F of the slag mixtures at 56 days almost exceeds that of the reference mixture.

4. An increase in compressive strength of slag concrete of 10% resulted in a fracture energy increase of around 18%. This raise rate is significantly higher than that previously found in high-strength concretes without slag, indicating that use of the finer slag can have a unique effect on the enhancement of the fracture resistance of concrete.

5. The related tendency of the critical stress intensity factor (K^S_{IC}) of the slag concretes is similar to that of the fracture energy. At early ages (14 days), the K^S_{IC} values of slag concrete are less than that of the reference concrete (R0) but exceed that of R0 after 56 days.

6. Concretes incorporating finer slag exhibit larger K^S_{IC}, and the K^S_{IC} increases in conjunction with the fineness level of the slag. This also implies an increase in the fracture resistance of the concrete.

Author Contributions: Conceptualization, C.-H.W. and T.Y.; Methodology, C.-H.W.; Validation, C.-H.W., C.-H.H. and S.-K.L.; Formal analysis, C.-H.W. and C.-H.H.; Investigation, C.-H.W. and C.-H.H.; Resources, C.-H.W.; Data curation, C.-H.W. and S.-K.L.; Writing—original draft preparation, C.-H.W.; Writing—review & editing, C.-H.W. and T.Y.; Visualization, C.-H.W. and S.-K.L.; Supervision, T.Y.; Project administration, C.-H.H.

Funding: This research received no external funding.

Conflicts of Interest: The authors declare no conflict of interest.

References

1. Babu, K.G.; Kumar, V.S.R. Efficiency of GGBS in Concrete. *Cem. Concr. Res.* **2000**, *30*, 1031–1036. [CrossRef]
2. Mehta, P.K. Pozzolanic and Cementitious By-products as Mineral Admixtures for Concrete—A Critical Review. *Int. Concr. Abstr. Portal* **1983**, *79*, 1–46.
3. Malhotra, V.M.; Mehta, P.K. Pozzolanic and Cementitious Materials. In *Advances in Concrete Technology*; Gordon and Breach: London, UK, 1996.
4. Babu, K.G.; Kumar, V.S.R. Performance of GGBS in Cementitious Composites. In Proceedings of the Sixth NCB International Seminar on Cement and Building Materials, New Delhi, India, 24–27 November 1998; p. XIII-76.
5. Tan, K.; Pu, X. Strengthening Effects of Finely Ground Fly Ash, Granulated Blast Furnace Slag, and Their Combination. *Cem. Concr. Res.* **1998**, *28*, 1819–1825. [CrossRef]
6. Mantel, D.G. Investigation into the Hydraulic Activity of Five Granulated Blast Furnace Slags with Eight Different Portland Cements. *ACI Mater. J.* **1994**, *91*, 471–477.
7. Yűksel, I.; Özkan, Ö.; Bilir, I. Use of Granulated Blast Furnace Slag in Concrete as Fine Aggregate. *ACI Mater. J.* **2006**, *103*, 203–208.
8. Duan, P.; Shui, S.; Chen, W.; Shen, C. Effects of Metakaolin, Silica Fume and Slag on Pore Structure, Interfacial Transition Zone and Compressive Strength of Concrete. *Constr. Build. Mater.* **2013**, *44*, 1–6. [CrossRef]
9. Gao, J.M.; Qian, C.X.; Lin, H.F.; Wang, B.; Li, L. ITZ Microstructure of Concrete containing GGBS. *Cem. Concr. Res.* **2005**, *35*, 1299–1304. [CrossRef]
10. Chinaprasirt, P.; Homwuttiwong, S.; Sirivivantananon, V. Influence of Fly Ash Fineness on Strength, Drying Shrinkage and Sulfate Resistance of Blended Cement Mortar. *Cem. Concr. Res.* **2004**, *34*, 1087–1092. [CrossRef]
11. Zhu, J.; Zhong, Q.; Chen, F.; Li, D. Effect of Particle Size of Blast Furnace Slag on Properties of Portland Cement. *Procedia Eng.* **2012**, *27*, 231–236. [CrossRef]
12. Nadeau, J.C. A Multiscale Model for Effective Moduli of Concrete Incorporating ITZ Water-Cement Ratio Gradients, Aggregate Size Distributions and Entrapped Voids. *Cem. Concr. Res.* **2003**, *33*, 103–113. [CrossRef]
13. Hillerborg, A.; Modeer, M.; Petersson, P.E. Analysis of Crack Formation and Crack Growth in Concrete by Means of Fracture Mechanics and Finite Elements. *Cem. Concr. Res.* **1976**, *6*, 773–782. [CrossRef]
14. Petersson, P.E. *Crack Growth and Development of Fracture Zones in Plain Concrete and Similar Materials*; Rep. TVBM-1006; Division of Building Materials, Lund Institute of Technology: Lund, Sweden, 1981.
15. Hillerborg, A. The Theoretical Basis of a Method to Determine the Fracture Energy G_F of Concrete. *Mater. Struct.* **1985**, *18*, 291–296. [CrossRef]
16. Rao, G.A.; Raghu Prasad, B.K. Size Effect and Fracture Properties of HPC. In Proceedings of the 14th Engineering Mechanics International Conference (ASCE), Austin, TX, USA, 21–24 May 2000; p. 104.
17. Rao, G.A. Fracture Energy and Softening Behavior of High-Strength Concrete. *Cem. Concr. Res.* **2001**, *32*, 247–252. [CrossRef]
18. Surendra, P.S. Determination of the Fracture Parameters (K^S_{IC} and CTODC) of Plain Concrete Using Three-Point Bend Test. *Mater. Struct.* **1990**, *23*, 457–460.
19. Jenq, Y.S.; Shah, S.P. Two Parameters Fracture Model for Concrete. *J. Eng. Mech. ASCE* **1985**, *111*, 1227–1241. [CrossRef]
20. Jensen, E.A.; Hansen, W. Fracture Energy Test for Highway Concrete—Determining the Effect of Coarse Aggregate on Crack Propagation Resistance. *Transp. Res. Rec.* **2001**, *1730*, 10–16. [CrossRef]
21. RILEM Technical Committee 50-FMC, Proposed RILEM Recommendation. Determination of The Fracture Energy of Mortar and Concrete by Means of Three-Point Bend Tests on Notched Beams. *Mater. Struct.* **1985**, *18*, 285–290.
22. Giaccio, G.; Rocco, C.; Zerbino, R. The Fracture Energy (G_F) of High-Strength Concretes. *Mater. Struct.* **1993**, *26*, 381–386. [CrossRef]

23. Gettu, R.; Bažant, Z.P.; Karr, M.E. Fracture Properties and Brittleness of High-Strength Concrete. *ACI Mater. J.* **1990**, *87*, 608–617.
24. Xie, J.; Elwi, A.E.; MacGregor, J.G. Mechanical Properties of Three High-Strength Concretes Containing Silica Fume. *ACI Mater. J.* **1995**, *92*, 135–145.

Article

Cracking Behavior of RC Beams Strengthened with Different Amounts and Layouts of CFRP

Muhammad Umair Saleem [1,*], **Hisham Jahangir Qureshi** [1], **Muhammad Nasir Amin** [1], **Kaffayatullah Khan** [1] and **Hassan Khurshid** [2]

[1] Department of Civil and Environmental Engineering, College of Engineering, King Faisal University, 31982 Hofuf, Alahsa, Saudi Arabia; hqureshi@kfu.edu.sa (H.J.Q.); mgadir@kfu.edu.sa (M.N.A.); kkhan@kfu.edu.sa (K.K.)

[2] Department of Mechanical Engineering, College of Engineering, King Faisal University, 31982 Hofuf, Alahsa, Saudi Arabia; hkhurshid@kfu.edu.sa

* Correspondence: mmsaleem@kfu.edu.sa; Tel.: +966-545-672946

Received: 4 February 2019; Accepted: 7 March 2019; Published: 12 March 2019

Featured Application: This study provides the valuable data to assess the failure mechanism of CFRP strengthened reinforced concrete beams and structural systems.

Abstract: The bending and shear behavior of RC beams strengthened with Carbon Fiber-Reinforced Polymers (CFRP) is the primary objective of this paper, which is focused on the failure mechanisms and on the moment-curvature response prior-to, and post, strengthening with different amounts and layouts of the CFRP reinforcement. Seven reinforced concrete beams were tested in 4-point bending, one without any CFRP reinforcement (control beam, Specimen C1), four with the same amount of CFRP in flexure but with different layouts of the reinforcement for shear (Specimens B1–B4), and two with extra reinforcement in bending, with and without reinforcement in shear (Specimens B6 and B5, respectively). During each test, the load and the mid-span deflection were monitored, as well as the crack pattern. The experimental results indicate that: (a) increasing the CFRP reinforcement above certain levels does not necessarily increase the bearing capacity; (b) the structural performance can be optimized through an appropriate combination of CFRP flexural and shear reinforcement; and (c) bond properties at the concrete–CFRP interface play a vital role, as the failure is very often triggered by the debonding of the CFRP strips. The experimental values were also verified analytically and a close agreement between the analytical and experimental values was achieved.

Keywords: concrete cracking; crack patterns; carbon fiber-reinforced polymers—CFRP; RC strengthening (in bending and shear); RC beams

---

1. Introduction

The Kingdom of Saudi Arabia, as well as rest of the world, contains a wide range of reinforced concrete infrastructure which ranges from small residential houses to multi-story buildings, towers, and pre-stressed concrete bridges. The majority of the infrastructure in Saudi Arabia was designed and constructed on old design codes, standards, and specifications, and as a result of that, it is currently experiencing weathering because of harsh climate, saltwater, acid attack, extreme temperature changes, and due to the inferior quality of building material. There is a necessity to strengthen old existing infrastructure in Saudi Arabia, which lacks in strength and stiffness. There are many different ways to strengthen reinforced concrete structures, and one of the most common methods is the application of CFRP (Carbon Fiber Reinforced Polymer) to the reinforced concrete member.

CFRP has been widely used for the purpose of strengthening reinforced concrete, masonry, and steel infrastructure. Over the past few decades, many researchers have proved the suitability of CFRP material for structures composed of different materials, stiffness, and ductility [1–6]. By adopting the correct retrofitting technique, CFRP can significantly increase the shear and flexure strength of concrete structures as compared to the normal concrete structure. CFRP gains its strength via epoxy, which is the glue that makes a bond of CFRP with the concrete surface. Flexure, shear, and compression cracks can be prevented in reinforced concrete beams with the help of CFRP application. The benefit of using CFRP retrofitting comes from its low density and its resistance to higher tensile forces, fatigue, and corrosion.

However, the debonding failure mechanism, structural ductility, and long-term durability are the main problems, which CFRP is facing due to poor bonding and reduced vapor pressure [7–9]. Fracture of CFRP concrete beams is mainly attributed due to debonding of CFRP from the concrete surface. Hence, flexural strength, failure behavior, and structural ductility are the most important parameters that are always under consideration in structural design, especially when the structure is located in the higher earthquake zone region. CFRP is a material which is brittle in nature and fails suddenly, and has a linear elastic behavior up until failure, i.e., the tensile failure strain of FRP ranges from 2% to 4% [6]. Once the failure strain is reached, CFRP shows no signs of warning, breaks suddenly, and results in the loss of its strength. This behavior of low ductility in reinforced concrete structures retrofitted with CFRP is not desirable, as it does not provide any kind of early warning before failure, thus resulting in the sudden collapse of the structure.

2. Literature Review

CFRP has been widely used in the past for repair and rehabilitation of civil infrastructure that was showing signs of deterioration and distress due to aging. If applied properly, CFRP can increase the service life of the structure. CFRP is a brittle material and usually fails (i.e., debonding or horizontal crack propagation) at a lower load level, hence the ultimate capacity of the reinforced concrete structural member is very hard to achieve [10]. In the past, many researchers conducted experiments on reinforced concrete beams retrofitted in flexure by CFRP and resulting failure patterns were observed.

Arduini and Nanni [11] conducted a parametric analysis to investigate the effects of CFRP reinforcement on serviceability, strength, and failure mechanisms of repaired RC beams. They presented the results of their analysis in terms of repaired/un-repaired strength and deflection ratios. They observed that a brittle failure mechanism can develop at loads much lower than expected when considering only flexural performance of the FRP Strengthened beams. Their research work also concluded that the application of CFRP reinforcement can considerably result in an increase in load-bearing capacity and can also limit deflection at the service level. Smith and Teng [12] conducted a comprehensive review of existing plate debonding strength models that were presented by many researchers in past. Each model was summarized and classified into one of the three categories based on the considered approach. Teng et al. [13] conducted extensive research that has been carried out in recent years on the use of fiber-reinforced polymer (FRP) composites for the purpose of strengthening of reinforced concrete (RC) structures. Their paper provides a concise review of existing research on the behavior and strength of FRP-strengthened RC structures, with a strong focus on those studies that contribute directly to the development of strength models. Topics that were covered in their research work includes flexural and shear strengthening of beams, flexural strengthening of slabs, and strengthening of columns subjected to both static and seismic loads. For each of the topics covered, the methods of strengthening were first explained, followed by a description of the common failure modes. Kotynia [14] conducted tests on reinforced concrete (RC) beams that were strengthened externally with CFRP strips in flexure only. The flexural behavior of the beams, as well as their failure modes, were discussed in detail. Teng and Chen [15] provided a summary of debonding failure modes of reinforced concrete beams that were strengthened externally with FRP reinforcement. Their paper addressed the following three issues: (a) classification of debonding failure modes; (b) mechanisms and processes of debonding failures; (c) and theoretical models for debonding failures.

Kang et al. [10] conducted a detailed review of previous research programs that were conducted by past researches in relation to debonding failure of FRP attached externally to the concrete surface. Li et al. [16] conducted a series of experimental tests on reinforced concrete beams that were strengthened with externally bonded CFRP sheets in flexure. They investigated debonding initiation and the allowable tensile strain of FRP sheets in flexurally-strengthened RC beams in comparison to different design code provisions. Hasnat et al. [17] conducted research on simply supported reinforced concrete (RC) beams that were strengthened by carbon-fiber-reinforced polymer wrap. A CFRP wrap resisted the premature cover debonding and acted as a U-clamp that results in an increase in the ultimate moment capacity. Mostafa and Razaqpur [18] conducted the experiment on reinforced concrete beams made of T-section. The load was applied and deflection responses were measured for each beam. The complete post-peak load/softening response of each beam was also captured.

Fu et al. [19] conducted research on the effectiveness of a U-jacketing system on delaying or preventing debonding failure. In their experimental research study, they tested eight large-scale RC beams in order to study and investigate the effects of different forms of FRP U-jacketing on debonding failure. Abid and Al-lami [20] conducted an extensive review of past research studies that emphasis the strength and durability of concrete beams that were externally bonded with FRP reinforcement. The focus of the research review was on the bond behavior, testing techniques, and models used to assess bond strength. Flexure, shear, and fatigue behaviors of different strengthening techniques were also reviewed and discussed in detail.

Wenwei and Guo conducted flexural testing on six reinforced concrete beams reinforced with external CFRP laminates in order to study the effects of initial load or load history on the ultimate strength [21]. Experimental results were explained in quantitate terms and for that purpose, a theoretical model for flexural behavior was developed. Lee and Moy carried out an experimental and an analytical study and developed a design-oriented expression to determine an effective laminate strain in bonded CFRP [22]. The developed expression was also used for predicting the deboning failures. Barros et al. conducted experimental and analytical research on reinforced concrete structures strengthened with CFRP systems [23]. Their experimental studies show that the CFRP debonding strain is dependent on the CFRP percentage and the longitudinal steel reinforcement ratio. Mansour and Mahmoud conducted an experimental and analytical study to predict the ultimate moment capacity of RC beams externally strengthened with CFRP [24]. They proposed prediction models to obtain the load capacities for rectangular as well as T beam sections. Pan et al. conducted experimental testing on eight reinforced concrete beams that were strengthened with FRP sheets [25]. They also proposed an analytical model that accounts for the opening of shear and flexural cracks along the beam. Later the results of the experimental data were compared with the proposed analytical model. Ghandour conducted three-point loading testing on seven half-scale reinforced concrete beams, which were strengthened with CFRP longitudinal sheets and U-wraps [26]. Kara and Ashour developed a numerical method for predicting the curvature, deflection, and moment capacity of reinforced concrete beams strengthened with FRP [27]. Later, the analytical results obtained from the numerical model were compared with the published experimental data. Pellegrino and Vasic [28] did the assessment on the design procedures that were available to predict the shear capacity of reinforced concrete beams externally strengthened with FRP composites. Their research work was based on a database, which was collected from recent literature and mainly focused on the basic codes for reinforced concrete structures (without strengthening) and current models for FRP structures strengthened in shear. Li et al. investigated debonding initiation and tensile strain of FRP laminates adhered externally to concrete beam surfaces [16]. Experimental testing was carried out and the allowable tensile strain in FRP sheets proposed by prevalent code provisions was assessed. Jung et al. presented both experimental and analytical research results to predict the flexural capacity of reinforced concrete beams strengthened in flexure with fabric reinforced cementitious matrix (FRCM) [29]. Six beams were strengthened in flexure with FRCM composite under four-point testing. A new bond strength model was proposed using a test database to predict the strengthening performance of the FRCM composite. Foster et al. did an experimental investigation on reinforced concrete T-beams of varied sizes to determine the effectiveness of unanchored and anchored externally bonded U-wrapped CFRP [30]. Beams were subjected to three-point bending with a span to depth ratio of 3.5. Dias and Barros did experimental research to understand the effectiveness of near surface mounted (NSM) technique by using CFRP laminates in shear [31]. T-Sections reinforced concrete beams were externally reinforced in shear with the help of CFRP laminates at 52- and 90-degree angles. Furthermore, they discovered that inclined laminates are more effective than vertical laminates. Osman et al. did experimental testing on seven CFRP strengthened reinforced concrete beams under four-point loading having different shear span-to-depth ratios [32]. A numerical analysis was also carried out on 27 reinforced concrete beams with and without CFRP sheets. The results achieved using ANSYS were close to the experimental results.

Keeping in mind the aforementioned literature review, it is found that there are very few researchers who studied the effect of different CFRP reinforcement amounts and layouts on beam moment capacity and on the failure behavior of the RC beams. The current study evaluated the effect of the CFRP reinforcement ratio on failure patterns of reinforced concrete beams. The CFRP reinforcement ratio was varied in the form of flexure and shear strips that resulted in different strengthening layouts and failure patterns of the concrete beams. The originality of this research work mainly lies in the determination of specific load levels at which debonding failure of the CFRP initiates and how different layouts of CFRP strengthening affect the shear and flexural strength and failure modes of the reinforced concrete beams. RC beams strengthened with different amounts and layouts of CFRP were tested under four-point bending. Moment-curvature behavior, failure loads, deflections, and failure modes were determined experimentally, which later were also compared with the analytical capacity evaluation models proposed by American Concrete Institute ACI 440.2R-08 [33] provisions.

3. Experimental Plan and Setup

In this research work, seven reinforced concrete beams were prepared and tested under four-point loading. The cross-sectional sizes of all seven reinforced concrete beams were kept constant and are selected equal to 100 mm × 200 mm with an overall equal length of 1200 mm. The beams were designed as per ACI 318-08 [34] for tension controlled failure. Two deformed rebars of 14 mm nominal diameter were used as a flexural steel reinforcement. Figures 1 and 2 give the details of the beams tested under the current experimental plan. The load is applied under a displacement control system at a constant strain rate of 0.03/min. The bending setup located in highway laboratory of King Faisal University was used to carry out the experimental research work as shown in Figure 3. The beams were tested under four-point loading, which gives a constant mid-span bending moment for a given increment of load between the two loading points. Effective span in all beams was kept equal to 1100 mm. The applied bending moment M of the beam is calculated by Equation (1). As the moment between loading point remains constant, the Sagital Method was used to calculate the curvature φ by using mid-span deflection δ of the beam.

$$M = \frac{P}{2} \times L_1 \tag{1}$$

where P is the load directly obtained from the Universal Testing Machine at each increment of displacement. L_1 is the distance between the support and point of application of load = 475 mm and L is the effective span of the beam = 1100 mm, as given in Figure 3b.

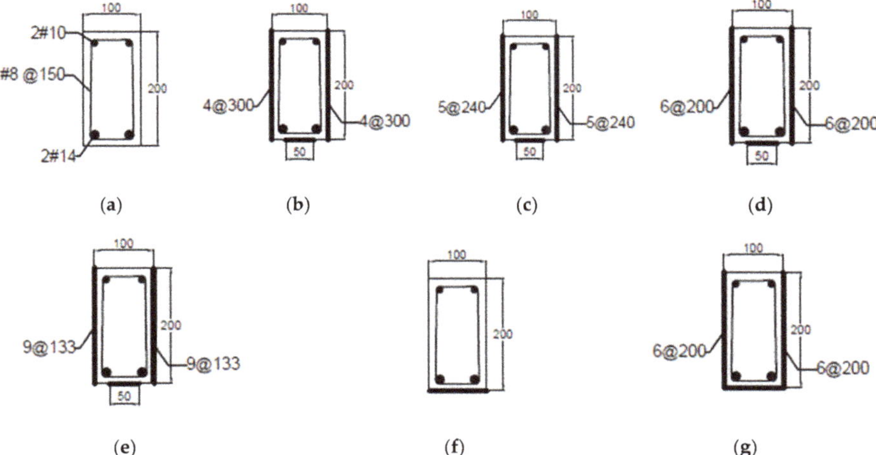

Figure 1. Details of beam specimens (**a**) Control specimen, (**b**) CFRP-B1, (**c**) CFRP-B2, (**d**) CFRP-B3, (**e**) CFRP-B4, (**f**) CFRP-B5, and (**g**) CFRP-B6.

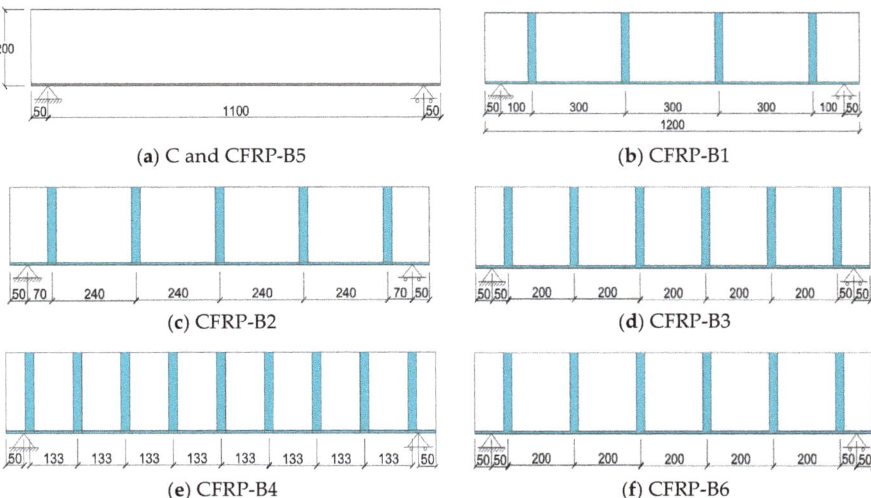

Figure 2. Lateral views of the beams strengthened with CFRP layers and strips.

One out of seven beams was selected as a control specimen, while the rest of the six beams were strengthened and retrofitted with external CFRP reinforcement ratios given in Table 1. The thickness of external flexure and shear CFRP strips was kept equal to 1.5 mm. For all beams, the cross-sectional sizes, internal reinforcement ratios, overall length, and thickness of CFRP strips were kept constant, while the external flexure and shear CFRP reinforcement ratios were varied. Table 1 shows the strengthening scheme of the beams, which were divided into two groups. Beams B1, B2, B3, and B4 were placed in group-1, where the external CFRP flexural reinforcement ratio was kept constant (0.38%). Beam B5 and B6 were placed in group-2, where the external CFRP flexural reinforcement ratio was doubled and kept equal to 0.75%. The flexural reinforcement ratio refers to the quantity of CFRP that is applied at the bottom surface of the beam. Reinforcement ratio is calculated by dividing the cross-sectional area of CFRP with the cross-sectional area of the concrete beam. In a similar way, the shear reinforcement ratio was calculated. Hence, the total CFRP reinforcement ratio is the sum of the flexural and shear CFRP reinforcement ratio. However, the external CFRP shear reinforcement ratio was also varied and was selected equal to 0 and 0.38% for Beam B5 and B6, as given in Table 1.

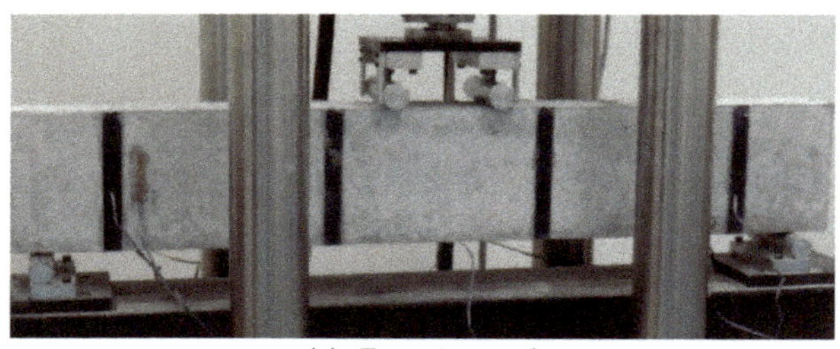

(a) Experimental

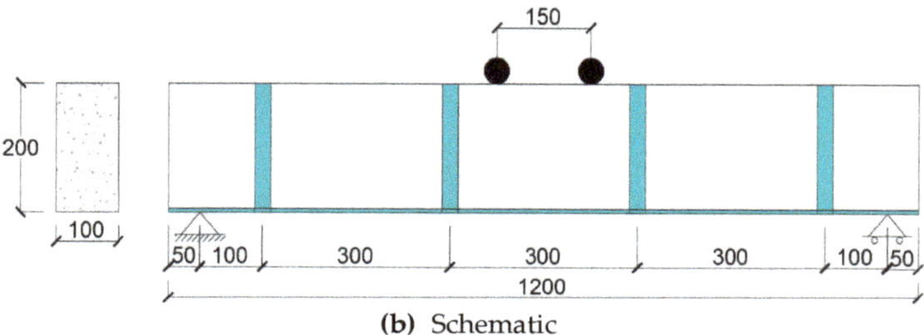

(b) Schematic

Figure 3. Test setup (a) and test layout for Beam CFRP-B1 (b).

Table 1. Testing specimen details.

Sr. No.	Specimen Designation	External CFRP Flexural Reinforcement Ratio (% Age)	External CFRP Shear Reinforcement Ratio	Total CFRP Reinforcement Ratio	Number of CFRP Shear Strips	Width of Flexural CFRP Strips (mm)
1	Control-C1	N/A	N/A	N/A	N/A	N/A
2	CFRP-B1	0.38	0.25	0.63	8	50
3	CFRP-B2	0.38	0.31	0.69	10	50
4	CFRP-B3	0.38	0.37	0.75	12	50
5	CFRP-B4	0.38	0.56	0.94	18	50
6	CFRP-B5	0.75	0.00	0.75	N/A	100
7	CFRP-B6	0.75	0.38	1.13	12	100

Material Properties

Concrete with a compressive strength of 28 MPa was used for the preparation of reinforced concrete beam specimens. The average compressive strength of concrete was determined by the procedure mentioned in ASTM C-39 [35] and was found equal to $f'_c = 28$ MPa. The modulus of elasticity and shear modulus of concrete were 24,870 MPa and 10,360 MPa, respectively. The modulus of elasticity of concrete was calculated using the ACI 318-08 [34] and shear modulus of elasticity was computed using poison ratio of 0.2. For quality control, the ready-mix concrete was delivered by a local supplier from a plant located in Saudi Arabia. For internal reinforcement, the yield strength of $f_y = 420$ MPa was selected for deformed rebar. The steel reinforcement has a tensile rupture strength of 620 MPa with elastic and shear modulus values of 200,000 MPa and 76,920 MPa, respectively.

Concrete beam specimens were cast and cured as per ASTM-C31 [36]. On completion of curing time, the external surfaces of the beam specimens were cleaned with a wet cloth and acetone. Epoxy (Sikadur 330) of uniform thickness was used to attach external CFRP strips to the concrete surface. After the application of CFRP strips on the surface of the concrete, the beams specimens were cured for three days by using a wet layup method. Figures 1 and 2 show the layout of beam specimens along with their external and internal reinforcement details. The CFRP has a tensile strength of 1600 MPa with the ultimate tensile elongation value of 1.8%. The tensile modulus of elasticity of CFRP was 120,000 MPa. The epoxy has tensile bond and compressive bond strength values of 9.6 MPa and 21 MPa, respectively. The aforementioned material properties of CFRP and epoxy were provided by their respective suppliers.

4. Results and Discussions

Figure 4 shows the mid-span bending moment-curvature behavior for the beam specimens C1, CFRP-B1, CFRP-B2, CFRP-B3, and B4, respectively. The control specimen has shown a peak moment value of 17.73 kN-m with the curvature value of 0.0031. Once the peak moment was reached, the control beam specimen C1 started losing its strength and finally, it reduced to 4.92 kN-m at a curvature value of 0.0057. The control specimen has shown a residual strength of 9.5 kN-m, which was approximately half of the peak strength (17.73 kN-m) of the control specimen. The residual strength of the concrete beam was mainly attributed to the cracked concrete section, which contains tensile reinforcement having a tensile strain greater than yield but lesser than the rupture strain. In the case of different CFRP strengthened beam specimens, higher peak moments were observed. The presence of CFRP strips has contributed to the flexural capacity and increased not only the peak strength but has also undergone higher displacements at the peak moment values. All of the beam specimens have shown peak strengths ranging from 27.7 kN-m to 26.6 kN-m. The maximum peak moment of 27.7 kN-m was observed in the case of beam CFRP-B4. All of the beams (CFRP-B1, B2, B3, and B4) carry a uniform external CFRP flexural reinforcement ratio of 0.38%, however, the shear reinforcement was gradually increased from 0. 25% to 0. 56%, respectively. The effect of this change in CFRP shear reinforcement ratio could be seen in the form of small variations in peak moment carrying capacities of the beams. For all of the beams, the relationship between curvature and moment capacities remained linear until the curvature value of 0.0017 was reached. After this value, the inelastic behavior of reinforced concrete became dominant and stiffness of the beam gradually started decreasing. All of the CFRP strengthened beams have shown a high residual strength compared to the control specimen C1. However, the highest residual strength was provided by specimen B3 and it was approximately 1.5 times the residual strength of the control specimen. Moreover, compared to the control specimen, the deformation capacities or curvature of the CFRP strengthened beams B1, B2, B3, and B4 were also higher. Beams B1, B2, B3, and B4 have shown a rise and fall in the moment values. This rise and fall behavior in the moment values represents the debonding of CFRP strips from the surface of the beam. After a small segment of CFRP strip was detached from the beam surface, a drop in strength occurred. With a further increase of load, the remaining bonded part of CFRP and concrete came into action and beams again started taking the load until the peak moment capacities of the beams was reached. Beam B1 has the minimum CFRP reinforcement compared to all other beams, and as a result, a sudden drop in strength was observed due to complete detachment of the CFRP strip from the concrete surface. As the beams were subjected to further load increments, the strength of beams B1, B2, B3, and B4 became equal to the control specimen. It showed that the contribution of CFRP reinforcement had completely gone and only reinforced concrete beam residual strength was in action.

The mid-span bending moment-curvature behaviors of beam CFRP-B5 and CFRP-B6 and the control specimen C1 are given in Figure 5. In the case of beam CFRP-B5 and CFRP-B6, the external CFRP flexural reinforcement was doubled compared to beam specimens CFRP-B1, B2, B3, and B4. Both beams B5 and B6 carry the same flexural reinforcement of 0.75%. However, beam B6 has a 0.38% shear CFRP reinforcement ratio, whereas beam B5 does not carry any shear reinforcement. Beam B5 has shown flexural strength of 26.52 kN-m that is 1.5 times higher than the peak strength of the control specimen (17.73 kN-m). On the other hand, beam B6 has shown a peak strength value of 33.71 kN-m, which is approximately double the strength of the control specimen. Adding 0.38% shear reinforcement ratio has increased the flexural strength from 26.62 to 33.71 kN-m. The presence of shear reinforcement hindered the diagonal shear crack propagation by intercepting their path, and in return, the moment carrying capacity of beam B6 has improved. The relation of moment-curvature remained linear prior to reaching the curvature values of 0.002. Beyond this value, a small rise and fall in the moment values was observed. These falls show the initiation of debonding failure between the CFRP and the concrete surface on the tension face of the beam.

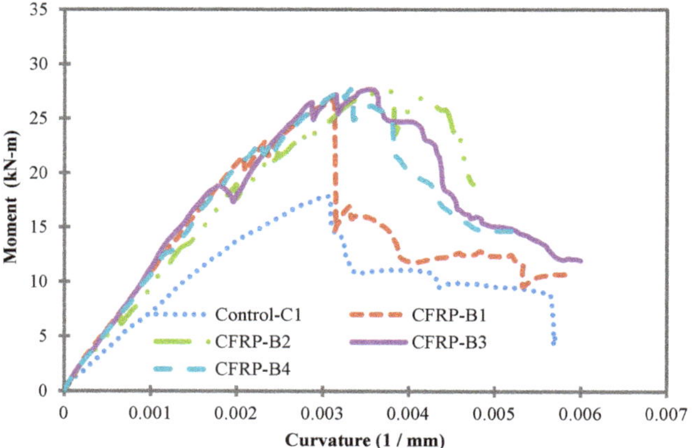

Figure 4. Moment-curvature relation for control specimen C1, CFRP-B1, CFRP-B2, CFRP-B3, and CFRP-B4.

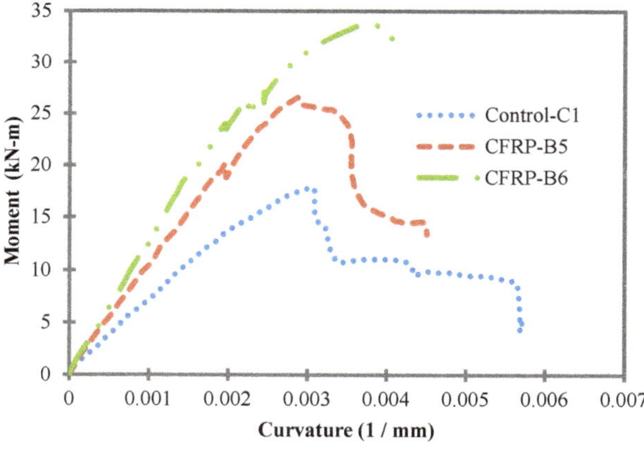

Figure 5. Moment-curvature relation for control specimen C1, CFRP-B5, and CFRP-B6.

Moreover, beam B6 has shown higher deformation or curvature compared to beam B5. For instance, the peak moment carrying capacity of the beam B5 occurred at a curvature value of 0.0028, whereas the peak strength of beam B6 was observed at a curvature value of 0.0038. It implies that the adding shear CFRP reinforcement in the presence of CFRP flexural reinforcement has not only increased the overall flexural strength of the beam but also the deformation capacity of the beam.

In order to understand the effect of shear reinforcement on the overall strength, a comparison of the mid-span bending moment-curvature behavior of beam CFRP-B3 and B5 is presented in Figure 6. Both beams CFRP-B3 and B5 carry an equal amount of overall (flexural plus shear) CFRP reinforcement ratio of 0.75%. However, in the case of beam B3, half (0.37%) of the total reinforcement (0.75%) was provided for flexure and the remainder was used for the shear. In the case of beam B5, the total reinforcement (0.75%) was provided as the flexural CFRP reinforcement, as given in Table 1. Beam B3 has shown peak strength of 27.7 kN-m compared to beam B5 that has a peak strength of 26.5 kN-m. Although a very small increase in overall strength was observed, beam B3 has shown better performance in terms of deformation capacity. Beam B5 has shown a curvature value 0.0028 at the peak moment of 26.5 kN-m and this was lower than the corresponding curvature value of 0.0035 for beam B3. For both beams, the debonding of CFRP from the tensile face started at the same curvature level, as indicated by a small fall in the moment at a curvature of 0.0019. However, the presence of shear strips provided resistance against shear cracks and improved the overall flexural behavior of beam B3. Moreover, beam B3 has shown better post-peak behavior compared to beam B5, as the residual strength of beam B3 was higher than beam B5.

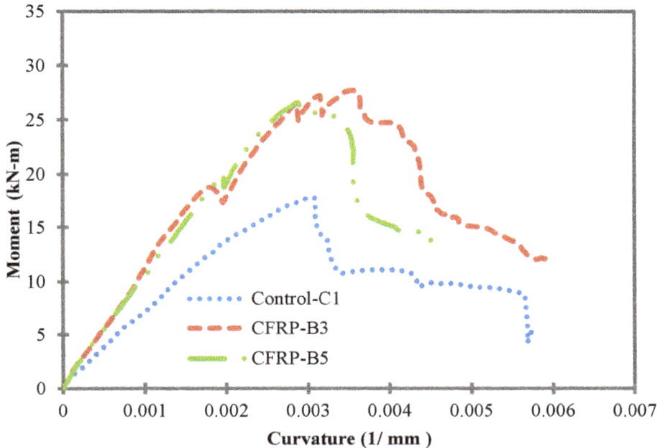

Figure 6. Moment-curvature relation for control specimen C1, CFRP-B3, and CFRP-B5.

The test results show that beams with higher CFRP shear reinforcement but with a lower amount of flexural CFRP reinforcement cannot contribute significantly in terms of the overall strength of the beam. For instance, group-1 beams had a lower flexural CFRP reinforcement ratio (0.38%) and the CFRP shear reinforcement was gradually increased from 0.25% to 0.56%. This increase in CFRP shear reinforcement ratio has increased the strength of beam B1 from 26.6 kN-m to 27.7 kN-m, which is not a significant contribution. On the other hand, when the flexural reinforcement doubled as in the case of the group-2 beam (B6), the moment carrying capacity increased to 33.8 kN-m.

5. Analytical Prediction of Beam Capacities

Several researchers compared the experimental and analytical flexural and shear capacities of different CFRP strengthened RC beams [21–32,37,38]. However, most of them [22,26–28,37] found close analytical values when ACI 440.2R-08 [33] was used. In the current study, ACI 440.2R-08 [33] was adopted to analytically obtain the flexural and shear capacities of the beams. The flexural strength of CFRP strengthened beams depends upon the governing mode of failure. Reinforced concrete beams strengthened with CFRP can experience three major failure modes: (i) tension failure of CFRP strip; (ii) debonding of CFRP from the concrete surface; and (iii) the crushing of concrete on the compression sides. ACI 440.2R-08 [33] considers all three failure modes to satisfy strain compatibility and force equilibrium conditions.

Figure 7 illustrates the internal strain and stress distribution of a CFRP strengthened rectangular RC beam under flexure. The ultimate nominal strength of beam (M_n) is obtained by satisfying strain compatibility and force equilibrium conditions and also considers the governing mode of failure. Equations (2)–(6) give the flexural or ultimate moment capacity of CFRP strengthened beams. In order to check the debonding failure of CFRP strengthened beams, the effective strain in the CFRP strips (ε_f) should be limited to debonding strain level (ε_{fd}), as given by Equation (4).

$$M_n = A_s f_y \left(d - \frac{a}{2} \right) + A_f f_f \left(d_f - \frac{a}{2} \right) \tag{2}$$

$$\varepsilon_{fd} = 0.41 \sqrt{\frac{f_c'}{n E_f t_f}} \le 0.9 \varepsilon_{fu} \tag{3}$$

$$\varepsilon_f = \varepsilon_{cu} \left(\frac{d_f - c}{c} \right) \le \varepsilon_{fd} \tag{4}$$

$$f_f = \varepsilon_f E_f \tag{5}$$

$$a = \beta_1 c \tag{6}$$

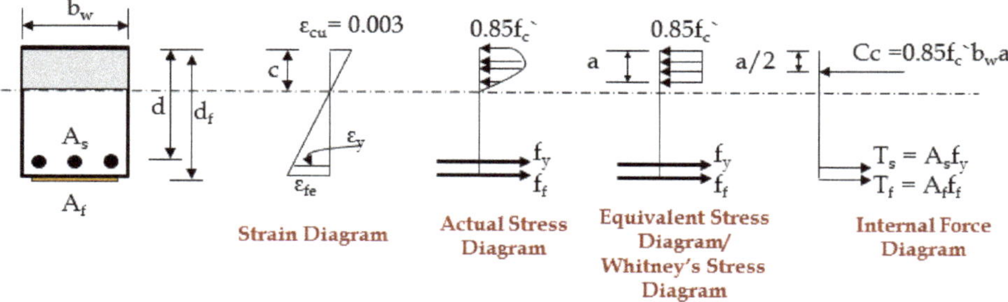

Figure 7. Stain, stress, and force distribution diagram of CFRP strengthened RC beam.

According to ACI 440.2R-08, the shear strength of CFRP strengthened beams can be computed by Equation (7), as follows.

$$V_n = V_c + V_s + \Psi_f V_f \tag{7}$$

where V_n is the nominal strength of a CFRP strengthened beam, V_s is shear strength provided by the steel and calculated by ACI 318-08 [34], as given by Equation (8), and V_c is the shear strength provided by the concrete, which can be computed by ACI 318-08 [34], as given by Equation (9).

$$V_s = \frac{A_v f_y d}{s} \tag{8}$$

$$V_c = 0.17 \sqrt{f_c'} b_w d \tag{9}$$

where V_f is the shear strength provided by the CFRP shear strips and $\Psi_f = 0.85$ is a reduction factor for CFRP shear strips applied on the sides of beams. ACI 440.2R-08 [33] recommends Equation (10) to calculate the shear strength provided by the FRP strips. The effective stress in FRP and area of CFRP in shear can be calculated by using Equations (11) and (12), respectively. The dimensional variables used in Equations (10) and (12) can be seen in Figure 8.

$$V_f = \frac{A_{fv} f_{fe} (\sin\alpha + \cos\alpha) d_{fv}}{s_f} \tag{10}$$

$$f_{fe} = \varepsilon_{fe} E_f \tag{11}$$

$$A_{fv} = 2 n t_f w_f \tag{12}$$

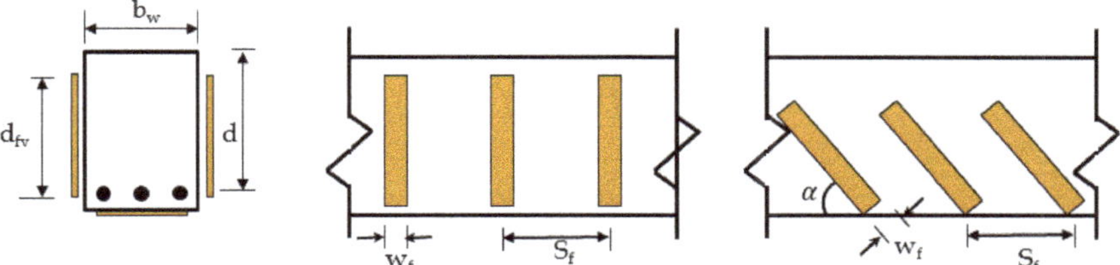

Figure 8. Dimensional variables used in shear-strengthening calculations.

For bonded face plies or CFRP strips, which is a strengthening system that does not enclose the entire section of the beam, failure of CFRP shear strips is mainly attributed to the delamination of CFRP from the concrete surface. Due to this reason, the bond stress was analyzed by Triantafillou [39] to determine the usefulness of these systems and proposed the effective strain level that can be achieved. The effective strain (ε_{fe}) is calculated using a bond-reduction coefficient k_v. Following Equations (13) and (14) gives the effective strain (ε_{fe}) and bond-reduction coefficient k_v. Equations (15)–(17) give the factors k_1, k_2, and effective length L_e used in Equation (14).

$$\varepsilon_{fe} = k_v \varepsilon_{fu} \leq 0.004 \tag{13}$$

$$k_v = \frac{k_1 k_2 L_e}{11,900\varepsilon_{fu}} \leq 0.75 \tag{14}$$

$$k_1 = \left(\frac{f'_c}{27}\right)^{2/3} \tag{15}$$

$$k_2 = \frac{d_{fv} - 2L_e}{d_{fv}} \tag{16}$$

$$L_e = \frac{23,300}{\left(nt_f E_f\right)^{0.58}} \tag{17}$$

Table 2 gives a detailed comparison of analytical and experimental load-carrying capacities of the beam specimens. The control beam did not carry any flexural and shear CFRP reinforcement and its flexural and shear capacities were computed by ACI 318-08 [34]. The flexural and shear capacities of the CFRP strengthened beams are calculated by Equations (2) and (7), as recommended by ACI 440.2R [33]. The flexural capacity of the beam is the main function of the amount of CFRP applied at the tensile face of the beam. As beams B1, B2, B3, and B4 carry the same amount of CFRP flexural reinforcement ratio, these beams had the same analytical moment capacity of 25.35 kN-m. Similarly, beam B5 and B6 have the same analytical moment capacity of 33.41 kN-m. On the other hand, the shear capacity of the CFRP strengthened beams is proportional to the amount of CFRP shear reinforcement. Due to the different amounts of CFRP shear reinforcement, specimens B1 to B5 have different analytical shear capacities. However, beam B3 and B6 have the same CFRP shear reinforcement ratio and as a result, these have the same analytical shear capacities. From the analytical flexural and shear strengths of beams, their corresponding peak loads P_m for moment and P_v for shear are calculated and presented in Table 2. For instance, the analytical moment capacity of control beam 17.28 kN-m will result in the corresponding load capacity of ($P_m = \frac{M_n \times 2}{L_1}$) 72.8 kN. The analytical peak load of each beam is evaluated as the lesser of P_m and P_v, depending upon which type of failure will govern the ultimate load carrying capacity of the beam. In order to understand the effect of CFRP flexural and shear reinforcement, the control beam was designed to have the same flexural and shear capacity. However, the P_m for control beam was slightly lesser than the P_v and it suggests that the dominant mode of failure in the beam is flexural rather than shear, or it indicates that the flexural failure will occur first, followed by the shear failure of the beam specimen. Similarly, in the case of beams B1 to B4 and B6, flexural failure governed the overall design, and in the case of beam B5 the shear failure governed the peak load of the beam, as given in Table 2. A comparison of experimental and analytical peak loads has also been provided in the last column of Table 2. A very close agreement between analytical and experimental load values was achieved for all of the tested beams. In the case of beam B5, the difference is slightly more compared to the other beams. The experimental peak load of beam B5 was slightly higher compared to the analytical load. In reality, the CFRP flexural strip also contributes to the shear resistance of the beam, so to get closer analytical values requires an accurate prediction of CFRP flexural strip contribution in shear.

Table 2. Comparison of Experimental and Analytical Failure Loads.

Specimen	Analytical M_n (kN-m)	P_m (kN)	Analytical V_n (kN-m)	P_v (kN)	Analytical Peak Load, P_a (kN)	Experimental Peak Load, P_e (kN)	P_a/P_e
C1	17.28	72.8	36.9	73.8	72.8	75.1	0.97
CFRP-B1	25.35	106.7	61.7	115.9	106.7	112	0.95
CFRP-B2	25.35	106.7	75.7	139.7	106.7	116	0.92
CFRP-B3	25.35	106.7	92.8	168.7	106.7	116.6	0.92
CFRP-B4	25.35	106.7	162.8	287.8	106.7	116.8	0.91
CFRP-B5	33.41	140.7	49.3	98.6	98.6	111.7	0.88
CFRP-B6	33.41	140.7	92.8	168.7	140.7	142	0.99

6. Cracking Behavior of Beams

6.1. Control Specimen C-1

Figure 9 shows the fracture behavior of control specimen C-1 at different load levels. Hairline cracks were observed prior to reaching a moment value of 8.7 kN-m and a displacement value of 3.5 mm. However, with an increase of moment from 8.7 kN-m to 13.1 kN-m, the crack started to propagate to the compression face of the beam, as shown in Figure 9a. With further increases in load, the crack further widened and became more prominent, as shown in Figure 9b. The cracks initially generated vertically at the bottom face of the beam and became inclined with an increase in the depth of the beam. It shows a flexural failure followed by the shear failure of the beam. A clear wide crack was observed prior to the failure of the beam at a moment and displacement level of 17.8 kN-m and 8.78 mm, respectively, as given in Figure 9c. After the peak moment capacity of 17.8 kN-m, the beam started losing its capacity and load values started decreasing, with a significant increase in the deformation capacity of the beam. No hairline cracks with uniform spacing were observed at the bottom section of the beam, which indicates the section was quite weak in strength. Figure 9d shows the failure pattern of a control specimen prior to the completion of the test. At the deflection level of 12.57 mm, a major wide crack was observed, which had split the beam into two distinct parts.

6.2. Beam CFRP B1

Failure patterns of beam CFRP B1 at different load levels are shown in Figure 10. In the case of beam B1, the initial minor cracks in the form of diagonal hairline cracks were spotted near the supports of the beam. These initial hairline cracks started to appear at a deflection value of 5.2 mm, as given in Figure 10a. With a further increase in the load, the beam started generating the new cracks and existing cracks were widened further, as shown in Figure 10b. As these cracks appear next to support in the form of diagonal movement, it confirms the shear dominant failure of the beams. The failure behavior of beam specimen B1 at the peak moment capacity of 26.6 kN-m and at a defection value of 8.9 mm is shown in Figure 10c. At a peak moment, cracks cross the FRP shear strip, however, due to higher shear forces, the existence of CFRP is not able to contribute much toward controlling the failure of the beam. Figure 10d presents the post-peak failure pattern of the beam at moment and deflection values of 10.73 kN-m and 17.3 mm, respectively. Beam B1 has only four CFRP shear strips on each face of the beam, with a higher spacing of CFRP shear strips. None of the CFRP shear strip intercepted the shear cracks initiated close to the right support of the beam. Therefore, the crack propagated diagonally to the compression face of the beam and resulted in the loss of beam strength. A significant amount of deflection and concrete crushing was observed close to the right support of the beam, which indicates that the bearing resistance of the beam was reduced and the concrete was finally crushed.

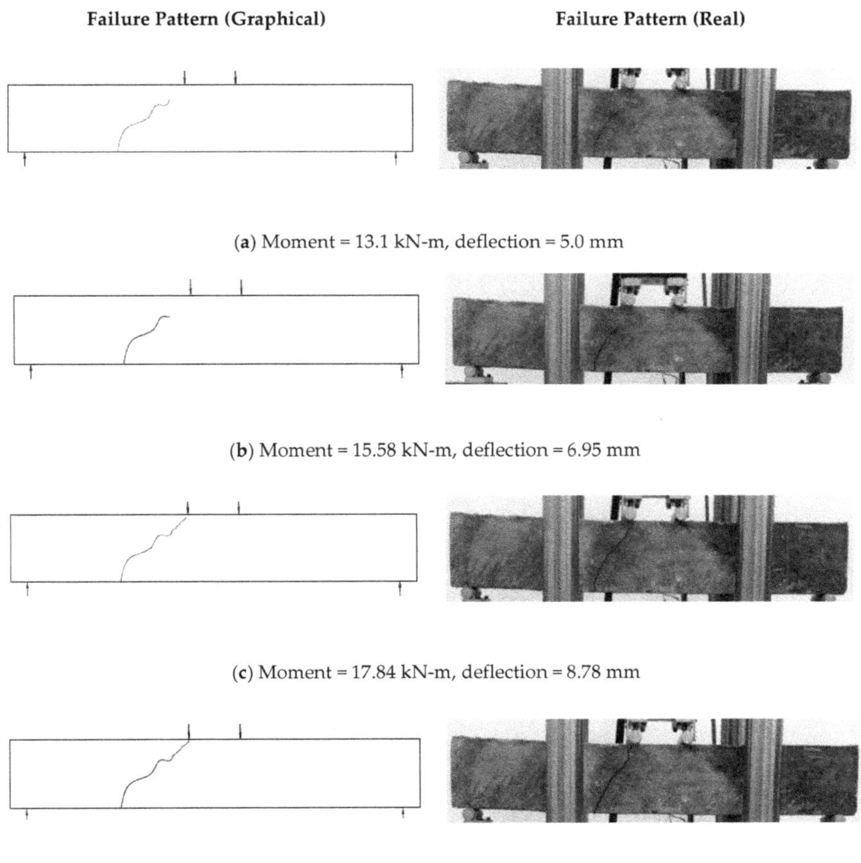

Failure Pattern (Graphical)   **Failure Pattern (Real)**

(**a**) Moment = 13.1 kN-m, deflection = 5.0 mm

(**b**) Moment = 15.58 kN-m, deflection = 6.95 mm

(**c**) Moment = 17.84 kN-m, deflection = 8.78 mm

(**d**) Moment = 9.55 kN-m, deflection = 12.57 mm

Figure 9. Failure pattern of control specimen C1 at different load levels.

6.3. Beam CFRP B2

Figure 11 shows the failure patterns of beam CFRP B2 at different load levels. Beam B2 carries a CFRP flexural reinforcement ratio of 0.38%, which was similar to beam B1, however, the CFRP shear reinforcement of beam B2 (0.31%) was slightly higher than beam B1 (0.25%). As a result, beam B2 carries five CFRP strips on each face of the beam, as shown in Figure 11. At a deflection value of 5.2 mm, no significant cracks were observed, however small hairline cracks were observed close to the left support of the beam, as shown in Figure 11a. A similar kind of pattern was noticed when the load on beam B2 was further increased until the mid-span deflection became equal to 7.7 mm, as given in Figure 11b. With a further increase in load, the cracks not only propagated diagonally but also became wide and visible to the naked eye. Figure 11c gives the cracking pattern of beam B2 at the peak moment capacity of 27.6 kN-m. Once the peak moment was reached, beam B2 started losing its capacity and cracks became more widespread and visible. In the case of beam B2, flexural mode of beam failure was dominant. Prior to complete failure, the diagonal crack reached the compression face of the beam and at that point, the beam had a residual moment carrying capacity of 18.9 kN-m, as shown in Figure 11d.

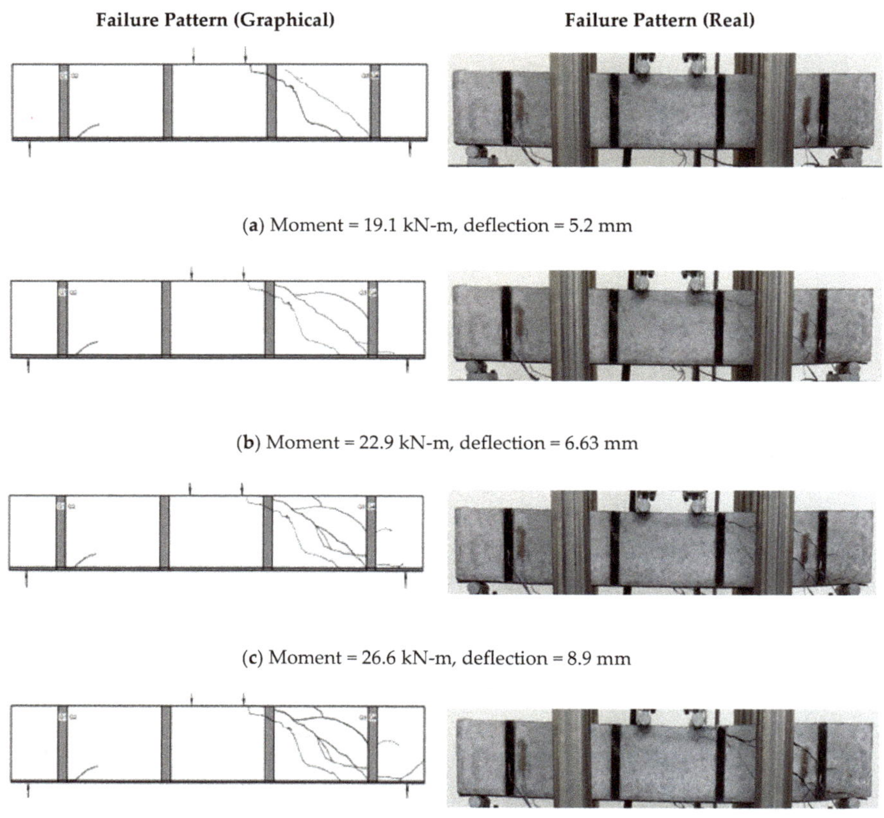

Failure Pattern (Graphical)   **Failure Pattern (Real)**

(**a**) Moment = 19.1 kN-m, deflection = 5.2 mm

(**b**) Moment = 22.9 kN-m, deflection = 6.63 mm

(**c**) Moment = 26.6 kN-m, deflection = 8.9 mm

(**d**) Moment = 10.73 kN-m, deflection = 17.3 mm

Figure 10. Failure Pattern of CFRP-B1 at different load levels.

6.4. Beam CFRP B3

The graphical and experimental failure patterns of beam B3 are shown in Figure 12. In the case of beam B3, the shear reinforcement has been increased compared to beam B2, while the external CFRP flexural reinforcement was kept similar to beam B1 and B2. In beam B3, six shear strips were provided on each face of the beam, as shown in Figure 12. At a deformation level of 5 mm, a moment capacity of 18.8 kN-m was observed, which was greater than the moment capacities of beam B1 and B2, respectively. No significant number of cracks were observed in the case of beam B3, as shown in Figure 12a. When the load was further increased, very small hairline cracks were observed on the beams, as given in Figure 12b. Beam B3 showed better behavior compared to beam B1 and B2 at the deformation level of 7.3 mm. Figure 12c shows the failure behavior of the beam at the peak moment carrying capacity of 27.7 kN-m. At a peak load, the beam underwent a significant level of cracking and slight crushing of the concrete was also observed. It shows a flexural failure due to the crushing of concrete. However, the crack remained confined between the two FRP strips. The major cracks were spotted near the left part of the beam. Figure 12d shows the post peak-cracking pattern of beam B3. The residual strength of beam B3 was close to beam B2; however, the cracks remained restricted between two CFRP shear strips.

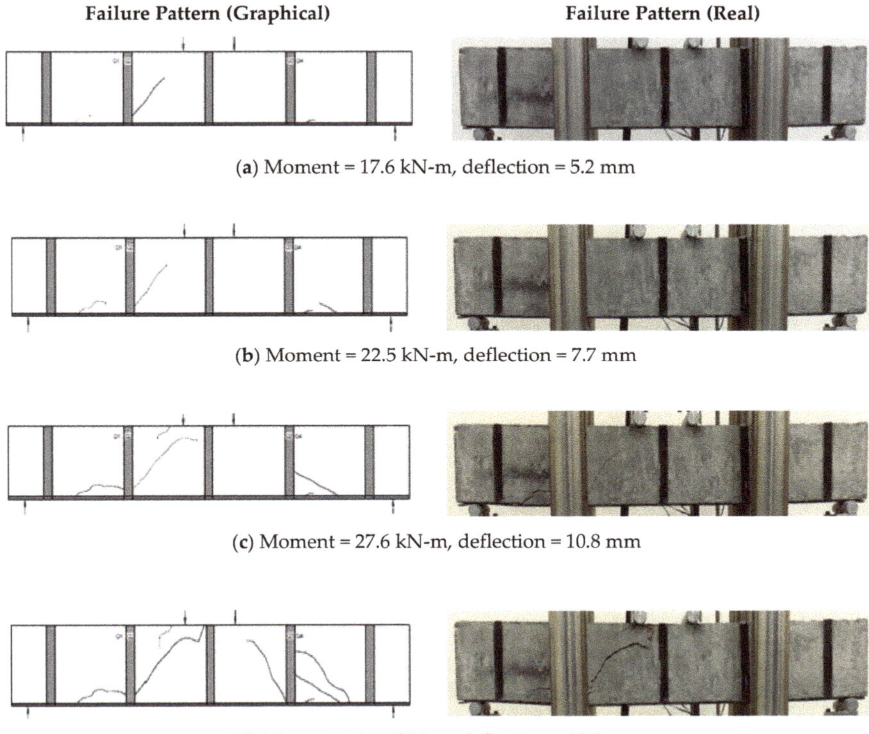

(a) Moment = 17.6 kN-m, deflection = 5.2 mm

(b) Moment = 22.5 kN-m, deflection = 7.7 mm

(c) Moment = 27.6 kN-m, deflection = 10.8 mm

(d) Moment = 18.9 kN-m, deflection = 13.8 mm

Figure 11. Failure pattern of CFRP-B2 at different load levels.

6.5. Beam CFRP B4

Figure 13 shows the graphical and experimental failure patterns of beam B4 at different load levels. In the case of beam B4, the CFRP shear reinforcement ratio was increased to 0.56% compared to beam B3 (0.38%), which resulted in a total of nine CFRP shear strips on each face of the beam. Figure 13a shows no significant cracks at a deflection value of 5.2 mm. However, when the load was slightly increased, it resulted in the formation of very small flexural and shear cracks near the mid-span of the beam, as given in Figure 13b. Beam B4 showed a peak moment capacity of 27.7 kN-m, which is similar to the moment capacity of beam B3. At the peak load level, the flexural cracks became more prominent and wider, as shown in Figure 13c. However, these cracks remained confined to the two CFRP shear strips. The cracks were also very steep compared to beam B1, B2, and B3, respectively. Figure 13d shows the cracking behavior of the beam at a deflection value of 15.4 mm. At this load level, the cracks passed through the CFRP shear strips and resulted in debonding of the shear strips from the concrete surface.

Failure Pattern (Graphical)   Failure Pattern (Real)

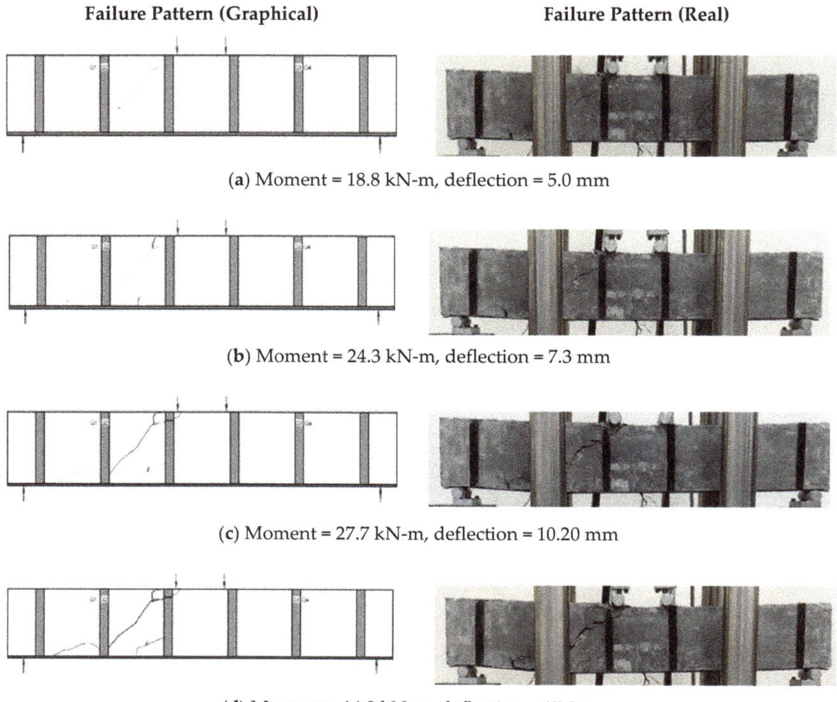

(**a**) Moment = 18.8 kN-m, deflection = 5.0 mm

(**b**) Moment = 24.3 kN-m, deflection = 7.3 mm

(**c**) Moment = 27.7 kN-m, deflection = 10.20 mm

(**d**) Moment = 14.3 kN-m, deflection = 15.8 mm

Figure 12. Failure pattern of CFRP-B3 at different load levels.

6.6. Beam CFRP B5

In beam B5, the amount of flexural reinforcement was doubled compared to beam B1, B2, B3, and B4. However, beam B5 did not contain any external CFRP shear reinforcement. Figure 14a–d presents the graphical and experimental cracking patterns of beam B5 at different load and deflection levels. A small number of cracks were observed at a deflection level of 5.2 mm (Figure 14a), which later on further increased in length and thickness when the deflection value reached 6.59 mm, as given in Figure 14b. These cracks were a mix of Mode-1 (flexural) and Mode-2 (shear) cracks, as they resulted from the interaction of bending and shear. Beam B5 has shown a peak moment capacity of 26.5 kN-m, which was greater than control beam but less than the other beams, such as B1, B2, B3, and B4, respectively. Prior to the peak moment capacity of 26.5 kN-m, a debonding type of failure of CFRP from the concrete surface was also observed. Figure 14c shows a significant number of cracks in beam B5, which were considerably wider in nature compared to the cracks that occurred at the deflection value of 6.59 mm. These cracks became even wider and longer when the deflection was increased and reached a value of 13.01 mm (Figure 14d).

6.7. Beam CFRP B6

Figure 15 shows the failure pattern of beam B6, which contains both flexural and shear CFRP reinforcement. The flexural reinforcement of beam B6 was similar to beam B5 (0.75%), with a shear reinforcement ratio of 0.38%. Like other beams, no significant cracks were observed at a deflection value of 5.0 mm, as shown in Figure 15a. With a further increase in load and deflection value, small hairline cracks started appearing closer to the beam support. Figure 15b shows that these cracks were shear in nature and occurred at a deflection value of 6.94 mm. Beam B6 showed a maximum peak strength moment capacity of 33.8 kN-m at a deflection value of 10.8 mm. At this peak point, some of the CFRP strips were de-bonded from the concrete surface and the cracks further moved diagonally. Although these cracks were intercepted by the shear CFRP strips, CFRP shear strips could not hinder their propagation in the diagonal direction, as given in Figure 15c. The final failure of the beam was mainly due to the crushing of concrete in the compression side of the beam. The post-peak cracking pattern of beam B6 is presented in Figure 15d. Beam B6 has shown a residual strength of 31.6 kN-m at a deflection level of 12.1 mm. At a deflection value of 12.1 mm, one of CFRP shear strips was completely de-bonded and fell apart.

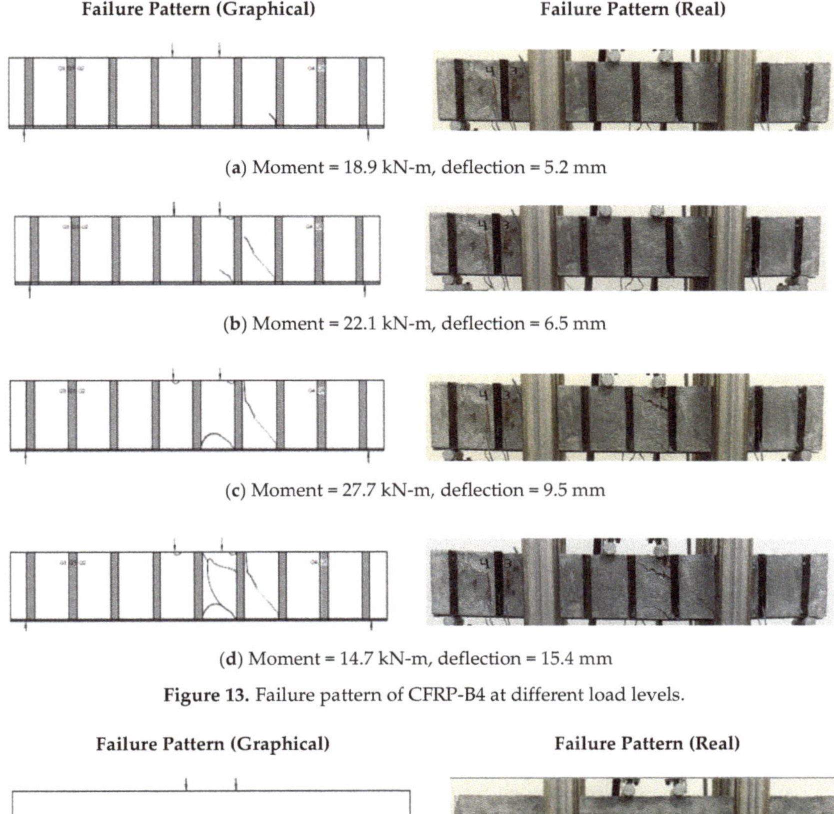

Failure Pattern (Graphical)   **Failure Pattern (Real)**

(**a**) Moment = 18.9 kN-m, deflection = 5.2 mm

(**b**) Moment = 22.1 kN-m, deflection = 6.5 mm

(**c**) Moment = 27.7 kN-m, deflection = 9.5 mm

(**d**) Moment = 14.7 kN-m, deflection = 15.4 mm

Figure 13. Failure pattern of CFRP-B4 at different load levels.

Failure Pattern (Graphical)   **Failure Pattern (Real)**

(**a**) Moment = 18.9 kN-m, deflection = 5.20 mm

(**b**) Moment = 22.9 kN-m, deflection = 6.59 mm

Figure 14. *Cont.*

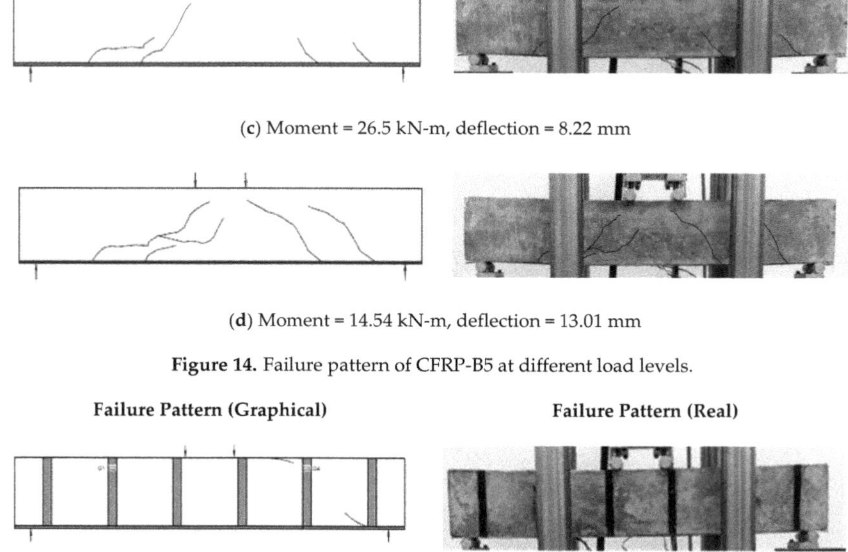

(**c**) Moment = 26.5 kN-m, deflection = 8.22 mm

(**d**) Moment = 14.54 kN-m, deflection = 13.01 mm

Figure 14. Failure pattern of CFRP-B5 at different load levels.

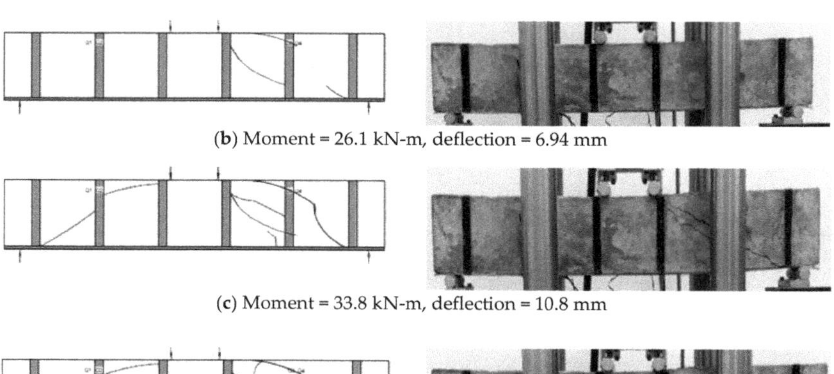

Failure Pattern (Graphical)   **Failure Pattern (Real)**

(**a**) Moment = 21.9 kN-m, deflection = 5.0 mm

(**b**) Moment = 26.1 kN-m, deflection = 6.94 mm

(**c**) Moment = 33.8 kN-m, deflection = 10.8 mm

(**d**) Moment = 31.6 kN-m, deflection = 12.1 mm

Figure 15. Failure pattern of CFRP-B6 at different load levels.

7. Conclusions

In this research study, a series of four-point bending tests were carried on reinforced concrete beams which were strengthened with different amounts and layouts of external CFRP flexural and shear reinforcement ratios. The layout of the CFRP application and the CFRP reinforcement ratio were the main parameters of the research study. Effects of these parameters were studied in terms of moment-curvature values, ultimate load carrying capacities, and failure patterns of the beams. A comparison between experimental and analytical ultimate load carrying capacities of the beams is also presented in this research study. The reinforced concrete beams were divided into two groups. Beams in group 1 had a CFRP flexural reinforcement ratio half of that of group 2 beams. However, the CFRP shear reinforcement was gradually increased in the reinforced concrete beams. Based on the results obtained from the experimental study, the following major conclusions are summarized.

- The choice of CFRP flexural reinforcement can influence the failure mode and strength gain of the RC beams. The ultimate load carrying capacities and the deformation capacities of the beams can increase to a certain level by increasing flexural or shear CFRP reinforcement. For this reason, a better understanding of beam governing mode of failure is necessary. For instance, a beam which is over-reinforced in flexure may result in the shear failure of the beam.
- Increasing the external CFRP shear reinforcement in reinforced concrete beams that have less flexural reinforcement did not result in a significant increase in strength. It was also observed that increasing CFRP flexural reinforcement in absence of external CFRP shear reinforcement did not result in a significant increase in strength.
- It was noted that the failure modes of the beams were highly dependent on the CFRP reinforcement layout. For the beams that had higher spacing between CFRP shear strips or no shear reinforcement, the failure occurred near the supports, in terms of diagonal shear cracks having a smaller angle of occurrence. On the other hand, for the beams that had smaller spacing between CFRP shear strips, the cracks that appeared were found to be steep and remained confined within the CFRP shear strips. The presence of CFRP shear strips hindered crack propagation, as well as resisted beam deformations.
- The flexural failure of the beams was initiated due to debonding of CFRP strips from the tensile face of the beams. The delamination of CFRP flexural strips started from the mid-span of the beam and moved laterally to the support, proving a bearing failure of the beam.
- The analytical procedure proposed by ACI R440.2-08 [33] was used to predict the moment and shear capacities of simply supported CFRP strengthened beams and a good agreement between the experimental and analytical ultimate load carrying capacities of the beam was obtained. Therefore, ACI 440.2R-08 [33] can be effectively used for the design and evaluation purposes of CFRP strengthened reinforced concrete beams.

Author Contributions: The most significant contributions in this research study were made by M.U.S. and H.J.Q. The experimental scheme of the research study was planned by M.U.S. and H.J.Q., M.N.A., K.K., and H.K. prepared the beam specimens and also cured the beams at later stages. M.U.S., H.J.Q., M.N.A., and K.K. tested the specimens and collected the data. Results were prepared and analyzed by M.U.S., H.J.Q and M.N.A. The original draft was prepared by M.U.S., H.J.Q., and H.K. The review and editing were later carried out by M.U.S. and H.J.Q.

Funding: This research was supported by the Deanship of Scientific Research (DSR) at King Faisal University (KFU) through its "Eighteenth annual research project no. 180061".

Acknowledgments: This research was supported by the Deanship of Scientific Research (DSR) at King Faisal University (KFU) through its "Eighteenth annual research project no. 180061". Authors are thankful to the Deanship of Scientific Research, King Faisal University for supporting this entire research study. In particular, authors are thankful to the Dean of Scientific Research, King Faisal University, who provided a lot of administrative support for the successful completion of this research project.

Conflicts of Interest: Authors reserves the complete right of the data and declare no conflict of interest with any private and public sector body.

Abbreviations

A_f	area of FRP external reinforcement, (mm^2)
A_{fv}	area of FRP shear reinforcement with spacing, (mm^2)
A_s	area of non-prestressed steel reinforcement, (mm^2)
a	depth of Whitney's stress diagram, (mm)
b_w	web width or diameter of circular section, (mm)
c	distance from extreme compression fiber to the neutral axis, (mm), $c = 3/8$ d for tension controlled failure of beams

d	distance from extreme compression fiber to centroid of tension reinforcement, (mm)
d_f	effective depth of FRP flexural reinforcement, (mm)
d_{fv}	effective depth of FRP shear reinforcement, (mm)
E_f	tensile modulus of elasticity of FRP, (MPa)
f_c'	specified compressive strength of concrete, (MPa)
f_f	effective flexural stress in the FRP; stress level attained at section failure, (MPa)
f_{fe}	effective shear stress in the FRP; stress level attained at section failure, (MPa)
f_y	specified yield strength of non-prestressed steel reinforcement, (MPa)
h	overall thickness or height of a member, (mm)
k_1	modification factor applied to κv to account for concrete strength
k_2	modification factor applied to κv to account for wrapping scheme
L_e	active bond length of FRP laminate, (mm)
M	applied moment using testing machine
M_n	nominal flexural strength, (N-mm)
n	number of or strips of FRP reinforcement
P_m	peak load of the beam calculated from nominal moment capacity of the beam
P_v	peak load of the beam calculated from nominal shear capacity
P_a	peak load of the beam determined analytically
P_e	peak load of the beam determined experimentally
t_f	nominal thickness of one ply of FRP reinforcement, (mm)
V_c	nominal shear strength provided by concrete with steel flexural reinforcement, (N)
V_f	nominal shear strength provided by FRP stirrups, (N)
V_n	nominal shear strength, lb (N)
V_s	nominal shear strength provided by steel stirrups, (N)
w_f	width of FRP reinforcing plies, (mm)
β_1	ratio of depth of equivalent rectangular stress block to depth of the neutral axis for $f_c' = 28$ MPa, $\beta_1 = 0.85$
ε_{cu}	ultimate axial strain of unconfined concrete corresponding to $0.85f_c'$ or maximum usable strain of unconfined concrete (mm/mm), which can occur at $0.85fc'$ or 0.003, depending on the obtained stress-strain curve
ε_{fd}	debonding strain of externally bonded FRP reinforcement, (mm/mm)
ε_f	Effective flexural strain level in FRP reinforcement attained at failure, (mm/mm)
ε_{fe}	effective shear strain level in FRP reinforcement attained at failure, (mm/mm)
ε_{fu}	design rupture strain of FRP reinforcement, (mm/mm)
κ_v	bond-dependent coefficient for shear
ψ_f	FRP strength reduction factor, 0.85 for shear (based on reliability analysis) for three-sided FRP U-wrap or two-sided strengthening schemes
δ	mid span deflection of beam
φ	curvature of the beam

References

1. Sulaimani, G.J.; Shariff, A.; Basanbul, I.A.; Baluch, M.H.; Ghaleb, B.N. Shear repair of reinforced concrete by fiber glass plate bonding. *Aci Struct. J.* **1994**, *91*, 458–464.
2. Chajes, M.J.; Jansuska, T.F.; Mertz, D.R.; Thomson, T.A.; Finch, J.R. Shear strength of RC beams using externally applied composite fabrics. *Aci Struct. J.* **1995**, *92*, 295–303.
3. Chen, J.F.; Teng, J.G. Shear capacity of FRP strengthened RC beams: FRP debonding. *Constr. Build. Mater.* **2003**, *17*, 27–41. [CrossRef]
4. Deniaud, C.; Cheng, J.J.R. Simplified shear design method for concrete beams strengthened with fiber reinforced polymer sheets. *J. Compos. Constr.* **2004**, *8*, 425–433. [CrossRef]
5. Eshwar, N.; Nanni, A.; Ibell, T.J. Performance of two anchor systems of externally bonded fiber-reinforced polymer laminates. *Aci Mater. J.* **2008**, *105*, 72–80.
6. Saleem, M.U.; Numada, M.; Amin, M.N.; Meguro, K. Shake Table Tests on FRP Retrofitted Masonry Building Models. *J. Compos. Constr.* **2016**, *20*, 1–19. [CrossRef]

7.   Takewaka, K.; Khin, M. Deterioration and stress rupture of FRP rods in alkaline solution simulating a concrete environment. In Proceedings of the 2nd International Conferences on Advanced Composite Materials in Bridges and Structures, Winnipeg, QC, Canada, 11–14 August 1996.

8.   Sen, R.; Shahawy, M.; Rosas, J.; Sukumar, S. Durability of aramid pretensioned elements in a marine environment. *Aci Struct. J.* **1998**, *95*, 578–587.

9.   Uomoto, T. Durability of FRP reinforcement as concrete reinforcement. In Proceedings of the International Conferences on FRP composites in Civil Engineering, Hong Kong, China, 12–15 December 2001.

10. Kang, T.H.K.; Howell, J.; Kim, S.; Lee, J. A state of the art review on debonding failures of FRP laminates externally adhered to concrete. *Int. J. Concr. Struct. Mater.* **2012**, *6*, 123–134. [CrossRef]

11. Arduini, M.; Nanni, A. Behaviour of precracked RC beams strengthened with carbon FRP sheets. *J. Compos. Constr.* **1997**, *1*, 63–73. [CrossRef]

12. Smith, S.T.; Teng, J.G. FRP strengthened RC beams: Assessment of debonding strength models. *Eng. Struct.* **2002**, *24*, 397–417. [CrossRef]

13. Teng, J.G.; Chen, J.; Smith, S.T.; Lam, L. Behaviour and strength of FRP strengthened RC structure: A state of the art review. *Struct. Build.* **2003**, *156*, 51–62. [CrossRef]

14. Kotynia, R. Debonding failures of RC beams strengthened with externally bonded strips. In Proceedings of the International Symposium on Bond Behavior of FRP in Structures, International institute for FRP in Constructions 2005, Hong Kong, China, 7–9 December 2005.

15. Teng, J.G.; Chen, J.F. Debonding failures of RC beams strengthened with externally bonded FRP reinforcement: Behavior and modelling. In Proceedings of the Asia pacific conference on FRP in Structures (APFIS 2007), Hong Kong, China, 12–14 December 2007.

16. Li, G.; Zhang, A.; Guo, Y. Debonding related strain limits for externally bonded FRP sheets in flexurally strengthened reinforced concrete beams. *Open Civ. Eng. J.* **2013**, *7*, 58–67. [CrossRef]

17. Hasnat, A.; Islam, M.M.; Amin, A.F.M.S. Enhancing the debonding strain limit for CFRP strengthened RC beams using U-clamps: Identification of design parameters. *J. Compos. Constr. ASCE* **2015**, *20*, 1–30. [CrossRef]

18. Mostafa, A.A.B.; Razaqpur, A.G. Finite element model for predicting post delamination Behavior in FRP-retrofitted beams in flexure. *Constr. Build. Mater.* **2017**, *131*, 195–204. [CrossRef]

19. Fu, B.; Chen, G.M.; Teng, J.G. Mitigation of intermediate crack debonding in FRP plated RC beams using FRP U-jackets. *Compos. Struct.* **2017**, *176*, 883–897. [CrossRef]

20. Abid, S.R.; Al-Lami, K. Critical review of strength and durability of concrete beams externally bonded with FRP. *Cogent Eng.* **2018**, *5*, 1–27. [CrossRef]

21. Wenwei, W.; Guo, L. Experimental study and analysis of RC beams strengthened with CFRP laminates under sustaining load. *Int. J. Solids Struct.* **2006**, *43*, 1372–1387. [CrossRef]

22. Lee, S.; Moy, S. A method for predicting the flexural strength of RC beams strengthened with carbon fiber reinforced polymer. *J. Reinf. Plast. Compos.* **2007**, *26*, 1383–1401. [CrossRef]

23. Barros, J.A.O.; Dias, S.J.E.; Lima, J.L.T. Analytical and numerical analysis of the behavior of RC beams flexural strengthened with CFRP. In Proceedings of the Fourth International Conference on FRP Composites in Civil Engineering, Zurich, Switzerland, 22–24 July 2008.

24. Mansour, M.H.; Mahmoud, K.A. Prediction of ultimate flexural capacity of reinforced concrete beams strengthened with CFRP laminate. *J. Eng. Sci. Assuit Univ.* **2008**, *36*, 1119–1132.

25. Pan, J.; Leung, C.K.Y.; Luo, M. Effect of multiple secondary cracks on FRP debonding from the substrate or reinforced concrete beams. *Constr. Build. Mater.* **2010**, *24*, 2507–2516. [CrossRef]

26. El-Ghandour, A.A. Experimental and analytical investigation of CFRP flexural and shear strengthening efficiencies of RC beams. *Constr. Build. Mater.* **2011**, *25*, 1419–1429. [CrossRef]

27. Kara, I.K.; Ashour, A.F. Flexural performance of FRP reinforced concrete beams. *Compos. Struct.* **2012**, *94*, 1616–1625. [CrossRef]

28. Pellegrino, C.; Vasic, M. Assessment of design procedures for the use of externally bonded FRP composites in shear strengthening of reinforced concrete beams. *Compos. Part B* **2013**, *45*, 727–741. [CrossRef]

29. Jung, K.; Hong, K.; Han, S.; Park, J.; Kim, J. Prediction of flexural capacity of RC beams strengthened in flexure with FRP fabric and cementitious matrix. *Int. J. Polym. Sci.* **2015**, *2015*. [CrossRef]

30. Foster, R.M.; Brindley, M.; Lees, J.M.; Ibell, T.J.; Morley, C.T.; Darby, A.P.; Evernden, M.C. Experimental investigation of reinforced concrete T-Beams strengthened in shear with externally bonded CFRP sheets. *J. Compos. Constr.* **2017**, *21*. [CrossRef]

31. Dias, S.J.E.; Barros, J.A.O. NSM shear strengthening technique with CFRP laminates applied in high T cross section RC beams. *Compos. Part B* **2017**, *114*, 256–267. [CrossRef]

32. Osman, B.H.; Wu, E.; Ji, B.; Abdulhameed, S.S. Effect of reinforcement ratios on shear behaviors of concrete beams strengthened with CFRP sheets. *HBRC J.* **2018**, *14*, 29–36. [CrossRef]

33. *Guide for the Design and Construction of Externally Bonded FRP System for Strengthening Concrete Structures*; American Concrete Institute: Farmington Hills, MI, USA, 2008; ACI 440.2R-08.

34. *Building Code Requirements for Structural Concrete and Commentary*; American Concrete Institute: Farmington Hills, MI, USA, 2008; ACI 318-08.

35. *Standard Test Method for Compressive Strength of Cylindrical Concrete Specimens*; ASTM International: West Conshohocken, PA, USA, 2018; ASTM C39-18.

36. *Standard Practice for Making and Curing Concrete Test Specimens in the Field*; ASTM International: West Conshohocken, PA, USA, 2018; ASTM C31-18.

37. Chellapandian, M.; Prakash, S.S.; Sharma, A. Experimental and finite element studies on the flexural behavior of reinforced concrete elements strengthened with hybrid FRP technique. *Compos. Struct.* **2019**, *208*, 466–478. [CrossRef]

38. Zhou, W.; Xie, X. Flexural response of continuous unbonded post- tensioned beams strengthened with CFRP laminates. *Compos. Struct.* **2019**, *211*, 455–468. [CrossRef]

39. Triantafillou, T.C. Shear Strengthening of Reinforced Concrete Beams Using Epoxy-Bonded FRP Composites. *ACI Struct J.* **1998**, *95*, 107–115.

Article

Determination of Fracture Properties of Concrete Using Size and Boundary Effect Models

Xiaofeng Gao [1], Chunfeng Liu [2], Yaosheng Tan [2], Ning Yang [2], Yu Qiao [2], Yu Hu [1,*], Qingbin Li [1,*], Georg Koval [3] and Cyrille Chazallon [3,4]

[1] State Key Laboratory of Hydroscience and Engineering, Tsinghua University, Beijing 100084, China; gaoxiaofeng@tsinghua.edu.cn

[2] China Three Gorges Projects Development Co., Ltd., Chengdu 610017, China; liu_chunfeng@ctg.com.cn (C.L.); tan_yaosheng@ctg.com.cn (Y.T.); yang_ning1@ctg.com.cn (N.Y.); qiao_yu@ctg.com.cn (Y.Q.)

[3] ICUBE Laboratory, CNRS, University of Strasbourg, 67084 Strasbourg, France; georg.koval@insa-strasbourg.fr (G.K.); cyrille.chazallon@insa-strasbourg.fr (C.C.)

[4] Shandong Provincial Key Laboratory of Road and Traffic Engineering, Shandong Jianzhu University, Jinan 250101, China

* Correspondence: yu-hu@tsinghua.edu.cn (Y.H.); qingbinli@tsinghua.edu.cn (Q.L.); Tel.: +86-010-62781161 (Y.H.); +86-010-62771015 (Q.L.)

Received: 26 February 2019; Accepted: 22 March 2019; Published: 29 March 2019

Abstract: Tensile strength and fracture toughness are two essential material parameters for the study of concrete fracture. The experimental procedures to measure these two fracture parameters might be complicated due to their dependence on the specimen size or test method. Alternatively, based on the fracture test results only, size and boundary effect models can determine both parameters simultaneously. In this study, different versions of boundary effect models developed by Hu et al. were summarized, and a modified Hu-Guan's boundary effect model with a more appropriate equivalent crack length definition is proposed. The proposed model can correctly combine the contributions of material strength and linear elastic fracture mechanics on the failure of concrete material with any maximum aggregate size. Another size and boundary model developed based on the local energy concept is also introduced, and its capability to predict the fracture parameters from the fracture test results of wedge-splitting and compact tension specimens is first validated. In addition, the classical Bažant's Type 2 size effect law is transformed to its boundary effect shape with the same equivalent crack length as Koval-Gao's size and boundary effect model. This improvement could extend the applicability of the model to infer the material parameters from the test results of different types of specimens, including the geometrically similar specimens with constant crack-length-to-height ratios and specimens with different initial crack-length-to-height ratios. The test results of different types of specimens are adopted to verify the applicability of different size and boundary effect models for the determination of fracture toughness and tensile strength of concrete material. The quality of the extrapolated fracture parameters of the different models are compared and discussed in detail, and the corresponding recommendations for predicting the fracture parameters for dam concrete are proposed.

Keywords: boundary effect; size effect; fracture toughness; tensile strength; concrete

---

1. Introduction

At present, several super-high arch dams have been constructed or are under construction in Southwest China [1–3]. To ensure the good working performance of concrete dams, one of the prerequisites is to prevent and control the formation and propagation of concrete cracks that may appear during the construction and operation periods. In fact, cracking is a very common, classical, but also complicated problem in concrete structures in real-life service conditions [4], which has been studied comprehensively from different perspectives by many researchers [5–9]. In terms of cracking risk analysis, the true fracture toughness K_c and tensile strength f_t of the concrete are the key material parameters that should be provided. The improper fracture properties may lead to a result with great deviation from the reality. Many test results showed that the fracture toughness K_c of concrete has obvious size effect, and tensile strength f_t usually depends on the specimen size and test method. To overcome the problem of size effect in fracture properties, it is often necessary to pour different sizes of concrete specimens to obtain the stable material parameter values. Thus, the test process is often complicated, especially for dam concrete, and the stable material parameters can only be measured from huge specimens [1,10]. In addition, the experimentally measured tensile strength of concrete varies for different test methods (uni-axial tension test, splitting test, flexure test, etc.) [11,12] due to the different fracture mechanisms. These size- or test method-dependent fracture parameters increase the difficulty of structure design and cracking risk analysis. Hence, it is crucial to find out a simple and relatively more accurate way to measure or predict the real fracture parameters of dam concrete.

Size effect laws (SEL) [13–17] and boundary effect models (BEM) [18,19] are two main asymptotic approaches to capture the size effect of concrete fracture. With these two kinds of models, the failure stress of structure with any specimen size or any crack size can be estimated, if the model parameters are fully known. On the contrary, if the test results of geometrically similar specimens with constant crack-length-to-height ratios or specimens with different crack to width ratios are available, size and boundary effect models can predict the material tensile strength f_t and fracture toughness K_c. The applicability of the different size and boundary effect models on the determination of fracture toughness and tensile strength have been carefully examined with many test results in the literature [20–23]. It is widely accepted that the size effect laws can provide the fracture toughness directly based on the geometrically similar fracture tests; however, in order to estimate the tensile strength, a material characteristic length l_{ch} must be assumed in advance [20]. The different versions of boundary effect models proposed by Hu et al. [19,21,23–25] can directly determine the tensile strength and fracture toughness. These models can be classified as the local and non-local models, according to the stress value adopted in the models to predict the failure due to the strength of material. The local models only use the point stress information at the crack tip, while an average stress information along a certain length emanated from the crack tip is needed for the non-local boundary effect models. The Koval-Gao's size and boundary effect model developed by Gao et al. [26,27] is also a local model, which adopts the derivative of energy release rate evaluated locally at the crack tip as the parameter for the study of fracture dominated by the strength of material. The Koval-Gao's size and boundary effect model can also predict simultaneously the fracture parameters from the fracture test results.

In this study, the local and non-local versions of boundary effect models developed by Hu et al. [19,21,23–25] are summarized, and a modified Hu-Guan's boundary effect model with a more appropriate equivalent crack length definition is proposed. The proposed modified Hu-Guan's boundary effect model considers the effect of maximum aggregate size on the equivalent crack length, thus changes the contributions of strength of material and linear elastic fracture mechanics (LEFM) on the specimen failure. This improvement is anticipated to be important when the maximum aggregate size of concrete getting larger. Another size and boundary effect model developed based on a local concept is also introduced [26,27], and its capability to predict the fracture parameters from the fracture test results of wedge-splitting and compact tension specimens is firstly validated. Besides, the classical Bažant's Type 2 size effect law is transformed to a shape similar to the boundary effect model by the authors, which could be easier to be used to determine the fracture parameters from fracture tests on specimens with any sample size or crack size . The test results of geometrically similar wedge-splitting and compact tension specimens with constant crack-length-to-height ratios [28–30] and specimens with different initial crack-length-to-height ratios [31–33] are adopted, to verify the applicability of the different boundary effect models on the determination of fracture toughness and tensile strength for concrete material.

2. Size and Boundary Effect Models

The size and boundary effects of quasi-brittle fracture have been studied systematically in recent decades [13,18,19,26,27], and some progresses have been achieved in recent years by considering the effect of aggregate size on the fracture behavior [21–24,34]. These models were developed initially for the nominal strength prediction of samples with various specimen size and crack size, and later have been extended to the determination of the real fracture parameters from fracture test results.

In this section, the local and non-local versions of boundary effect models developed by Hu et al. [19,21,23–25] are summarized. Then, a modified Hu-Guan's boundary effect model with a new equivalent crack length definition is proposed. Following by the brief introductions of Koval-Gao's size and boundary effect models and Bažant's Type 2 size effect law. All the models are written as their own linear forms, which can be used directly to extrapolate experimental results obtained from laboratory size specimens. The equivalent crack lengths of different models are compared in detail and the method for the determination of fracture parameters is presented at the end of this section.

2.1. Hu et al. Boundary Effect Models

2.1.1. Hu-Duan's Boundary Effect Model

The boundary effect models can characterize the effect of crack length on the failure load. Figure 1a shows a wedge-splitting (or compact tension) specimen with an initial crack length a_0 under its failure load P_{max}, and Figure 1b presents the two nominal stresses in the fracture analysis. The nominal strength σ_N of the specimen is defined without considering the existence of the initial crack a_0, which can be calculated by the equilibrium conditions of the forces and moments [23]:

$$\sigma_N = \frac{4P_{max}}{Wt},\tag{1}$$

where W is the sample size, t is the thickness of the specimen.

σ_{n1} in Figure 1b represents the stress at the crack tip without considering the stress singularity, and assuming a linear distribution along the ligament, which reads [23]:

$$\sigma_{n1} = \frac{P_{max}(4W + 2a_0)}{t(W - a_0)^2},\tag{2}$$

where a_0 is the initial crack length.

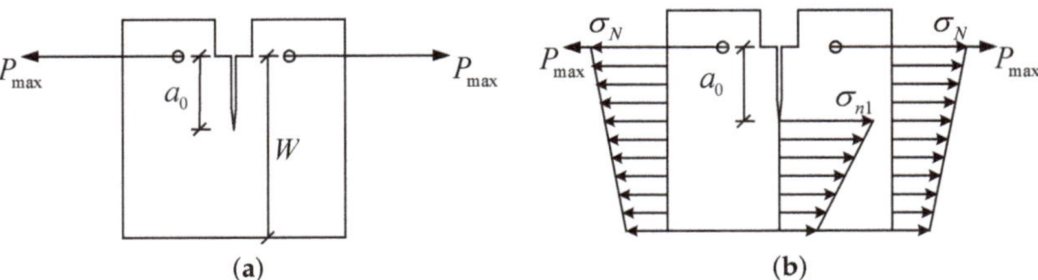

Figure 1. (a) A wedge-splitting specimen under its peak load and (b) the definition of two nominal stresses in Hu-Duan's boundary effect model.

In Hu-Duan's boundary effect model, the ratio of two nominal stresses σ_N and σ_{n1} is a dimensionless factor $A_1(\alpha)$ depending only on the initial crack length to sample size ratio α ($\alpha = a/W$).

For wedge-splitting (compact tension) specimen, $A_1(\alpha)$ can be calculated by relating Equations (1) and (2).

$$A_1(\alpha) = \frac{\sigma_N}{\sigma_{n1}} = \frac{2(1-\alpha)^2}{2+\alpha}, \tag{3}$$

The stress criterion adopted in Hu-Duan's boundary effect model assumes failure happens when σ_{n1} reaches its critical value, the material tensile strength f_t. Thus, one can obtain the nominal strength σ_N predicted by the stress criterion by the following expression:

$$\sigma_N = A_1(\alpha) f_t, \tag{4}$$

The criterion of LEFM is directly adopted as the energy part of Hu-Duan's boundary effect model. The nominal strength predicted by the LEFM criterion can be written as a function of crack length a_0:

$$\sigma_N = \frac{K_c}{Y(\alpha)\sqrt{\pi a_0}}, \tag{5}$$

where $Y(\alpha)$ is a geometrical factor. For wedge-splitting and compact tension specimen, the expression of $Y(\alpha)$ is identical [35]. This is because the geometry and loading conditions are the same for compact tension and wedge-splitting fracture tests [36]. It is given by:

$$Y(\alpha) = \frac{(2+\alpha)(0.886 + 4.64\alpha - 13.32\alpha^2 + 14.72\alpha^3 - 5.6\alpha^4)}{4\sqrt{\pi a}(1-\alpha)^{3/2}}. \tag{6}$$

To link Equations (4) and (5), a transition (or reference) crack length $a_\infty^\star$ needs to be introduced, which reads:

$$a_\infty^\star = \frac{1}{1.12^2 \pi} \left(\frac{K_c}{f_t} \right)^2. \tag{7}$$

The transition crack length $a_\infty^\star$ is a material constant proportional to material's characteristic length l_{ch} [37]. Thus, Equation (5) can be rewritten as:

$$\sigma_N = \frac{f_t}{\sqrt{[Y(\alpha)/1.12]^2 a_0 / a_\infty^\star}}, \tag{8}$$

The Hu-Duan's boundary effect model for the nominal strength prediction of structure with any crack length, can be obtained by relating Equations (4) and (8):

$$\sigma_N = \frac{A_1(\alpha) f_t}{\sqrt{1 + [A_1(\alpha) \times Y(\alpha)/1.12]^2 a_0 / a_\infty^\star}} = \frac{A_1(\alpha) f_t}{\sqrt{1 + a_{e1}/a_\infty^\star}}, \tag{9}$$

or

$$\sigma_{n1} = \frac{f_t}{\sqrt{1 + a_{e1}/a_\infty^\star}}, \tag{10}$$

where a_{e1} is the equivalent crack length and depends purely on the geometrical information of the specimen, which can be calculated by:

$$a_{e1} = \left[\frac{A_1(\alpha) \times Y(\alpha)}{1.12} \right]^2 a_0. \tag{11}$$

Equation (10) can be further rearranged as follows:

$$\frac{1}{\sigma_{n1}^2} = \frac{1}{f_t^2} + \frac{1.12^2 \pi}{K_c^2} a_{e1}, \tag{12}$$

Once the peak loads are recorded after the fracture tests, the equivalent crack length a_{e1} and nominal strength σ_{n1} are known accordingly. hence, the tensile strength f_t and fracture toughness K_c can be calculated from the intercept and slope of the linear regression.

2.1.2. Hu-Guan's (Hu-Wang's) Boundary Effect Model

The Hu-Guan's and Hu-Wang's boundary effect models [21,23,34] are developed for wedge-splitting and three-point bending tests respectively, which all assume a constant crack-bridging stress σ_{n2} within the partially developed fracture process zone or Δa_{fic} (see Figure 2a). In contrast to the local point stress value σ_{n1} (see Figure 1b) defined in Hu-Duan's boundary effect model, σ_{n2} (see Figure 2a) is associated with the fictitious crack length Δa_{fic}, and its value is affected by the choice of this length, thus becomes a non-local stress parameter. The strain condition along the crack plane, the equilibrium conditions of bending stress and bending moment were considered to determine this constant stress σ_{n2} [21,34]. It is given by [23]:

$$\sigma_{n2} = \frac{P_{max}(3W_2 + W_1)}{6t} \Big/ \left(\frac{W_1^2}{6} + \frac{\Delta a_{fic}}{6}W_1 + \frac{W - a_0}{2}\Delta a_{fic} \right), \tag{13}$$

with $W_1 = W - a_0 - \Delta a_{fic}$ and $W_2 = W + a_0 + \Delta a_{fic}$.

The advancement from Hu-Duan's boundary effect model to Hu-Guan's boundary effect model is the non-local nominal stress σ_{n2} used in the stress criterion. Thus, the local version of the boundary effect model was improved as a non-local version. Following the same idea of Hu-Duan's boundary effect model, Guan et al. proposed in [23] the way of calculating the tensile strength and fracture toughness by the following linear relation:

$$\frac{1}{\sigma_{n2}^2} = \frac{1}{f_t^2} + \frac{1.12^2\pi}{K_c^2}a_{e1}. \tag{14}$$

The local stress σ_{n1} is replaced by the non-local stress σ_{n2}, and a_{e1} is the same as the one in Hu-Duan's boundary effect model. It is admitted that Hu-Duan's boundary effect model ($\Delta a_{fic} = 0$) may overestimate the tensile strength of the material, and Hu-Guan's boundary effect model can lower the predicted tensile strength when Δa_{fic} increases [22]. This fictitious crack length Δa_{fic} was assumed to be proportional to the maximum aggregate size d_{max} of concrete [23]:

$$\Delta a_{fic} = \beta_1 d_{max}. \tag{15}$$

It is concluded in [23] that $\beta_1 = 1$ is a good approximation when the maximum aggregate size d_{max} is the dominant aggregate size of a concrete mix and the W/d_{max} ratio is below 20 (or even 50).

2.1.3. Hu-Zhang's Boundary Effect Model

Hu-Zhang's boundary effect model [24,25] added two more assumptions about the length scale than Hu-Guan's boundary effect model. That are, the fictitious crack length $\Delta a_{fic} = \beta_2 d_{av} \approx 1.5 d_{av}$ (see Figure 2b, d_{av} is the average aggregate size) is accurate enough when the ratio of the sample size and the average aggregate size varies in a certain range, and the transition crack length $a_\infty^\star \approx 3 d_{av}$ is appropriate to consider the micro-structure influence of granite and fine-grained polycrystalline ceramics studied in [24,25]. Based on these two assumptions, if the average aggregate size d_{av} (transition crack length $a_\infty^\star$) and tensile strength f_t are decided, then one can directly estimate the fracture toughness K_c by Equation (7).

Since the average aggregate sizes are not known for the concrete fracture tests adopted in this study, the Hu-Zhang's boundary effect model will not be used to estimate the tensile strength and fracture toughness in the next sections.

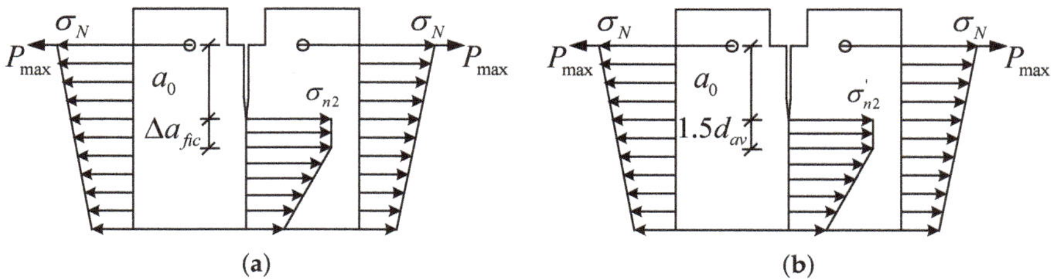

Figure 2. The definition of two nominal stresses in (**a**) Hu-Guan's boundary effect model and (**b**) Hu-Zhang's boundary effect model.

2.1.4. Proposed Modified Hu-Guan's Boundary Effect Model

Based on Equation (1) and Equation (13) in Hu-Guan's boundary effect model, the nominal strength σ_N can be rewritten as a function of σ_{n2}:

$$\sigma_N = \sigma_{n2} \times \left[\frac{2(1-\alpha)(1-\alpha+2\Delta\alpha_{fic})}{2+\alpha+\Delta\alpha_{fic}} \right], \tag{16}$$

with

$$A_2(\alpha,\Delta\alpha_{fic}) = \frac{2(1-\alpha)(1-\alpha+2\Delta a_{fic})}{2+\alpha+\Delta\alpha_{fic}}. \tag{17}$$

Unlike $A_1(\alpha)$ in Equation (3), $A_2(\alpha,\Delta\alpha_{fic})$ is no longer just geometry-related, but also material-related by the fictitious crack length Δa_{fic}, which is assumed to be proportional to the maximum aggregate size d_{max}. In the equivalent crack length calculation, to be more appropriate, it is recommended in this study to use $A_2(\alpha,\Delta\alpha_{fic})$ instead of $A_1(\alpha)$, thus, the modified equivalent crack length a_{e2} would be:

$$a_{e2} = \left[\frac{A_2(\alpha,\Delta\alpha_{fic}) \times Y(\alpha)}{1.12} \right]^2 a_0. \tag{18}$$

In the linear regression of tensile strength f_t and fracture toughness K_c calculation, a_{e1} should be replaced by a_{e2} in this modified Hu-Guan's boundary effect model:

$$\frac{1}{\sigma_{n2}^2} = \frac{1}{f_t^2} + \frac{1.12^2\pi}{K_c^2}a_{e2}. \tag{19}$$

2.2. Koval-Gao's Size and Boundary Effect Model

The Koval-Gao's size and boundary effect model [26,27,38] has two different forms. The second form is similar to the traditional Type 2 size effect law [14], which can be used to predict the size effect induced by the specimen size. The first form is close to the boundary effect model proposed by Hu et al., which can predict simultaneously the tensile strength and fracture toughness from the fracture test results. The Koval-Gao's size and boundary effect model was developed initially for cracked samples [26] by relating the contribution of the tensile strength and fracture energy. The contribution of the tensile strength is obtained from the derivative of the energy release rate, which is evaluated locally at the crack tip. While the contribution of the fracture energy is simply quantified by the LEFM criterion, same as the boundary effect models developed by Hu et al. The Koval-Gao's size and boundary effect model was further improved as a local fracture criterion in [27,38], which can be easily applied for any geometry. The first form of Koval-Gao's size and boundary effect model reads:

$$\sigma_N = \frac{1.12 f_t}{H(\alpha)\sqrt{1 + a_{e3}/a_\infty^\star}}, \tag{20}$$

where $H(\alpha) = \sqrt{Y^2(\alpha) + 2Y(\alpha) \times dY(\alpha)/d\alpha \times \alpha}$, and the equivalent crack length a_{e3} is defined as:

$$a_{e3} = \frac{Y^2(\alpha)}{H^2(\alpha)} \times a_0. \tag{21}$$

This model can also be written as a linear form:

$$\frac{1}{\sigma_{n3}^2} = \frac{1}{f_t^2} + \frac{1.12^2 \pi}{K_c^2} a_{e3}, \tag{22}$$

with

$$\sigma_{n3} = \frac{H(\alpha)}{1.12} \sigma_N. \tag{23}$$

So, the material tensile strength f_t and fracture toughness K_c can be calculated from the intercept and slope of the linear regression.

2.3. Bažant's Type 2 Size Effect Law

Bažant's Type 2 size effect law was originally developed for the size effect study of geometrically similar specimens [13]. The law reads

$$\sigma_N = \hat{B} f_t \left(1 + \frac{W}{W_0}\right)^{-1/2}, \tag{24}$$

where $\hat{B}$ is a geometrical dimensionless constant; W_0 is a material constant proportional to l_{ch}. $\hat{B} f_t$ and W_0 were given as follows in [15]:

$$\hat{B} f_t = \sqrt{\frac{E G_c}{g'(\alpha) c_f}}, \tag{25}$$

$$W_0 = \frac{c_f g'(\alpha)}{g(\alpha)}, \tag{26}$$

where $g(\alpha) = K_I^2(\alpha) W(t/P)^2 = Y^2(\alpha) \pi \alpha$ is defined as the dimensionless energy release function of LEFM; the length parameter c_f is proportional to l_{ch}, which is the effective size of fracture process zone [15]. $g'(\alpha_0)$ can be written as a function of the geometrical correction factor $H(\alpha)$ presented in Equation (20):

$$g'(\alpha) = H^2(\alpha) \pi. \tag{27}$$

After the substitution of Equations (7), (25)–(27) into Equation (24), the type 2 size effect law can be finally transformed to a shape similar to the boundary effect model:

$$\sigma_N = \frac{1.12 f_t \sqrt{a_\infty^\star / c_f}}{H(\alpha)} \left(1 + \frac{a_{e3}}{c_f}\right)^{-1/2}, \tag{28}$$

where a_{e3} shares the same definition as the one presented in Koval-Gao's size and boundary effect model. This form of the Bažant's Type 2 size effect law can be further written as:

$$\frac{1}{\sigma_{n4}^2} = \frac{c_f}{K_c^2} + \frac{1}{K_c^2} a_{e3}, \tag{29}$$

with

$$\sigma_{n4} = \sqrt{\pi}H(\alpha)\sigma_N. \tag{30}$$

Thus, the fracture toughness K_c and the effective size of the fracture process zone c_f can be identified from the size effect or boundary effect fracture test results. To obtain the tensile strength of the material, a ratio γ_1 ($\gamma_1 = c_f/l_{ch}$) between c_f and the material characteristic length l_{ch} should be assumed. It is reported in [20,39,40] that γ_1 can be 0.28 or 0.29 for concrete material. Since $l_{ch} = 1.12^2 \pi a_\infty^\star$, another ratio $\gamma_2 = c_f/a_\infty^\star$ can also be obtained, which is close to 1 when $\gamma_1 \approx 0.28 \; or \; 0.29$ ($\gamma_2 \approx 1.10 \; or \; 1.14$), means the effective size of the fracture process zone may be roughly equal to the transition crack size.

It is interesting to notice that the effective size of the fracture process zone c_f in Bažant's Type 2 size effect law might be able to link to the fictitious crack length Δa_{fic} proposed in Hu-Guan' (Hu-Wang') boundary effect model (see Section 2.1.2). A reasonable choice of γ_1 or γ_2 would provide a better estimation of material tensile strength f_t. The theoretical and experimental studies can be done in the future work to find out the approximate relation between c_f and maximum aggregate size d_{max} or average aggregate size d_{av}.

2.4. Comparison of Different Equivalent Crack Lengths

Different size and boundary effect models have adopted different expressions for the calculation of equivalent crack lengths. a_{e1} (Equation (11)) in Hu-Guan's boundary effect model depends only on the crack-length-to-height ratio α, while a_{e2} (Equation (18)) proposed in modified Hu-Guan's boundary effect model believes the equivalent crack length depends on both the crack-length-to-height ratio α and fictitious crack-length-to-height ratio $\Delta\alpha_{fic}$, thus becomes geometry- and material-related. The expression a_{e3} (Equation (21)) in Koval-Gao's size and boundary effect model, and Bažant's Type 2 size effect law is identical, which is only geometry-dependent. To better compare the different equivalent crack lengths, a wedge-splitting specimen with $W = 1000$ mm (see Figure 1a) is taken as an example, and its initial crack length varies from 50 mm to 950 mm. The variations of the equivalent crack length for different maximum aggregate size d_{max} with fixed discrete number β_1, and variable β_1 with fixed d_{max} are plotted respectively in Figures 3 and 4.

In different constructions, the maximum aggregate size d_{max} can be very different. For concrete used in Wudongde and Baihetan super-high arch dams, $d_{max} = 40$ mm, 80 mm or 150 mm for different parts of the dam. Meanwhile for the relatively smaller structures such as piers or beams, the frequently used d_{max} is only 10 mm or 20 mm. These possible d_{max} are all considered in this section, and the corresponding a_{e2} are plotted in Figure 3, together with the material (d_{max}) independent equivalent crack lengths a_{e1} and a_{e3}. For the original Hu-Guan's boundary effect model, a_{e1} reaches its maximum value for any d_{max} when the crack-length-to-height ratio α is around 0.2, means for this ratio, the LEFM contributes the most to the failure of the specimen, thus should be recommended as the initial ratio for the fracture test on wedge-splitting specimen. This recommended ratio for Koval-Gao's size and boundary effect model, and Bažant's Type 2 size effect law is 0.3 to 0.4, which is closer to the usual initial crack-length-to-height ratio reported in the references [1,10]. In terms of a_{e2} proposed in this study, it is obvious that it is different from a_{e1}, and this difference becomes larger as the increase of d_{max}. Hence, the proposed modified Hu-Guan's boundary effect model would be more appropriate than its original version, due to the fact that the two failure mechanisms are correctly combined even for the fracture analysis of dam concrete with maximum aggregate size reaches 150 mm. However, it should be admitted that for d_{max} such as 10 mm or 20 mm, the difference in a_{e2} and a_{e1} may has neglectable effect on the fracture properties estimated by the original and modified Hu-Guan's boundary effect model. In addition, the increase of a_{e2} at the tail part can be observed in Figure 3. This trend appears naturally due to its definition. An increase a_{e2} could maintain the validity of the failure mechanism of LEFM for all the possible crack-length-to-height ratios.

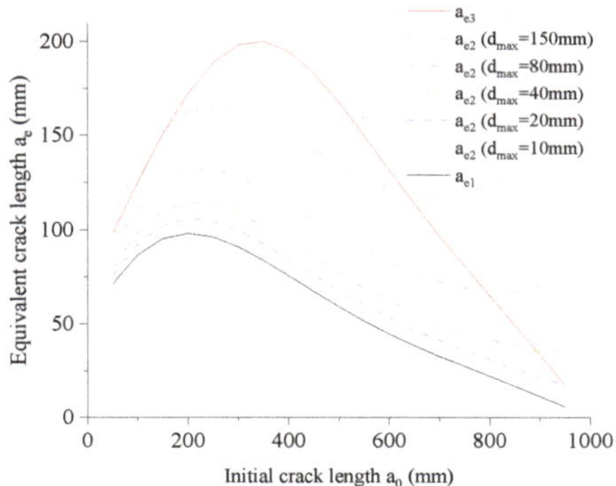

Figure 3. Equivalent crack length a_e versus initial crack length a_0 for different maximum aggregate size d_{max} when $\beta_1 = 1$.

Figure 4 shows the equivalent crack length a_e versus initial crack length a_0 for different discrete number β_1 when $d_{max} = 150$ mm. a_{e1} and a_{e3} are the same as the ones plotted in Figure 3 since they are independent on the fictitious crack length. β_1 ranges from 0.25 to 2 is considered, which is a reasonable range for concrete material [23]. As shown in Figure 4, a_{e2} is strongly affected by the choice of β_1 when $d_{max} = 150$ mm, thus may provide very different fracture parameters for different boundary effect models. It should be noted that according to the definition of σ_{n2} (see Figure 2a) in original and modified Hu-Guan's boundary effect model, the fictitious crack length Δa_{fic} should be smaller than the length of the ligament, so the Hu-Guan's boundary effect model cannot be applied directly to some special cases (the tail part of a_{e2} in Figure 4) when the ligament is not enough to distribute the stress. This might be a shortcoming of the non-local models, but in fact, can be rarely encountered in the analysis of usual fracture test results.

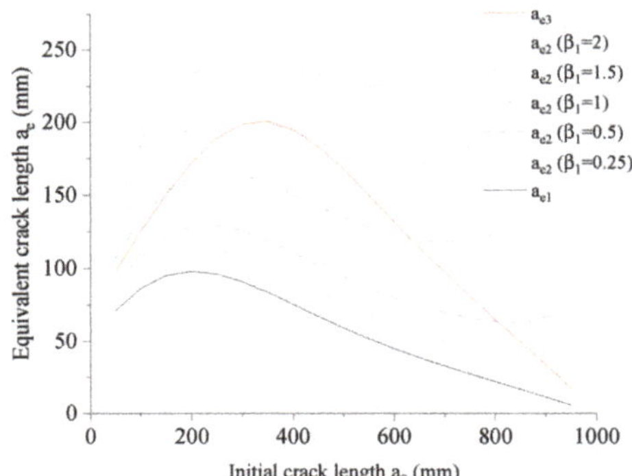

Figure 4. Equivalent crack length a_e versus initial crack length a_0 for different discrete number β_1 when $d_{max} = 150$ mm.

2.5. Method for the Determination of Fracture Parameters

The fracture tests on geometrically similar specimens with constant crack-length-to-height ratios, and specimens with different initial crack-length-to-height ratios are frequently performed to study the dependence of fracture behaviors on specimen sizes and initial crack lengths. The test results obtained from these two types of specimens can be used to extrapolate the tensile strength f_t and fracture toughness K_c by the size and boundary effect models.

The linear forms of original Hu-Guan's boundary effect model (Equation (14)), modified Hu-Guan's boundary effect model (Equation (19)), Koval-Gao's size and boundary effect model (Equation (22)), and Bažant's Type 2 size effect law (Equation (29)) have presented previously in this section. Only the maximum loads and geometrical information of the tested specimens are needed for the linear regression analysis. The nominal stresses in the different linear models can be calculated by Equation (13) (original and modified Hu-Guan's boundary effect model), Equation (23) (Koval-Gao's size and boundary effect model), and Equation (30) (Bažant's Type 2 size effect law). In addition, the different equivalent crack lengths can be calculated by Equation (11) (original Hu-Guan's boundary effect model), Equation (18) (modified Hu-Guan's boundary effect model), and Equation (21) (Koval-Gao's size and boundary effect model and Bažant's Type 2 size effect law). With these two known values from the test results and specimens' geometrical information, one can do the linear regression analysis by the different models. Thus, the intercept and slope obtained from the best linear fit can be used to predict the material properties.

It should be noted that for the original and modified Hu-Guan's boundary effect models, the extrapolated fracture parameters vary together as the adjustment of the discrete number β_1. Thus, special attention should be paid to the regression analysis, to ensure the predicted tensile strength and fracture toughness are all acceptable. For Bažant's Type 2 size effect law, as introduced in Section 2.3, only the fracture toughness K_c and the effective size of the fracture process zone c_f can be obtained from the linear fit. To estimate the tensile strength f_t of the material, a ratio γ_1 ($\gamma_1 = c_f / l_{ch}$) between c_f and the material characteristic length l_{ch} should be assumed. Then f_t can be calculated as follows:

$$f_t = \sqrt{\frac{\gamma_1 K_c^2}{c_f}}. \tag{31}$$

Koval-Gao's size and boundary effect model is less flexible than the other two models, because the liner fit is unique for one set of the test results, and one can calculate the tensile strength from the intercept and fracture toughness from the slope, respectively.

3. Model Verification and Comparison

The test results of geometrically similar wedge-splitting [28,29] and compact tension specimens [30] with constant crack-length-to-height ratios, and wedge-splitting specimens with different initial crack-length-to-height ratios [31–33] are adopted, to verify the applicability of the different size and boundary effect models on the determination of fracture properties for concrete material.

3.1. Tests of Geometrically Similar Specimens with Constant Crack-Length-to-Height Ratios

3.1.1. Zhang's Experiments

Zhang et al. [28,29] carried out a series of experiments with geometrically similar wedge-splitting specimens, to investigate the size effects of concrete with small aggregate size. The height of the specimens ranges from 150 mm to 1200 mm, with the maximum aggregate size of 10 mm. There were 24 specimens in total, which were divided into 7 groups according to their heights, and each group had at least 3 samples. The compressive strength at 60 days was 29.56 MPa.

The sample dimensions, test results, equivalent crack lengths and nominal stresses in different boundary effect models are listed in Table 1. It should be noted that the dimensions of different samples in one group were slightly different due to the size deviations of the molds.

Table 1. Geometry, test results and equivalent crack length of wedge-splitting specimens [28,29].

Label	W (mm)	a_0/W	t (mm)	P_{max} (kN)	a_{e1} (mm)	a_{e2} (mm)	a_{e3} (mm)	σ_N (Mpa)	σ_{n2} (Mpa)
WS150-1	150	0.4638	200	4.73405	9.75	14.41	26.89	0.631	2.225
WS150-2	150	0.477	193	5.59249	9.43	14.08	26.27	0.773	2.863
WS150-3	150	0.4868	200	5.22936	9.19	13.84	25.8	0.697	2.683
WS200-1s	200	0.4579	200	7.08092	13.19	17.78	36.2	0.708	2.551
WS200-2s	200	0.4703	200	6.885	12.79	17.36	35.45	0.689	2.601
WS200-3d	200	0.4703	200	6.61111	12.79	17.36	35.45	0.661	2.498
WS400-1	400	0.4527	200	11.37149	26.73	31.19	72.99	0.569	2.155
WS400-2	400	0.4602	198	14.27358	26.23	30.69	72.12	0.721	2.814
WS400-3	400	0.4869	199	12.00834	24.5	28.93	68.78	0.603	2.623
WS600-1	600	0.4618	193	17.82212	39.19	43.6	107.9	0.616	2.48
WS600-2	599	0.4551	200	16.97275	39.78	44.21	108.9	0.567	2.223
WS600-3	600	0.4551	193	16.68785	39.85	44.27	109.08	0.576	2.261
WS600-5	599	0.4551	200	18.22867	39.78	44.21	108.9	0.609	2.387
WS800-1	799	0.4624	196	18.65977	52.11	56.5	143.54	0.477	1.95
WS800-2	800	0.4624	194	22.60376	52.17	56.56	143.72	0.583	2.384
WS800-4	798	0.4539	200	21.62546	53.16	57.56	145.35	0.542	2.143
WS800-5	801	0.4577	200	21.98917	52.86	57.25	145.02	0.549	2.204
WS1000-1	997	0.4521	200	22.14598	66.71	71.1	182.11	0.444	1.757
WS1000-3	997	0.4531	200	23.33836	66.55	70.94	181.82	0.468	1.86
WS1000-4	999	0.4522	196	25.81247	66.83	71.22	182.44	0.527	2.087
WS1000-5	1000	0.4506	200	24.4032	67.16	71.56	183.08	0.488	1.919
WS1200-0	1198	0.4597	200	24.7045	78.66	83.03	216.19	0.412	1.691
WS1200-1	1200	0.4554	201	24.2143	79.64	84.02	218.05	0.402	1.618
WS1200-2	1200	0.4545	200	28.2619	79.82	84.2	218.36	0.471	1.891

To obtain simultaneously the tensile strength and fracture toughness, the linear regressions of original Hu-Guan's boundary effect model ($\Delta a_{fic} = d_{max}$), modified Hu-Guan's boundary effect model ($\Delta a_{fic} = d_{max}$), Koval-Gao's size and boundary effect model, and Bažant's Type 2 size effect law are plotted respectively in Figure 5. The estimated materials parameters (see Figure 5a) for original Hu-Guan's boundary effect model ($\beta_1 = 1$) are: $f_t = 3.05$ MPa, $K_c = 1.25$ MPa$\sqrt{m}$, $a_\infty^\star = 42.9$ mm and $l_{ch} = 168.9$ mm. It should be noticed that the results of Hu-Guan's boundary effect model are slightly different from the results presented in [23], because in this analysis, the more accurate sample dimensions and peak loads are adopted from [28,29]. For modified Hu-Guan's boundary effect model (see Figure 5b), when $\beta_1 = 1$ ($\Delta a_{fic} = d_{max}$), one can obtain $f_t = 3.23$ MPa, $K_c = 1.25$ MPa$\sqrt{m}$, $a_\infty^\star = 38.1$ mm and $l_{ch} = 150.0$ mm. For Koval-Gao's size and boundary effect model, $f_t = 2.36$ MPa, $K_c = 1.16$ MPa$\sqrt{m}$, $a_\infty^\star = 61.6$ mm and $l_{ch} = 242.9$ mm can be calculated from the intercept and slope shown in Figure 5c. Bažant's Type 2 size effect law (see Figure 5d) gives $K_c = 1.16$ MPa$\sqrt{m}$ and $c_f = 61.6$ mm (Equation (29)). As already introduced previously in Sections 2.3 and 2.5, in order to obtain the tensile strength f_t of the material, a ratio γ_1 should be assumed, then f_t can be calculated by Equation (31). It is obvious that mathematically f_t would increase monotonically as the increase of γ_1. For the test results analyzed in this section, $\gamma_1 = 0.1$ and 1 gives $f_t = 1.48$ MPa and 4.69 MPa, respectively. If $\gamma_1 = 0.28$ or 0.29, the corresponding $f_t = 2.48$ *or* 2.52 MPa, which is close to the estimation of Koval-Gao's size and boundary effect model. The transition crack length $a_\infty^\star = 55.86$ mm and material characteristic length $l_{ch} = 220.13$ mm can also be calculated for the chosen $\gamma_1 = 0.28$.

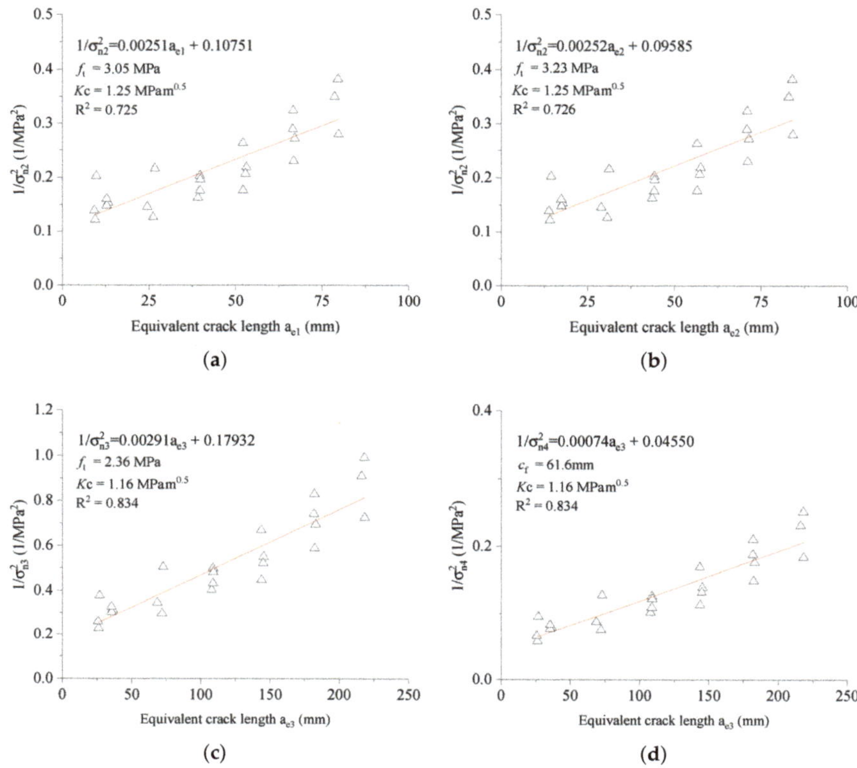

Figure 5. Test results in [28,29] and the corresponding fitted curves of (**a**) Hu–Guan's boundary effect model with $\beta_1 = 1$, (**b**) modified Hu–Guan's boundary effect model with $\beta_1 = 1$, (**c**) Koval-Gao's size and boundary effect model and (**d**) Bažant's Type 2 size effect law.

The tensile strength was not measured in [29], alternatively, we can estimate it from the measured compressive strength by $f_t = 0.24 f_{cu}^{3/2} = 2.29$ MPa [31], which is only 2.9% difference with the estimation of Koval-Gao's size and boundary effect model. The Hu–Guan's boundary effect model and its modified version still somehow overestimate the material tensile strength with $\beta_1 = 1$. However, as the increase of Δa_{fic}, the estimated f_t would decrease accordingly, as what has been done in [23], to test the effect of different Δa_{fic} on the predicted tensile strength.

For fracture toughness K_c, when the sample size is larger than 600 mm, the nominal fracture toughness K_c tends to be stable, and its value ranges from 0.93 to 1.13 MPa$\sqrt{m}$, hence, $K_c = 1.16$ MPa$\sqrt{m}$ estimated by the Koval-Gao's model and Bažant model is acceptable. While for the original Hu–Guan model and its modified version, the estimated $K_c = 1.25$ MPa$\sqrt{m}$ when $\Delta a_{fic} = d_{max}$ seems slightly higher, since the largest measured K_c is only 1.09 MPa$\sqrt{m}$ for the largest specimen with $W = 1200$ mm, whose failure can be assumed as totally LEFM control. Nevertheless, as the variation of the value of Δa_{fic} (or β_1), the best estimations of f_t and K_c given by Hu–Guan's boundary effect model can be achieved.

It is worthwhile to mention that the adjustment of β_1 in original and modified Hu–Guan's boundary effect models would alter the predicted f_t and K_c values simultaneously, to be more exact, as the increase of β_1, the estimated f_t would decrease and K_c would increase. This is simply due to the model assumption about the competition between two failure mechanisms (strength of material and

LEFM). A larger fictitious crack length gives a lower non-local stress σ_{n2} (see Figure 2a) distributed along the fictitious crack, thus reduces the contribution of strength of material on the specimen failure, and finally leads to the different combination of predicted fracture parameters. This feature may introduce some difficulties to directly adopt the extrapolated fracture parameters to the numerical simulation of a real structure, since one may suspend the veracity of the estimated parameters. Therefore, before using the Hu-Guan's boundary effect model to extrapolate the fracture parameters, it is recommended to fix one parameter by tests or give a narrow range to it in advance. In this way, the other parameter can be more reliable, and a reasonable β_1 is naturally obtained.

In terms of fracture parameters extrapolation, Bažant's Type 2 size effect law is also aided by a length assumption ($c_f = \gamma_1 l_{ch}$). However, the different choice of γ_1 would change the prediction of f_t only. It is reported in [20,39,40] that γ_1 is about 0.28 for concrete material, but as the increase of maximum aggregate size, especially for the dam concrete with $d_{max} = 150$ mm, $\gamma_1 \approx 0.28$ should be better verified by more experiential results. The third model, Koval-Gao's size and boundary effect model only provides one set of fracture parameters, which seems to be less flexible than the other models, but its estimated parameters are acceptable for this case.

3.1.2. Wittmann's Experiments

Wittmann et al. [30] performed the fracture tests on geometrically similar compact tension specimens. The heights of the specimens were 300 mm, 600 mm and 1200 mm, with 6 samples for each specimen size and the maximum aggregate size was 16 mm. The measured compressive strength at 28 days was 42.9 MPa. The sample dimensions, test results and model parameters are listed in Table 2. Only the mean value of the maximum loads of each group (6 samples) has been reported in [30].

Table 2. Geometry, test results and equivalent crack length of compact tension specimens [30].

Series	W (mm)	a_0/W	t (mm)	P_{max} (kN)	a_{e1} (mm)	a_{e2} (mm)	a_{e3} (mm)	σ_N (MPa)	σ_{n2} (MPa)
Small	300	0.5	120	7.30	17.76	42.18	50.27	0.811	2.631
Medium	600	0.5	120	12.60	35.51	58.10	100.54	0.700	2.736
Large	1200	0.5	120	20.70	71.02	92.58	201.09	0.575	2.518

The linear regressions of different size and boundary effect models are plotted in Figure 6. When $\beta_1 = 1$, the estimated materials parameters (see Figure 6a) for original Hu-Guan's boundary effect model are: $f_t = 3.83$ MPa, $K_c = 2.10$ MPa$\sqrt{m}$, $a_\infty^\star = 76.2$ mm, $l_{ch} = 300.3$ mm, and for modified Hu-Guan's boundary effect model (see Figure 6b), the extrapolated parameters are: $f_t = 4.03$ MPa, $K_c = 2.09$ MPa$\sqrt{m}$, $a_\infty^\star = 68.3$ mm and $l_{ch} = 269.3$ mm. Koval-Gao's size and boundary effect model gives $f_t = 2.93$ MPa, $K_c = 1.87$ MPa$\sqrt{m}$, $a_\infty^\star = 103.3$ mm and $l_{ch} = 407.2$ mm (see Figure 6c). Bažant's Type 2 size effect law (see Figure 6d) gives $K_c = 1.87$ MPa$\sqrt{m}$ and $c_f = 103.34$ mm. In addition, the tensile strength $f_t = 3.09$ MPa is obtained for $\gamma_1 = 0.28$.

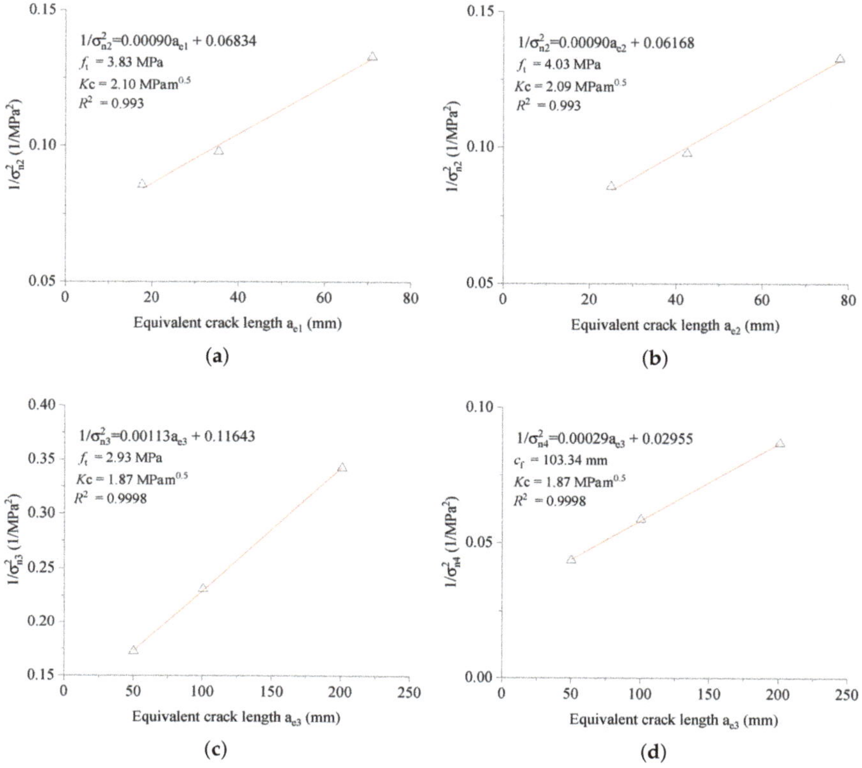

Figure 6. Test results in [30] and the corresponding fitted curves of (**a**) Hu-Guan's boundary effect model with $\beta_1 = 1$, (**b**) modified Hu-Guan's boundary effect model with $\beta_1 = 1$, (**c**) Koval-Gao's size and boundary effect model and (**d**) Bažant's Type 2 size effect law.

The tensile strength can be estimated from the compressive strength by $f_t = 0.24 f_{cu}^{3/2} = 2.94$ MPa [31]. Similar to the estimated results of Zhang's experiments (see Section 3.1.1), the extrapolated tensile strengths of Koval-Gao's model and Bažant's model seem to be more reasonable than the values given by the original and modified Hu-Guan's boundary effect models. In terms of fracture toughness, the difference of the values given by the different models is within 15%, therefore, the model predictions are all acceptable. However, for the largest tested specimen ($W = 1200$ mm), the measured K_c is only 1.521 MPa$\sqrt{m}$. According to the experimental size effect study [1,10], for such a relatively huge specimen with ligament length ($0.5W$) to maximum aggregate size (d_{max}) ratio equals 37.5, the measured fracture toughness should be close to the real material fracture toughness already. Therefore, the lower fracture toughness prediction given by Koval-Gao's model and Bažant's model might be more appropriate. As already mentioned in Section 3.1.1, a choice of a smaller β_1 could lower the prediction of fracture toughness given by the Hu-Guan's model, but at the same time, an unrealistic larger tensile strength would be obtained. This issue of the Hu-Guan's boundary effect model should be further studied.

3.2. Tests of Specimens with Different Initial Crack-Length-to-Height Ratios

3.2.1. Xu's Experiments

Xu et al. [31] performed the wedge-splitting tests made of concrete with different initial crack-length-to-height ratios. The maximum aggregate size used was 20 mm. The height and thickness were fixed for all the specimens ($W = 170$ mm, $t = 200$ mm). There were three different initial crack lengths a_0, which were 60 mm, 80 mm and 100 mm, respectively. The measured compressive strength f_{cu} is 47.96 MPa, and the tensile strength f_t was estimated as $f_t = 0.24 f_{cu}^{3/2} = 3.17$ MPa by Xu et al. [31]. Details of the specimen dimensions, test results and equivalent crack lengths are listed in Table 3.

Table 3. Geometry, test results and equivalent crack length of wedge-splitting specimens [31].

Label	W (mm)	a_0/W	t (mm)	P_{max} (kN)	a_{e1} (mm)	a_{e2} (mm)	a_{e3} (mm)	σ_N (MPa)	σ_{n2} (MPa)
WS70-1	170	0.353	200	10.71	14.16	23.89	34.05	1.260	2.727
WS70-2	170	0.353	200	9.65	14.16	23.89	34.05	1.135	2.457
WS70-3	170	0.353	200	11.02	14.16	23.89	34.05	1.296	2.806
WS90-1	170	0.471	200	6.56	10.85	20.63	30.1	0.772	2.471
WS90-2	170	0.471	200	6.77	10.85	20.63	30.1	0.796	2.550
WS90-3	170	0.471	200	6.96	10.85	20.63	30.1	0.819	2.621
WS110-1	170	0.588	200	4.09	7.88	17.81	23.11	0.481	2.441
WS110-2	170	0.588	198	4.36	7.88	17.81	23.11	0.513	2.602
WS110-3	170	0.588	199	4.94	7.88	17.81	23.11	0.581	2.948
WS110-4	170	0.588	193	3.98	7.88	17.81	23.11	0.468	2.375
WS110-5	170	0.588	200	4.48	7.88	17.81	23.11	0.527	2.674
WS110-6	170	0.588	193	4.94	7.88	17.81	23.11	0.581	2.948
WS110-7	170	0.588	200	4.97	7.88	17.81	23.11	0.585	2.966
WS110-8	170	0.588	196	4.82	7.88	17.81	23.11	0.567	2.877

Figure 7 presents the test results in [31] and the fitted curves of the different size and boundary effect models. It should be noted that the low coefficients of determination (R^2) are simply due to the fluctuation of the test results. As shown in Figure 7, the linear trend of the test results is very clear and therefore, can be used to the prediction of fracture properties. For the original Hu-Guan's boundary effect model (see Figure 7a), it is provided in [23] that $f_t = 2.78$ MPa, $K_c = 1.73$ MPa$\sqrt{m}$, $a_\infty^\star = 97.65$ mm and $l_{ch} = 384.84$ mm when $\beta_1 = 1$. The modified Hu-Guan's boundary effect model (see Figure 7b) gives $f_t = 2.94$ MPa, $K_c = 1.71$ MPa$\sqrt{m}$, $a_\infty^\star = 85.63$ mm and $l_{ch} = 337.45$ mm when $\beta_1 = 1$. For Koval-Gao's size and boundary effect model, as shown in Figure 7c, one can obtain $f_t = 3.08$ MPa, $K_c = 1.10$ MPa$\sqrt{m}$, $a_\infty^\star = 32.25$ mm and $l_{ch} = 127.09$ mm. The estimated values for Bažant's Type 2 size effect law (see Figure 7d) are: $K_c = 1.10$ MPa$\sqrt{m}$ and $c_f = 32.25$ mm. When $\gamma_1 = 0.28$, the material tensile strength $f_t = 3.24$ MPa can be obtained by Equation (31), and the corresponding $a_\infty^\star = 29.23$ mm and $l_{ch} = 115.18$ mm.

The tensile strength calculated by Koval-Gao's size and boundary effect model is again within 5% difference with the estimated tensile strength $f_t = 0.24 f_{cu}^{3/2} = 3.17$ MPa by Xu et al. [31]. With the same discrete number β_1, the original Hu-Guan's boundary effect model gives a lower tensile strength than its modified one. In terms of fracture toughness, the estimation of Koval-Gao's size and boundary effect model is the same as the Bažant's Type 2 size effect law, and lower than the values given by the original and modified Hu-Guan's boundary effect models. However, under current conditions, it is difficult to decide which model provides the more appropriate fracture toughness. This is because the tested specimen size $W = 170$ mm, which is only 8.5 times bigger than the maximum aggregate size. For such a W/d_{max} ratio, the failure of the pre-cracked specimen is certainly controlled by the non-LEFM, and the real fracture toughness of the material should be evaluated by a larger specimen or estimated by the size and boundary effect models.

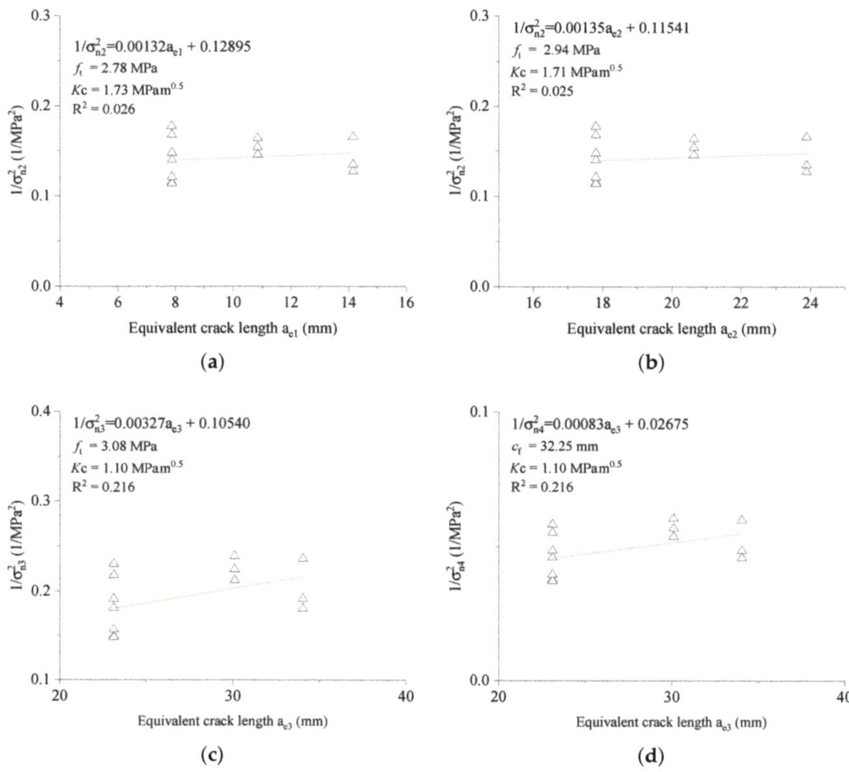

Figure 7. Test results in [31] and the corresponding fitted curves of (**a**) Hu-Guan's boundary effect model with $\beta_1 = 1$, (**b**) modified Hu-Guan's boundary effect model with $\beta_1 = 1$, (**c**) Koval-Gao's size and boundary effect model and (**d**) Bažant's Type 2 size effect law.

3.2.2. Wu's Experiments

Another series of wedge-splitting tests with different initial crack-length-to-height ratios were carried out by Wu et al. [32,33]. The maximum aggregate size was 20 mm. The height and thickness of the specimens were $W = 170$ mm and $t = 200$ mm, respectively. The samples were classified as 5 groups according to the different initial crack-length-to-height ratios ranging from 0.2 to 0.8. The measured compressive strength f_{cu} was 38.8 MPa, and splitting tensile strength was 3 MPa. Details of the tests are listed in Table 4, the failure loads P_{max} were the mean values of the different groups provided in [32,33].

Table 4. Geometry, test results and equivalent crack length of wedge-splitting specimens [32,33].

Label	W (mm)	a_0/W	t (mm)	P_{max} (kN)	a_{e1} (mm)	a_{e2} (mm)	a_{e3} (mm)	σ_N (MPa)	σ_{n2} (MPa)
WS1	400	0.2	200	34.20	39.22	47.46	68.95	1.710	2.672
WS2	400	0.4	200	25.20	30.25	39.51	77.94	1.260	3.675
WS3	400	0.5	200	16.00	23.67	32.77	67.03	0.800	3.400
WS4	400	0.6	200	10.80	17.92	26.95	52.63	0.540	3.578
WS5	400	0.8	200	3.20	8.79	19.09	25.94	0.160	3.800

The linear regressions of different models are plotted in Figure 8. When $\beta_1 = 1$, the estimated materials parameters (see Figure 8a) for original Hu-Guan's boundary effect model are: $f_t = 4.83$ MPa, $K_c = 1.42$ MPa $\sqrt{m}$, $a_\infty^\star = 22.0$ mm, $l_{ch} = 86.6$ mm, and for modified Hu-Guan's boundary effect model (see Figure 6b), the extrapolated parameters are: $f_t = 6.74$ MPa, $K_c = 1.39$ MPa $\sqrt{m}$, $a_\infty^\star = 10.8$ mm and $l_{ch} = 42.5$ mm. Koval-Gao's size and boundary effect model gives $f_t = 3.90$ MPa, $K_c = 1.61$ MPa $\sqrt{m}$, $a_\infty^\star = 43.1$ mm and $l_{ch} = 169.9$ mm (see Figure 6c). Bažant's Type 2 size effect law (see Figure 6d) gives $K_c = 1.61$ MPa $\sqrt{m}$ and $c_f = 43.13$ mm. In addition, the tensile strength $f_t = 4.10$ MPa is obtained for $\gamma_1 = 0.28$.

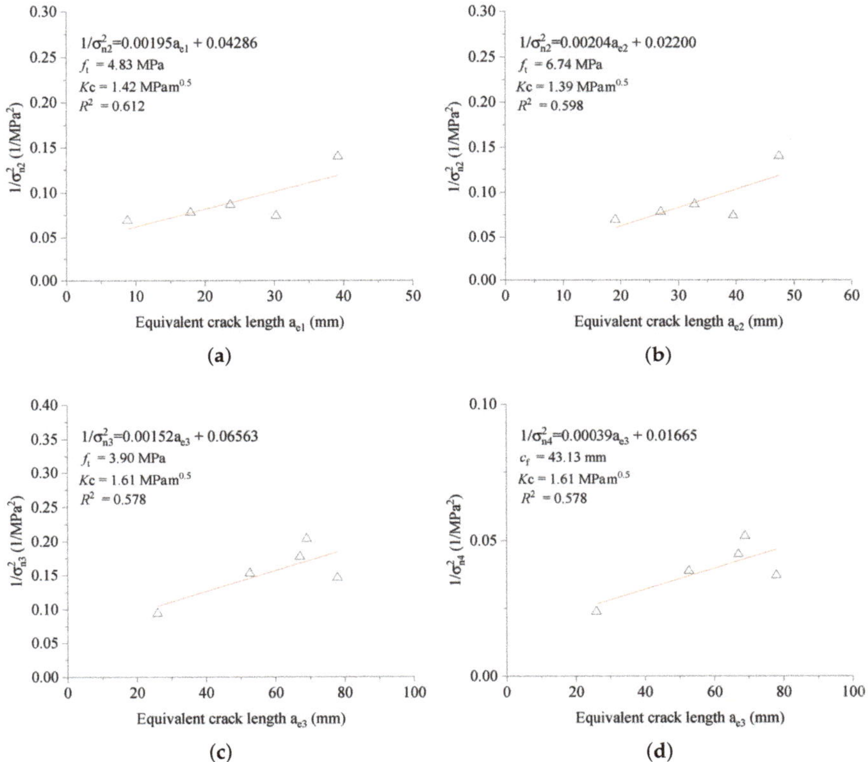

Figure 8. Test results in [32,33] and the corresponding fitted curves of (**a**) Hu-Guan's boundary effect model with $\beta_1 = 1$, (**b**) modified Hu-Guan's boundary effect model with $\beta_1 = 1$, (**c**) Koval-Gao's size and boundary effect model and (**d**) Bažant's Type 2 size effect law.

For this example, the extrapolated tensile strengths of all the models are larger than the measured splitting tensile strength (3 MPa), this is because the different fracture mechanisms of the splitting and wedge-splitting tests. The latter one's failure mechanism is close to the flexure test. Therefore, the extrapolated tensile strength from the wedge-splitting tests should be close to the tensile strength measured by the flexure test. Raphael [11,12] concluded from plenty of the experimental results that the tensile strength measured by the flexure test is around 35% higher than the one measured by the splitting test. Hence, the extrapolated tensile strengths by the different models are still reasonable.

The fracture toughness values given by the Koval-Gao's model and Bažant's model ($K_c = 1.61$ MPa $\sqrt{m}$) are higher than the values obtained from the Hu-Guan's boundary effect model

($K_c = 1.42$ MPa$\sqrt{m}$) when $\beta_1 = 1$. Since the measured nominal fracture toughness for the specimen with initial crack-length-to-height ratio equals to 0.4 was 1.45 MPa$\sqrt{m}$, it is recommended that a larger value of β_1 should be adopted, to obtain a more realistic fracture toughness. For instance, one can get $f_t = 3.89$ MPa and $K_c = 1.58$ MPa$\sqrt{m}$ by Hu-Guan's boundary effect model when $\beta_1 = 1.5$. This set of parameters is close to Koval-Gao's and Bažant's models.

4. Conclusions

The local and non-local versions of boundary effect models developed by Hu et al., the modified Hu-Guan's boundary effect model proposed in this study, the Koval-Gao's size and boundary model, and Bažant's Type 2 size effect law are all able to capture the effects of crack length and sample size on the fracture behavior of wedge-splitting and compact tension specimens. The proposed modified Hu-Guan's boundary effect model provides a more appropriate definition of equivalent crack length. This proposed model can correctly combine the contributions of strength of material and LEFM on the material failure for concrete with any maximum aggregate size. The boundary effect shape of Bažant's Type 2 size effect law shares the same equivalent crack length as Koval-Gao's size and boundary effect model. This improvement could extend the applicability of the model to extrapolate the material parameters by the test results obtained from both the geometrically similar specimens with constant crack-length-to-height ratios and specimens with different initial crack-length-to-height ratios.

The applicability of the different size and boundary effect models on the determination of fracture toughness and tensile strength for concrete material are verified and compared by test results reported in the literature. The original and modified Hu-Guan's boundary effect model is more flexible, because the two fracture parameters vary together as the adjustment of the discrete number β_1. However, there exists the risk that the reasonable fracture parameters cannot be obtained with one single β_1. Therefore, before using the Hu-Guan's boundary effect model to extrapolate the fracture parameters, it is recommended to fix one parameter by tests or give a narrow range to it in advance. In this way, the other parameter can be more reliable, and a reasonable β_1 is naturally obtained. Besides, the proposed modified Hu-Guan's boundary effect model would be more appropriate than its original version, due to the fact that the two failure mechanisms are correctly combined even for the fracture analysis of dam concrete with maximum aggregate size reaches 150 mm. Nevertheless, for d_{max} such as 10 mm or 20 mm, this improvement is neglectable. Bažant's Type 2 size effect law is less flexible than Hu-Guan's boundary effect model, because once the test results are given, the estimated fracture toughness is decided. The variation of the length scale or γ_1 would change the estimated value of tensile strength only. In terms of Koval-Gao's size and boundary effect model, only one set of fracture parameters can be extrapolated from the test results. However, for the test results adopted in this study, the predictions of the fracture properties given by Koval-Gao's size and boundary effect model are all acceptable.

For dam concrete, the specimen size needed to measure the stable tensile strength and fracture toughness can be huge. Hence, it is recommended to use the size and boundary effect to determine the two fracture parameters simultaneously. The fracture tests on at least three different sizes of geometrically similar specimens (e.g., $W = 750$ mm, 1500 mm, 2250 mm with $\alpha_0 = 0.4$) and one specimen size with three different initial crack-length-to-height ratios (e.g., $W = 1125$ mm with $\alpha_0 = 0.2, 0.4, 0.6$) are encouraged to be performed, for the purpose of the extrapolation of fracture parameters by the theoretical models. More experimental findings will be reported in the authors' further coming publications. Comprehensive experimental studies on the wedge-splitting specimens made of fully graded concrete are currently undergoing. All the specimens are casted in the same time at the construction sites of Wudongde and Baihetan super-high arch dams. The mechanical and fracture tests are performed simultaneously, to obtain the reliable and comprehensive test results for the dam concrete. The further coming experimental evidence will improve the capability of the different size and boundary effect models on the determination of fracture toughness and tensile strength for concrete material, especially the fully graded concrete, whose fracture parameters are crucial to the dam construction and operation, but not easy to be obtained.

Author Contributions: Conceptualization, X.G.; Formal analysis, X.G., Y.H., Q.L. and G.K.; Funding acquisition, Q.L. and X.G.; Supervision, Y.H., Q.L., G.K. and C.C.; Validation, Y.T., N.Y., Y.Q., C.L. and C.C.; Writing—original draft, X.G.; Writing—review and editing, X.G., Y.H., Q.L., Y.T., N.Y., Y.Q., C.L., G.K. and C.C.

Funding: This research was funded by the Research Projects of China Three Gorges Corporation (Contract numbers: BHT/0806 and WDD/0427) and China Postdoctoral Science Foundation grant number 2018M631478.

Conflicts of Interest: The authors declare no conflict of interest.

Abbreviations

The following abbreviations are used in this manuscript:

BEM   Boundary effect model
SEL   Size effect law
LEFM   Linear elastic fracture mechanics

References

1. Li, Q.; Guan, J.; Wu, Z.; Dong, W.; Zhou, S. Fracture Behavior of Site-Casting Dam Concrete. *ACI Mater. J.* **2015**, *112*, 11–20. [CrossRef]
2. Hu, Y.; Liang, G.; Li, Q.; Zuo, Z. A monitoring-mining-modeling system and its application to the temperature status of the Xiluodu arch dam. *Adv. Struct. Eng.* **2017**, *20*, 235–244. [CrossRef]
3. Lin, P.; Wei, P.; Wang, W.; Huang, H. Cracking Risk and Overall Stability Analysis of Xulong High Arch Dam: A Case Study. *Appl. Sci.* **2018**, *8*, 2555. [CrossRef]
4. Safiuddin, M.; Kaish, A.; Woon, C.; Raman, S. Early-Age Cracking in Concrete: Causes, Consequences, Remedial Measures, and Recommendations. *Appl. Sci.* **2018**, *8*, 1730. [CrossRef]
5. Mi, Z.; Hu, Y.; Li, Q.; An, Z. Effect of curing humidity on the fracture properties of concrete. *Constr. Build. Mater.* **2018**, *169*, 403–413. [CrossRef]
6. Mi, Z.; Hu, Y.; Li, Q.; Zhu, H. Elevated temperature inversion phenomenon in fracture properties of concrete and its application to maturity model. *Eng. Fract. Mech.* **2018**, *199*, 294–307. [CrossRef]
7. Liu, L.; Ouyang, J.; Li, F.; Xin, J.; Huang, D.; Gao, S. Research on the Crack Risk of Early-Age Concrete under the Temperature Stress Test Machine. *Materials* **2018**, *11*, 1822. [CrossRef] [PubMed]
8. Mi, Z.; Hu, Y.; Li, Q.; Gao, X.; Yin, T. Maturity model for fracture properties of concrete considering coupling effect of curing temperature and humidity. *Constr. Build. Mater.* **2019**, *196*, 1–13. [CrossRef]
9. Li, J.; Gao, X.; Fu, X.; Wu, C.; Lin, G. A Nonlinear Crack Model for Concrete Structure Based on an Extended Scaled Boundary Finite Element Method. *Appl. Sci.* **2018**, *8*, 1067. [CrossRef]
10. Guan, J.; Li, Q.; Wu, Z.; Zhao, S.; Dong, W.; Zhou, S. Minimum specimen size for fracture parameters of site-casting dam concrete. *Constr. Build. Mater.* **2015**, *93*, 973–982. [CrossRef]
11. Raphael, J. Tensile Strength of Concrete. *J. Am. Concrete Inst.* **1984**, *81*, 158–165.
12. Wu, M.; Zang, C.; Chen, Z. Study on the direct tensile, splitting and flexure strengths of concrete. *J. Hydraul. Eng.* **2015**, *46*, 981–988.
13. Bažant, Z. Size effect in blunt fracture: Concrete, rock, metal. *J. Eng. Mech.* **1984**, *110*, 518–535. [CrossRef]
14. Bažant, Z. Scaling theory for quasibrittle structural failure. *Proc. Natl. Acad. Sci. USA* **2004**, *101*, 13400–13407. [CrossRef]
15. Bažant, Z. *Scaling of Structural Strength*; Butterworth-Heinemann: Oxford, UK, 2005.
16. Bažant, Z.; Kazemi, M. Size dependence of concrete fracture energy determined by RILEM work-of-fracture method. *Int. J. Fract.* **1991**, *51*, 121–138.
17. Maimí, P.; González, E.; Gascons, N.; Ripoll, L. Size effect law and critical distance theories to predict the nominal strength of quasibrittle structures. *Appl. Mech. Rev.* **2013**, *65*, 020803. [CrossRef]
18. Hu, X. An asymptotic approach to size effect on fracture toughness and fracture energy of composites. *Eng. Fract. Mech.* **2002**, *69*, 555–564. [CrossRef]
19. Duan, K.; Hu, X.; Wittmann, F. Scaling of quasi-brittle fracture: Boundary and size effect. *Mech. Mater.* **2006**, *38*, 128–141. [CrossRef]

20. Hoover, C.; Bažant, Z. Comparison of the Hu-Duan boundary effect model with the size-shape effect law for quasi-brittle fracture based on new comprehensive fracture tests. *J. Eng. Mech.* **2014**, *140*, 480–486. [CrossRef]
21. Wang, Y.; Hu, X. Determination of tensile strength and fracture toughness of granite using notched three-point-bend samples. *Rock Mech. Rock Eng.* **2017**, *50*, 17–28. [CrossRef]
22. Hu, X.; Guan, J.; Wang, Y.; Keating, A.; Yang, S. Comparison of boundary and size effect models based on new developments. *Eng. Fract. Mech.* **2017**, *175*, 146–167. [CrossRef]
23. Guan, J.; Hu, X.; Xie, C.; Li, Q.; Wu, Z. Wedge-splitting tests for tensile strength and fracture toughness of concrete. *Theor. Appl. Fract. Mech.* **2018**, *93*, 263–275. [CrossRef]
24. Zhang, C.; Hu, X.; Wu, Z.; Li, Q. Influence of grain size on granite strength and toughness with reliability specified by normal distribution. *Theor. Appl. Fract. Mech.* **2018**, *96*, 534–544. [CrossRef]
25. Zhang, C.; Hu, X.; Sercombe, T.; Li, Q.; Wu, Z.; Lu, P. Prediction of ceramic fracture with normal distribution pertinent to grain size. *Acta Mater.* **2018**, *145*, 41–48. [CrossRef]
26. Gao, X.; Koval, G.; Chazallon, C. A Size and Boundary Effects Model for Quasi-Brittle Fracture. *Materials* **2016**, *9*, 1030. [CrossRef]
27. Gao, X.; Koval, G.; Chazallon, C. Energetical formulation of size effect law for quasi-brittle fracture. *Eng. Fract. Mech.* **2017**, *175*, 279–292. [CrossRef]
28. Zhang, X.; Xu, S.; Gao, H. Superposition calculation of double-K fracture parameters of concrete using wedge splitting geometry and boundary effect. *J. Dalian Univ. Technol.* **2006**, *46*, 868–874.
29. Zhang, X.; Xu, S.; Zheng, S. Experimental measurement of double-K fracture parameters of concrete with small-size aggregates. *Front. Archit. Civ. Eng. China* **2007**, *1*, 448–457. [CrossRef]
30. Wittmann, F.; Rokugo, K.; Brühwiler, E.; Mihashi, H.; Simonin, P. Fracture energy and strain softening of concrete as determined by means of compact tension specimens. *Mater. Struct.* **1988**, *21*, 21–32. [CrossRef]
31. Xu, S.; Zhao, Y.; Wu, Z. Study on the average fracture energy for crack propagation in concrete. *J. Mater. Civ. Eng.* **2006**, *18*, 817–824. [CrossRef]
32. Wu, Z.; Zhao, G. Influences of relative crack length a_0/h of concrete specimens on crack propagation process and fracture toughness. *J. Basic Sci. Eng.* **1995**, *3*, 126–130.
33. Wu, Z.; Xu, S.; Lu, J.; Liu, Y. Influence of specimen initial crack length on double-K fracture parameter of concrete. *J. Hydraul. Eng.* **2000**, *31*, 35–39.
34. Wang, Y.; Hu, X.; Liang, L.; Zhu, W. Determination of tensile strength and fracture toughness of concrete using notched 3-p-b specimens. *Eng. Fract. Mech.* **2016**, *160*, 67–77. [CrossRef]
35. Tada, H.; Paris, P.; Irwin, G. *The Stress Analysis of Cracks Handbook*; Wiley: Hoboken, NJ, USA, 2000.
36. Xu, S.; Reinhardt, H. Determination of double-K criterion for crack propagation in quasi-brittle fracture, Part III: Compact tension specimens and wedge splitting specimens. *Int. J. Fract.* **1999**, *98*, 179–193. [CrossRef]
37. Irwin, G. Analysis of stresses and strains near the end of a crack traversing a plate. *J. Appl. Mech.* **1957**, *24*, 361–364.
38. Gao, X. Modelling of Nominal Strength Prediction for Quasi-Brittle Materials. Application to Discrete Element Modelling of Damage and Fracture of Asphalt Concrete Under Fatigue Loading. Ph.D. Thesis, Université de Strasbourg, Strasbourg, France, 2017.
39. Cusatis, G.; Schauffert, E. Cohesive crack analysis of size effect. *Eng. Fract. Mech.* **2009**, *76*, 2163–2173. [CrossRef]
40. Yu, Q.; Le, J.; Hoover, C.; Bažant, Z. Problems with Hu-Duan boundary effect model and its comparison to size-shape effect law for quasi-brittle fracture. *J. Eng. Mech.* **2009**, *136*, 40–50. [CrossRef]

applied sciences

MDPI

Article

Effects of Fineness and Dosage of Fly Ash on the Fracture Properties and Strength of Concrete

Chung-Hao Wu [1], Chung-Ho Huang [2,*], Yu-Cheng Kan [3] and Tsong Yen [4]

[1] Department of Civil Engineering, Chienkuo Technology University, No.1, Chiehshou North Road, Chunghau City 500, Taiwan; chw@ctu.edu.tw
[2] Department of Civil Engineering, National Taipei University of Technology, No.1, Sec. 3, Zhongxiao E. Rd., Da'an Dist., Taipei City 10668, Taiwan
[3] Department of Construction Engineering, Chaoyang University of Technology, 168, Jifeng E. Rd., Wufeng District, Taichung 41349, Taiwan; yckan@cyut.edu.tw
[4] Department of Civil Engineering, National Chung-Hsing University, No.250, Kuo Kuang Road, Taichung 402, Taiwan; tyen@dragon.nchu.edu.tw
* Correspondence: cdewsx.hch@gmail.com; Tel.: +886-911777424

Received: 16 April 2019; Accepted: 28 May 2019; Published: 31 May 2019

Abstract: This study focuses on evaluating the effects of the fineness of fly ash on the strength, fracture toughness, and fracture resistance of concrete. Three fineness levels of fly ash that respectively pass sieves—no. 175, no. 250, and no. 32—were used. In addition to the control concrete mixture without fly ash, two fly ash replacement levels of 10% and 20% by weight of the cementitious material were selected for the concrete mixture. The experimental results indicate that the compressive strength of the fly ash concrete decreases with the increase in the replacement ratio of fly ash but increases in conjunction with the fineness level of fly ash. The presence of finer fly ash can have beneficial effects on the fracture energy (G_F) of concrete at an early age (14 days) and attain a higher increment of G_F at a later age (56 days). The concrete containing finer fly ash was found to present larger critical stress intensity factors (K^S_{IC}) at various ages, and the K^S_{IC} also increases in conjunction with the fineness levels of fly ash.

Keywords: fly ash; fineness; compressive strength; fracture energy; critical stress intensity factor

---

1. Introduction

Incorporating fly ash in concrete may affect the rate of hydration reaction and improve its microstructure [1,2]. The improvement in microstructure occurs due to grain and pore refinements, especially in the interfacial transition zone (ITZ) [3]. The improved microstructure can cause enhancement in the mechanical and durability-related properties of concrete. For the replacement amounts normally used, most fly ashes tend to reduce the early strength up to 28 days but increase the long term strength [4].

It is generally recognized that the use of finer fly ash may increase the properties of concrete [5] Mehta [6] reported that the majority of the reactive particles in fly ash are actually less than 10 micrometers in diameter. Studies [3–6] showed that increasing the fineness of fly ash by grinding improves the reactivity to some extent and increases the strength of the concrete, but eventually leads to increased water demand. Obla et al. [7] tested ultra-fine fly ash with a mean particle diameter of about 3 µm and found that significant improvements in the concrete strength and durability without loss in workability could be achieved with a commercially available ultra-fine fly ash. It could be primarily concluded that increasing the fineness of fly ash may increase the strength of concrete attributed to the micro-cracking resistance of cement paste, particularly in the ITZ.

The interfacial transition zone in concrete, which is often considered to be a three-phase material, is represented as the third phase [8,9]. Existing as a thin shell around the larger aggregate, the ITZ is generally weaker than either of the two main components of the concrete, which are the hydrated cement paste and the aggregate. The ITZ has less crack resistance, leading to the preferential occurrence of cracking. A major factor responsible for the poor strength of the ITZ is the presence of micro-cracks. Much of the physical nature of the response of concrete under loading can be described in terms of micro-cracks that can be observed at relatively low magnification. Cracking in concrete is mostly due to the tensile stress that occurs under load or environmental changes. As such, the failure of concrete in tension is governed by micro-cracks, which are associated particularly with the ITZ [10,11].

The tensile strength of concrete is a very essential property. Not only the tensile strength but also the behavior at the tensile fracture is of importance, particularly the fracture toughness. When concrete fails in tension, its behavior is characterized by both the peak stress and the energy required to fully generate a crack. Fracture mechanics can, in principle, be a suitable basis for analyzing the fracture toughness of concrete [12].

The defects and stress concentrations are the main factors causing failure. Concrete with defects or stress concentrations is more prone to cracking, which results in reduced strength. A structure made with high-strength concrete may fail under critical stress when small cracks simultaneously occur. It is essential, therefore, to determine fracture behavior and toughness for the widely used high-strength high-performance concretes that normally contain fly ash or furnace slag [13,14].

Due to the non-homogeneous characteristics of concrete, its fracture behavior is quite complicated. It has been proved that the methods developed in conventional fracture mechanics are unsuitable for the analysis of the influence of fracture toughness on the behavior of concrete structures [15,16]. Many theoretical models have been developed to make such an analysis possible. Among them, three well-known models are the fictitious crack model, the crack band model, and the two-parameter fracture model. By means of numerical techniques, for example, the finite element method, it is possible to use these models to make a theoretical analysis for the development of the damage zone and the complete behavior of the structure. As for the fracture mechanics of concrete, the International Union of Laboratories and Experts in Construction Materials, Systems and Structures (RILEM, from the name in French) [17] has provided a proposal of several methods for determining the fracture properties and parameters of concrete with a three-point bend test on a notched beam. Hillerborg [12] determined the fracture energy according to the fictitious crack model. Shah and Jenq [18] used the two-parameter fracture model to determine the critical stress intensity factor.

To investigate the effects of the fineness of fly ash on the strength and fracture toughness of concrete, this study adopted the three-point bend test to analyze the fracture energy and the critical stress intensity factor for evaluating the fracture toughness and fracture resistance of fly ash concrete.

2. Research Significance

Laboratory investigations have shown that when the fly ash particle size is reduced, its mechanical performance in concrete is improved. This study additionally confirms the comprehensive results that significant improvements in concrete strength and fracture toughness without loss in workability can be achieved with fine fly ashes, whose particle size passes through no. 250 and no. 325 sieves. Furthermore, the finer the fly ash is, the higher the enhancement of the strength, fracture energy, and critical stress intensity factor of the concrete will be.

3. Experimental Details

Concrete beam specimens were tested to determine the relationship between compressive strength, fracture energy, and stress intensity factor as a function of the cement replacement ratio of fly ash and the fineness of fly ash. The water-cementitious ratio (w/cm) used was 0.35 for all concrete mixtures. Three cement replacement ratios of 0%, 10%, and 20%, and three particle sizes of fly ash that pass through no. 175 (90 μm), no. 250 (63 μm), and no. 325 (45 μm) sieves were selected for the mixtures. The three-point bend test on notched beams was used to determine the load-deflection relations, and the maximum load for calculating the fracture energy (G_F) and the critical stress intensity factor (K^S_{IC}).

3.1. Materials

Materials used for preparing the concrete mixtures included a Type I Portland cement (Table 1), river sand, crushed coarse aggregate with a maximum size of 10 mm, a naphthalene-sulfurate based superplasticizer, and fly ashes with three levels of fineness. Class F fly ash from the Taichung Power Station in Taiwan was used. The fly ash complied with the requirement of ASTM C168. Table 1 lists its chemical and physical properties. The fly ash was classified using the method of sieving and consisted of three lots:

1. F175: fly ash passed sieve no. 175 (90 µm);
2. F250: fly ash passed sieve no. 250 (63 µm);
3. F325: fly ash passed sieve no. 325 (45 µm).

Table 1. Mixture proportions of fresh concrete.

Oxide (%)	Cement	Fly Ash
SiO_2	21.04	56
Fe_2O_3	5.46	24.81
Al_2O_3	2.98	5.3
CaO	63.56	4.8
MgO	2.52	1.48
SO_3	2.01	0.36
Loss on ignition (%)	0.92	4.12
Specific gravity	3.15	2.31

3.2. Mixture Proportions

In the experimental program, the control mixture (R0) without adding fly ash was designed to have a target compressive strength at 28 days of 63 MPa. The fly ash concrete mixtures designated as F1R1, F1R2, F2R1, F2R2, F3R1, and F3R2 were prepared using varying amounts of F175, F250, and F325 fly ash. F1, F2, and F3 referred to F175, F250, and F325, respectively, and R1 and R2 referred to the replacement ratio of 10% and 20%, respectively. The dosage of the superplasticizer was adjusted to produce the desired slump of 250 ± 25 mm and a slump-flow of 600 ± 50 mm. The mixture proportions and the slump test results are shown in Tables 2 and 3, respectively.

Table 2. Mixture proportions of concrete.

Mixture no. [1]	w/cm [2]		Batch Quantities (kg/m^3)				
		Water	Cement	Fly Ash	Sand	Coarse Aggregate	SP [3]
R0			514	0	853	821	8.2
F1R1			463	51	823	834	8.0
F1R2			411	103	822	819	8.8
F2R1	0.35	180	463	51	823	834	8.4
F2R2			411	103	822	819	9.4
F3R1			463	51	823	834	9.8
F3R2			411	103	822	819	10.5

[1] F1, F2, F3 = Fly ash of F175, F250, and F325, respectively; R1, R2 = Fly ash replacement ratio of 10% and 20%, respectively; R0 = concrete without fly ash. [2] w/cm = water/(cement + fly ash) by weight. [3] SP = Superplasticizer.

Table 3. The slump test results of fresh concrete.

Mixture no.	w/cm	Slump (mm)	Slump-Flow (mm)
R0		250	570
F1R1		255	640
F1R2		240	560
F2R1	0.35	260	610
F2R2		250	570
F3R1		248	580
F3R2		250	620

3.3. Preparation and Casting of Test Specimens

Cylinder specimens (100 × 200 mm) were cast from each mixture for compression tests, three cylinders each for testing at five ages. Flexure specimens with dimensions of 50 × 50 × 650 mm and 80 × 150 × 700 mm were cast for the determination of G_F and K^S_{IC}, respectively. Four specimens were fabricated for each mixture for the testing of each two specimens at two stages of 14 and 56 days. The specimens were prepared in accordance with ASTM C192. After casting, test specimens were covered with a plastic sheet for 24 hours. Then, they were demolded and put into a 100% relative humidity (RH) moist-curing room at about 23 °C until the time of testing. At least 24 hours before the specimens were tested, the fracture energy specimens and the critical stress intensity factor specimens, as shown in Figures 1 and 2, were prepared by cutting a 25 mm deep by 2 mm width notch and a 50 mm by 2 mm notch, respectively, on one side at midspan.

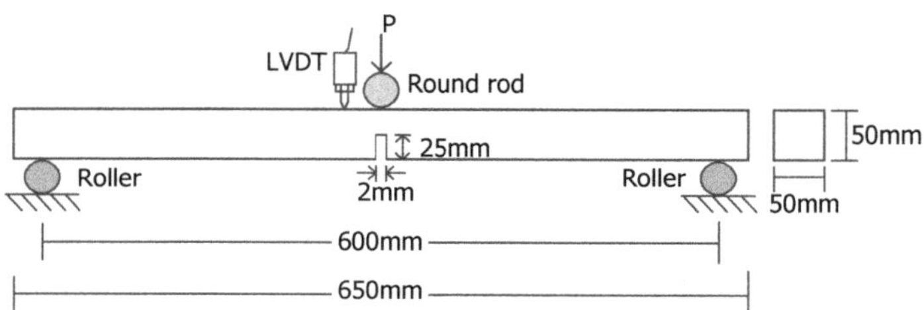

Figure 1. Set-up of the three-point bending test for the determination of the fracture energy of concrete. LVDT is the linear variable differential transformer and P is the load.

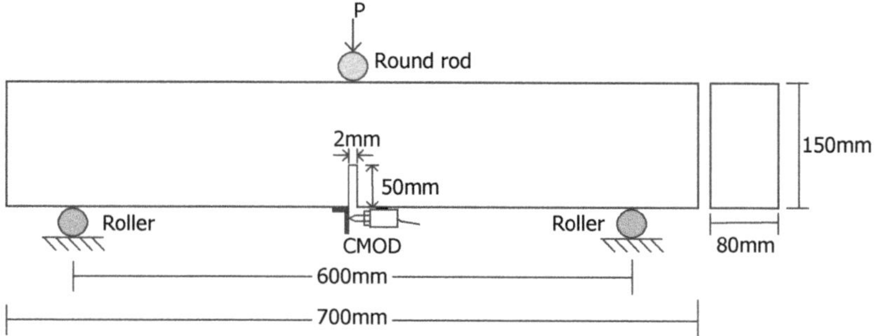

Figure 2. Set-up of the three-point bending test for the determination of the critical stress intensity factor of concrete. CMOD is the crack mouth opening displacement.

3.4. Testing of Specimens

A compression test was carried out in accordance with ASTM C39. The three-point bend test was performed on the notched beam as follows to determine the G_F and K^S_{IC}:

1. Test for determining G_F.

The fracture energy test followed the guidelines established by RILEM [17] using a closed-loop testing machine. The testing arrangement is shown in Figure 1. A linear variable differential transformer (LVDT) was installed at midspan of the beam to measure the deflection. The loading velocity was chosen so that the maximum load was reached 30 s after the loading started. The loading rate selected was 0.25 mm/min. A load-deflection (F-δ) curve was then plotted, with the energy "W_0" representing the area under the curve. The fracture energy (G_F) according to this work of fracture method can be calculated as [12,18]:

$$G_F = \frac{W_0 + mg\delta_0}{A} \ (N/m)$$

(1)

where W_0 = area under the load-deflection curve (N/m); mg = weight of the specimen (kg); g = acceleration due to gravity; δ_0 = maximum deflection recorded (m); and A = cross-sectional area of the beam above the notch (m^2).

2. Test for determining K^S_{IC}.

The test for the determination of K^S_{IC} followed the test method suggested by Shah and Jenq [18] using a closed-loop testing machine under the crack mouth opening displacement (CMOD) control. The test set-up is shown in Figure 2. A clip gauge was used to measure the CMOD. The CMOD and the applied load were recorded continuously during the test. The test procedure included loading and unloading. The test was started from loading until the peak load was reached, and then the applied load was manually reduced (also termed unloading) when the load passed the maximum load and was at about 95% of the peak load. When the applied load was reduced to zero, the specimen was reloaded. Each loading and unloading cycle was finished in about one min. Only one cycle of loading-unloading was required for the test. The K^S_{IC} was calculated using the following Equation:

$$K^S_{IC} = 3(P_{max} + 0.5W)\frac{S(\pi a_c)^{1/2}F(\alpha)}{2d^2b}\left(Nm^{-3/2}\right)$$

(2)

in which

$$W = W' \times \frac{S}{L}$$

(3)

$$F(\alpha) = \frac{1.99 - \alpha(1-\alpha)\left(2.15 - 3.93\alpha + 2.7\alpha^2\right)}{\sqrt{\pi}(1+2\alpha)(1-\alpha)^{3/2}}$$

(4)

where P_{max} = the measured maximum load (N); S = the span of the beam (m); b = beam width (m); d = beam depth (m); L = beam length (m); W' = self-weight of the beam (N); α_c = critical effective crack length (m); and $\alpha = \alpha_c/d$.

4. Results and Discussion

4.1. Compressive Strength

Compressive strengths were determined at five ages of 7, 14, 28, 56, and 91 days on concrete stored under moist-curing conditions. The results are given in Table 4. As can be seen in Table 3, the slump values of all concrete mixtures were in the range of 240–260 mm, and similarly, the slump-flow fell in the range of 550–640 mm. This indicated that all mixtures had quite similar workability. In addition, all mixtures were proportioned to have an equivalent target strength of 63 MPa at the age of 28 days. This was generally achieved, except for mixture F1R2. In this case, the strength equivalence was achieved later, at the age of 56 days.

Table 4. Compressive strength of concrete cylinders.

Mixture no.	w/cm	Compressive Strength, MPa *				
		7 Days	14 Days	28 Days	56 Days	91 Days
R0		56.5 ± 0.26 (100%)	62.1 ± 0.29 (110%)	66.1 ± 0.28 (117%)	70.3 ± 0.20 (124%)	71.2 ± 0.38 (126%)
F1R1		53.0 ± 0.17 (100%)	57.3 ± 0.33 (108%)	63.0 ± 0.26 (119%)	69.8 ± 0.31 (132%)	72.4 ± 0.41 (137%)
F1R2		52.2 ± 0.20 (100%)	56.0 ± 0.34 (107%)	61.4 ± 0.24 (118%)	68.7 ± 0.36 (132%)	71.7 ± 0.45 (137%)
F2R1	0.35	54.6 ± 0.12 (100%)	61.9 ± 0.29 (113%)	68.2 ± 0.26 (125%)	74.2 ± 0.25 (136%)	75.8 ± 0.34 (139%)
F2R2		53.3 ± 0.15 (100%)	59.2 ± 0.19 (111%)	66.8 ± 0.36 (125%)	73.1 ± 0.35 (137%)	74.6 ± 0.41 (140%)
F3R1		55.7 ± 0.23 (100%)	63.5 ± 0.35 (114%)	70.3 ± 0.35 (126%)	75.8 ± 0.29 (136%)	77.9 ± 0.34 (140%)
F3R2		53.9 ± 0.14 (100%)	61.5 ± 0.29 (114%)	69.1 ± 0.19 (128%)	74.6 ± 0.23 (138%)	76.5 ± 0.28 (142%)

* Average value of three specimens.

Figure 3 shows the strength development for each concrete mixture. At an early age (7 days), as expected, the control mixture (R0) without fly ash displayed higher strength than the other fly ash mixtures. The early strength gain of the control mixture was superior to that of the fly ash mixtures, indicating that replacing any amount of cement with fly ash of various fineness may reduce the strength gain of the concrete. In addition, the concrete incorporating more fly ash (20%) presented lower strength than that with lower fly ash content (10%). Nevertheless, the curve slope of the fly ash mixtures in Figure 3 tended to be steeper than that of the control mixture at later ages. In other words, the strength increment versus age for the fly ash mixture was larger than that of the control mixture. Consequently, the strength of each fly ash mixture exceeded that of the control mixture at later ages.

On the other hand, it can be seen in Table 4 that for the same replacement ratio (20%) of fly ash, mixture F3R2, had the highest strength at an early age (7 days), followed by F2R2, and the F1R2 had the lowest. This revealed that incorporating finer fly ash in concrete may fill the micro-voids much better and produce higher packing density, resulting in a larger strength enhancement. In addition, a beneficial effect of the fineness of fly ash on the concrete strength can be seen in Figure 4; the compressive strength of fly ash concrete increased in accordance with the enhancement of the fineness level of fly ash for each age and various fly ash replacement ratios.

Moreover, as seen in Table 4, there was a significant strength gain from 7 to 28 days, and again from 28 days to 91 days for the mixtures containing finer fly ash. For example, the strength increment of the mixtures F2R1 and F3R1, which contained finer fly ash, displayed an increased rate from 100% at 7 days to 125% and 126% at 28 days, respectively, and 139% and 140% at 91 days, respectively, which were obviously larger than those of the corresponding mixtures F1R1 that increased from 100% at 7 days to 119% at 28 days and 137% at 91 days. At 28 days, the strengths of the finer fly ash mixtures of F2R1, F2R2, F3R1, and F3R2 were shown to consistently exceed that of the control mixture. This was particularly significant considering that the cement replacement of finer fly ash in each case was only 10–20%. In addition, the highest strength achieved at 91 days was 76.5 MPa for the mixture F3R2 with 20% F325 fly ash replacement, while the strength obtained for the corresponding control concrete (R0) was the relatively lower value of 71.2 MPa.

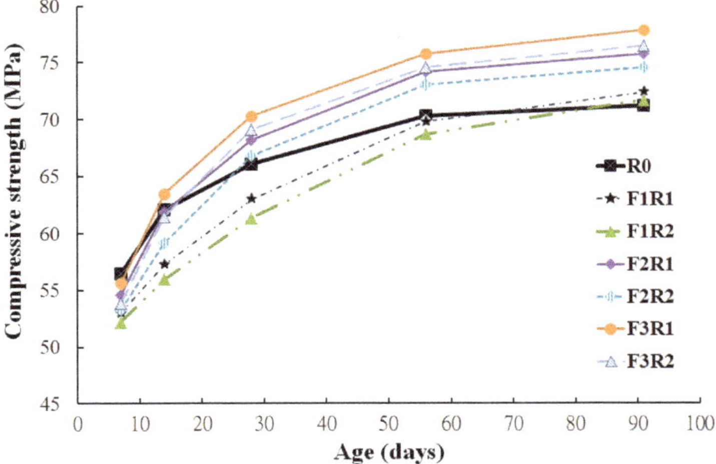

Figure 3. Compressive strength development for cylindrical specimens.

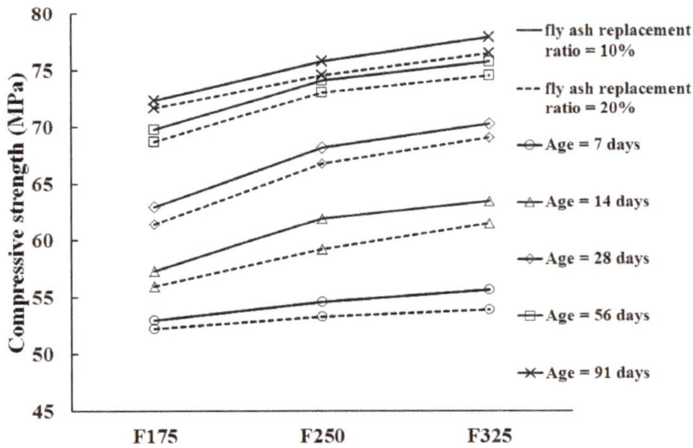

Figure 4. Compressive strength of fly ash concrete versus the particle size of fly ash.

4.2. Fracture Energy

Fracture energy (G_F) is the energy dissipated per unit area during the formation of a crack. In this research, G_F was determined using a notched beam in a three-point bending test. Load-deflection curves were plotted from the tests. The energy absorbed in the test to the point of failure was represented by the area below the load-deflection curve for the specimen. This area represented the fracture energy per unit area of the fracture surface. The G_F was calculated using Equation (1). Concrete specimens were tested at 14 and 56 days. Table 5 lists the test results. It shows that, at early ages, the G_F values of all fly ash concrete were lower than that of control concrete without fly ash (R0), except for the fly ash mixture F3R2. Nevertheless, the presence of the finer fly ash had a beneficial effect on the G_F of concrete. As shown in Figure 5, when the fineness level of the fly ash increased, the G_F value of the concrete increased. Even at the earlier age of 14 days, the G_F value of the mixture with finer fly ash (F3R2) was in turn higher than those of the mixtures F2R2 and F1R2 for the similar fly ash replacement ratio of 20%; furthermore, the G_F of the mixture F3R1 was higher than those of F2R1 and F1R1. This implies that although the pozzolanic reaction had not yet occurred at the early ages, the superior filling effects of the finer fly ash could increase the denseness of the concrete, resulting in increased fracture toughness. After 56 days, the G_F values of the fly ash concrete mixtures all exceeded that of the control mixture, except for the mixture F1R2. In addition, it was also found that the increment of the fracture energy for each fly ash concrete from 14 days to 56 days was attained by 21–31%, which was much higher than that of the control concrete (13.6%). This signifies that, during this period, the pozzolanic reaction of fly ash became active, which together with the filling effect, enhanced the fracture toughness of the fly ash concrete.

Table 5. Fracture energy of concrete.

Mixture no.	Compressive Strength (MPa)		Fracture Energy * (N/m)		
	14 Days	56 Days	14 Days	56 Days	Increment N/m (%)
R0	62.1 ± 0.29	70.3 ± 0.20	97.1 ± 2.21	110.3 ± 2.51	13.2 (13.6)
F1R1	57.3 ± 0.33	69.8 ± 0.31	84.3 ± 1.69	109.4 ± 2.44	25.1 (29.8)
F1R2	56.0 ± 0.34	68.7 ± 0.36	78.5 ± 1.88	103.5 ± 2.10	25.0 (31.8)
F2R1	61.9 ± 0.29	74.2 ± 0.25	95.7 ± 2.46	115.8 ± 1.98	20.1 (21.0)
F2R2	59.2 ± 0.19	73.1 ± 0.35	88.5 ± 2.21	114.9 ± 2.56	26.4 (29.8)
F3R1	63.5 ± 0.35	75.8 ± 0.29	95.8 ± 3.10	121.4 ± 2.88	25.6 (26.7)
F3R2	61.5 ± 0.29	74.6 ± 0.23	98.1 ± 2.66	123.5 ± 3.11	25.4 (25.9)

* Average value of two specimens.

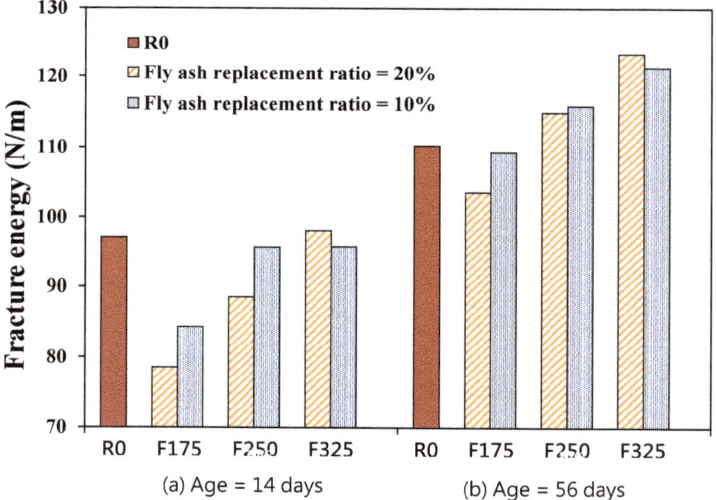

Figure 5. Fracture energy of concretes with various fineness levels and the replacement ratios of fly ash.

Figure 6 compares the fracture energy, and the compressive strength of all fly ash concretes in the study. The G_F was found to increase in conjunction with the compressive strength. This trend compares favorably with the research of Gettu [19], Giaccio [20], and Xie [21]. Nevertheless, these researchers observed smaller increases in G_F with increases in compressive strength of high strength concrete. Gettu et al. [19] found that an increase in compressive strength of 160% resulted in an increase in G_F of only 12%. Giaccio et al. [20] observed that G_F increased as compressive strength increased, but only at a fraction of the rate. Xie et al. [21] found that increases in compressive strength of 29–35% resulted in a G_F increase of only 11–13%. In this study, an increase in compressive strength of 10% resulted in a larger increase in G_F of around 14% (~13–15%) when the compressive strength varied between 50 and 80 MPa. This demonstrates the unique effects of using finer fly ash on the enhancement of the fracture toughness of high-strength concrete.

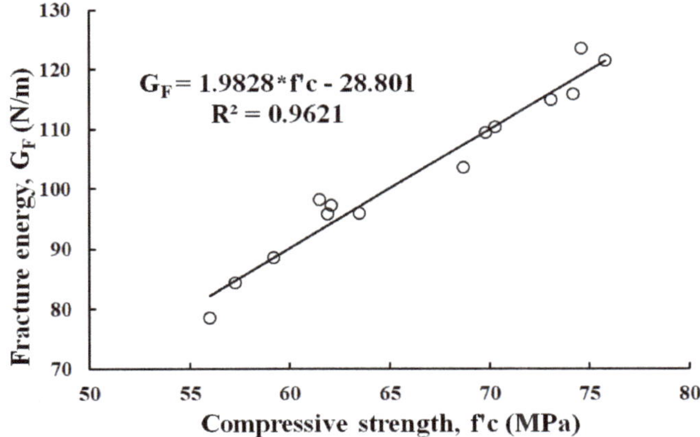

Figure 6. Fracture energy versus compressive strength of fly ash concrete.

4.3. Critical Stress Intensity Factor

The critical stress intensity factor (K^S_{IC}) is the stress intensity factor calculated at the critical effective crack tip, using the measured maximum load. In this research, the K^S_{IC} was determined with a notched beam in a three-point bend test, using a closed-loop testing machine under the crack mouth opening displacement control. Load-CMOD curves were plotted from the tests. The K^S_{IC} was calculated using Equation (2). Concrete specimens were tested at 14 and 56 days. As seen in Table 6, results showed that the related tendency of the K^S_{IC} of concrete was similar to that of the G_F. At an early age (14 days), most of the K^S_{IC} values of the fly ash concrete were lower than that of the control concrete (R0) but exceeded that of R0 after 56 days. This implies that the additional pozzolanic reaction of fly ash with $Ca(OH)_2$ in the concrete became active during the period between 14 and 56 days, which led to the enhancement of K^S_{IC} of the fly ash concrete.

Table 6. Critical stress intensity factor of concrete.

Mixture no.	Compressive Strength (MPa)		Critical Stress Intensity Factor ($N \times m^{-3/2}$) *	
	14 Days	56 Days	14 Days	56 Days
R0	62.1 ± 0.29	70.3 ± 0.20	1.030 ± 0.0127	1.231 ± 0.0212
F1R1	57.3 ± 0.33	69.8 ± 0.31	0.964 ± 0.0121	1.289 ± 0.0189
F1R2	56.0 ± 0.34	68.7 ± 0.36	0.908 ± 0.0136	1.218 ± 0.0210
F2R1	61.9 ± 0.29	74.2 ± 0.25	1.025 ± 0.0135	1.342 ± 0.0250
F2R2	59.2 ± 0.19	73.1 ± 0.35	0.999 ± 0.0128	1.310 ± 0.0244
F3R1	63.5 ± 0.35	75.8 ± 0.29	1.135 ± 0.0131	1.402 ± 0.0260
F3R2	61.5 ± 0.29	74.6 ± 0.23	1.085 ± 0.0120	1.389 ± 0.0272

* Average value of two specimens.

It was also found that concrete containing finer fly ash presented larger K^S_{IC} values. This can be observed in Figure 7, which shows that the increase in the fineness level of fly ash led to an increase in the K^S_{IC} values of concrete at various ages. In other words, the K^S_{IC} of the mixture F3R2 with finer fly ash was in turn higher than that of F2R2 and F1R2, for the same fly ash replacement ratio of 20%. Furthermore, the K^S_{IC} of F3R1 > F2R1 > F1R1 for a similar fly ash replacement ratio of 10%. The reason for attaining this result is that the finer fly ash particle had a larger surface area, exhibiting a more active pozzolanic reaction, and thus resulting in an increase in the strength and fracture resistance (K^S_{IC}).

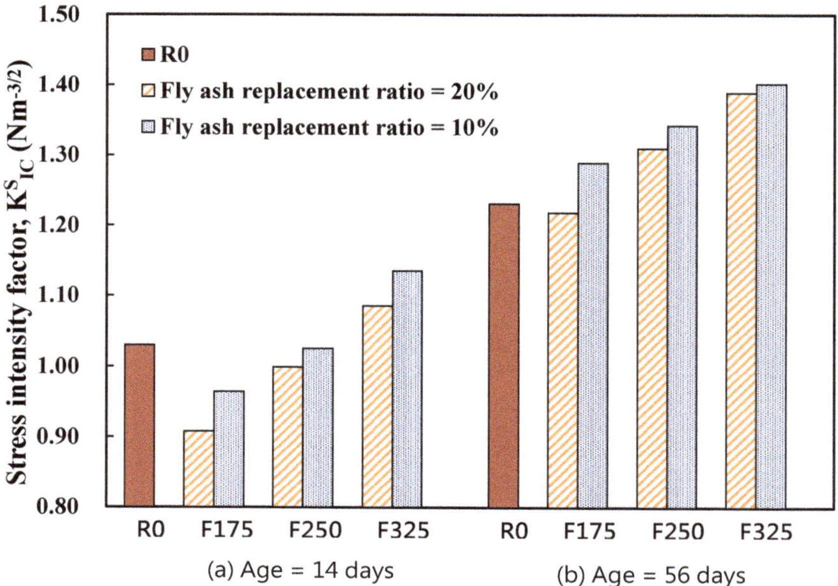

Figure 7. Stress intensity factor of concrete with various fineness levels, and the replacement ratio of fly ash.

The stress intensity factors of fly ash concrete in the fracture test were compared to compressive strength, as seen in Figure 8. As shown in the figure, the two values of stress were nearly linearly related. The relationships shown in Figure 8 are of some importance because based on the close relationship between the particle size of fly ash and the compressive strength (Figure 4), the stress intensity factor increases with the increase of the fineness level of fly ash.

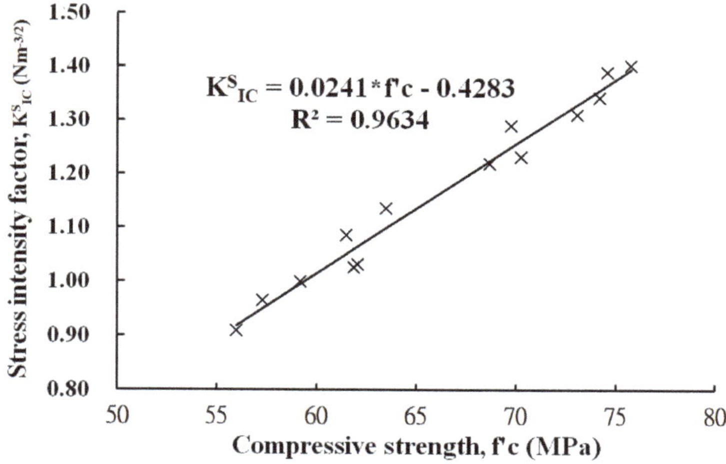

Figure 8. Stress intensity factor versus compressive strength of fly ash concrete.

5. Conclusions

Based on the tests and evaluations presented in this study, the following conclusion can be reached:

1. Incorporating finer fly ash into concrete may fill the micro-voids much better and produce higher packing density, resulting in larger strength enhancement. The compressive strength of fly ash concrete increases in conjunction with the fineness level of fly ash presented in the mixtures for each age and various fly ash replacement ratios.
2. Use of finer fly ash has beneficial effects on the fracture energy (G_F) of concrete, even at an early age (14 days). The finer fly ash used, the greater the fracture energy of the concrete exhibited.
3. The increment of the fracture energy of all the fly ash concrete measured in this study from 14 days to 56 days was attained by 21–31%, which was much higher than that of control concrete (13.6%), such that the G_F of the fly ash mixtures at 56 days almost exceeded that of the control concrete.
4. An increase in fly ash concrete compressive strength of 10% resulted in an increase in the G_F value of around 14% when the compressive strength varied between 50 and 80 MPa. This rate of increase is significantly higher than that found in normal concrete, indicating that the use of finer fly ash can have unique effects on the enhancement of the fracture toughness of high-strength concrete.
5. Similar to the trend of development of the fracture energy, at an early age (14 days), most of the stress intensity factor (K^S_{IC}) of the fly ash concrete was lower than that of the control concrete (R0), but exceeded that of the R0 after 56 days.
6. Concrete containing finer fly ash exhibited larger K^S_{IC} value for various ages, and the stress intensity factor increased in conjunction with the fineness level of fly ash. This also implicates an increase in the fracture resistance of concrete.

Author Contributions: Conceptualization—C.-H.H. and T.Y.; Methodology—C.-H.H.; Validation—C.-H.W., C.-H.H., and Y.-C.K.; Formal Analysis—C.-H.W. and C.-H.H.; Investigation—C.-H.W. and C.-H.H.; Resources—C.-H.H.; Data Curation—C.-H.W. and C.-H.H.; Writing and Original Draft Preparation—C.-H.W., C.-H.H. and T.Y.; Writing, Review, and Editing—C.-H.H. and T.Y.; Visualization—C.-H.H. and Y.-C.K.; Supervision—T.Y.; Project Administration—C.-H.H.

Funding: This research received no external funding.

Conflicts of Interest: The authors declare no conflict of interest.

References

1. Pejman, K.D.; Savas, E.; Marva, A.B. Physico-mechanical, microstructural and dynamic properties of newly developed artificial fly ash based lightweight aggregate—Rubber concrete composite. *Compos. Part B Eng.* **2015**, *79*, 451–455.
2. Sukumar, B.; Rani, M.U.; Silvac, P.J. Micro-structural characteristics of self-compacting concrete with high volume fly ash. *Int. J. Appl. Eng. Res.* **2015**, *10*, 10265–10277.
3. Rahimah, E.; Andri, K.; Nasir, S.; Muhd, F.N. Strength and microstructural properties of fly ash based geopolymer concrete containing high-calcium and water-absorptive aggregate. *J. Clean. Prod.* **2016**, *112*, 816–822.
4. Huang, C.H.; Lin, S.K.; Chang, C.S.; Chen, H.J. Mix proportions and mechanical properties of concrete containing very high-volume of Class F fly ash. *Constr. Build. Mater.* **2013**, *46*, 71–78. [CrossRef]
5. Bagheri, A.; Zanganeh, H.; Alizadeh, H.; Shakerinia, M.; Marian, M.A.S. Comparing the performance of fine fly ash and silica fume in enhancing the properties of concretes containing fly ash. *Constr. Build. Mater.* **2013**, *47*, 1402–1408. [CrossRef]
6. Mehta, P.K. Influence of Fly Ash Characteristics on the Strength of Portland-Fly Ash Mixtures. *Cem. Concr. Res.* **1985**, *15*, 669–674. [CrossRef]
7. Obla, K.H.; Hill, R.L.; Thomas, M.D.A.; Shashiprakash, S.G.; Perebatova, O. Properties of Concrete Containing Ultra-Fine Fly Ash. *ACI Mater. J.* **2003**, *100*, 426–433.
8. Xiao, J.H.; Li, W.; Corr, D.J.; Shah, S.P. Effects of interfacial transition zones on the stress–strain behavior of modeled recycled aggregate concrete. *Cem. Concr. Res.* **2013**, *52*, 82–99. [CrossRef]

9. Xie, Y.; Corr, D.J.; Jin, F.; Zhou, H.; Shah, S.P. Experimental study of the interfacial transition zone (ITZ) of model rock-filled concrete (RFC). *Cem. Concr. Compos.* **2015**, *55*, 223–231. [CrossRef]

10. Alberti, M.G.; Enfedaque, A.; Gálvez, J.C. Fracture mechanics of polyolefin fibre reinforced concrete: Study of the influence of the concrete properties, casting procedures, the fibre length and specimen size. *Eng. Fract. Mech.* **2016**, *154*, 225–244. [CrossRef]

11. Rezaie, F.; Farnam, S.M. Fracture mechanics analysis of pre-stressed concrete sleepers via investigating crack initiation length. *Eng. Fail. Anal.* **2015**, *58*, 267–280. [CrossRef]

12. Hillerborg, A. The Theoretical Basis of a Method to Determine the Fracture Energy G_F of Concrete. *Mater. Struct.* **1985**, *18*, 291–296. [CrossRef]

13. Rao, G.A.; Raghu Prasad, B.K. Size effect and Fracture Properties of HPC. In Proceedings of the 14th Engineering Mechanics International Conference (ASCE), Austin, TX, USA, 21–24 May 2000; Volume 104, pp. 21–24.

14. Appa Rao, G.; Raghu Prasad, B.K. Fracture energy and softening behavior of high-strength concrete. *Cem. Concr. Res.* **2002**, *32*, 247–252.

15. Golewski, G.L.; Sadowski, T. An analysis of shear fracture toughness KIIc and microstructure in concretes containing fly-ash. *Constr. Build. Mater.* **2014**, *51*, 207–214. [CrossRef]

16. Teng, S.; Liu, Y.; Yang, T.; Lim, D. Determination of fracture energy of ultra-high strength concrete. *Eng. Fract. Mech.* **2014**, *131*, 602–615. [CrossRef]

17. RILEM Technical Committee 50-FMC, Proposed RILEM Recommendation. Determination of The Fracture Energy of Mortar and Concrete by Means of Three-Point Bend Tests on Notched Beams. *Mater. Struct.* **1985**, *18*, 285–290.

18. Shah, S.P.; Jenq, Y.S. Determination of Fracture Parameters (K^S_{IC} and CTODC) of Plain Concrete Using Three-Point Bend Tests. Final Report of RILEM Committee 89-FMT Fracture Mechanics of Concrete: Test Method. *Mater. Struct.* **1988**, *23*, 457–460. [CrossRef]

19. Gettu, R.; Bazant, Z.P.; Karr, M.E. Fracture properties and brittleness of high-strength concrete. *ACI Mater. J.* **1990**, *87*, 608–617.

20. Giaccio, G.; Rocco, C.; Zerbino, R. The fracture energy of high strength concretes. *Mater. Struct.* **1993**, *26*, 381–386. [CrossRef]

21. Xie, J.; Elwi, A.E.; MacGregor, J.G. Mechanical properties of three high-strength concretes containing silica fume. *ACI Mater. J.* **1995**, *92*, 135–145.

Article

Meso-Scale Simulation of Concrete Based on Fracture and Interaction Behavior

Xueyu Xiong [1,2,*] and Qisheng Xiao [2]

[1] Department of Structural Engineering, College of Civil Engineering, Tongji University, Siping Lu 1239 hao, Yangpu Qu, Shanghai 200092, China
[2] Key Laboratory of Ministry of Education for Advanced Civil Engineering and Materials, Tongji University, Siping Lu 1239 hao, Yangpu Qu, Shanghai 200092, China
* Correspondence: xueyu@tongji.edu.cn; Tel.: +86-137-0191-8688

Received: 27 June 2019; Accepted: 22 July 2019; Published: 25 July 2019

Abstract: Based on the cohesive zone model, a meso-scale model is developed for numerical studies of three-phase concrete under tension and compression. The model is characterized by adopting mixed-mode fracture and interaction behavior to describe fracture, friction and collision in tension and compression processes. The simulation results match satisfactorily with the experimental results in both mechanical characteristics and failure mode. Whole deformation and crack propagation process analyses are conducted to reveal damage evolution of concrete. The analyses also set a foundation for the following parametric studies in which mode II fracture energy, material parameter, frictional angle and aggregates' mechanical characteristics are considered as variables. It shows that the mixed-mode fracture accounts for a considerable proportion, even in tension failure. Under compression, the frictional stress can constrain crack propagation at the beginning of the damage and reestablish loading path during the softening stage. Aggregates' mechanical characteristics mainly affect concrete's performance in the mid-and-late softening stage.

Keywords: mesostructure; finite element analysis; cohesive zone model; crack propagation

1. Introduction

It is widely accepted that the dominant failure mode of concrete is quasi-brittle fracture, which is mainly due to the expanding and connecting of initial defects and microcracks. To explain the behavior of concrete's fracture, concepts of fracture mechanics were introduced more than 50 years ago [1]. Although it is suitable for applying fracture mechanics theories to the analysis of mass concrete structures, it is inappropriate for application when microcrack and subcritical crack are not ignorable [2]. The sizes of microcracks and subcritical cracks of concrete are several orders of magnitude larger than that of metal. Meanwhile, concrete is a kind of multiphase and heterogeneous material, so the COD (crack opening displacement) and J-integral in elastic-plastic fracture mechanics are inappropriate for application [3].

The macro-mechanic response of concrete is the reflection of its meso-mechanical and micro-mechanic characteristics [2]. According to this idea, scholars have attempted to use finite element methods to research fractures in concrete, similar to the research of metal fractures [4]. Due to the advance of multi-processors' parallel processing capability, finite element simulation of composite material's meso-mechanical characteristics has progressed [5]. Through simulation, it is possible to predict the macro performance of concrete with its specific meso-scale structure, material and internal property [6].

Based on the macroscopic experimental phenomena of concrete, several concrete crack models have been proposed to simulate the fracture behavior of concrete. In the 1960s, two basic models were proposed: the discrete crack model [7] and smeared crack model [8]. The former simulates cracking by the definition of nodes' split criterion, which can describe the geometric discontinuity at cracks. The latter simulates cracking by the definition of elements' damage, and the model is maintained to be geometrically continuous. Representative discontinuous models include the lattice model [9], rigid bodies-spring model [10], and interface constitutive model [11]. Although the previously mentioned discrete crack model can provide intuitionistic and realistic simulations of crack behavior, their calculation results are sensitive to mesh distortion. In recent years, simulating discrete crack by cohesive zone model [12] and extended finite element method [13] are widely recognized as reliable approaches for their low mesh sensitivity and satisfactory convergent property [14].

The cohesive zone model simulates fractures by inserting cohesive elements at the potential crack propagation paths. The method breaks through the limit of using stress fields and stress intensity factor to describe concrete fracture. Unlike traditional investigation of fracture mechanics, the method uses fracture energy to define the constitutive relation between element deformation and damage on a much grander scale. As mentioned above, the special fracture behavior caused by microcracks and subcritical cracking can be described by elements' fracture energy. Ongoing studies have been being conducted on concrete fracture energy [15–20], and corresponding testing methods of different fracture types have been proposed [21,22].

The cohesive zone model has been frequently adopted in the simulation of the concrete failure problem and proved to be feasible in concrete tension fracture simulation [23–28]. However, very few studies report its application in concrete compression simulation, which has a more complex damage evolution in the fracture process [29].

Although it is widely accepted that the main concrete failure process is fracturing [30] and many concrete crack models have been proposed to describe this process, there is still a need of a unified approach to simulate the mechanical behavior and appearance of concrete under different loading conditions. The objective of this research is to extend the application range of the cohesive zone model, making the model suitable for concrete fracture simulation both under tension and compression. Quantifying the effects of the different material and interaction parameters studied in this paper contributes to revealing the relationship between meso-scale fracture and macroscopic behavior of concrete.

2. Finite Element Modeling Method

The explicit method in the Abaqus/Explicit module is used in this paper to simulate concrete tension and the compression loading process. Since friction and collision of particles play a significant role in concrete's compression softening stage, interaction behavior is defined to describe the contact behavior. The general contact algorithm in Abaqus/Explicit can only be used with three-dimensional surfaces [31], so three-dimensional models are used for simulation.

The models are built to simulate concrete uniaxial tension and compression experiments conducted by Ren et al. [32]. The specimens used for experiments are cuboids with dimensions of $150 \times 150 \times 50$ mm, which were cut from large plate specimens with dimensions of $500 \times 500 \times 50$ mm. The mean tension and compression elastic modules are 42.0 GPa and 37.5 GPa, respectively. The mean tension and compression strengths are 3.24 MPa and 37.5 MPa, respectively. The volume proportion of aggregates calculated by the mixture ratio is 45.1%. For computing efficiency, aggregates whose particle diameters are greater than 2.5 mm are modeled. The fine aggregates and cement matrix are treated as mortar whose mechanical behavior is simulated by mortar particle elements and internal interface elements.

As incomputable microcracks occur during the fracture process, it is difficult to apply the fracture criterion to simulate concrete fracture at the micro-scale. However, it is appropriate to use the traction-separation law to describe the meso-scale damage and fracture caused by the microcracks. For this reason, cohesive elements are used to simulate damage evolution and energy dissipation of concrete. In the main study of this paper (Section 6.4), the cohesive elements are only inserted between the elements of aggregates and mortar (interfacial transitional zones, ITZs), and inside the mortar elements (mortar internal interface, MII), as shown in Figure 1. In the study of Section 6.4, the cohesive elements are also inserted inside the aggregate elements (aggregate internal interface, AII), to study the effects of aggregates' mechanical characteristics. In this paper, the study adopts the simplifying assumption that the interface properties of ITZs, MII, AII are the same.

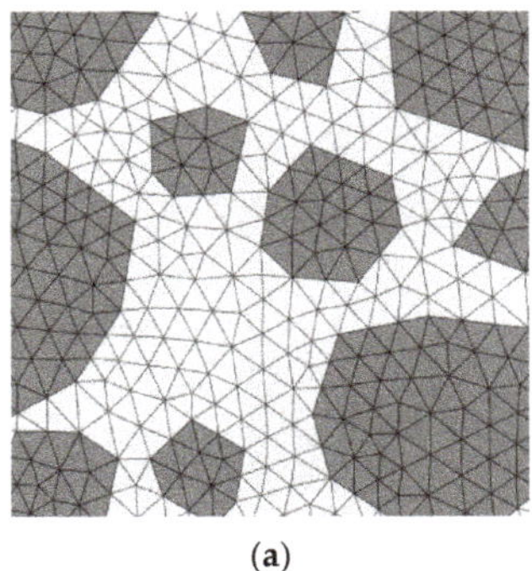

(a)

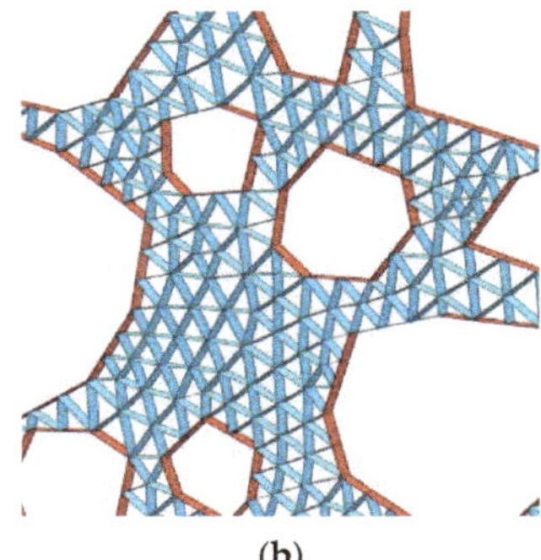

(b)

Figure 1. Element mesh of model: (**a**) bulk elements of aggregate (grey colored) and mortar particle (white colored); (**b**) cohesive elements of ITZs (interfacial transitional zones, red colored) and mortar internal interface (blue coloured).

The models are meshed by irregular triangle elements in the plane (Figure 1). Compared with other polygon elements, triangle elements provide the maximum number of interfaces with the same nodes in the plane. Cohesive elements are inserted into the interfaces to simulate damage and fracture behaviors.

The displacement loading control is adopted to obtain the complete tension and compression process data. The vertical motions of the nodes at the bottom boundary are constrained. The load is applied by controlling the motions of the nodes at the top boundary. All nodes' motions in the direction normal to the plane are constrained. In the tension and compression simulations, the final vertical displacements of the top nodes are 0.09 mm and 0.9 mm, respectively. The total time of loading is 100 s. In the first 20 s and last 80 s, the vertical motions of the top nodes are uniformly accelerated and uniform, respectively.

2.1. Mortar Particle and Aggregate Material Model

The linear elastic model is adopted for modeling the behavior of the mortar particles and aggregates. The mortar particles and aggregates are modeled by the six-node linear triangular prism (C3D6) elements. Although the mortar particles and aggregates are assumed to be isotropic in the two-dimensional plane, they need to be defined as anisotropic to eliminate the Poisson effect in the direction normal to the plane. The Poisson ratios normal to the plane are set to zero.

Referring to previous material test results, different elastic moduli are set to the mortar particles and aggregates to simulate the crack propagation caused by local deformation difference. The macroscopic modulus (E_m) of specimens can be estimated by the following formula:

$$E_m = \frac{(E_a/P_a) \cdot (E_m/P_m)}{E_a/P_a + E_m/P_m} \tag{1}$$

where E_a and E_m are the moduli of aggregates and mortar, and P_a and P_m are the volume proportions of aggregates and mortar. The formula is obtained through a short derivation. In the derivation, it is assumed that mortar and aggregates are connected in series. Based on the volume proportions and moduli, Equation (1) was derived.

2.2. Aggregate Generation

The aggregates are generated through modeling arbitrary polygon aggregates based on a pre-generated mesh [33]. The plane is meshed by triangular grids. The nodes' coordinate information is extracted as the basis to model aggregates, and check conflicts and overlaps. The distributions of aggregates accord with Fuller grading cumulative distribution. The function of Fuller grading curve can be decomposed into two dimensions [34] as:

$$P_c(D < D_0) = \qquad P_k\big(1.065D_0{}^{0.5}D_{max}{}^{-0.5} - 0.053D_0{}^4 D_{max}{}^{-4} \\ -0.012D_0{}^6 D_{max}{}^{-6} - 0.0045D_0{}^8 D_{max}{}^{-8} - 0.0025D_0{}^{10}D_{max}{}^{-10}\big) \tag{2}$$

where $P_c(D < D_0)$ denotes the proportion of aggregates whose particle diameter D are smaller than the threshold value D_0, P_k is the ratio of the total volume of aggregates to the concrete volume, and D_{max} is the diameter of the largest aggregate particle.

The geometrical shapes of aggregates are formed by the polygons' vertices. A random function expressed in terms of the polar coordinates r and θ is used to generate the vertices, as shown in Figure 2. To enforce randomicity and ergodicity, random numbers are introduced while generating the radius and angle. The vertices location (r, θ) is defined as follows:

$$\begin{cases} r = \frac{D}{2} \times [1 + rand(-1,1) \times f_r], \; 0 < f_r < 1 \\ \theta = \frac{2\pi}{\alpha} \times [i + rand(-1,1) \times f_\theta], \; 0 < f_\theta < 0.5, \; i = 0,1,2,\ldots\ldots,\alpha - 1 \end{cases} \tag{3}$$

where $rand(-1,1)$ is a random number less than 1 and greater than -1, α is the edge number, and f_r and f_θ are the radius and angle fluctuations respectively. f_r and f_θ are used to control the irregularity of aggregates. The greater f_r and f_θ, the more irregular the generated aggregates will be. If $f_r = 0$ and $f_\theta = 0$, the generated aggregates are regular polygons.

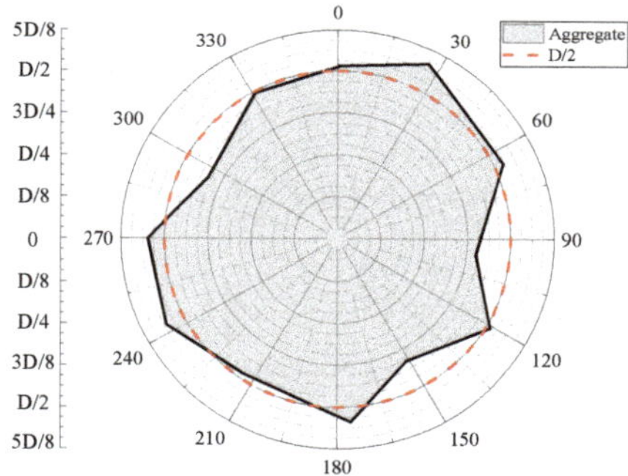

Figure 2. Aggregate generation in polar coordinates.

2.3. Interfacial Transitional Zones (ITZs) and Mortar Internal Interface (MII) Model

The ITZs and MII are modeled by eight-node three-dimensional cohesive elements (CH3D8). Cohesive elements only allow separation in one specific direction (Figure 3). Therefore, cracks can only propagate along the boundaries of particle elements, which may lead to mesh sensitivity [35]. In the case of control computing cost, the fine and irregular mesh is generated to provide more path choices for the crack propagation and to reduce the mesh sensitivity to some extent [36]. The geometrical (constitutive) thickness of cohesive elements is set to zero to maintain the original geometrical partition. The quadratic nominal stress criterion is adopted to define the beginning of an element's degradation, which is presented as follows:

$$\left(\frac{t_n}{t_n^0}\right)^2 + \left(\frac{t_s}{t_s^0}\right)^2 + \left(\frac{t_t}{t_t^0}\right)^2 = 1 \tag{4}$$

where t_n^0, t_s^0 and t_t^0 represent the maximum allowable nominal stresses in normal and two shear directions, respectively. Likewise, t_n, t_s and t_t represent the values of the nominal tensile stress and nominal shear stresses in two directions, respectively.

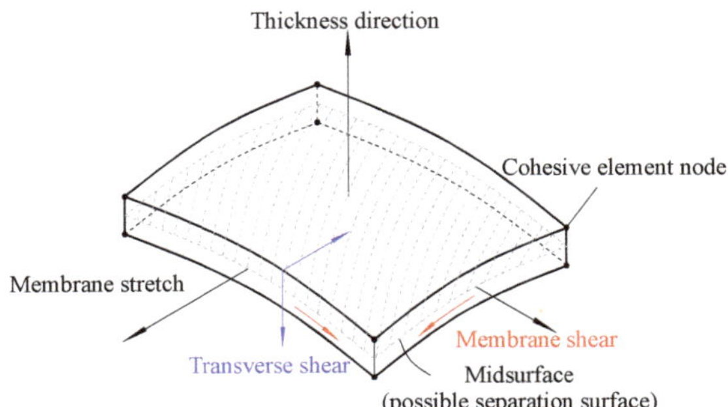

Figure 3. Spatial representation of CH3D8 (eight-node three-dimensional) cohesive element.

In this paper, linear damage evolution is selected to describe the traction-separation law, for its excellent computing efficiency and high accuracy [37].

Damage evolution is defined based on the mode I and II fracture energies, which are obtainable from the experiments. Since the model is two-dimensional, the out-of-plane shear failure mode will not happen, and the fracture energy in mode III is not taken into account. Different types of concrete shear experiments have shown that the mode II fracture energy of concrete is about 20 to 25 times larger than mode I fracture energy [22,38,39]. The findings are used in the present research.

Since crack propagation paths are meandering curves at the meso-scale, mixed-mode (I + II) fractures are more general than pure mode I or II fractures in concrete failure. The Benzeggagh–Kenane (B–K) fracture criterion [40] is applied to define the mixed-mode fracture energy. It is given by:

$$G^C = G_n^C + \left(G_s^C - G_n^C\right)\left\{\frac{G_s}{G_n + G_s}\right\}^{\eta} \tag{5}$$

where G^C is mixed-mode fracture energy, G_n^C and G_s^C are mode I and II fracture energies, G_s is the sum of shear strain energies in two directions, G_n is strain energy in the normal direction, and η is the material parameter. Since tearing mode fracture will not happen in the 2D models, G_s is equal to in-plane shear energy. G_n and G_s of a cohesive element are computed based on the deformation history at integration points in computation, which also relates to the damage evolution of the element [31]. The quadratic nominal stress criterion and mixed-mode fracture criterion are shown in Figure 4.

2.4. Interaction

Since particles bump and grind during compression, it is necessary to define the interaction characteristics to simulate the collision and friction. It is assumed that when the particle surfaces are in contact, any contact pressure can be transmitted. The 'hard' contact is adopted to describe the normal behavior of the interaction. According to the Coulomb friction mechanism, the friction behavior influences ultimate compressive strength and concrete performance in the softening stage [41]. The Coulomb friction model is adopted to describe the tangential behavior of the interaction. As with geomaterial, the friction behavior of concrete can be depicted by the friction angle φ [42]. The angle can be measured by the direct shear test and uniaxial compression test [43]. With the increase of concrete strength, the angle grows slightly. The friction angle ranges from 50 to 60 degrees. In the modeling, the friction angle φ is transformed into the friction coefficient μ. The transformation formula is shown as follows:

$$\mu = \tan \varphi \tag{6}$$

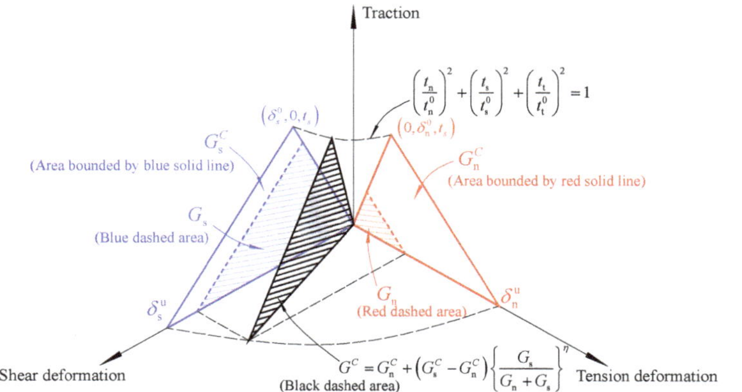

Figure 4. Illustration of quadratic nominal stress criterion and mixed-mode fracture criterion.

3. Calibration of Numerical Simulation

Modeling concrete, which contains multiple materials and interfaces at the meso-scale, requires a large number of parameters. Few experiments can provide the corresponding parameters that can one-to-one match with those used in the numerical simulation presented in this paper. Moreover, the cohesive zone model has a mesh sensitivity problem, so calibration is necessary for obtaining a suitable set of parameters and mesh precision.

In the calibration, it is found that the specimens with the same grading cumulative distribution but different aggregate distributions present different compressive properties, especially during the softening stage. Therefore, it is inadequate to include a single specimen to study the entire concrete performances. In order to conduct universal studies, ten specimens with different aggregate distributions are modeled randomly according to Equation (2) and (3). The number and proportion of aggregates included in the specimens are summarized in Table 1.

Table 1. Aggregate gradation and particle number.

Aggregate Particle Diameter D (mm)	Proportion	Particle Number
12 < D ≤ 15	5.36%	8–11
9 < D ≤ 12	8.83%	20–25
6 < D ≤ 9	11.74%	65–72
2.5 < D ≤ 6	19.18%	291–308
D ≤ 2.5	35.69%	/

To check the mesh dependence and computing efficiency, three models with the same aggregate distribution and different mesh precisions are tested. The approximate element sizes of the models are 1.5 mm, 1.0 mm, 0.5 mm (Figure 5). The number of bulk elements and cohesive elements are listed in Table 2. The time consumption of the models with element size of 1.0 mm and 0.5 mm are 2.5 and 11.6 times that of the model with 1.5 mm element size, respectively.

Table 2. The bulk and cohesive elements number of the model with different mesh precision.

Approximate Elements Size (mm)	Bulk Element	Cohesive Element
1.5	22,076	19,877
1.0	50,890	44,357
0.5	199,038	166,548

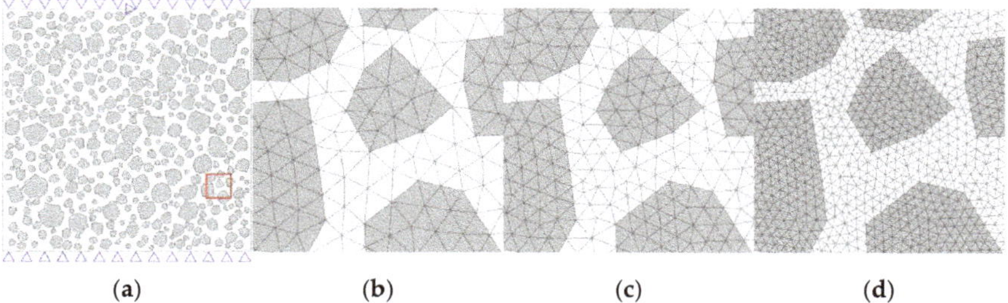

Figure 5. The meshes with different approximate element sizes: (**a**) boundary constraint condition marked by the blue triangles and the zoom area marked by the red square; (**b**) approximate size of 1.5 mm; (**c**) approximate size of 1.0 mm; (**d**) approximate size of 0.5 mm.

Figure 6 shows the stress-strain curves of the specimens with different element sizes. Comparing the tension and compression simulation results, it is shown that the compressive performance is more sensitive to mesh precision. It is indicated that the damage and fracture of cohesive elements lead to stress redistribution. With the increase of plastic deformation ability in the local area, the stress distribution becomes more uniform, which can make the particles stay at a higher average stress level. Hence, the increase of cohesive element number can contribute to higher ultimate compressive strength and higher residual strength in the softening stage. The mesh size of 1.0 mm is chosen for further experiments due to its good simulation accuracy and acceptable time consumption.

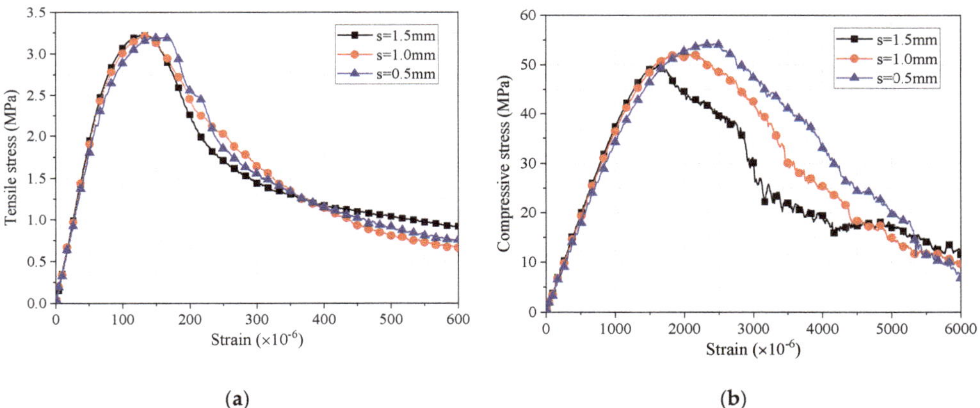

Figure 6. Effect of element size on global stress-strain relation under (**a**) tension and (**b**) compression.

Based on the outcomes of previous research and the simulation technique discussed in Section 2, the models are calibrated by considering interaction relations, material and interface parameters. Through the calibration, the frictional angle of 54.4° and the material parameters listed in Table 3 can make the models achieve satisfactory results, which are in close agreement with experimental results in both mechanical response and failure mode. In the following, the standard models used for analysis and parametric study are the models with ten different aggregate distributions and the calibrated parameters.

Table 3. Material properties of aggregate particles, mortar particles, interfacial transition zones (ITZs), and mortar internal interface (MII).

Parameter	Aggregate	Mortar	ITZs	MII
Elastic Modulus, E (GPa)	72	28	24	26
Poisson's ratio in plane, v_{12}	0.16	0.2	-	-
Poisson's ratio normal to plane, v_{13} and v_{23}	0	0	-	-
Shear modulus G (GPa)	31.0	11.7	10.0	10.8
Maximum nominal stress in normal direction, t_n^0 (MPa)	-	-	2.6	4
Maximum nominal stress in shear direction, t_s^0 and t_t^0 (MPa)	-	-	10	30
Normal Mode fracture energy, G_n^C (N/mm)	-	-	0.025	0.1
Shear mode Fracture energy, G_s^C (N/mm)	-	-	0.625	2.5
B-K criterion material parameter, η	-	-	1.2	1.2

4. Tension Simulation Result

The complete tensile stress-strain curves of the specimens are shown in Figure 7. A mean curve is plotted to show the integral trend of the curves. Assume that $\sigma_i(\varepsilon)$ is the stress of i-th specimen at the global strain ε. The points $(\overline{\sigma}, \varepsilon)$ on the mean curve can be calculated as follows:

$$\overline{\sigma} = \frac{1}{N}\sum_{i=1}^{N}\sigma_i(\varepsilon) \tag{7}$$

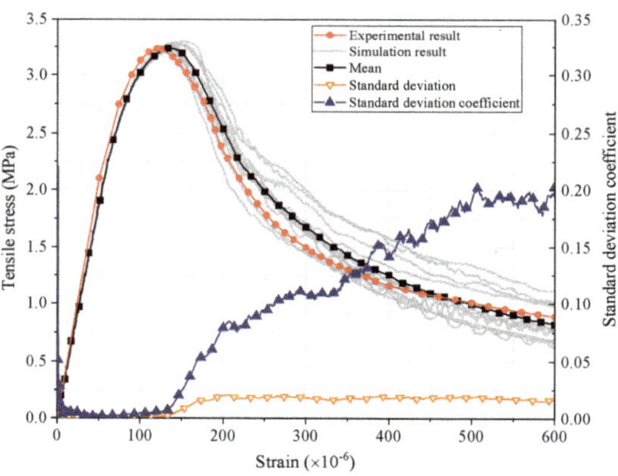

Figure 7. Statistical characteristic of global stress-strain curve under tension.

Through the calibration, the mean curve agrees well with the curve of the experimental result [32]. The curves of standard deviation and standard deviation coefficient that are calculated with simulation data are also plotted in Figure 7. The curves represent the absolute and relative errors of the simulation data, respectively. It is worth noting that the standard deviation almost remains constant in the softening stage. The stable deviation indicates that the specimens have a close damage evolution rate at the same global deformation level in spite of their different aggregate distributions.

As shown in Figure 8, there are two typical tension failure modes: the specimen fracture with one or two dominant cracks perpendicular to the tension direction. In the failure mode with only one dominant crack, the crack extends through the whole cross-section as the experimental result. The majority of the tension failures are this mode. In the other failure mode, the two dominant cracks at different horizontal positions extend from both sides. With the crack propagation, the horizontal projections of the two cracks intersect each other. Afterward, one of the crack's propagation rates declines, and the other crack extends through the whole cross-section. Figure 9 shows the strain-stress curves of the specimens with one or two dominant cracks. The strength and energy dissipation of the specimen with two dominant cracks is higher than that of the specimen with one dominant crack

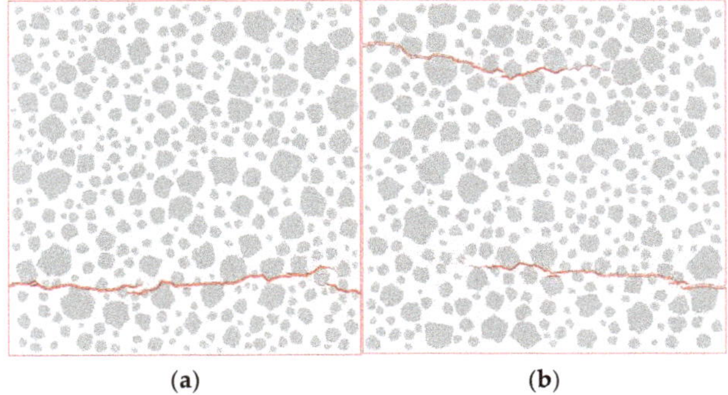

(a)   (b)

Figure 8. Failure modes of the specimens under tension: (**a**) simulation result (with one dominant crack); (**b**) simulation result (with two dominant cracks).

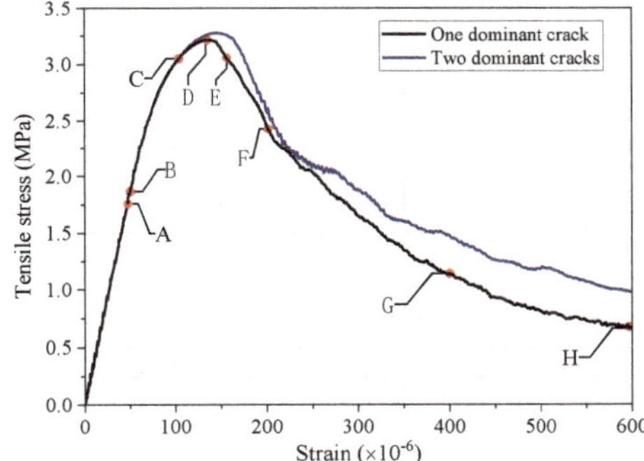

Figure 9. Global stress-strain curve of the specimens with one and two dominant cracks.

The specimen with one dominant crack in Figure 8 is selected for the deformation process analysis. In order to observe the whole deformation process, the deformation of the specimen is enlarged geometrically, as shown in Figure 10. The specimen's original deformation in the direction of tension is multiplied by a deformation scale factor D_V to make the aspect ratio become 2:1 from 1:1. Eight characteristic states are selected to describe the deformation process. The corresponding points of the states are plotted on the stress-strain curve in Figure 9.

After the specimen is loaded, before state A, the deformation evenly distributes at the aggregates and mortar, as shown in Figure 10a. The obvious plastic deformation appears since state B. Although the global stress hasn't reached the damage initial stress of the ITZ elements, a few ITZ elements begin sustaining damage due to the stress concentration caused by the deformation difference between mortar and aggregates.

It is obvious from Figure 10c that the deformation concentrates at ITZ elements and nearby MII elements at state C. Many deformation concentrated areas are formed as spindle-shaped, which are comprised by ITZ elements at mid-piece and MII elements at both ends. When the specimen attained the ultimate tensile strength state, two potential fracture zones (FZ1 and FZ2) are formed by the deformation concentrated areas, as shown in Figure 10d.

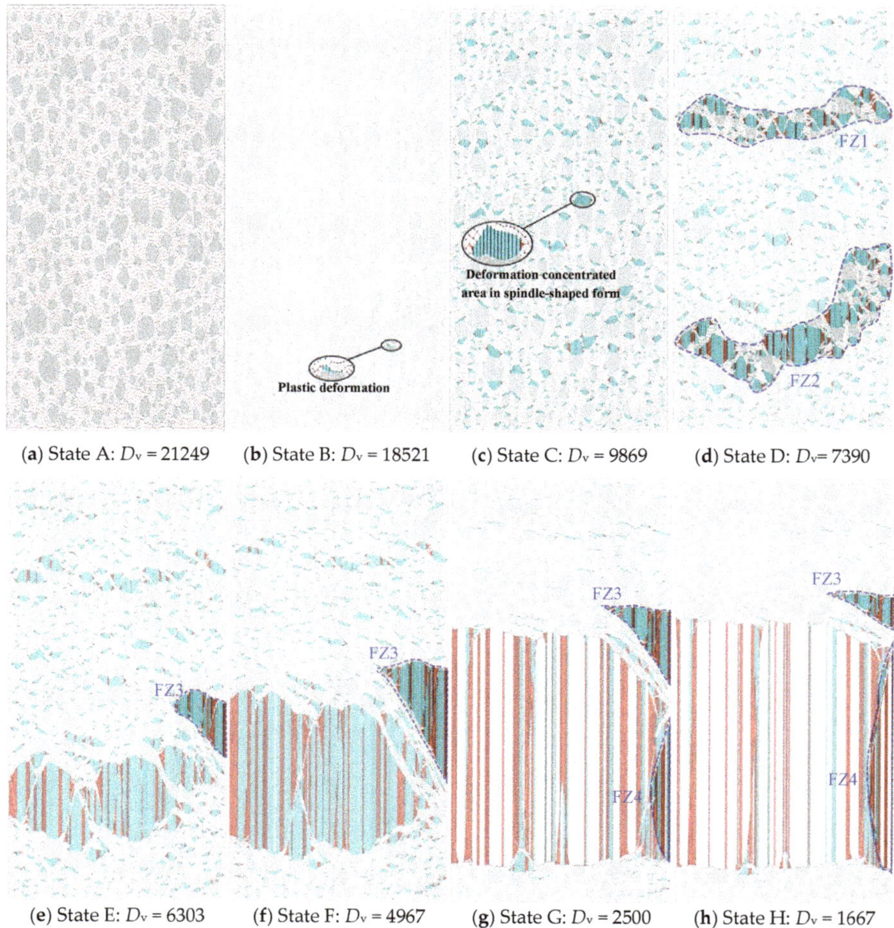

(**a**) State A: $D_v = 21249$ (**b**) State B: $D_v = 18521$ (**c**) State C: $D_v = 9869$ (**d**) State D: $D_v = 7390$

(**e**) State E: $D_v = 6303$ (**f**) State F: $D_v = 4967$ (**g**) State G: $D_v = 2500$ (**h**) State H: $D_v = 1667$

Figure 10. Deformation magnification in loading direction of a specimen in eight states. Cohesive elements of ITZs (blue coloured) and mortar internal interface (red coloured).

At state E, one of the potential fracture zones (FZ1) degrades, and another one (FZ2) extends unstably. At state F, the first fracture initiates at the center-left. The deformation of MII elements is very close to that of the nearby ITZ elements. The original spindle-shaped characteristic of the deformation areas has almost disappeared.

Although the cracks develop and extend quickly afterward, the drop rate of bearing capacity slows down, and the curve gradually flattens. At state G, a large number of ITZ elements are eliminated. As can be noticed, a new deformation concentrated area (FZ4) has been generated at the bottom right corner of the potential fracture zone during the stage between state F and G, while the potential fracture zone (FZ3), which was generated since state D, begins contracting. At state H, fracture happens in this new deformation concentrated area (FZ4). This illustrates that although the tension crack generally propagates through the main deformation concentrated areas, the crack would bypass parts of areas and propagate through a new fracture path due to the force redistribution.

5. Compression Simulation Result

The aggregate distribution has a substantial influence on the concrete compressive property, especially in the softening stage. Compared with the tension simulation result, the result of compression simulation shows greater discreteness. The tensile behavior is determined by the weakest section, while compressive behavior is determined by the entire specimen [41].

The crack initiation is caused by the localized shear failure, as shown in Figure 11. There are two to four main shear bands crossing the whole section in the middle region accompanied by several small bands on both sides. The angles between the shear bands and the direction of compression range from 5 to 15 degrees. The specimens are divided into several wedge-shaped blocks by the shear bands as the experimental result [32].

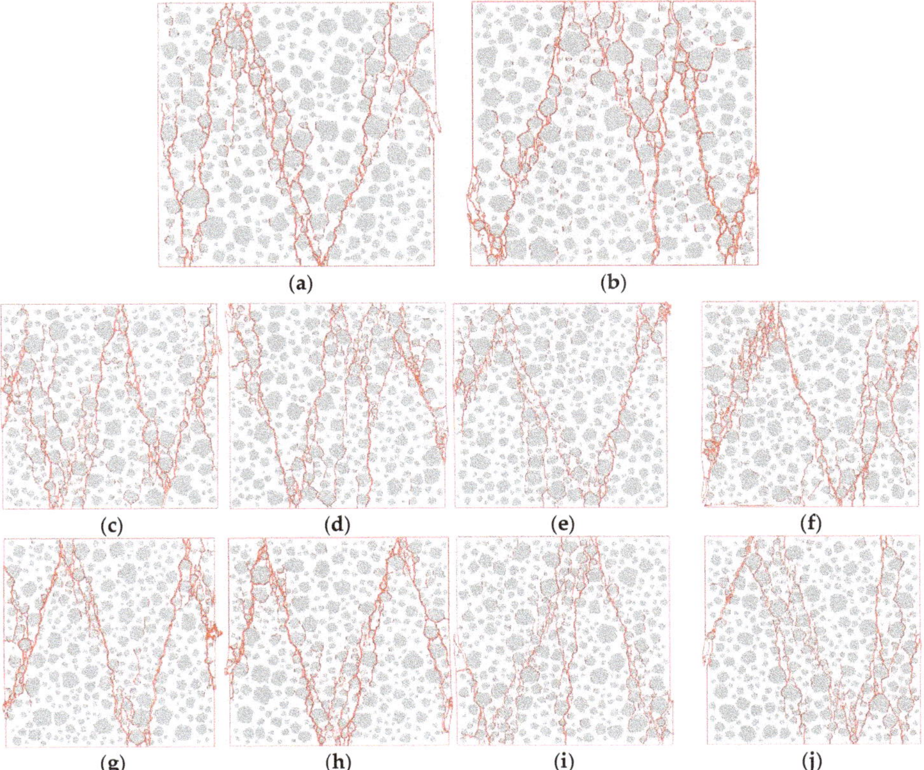

Figure 11. Failure modes of the specimens with different aggregate distributions under compression

The complete compressive stress-strain curves are shown in Figure 12. As with the statistical characteristics of tension stress-curves, the datum maintains a low dispersion before the mean compressive stress reaches its peak. Remarkably, both the curves of standard deviation and standard deviation coefficient have a 'valley' at the specimens' softening stage. In the descending branch, the stress-strain curves of the specimens are pinched together at a certain state. The 'valley' symbolizes that there is a relatively steady state in the softening stage, at which the specimens have a close residual bearing capacity.

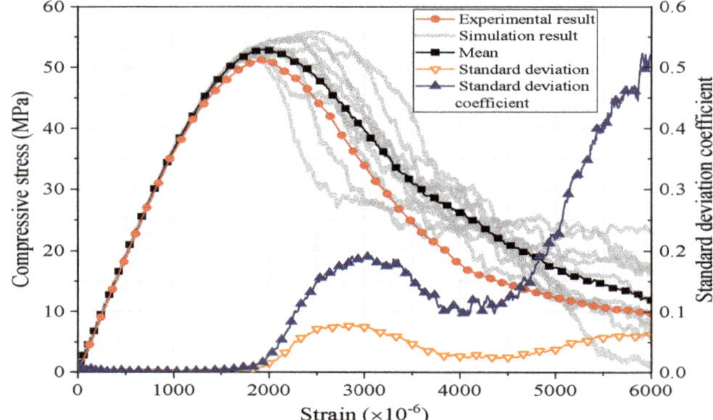

Figure 12. Statistical characteristic of global stress-strain curve under compression.

Figure 13 shows the complete compressive stress-strain curve of the model with the same aggregate distribution as the one analyzed in Section 4. Seven states labeled from 'A' to 'G' on the curve are selected to describe the crack propagation process as shown in Figure 14. The cracks newly appearing in each state are differentiated by colors.

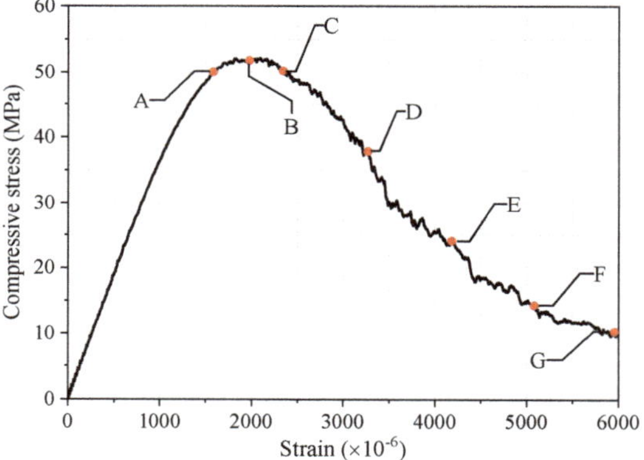

Figure 13. Global stress-strain curve of a specimen under compression.

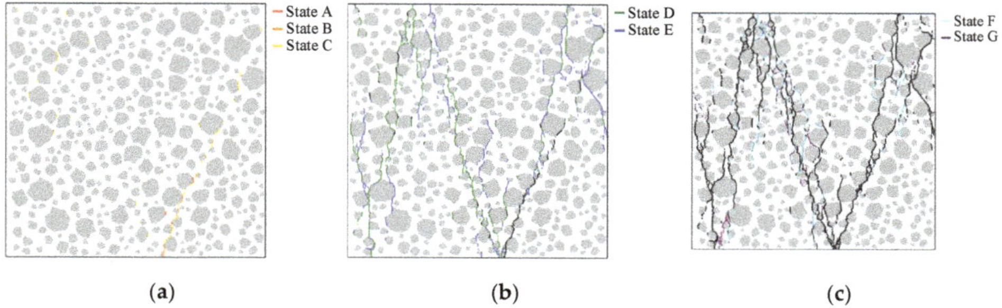

(**a**)　　　　　　　　　　(**b**)　　　　　　　　　　(**c**)

Figure 14. Crack propagation process of a specimen: (**a**) State A, B, C; (**b**) State D, E; (**c**) State F, G.

The sums of the lengths of ITZ and MII elements with different damage degrees are accumulated, as shown in Figure 15. SDEG (overall value of the scalar damage variable) values of the elements are extracted from the result file to determine the damage degree. When an element's SDEG value equals 1, the element would be eliminated to simulate visible fracture. The lengths of the elements are calculated by the coordinates of the elements' nodes. Since the nonlinear behavior of the specimen is simulated by the damage of ITZ and MII elements, the total length of the elements represents a specimen's damage condition.

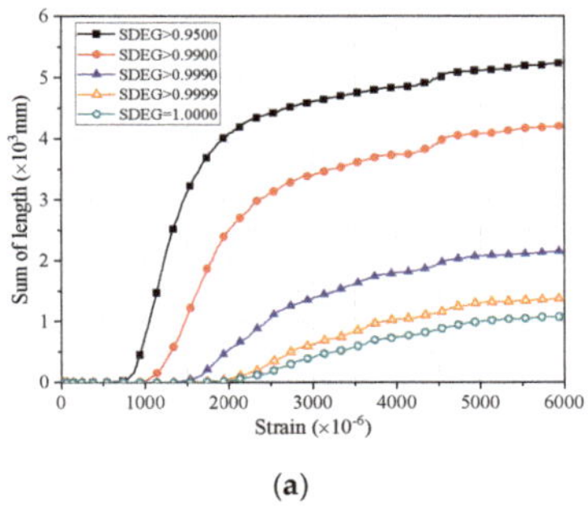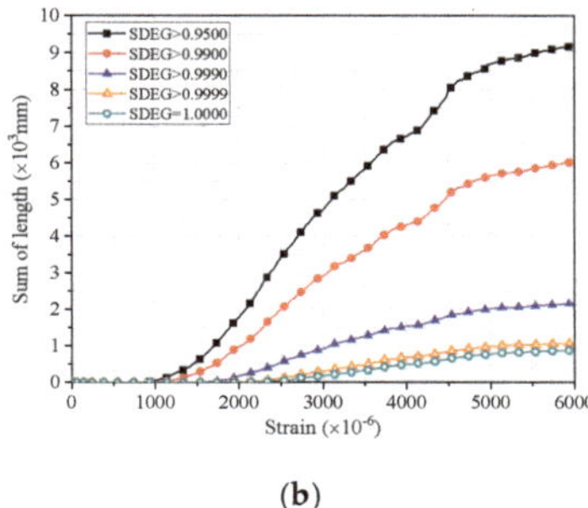

(a)

(b)

Figure 15. The relationship between global strain and the sum of length of (**a**) ITZ elements and (**b**) MII elements with different damage degree.

The specimen is linear-elastic until the first damage occurs. Damage firstly occurs at ITZ elements, and then MII elements. The damage evolution rate of the specimen gradually accelerates and achieves a constant speed with the deformation increased. Although the ultimate bearing capacity hasn't been reached, the crack initiates at state A. The crack nucleates at the ITZ elements parallel to the loading direction and propagates slowly before the state B.

Afterwards, several small cracks appear from the bottom middle-right to the top-right corner during the stage between state B and C. The bearing capacity gradually declines at this stage. A rudiment of a shear band seems to be formed by these small cracks.

The residual bearing capacity descends rapidly during the stage between states C and E, accompanied by the appearance of main shear bands. The MII elements' fracture leads to the branching and coalescence of the crack. In this stage, the damage evolution rate is fast but steady.

It is interesting that the valleys of the standard deviation curve and standard deviation coefficient curve occur when the specimen is at state E, which indicates that the relatively steady state is attributed to the formation of main shear bands. At this state, the residual capacity depends on the friction action of the blocks divided by the shear bands. In the following loading process, the crack propagation rate slows down. The specimen's deformation gradually evolves into the slipping of the wedge-shaped blocks.

6. Parametric Study

6.1. The Effect of Mode II Fracture Energy

Since mode I fracture was considered as the dominant form in tension damage, single fracture energy was adopted to describe the interface fracture behavior in the previous concrete tension simulation. In this method, the fracture energies of the interface elements are fixed values regardless of fracture type. Moreover, since the compression failure of concrete is mostly due to shear fracture [41], modeling with one specific fracture energy is unable to simulate the fracture behavior of concrete both under tension and compression. Due to the multiphase and heterogeneous property of concrete, the mixed-mode fracture would occur inevitably in concrete failure, even under uniaxial tension conditions. Mixed-mode fracture energy is determined by mode I and mode II fracture energy, and the two pure-mode fracture energies both have influences on concrete tensile and compressive behavior.

In this section, the effect of mode II fracture energy on the performance of concrete under tension and compression is studied.

As mentioned in Section 2, the experiment results show that there is a relationship between mode I and II fracture energy. In this section, mode I fracture energy is fixed. The ratio of mode II and mode I fracture energy (ξ) is taken as the variable to investigate the effect of mode II fracture energy. The ratio ξ is calculated by the following equation:

$$\xi = G_s^C / G_n^C \tag{8}$$

The tension simulation result of varying ξ is shown in Figure 16a. The influence on the ultimate tensile strength is negligible. It is noteworthy that the mode II fracture energy influences the specimen's tensile behavior in the softening stage. The mixed-mode fracture accounts for a large proportion. In the case that the mode II fracture energy equals the mode I fracture energy ($\xi = 1$), the total fracture energy is 48.2% lower than that of the standard model ($\xi = 25$). It can be inferred that the mixed-mode fracture energy caused by the difference of the two pure-mode fracture energy accounts for about half of the total fracture energy.

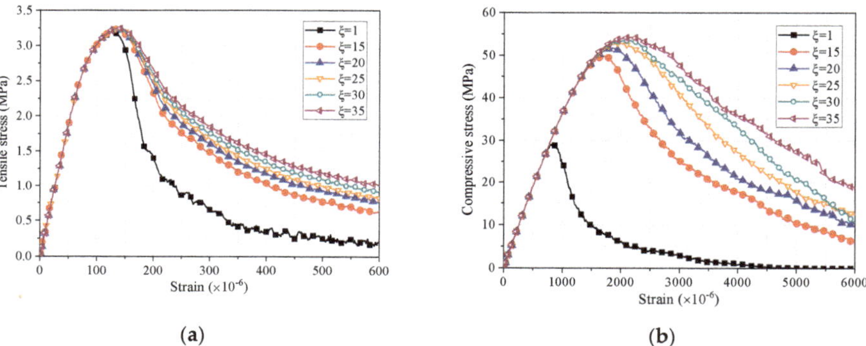

Figure 16. Effect of mode II fracture energy on global stress-strain relationship under tension (**a**) and compression (**b**).

The compression simulation result of varying the multiple is shown in Figure 16b. The ultimate compressive strength is significantly influenced by the mode II fracture energy, which is different from tensile strength. When the mode II fracture energy equals the mode I fracture energy ($\xi = 1$), the compressive strength and total fracture energy is 44.7% and 82.2% lower than that of the standard model ($\xi = 25$), respectively. As described in Sections 3 and 4, the crack initiates before and after the specimen reaches the ultimate bearing capacity state under compression and tension, respectively. Low mode II fracture energy will limit interface elements' shear deformation under compression, while it has a relatively low influence on the interface elements' tensile deformation under tension. It can be noticed that with the increase of mode II fracture energy, the growth of compressive strength decreases. The limit of compressive strength gradually translates from deformation ability to the shear strength of interface elements.

6.2. The Effect of Mixed-Mode Fracture Criterion

The concrete's fracture is generally mixed-mode, especially under compression. The Benzeggagh–Kenane (B–K) fracture criterion with the material parameter η is adopted to define the damage evolution of mixed-mode fracture. It is accepted that the B–K criterion can adequately specify the mixed-mode fracture energy for a wide range of composite material [44].

In this section, the material parameter η is taken as the variable to investigate the effect of mixed-mode fracture criterion. The other parameters are held the same, as listed in Table 3. Since there are limited studies of the material parameter η of concrete, the parameter value of brittle epoxy resin is taken as a reference, which ranges from 1.0 to 2.0. Through calibration, it seems that 1.2 is an appropriate value for the parameter η under both compression and tension conditions.

Figure 17 shows the complete tensile and compressive stress-strain mean curves for varying the parameter η. The results indicate that the material parameter mainly affects the concrete performance in the softening stage under both tension and compression. The ultimate tensile and compressive strengths are slightly affected. It seems that the curves separate at the peak and gradually tend to decrease in parallel with the increase of global strain. For the specimens with the same aggregate distribution, the crack propagation paths are similar under tension, as are the failure modes and the sum of crack length under compression, as shown in Figure 8. According to the B–K criterion expression (Equation (5)), if the ratio of strain energy at a tangent to that in the normal direction is constant, a decrease of material parameter η leads to the increase of total fracture energy G^C. Due to the similarity in failure modes, the bearing capacities of the specimens are mainly determined by the fracture energy.

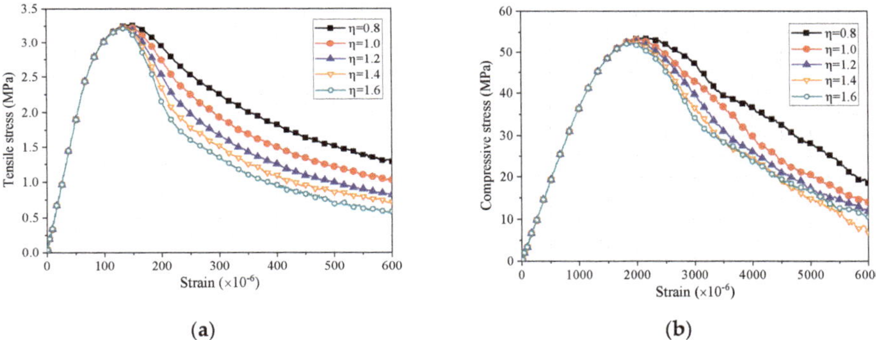

(**a**)　　　　　　　　　　　(**b**)

Figure 17. Effect of material parameter on global stress-strain relationship under (**a**) tension and (**b**) compression.

6.3. The Effect of Frictional Behavior

Coulombic friction is considered a key factor that influences the ultimate strength and residual bearing capacity of concrete. The results of the direct shear test and uniaxial compression test showed that the concrete with higher compressive strength has a greater internal friction angle. Although a desirable strength of concrete specimens can be achieved in experiments by adjusting the material types and aggregate gradation, it is hard to control the maximum allowable frictional stress. Therefore, the frictional angle α is taken as the variable to investigate the effect of frictional behavior.

As shown in Figure 18, the internal frictional behavior of concrete affects the final failure mode. With the increase of frictional angle, the inclination angles of main shear bands decrease. The local cracks tend to propagate dispersedly into concrete blocks rather than in the direction of the shear bands. The stress-strain curves of varying frictional angle α are shown in Figure 19. As expected, the increase of frictional angle value has a positive influence on the concrete bearing capacity in the whole loading process. The influence strengthens once the plastic deformation appears and weakens when the specimens' bearing capacity declines by about a third. The sum of the crack length of the specimens is calculated as shown in Figure 20.

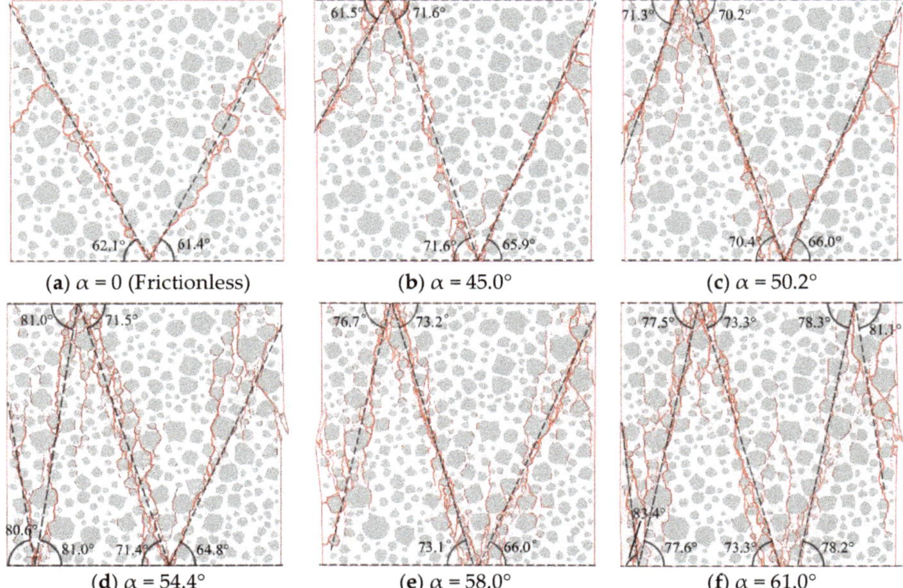

Figure 18. Failure modes of a specimen under the different frictional angles.

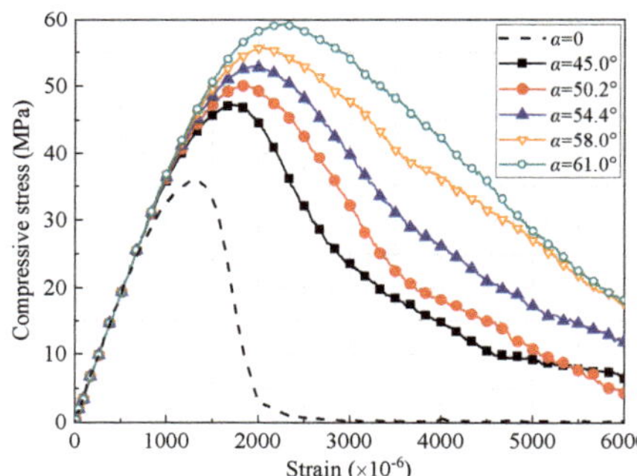

Figure 19. Effect of frictional angle on global stress-strain relationship under compression.

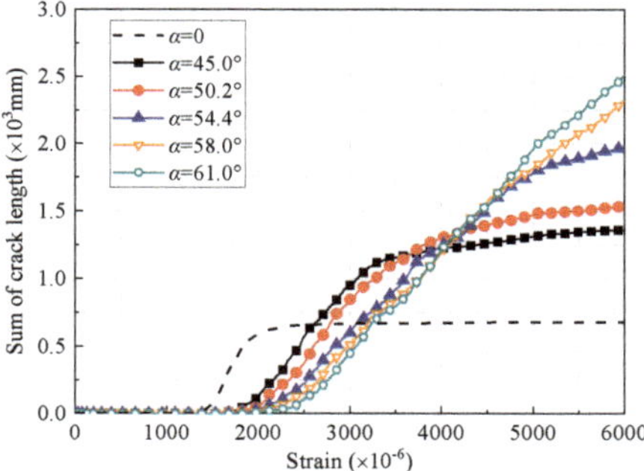

Figure 20. Effect of frictional angle on crack propagation under compression.

When the tangential interaction is frictionless ($\alpha = 0$), the compressive strength and energy dissipations are 35.9% and 77.1% lower than that of the standard model ($\alpha = 54.4°$), respectively. It is noticed that the crack propagates more sufficiently and the inclination angle of the shear bands decreases with the increase of frictional angle. Although the cracks' local propagation paths are different, the main shear bands of the specimens are similar. The friction stress could constrain the crack propagation at the beginning of the damage, which benefits the ultimate bearing capacity. In the softening stage, the friction stress could help reestablish the loading path and make more blocks participate in bearing load, which may cause more cracks.

6.4. The Effect of Aggregates' Mechanical Characteristics

Although the tensile behavior of concrete is dominated by the properties of ITZs and mortar, damage and fracture may happen in aggregates during compression [45]. Different kinds of aggregates have differences in mechanical characteristics, geometrical shape, and corresponding ITZ properties [46]. To study the effect of aggregates' mechanical characteristics on concrete compressive performance, based on standard models, cohesive elements are inserted into aggregates to simulate aggregate internal interfaces (AII). The material parameters of AII elements are set by referring to the material tests of the following common aggregate rocks: limestone [47], marble [48], granite [49], and basalt [50]. The material parameters of AII elements are listed in Table 4.

Table 4. Material properties of aggregate internal interfaces (AIIs).

Parameter	Limestone	Marble	Granite	Basalt
Maximum nominal stress in normal direction, t_n^0 (MPa)	6.0	4.5	10.0	15.0
Maximum nominal stress in shear direction, t_s^0 and t_t^0 (MPa)	45.0	40.0	90.0	120.0
Normal Mode fracture energy, G_n^C (N/mm)	60	160	400	1400
Shear mode Fracture energy, G_s^C (N/mm)	1500	4000	10,000	35,000
B-K criterion material parameter, η	1.2	1.2	1.2	1.2

The stress-strain curves of varying aggregates' mechanical characteristics are shown in Figure 21. Compared with the ultimate bearing capacity of the models without considering aggregate damage and fracture, the ultimate bearing capacities of the models with limestone, marble, granite, basalt aggregates fall by averages of 1.1%, 1.0%, 0.23%, 0.18%, respectively. The effect of aggregate type on the stress-strain relationship is mainly reflected in the softening stage. As shown in Figure 22, damage evolution of aggregates accelerates in the early softening stage. Since marble aggregates have the lowest tensile and shear strength, they are more likely to be damaged during the compression process. The energy dissipation of aggregates can lead to stress redistribution at the local area and enhance residual strength.

Figure 23 shows the failure modes of the models with the same aggregate distribution and different aggregates' mechanical characteristics. It can be seen that the fracture of aggregates only constitutes a minority, but the crack propagation paths at the models' left half are obviously influenced by local crack propagation differences caused by aggregates' fracture. Since the shear bands at the models' right half form earlier than those at left half, the models have very similar failure modes (FZ5–FZ9) at the right half. In view of these observations, it can be concluded that the influence of aggregates' mechanical characteristics on concrete's mechanical characteristics and failure modes is mainly concentrated in the mid-and-late softening stage.

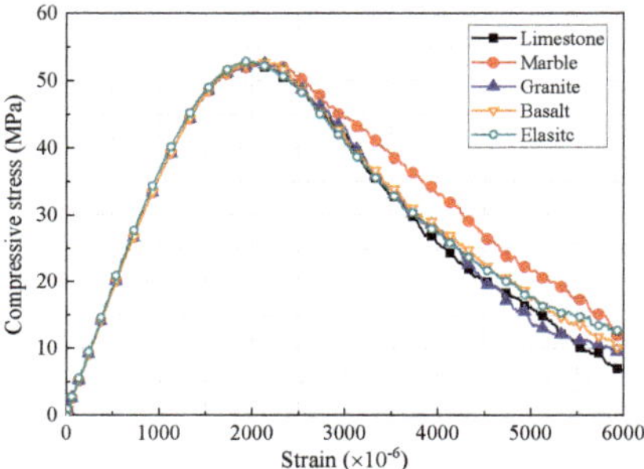

Figure 21. Effect of aggregates' mechanical characteristics on global stress-strain relationship under compression.

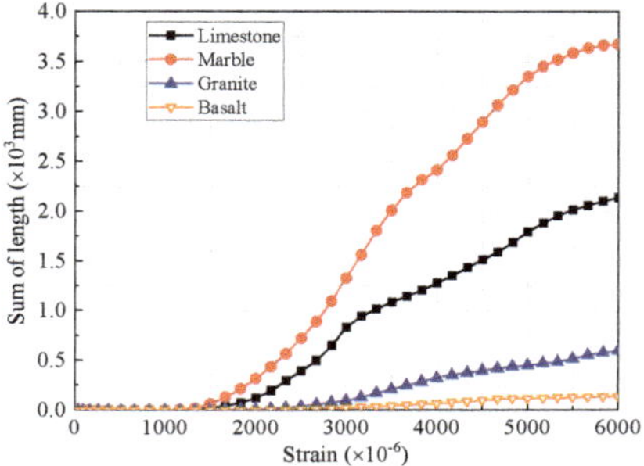

Figure 22. The relationship between strain and the sum of length of AII elements whose SDEG (overall value of the scalar damage variable) values are higher than 0.95.

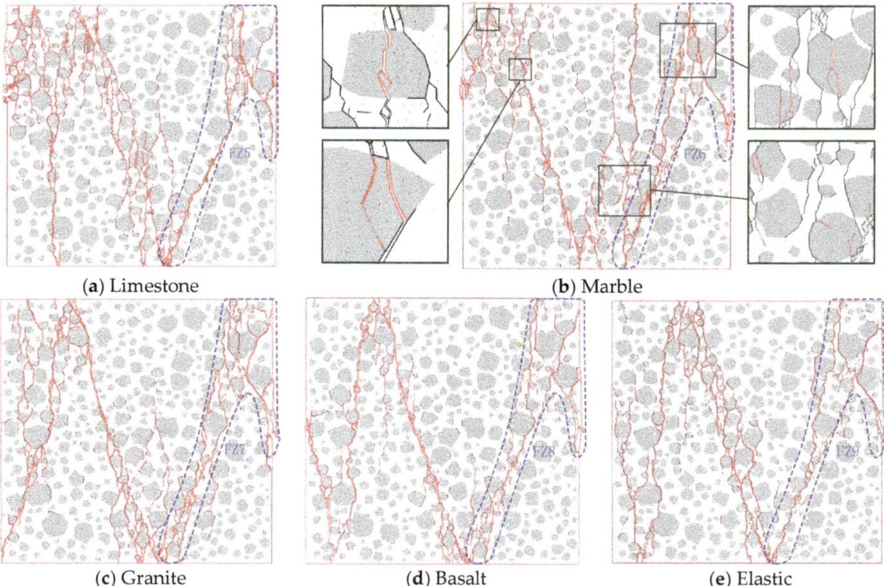

(**a**) Limestone   (**b**) Marble

(**c**) Granite   (**d**) Basalt   (**e**) Elastic

Figure 23. Failure modes of a specimen with different aggregate types; in the zoom areas of Figure 23b, the cracks through aggregates are marked by red, the others are marked by black.

7. Conclusions

The application range of the cohesive zone model for simulation of three-phase concrete (i.e., aggregate, mortar, and ITZs) has been extended. Through the definition of mixed-mode fracture and interaction behavior, the model is suitable for simulation both under tension and compression. The modeling method and relevant material parameters are discussed based on the simulation technique and previous material experiment results. A balance between mesh sensitivity and computing efficiency has been achieved through the calibration of mesh precision. The simulation results have good agreement with the experiments in both mechanical characteristics and failure modes. The whole process analyses and parametric studies are conducted to investigate the failure mechanism of concrete under tension and compression. The following conclusions can be drawn from this study.

The ITZ and MII elements modeled by cohesive elements are appropriate to simulate concrete meso-scale fracture behavior and damage accumulation at a smaller scale. The mesh precision level in which the approximate element size is 1/150 of the size of the model can provide a relatively high simulation accuracy and calculation efficiency.

The discreteness of compressive behavior caused by aggregate distribution on the stress-strain curve is larger than that of tension behavior. In the tension simulation, the specimens with different aggregate distributions have similar damage evolution rates at the same global deformation level. The failure modes of concrete can be separated into two types: one or two dominant cracks perpendicular to the tension direction.

In the concrete compression simulation, the inclination angles of shear bands range from 5 to 15 degrees. A valley in the standard deviation coefficient curve is formed, which symbolizes a relatively steady state in the softening stage when main shear bands form.

In the parametric study, it is found that the mixed-mode fracture accounts for a considerable proportion both in tension and compression failure. The fracture energy caused by the difference of the two pure-mode fracture energy occupies 48.2% and 82.2% of the total fracture energy under tension and compression, respectively. Mode II fracture energy has a significant impact on both ultimate and residual compressive strength of concrete. In the case that mode II fracture energy is equal to mode I fracture energy, the ultimate compressive strength reduces by 44.7%.

The mixed-mode fracture criterion mainly affects concrete performance in the softening stage under both tension and compression, while the effects on failure mode, tensile and compressive strength are negligible.

Frictional behavior mainly affects concrete compressive performance. In the case that there is no internal frictional behavior, the ultimate compressive strength and energy dissipation reduce by 35.9% and 77.1%, respectively. The frictional stress can constrain crack propagation at the beginning of damage and reestablish the loading path in the softening stage.

If the influence of aggregates' geometrical shapes and ITZ properties are subtracted, aggregates' mechanical characteristics affect concrete's mechanical characteristics and failure modes, mainly concentrated in the mid-and-late softening stage.

Author Contributions: Conceptualization, X.X. and Q.X.; methodology, X.X. and Q.X.; software, Q.X.; validation, X.X.; formal analysis, Q.X.; investigation, X.X. and Q.X.; resources, X.X.; data curation, Q.X.; writing—original draft preparation, Q.X.; writing—review and editing, X.X. and Q.X.; visualization, X.X. and Q.X; supervision, X.X.; project administration, X.X.; funding acquisition, X.X.

Funding: This research was funded by the National Natural Science Foundation of China (Grant Nos. 51578400, 51178328), Project of Science and Technology Commission of Shanghai Municipality (Grant No. 16DZ1201602).

References

1. Kaplan, M.F. Crack propagation and the fracture of concrete. *J. Am. Concr. Inst.* **1961**, *58*, 591–610.
2. Wittmann, F.H. (Ed.) Structure of concrete with respect to crack formation. In *Fracture Mechanics of Concrete*; Elsevier: London, UK, 1983; pp. 43–74.
3. Yu, X. (Ed.) Nonlinear fracture mechanics of concrete. In *Fracture Mechanics Rock Concrete*; Central South University of Technology Press: Changsha, China, 1991; pp. 335–381.
4. Kaczmarczyk, J.; Kozłowska, A.; Grajcar, A.; Sławski, S. Modelling and microstructural aspects of ultra-thin sheet metal bundle cutting. *Metals* **2019**, *9*, 162. [CrossRef]
5. Baskes, M.; Hoagland, R. Summary report: Computational issues in the mechanical behavior of metals and intermetallics. *Mater. Sci. Eng.* **1992**, *159*, 1–34. [CrossRef]
6. Dawson, P.R.; Needleman, A.; Suresh, S. Issues in the finite-element modeling of polyphase plasticity. *Mater. Sci. Eng. A-Structural Mater. Prop. Microstruct. Process.* **1994**, *175*, 43–48. [CrossRef]
7. Ngo, D.; Scordelis, C. Finite element analysis of reinforced concrete beams. *ACI* **1967**, *64*, 152–163.
8. Rashid, Y.R. Ultimate strength analysis of prestressed concrete pressure vessels. *Nucl. Eng. Des.* **1968**, *7*, 334–344. [CrossRef]
9. Herrmann, H.J.; Hansen, A.; Roux, S. Fracture of disordered, elastic lattices in two dimensions. *Phys. Rev. B* **1989**, *39*, 637–648. [CrossRef]
10. Kikuchi, A.; Kawai, T.; Suzuki, N. The rigid bodies-spring models and their applications to three-dimensional crack problems. *Comput. Struct.* **1992**, *44*, 469–480. [CrossRef]
11. Carol, I.; López, C.M.; Roa, O. Micromechanical analysis of quasi-brittle materials using fracture-based interface elements. *Int. J. Numer. Methods Eng.* **2001**, *52*, 193–215. [CrossRef]
12. Hillerborg, A.; Modeer, M.; Petersson, P. Analysis of crack formation and crack growth in concrete by means of fracture mechanics and finite elements Hillerborg. *Cem. Concr. Res.* **1976**, *6*, 773–782. [CrossRef]
13. Belytscho, T.; Black, T. Elastic crack growth with minimal remeshing. *Int. J. Numer. Methods Eng.* **1999**, *45*, 610–620.
14. De Cicco, D.; Taheri, F. Delamination buckling and crack propagation simulations in fiber-metal laminates using xfem and cohesive elements. *Appl. Sci.* **2018**, *8*, 2440. [CrossRef]
15. Bažant, B. Oh, crack band theory of concrete. *Mater. Struct.* **1983**, *16*, 155–177. [CrossRef]
16. Duan, K.; Hu, X.; Wittmann, F.H. Boundary effect on concrete fracture and non-constant fracture energy distribution. *Eng. Fract. Mech.* **2003**, *70*, 2257–2268. [CrossRef]
17. Cifuentes, H.; Alcalde, M.; Medina, F. Measuring the size-independent fracture energy of concrete. *Strain* **2013**, *49*, 54–59. [CrossRef]
18. Vishalakshi, K.P.; Revathi, V.; Sivamurthy Reddy, S. Effect of type of coarse aggregate on the strength properties and fracture energy of normal and high strength concrete. *Eng. Fract. Mech.* **2018**, *194*, 52–60. [CrossRef]
19. Merta, I.; Tschegg, E.K. Fracture energy of natural fibre reinforced concrete. *Constr. Build. Mater.* **2013**, *40*, 991–997. [CrossRef]
20. Vydra, V.; Trtík, K.; Vodák, F. Size independent fracture energy of concrete. *Constr. Build. Mater.* **2012**, *26*, 357–361. [CrossRef]
21. RILEM Technical Committee 50. Determination of fracture energy of mortar and concrete by means of three point bend tests on notched beams. *Mater. Struct.* **1985**, *18*, 285–290.
22. Bažant, Z.P.; Pfeiffer, P.A. Shear fracture tests of concrete. *Mater. Struct.* **1986**, *19*, 111–121. [CrossRef]
23. Kim, Y.R.; de Freitas, F.A.C.; Jung, J.S.; Sim, Y. Characterization of bitumen fracture using tensile tests incorporated with viscoelastic cohesive zone model. *Constr. Build. Mater.* **2015**, *88*, 1–9. [CrossRef]
24. Kim, S.M.; Al-Rub, R.K.A. Meso-scale computational modeling of the plastic-damage response of cementitious composites. *Cem. Concr. Res.* **2011**, *41*, 339–358. [CrossRef]
25. Trawiński, W.; Bobiński, J.; Tejchman, J. Two-dimensional simulations of concrete fracture at aggregate level with cohesive elements based on X-ray μCT images. *Eng. Fract. Mech.* **2016**, *168*, 204–226. [CrossRef]
26. Liang, S.; Li, J.; Yu, F. A Multi-scale analysis-based stochastic damage model of concrete. *J. Tongji Univ.* **2017**, *45*, 1249–1257.
27. Yılmaz, O.; Molinari, J.-F. A mesoscale fracture model for concrete. *Cem. Concr. Res.* **2017**, *97*, 84–94. [CrossRef]

28. Trawiński, W.; Tejchman, J.; Bobiński, J. A three-dimensional meso-scale modelling of concrete fracture, based on cohesive elements and X-ray μCT images. *Eng. Fract. Mech.* **2018**, *189*, 27–50. [CrossRef]

29. Unger, J.F.; Eckardt, S.; Koenke, C. A mesoscale model for concrete to simulate mechanical failure. *Comput. Concr.* **2011**, *8*, 401–423. [CrossRef]

30. Gao, X.; Liu, C.; Tan, Y.; Yang, N.; Qiao, Y.; Hu, Y.; Li, Q.; Koval, G.; Chazallon, C. Determination of fracture properties of concrete using size and boundary effect models. *Appl. Sci.* **2019**, *9*, 1337. [CrossRef]

31. Dassault Systèmes, D.S. *Abaqus Analysis User's Guide*; Technical Report Abaqus 6.14 Documentation. Simulia Corp., 2016. Available online: http://abaqus.software.polimi.it/v6.14/index.html (accessed on 25 July 2019).

32. Ren, X.; Yang, W.; Zhou, Y.; Li, J. Behavior of high-performance concrete under uniaxial and biaxial loading. *ACI Mater. J.* **2008**, *105*, 548–558.

33. Wang, B.; Hong, W.; Zhang, Z.Q.; Zhou, M.J. Mesoscopic modeling method of concrete aggregates with arbitrary shapes based on mesh generation. *Chin. J. Comput. Mech.* **2017**, *34*, 591–596.

34. Walraven, J.; Reinhardt, H. Theory and experiments on the mechanical behaviour of cracks in plan and reinforced concrete subjected to shear loading. *Heron J.* **1981**, *26*, 5–68.

35. Bažant, Z.P.; Jirasek, M. Nonlocal integral formulations of plasticity and damage: Survey of progress. *J. Eng. Mech.* **2016**, *128*, 1119–1149. [CrossRef]

36. Yang, Z.; Xu, X.F. A heterogeneous cohesive model for quasi-brittle materials considering spatially varying random fracture properties. *Comput. Methods Appl. Mech. Eng.* **2008**, *197*, 4027–4039. [CrossRef]

37. Alfano, G. On the influence of the shape of the interface law on the application of cohesive-zone models. *Compos. Sci. Technol.* **2006**, *66*, 723–730. [CrossRef]

38. Reinhardt, H.W.; Xu, S. A practical testing approach to determine mode II fracture energy GIIF for concrete. *Int. J. Fract.* **2000**, *105*, 107–125. [CrossRef]

39. Kumar, C.N.S.; Rao, T.D.G. Punching shear resistance of concrete slabs using mode-II fracture energy. *Eng. Fract. Mech.* **2012**, *83*, 75–85. [CrossRef]

40. Benzeggagh, M.L.; Kenane, M.; Aldcb, A. Measurement of mixed-mode delamination fracture toughness of unidirectional glass/epoxy composites with mixed-mode bending apparatus. *Compos. Sci. Technol.* **1996**, *56*, 439–449. [CrossRef]

41. Van Mier, J.G.M. Mode II fracture localization in concrete loaded in compression. *J. Eng. Mech.* **2009**, *135*, 1–8. [CrossRef]

42. Valsangkar, A.J.; Holm, T.A. Friction angle between expanded shale aggregate construction materials. *Geotech. Test. J.* **1997**, *20*, 252–257.

43. Cong, Y.; Kong, L.; Zheng, Y.; Erdi, A.B.; Wang, Z. Experimental study on shear strength of concrete. *Concrete* **2015**, *5*, 40–45.

44. Bui, Q.V. A modified Benzeggagh-Kenane fracture criterion for mixed-mode delamination. *J. Compos. Mater.* **2011**, *45*, 389–413. [CrossRef]

45. Aïtcin, P.-C.; Mehta, P.K. Effect of coarse aggregate characteristics on mechanical properties of high-strength concrete. *Mater. J.* **1990**, *87*, 103–107.

46. Rocco, C.G.; Elices, M. Effect of aggregate shape on the mechanical properties of a simple concrete. *Eng. Fract. Mech.* **2009**, *76*, 286–298. [CrossRef]

47. Bažant, Z.P.; Gettu, R.; Kazemi, M.T. Identification of nonlinear fracture properties from size effect tests and structural analysis based on geometry-dependent R-curves. *Int. J. Rock Mech. Min. Sci. Geomech. Abstr.* **1991**, *28*, 43–51. [CrossRef]

48. Shaofeng, Y.; Zhang, Z.; Ge, X.; Qiu, Y.; Xu, J. Correlation between fracture energy and geometrical characteristic of mesostructure of marble. *Rock Soil Mech.* **2016**, *37*, 2341–2346.

49. Ying, X.; Junchen, Z.; Wei, Y.; Kaiwen, X. Experimental study of dynamic fracture energy anisotropy of granitic rocks. *Chin. J. Rock Mech. Eng.* **2018**, *37*, 3231–3238.

50. Jianping, Z.; Nengbin, C.; Hongwei, Z. Investigation on failure behavior of basalt from different depth based on three-point bending meso-experiments. *Chin. J. Rock Mech. Eng.* **2013**, *32*, 689–695.

applied sciences

Article

Three-Dimensional Physical and Numerical Modelling of Fracturing and Deformation Behaviour of Mining-Induced Rock Slopes

Guoxiang Yang [1], Anthony K. Leung [2], Nengxiong Xu [1,*], Kunxiang Zhang [1] and Kunpeng Gao [1]

[1]   School of Engineering and Technology, China University of Geosciences, Beijing 100083, China;
       yanggx@cugb.edu.cn (G.Y.); 2002170033@cugb.edu.cn (K.Z.); gao.kunpeng@foxmail.com (K.G.)
[2]   Department of Civil and Environmental Engineering, Hong Kong University of Science and Technology,
       Hong Kong SAR 999077, China; ceanthony@ust.hk
*   Correspondence: xunengxiong@cugb.edu.cn; Tel.: +86-10-8232-3010

Received: 27 February 2019; Accepted: 28 March 2019; Published: 31 March 2019

Abstract: Fracturing behaviour of jointed rock mass subjected to mining can significantly affect the stability of the rock structures and rock slopes. Ore mining within an open-pit final slope would lead to large-scale strata and surface movement of the rock slope. Rock mass structure, or more specifically, the strength, spacing and distribution of rock joints, are the controlling factors that govern the failure and deformation mechanisms of the final slope. Two-dimensional (2-D) physical modelling tests have been conducted in the literature, but in general, most of them have simplified the geological conditions and neglected some key features of rock mass structure in the field. In this study, new three-dimensional (3-D) physical modelling methods are introduced, with realistic modelling of mechanical behaviour of rock mass as well as identified properties of predominant rock joint sets. A case study of Yanqianshan iron mine is considered and the corresponding 1:200 model rock slope was created for studying the rock joint effects on the strata movement and the subsidence mechanism of the slope. The physical model test results are subsequently verified with 3-D discrete element numerical modelling. Due to the presence of the predominant joints, the observed well-shaped strata subsidence in Yanqianshan iron mine was successfully reproduced in the 3-D physical model. The failure mechanism of rock slopes differs from the trumpet-shaped subsidence observed in unconsolidated soil. Due to the formation of an arching mechanism within the rock mass, the strata deformation transferred gradually from the roof of the goaf to the slope surface.

Keywords: physical modelling test; rock structure; fracture; deformation; mining

1. Introduction

The stability of rocks is an important subject in rock and slope engineering [1]. Strata and surface movements induced by mining activities in an open-pit final rock slope could trigger slope failure and surface subsidence, creating safety risks to the mining workers. It is thus important to study mining-induced strata and surface movement for better improving the understanding of the stress transfer mechanism in the rock slopes, so as to apply appropriate engineering mitigation measures to prevent and reduce disasters associated with the mining-induced slope failures. In the literature, the methods of investigation can be broadly categorised as theoretical analysis, numerical simulation and physical modelling. Most of the existing theoretical analysis idealises the strata as a beam or slab, which is then analysed by various mechanical analysis methods. A commonly-used mechanical model is, namely, the compressive arch theory. For example, He and Zhang applied the discontinuous deformation analysis to investigate the formation of pressure arch [2]. Wang, Jing et al. used the compressive arch theory to predict collapse of deep-buried tunnel [3]. On the other hand, Chen et al. used the cantilever hypothesis to analysis the mechanism of strata movement and surface deformation in an iron mine [4]. Tu et al. conducted a research on the failure of a gate road system based on the cantilever hypothesis [5]. Li et al. determined the static stress within fault-pillars using the Voussoir beam theory [6]. Ju and Xu found and defined three kinds of structural model affected by the key strata's position in super great mining height long wall face [7]. Although the mechanism of mining-induced strata movements can be reasonably captured by using the existing theoretical methods, idealising the strata to be a simple solid beam or slab would lead to an inaccurate estimation of rock mass deformation, especially under complicated geological settings, where a simple beam or slab geometry would be insufficient. With the development of the numerical simulation method, it has been widely used to study underground mining-induced strata and surface movement [8–14]. By numerical simulation, the stress, strain, and displacement of strata could be conveniently analysed, but on account of the constitutive relation and mechanical parameters of rock mass are difficult to be defined accurately.

Physical modelling, on the other hand, has a major advantage over the theoretical method, as a carefully designed physical model can reproduce the deformation and failure mechanisms of the strata induced by mining under a more realistic geological conditions [15,16]. Two-dimensional (2-D) physical models with simplified geological condition have been constructed [17], but there are only a few attempts to model the strata behaviour three-dimensionally, especially when the strata in the prototype has complex joint systems. Existing 2-D or three-dimensional (3-D) model tests considered only the major structural planes, such as the fault plane and bedding plane. Any widely-distributed joints within the rock mass are normally ignored or highly simplified [18–22]. Indeed, the presence of rock joints and their complex distribution typically control the mechanisms of rock mass deformation and strata movement caused by mining. When the rock mass is free from joints or relatively intact, the failure mode of the strata caused by mining is usually collapse or topographic avalanche [23]. On the contrary, deep subsidence pit (up to 100 m) was formed when rock mass is composed of complex joint systems, and a series of ground fractures were also observed at the upper part of eastern final slope (see Figure 1 for an example found at the open-pit slope of Yanqianshan iron mine). Thus, a physical model that could properly and realistically capture the rock joint distribution and orientation is necessary to more correctly study the effects of rock joints on strata and surface movement.

Figure 1. Subsidence pit observed at eastern final slope in Yanqianshan iron mine.

Physical modelling is an important research method but has difficulties and limitations when used in quantitative research. This method can directly reflect the actual process of strata movement expected in the field. The 2-D physical model is normally constructed using simplified geological conditions, without considering the important influence of wide spread joints in the actual rock mass. The absence of rock joints in a physical model would lead to large differences between the experimental results and field observations, which is one of the current challenges in physical modelling tests. This paper aims to conduct 3-D physical modelling tests which can consider the effects of multiple rock joint orientations on strata and surface movements induced by mining processes. The eastern final slope in Yanqianshan iron mine, Liaoning, China was chosen as a case study. The rock joint spacing used in the physical model tests was carefully designed via a series of discrete element analyses. The numerical outcome determined a representative rock joint spacing, which was then adopted for detailed testing and investigation. The observed deformation and failure mechanism of the model slope was subsequently analysed by comparing it with the discrete element calculations.

2. Geological Settings

The Yanqianshan iron mine is located in Anshan City, Liaoning Province, China. The basic rock structure of the mining area is a steep monoclinic structure trending toward 270°~300°, with a dip angle of 70°~88° in the northeast or southeast direction. In the area of the eastern final slope, the iron ore body is located in the middle of the area, strikes nearly east¬east, and dips in the northeast direction at approximately 70°. The ore body of the eastern final slope has a length of 300 m~550 m and an average thickness of 80 m. The eastern final slope is located at the east of the XIV prospecting line three sets of mutually intersecting dominant joints are found in the rock mass. One of the joint sets is a strata layer and the other two that crosscut each other intersect this layer. To facilitate the model construction in this study, the eastern final slope was divided into two parts along the iron vein axis. The red shaded area shown in Figure 2 was selected as the prototype for model simulation. Laboratory element testing including the uni-axial compression tests, splitting tests and direct shear tests have been conducted on the field samples to obtain the mechanical parameters. The typical rock types and the associated mechanical parameters of the rock mass found in the Yanqianshan iron mine are listed in Table 1.

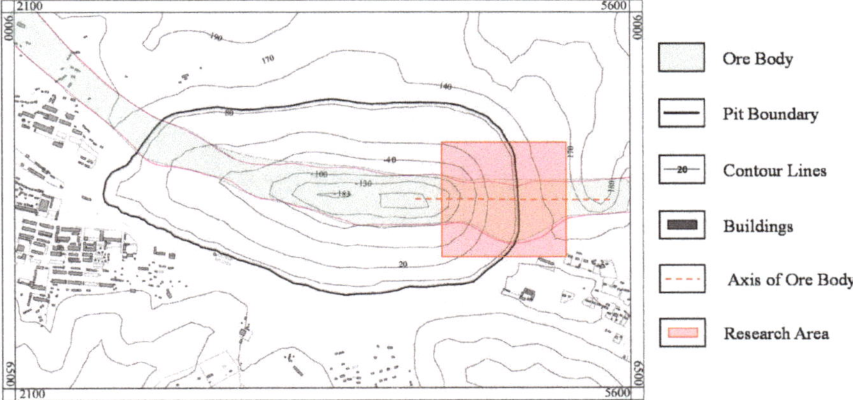

Figure 2. Overview of the study area. Shaded region represents the eastern final slope investigated in this study (Xu et al. 2016).

Table 1. Mechanical parameters of the major rock masses found in the Yanqianshan iron mine.

Rock Mass	Compressive Strength (MPa)	Deformation Modulus (GPa)	Cohesion (MPa)	Internal Friction Angle (°)
Mixed rock	164.34	3~5	40~50	38~40
Diorite	181.47	2	55~60	40~42
Carbonaceous phyllite	44.52	1.5	35~38	35~38
Chlorite quartz schist	98.56	1.5~2	40~45	38~40

3. Physical Model Tests

3.1. Model Contaniner

In this study, a large-scale 3-D container was created. The container was 4.3 m long, 2.3 m wide and 3.6 m tall (Figure 3). Face A of the container is open for researcher's access, while face B is closed. Faces C and D are made of high-strength plexiglass, through which the in-plane deformation of a model slope can be imaged by 3-D laser scanner and photogrammeter. Face C of the model container is a vertical section along the centre of the veins (refer to Figure 1), and the strata and surface movement in this section can be observed. To track the deformation, observation points can be marked on the plexiglass. Then, a high-resolution camera can be used to obtain the location of each of these points during the simulation of mining process. Hence, a displacement field of this section can be calculated for interpretation.

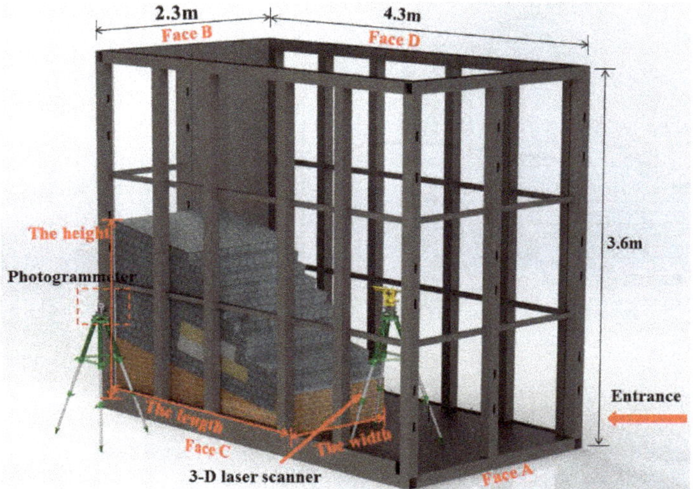

Figure 3. The model box and monitoring program.

The model slopes used for testing were created according to the geological and geometric characteristics of the eastern final slope at the Yanqianshan iron mine. The study area was scaled down according to a geometric similarity ratio of 200:1, with due consideration of the space constraints in the laboratory. The model has a length of 2.3 m, a width of 1.2 m and a height of 2.0 m (Figure 3), the side view of the completed model is shown in Figure 4.

3.2. Modelling of Rock Mass

In this study, the geometric and material similarities of the tested slope were obtained based on based on the similarity theory [24], and the similarity relationships are summarized in Table 2. In order to correctly simulate the mechanical properties of the rock mass found in the mine area, a mixture of cement, quartz sand, barite, iron powder, gypsum, and water was used to produce cubic blocks of model rock mass. The final mass proportion of the model material used was determined through mass proportion test [25,26]. The mass proportion and the mechanical parameters of the material are given in Tables 3 and 4. The length of each cubic rock block was selected to be 7.5 cm, based on the choice of rock joint spacing adopted in the physical model. More discussion on the choice of block size is given later when discussing the joint spacing.

Figure 4. The side view and the completed model.

Table 2. Similarity relationships of the main controlling parameters of the physical model.

Parameters	Similarity Relationship	Similarity Ratio
Geometry (C_L)	——	200
Bulk Density (C_r)	——	1
Stress (C_ε)	——	1
Poisson's Ratio (C_μ)	——	1
Friction Angle (C_φ)	——	1
Strain (C_σ)	$C_\sigma = C_r \times C_L$	200
Elastic Modulus	$C_E = \dfrac{C_\sigma}{C_\varepsilon}$	200

Table 3. The mass proportion of the model rock mass.

Cement	Quartz Sand	Barite	Iron Powder	Gypsum	Water
1	28	28	6.67	3	7.07

Table 4. Mechanical properties of the model rock mass.

Density (g/cm^3)	Uniaxial Compressive Strength (MPa)	Deformation Modulus (MPa)	Cohesion (MPa)	Friction Angle (°)
2.56	0.80	200.61	0.1735	38.94

3.3. Modelling of Rock Joints

In reality, a rock mass has a complex rock joint system. However, it is often impractical to consider and model all the joint systems in a reduced-scaled physical model. In this study, only the joints that have frequent occurrences and that would potentially affect the structure, strength and deformation of rock mass were considered when constructing a physical model. Before construction, the predominant joints in a prototype rock mass were classified into several sets based on the frequency of their occurrences. Each set of joint surfaces was simulated using parallel planes, whereas the block geometry for the model construction was determined according to the mutual intersection of the actual predominant joints. For the case of the Yanqianshan iron mine, the rock joints were simplified into three predominant sets, of which any two sets would be orthogonal to each other (Figure 5). Hence, three idealised sets of predominant joints were created in the model at $90°\angle10°$, $0°\angle90°$ and $270°\angle80°$.

Each block was cubic. The contact surface between blocks was the joint surface, while the side length of each cubic block was the joint spacing.

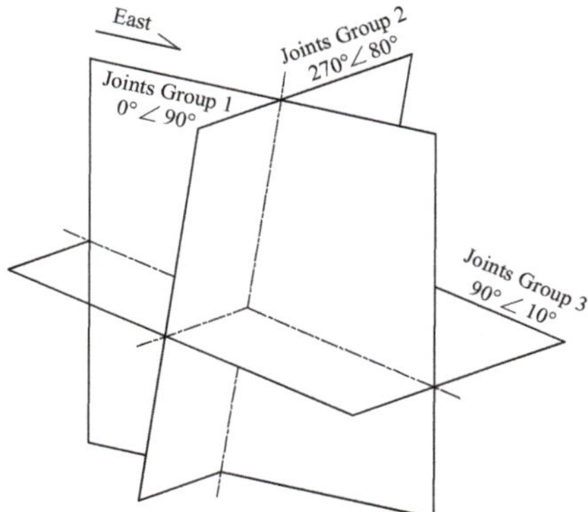

Figure 5. Three predominant sets of rock joints idealised from the Yanqianshan iron mine for physical model testing.

According to the similarity relationships and geometric similarity ratio shown in Table 2, the joint spacing of the prototype was divided by the geometric ratio to get the appropriate joint spacing required in the test. For the case of the Yanqianshan iron mine, the average joint spacing in the prototype is about 30–40 cm. Using the scaling factor of 200, the model joint spacing would be only 1.5–2 mm. When using cubic blocks with a length of 1.5–2 mm to construct a model with a total length of 2 m, approximately 580 million cubic blocks are required. The very small joint spacing and the huge number of blocks required at the 200th scale created practical difficulties for the construction. To maintain the practicality of model construction while not compromising the deformation and failure characteristics of the mine in the prototype scale, discrete element modelling (DEM) analyses were conducted to determine a representative joint spacing used for physical model testing. More details of the numerical modelling are given in the next section.

On the other hand, joint strength needs to be properly scaled because it is an important parameter that directly controls the deformation characteristics and failure modes of a rock mass. In this study, the equivalent discontinuous modelling method of jointed rock mass proposed by Xu and Bayisa was adopted to determine the model joint strength [27], and the relationship between joint spacing and joint mechanical parameters was built by them. In this study, the interfaces between the blocks represented rock joints in prototype. Thus, the friction between the blocks was simulated to follow the prototype joint strength. In this physical modelling work, an adhesive with a strength comparable to rock joint strength was used to fill the block interfaces. The adhesive was a mixture of barite, quartz sand, gypsum, and white latex. The strength of the adhesive was determined by the mass proportion of the various components. By changing the mass proportion of these components (see Table 5), adhesives with different strength were obtained. In this study, the adhesive strength was obtained by uniaxial compression tests, splitting tests and direct shear tests. The test results are summarized in Table 6.

Table 5. Mass proportion of the adhesive used.

Barite	Quartz Sand	Gypsum	White Latex
3.5	4.8	0.9	1

Table 6. Mechanical properties of rock joint strength (model scale).

	Cohesion (MPa)	Friction Angle (°)	Tensile Strength (MPa)
Joint strength of model slope	0.164	20.45	0.00698
Joint strength of field rock	33	21	1.5

3.4. Modelling of Mining Processes

The instability of the model slope was introduced by modelling sequences of mining processes. Based on the site information, the ore body was approximately located at 0.75 m away from the top surface of the slope, and it had a length of 1 m, a width of 0.4 m and a thickness of about 0.2 m (Figures 2 and 6). The ore body was composed of four individual sandbags. The simulation of mining process was divided into four steps by sequentially removing the sand from each bag following the order shown in Figure 6. This is a new modelling approach that is more advantageous over the existing methods, where blocks or PVC pipes were often used to replace an ore body and the mining process was simulated by an extraction of these blocks or pipes at one time [28–32]. The whole process of deformation and failure mode of the strata could be identified and investigated by the new method.

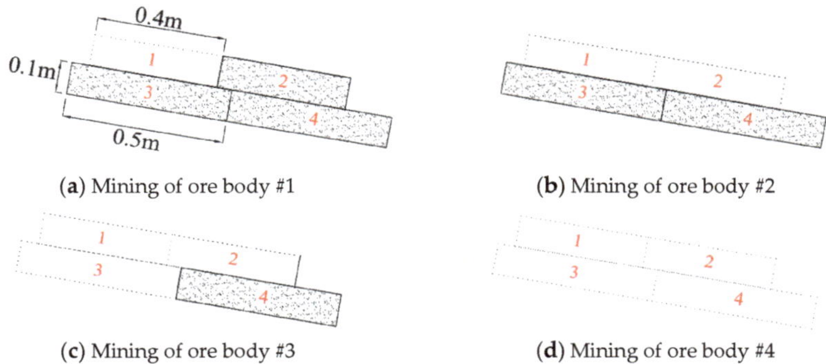

(**a**) Mining of ore body #1   (**b**) Mining of ore body #2

(**c**) Mining of ore body #3   (**d**) Mining of ore body #4

Figure 6. Illustration of the mining process.

4. Discrete Element Modelling

Numerical modelling of the behaviour of the rock slope was carried out using discrete element method (DEM), for two purposes, the first one was to perform analysis to influence of joint spacing on rock slope failure induced by mining, and to determine a representative joint spacing for informing the model design of the reduced-scale rock slope models. The second purpose was to back-analyse the physical model tests to improve the understanding of the rock mass deformation and failure characteristics upon mining. In this paper, the DEM software, Three-dimensional Distinct Element Code (3DEC) [33], was adopted in all numerical simulations. 3DEC can simulate the mechanical behaviour of a discountinuum material, such as a jointed rock mass. The material is represented as a collection of three-dimensional blocks. The discontinuities which bound the blocks are treated as boundary conditions, large displacements are permitted along the boundary [34]. 3DEC simulated mining by the "null" blocks. Using numerical modelling method to study the rock slope behaviour overcame the difficulties encountered in the physical tests, whereby once large deformation of rock mass occurs, any sensor installed within the slope mass would be displaced or even destroyed. Through numerical back-analysis, it is possible to determine the stress and strain induced in each individual block and hence to investigate the stress transfer mechanisms upon mining-induced unloading, providing new insights into the deformation and failure mechanisms of the rock slope.

With regard to the first objective, analysis to the influence of joint spacing on rock slope failure induced by mining was conducted to determine a critical joint spacing (l_{cr}), which would be practical for modelling in the model container, while not compromising the deformation and failure characteristics of the mine in prototype scale. Eleven rock slope models that have the same geometry to the physical models were created in the software (Figure 7). The dimension of the numerical model is also with a length of 2.3 m, width of 1.5 m and height of 2.0 m, the numerical model is composed of 8747 elements and 10373 nodes. In all the analyses, the constitutive model used is Mohr-Coulomb failure criterion model. The rigid stress-strain constitutive criterion is applied to the materials. A range of joint spacing between 5 cm and 10 cm (0.5 cm interval) was examined to investigate its effects on discrete element modelling (DEM) analyses were conducted to determine a representative joint spacing used for physical model tests. In this analysis, different numerical slope models having different joint spacings were constructed in the DEM software, 3DEC, for predicting the slope corresponding strata and surface movements when having different values of joint spacing. The input parameters of rock mass and rock joints are summarised in Tables 4 and 6, respectively. The horizontal direction (X) of the eastern and western boundaries, as well as the horizontal direction (Y) of the southern and northern boundaries of the model slopes were all fixed boundaries. On the other hand, the bottom boundary was also set fixed in the vertical direction (Z). No mechanical constraints were applied to the surface of the model (i.e., free to deform). In each analysis, the slope was subjected to the identical mining methods and procedures as in the physical model tests. Four blocks having the same size of the sandbag were set void sequentially (following Figure 6), to simulate the process of sand removal and hence unloading.

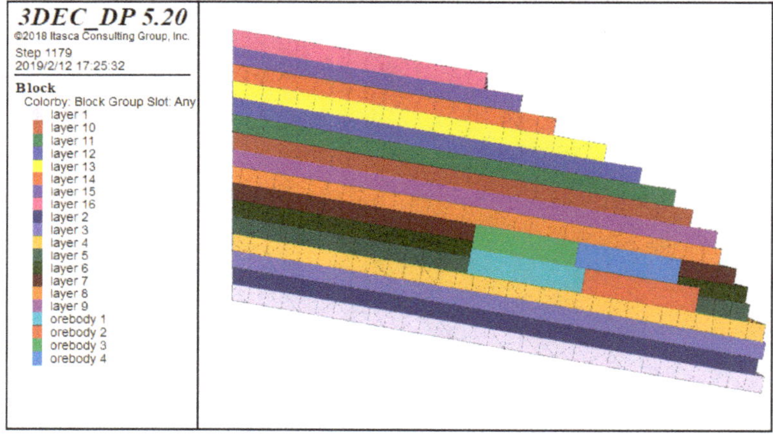

Figure 7. Elevation view of the 3-D numerical model.

The outcomes of the numerical analyses about influence of joint spacing on strata movement are given in Figure 8. It can be seen that the deformation of the rock slopes that have a joint spacing between 8 to 10 cm was almost identical, whereas while the joint spacing changes from 7.5 cm to 5 cm the slope deformation is very distinctive, which indicates that the influence of joint spacing on the slope behaviour should not be neglected. Hence, the critical joint spacing l_{cr} was found to be 7.5 cm, and this critical value was adopted in the physical model tests.

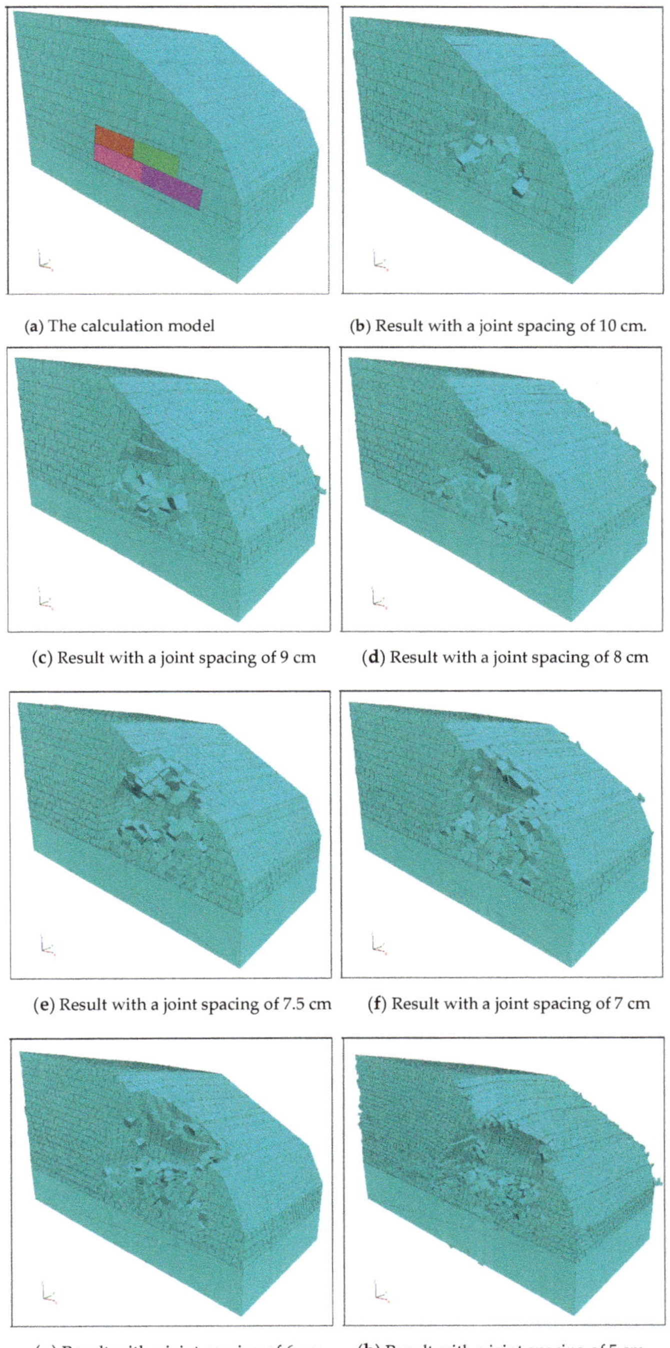

(**a**) The calculation model

(**b**) Result with a joint spacing of 10 cm.

(**c**) Result with a joint spacing of 9 cm

(**d**) Result with a joint spacing of 8 cm

(**e**) Result with a joint spacing of 7.5 cm

(**f**) Result with a joint spacing of 7 cm

(**g**) Result with a joint spacing of 6 cm

(**h**) Result with a joint spacing of 5 cm

Figure 8. Results of numerical analysis at different joint spacing; (**a**) model setup; (**b**) 10 cm; (**c**) 9 cm; (**d**) 8 cm; (**e**) 7.5 cm; (**f**) 7 cm; (**g**) 6 cm; and (**h**) 5 cm.

5. Physical Test Results

The observed deformation and displacement vector after each stage of mining simulation are shown in Figures 9–12. After mining the ore body #1 (see Figure 9), the blocks that were right above the mining area collapsed and fell off, as expected, due to the loss of support. Then, the strata overlying this layer of blocks underwent a substantial downward movement, which caused separation of the overlaying strata along the flat joints. There was no change in the shape of slope surface possibly due to the formation of arching mechanism within blocks (Figure 9a), Figure 9b also shows that the strata deformation has not transferred to the slope surface. After mining the ore body #2 (Figure 10), the strata deformed further towards the open area created by the previous step of mining process. The strata right above this area of mining collapsed almost vertically, introducing significant subsidence pit on the slope surface. This mining process did not introduce any surface subsidence near the crest of the model slope, but the strata deformation transferred to the slope surface (Figure 10b). After mining the ore body #3 (Figure 11), the strata right above the mined area was extensively fractured and more blocks right above collapsed. An arch was formed to support the blocks away from the mined area (Figure 11a). Although there was little or no surface subsidence near the crest of the slope, two rock joints were expanded laterally in size, forming two prominent vertical cracks in this slope section. A toppling avalanche was found near the toe of the slope. The surrounding blocks displaced toward the mined area, and local sliding occurred along the slope surface right above the ore body #3. Finally, after mining the ore body #4 (Figure 12), the strata above the mined area collapsed further, though interestingly, the arch formed in the previous stage of mining remained intact and was apparently unaffected by this last stage of mining (Figure 12a). Substantial amount of rock mass was fallen into the mined area, resulting in large slope surface subsidence at middle of the slope. Blocks near the subsidence pit experienced intensive disturbance and started to slip along the flat joint surface. At the end of the mining processes, a subsidence pit was formed right above the mining area (Figure 13) due to the significant vertical downward movement of strata. Due to the existence of predominant joints, the observed failure mode differs from the trumpet-shaped subsidence experienced in slopes made of unconsolidated soil [30].

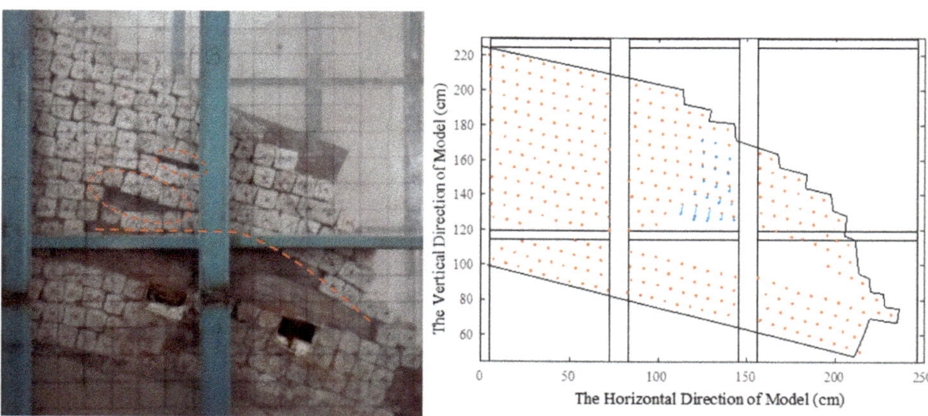

(**a**) Deformation after mining ore body #1   (**b**) Displacement vector after mining ore body #1

Figure 9. Slope deformation and displacement vector after mining ore body #1.

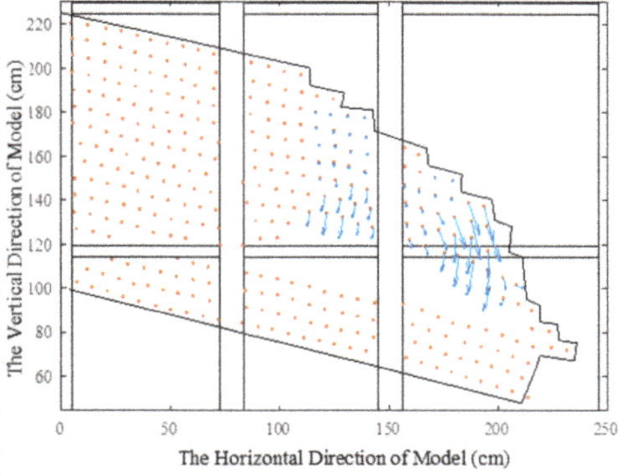

(**a**) Deformation after mining ore body #2   (**b**) Displacement vector after mining ore body #2

Figure 10. Slope deformation and displacement vector after mining ore body #2.

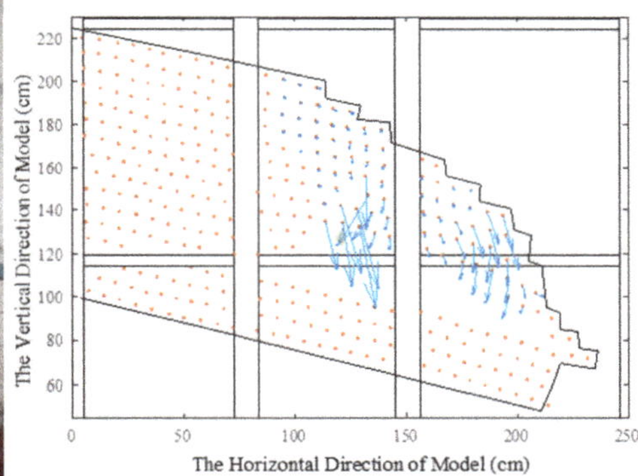

(**a**) Deformation after mining ore body #3   (**b**) Displacement vector after mining ore body #3

Figure 11. Slope deformation and displacement vector after mining ore body #3.

(**a**) Deformation after mining ore body #4   (**b**) Displacement vector after mining ore body #4

Figure 12. Slope deformation and displacement vector after mining ore body #4.

Figure 13. Front view of the model slope after four stages of mining. Subsidence pit was formed above the mining area.

6. Discussion

6.1. Comparative Analysis of Physical Modelling Result and Numerical Simulation Result

Figure 14 illustrates the process of mining induced strata movement obtained by numerical simulation, which is basically consistent with the physical modelling test result. The main deformation phenomenon reappeared in numerical simulation, though the internal displacement of the rock mass is not obtained by physical modelling, the displacement obtained by numerical simulation is consistent with the displacement vector graph in the physical modelling test and can support the phenomenon of strata movement obtained by physical modelling. Displacement monitoring points are set in ten strata from the first strata directly above the mined area to the slope surface (layer 7, 8, 10, 12, 14 and 16 illustrated in Figure 7), and the horizontal spacing of these monitoring points is 15 cm. Figure 15 shows the vertical displacement of the strata directly above the mined-out area, and the displacement value is built up with the mining. The maximum displacement will always appear in the strata that are directly above the mined area, which indicates that mining-induced unloading triggered the strata movement. All the displacement curves of the four mining steps tell us that a significant increase in displacement began to appear near the boundary between the mined area and surrounding rock, which indicates that due to the existence of steep joints, the rock mass deformation transferred mainly upward to the slope surface. The final well-shaped displacement curve also indicates a well-shaped rock mass subsidence.

In this study, the mining-induced strata movements in the Yanqianshan iron mine can be summarised as follows.

I. The main deformation modes were: (a) the strata separated from the flat joint surface due to the losing of support and the overburden pressure provided by the strata, (b) the arch was formed by the collapsed rock mass, and (c) the subsidence pit was formed by downward strata movement along the steep joint surface and ground cracks formed by extension of steep joints.

II. Transfer of the strata movements. The strata movement transferred gradually from the roof of the mining area to the ground surface. The mining depth or the depth of overlying strata were of significant influence on the depth of subsidence pit formed. Both the displacement vector in Figure 12b and the subsidence curve in Figure 15d showed that the mining-induced deformation in the rock mass that has a realistic joint system (Figure 5) was well-shaped, and the boundary of the subsidence pit was controlled by the steep joints.

III. Surface deformation. Ground surface cracks formed by the extension of steep joints experienced near the slope shoulder, the subsidence pit at slope surface resulted in tensile force in horizontal direction, the steep joint extended under the tensile force and the cracks formed in a certain depth from the slope surface to the internal.

The results from the numerical simulation and physical modelling were consistent with each other. However, several issues need to be highlighted. First, the strata movement process at the macro level was simulated in the numerical modelling, while the strata movement at the micro level was considered in the physical modelling. Therefore, the phenomenon investigated and obtained by the two methods are not completely consistent, especially in Figures 10a and 14c. Second, the simulation of the mining process between two modelling methods was also different. The mining process in the physical modelling was achieved by removing the sand from the sandbag, yet little sand was left in the sandbag. No such issue occurs in the numerical simulation. In addition, the physical modelling can reflect the rock joint influence on the mining induced strata movement in detail and reproduce the whole process of slope failure microscopically. Although there are several differences between the numerical and physical modelling, the two methods that are complementary to each other should be interpreted holistically.

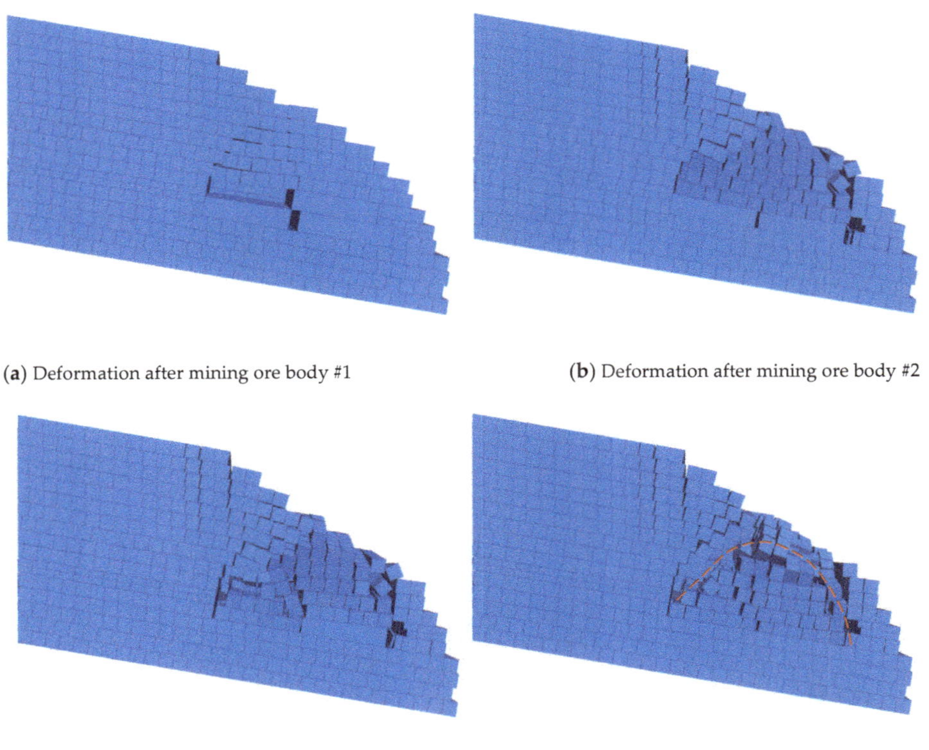

(**a**) Deformation after mining ore body #1 (**b**) Deformation after mining ore body #2

(**c**) Deformation after mining ore body #3 (**d**) Deformation after mining ore body #4

Figure 14. Process of strata movement by numerical simulation.

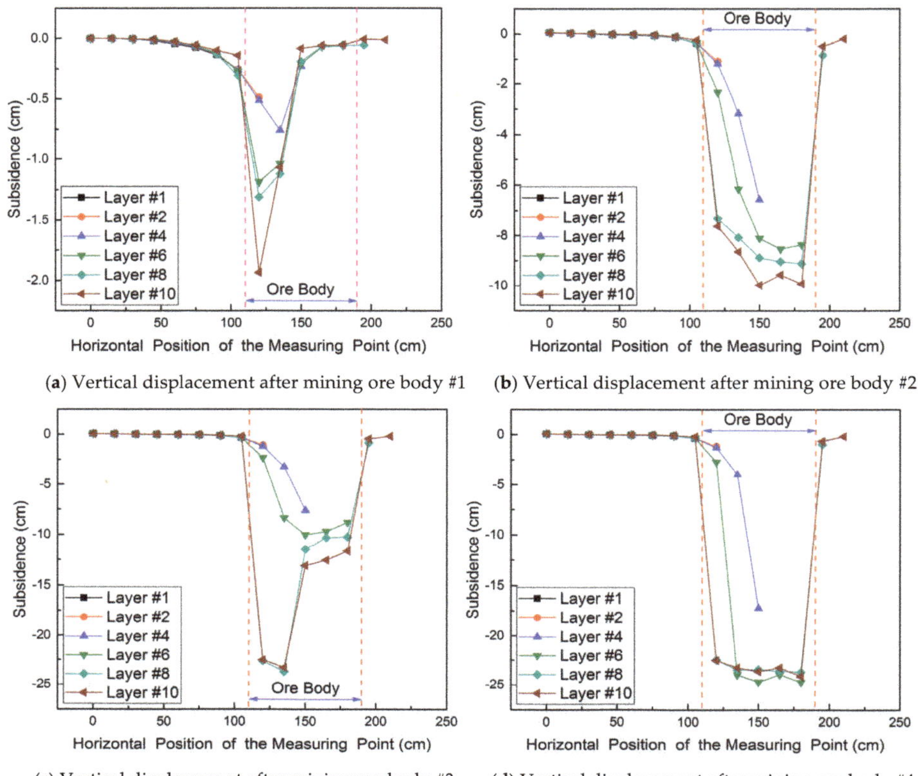

(**a**) Vertical displacement after mining ore body #1

(**b**) Vertical displacement after mining ore body #2

(**c**) Vertical displacement after mining ore body #3

(**d**) Vertical displacement after mining ore body #4

Figure 15. Vertical displacement by numerical simulation.

6.2. Analysis of the Typical Strata Movement Characteristics

The strata movement and deformation observed in the physical model tests and the numerical simulation demonstrate that the presence of dominant joints has a significant influence on the strata movement and slope failure. The vertical cracks that appeared near the slope shoulder are mainly along the steep joints. The angle between the vertical boundary of the mining area and the cracks were small. Thus, the whole deformation zone observed in the physical model was well-shaped and could be a referenced to support the strata and surface movement observed in the field. Additionally, it is interesting to reveal that the arch formed during mining was an asymmetrical pressure arch. Although this arch can remain stable over a short period of time right after mining, the arch finally collapsed due to further mining and other impacts, including traffic loading, mining disturbance, rainfall and so on. The subsequent mining and activities may explain the field observations, where mining under eastern final slope in the Yanqianshan iron mine was completed in May 2014. No large-scale deformation was observed until March 2015; at that time, a subsidence pit appeared and the road access was destroyed. This phenomenon could be another reference for the prediction and supporting design of the mining-induced strata and surface movement.

7. Conclusions

In this study, a 1:200 slope model was created according to the field observation made from the Yanqianshan iron mine. Various 3-D physical model tests were performed to investigate the effects of predominant joints on the failure and deformation of rock slope when subjected to sequential excavation due to mining. The main benefits of the new 3-D modelling method are threefold: (i) it overcame the limitations of most existing 2-D physical models, where simplified geological conditions were often considered without a detailed consideration and modelling of rock joint distribution (3-D modelling, however, can realistically capture the mechanical properties of rock mass and the properties of rock joint including strengths and distribution in 3D space); (ii) the use of a 3-D model can (a) reproduce the whole process of strata deformation and movement under realistic geological conditions, and (b) record and measure the slope deformation characteristics during a test, which is important for the post-test analysis of strata movement mechanism; and (iii) with the aid of 3-D discrete element modelling, an equivalent joint spacing could be determined and was then used in the physical tests, usefully by passing the practical difficulties of rock blocks in an experiment.

By using the new 3-D physical model test and the complementary numerical modelling, the strata movement process was recorded and the mechanism of well-shaped subsidence in jointed rock mass was obtained. The new modeling method has provided a new effective means to study mining-induced strata and surface movement in jointed rock masses. Nevertheless, the observed deformation mechanism of the physical model was qualitatively consistent with what was observed in the field. Further study is necessary to improve the physical model to more quantitatively capture the field observation such as the displacement and failure scope.

Author Contributions: Conceptualization, G.Y.; Physical modelling, G.Y., N.X. and K.G.; Numerical simulation, N.X., A.K.L. and K.Z.; Funding acquisition, N.X. and G.Y.; Data analysis, G.Y., A.K.L., K.Z. and K.G.; Writing—original draft, G.Y.; Writing—review and editing, G.Y., A.K.L., N.X., K.Z. and K.G.

Funding: This research was funded by the National Natural Science Foundation of China, grant No. 41772326.

References

1. Fu, H.; Wang, S.; Pei, X.; Chen, W. Indices to Determine the Reliability of Rocks under Fatigue Load Based on Strain Energy Method. *Appl. Sci.* **2019**, *9*, 360. [CrossRef]

2. He, L.; Zhang, Q.B. Numerical investigation of arcing mechanism to underground mining in jointed rock mass. *Tunn. Undergr. Space Technol.* **2005**, *50*, 54–67. [CrossRef]

3. Wang, Y.; Jing, H.; Zhang, Q.; Luo, N.; Yin, X. Prediction of Collapse Scope of Deep-Buried Tunnels Using Pressure Arch Theory. *Math. Probl. Eng.* **2016**, *3–4*, 1–10. [CrossRef]

4. Cheng, G.; Chen, C.; Ma, T.; Liu, H.; Tang, C. A Case Study on the Strata Movement Mechanism and Surface Deformation Regulation in Underground Iron Mine. *Rock Mech. Rock Eng.* **2016**, *50*, 1011–1032. [CrossRef]

5. Bai, Q.; Tu, S.; Wang, F.; Zhang, C. Field and numerical investigations of gate road system failure induced by hard roofs in a longwall top coal caving face. *Int. J. Coal Geol.* **2017**, *173*, 176–199. [CrossRef]

6. Li, Z.; Dou, L.; Cai, W.; Wang, G.; Ding, Y.; Kong, Y. Mechanical Analysis of Static Stress with in Fault-Pillars Based on a Voussoir Beam Structure. *Rock Mech. Rock. Eng.* **2015**, *49*, 1097–1105. [CrossRef]

7. Ju, J.; Xu, J. Structural characteristics of key strata and strata behaviour of a fully mechanized longwall face with 7.0 m height chocks. *Int. J. Rock Mech. Min.* **2013**, *58*, 46–54. [CrossRef]

8. Unver, B.; Yasitli, N. Modelling of strata movement with a special reference to caving mechanism in thick seam coal mining. *Int. J. Coal Geol.* **2016**, *66*, 227–252. [CrossRef]

9. Guo, H.; Yuan, L.; Shen, B.; Qu, Q.; Xue, J. Mining-induced strata stress changes, fractures and gas flow dynamics in multi-seam longwall mining. *Int. J. Rock Mech. Min.* **2012**, *54*, 129–139. [CrossRef]

10. Ye, Q.; Wang, W.; Wang, G.; Jia, Z. Numerical simulation on tendency mining fracture evolution characteristics of overlying strata and coal seams above working face with large inclination angle and mining depth. *Arab. J. Geosci.* **2017**, *10*, 82. [CrossRef]

11. Wang, X.; Kulatilake, P.H.S.W.; Song, W. Stability investigations around a mine tunnel through three-dimensional discontinuum and continuum stress analyses. *Tunn. Undergr. Space Technol.* **2012**, *32*, 98–112. [CrossRef]

12. Gao, F.; Stead, D. Discrete element modelling of cutter roof failure in coal mine roadways. *Int. J. Coal Geol.* **2013**, *116–117*, 158–171. [CrossRef]

13. Xu, N.; Zhang, J.; Tian, H.; Mei, G.; Ge, Q. Discrete element modelling of strata and surface movement induced by mining under open-pit final slope. *Int. J. Rock Mech. Min.* **2016**, *88*, 61–76. [CrossRef]
14. Vyazmensky, A.; Stead, D.; Elmo, D.; Moss, A. Numerical Analysis of Block Caving-Induced Instability in Large Open Pit Slopes: A Finite Element/Discrete Element Approach. *Rock Mech. Rock Eng.* **2009**, *43*, 21–39. [CrossRef]
15. Guo, S.; Qi, S.; Yang, G.; Zhang, S.; Saroglou, C. An Analytical Solution for Block Toppling Failure of Rock Slopes during an Earthquake. *Appl. Sci.* **2017**, *7*, 1008. [CrossRef]
16. Zhan, Z.; Qi, S. Numerical Study on Dynamic Response of a Horizontal Layered-Structure Rock Slope under a Normally Incident Sv Wave. *Appl. Sci.* **2017**, *7*, 716. [CrossRef]
17. Ghabraie, B.; Ren, G.; Smith, J.V. Characterising the multi-seam subsidence due to varying mining configuration, insights from physical modelling. *Int. J. Rock Mech. Min.* **2017**, *93*, 269–279. [CrossRef]
18. Ren, W.; Guo, C.; Peng, Z.; Wang, Y. Model experimental research on deformation and subsidence characteristics of ground and wall rock due to mining under thick overlying terrane. *Int. J. Rock Mech. Min.* **2010**, *47*, 614–624. [CrossRef]
19. Zhu, W.; Li, Y.; Li, S.; Wang, S.; Zhang, Q. Quasi-three-dimensional physical model tests on a cavern complex under high in situ stresses. *Int. J. Rock Mech. Min.* **2011**, *48*, 199–209.
20. Li, S.C.; Wang, Q.; Wang, H.T.; Jiang, D.C.; Wang, B.; Zhang, Y.; Li, G.; Ruan, Q. Model test study on surrounding rock deformation and failure mechanisms of deep roadways with thick top coal. *Tunn. Undergr. Space Technol.* **2015**, *47*, 52–63. [CrossRef]
21. Fang, Y.; Xu, C.; Cui, G.; Kenneally, B. Scale model test of highway tunnel construction underlying mined-out thin coal seam. *Tunn. Undergr. Space Technol.* **2016**, *56*, 105–116. [CrossRef]
22. Ju, M.; Li, X.; Yao, Q.; Liu, S.; Liang, S.; Wang, X. Effect of sand grain size on simulated mining-induced overburden failure in physical model tests. *Eng. Geol.* **2017**, *226*, 93–106. [CrossRef]
23. Xia, K.; Chen, C.; Deng, Y.; Xiao, G.; Zheng, Y.; Liu, X.; Fu, H.; Song, X.; Chen, L. In situ monitoring and analysis of the mining-induced deep ground movement in a metal mine. *Int. J. Rock Mech. Min.* **2018**, *109*, 32–51. [CrossRef]
24. Luo, X.; Ge, X. *Theory and Application of Model Test on Landslide*; China Waterpower Press: Beijing, China, 2008.
25. Wang, H.; Li, S. Development of a new geomechanical similar material. *Chin. J. Rock Mech. Eng.* **2006**, *25*, 1842–1847.
26. Wang, H.; Li, S.; Zheng, X. Research progress of geomechanical model test with new technology and its engineering application. *Chin. J. Rock Mech. Eng.* **2009**, *28*, 2765–2771.
27. Regassaa, B.; Xu, N.; Mei, G. An Equivalent Discontinuous Modelling Method of Jointed Rock Masses for DEM Simulation of Mining-induced Rock Movements. *Int. J. Rock Mech. Min.* **2018**, *108*, 1–14. [CrossRef]
28. Huayang, D.; Xugang, L.; Jiyan, L.; Yixin, L.; Yameng, Z.; Weinan, D.; Yinfei, C. Model study of deformation induced by fully mechanized caving below a thick loess layer. *Int. J. Rock Mech. Min.* **2010**, *47*, 1027–1033. [CrossRef]
29. Ghabraie, B.; Ren, G.; Smith, J.; Holden, L. Application of 3D laser scanner, optical transducers and digital image processing techniques in physical modelling of mining-related strata movement. *Int. J. Rock Mech. Min.* **2015**, *80*, 219–230. [CrossRef]
30. Xu, Y.; Wu, K.; Li, L.; Zhou, D.; Hu, Z. Ground cracks development and characteristics of strata movement under fast mining: A case study at Bulianta coal mine, China. *Bull. Eng. Geol. Environ.* **2017**, 1–16. [CrossRef]
31. Sun, G. *Rockmass Structural Mechanism*; Science Press: Beijing, China, 1988.
32. Guo, Q.; Guo, G.; Lv, X.; Zhang, W.; Lin, Y.; Qin, S. Strata movement and surface subsidence prediction model of dense solid backfilling mining. *Environ. Earth Sci.* **2016**, *75*. [CrossRef]
33. Itasca Consulting Group. *3DEC District Element Code in 3 Dimensional, Version 5.0. (a) User's Guide and (b) Theory and Background*; Itasca Consulting Group: Minneapolis, MN, USA, 2013.
34. Brideau, M.A.; Stead, D. Controls on Block Toppling Using a Three-Dimensional Distinct Element Approach. *Rock Mech. Rock Eng.* **2010**, *43*. [CrossRef]

Article

Experimental and DEM Analysis on Secondary Crack Types of Rock-Like Material Containing Multiple Flaws Under Uniaxial Compression

Yong Li [1,2,*], Weibing Cai [1,2], Xiaojing Li [3,4,*], Weishen Zhu [2], Qiangyong Zhang [2] and Shugang Wang [2]

[1] School of Qilu Transportation, Shandong University, Jinan 250061, China; 201714552@mail.sdu.edu.cn
[2] Geotechnical & Structural Engineering Research Center, Shandong University, Jinan 250061, China; zhuw@sdu.edu.cn (W.Z.); qiangyongz@sdu.edu.cn (Q.Z.); sdgeowsg@gmail.com (S.W.)
[3] School of Civil Engineering, Shandong Jianzhu University, Jinan 250101, China
[4] State Key Laboratory of Water Resources and Hydropower Engineering Science, Wuhan University, Wuhan 430072, China
* Correspondence: yongli@sdu.edu.cn (Y.L.); li8021@163.com (X.L.); Tel.: +86-13606404829 (Y.L.); +86-15954128108 (X.L.)

Received: 8 March 2019; Accepted: 23 April 2019; Published: 27 April 2019

Abstract: To better understand the evolution of crack propagation in brittle rock mass, the particle velocity field evolution on both sides of secondary crack in rock-like materials (cement mortar specimens) with pre-existing parallel double flaws under uniaxial compression is analyzed based on the discrete element theory. By bringing in strain rate tensor, a new technique is proposed for quantifying the failure mechanism of cracks to distinguish the types and mechanical behaviors of secondary cracks between pre-existing parallel flaws. The research results show that the types and mechanical behaviors of secondary cracks are distinct at different axial loading stages and can be directly identified and captured through the presented approach. The relative motion trend between particles determines the types and mechanical behaviors of secondary cracks. Based on particles movement on both sides of secondary cracks between cracks, the velocity fields of particles can be divided into four types to further analyze the causes of different types of cracks. In different axial loading stages, the velocity field types of particles on both sides of cracks are continuously evolving. According to the particle velocity field analysis and the proposed novel way, the types of macroscopic cracks are not directly determined by the types of dominated micro-cracks. Under uniaxial compression, the particles between secondary cracks and pre-existing parallel flaws form a confined compressive member. Under the confinement of lateral particles, secondary cracks appear as shear cracks between pre-existing parallel flaws at the beginning stage of crack initiation.

Keywords: rock-like material; crack propagation; discrete element; strain rate tensor; velocity field

---

1. Introduction

Fractured rock mass is one of the most significant construction objects encountered in geotechnical engineering. Under high in-situ stress, crack propagation, and coalescence in fractured rock mass could result in local damage or even failure, which could eventually threaten the stability and safety of rock engineering projects [1–5]. Therefore, a thorough understanding of cracking propagation emanating from existing flaws in fractured rock mass can benefit geotechnical engineering design and implementation, and relevant research has had widespread attention.

A large number of experimental works are available on crack propagation and failure mode from pre-cracked brittle rock-like materials under uniaxial compression [6–10]. It is generally accepted and confirmed that wing cracks in brittle rock materials under compression are mostly tensile cracks [11–14]. The initiation of secondary cracks is often related to the stress field at or near the tips of the pre-existing flaws, but the propagation direction is distinct from the wing crack. Bobet [15] observed in the laboratory that two initiation directions are possible: one coplanar or quasi-coplanar to the flaw, and the other one parallel to the wing cracks but in the opposite direction. Cao et al. [16] found that the wing cracks propagate to a certain length and then stop through prefabricating the cracks in the cement mortar material. Then, as the load is increased, the secondary cracks begin to initiate in large amount. Compared with wing cracks, secondary cracks often initiate with a large quantity, appearing to be a rather complicated process, which are often difficult to distinguish in the laboratory without advanced technology. Numerous studies [17–20] have confirmed that the relative shear results in the initiation of secondary cracks, suggesting that the secondary cracks are shear in essence. However, Wong and Einstein [21,22] conducted a series of laboratory experiments to find that not only is the secondary crack made up with shear cracks, but also contains tensile cracks. Wong et al. [23] also found that shear bands contain a large number of tensile micro-cracks in the rock bridge area, indicating that previous understanding of secondary cracks is not profound. In addition, the propagation direction of these micro-cracks is almost parallel to the most compressive direction, which demonstrates that the current description of the crack nature is not accurate. Consequently, the urgent demanding for identifying secondary crack is of strong interest to scholars.

Due to the rapidity and convenience of numerical methods, numerical simulation has become a widely used method to study the deformation and failure mechanism of materials. The discrete element method proposed by Cundall [24] is very effective for analyzing the crack propagation process and explaining the types of cracks observed in the previous physical experiments. The parallel bond model based on the discrete element theory has been widely used in rock damage analysis for decades [25–29], and the numerical simulation results are generally in good agreement with the laboratory results. However, due to the lack of effective approaches, the current numerical model based on parallel bonding is not proficient in distinguishing the mechanical behaviors and types of cracked shear bands in the process of crack propagation and coalescence for several years. Hazzard [30] presented a technique which is described for quantifying the seismic source mechanisms of the modelled events to investigate the failure mechanism in rock, providing insight into understanding of crack nature. Based on the moment tensor inversion analysis, Zhang et al. [31] found that a large number of tensile micro-cracks appeared in the rock bridge area at the initial loading stage, and an obvious shear band formed due to the relative slip between the particles, indicating that the macroscopic shear fracture is not completely composed of shear micro-cracks. Strain rate tensor and velocity field analysis have been widely performed in the analysis of the deformation mechanism of earthquake fault [32,33]. Ge et al. [34] employed tiny blocks to reveal the crustal movement deformation mode by analyzing the velocity field and strain rate field in the region. Kostrov et al. [35] proposed that the average strain rate tensor caused is equal to the sum of the moment tensors of all earthquakes occurring in a unit volume. Compared to the moment tensor analysis, the strain rate tensor is more suitable for distinguishing the mechanical behaviors and types of secondary cracks since strain rate tensor can comprehensively characterize the source evolution mechanism of faults within a given volume without further analysis of the moment tensors one by one. Additionally, the moment tensor analysis is required to compile complex codes with time consuming. However, based on the discrete element theory, we can easily obtain the strain rate tensor by arranging the measurement circle between pre-existing parallel flaws, and accurately define two variables to quantify the crack failure mechanism, providing a more efficient method to distinguish secondary crack types.

To further gain insight into the mechanism of secondary crack propagation, we proposed a technique to quantify the failure mechanism of secondary crack at different loading stages by means of adopting strain rate tensor analysis. Furthermore, according to the relative motion trend between particles at various stress levels, the velocity fields of both sides of secondary cracks are classified for better explanation of crack types. Combined with particle velocity field analysis and values of variable R, it is convenient for us to distinguish crack types and reveal essential crack characteristics.

2. Mesoscopic Parameter Calibration and Basic Theory

2.1. Specimen Preparation and Mesoscopic Parameter Calibration

The cement mortar, as a typical rock-like material for laboratory uniaxial compression tests, is made from a mixture of 42.5R ordinary Portland cement, quartz sand, and water, with a mass ratio of 1:2.34:1.35, respectively. In order to promote the fluidity of cement mortar, a small amount of water reducing agent was added during the preparation of specimens. The physico-mechanical parameters of this rock-like material are listed in Table 1. The physico-mechanical properties are similar to those of typical rock materials such as sandstone, and the ratio of tensile strength to compressive strength is close to 1:10, indicating that it is a comparatively ideal rock-like material with high brittleness. Therefore, it can be used as a rock-like material to study the evolution of secondary crack propagation in brittle rocks.

Table 1. Physico-mechanical parameters of cement mortar and Sandstone.

Material	Compressive Strength σ_c (MPa)	Tensile Strength σ_t (MPa)	Young's Modulus E (GPa)	Poisson's Ratio v	Density ρ (g/cm^3)
Cement mortar	58.25	5.62	11.63	0.20	2.38
Sandstone	20~170	4~25	3~35	0.02~0.25	2.10~2.40

The specimens used in uniaxial test series are cuboid blocks with dimensions of 140 mm in height, 70 mm in width, and 40 mm in thickness. Prior to casting the cement mortar specimens, the iron piece, fixed in the mold, is smeared with a little epoxy resin on the surface. After meticulous maintenance in the mold for 24 h, two flaws, created by pulling out the thin iron pieces, are formed through the thickness of the specimens during casting in such a way that the plane of the flaws is perpendicular to the faces of the specimens. Prefabricated cracks are open cracks with a certain degree of openness, so the internal faces do not touch each other during fabrication and loading. Two flaws are always parallel to each other, and have a constant length of 12 mm. The thickness of the flaws is 1.2 mm approximately. To study the influence of the crack inclination angle on the mechanical properties and failure process of the rock mass, three flaw inclination angles of 30°, 45°, and 60° are used, and the spacing of pre-existing parallel flaws is 15 mm. Two flaws are located at the center of the specimen. Six cement mortar specimens (a total of 18 specimens) were prepared with the same flaw inclination angle, and the average values of the test data are used for analysis.

The calculations performed in PFC2D (Particle Flow Code in 2 dimensions) is based on Newton's second law and a force-displacement law at the contacts. Newton's second law is used to determine the motion of each particle arising from the contact and body forces acting upon it, while the force-displacement law is used to update the contact forces arising from the relative motion at each contact [36]. Particle Flow Code (PFC) has great advantage in simulating the micro-mechanical behavior and investigating the mechanism of crack propagation in brittle materials. However, the straightforward adoption of circular (or spherical) particles cannot fully capture the behavior of complex shaped and highly interlocked grain structures [37]. Further, PFC fails to simulate the mechanical properties of brittle rock with higher internal friction angle. A parallel bond model is adopted as the numerical model of specimens, and the model size and crack layout are shown in Figure 1.

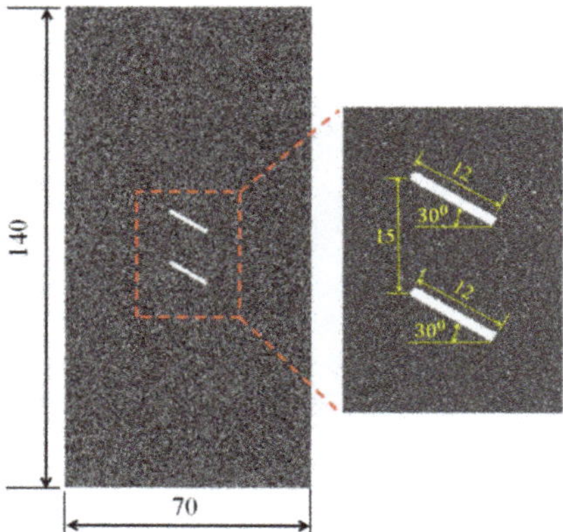

Figure 1. Numerical model for pre-existing parallel flaws layout ($\alpha = 30°$, Unit: mm).

The loading stops when the axial stress drops to 50% of the peak strength. Zhang et al. [38] studied the effect of loading rate on the crack propagation and failure modes of the specimen under uniaxial loading. To ensure that the numerical model maintains static equilibrium during the loading process, the displacement loading rate of the numerical model is taken as 0.08 m/s. In PFC, it is a crucial step to calibrate the mesoscopic parameters by performing a laboratory compression test on the standard specimens. The macroscopic mechanical properties of specimens are determined by the values of the mesoscopic parameters between the particles. A few references [39,40] reveal that the value of the compressive strength ($\overline{\sigma_c}$) and the tensile strength ($\overline{\tau_c}$) will affect the failure mode and the type of micro-crack, as such the friction coefficient between the particles has less influence on the significant parameters such as initiation stress, peak strength, and elastic modulus of specimens. According to the characteristics of cement mortar material, the parameters obtained by numerical simulation are consistent with the physical experimental parameters of the complete standard specimen by adjusting the mesoscopic parameters. The specific parameters are listed in Table 2. Since cement mortar is a brittle material inducing complex crack types under compressive loads, so it is essential to control the type of micro-crack by continuously adjusting the values of $\overline{\sigma_c}/\overline{\tau_c}$, to make sure that the failure modes of the specimens obtained by numerical simulation are in a good agreement with the laboratory test results. The specific failure modes are shown in Figure 2, and the final mesoscopic calibration parameters are given in Table 3.

Table 2. Physico-mechanical parameters of intact cement mortar specimens for laboratory tests and numerical simulations.

Properties	Specimens for Laboratory Tests	Specimens for Numerical Simulation
Density ρ (g/cm^3)	2.38	2.38
Young's modulus E (GPa)	11.63	11.95
Poisson's ratio ν	0.20	0.21
Uniaxial compressive strength σ_c (MPa)	58.25	57.30

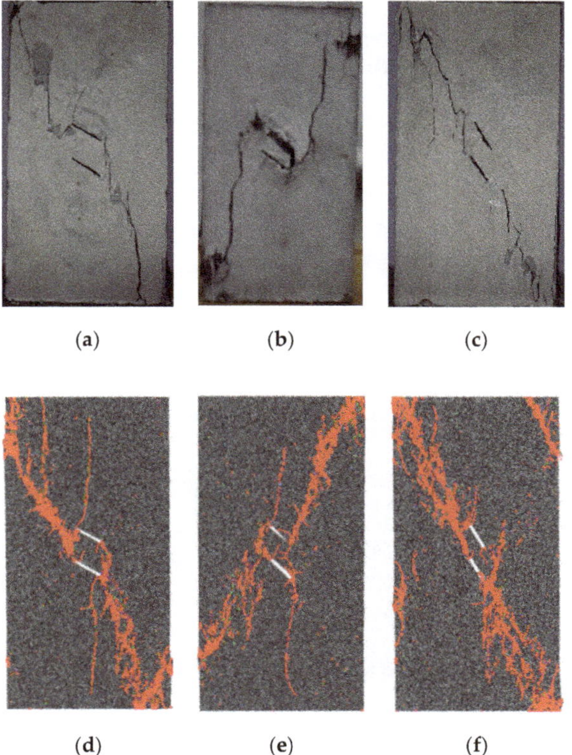

Figure 2. Comparison of failure modes between laboratory experiment and numerical simulation. (**a**) $\alpha = 30°$; (**b**) $\alpha = 45°$; (**c**) $\alpha = 60°$; (**d**) $\alpha = 30°$; (**e**) $\alpha = 45°$; (**f**) $\alpha = 60°$. (Pictures **a**, **b** and **c** are experimental results, and pictures **d**, **e** and **f** are numerical results).

Table 3. Microscopic parameters used in the numerical model.

Particle Parameters	Values
Minimum radius R_{min} (mm)	0.18
Particle radius ratio R_{max}/R_{min}	1.66
Density ρ (g/cm^3)	2.38
Friction μ	0.55
Effective modulus E_c (GPa)	5.5
Normal/shear stiffness ratio k_n/k_s	2.0
Tensile strength $\overline{\sigma}_c$ (MPa)	22.5 ± 2.0
Cohesion $\overline{c}$ (MPa)	19.5 ± 2.0
Angle of internal friction $\overline{\varphi}$ (0)	35
Bond effective modulus $\overline{E}_c$ (GPa)	5.5
Bond normal/shear stiffness ratio $\overline{k}_n/\overline{k}_s$	2.0

2.2. Strain Rate Tensor

Based on acoustic emission technology, the moment tensor inversion analysis [41–43] has been widely used to distinguish rock fracture types. Ming et al. [44] proposed a criterion for judging rock rupture through reasonably decomposing the variability tensor. The strain rate tensor can be calculated through the moment tensor inversion, but it is still inadequate to adopt strain rate tensor to judge the crack type and analyze the crack propagation law. The PFC software developed based on discrete element theory can not only simulate the propagation and evolution of micro-cracks in rock masses, but also consistently monitor the strain rate tensor changes in specific regions through the measurement circle. The schematic diagram of the main strain rate tensor when micro-cracks appear is shown in Figure 3. When the contact between the particles breaks, the velocity of the particles will instantly change, triggering the variation of the magnitude and direction of the strain rate tensor in the monitored region. Therefore, the evolution law of crack propagation can be accurately reflected by strain rate tensor.

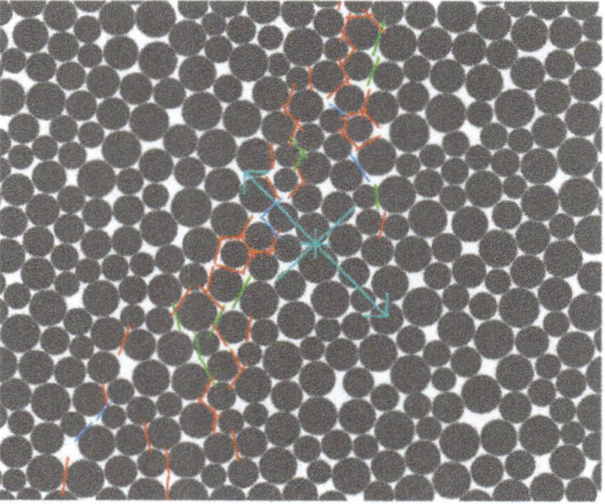

Figure 3. Schematic diagram of the principal strain rate tensor. (The red, green and blue micro-cracks represent tensile micro-cracks, tensile-shear micro-cracks, and compressive-shear micro-cracks, respectively, and the principal strain tensor is denoted by two sets of light blue lines with arrows).

In PFC, when the absolute value of the difference between the actual velocity of the particles in the measurement circle and the calculated velocity is minimized, the strain rate tensor in the measurement circle can be obtained by establishing a system of equations [36].

Assuming that there are N particles in the measurement circle, the particle translation speed and the centroid position are V_i and x_i respectively, and the average velocity $\overline{V}_i$ and average position $\overline{x}_i$ can be expressed as:

$$\overline{V}_i = \frac{\sum\limits_N V_i}{N} \tag{1}$$

$$\overline{x}_i = \frac{\sum\limits_N x_i}{N} \tag{2}$$

The actual relative velocity $\widetilde{V}_i$ and relative position of the particles $\widetilde{x}_i$ are:

$$\widetilde{V}_i = V_i - \overline{V}_i \tag{3}$$

$$\widetilde{x}_i = x_i - \overline{x}_i \tag{4}$$

Assuming that the particles move from point x_i to point x_j during infinitesimal time, the average speed difference between the two points is:

$$dv_i = \dot{\alpha}_{ij} dx_j \tag{5}$$

If the velocity gradient tensor $\dot{\alpha}_{ij}$ is known, the relative velocity can be calculated as:

$$\overline{v}_i = \dot{\alpha}_{ij} \overline{x}_j \tag{6}$$

Then the square of the absolute value of the difference between the relative velocity calculated in the circle and the actual relative velocity is:

$$z = \sum_N \left| \overline{v}_i - \widetilde{V}_i \right|^2 \tag{7}$$

When z takes the minimum value, namely

$$\frac{\partial z}{\partial \dot{\alpha}_{ij}} = 0 \tag{8}$$

Here, the velocity gradient tensor can be solved by the following equation

$$\begin{bmatrix} \sum_N \widetilde{x}_1 \widetilde{x}_1 & \sum_N \widetilde{x}_2 \widetilde{x}_1 \\ \sum_N \widetilde{x}_1 \widetilde{x}_2 & \sum_N \widetilde{x}_2 \widetilde{x}_2 \end{bmatrix} \begin{pmatrix} \dot{\alpha}_{i1} \\ \dot{\alpha}_{i2} \end{pmatrix} = \begin{pmatrix} \sum_N \widetilde{V}_i \widetilde{x}_1 \\ \sum_N \widetilde{V}_i \widetilde{x}_2 \end{pmatrix} \tag{9}$$

In PFC, the velocity gradient tensor can characterize the strain rate tensor, and the principal strain rate tensor calculated by the strain rate tensor can reflect the crack propagation processes and distinguish the crack types. If the value of the strain rate in the measurement circle is zero, it means no micro cracks appear in this area. Once the secondary crack starts to initiate, the velocity field of the fracture region of the particle will vary, which consequently results in the changes of strain rate magnitudes and directions. The magnitudes and directions of the principal strain rate represent the number of micro-cracks and deformation characteristics of the crack, respectively. To better understand on the variation law of the strain rate tensor of secondary crack between parallel pre-existing flaws, the layout of the measurement circle is shown in Figure 4.

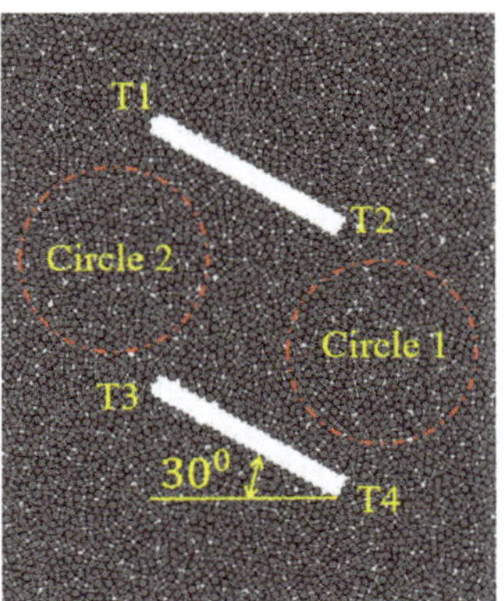

Figure 4. Layout of measuring circles 1 and 2. (T1, T2, T3 and T4 denote the four crack tips.).

3. Analysis of Numerical Simulation Results

In this paper, the cracks initiate between parallel pre-existing flaws before the axial stress peak strength are defined as secondary cracks. In PFC numerical simulation, the red, green, and blue micro-cracks represent tensile micro-cracks, tensile-shear micro-cracks, and compressive-shear micro-cracks, respectively.

3.1. Research on Crack Propagation Mechanism

Feignier et al. [45] suggested that the ratio of isotropic and deviatoric components of the moment tensor can be effectively used to quantify the failure mechanisms of events and distinguish the types of cracks. The average strain rate tensor caused by crack propagation is equal to the sum of the moment tensors in a unit area. Therefore, we define the variable R based on the strain rate tensor to analyze the crack initiation mechanism and judge the crack types. The variable R is given as:

$$R = \frac{tr(\dot{\alpha}_{ij}) * 100}{\left|tr(\dot{\alpha}_{ij})\right| + \Sigma\left|m_i^*\right|} \tag{10}$$

where $tr(\dot{\alpha}_{ij})$ is the trace of the moment tensor, which can be expressed as $\dot{\alpha}_{ij} = m_1 + m_2 + m_3$ and m_i (i = 1, 2, 3) are the eigenvalues of the moment tensor obtained by the calculation described in Equation (10). m_i^* is the deviatoric eigenvalue, which can be expressed as $m_i^* = m_i - tr(\dot{\alpha}_{ij})/3$. The ratio ($R$) ranges between 100 and −100.

In the breakage process of parallel bond contact, the propagation law, and types of cracks are determined to some extent by the velocity and movement tendency of the adjacent particles, indicating that the continuous evolution process of the crack is essentially the evolution process of the particle velocity. Therefore, according to the velocity and relative motion trend of the particles on both sides of the crack, the velocity field of the particles can be obviously divided into four types, namely Types I, II, III, and IV, as shown in Figure 5. The definitions of the four types of cracks are described as below.

(1) The typical characteristics of Type I are described as follows. The directions of horizontal component of particle velocity on both sides of the crack are opposite, the vertical component is in the same direction and the vertical component of the velocity is almost no different or zero. It can be obviously seen that the relative motion tendency of the particle is mainly controlled by the horizontal velocity component.

(2) For Type II, the directions of particle velocity on both sides of the crack are almost the same, and the values have no difference. In this case, the motion trend between the particles has certain inhibitory effects on the crack initiation and propagation.

(3) For Type III, the horizontal and vertical components of the particle velocity on both sides of the crack are the same, but the values are different.

(4) For Type IV, the directions of particle velocity on both sides of the crack are opposite.

For the convenience, the case of the dip angle of 30° is taken as an example to intensively study the evolution law of secondary crack propagation between parallel pre-existing double flaws. Since secondary cracks are characterized by a large quantity and instant with complicated initiation and coalescence mechanism, which include various kinds of micro-cracks, such as tensile micro-cracks, tensile-shear micro-cracks, and compressive-shear micro-cracks. For a better understanding of this phenomenon, different loading stages (a, b, c, d, e, and f) are selected for the main strain rate tensor analysis, as shown in Figure 6, and the axial stresses corresponding to each loading stage are 35.12, 36.96, 37.24, 43.40, 43.51, and 45.61 MPa, respectively.

The velocity fields and the strain rate tensors in the measurement circle at the different loading stages are shown in Figures 7 and 8. The velocity of the particles in Figure 7 is denoted by the black line with arrows and the thick black arrow represents the relative motion of the particles near the crack. The main strain rate tensor in Figure 8 is denoted by two sets of light green lines with arrows, with the direction and length of the arrows indicating the direction and relative size of the main strain rate, respectively. Table 4 shows the values of variable R in the measurement circle at different loading stages. The corresponding R values in the measurement circle ① and ② are recorded as R_1 and R_2, and the short dash line in Table 4 represents no secondary cracks at different loading stages.

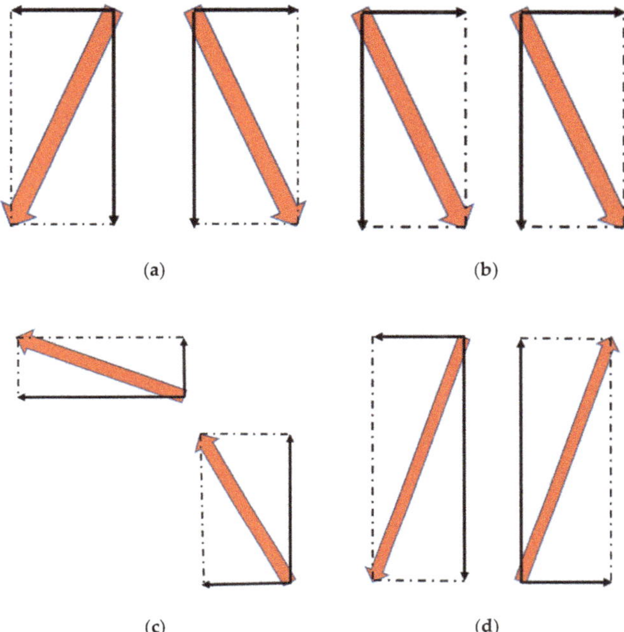

Figure 5. Schematic diagram of velocity fields for different types of particles. (**a**) Type I; (**b**) Type II; (**c**) Type III; (**d**) Type IV.

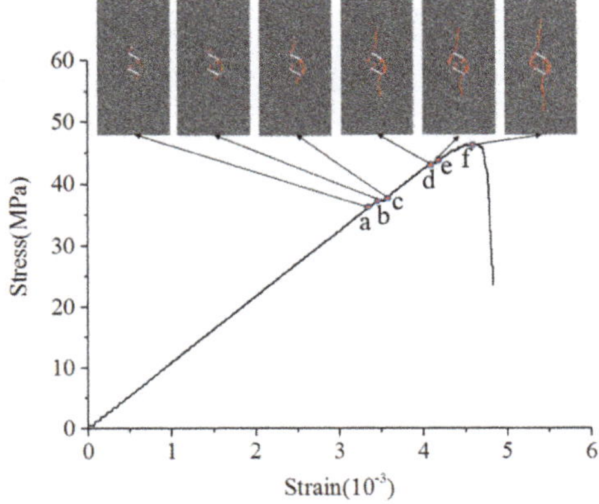

Figure 6. The stress-strain curve of a specimen under uniaxial compression. (Here, the dip angle of the flaws is $\alpha = 30°$. The points **a**, **b**, **c**, **d**, **e** and **f** correspond to different loading stages respectively.).

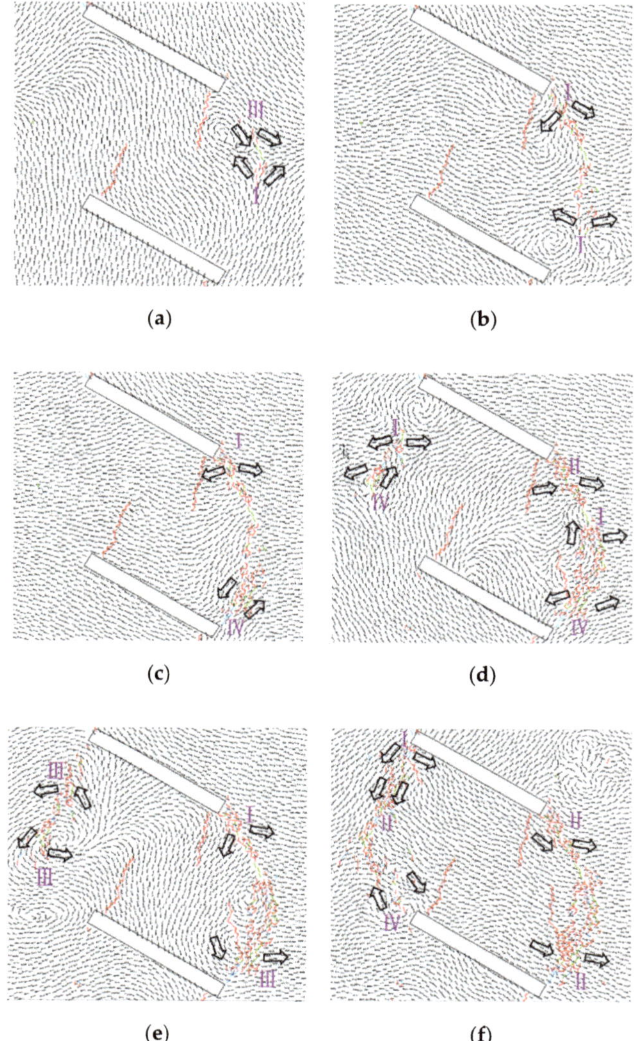

Figure 7. Particle velocity fields at different loading stages. (Pictures **a**, **b**, **c**, **d**, **e** and **f** respectively show particle velocity fields corresponding to different loading stages in Figure 6 when secondary cracks propagate.).

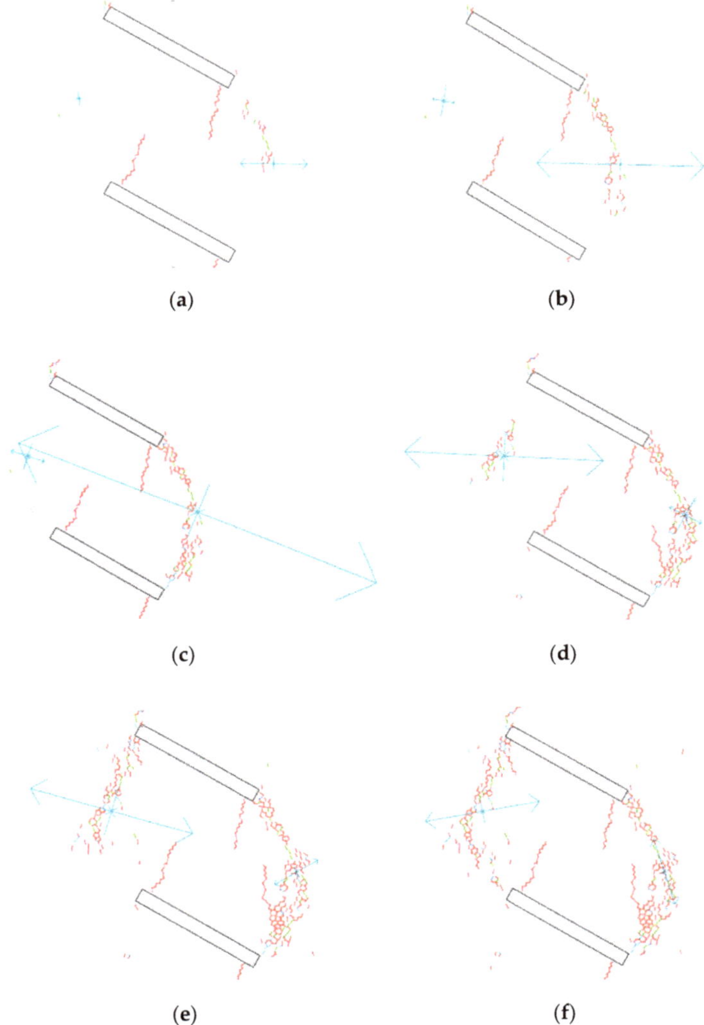

Figure 8. Principal strain rate tensors at different loading stages. (Pictures **a**, **b**, **c**, **d**, **e**, and **f** respectively show principal strain rate tensors in the two measurement circles corresponding to different loading stages in Figure 6 when secondary cracks propagate.).

Table 4. The values of variable R in the two measurement circles at different loading stages.

Loading Stages	Loading Stresses (MPa)	R_1	R_2
a	35.12	23.52	-
b	36.96	37.77	-
c	37.24	35.75	-
d	43.40	-	27.53
e	43.61	-	41.69
f	45.61	−45.60	23.02

When the secondary cracks between parallel pre-existing flaws instantaneously appear in large quantities, the values of the variable R can directly discriminate the types of newly generated cracks. 'Shear' is considered to occur for R varying between -30 and 30. 'Tensile' is considered to occur for an event with an R larger than 30. 'Compressive-shear' is considered to occur for an event with R smaller than -30. Therefore, the failure mechanisms of events can be exactly determined according to the values of variable R. Here, it should be noted that the time step adopted by PFC is not infinitely small. Even in the region where no crack initiates, the value of the strain rate tensor still exists, leading to the existence of the value of the variable R without no crack initiating. Consequently, we only need to pay more attentions to the strain rate tensor and the value of variable R at the time of secondary crack initiation.

For better analyzing crack propagation, the secondary cracks in the measurement circle ① and ② are named as SC-1 and SC-2 respectively. It can be seen from the analysis of Figures 7 and 8 that when the axial stress is increased to 35.12 MPa, SC-1 initiates near the tip T2 of the flaw, and the velocity fields of the upper and lower half of the newly generated cracks are Types III and I, which represent composite tensile-shear crack and tensile crack, respectively. At this time, the direction of the maximum principal strain rate is close to the horizontal direction. The ratio ($R_1 = 22.52$) indicates that the newly generated cracks are shear cracks in essence, where the relative motion of the particles are consistent with the direction of the maximum principal strain rate. As the axial stress is continuously increased to 36.96 MPa, one end of SC-1 extends to the tip T2 of the flaw, and the other end extends toward the tip T4 of the flaw. The velocity field on both sides of the newly generated cracks is Type I, and the value of R_1 is 37.77, indicating that the newly generated cracks are essentially tensile cracks. When the axial stress is increased to 37.24 MPa, SC-1 continues to extend to the crack tip T4, which finally generates an obvious crack zone to connect the flaw tip T2 with T4. Here the particle velocity field on both sides of the newly generated cracks near the flaw tip T4 are Type IV. Since relative shearing trend appears between the particles, the particle velocity field on both sides of the newly generated cracks near the flaw tip T2 are Type I. However, the variable ($R_1 = 45.61$) indicates that the newly generated crack are essentially tensile cracks.

As the strain increases and the axial stress is slowly increased up to 43.40 MPa, SC-2 suddenly initiates between the crack tip T1 and T3. The particle velocity fields on both sides of the upper and lower part of the newly generated cracks are Types I and IV, which represents that the corresponding newly generated cracks are tensile and shear crack, respectively. Here, the variable ($R_2 = 27.53$) indicates that that the newly generated cracks are shear cracks. As the axial stress is slightly increased up to 43.61 MPa, one end of SC-2 extends to the tip T1 of the flaw, and the other end extends toward the tip T3 of the flaw. The velocity fields on both sides of the newly initiated cracks are Type III and the variable ($R_2 = 41.69$) indicates that the newly generated cracks are tensile cracks. When the axial stress is increased up to 43.61 MPa, the axial stress approaches to the peak strength at this time and SC-2 continues to extend to the crack tip T3. Eventually an obvious crack zone is generated between the flaw tip T1 and T3. Since the relative shearing tendency appears between the particles, the particle velocity fields on both sides of the newly generated cracks near the crack tip T3 are Type IV, while those near the flaw tip T2 are mixed Types I and II. The variable ($R_2 = 23.02$) indicates that the newly generated cracks are shear cracks. Meanwhile, SC-1 continues to propagate, and eventually the particle velocity fields on both sides of the new crack turn into Type II. The variable ($R_2 = -45.60$) indicates that the newly generated cracks that continue to propagate on a basis of SC-1 are essentially composite compressive-shear cracks.

3.2. Analysis and Discussions

It can be seen from the above simulation results that SC-1 and SC-2 appear as shear cracks between pre-existing parallel flaws at the beginning stage of secondary crack initiation. However, from the velocity fields analysis of the particles, the upper part of the SC-1 are composite tensile-shear cracks and the lower part are tensile cracks; the upper part of the SC-2 are tensile cracks, and the lower part are shear cracks. In addition, the newly generated cracks contain tensile micro-cracks and tensile-shear micro-cracks, in which the number of tensile micro-cracks is dominant. As the axial stress is increased, SC-1 continues to extend to the tip of T4, and the newly generated cracks are tensile cracks. When SC-2 continues to extend to the tip of T3, the newly generated cracks are shear cracks. However, based on the velocity field analysis, it can be seen that the newly generated cracks near the flaws T3 and T4 tips are shear cracks, indicating that the types of macro-cracks are not totally determined by the types of partial cracks. Meanwhile, the type of macro-crack does not depend on the dominant type of micro-cracks. Since the tensile strength of the particles is much smaller than the shear strength, when the tensile micro-cracks initiate, a shear band is gradually formed under the action of the shearing force. Similarly, after the particles suffer the shear failure in the compressive zone, the stress concentration effect promotes the initiation of tensile cracks. Therefore, it can be concluded that macro tensile cracks and shear cracks partially contain shear cracks and tensile cracks respectively. SC-1 and SC-2 appear as shear cracks between pre-existing parallel flaws at the beginning stage of secondary crack initiation, and then extend up and down toward the tip of the flaw, and finally T2 and T4, as well as T1 and T3, are connected in the form of arcs. The particles between the pre-existing flaws form a confined compressive member under uniaxial compression. The type of particle velocity on both sides of the secondary crack between the fractures is most complicated, so the velocity field types of the particles are distinctive at different stress loading stages. For example, the velocity fields of the particles near the crack on the T4 tip of the flaw evolve from the initial Type IV into Type III, and finally remain in Type II. Thereafter, the cracks will not propagate any more, indicating that the type of newly generated cracks near the flaw T4 gradually evolves from shear cracks into composite tensile-shear cracks. Finally, because the velocity fields of the particles on both sides of the crack are Type II, the velocity direction and value of the particles are not much different, which can suppress the crack propagation to some extent.

4. Conclusions

In this study, the analysis of strain rate tensors and particle velocity fields are utilized to distinguish crack types and study the mechanical behaviors in the region between pre-existing flaws under uniaxial compression. The following conclusions have been drawn.

(1) By defining a variable R to quantify the crack failure mechanism, the types and mechanical behaviors of the secondary cracks between the flaws can be effectively distinguished. The initiation mechanism of secondary cracks between flaws is most complicated, and the types and mechanical behaviors of newly generated cracks are distinctive in different stress loading stages. According to the value of variable R, we can directly understand the types mechanical behaviors of secondary cracks.

(2) According to the velocity and relative motion trend of the particles on both sides of the crack, the velocity field of the particles can be obviously divided into four types. The type of particle velocity field on both sides of the newly generated cracks determines the type of crack in the measurement region. At different stress loading stages, the velocity field types of the particles on both sides of the crack are constantly evolving and complicated.

(3) Combined with the particle velocity field analysis and the value of the variable R, it can be seen that the macro tensile crack contains partial shear cracks, and the macro shear crack contains partial tensile cracks, indicating that the type of macro crack is not totally determined by the type of partial cracks. The secondary cracks contain tensile micro-cracks, shear micro-cracks, and compressive-shear micro-cracks, and the number of tensile micro-cracks is the largest. However, when the axial stress is reached to 35.12 MPa, the ratio ($R_1 = 22.52$) indicates that the newly generated cracks are shear cracks in essence. Therefore, we can see that the type of macro-crack does not depend on the dominant type of micro-cracks.

(4) At the beginning stage of secondary crack initiation, SC-1 and SC-2 appear as shear cracks between pre-existing parallel flaws, then extend up and down toward the tip of the flaws, and finally

connect the tips of T2 and T4, as well as T1 and T3, in the form of arcs. The particles between pre-existing parallel flaws form a confined compressive member under uniaxial compression. Under the confinement of lateral particles, the contacts between particles are broken owing to the combined compressive and shear actions, and eventually the shear cracks are successively formed.

Author Contributions: Conceptualization, Y.L. and X.L.; Methodology, W.C.; Software, W.C.; Validation, Y.L. and Q.Z.; Formal analysis; Writing—review and editing, W.Z. and Y.L.; Supervision, S.W. and W.C.

Funding: This research was funded by National Natural Science Foundation of China (Grant Nos. 51879149, 51779134 and 51579142), Taishan Scholars Project of Shandong Province and Open Research Fund Program of State Key Laboratory of Water Resources and Hydropower Engineering Science (Grant No. 2018SGG01).

Acknowledgments: The authors would like to thank the anonymous reviewers for their constructive suggestions to improve the quality of the paper.

Conflicts of Interest: The authors declare no conflict of interest.

References

1. Zhu, W.; Li, Y.; Li, S.; Wang, S.; Zhang, Q. Quasi-three-dimensional physical model tests on a cavern complex under high in-situ stresses. *Int. J. Rock Mech. Min.* **2011**, *48*, 199–209.
2. Li, Y.; Zhu, W.; Fu, J.; Guo, Y.; Qi, Y. A damage rheology model applied to analysis of splitting failure in underground caverns of Jinping I hydropower station. *Int. J. Rock Mech. Min.* **2014**, *71*, 224–234. [CrossRef]
3. Li, Y.; Zhou, H.; Zhang, L.; Zhu, W.; Li, S.; Liu, J. Experimental and numerical investigations on mechanical property and reinforcement effect of bolted jointed rock mass. *Constr. Build. Mater.* **2016**, *126*, 843–856. [CrossRef]
4. Li, Y.; Li, C.; Zhang, L.; Zhu, W.; Li, S.; Liu, J. An experimental investigation on mechanical property and anchorage effect of bolted jointed rock mass. *Geosci. J.* **2017**, *21*, 253–265.
5. Li, Y.; Zhou, H.; Dong, Z.; Zhu, W.; Li, S.; Wang, S. Numerical investigations on stability evaluation of a jointed rock slope during excavation using an optimized DDARF method. *Geomech. Eng.* **2018**, *14*, 271–281.
6. Zhou, X.; Wang, Y.; Zhang, J.; Liu, F. Fracturing behavior study of Three-Flawed specimens by uniaxial compression and 3D digital image correlation: Sensitivity to brittleness. *Rock Mech. Rock Eng.* **2019**, *52*, 691–718. [CrossRef]
7. Ma, G.W.; Dong, Q.Q.; Fan, L.F.; Gao, J.W. An investigation of non-straight fissures cracking under uniaxial compression. *Eng. Fract. Mech.* **2018**, *191*, 300–310. [CrossRef]
8. Tang, C.A.; Kou, S.Q. Crack propagation and coalescence in brittle materials under compression. *Eng. Fract. Mech.* **1998**, *61*, 311–324. [CrossRef]
9. Park, C.H.; Bobet, A. Crack initiation, propagation and coalescence from frictional flaws in uniaxial compression. *Eng. Fract. Mech.* **2010**, *77*, 2727–2748. [CrossRef]
10. Yang, S.Q.; Huang, Y.H.; Tian, W.L.; Zhu, J.B. Erratum to: An experimental investigation on strength, deformation and crack evolution behavior of sandstone containing two oval flaws under uniaxial compression. *Eng. Geol.* **2017**, *217*, 35–48. [CrossRef]
11. Yang, S.Q.; Ranjith, P.G.; Jing, H.W.; Tian, W.L.; Ju, Y. An experimental investigation on thermal damage and failure mechanical behavior of granite after exposure to different high temperature treatments. *Geothermics* **2017**, *65*, 180–197. [CrossRef]
12. Zhou, X.P.; Bi, J.; Qian, Q.H. Numerical simulation of crack growth and coalescence in Rock-Like materials containing multiple pre-existing flaws. *Rock Mech. Rock Eng.* **2015**, *48*, 1097–1114. [CrossRef]
13. Chen, W.; Li, S.; Zhu, W.; Qiu, X. Experimental and numerical research on crack propagation in rock under compression. *Chin. J. Rock Mech. Eng.* **2003**, *22*, 18–23.
14. Wu, Z.; Wong, L.N.Y. Frictional crack initiation and propagation analysis using the numerical manifold method. *Comput. Geotech.* **2012**, *39*, 38–53. [CrossRef]
15. Bobet, A. The initiation of secondary cracks in compression. *Eng. Fract. Mech.* **2000**, *66*, 187–219. [CrossRef]
16. Cao, P.; Liu, T.; Pu, C.; Lin, H. Crack propagation and coalescence of brittle rock-like specimens with pre-existing cracks in compression. *Eng. Geol.* **2015**, *187*, 113–121. [CrossRef]
17. Shen, B.; Stephansson, O.; Einstein, H.H.; Ghahreman, B. Coalescence of fractures under shear stress experiment. *J. Geophys. Res.* **1995**, *100*, 5975–5990. [CrossRef]

18. Wong, R.H.C.; Chau, K.T. Crack coalescence in a rock-like material containing two cracks. *Int. J. Rock Mech. Min.* **1998**, *35*, 147–164. [CrossRef]

19. Cheng, H.; Zhou, X.; Zhu, J.; Qian, Q. The effects of crack openings on crack initiation, propagation and coalescence behavior in Rock-Like materials under uniaxial compression. *Rock Mech. Rock Eng.* **2016**, *49*, 3481–3494. [CrossRef]

20. Bobet, A.; Einstein, H.H. Fracture coalescence in rock-type materials under uniaxial and biaxial compression. *Int. J. Rock Mech. Min.* **1998**, *35*, 863–888. [CrossRef]

21. Wong, L.N.Y.; Einstein, H.H. Crack coalescence in molded gypsum and carrara marble: Part 1. Macroscopic observations and interpretation. *Rock Mech. Rock Eng.* **2009**, *42*, 475–511. [CrossRef]

22. Wong, L.N.Y.; Einstein, H.H. Crack coalescence in molded gypsum and carrara marble: Part 2-Microscopic observations and interpretation. *Rock Mech. Rock Eng.* **2009**, *42*, 513–545. [CrossRef]

23. Wong, L.N.Y.; Einstein, H.H. Systematic evaluation of cracking behavior in specimens containing single flaws under uniaxial compression. *Int. J. Rock Mech. Min.* **2009**, *46*, 239–249. [CrossRef]

24. Cundall, P.A.; Strack, O. Discrete numerical-model for granular assemblies. *Geotechnique* **1979**, *29*, 47–65. [CrossRef]

25. Liu, J.; Wang, J.; Wan, W. Numerical study of crack propagation in an indented rock specimen. *Comput. Geotech.* **2018**, *96*, 1–11. [CrossRef]

26. Cao, R.H.; Cao, P.; Lin, H.; Ma, G.W.; Zhang, C.Y.; Jiang, C. Failure characteristics of jointed rock-like material containing multi-joints under a compressive-shear test: Experimental and numerical analyses. *Arch. Civ. Mech. Eng.* **2018**, *18*, 784–798. [CrossRef]

27. Park, B.; Min, K.; Thompson, N.; Horsrud, P. Three-dimensional bonded-particle discrete element modeling of mechanical behavior of transversely isotropic rock. *Int. J. Rock Mech. Min.* **2018**, *110*, 120–132. [CrossRef]

28. Cao, R.H.; Cao, P.; Lin, H.; Ma, G.W.; Fan, X.; Xiong, X.G. Mechanical behavior of an opening in a jointed rock-like specimen under uniaxial loading: Experimental studies and particle mechanics approach. *Arch. Civ. Mech. Eng.* **2018**, *18*, 198–214. [CrossRef]

29. Cao, R.; Lin, H.; Cao, P. Strength and failure characteristics of brittle jointed rock-like specimens under uniaxial compression: Digital speckle technology and a particle mechanics approach. *Int. J. Min. Sci. Technol.* **2018**, *28*, 669–677. [CrossRef]

30. Hazzard, J.F.; Young, R.P. Moment tensors and micromechanical models. *Tectonophysics* **2002**, *356*, 181–197. [CrossRef]

31. Zhang, X.; Zhang, Q. Distinction of crack nature in brittle Rock-Like materials: A numerical study based on moment tensors. *Rock Mech. Rock Eng.* **2017**, *50*, 2837–2845. [CrossRef]

32. Boyd, O.S.; Dreger, D.S.; Gritto, R.; Garcia, J. Analysis of seismic moment tensors and in situ stress during Enhanced Geothermal System development at the Geysers geothermal field, California. *Geophys. J. Int.* **2018**, *215*, 1483–1500. [CrossRef]

33. Stroujkova, A. Relative moment tensor inversion with application to shallow underground explosions and earthquakes. *Bull. Seismol. Soc. Am.* **2018**, *108*, 2724–2738. [CrossRef]

34. Ge, W.; Wang, M.; Shen, Z.; Yuan, D.; Zheng, W. Intersiesmic kinematics and defromation patterns on the upper crust of Qaidam-Qilianshan block. *Chin. J. Geophys. Chin.* **2013**, *56*, 2994–3010.

35. Kostrov, B.V.; Das, S. *Structural Earthquake Source Mechanics*; Science Press: Beijing, China, 1994.

36. Itasca Consulting Group Inc. *Users' Manual for Particle Flow Code (PFC)*; Version 5.0; Itasca Consulting Group Inc.: Minneapolis, MN, USA, 2014.

37. Lisjak, A.; Grasselli, G. A review of discrete modeling techniques for fracturing processes in discontinuous rock masses. *J. Rock Mech. Geotec. Eng.* **2014**, *6*, 301–314. [CrossRef]

38. Zhang, X.; Wong, L.N.Y. Loading rate effects on cracking behavior of flaw-contained specimens under uniaxial compression. *Int. J. Fracture* **2013**, *180*, 93–110. [CrossRef]

39. Yoon, J. Application of experimental design and optimization to PFC model calibration in uniaxial compression simulation. *Int. J. Rock Mech. Min.* **2007**, *44*, 871–889. [CrossRef]

40. Cho, N.; Martin, C.D.; Sego, D.C. A clumped particle model for rock. *Int. J. Rock Mech. Min.* **2007**, *44*, 997–1010. [CrossRef]

41. Ghazvinian, A.; Sarfarazi, V.; Schubert, W.; Blumel, M. A study of the failure mechanism of planar Non-Persistent open joints using PFC2D. *Rock Mech. Rock Eng.* **2012**, *45*, 677–693. [CrossRef]

42. Baker, C.; Young, R.P. Evidence for extensile crack initiation in point source time-dependent moment tensor solutions. *Bull. Seismol. Soc. Am.* **1997**, *87*, 1442–1453.
43. Martin, C.D.; Chandler, N.A. The progressive fracture of lac du bonnet granite. *Int. J. Rock Mech. Min. Sci. Geomech. Abstr.* **1994**, *31*, 643–659. [CrossRef]
44. Ming, H.J.; Feng, X.T.; Chen, B.R.; Zhang, C.Q. Analysis of rockburst mechanism for deep tunnel based on moment tensor. *Rock Soil Mech.* **2013**, *34*, 163–172.
45. Feignier, B.; Young, R.P. Moment tensor inversion of induced microseisnmic events: Evidence of non-shear failures in the $-4 < m < -2$ moment magnitude range. *Geophys. Res. Lett.* **1992**, *19*, 1503–1506.

Article

Numerical Analysis of the Mechanical Behaviors of Various Jointed Rocks under Uniaxial Tension Loading

Jiaming Shu, Lishuai Jiang *, Peng Kong and Qingbiao Wang

State Key Laboratory of Mining Disaster Prevention and Control, Shandong University of Science and Technology, Qingdao 266590, China; 15621565102@163.com (J.S.); 17854859770@163.com (P.K.); 18805381111@139.com (Q.W.)
* Correspondence: lsjiang@sdust.edu.cn

Received: 1 April 2019; Accepted: 27 April 2019; Published: 1 May 2019

Abstract: In a complex stress field of underground mining or geotechnical practice, tension damage/failure in rock masses is easily triggered and dominant. Unlike metals, rocks are generally bi-modularity materials with different mechanical properties (Young's modulus, etc.) in compression and tension. It is well established that the Young's modulus of a rock mass is directly related to the presence of the fracture or joint, and the Young's modulus estimation for jointed rocks and rock masses is essential for stability analysis. In this paper, the tensile properties in joint rocks were investigated by using numerical simulations based on the discrete element method. Four influencing parameters relating to the tensile properties (joint dip angle, joint spacing, joint intersection angle, and joint density) were studied. The numerical results show that there is an approximately linear relationship between the joint dip angle (α) and the joint intersection angle (β) with the tensile strength (σ_t), however, the changes in α and β have less influence on the Young's modulus in tension (E_t). With respect to joint spacing, the simulations show that the effects of joint spacing on σ_t and E_t are negligible. In relation to the joint density, the numerical results reveal that the joint intensity of rock mass has great effect on E_t but insignificant effect on σ_t.

Keywords: jointed rock; uniaxial tension loading; numerical analysis; discrete element method

---

1. Introduction

For brittle materials such as rock, tension failure is one of the most significant failure modes [1]. In a complex stress field of underground mining or geotechnical practice, tension damage/failure in rock masses is easily triggered and dominant because: (1) The tensile strength (σ_t) of rocks is much lower than their compressive strength, and (2) joints and fractures of rock mass can offer little resistance to tensile stress [2]. After an excavation is taken underground, tension failure and its induced fractures will occur in surrounding rock masses near the opening [3–8]. The initiated fractures will develop/propagate and bring weakening effects on the rock masses under the effect of time and constant stress disturbance (development of entry or shaft, mining activities, etc.). As it is well established that the Young's modulus of a rock mass is directly related to the fracture or joint intensity [9–13], the aforementioned fracture development can be described as a Young's modulus degradation process in the view of an equivalent material method in continuum mechanics.

The macroscopic mechanical behaviors of intact and jointed rocks have been widely investigated, most of which has been focused on the material behaviors in compression [14–16]. However, unlike metals or other materials, rocks are generally bi-modularity materials with different Young's moduli and Poisson's ratios in compression and tension. Previous studies done by direct tension tests showed that the Young's modulus in tension (E_t) was always no larger than the compressive Young's modulus

(E_c) [17–21]. Hawkes et al. [17] conducted direct tension tests on different kinds of rocks, and the results showed that the ratio of E_t/E_c was 1/9 for Barre sandstone and 0.5 for Barre granite. Stimpson and Chen [20] acquired the ratios to be 0.5, 1.0, 0.7, and 0.3–0.4 for four different rocks from cyclic loading uniaxial tension and compression tests. Similar results were obtained by Yu et al. [22] with an originally developed loading frame for direct tension. It has been addressed by researchers that the improper assumption that E_t equals to E_c may lead to errors in calculating stress distributions around underground openings by means of analytical or numerical analysis, as well as in determining the tensile strength of rocks with Brazilian tests [18,20,22,23].

To date, it is still quite challenging to conduct direct uniaxial tension tests on intact rock specimens in laboratories due to the difficulties in avoiding: 1) Unfavorable stress concentration over the grip and 2) bending moments due to the non-coaxial gripping and curvature of the specimen. Various attempts have been made in this regard. Stimpson and Chen [20] conducted their tests on rock samples with a special hollow cylinder geometry. Okubo and Fukui [24] glued the rock samples directly to the loading platen to carry out direct tension tests. Fahimifar and Malekpour [25] developed a direct tension apparatus with hard steel tension jaws to ensure the connection between samples and the apparatus and that the applied load is pure tension. Fuenkajorn and Klanphumeesri [26] developed a compression-to-tension load converter to determine the tensile strength and Young's modulus from dog-bone-shaped specimens.

As described above, the Young's modulus estimation for jointed rocks and rock masses is essential for stability analysis. However, rock specimens with joints or fractures are difficult to be prepared, and simplifications have to be made with rock-like materials (such as gypsum) or simple notches. Furthermore, the uniaxial tension tests rely on specially made, non-universal apparatuses.

In recent years, numerical and computational resources have taken significant leaps forward, and numerical methods have become strong and efficient research tools by overcoming the limitations in laboratory and physical tests. Some previous studies have been done with numerical modelling on the direct tension testing of rock specimens. Tang et al. [1] studied the growth of micro-fractures in a specimen under uniaxial tension and the influence of heterogeneity on rock strength with a self-developed finite element code (RFPA2D). Wang et al. [27] simulated crack initiation, propagation, and coalescence for intact, single-notched, and double-notched rock specimens with RFPA3D under uniaxial tension, and the effects of the separation distance and overlapping distance of the two notches were investigated. Hamdi et al. [28] studied the model I fracture propagation in brittle rock under direct tension by means of a discrete element method. Guo et al. [29] investigated fracture patterns in layered rocks under direct tension with a discrete element method.

According to the previous studies [1,27–32], the numerical studies on rock specimens under uniaxial or direct tension loads mainly focus on fracture initiation and propagation within the specimens. However, the effects on the mechanical behaviors of strength and deformability, i.e., tensile strength and Young's modulus in tension, of jointed rock specimens subjected to uniaxial tension load are barely discussed. In this study, rock specimens with various joint conditions are modeled with 3DEC (a 3-Dimension Distinct Element Code by Itasca) due to its capability in simulating the behaviors of joints (slip, separation, deformation, etc.) under mechanical loading, and the effects of joints on the tensile properties are studied by means of uniaxial tension numerical tests. Tests of rock samples with single joint, and multiple joints with different joint geometry parameters are designed, and their effects on the tensile properties of rock specimens are investigated. The goal of this study is to understand the effects of joints on the tensile properties of rocks.

2. Numerical Modelling

The numerical model is constructed by means of 3DEC, which is a three-dimensional numerical program employing the distinct element method for discontinuum modelling. In the following analysis, rock specimen models with different joint conditions (single, multiple parallel, and intersecting joints) are built, and the effects of joints on the tensile properties are investigated by applying a uniaxial tension load. The specimen model is a cylinder with 50 mm in diameter and 100 mm in height, as suggested by the International Society of Rock Mechanics (ISRM) [33]. All joints are considered to cut through the specimen. A constant rate of 0.005 mm/step is applied to the upper and lower boundary to simulate the uniaxial tension load. After the specimens fail in tension, 1000 extra steps are calculated to capture the after-peak behavior. With the obtained stress–strain curves, σ_t and E_t under different joint conditions are calculated, and the effects of joint condition on the tensile strength (σ_t) and E_t are hereby investigated.

In 3DEC, joints or fractures are considered as elements with mechanical properties. The rock and joint properties [34] (listed in Tables 1 and 2) are applied according to a study on tunnel stability analysis of an underground mine using 3DEC.

Table 1. Rock mass properties.

Lithology	Density (kg/m^3)	Bulk modulus K (GPa)	Shear modulus G (GPa)	Cohesion C (MPa)	Internal friction angle φ (°)	Tensile strength (MPa)
Sandstone	2630	26.49	19.05	3.75	25.9	2.25

Table 2. Joint properties.

Lithology	Cohesion C (MPa)	JKN (GPa/M)	JKS (GPa/M)	Joint friction angle (°)	Tensile strength (MPa)
Sandstone	3.67	26.49	19.05	25.9	1.88

To study the effects of joints on the tensile properties of rocks, the rock specimens with various joint conditions were described and modelled with four geometry parameters of joints: α (the angle between the joint and the horizontal direction), d (the perpendicular distance between the two joints), β (the angle at which the two joints intersect), and n (the joint density).

In cases of rock specimens with two joints, the effects of the perpendicular distance between the two joints (d), the angle between the joint and the horizontal direction (α) of parallel joints, and the joint density (n) on the complete stress–strain response were studied by numerical simulation. When d was kept constant at 10, 20, 30, and 40 mm, four different homogeneity indexes (α) were considered, which were 20, 30, 40 and 50°, respectively. Similarly, when α were kept constant, d was varied as 10, 20, 30, and 40 mm. Moreover, when d and α were kept constant, n was changed from 2 to 6.

In cases of rock specimens with multiple joints, the effects of the perpendicular distance between the two cracks (d), the angle at which the two joints intersected (β), and the joint density (n) of intersecting joints on the complete stress–strain response were studied by numerical simulation. When d and n were kept constant as 0 mm and 2, there were six different homogeneity indexes (β) of 20, 40, 60, 80, and 100°. When β and n were kept constant, d was varied as 10, 20, 30, and 40mm. As a comparison, when d and n were kept constant, β was separately varied as 40, 60, 80, and 100°. When d and β were kept constant, n was changed from 4 to 12.

3. Effects of Joints on the Tensile Properties

3.1. Effect of the Dip Angle of A Single Joint

In this study, all joints are assumed to cut through the specimen. Seven models with different dip angles (α = 0, 10, 20, 30, 40, and 50°) were built, as illustrated in Figure 1. The stress–strain curves and the tensile properties with respect to dip angles are shown in Figure 2 and Table 3, respectively.

As can be seen, the general shapes of the stress–strain curves are similar. The stresses of specimens linearly increase after the tension loading starts, and undergo a sudden drop after failure, due to the brittle nature of rocks. The dip angle of the joint clearly has a notable effect on the tensile properties of the rock specimens. The peaks (σ_t) of the pre-peak curves vary monotonically increases with the increase of α.

When $\alpha = 0°$, i.e., a horizontal joint, the σ_t and E_t of the specimen are 0.78 MPa and 0.80 GPa, respectively. When α reaches 50°, the σ_t and E_t increase to 1.05 MPa and 0.81 GPa, which are approximately 1.35 and 1.01 times the values in the case of $\alpha = 0°$, respectively. In comparison, the changes in α have a greater influence on σ_t than E_t. In reality, joint surfaces are usually rough, and the asperities provide resistance to shear stress, and the joint roughness is positively correlated to the shear strength [35]. The joint roughness is described with normal stiffness, shear stiffness, cohesion, etc. in 3DEC. When α is not 0°, the shear strength of the joint will provide extra resistance to the tensile load on rock failure and deformability, which leads to the high σ_t and E_t when is α higher. However, when $\alpha = 0°$, the joint is perpendicular to the tensile load, so no extra resistance can be provided from the shear behavior of the joint.

The relationship between α and σ_t after fitting is shown in Figure 3. As can be seen, σ_t is linearly and positively correlated to the dip angle of the joint, and the fitting formula is shown in Figure 3.

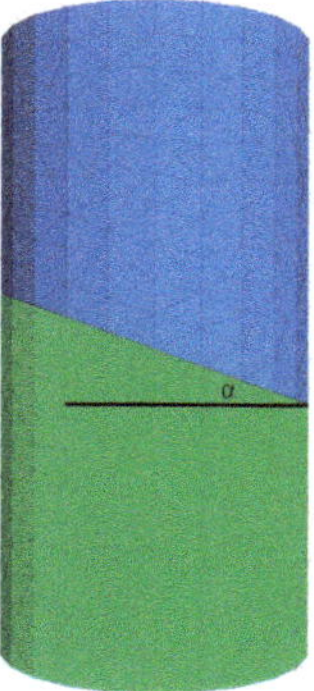

Figure 1. Rock specimen model with a single joint.

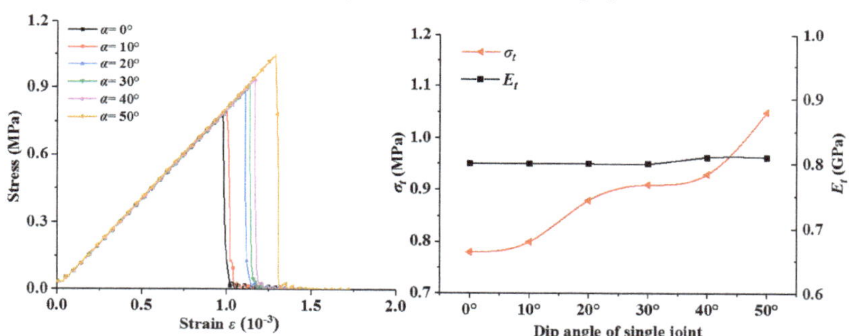

Figure 2. Mechanical behaviors of single-jointed rock specimens with different dip angles.

Table 3. Tensile strengths (σ_t) and Young's moduli in tension (E_t) of single-jointed rock specimens with different dip angles.

α	σ_t (MPa)	E_t (GPa)
$\alpha = 0°$	0.78	0.80
$\alpha = 10°$	0.80	0.80
$\alpha = 20°$	0.88	0.80
$\alpha = 30°$	0.91	0.80
$\alpha = 40°$	0.93	0.81
$\alpha = 50°$	1.05	0.81

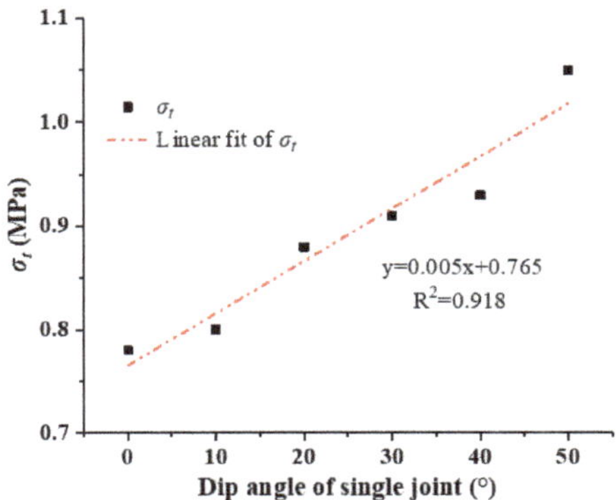

Figure 3. The relationship between joint dip angle α and σ_t.

3.2. Effects of Parallel Joints on the Tensile Properties

Due to the existence of weak planes in rocks or rock masses, joints are the initiation points of rock and rock mass failure in most cases. Rock masses with multiple joints are common in engineering practice.

3.2.1. Effects of Joint Spacing and Dip Angle of Two Parallel Joints

Joint spacing is the perpendicular distance between adjacent joints, and usually determines the sizes of the blocks making up the rock mass [2]. To investigate the effects of joint spacing (d) and dip angle (α) of two parallel joints on σ_t and E_t, four sets of models with different dip angles (α = 20, 30, 40, and 50°) were built (as illustrated in Figure 4) under the condition of n = 2, and for each set of the model, four subsets with different joint spacing (d = 10, 20, 30, and 40 mm) were analyzed. The test program specifics are shown in Table 4, and the stress–strain curves under uniaxial tensile load are shown in Figure 5. The results are shown in Table 5.

Table 4. Model designs for rock specimens with two parallel joints.

Sets of models	α	Subsets with respect to d
1	20°	10 mm 20 mm 30 mm 40 mm
2	30°	10 mm 20 mm 30 mm 40 mm
3	40°	10 mm 20 mm 30 mm 40 mm
4	50°	10 mm 20 mm 30 mm 40 mm

As can be seen, the slopes and peaks of the curves, with same dip angle but different joint spacing, are almost the same, which means that the effect of joint spacing on σ_t and E_t are negligible. When d decreases from 40 to 10 mm, σ_t and E_t merely decrease by 0.003 MPa and 0.002 GPa, respectively. This result is obtained under the premise that no joint development is considered. In practice, joints will develop, and adjacent joints will connect and merge under mechanical loading. However, this study focuses on the effects of existing joints on the tensile properties. Similar to the results from Section 3.1, σ_t and E_t are affected by the dip angle of the two parallel joints, but the effects vary. The dip angle has a greater effect on σ_t than E_t. As the dip angle increases from 20 to 50°, σ_t increases from 0.88 to 1.05 MPa, while E_t lightly increases from 0.71 to 0.73 GPa. When comparing with Table 3, it can be noticed that the value of σ_t is identical in the cases of single-jointed and double-jointed models with the same dip angle.

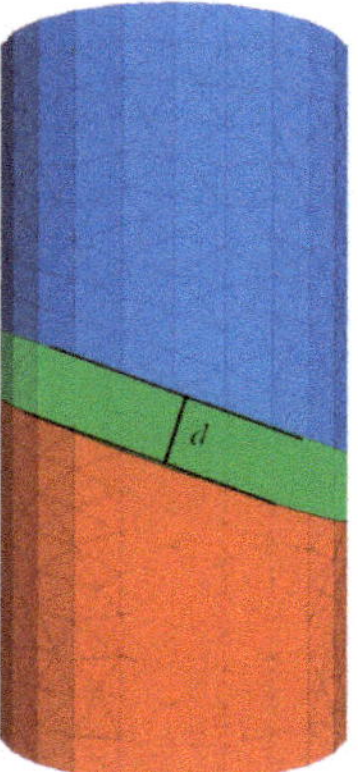

Figure 4. Rock specimen model with parallel joints.

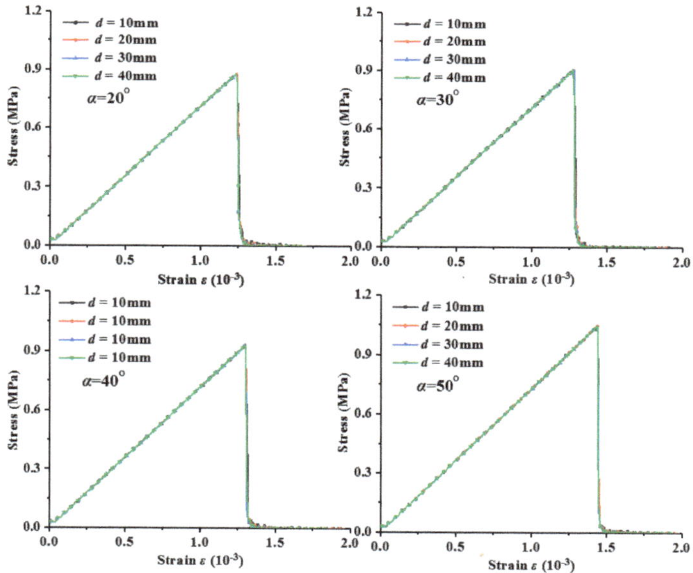

Figure 5. Tensile behaviors of double-jointed rock specimens with different joint dip angles.

Table 5. Tensile properties of double-jointed rock specimens with different joint dip angles.

α	σ_t (MPa)	E_t (GPa)
20°	0.88	0.71
30°	0.91	0.71
40°	0.93	0.73
50°	1.05	0.73

3.2.2. Effect of Joint Density of Parallel Joints

In a certain unit of rock mass, the number of joints is often referred to as the fracture intensity [36,37]. It is well addressed that the mechanism of deformation and failure of rock masses varies with fracture intensity, as well as the engineering properties such as cavability, permeability, and fragmentation characteristics. In this study, the effect of joint intensity on σ_t and E_t is investigated by means of the number of joints (n) inside the specimen. Four sets of models with different dip angles (α = 20, 30, 40, 50°) were built (as illustrated in Figure 6) under the condition of d = 5 mm, and for each set of the model, six subsets with different numbers of joints (n = 1, 2, 3, 4, 5, 6) were analyzed. The test program specifics are shown in Table 6 and the stress–strain curves under uniaxial tensile load are shown in Figure 7. The results are shown in Table 7.

Table 6. Model designs for rock specimens with multiple parallel joints.

Sets of models	α	Subsets with respect to n
1	20°	$n = 2$ $n = 3$ $n = 4$ $n = 5$
2	30°	$n = 2$ $n = 3$ $n = 4$ $n = 5$
3	40°	$n = 2$ $n = 3$ $n = 4$ $n = 5$
4	50°	$n = 2$ $n = 3$ $n = 4$ $n = 5$

As can be seen, for each subset with the same α, the peaks of the curves are almost identical, while the slopes are notably effected by n. Taking the model set with α = 40° as an example, as listed in Figure 7, when n increases from 1 to 6, E_t significantly drops from 0.81 to 0.52 GPa, while σ_t remains constant. The influencing pattern of n on E_t is identical for different values of α. These results indicate that the fracture intensity of a rock mass has a great effect on E_t but a negligible effect on σ_t.

The relationship between n and E_t with different values of α after fitting is shown in Figure 8. As can be seen, E_t is negatively correlated to n, and the relationship varies with different values of α. The change in E_t with n will be less significant in cases of higher values of α.

Figure 6. Rock specimen model with multiple parallel joints.

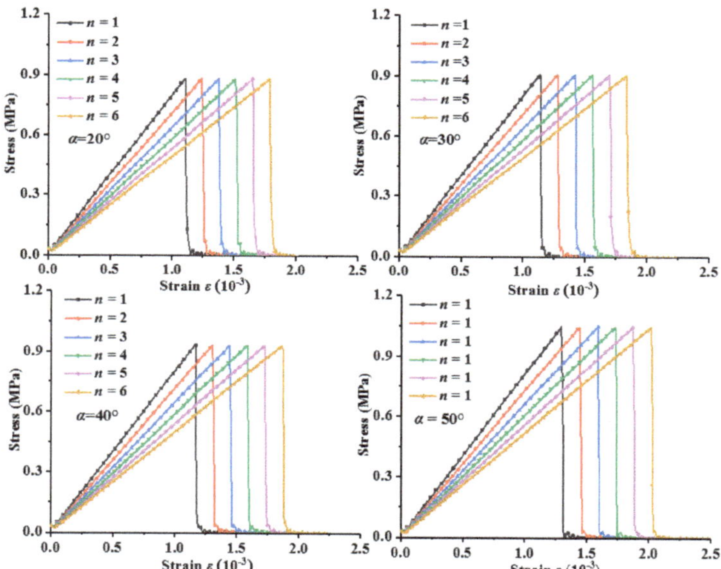

Figure 7. Tensile behaviors of rock specimens with different numbers of joints.

Table 7. Effect of joint density of parallel joints on tensile properties.

n	σ_t **(MPa)**	E_t **(GPa)**
$n = 1$	0.93	0.81
$n = 2$	0.93	0.73
$n = 3$	0.93	0.66
$n = 4$	0.93	0.61
$n = 5$	0.93	0.56
$n = 6$	0.93	0.52

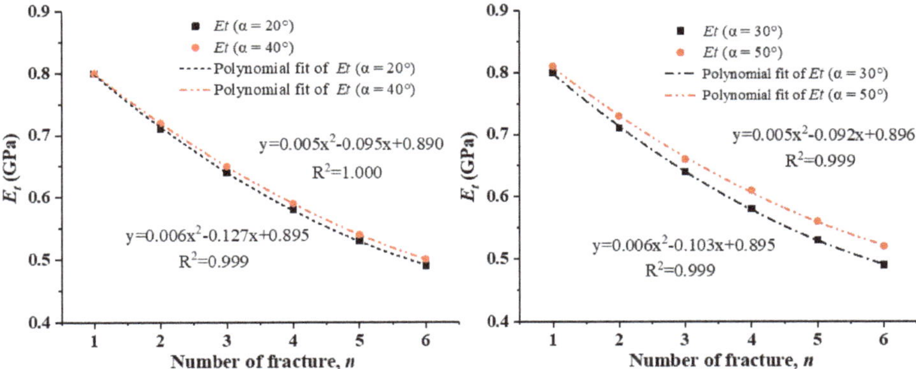

Figure 8. The relationship between the joint density (n) and E_t under the different values of α.

3.3. Effects of Intersecting Joints on the Tensile Properties

In actual rock engineering activities, the existence of fractures is usually very complicated. One or more intersecting joint sets, usually referred to as a joint system, are common in heavily jointed rock masses. In this section, the effects of intersecting joints on the tensile properties of rock specimens are analyzed.

3.3.1. Effect of the Intersection Angle of an Intersecting Joints Set

Seven models with symmetric intersecting joints, which have different intersection angles ($\beta = 20$, 40, 60, 80 and 100°), were built, as illustrated in Figure 9. The stress–strain curves and the tensile properties with respect to intersection angles are shown in Figure 10 and Table 8, respectively.

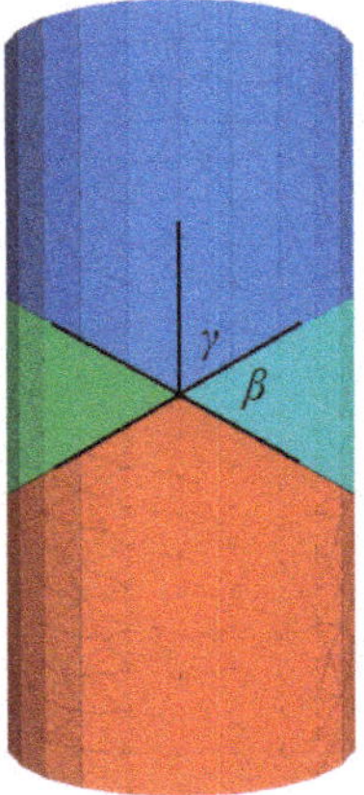

Figure 9. Rock specimen model with intersection joints.

As can be seen, the general patterns of the stress–strain curves are analogous. The stresses of specimens linearly increase after the tension loading starts, and undergo a sudden drop after failure, exhibiting the brittle behavior of rocks. The intersection angle of the joint clearly has a remarkable effect on the tensile properties of the rock specimens. The peaks (σ_t) and the slopes (E_t) of the pre-peak curves vary and $_t$ monotonically increase with the increase of β.

When $\beta = 20°$, the σ_t and E_t of the specimen are 0.80 MPa and 0.71 GPa, respectively. When β reaches 100°, the σ_t and E_t increase to 1.05 MPa and 0.73 GPa, which are approximately 1.3 and

1.03 times the values in the case of $\beta = 20°$, respectively. In contrast, the changes in β have a greater influence on σ_t than E_t.

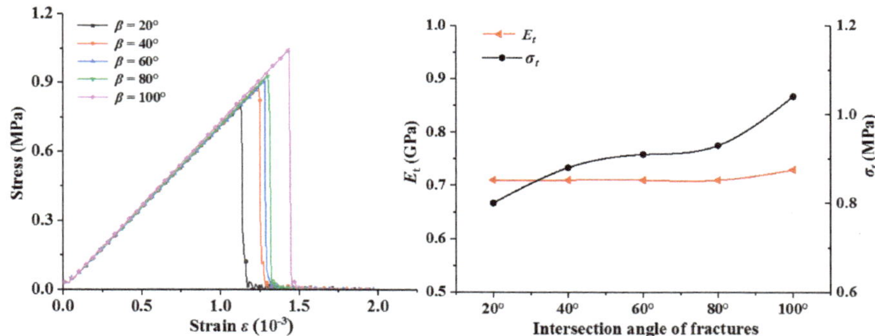

Figure 10. Tensile behaviors of intersection-jointed rock specimens with different joint angles.

Table 8. σ_t and E_t of intersection-jointed rock specimens with different joint angles.

α	σ_t (MPa)	E_t (GPa)
$\beta = 20°$	0.80	0.71
$\beta = 40°$	0.88	0.71
$\beta = 60°$	0.91	0.71
$\beta = 80°$	0.93	0.73
$\beta = 100°$	1.05	0.73

The relationship between β and σ_t after fitting is shown in Figure 11. As can be seen, σ_t is linearly and positively correlated to the intersection angle. The fitting formula is shown in Figure 11.

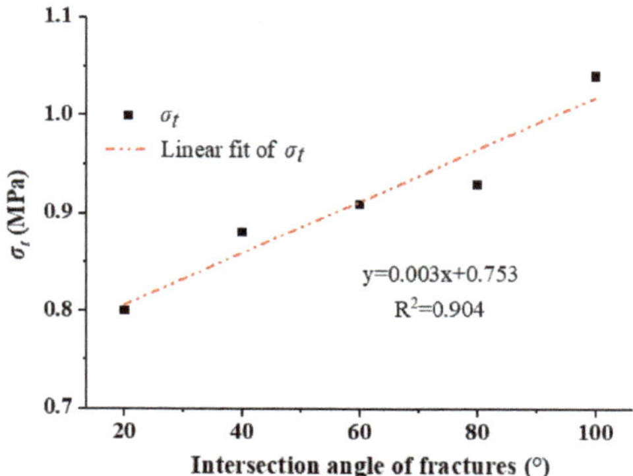

Figure 11. The relationship between the joint intersection angle (β) and σ_t.

3.3.2. Effects of Joint Spacing and Intersection Angle of a Joints Set

To investigate the effects of joint spacing (d) and intersection angle (β) of two intersection joints on σ_t and E_t, four sets of models with different dip angels ($\beta = 20, 40, 60,$ and $100°$) were built (as illustrated in Figure 12), and for each set of the model, four subsets with different joint spacing ($d = 10, 20, 30,$ and 40 mm) were analyzed. The test program specifics are shown in Table 9. The stress–strain curves under uniaxial tensile load are shown in Figure 13 and the results are shown in Table 10.

Figure 12. Rock specimen model with two intersection joints sets .

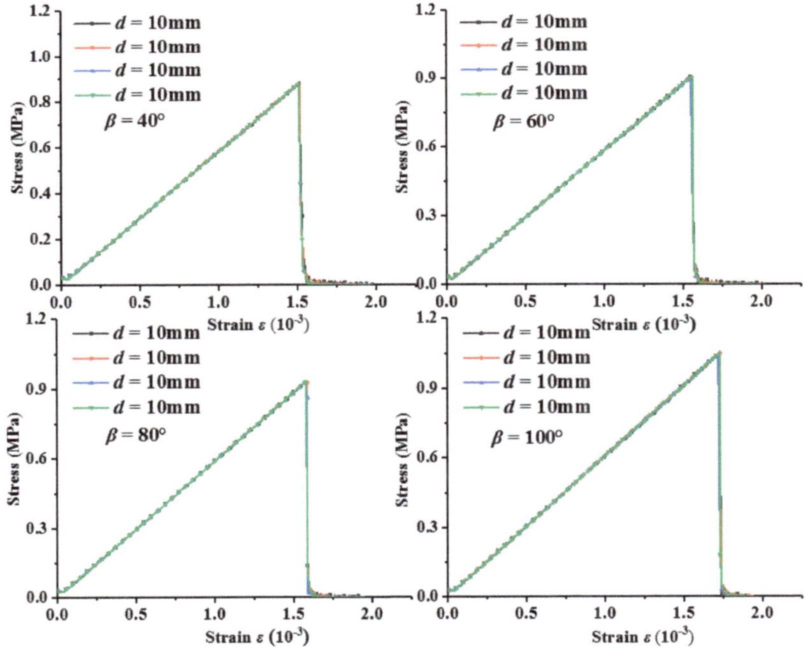

Figure 13. Tensile behaviors of intersection-jointed rock specimens with different intersection angles.

As can be seen, the slopes and peaks of the curves with same intersection angle but different joint spacing are almost the same, which means that the effects of joint spacing on σ_t and E_t are trivial. When d increases from 10 to 40 mm, σ_t and E_t merely increase by 0.0004 MPa and 0.0008 GPa, respectively. Similar to the results from Section 3.3.1, σ_t and E_t are affected by the joint spacing of the intersection joints, but the effect varies. The intersection angle has a greater effect on σ_t than E_t. As the intersection angle increases from 40 to 100°, σ_t doubles from 0.88 to 1.05 MPa, while E_t lightly increases from 0.58 to 0.61 GPa. It can be noticed that the value of σ_t is identical in the cases of single-jointed and double-jointed models with identical β.

Table 9. Model designs for rock specimens with two intersecting joint sets.

Sets of Models	d	Subsets with Respect to β
1	$d = 10$ mm	$\beta = 40°$ $\beta = 60°$ $\beta = 80°$ $\beta = 100°$
2	$d = 20$ mm	$\beta = 40°$ $\beta = 60°$ $\beta = 80°$ $\beta = 100°$
3	$d = 30$ mm	$\beta = 40°$ $\beta = 60°$ $\beta = 80°$ $\beta = 100°$
4	$d = 40$ mm	$\beta = 40°$ $\beta = 60°$ $\beta = 80°$ $\beta = 100°$

Table 10. σ_t and E_t of intersection-jointed rock specimens with different intersection angles.

β.	σ_t (MPa)	E_t (GPa)
$\beta = 40°$	0.88	0.58
$\beta = 60°$	0.91	0.58
$\beta = 80°$	0.93	0.58
$\beta = 100°$	1.05	0.61

3.3.3. Effect of Joint Density of Intersection Joints

In this study, the effects of joint intensity on σ_t and E_t are investigated by means of the number of joints (n) inside the specimen. Four sets of models with different dip angles ($\beta = 40, 60, 80,$ and $100°$) were built (as illustrated in Figure 14) under the condition of $d = 5$ mm, and for each set of the model, six subsets with different numbers of joints ($n = 4, 6, 8, 10,$ and 12) were analyzed. The test program specifics are shown in Table 11 and the stress–strain curves under uniaxial tensile load are shown in Figure 15. The results are shown in Table 12

Figure 14. Rock specimen model with intersecting joints.

Table 11. Model designs for rock specimens with two parallel joints.

Sets of Models	β	Subsets with Respect to n
1	40°	$n = 4$ $n = 6$ $n = 8$ $n = 10$ $n = 12$
2	60°	$n = 4$ $n = 6$ $n = 8$ $n = 10$ $n = 12$
3	80°	$n = 4$ $n = 6$ $n = 8$ $n = 10$ $n = 12$
4	100°	$n = 4$ $n = 6$ $n = 8$ $n = 10$ $n = 12$

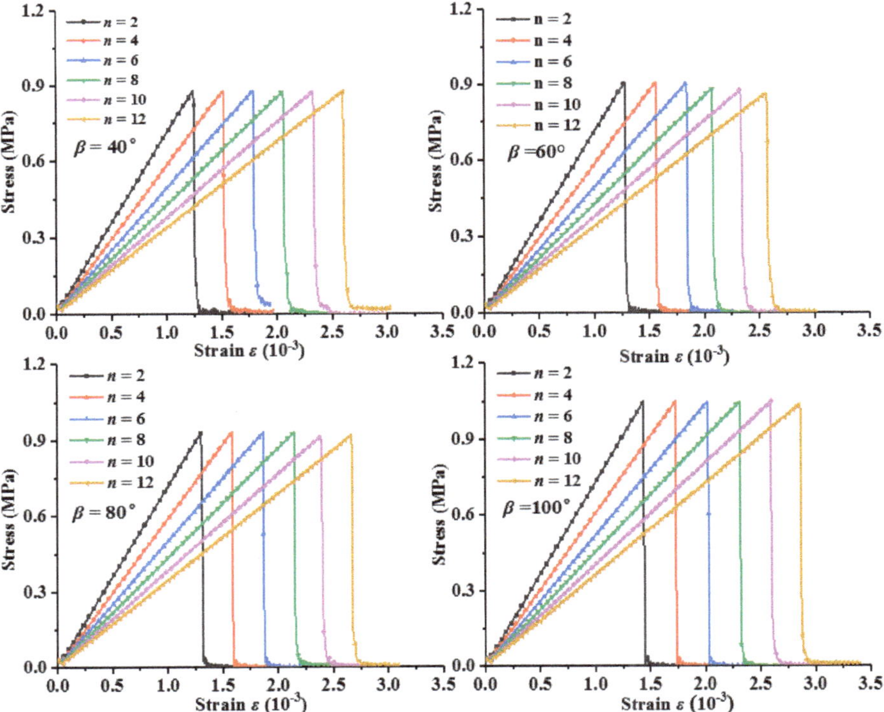

Figure 15. Tensile behaviors of intersection-jointed rock specimens with different numbers of intersection joints.

Table 12. Mechanical behaviors of rock specimens with different numbers of intersection joints.

n	σ_t (MPa)	E_t (GPa)
$n = 2$	0.93	0.73
$n = 4$	0.93	0.58
$n = 6$	0.93	0.50
$n = 8$	0.93	0.44
$n = 10$	0.93	0.38
$n = 12$	0.93	0.34

As can be seen, for each subset with the same α, the peaks of the curves are almost identical while the slopes are notably effected by n. Taking the model set with $\beta = 80°$ as an example, as listed in Figure 15, when n increases from 2 to 12, E_t significantly drops from 0.73 to 0.34 GPa, while σ_t remains constant. The influencing pattern of n on E_t is identical for different values of α. These results indicate that the fracture intensity of a rock mass has a great effect on E_t but a negligible effect on σ_t.

The relationship between n and E_t with different values of β after fitting is shown in Figure 16. As can be seen, E_t is negatively correlated to n, and the relationship varies with different values of β. The change in E_t with n will be less significant in case of a higher value of β.

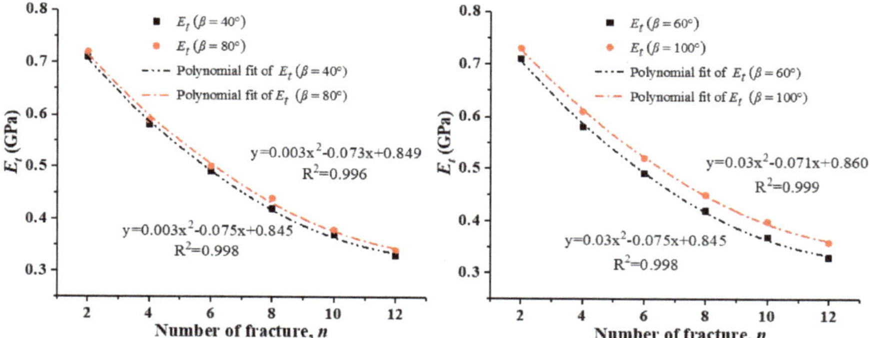

Figure 16. The relationship between n and E_t under the different values of β.

4. Discussion

The aforementioned results were obtained under the premise that joint development is not considered. In practice, joints will develop, and adjacent joints will connect and merge under mechanical loading. However, this study focuses on the effect of existing joints on the tensile properties.

The change in the number of joints can be considered as a change in the degree of development of fractures of the rock mass. Then, the geological strength index (*GSI*) is introduced [10]. *GSI* is generally used to calculate the strength and Young's modulus of a rock mass through laboratory rock mechanics properties and rock mass structural surface parameters. Figure 17 [38] is a *GSI* quantization table based on rock mass structures and surface features of the structures.

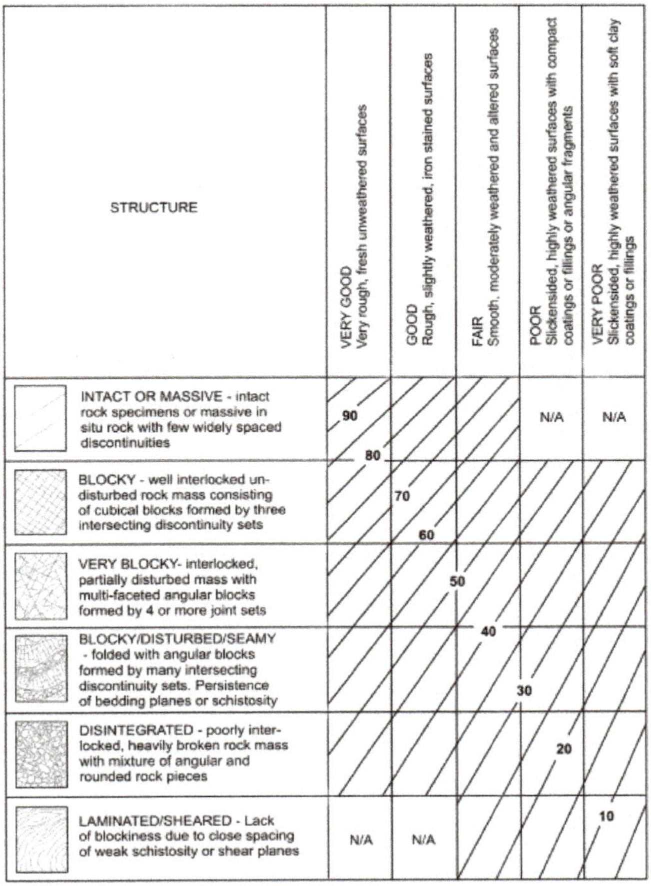

STRUCTURE	VERY GOOD Very rough, fresh unweathered surfaces	GOOD Rough, slightly weathered, iron stained surfaces	FAIR Smooth, moderately weathered and altered surfaces	POOR Slickensided, highly weathered surfaces with compact coatings or fillings or angular fragments	VERY POOR Slickensided, highly weathered surfaces with soft clay coatings or fillings
INTACT OR MASSIVE - intact rock specimens or massive in situ rock with few widely spaced discontinuities	90 / 80			N/A	N/A
BLOCKY - well interlocked un-disturbed rock mass consisting of cubical blocks formed by three intersecting discontinuity sets		70 / 60			
VERY BLOCKY- interlocked, partially disturbed mass with multi-faceted angular blocks formed by 4 or more joint sets			50		
BLOCKY/DISTURBED/SEAMY - folded with angular blocks formed by many intersecting discontinuity sets. Persistence of bedding planes or schistosity			40	30	
DISINTEGRATED - poorly inter-locked, heavily broken rock mass with mixture of angular and rounded rock pieces				20	
LAMINATED/SHEARED - Lack of blockiness due to close spacing of weak schistosity or shear planes	N/A	N/A			10

Figure 17. Quantitative chart of the geological strength index (*GSI*).

For rocks with a uniaxial compressive strength (σ_{ci}) < 100 MPa, the elastic modulus of a rock mass (E_t) is estimated from the following equation [39]:

$$E_t = \left(1 - \frac{D}{2}\right)\sqrt{\frac{\sigma_{ci}}{100}} \times 10^{\left(\frac{GSI-10}{40}\right)} \tag{1}$$

where σ_{ci} is the uniaxial compressive strength of intact rock, and D is a factor that depends on the degree of disturbance to which the rock mass has been subjected by blast damage and stress relaxation. When $D = 0$, the relationships between E_t and *GSI* under the conditions of different values of σ_{ci} are shown in Figure 18a. With the increase of the development of rock mass fissures, the Young's moduli decline notably. The degradation patterns with the number of parallel and intersection joints are illustrated in Figures 19a and 20a, respectively, which agree with the numerical results presented in the previous sections.

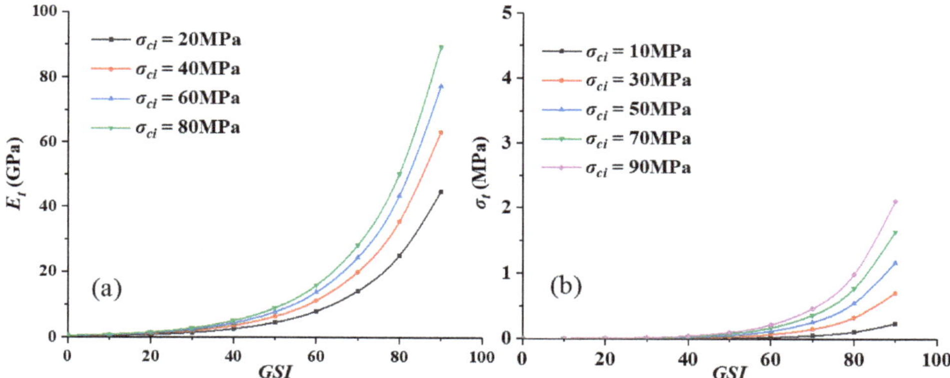

Figure 18. The effect of *GSI* on E_t, and σ_t of rock mass (**a**) Relationship between *GSI* and E_t, (**b**) Relationship between *GSI* and σ_t.

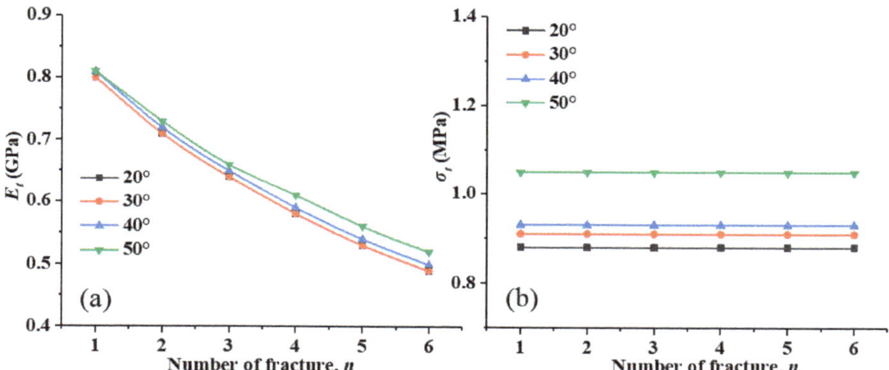

Figure 19. Effect of the number of parallel fractures on E_t and σ_t. (**a**) Effect of the number of fractures on E_t, (**b**) Effect of the number of fractures on σ_t.

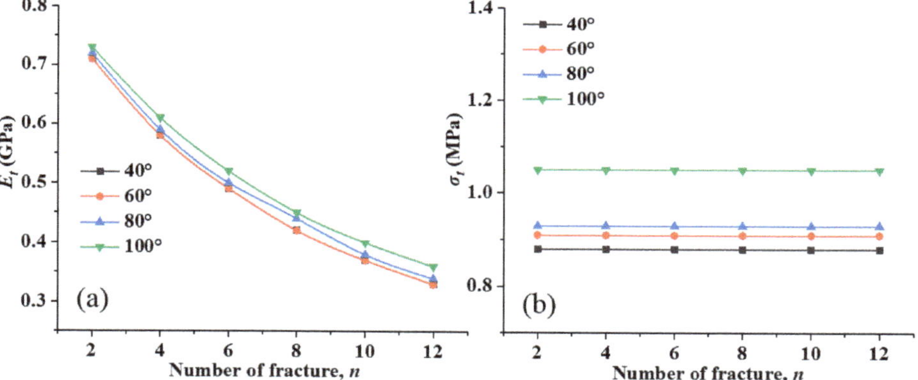

Figure 20. Effect of the number of intersecting parallel fractures on E_t and σ_t. (**a**) Effect of the number of fractures on E_t, (**b**) Effect of the number of fractures on σ_t.

According to the generalized Hoek–Brown peak strength criterion [40], the tensile strength of a rock mass is:

$$\sigma_t = -s\sigma_{ci}/m_b \tag{2}$$

where m_b is the reduced value of the material constant m_i for the rock mass, and is given by:

$$.m_b = m_i \exp\{(GSI - 100)/(28 - 14D)\} \tag{3}$$

and s is the rock mass constants, given by:

$$.s = \exp\{(GSI - 100)/(9 - 3D)\} \tag{4}$$

When $D = 0$, the relationships between σ_t and GSI under the conditions of different values of σ_{ci} are shown in Figure 18b. In a certain interval, the influence on the tensile strength is negligible with the increase of the development of rock mass fissures. Then, the rationalities of the conclusion in this paper were proved, as illustrated in Figures 19b and 20b.

5. Conclusions

According to the aforementioned analysis, the following conclusions can be obtained:

(1) For rock specimens with a single joint, the tensile strength (σ_t) is positively related to the joint angle (α). However, the Young's modulus in tension (E_t) is barely influenced by α.

(2) For rock specimens with parallel joints, the perpendicular distance (d) between the joints has negligible effects on σ_t and E_t. The influencing pattern is similar to the tests with single joints by changing α while keeping d constant. The number of parallel joints, or joint density, has notable effects on E_t but few effects on σ_t.

(3) For rock specimens with intersecting joints, the number of intersecting joints has notable effects on E_t but few effects on σ_t. For a given number of intersecting joints, σ_t is positively correlated to the angle between two interesting joints (β).

Author Contributions: Writing—original draft preparation: J.S.; writing-review and editing: L.J.; methodology: P.K. and Q.W.

Funding: The research of this study was sponsored by the National Key R&D Program of China (2018YFC0604703), the National Natural Science Foundation of China (51704182, 51774194), the Natural Science Foundation of the Shandong Province (ZR2017BEE050), and the Shandong University of Science and Technology.

Conflicts of Interest: The authors declare no conflict of interest.

References

1. Tang, C.A.; Tham, L.G.; Wang, S.H.; Liu, H.; Li, W.H. A numerical study of the influence of heterogeneity on the strength characterization of rock under uniaxial tension. *Mech. Mater.* **2007**, *39*, 326–339. [CrossRef]

2. Brady, B.H.G.; Brown, E.T. *Rock mechanics for underground mining*; Springer Science & Business Media: Berlin, Germany, 2004.

3. Shen, B. Development and applications of rock fracture mechanics modelling with FRACOD: A general review. *Geosyst. Eng.* **2014**, *17*, 235–252. [CrossRef]

4. Jiang, L.; Sainoki, A.; Mitri, H.S.; Ma, N.; Liu, H.; Hao, Z. Influence of fracture-induced weakening on coal mine gateroad stability. *Int. J. Rock Mech. Min.* **2016**, *88*, 307–317. [CrossRef]

5. Jiang, L.; Kong, P.; Shu, J.; Fan, K. Numerical analysis of support designs based on a case study of a longwall entry. *Rock Mech. Rock Eng.* **2019**, 1–12. [CrossRef]

6. Jiang, L.; Wu, Q.S.; Wu, Q.L.; Wang, P.; Xue, Y.; Kong, P.; Gong, B. Fracture failure analysis of hard and thick key layer and its dynamic response characteristics. *Eng. Fail. Anal.* **2019**, *98*, 118–130. [CrossRef]

7. Ni, G.; Xie, H.; Li, S.; Sun, Q.; Huang, D.; Cheng, Y.; Wang, N. The effect of anionic surfactant (SDS) on pore-fracture evolution of acidified coal and its significance for coalbed methane extraction. *Adv. Powder Technol.* **2019**, *30*, 940–951. [CrossRef]

8. Jiang, L.; Wang, P.; Zheng, P.; Luan, H.; Zhang, C. Influence of Different Advancing Directions on Mining Effect Caused by a Fault. *Adv. Civ. Eng.* **2019**, *2019*, 1–10. [CrossRef]

9. Cai, M.; Kaiser, P.K.; Martin, C.D. Quantification of rock mass damage in underground excavations from microseismic event monitoring. *Int. J. Rock Mech. Min.* **2001**, *38*, 1135–1145. [CrossRef]

10. Tajduś, K. New method for determining the elastic parameters of rock mass layers in the region of underground mining influence. *Int. J. Rock Mech. Min.* **2009**, *46*, 1296–1305. [CrossRef]

11. Hoek, E.; Carranza-Torres, C.; Corkum, B. Hoek-Brown failure criterion-2002 edition. *Proc. NARMS-Tac* **2002**, *1*, 267–273.

12. Mitri, H.S.; Edrissi, R.; Henning, J.G. Finite-element modeling of cable-bolted stopes in hard-rock underground mines. *SME Trans.* **1995**, *298*, 1897–1902.

13. Golsanami, N.; Sun, J.; Liu, Y.; Yan, W.; Lianjun, C.; Jiang, L.; Dong, H.; Zong, C.; Wang, H. Distinguishing fractures from matrix pores based on the practical application of rock physics inversion and NMR data: A case study from an unconventional coal reservoir in China. *J. Nat. Gas Sci. Eng.* **2019**, *65*, 145–167. [CrossRef]

14. Hoek, E.; Bieniawski, Z.T. Brittle Fracture Propagation in Rock Under Compression. *Int. J. Fracture* **1965**, *1*, 137–155. [CrossRef]

15. Paliwal, B.; Ramesh, K.T. An interacting micro-crack damage model for failure of brittle materials under compression. *J. Mech. Phys. Solids* **2008**, *56*, 896–923. [CrossRef]

16. Ashby, M.F.; Hallam, S.D. The failure of brittle solids containing small cracks under compressive stress states. *Acta Metall.* **1986**, *34*, 497–510. [CrossRef]

17. Hawkes, I.; Mellor, M.; Gariepy, S. Deformation of rocks under uniaxial tension. *Int. J. Rock Mech. Min. Sci. Geomech. Abstr.* **1973**, *10*, 493–507. [CrossRef]

18. Haimson, B.C.; Tharp, T.M. Stresses around boreholes in bilinear elastic rock. *Soc. Petrol. Engrs. J.* **1974**, *14*, 145–151. [CrossRef]

19. Barla, G.; Goffi, L. Direct tensile testing of anisotropic rocks, Advances in Rock Mechanics. In Proceedings of the 3rd. Congress of International Society for Rock Mechanics, Denver, CO, USA, 1–7 September 1974.

20. Chen, R.; Stimpson, B. Interpretation of indirect tensile strength tests when moduli of deformation in compression and in tension are different. *Rock Mech. Rock Eng.* **1993**, *26*, 183–189. [CrossRef]

21. Fuenkajorn, K.; Klanphumeesri, S. Laboratory Determination of Direct Tensile Strength and Deformability of Intact Rocks. *Geotech. Test. J.* **2010**, *34*, 103–134.

22. Yu, X.B.; Xie, Q.; Li, X.Y.; Na, Y.K.; Song, Z.P. Cycle loading tests of rock samples under direct tension and compression and bi-modular constitutive model. *Chin. J. Geotech. Eng.* **2005**, *9*, 988–993.

23. Sundaram, P.N.; Corrales, J.M. Brazilian tensile strength of rocks with different elastic properties in tension and compression. *Int. J. Rock Mech. Min. Sci. Geomech. Abstr.* **1980**, *17*, 131–133. [CrossRef]

24. Okubo, S.; Fukui, K. Complete stress-strain curves for various rock types in uniaxial tension. *Int. J. Rock Mech. Min. Sci. Geomech. Abstr.* **1996**, *33*, 549–556. [CrossRef]

25. Fahimifar, A.; Malekpour, M. Experimental and numerical analysis of indirect and direct tensile strength using fracture mechanics concepts. *B. Eng. Geol. Environ.* **2012**, *71*, 269–283. [CrossRef]

26. Fuenkajorn, K.; Kenkhunthod, N. Influence of Loading Rate on Deformability and Compressive Strength of Three Thai Sandstones. *Geotech. Geol. Eng.* **2010**, *28*, 707–715. [CrossRef]

27. Wang, S.Y.; Sloan, S.W.; Sheng, D.C.; Tang, C.A. 3D numerical analysis of crack propagation of heterogeneous notched rock under uniaxial tension. *Tectonophysics* **2016**, *677–678*, 45–67. [CrossRef]

28. Hamdi, J.; Souley, M.; Scholtès, L.; Heib, M.A.; Gunzburger, Y. Assessment of the Energy Balance of Rock Masses through Discrete Element Modelling. *Procedia Eng.* **2017**, *191*, 442–450. [CrossRef]

29. Guo, L.; Latham, J.P.; Xiang, J. A numerical study of fracture spacing and through-going fracture formation in layered rocks. *Int. J. Solids Struct.* **2017**, *110–111*, 44–57. [CrossRef]

30. Zhou, J.; Li, X.; Mitri, H.S. Evaluation Method of Rockburst: State-of-the-art Literature Review. *Tunn. Undergr. Space Technol.* **2018**, *81*, 632–659. [CrossRef]

31. Chen, L.; Li, P.; Cheng, W.; Liu, Z. Development of cement dust suppression technology during shotcrete in mine of China-A review. *J. Loss Prevent. Process Ind.* **2018**, *55*, 232–242. [CrossRef]

32. Jiang, L.; Zhang, P.; Chen, L.; Hao, Z.; Sainoki, A.; Mitri, H.S.; Wang, Q. Numerical Approach for Goaf-Side Entry Layout and Yield Pillar Design in Fractured Ground Conditions. *Rock Mech. Rock Eng.* **2017**, *50*, 3049–3071. [CrossRef]

33. ISRM. *The Complete ISRM Suggested Methods for Rock Characterization, Testing and Monitoring:1974-2006*; Ulusay, R., Hudson, J.A., Eds.; Suggested Methods Prepared by the Commission on Testing Methods, International Society for Rock Mechanics, Compilation Arranged by the ISRM Turkish National Group: Ankara, Turkey, 2007.

34. Wang, X.; Kulatilake, P.H.S.W.; Song, W. Stability investigations around a mine tunnel through three-dimensional discontinuum and continuum stress analyses. *Tunn. Undergr. Space Tech.* **2012**, *32*, 98–112. [CrossRef]

35. Barton, N. The shear strength of rock and rock joints. *Int. J. Rock Mech. Min. Sci. Geomech. Abstr.* **1976**, *13*, 255–279. [CrossRef]

36. Haftani, M.; Chehreh, H.A.; Mehinrad, A.; Binazadeh, K. Practical Investigations on Use of Weighted Joint Density to Decrease the Limitations of RQD Measurements. *Rock Mech. Rock Eng.* **2016**, *49*, 1551–1558. [CrossRef]

37. Yang, W.; Zhang, Q.; Ranjith, P.G.; Yu, R.; Luo, G.; Huang, C.; Wang, G. A damage mechanical model applied to analysis of mechanical properties of jointed rock masses. *Tunn. Undergr. Space Tech.* **2019**, *84*, 113–128. [CrossRef]

38. Hoek, E.; Marinos, P.G. Predicting tunnel squeezing problems in weak heterogeneous rock masses. *Tunn. Tunn. Int.* **2000**, *132*, 45–51.

39. Wang, P.; Jiang, L.; Li, X.; Qin, G.; Wang, E. Physical simulation of mining effect caused by a fault tectonic. *Arab. J. Geosci.* **2018**, *11*. [CrossRef]

40. Hoek, E.; Brown, E.T. Practical estimates of rock mass strength. *Int. Rock Mech. Min.* **1997**, *34*, 1165–1186. [CrossRef]

applied
sciences

Article

Numerical Modeling Approach on Mining-Induced Strata Structural Behavior by Considering the Fracture-Weakening Effect on Rock Mass

Jiaming Shu, Lishuai Jiang *, Peng Kong *, Pu Wang and Peipeng Zhang

State Key Laboratory of Mining Disaster Prevention and Control, Shandong University of Science and Technology, Qingdao 266590, China; 15621565102@163.com (J.S.); 15854848872@163.com (P.W.); zppsk2008@163.com (P.Z.)
* Correspondence: lsjiang@sdust.edu.cn (L.J.); 17854859770@163.com (P.K.)

Received: 3 April 2019; Accepted: 1 May 2019; Published: 3 May 2019

Abstract: By employing the longwall mining method, a series of intensive strata structure responses and activities will be induced including stress redistribution, fracture extension and strata movement. Due to the geological stratification feature of coal mine strata, tensile failure and tension-induced fracturing play dominant roles in the strata of the fractured zone. These responses induced in the strata require the consideration of the weakening effect on the rock mass behavior due to failure and fracturing in tension. In this study, a numerical modeling approach on mining-induced strata structural behaviors was proposed by considering the mechanical behaviors of the caved zone consolidation and tension-induced weakening in the fractured zone. Based on a numerical model built according to a study site, a parametric study with respect to different fracturing intensity parameters was performed to investigate the fracturing weakening effect on the mining-induced stress redistribution and strata movement. The numerical results showed that the tensile fracture intensity had a notable effect on the mining-induced stress distribution in two aspects: (1) Increase in peak and area of the front abutment stress; (2) variation in the patterns of stress recovery in the goaf. The stress data obtained from numerical simulation represent and help to back-analyze the structural behaviors (failure, movement) of the overlying strata. The high stress on the coal seam indicated that the strata lay on and transferred loads to the seam, while the low stress indicated the detachment between the seam and the suspending strata. With the increase in fracture intensity, the roof strata were more prone to breaking and caving, and the suspending length of the roof beam decreased, which made the strata sufficiently break, cave and transfer the overburden load to loose rock in the goaf; caving along the strike direction of the panel became the dominant overlying strata structure movement, while the dominant movement caved along the dip direction in the case of strong and intact overlying strata with few tensile fractures. Thus, the tensile fracturing intensity should not be ignored in studies related to the behaviors of the overlying strata. Validated by analytical studies, this study presents a novel numerical modeling approach for this topic and can be utilized for multiple studies based on proper roof fracturing estimation or back analysis.

Keywords: strata structural behavior; numerical simulation; tension weakening; fractures; goaf consolidation

---

1. Introduction

The longwall mining method in underground coal mines has been used worldwide for decades. After the coal in a longwall panel is extracted, the overlying strata tend to sag and fill the goaf (voids created by the extracted coal). Such an operation process will induce a series of intensive strata responses and activities, including massive stress redistribution, strata movement and fracture extension. The overlying strata structure, as well as stress distribution, are disturbed by the mining operation in the order of severity from the immediate roof upward to the ground surface. As illustrated in Figure 1, the overlying strata can be generalized into three different zones of disturbance in response to the longwall mining [1].

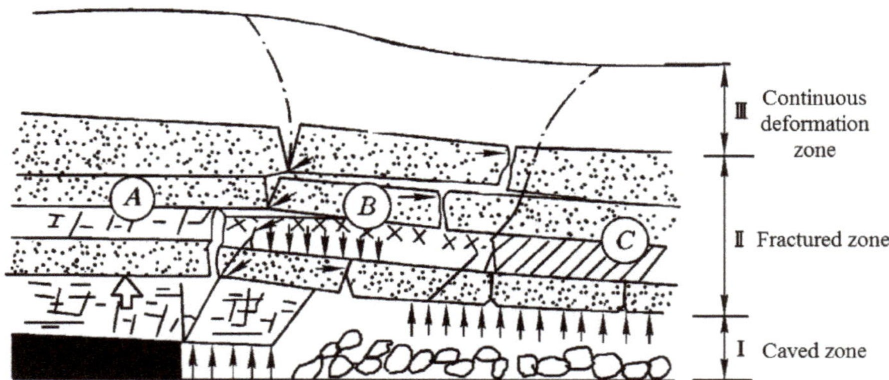

Figure 1. Zones of disturbance in response to longwall mining (Peng 2008).

The mining-induced behaviors of the three zones are of paramount importance for studies related to stress redistribution and strata structure movement. For example, the abutment pressure ahead of a longwall face and periodic weighting of the main roof induced by the mining operation are essential parameters in the design of shield support in longwall faces [2] and studies on the rock burst and gas outburst control [3,4].

The strata in the fractured zone (FZ), which is located in the roof above the caved zone (CZ), is broken into blocks by vertical and horizontal fractures because of the bending and bed separations between layers. Because of the fracture propagation and its water-inflow capacity, the FZ is considered a major research target if there is underground water in the roof strata. Some research attempts aimed at roof water inrush prevention, for instance Song et al. [5] and Wang et al. [6], have focused on the extent and development in height of the FZ induced by mining. Advances have been made in forecasting and monitoring the FZ development under various geological and engineering conditions [7–9] as well as analytical and numerical modeling [10,11]. The rock mass in the FZ is heavily fractured, and it is well established that the properties of rock masses are directly related to the occurrence of fractures [12–14]. Therefore, the fracturing intensity of multiple strata in the FZ varies and directly affects the strata structural behavior (movement, failure, etc.) and stress distribution, so it is a factor that should not be ignored in addition to the extent and height of the FZs. However, the fracturing intensity of strata due to longwall mining, and most importantly, its induced weakening effect on stress readjustment and strata structural behavior are inadequately discussed.

In the presented study, a numerical modeling approach was proposed by considering the mechanical behaviors of loose rock consolidation in the CZ and tension-induced weakening in the FZ. Based on a numerical model built according to a case study, a parametric study with respect to the fracturing intensity was performed to investigate the fracturing weakening effect on the mining-induced stress redistribution and strata behavior.

2. Mechanical Behavior of the Caved and Fractured Zones

The mechanical behavior of the overlying strata in response to mining activities can be described by two processes: Consolidation of the CZ and development of the FZ.

2.1. Extent of Caved Zone and Fractured Zone

It is generally believed that the extent of the CZ and FZ depends on the geological and geotechnical factors of a panel. The geological factors involve the lithology, thickness, mechanical properties (strength, Young's modulus, friction angle, etc.), etc. [1], of the coal seam and its overlying strata and mainly refer to the mining height and length of each mining cycle.

The CZ and FZ have been estimated using empirical and analytical methods. Among the methods, an analytical method proposed by Bai et al. [15] is widely used, which is based on the numbers of field investigations in Chinese coalfields. In this method, the vertical extents of the CZ and FZ were estimated as follows:

$$\begin{cases} H_c = \frac{100h}{c_1 h + c_2} \\ H_f = \frac{100h}{c_3 h + c_4} \end{cases} \tag{1}$$

where H_c is the height of the CZ, H_f is the height of the FZ, h is the mining height, and $c_1 \sim c_4$ are the coefficients that depend on the strata lithology [15].

2.2. Mechanical Behavior of Consolidation of the Caved Zone

After the coal extraction of each mining cycle, if no backfill is applied, the goaf is filled by the caved rock blocks from the roof. After the immediate roof collapse, rocks cave and fall into the FZ and are naturally and loosely placed in the CZ. By considering its mechanical behaviors, the CZ in this stage can be considered as a loose material with a lower strength. Then, the unsupported overlying strata will collapse or continuously cave, and they eventually fall and lie on the caved rocks in the CZ. The overburden load transfers to the caved rocks and compresses them into a zone of stiffer and denser material. When the rock mass in the CZ stiffens, it begins to provide increasing support to resist the overlying strata from sagging. This mechanical behavior is called goaf consolidation, which is essential to studies on overlying strata structure movement and stress distribution. A stress-strain relationship of the CZ material was first developed by Salamon [16] shown in Equation (2), and further developed and applied to simulate the goaf consolidation process of longwall mining [3,17,18] as Equation (3).

$$\sigma = \frac{E_0 \varepsilon}{1 - \varepsilon / \varepsilon_m} \tag{2}$$

$$\sigma = \frac{10.39 \sigma_c^{1.042}}{b^{7.7}} \times \frac{\varepsilon}{1 - \frac{b}{b-1}\varepsilon} \tag{3}$$

where σ is the uniaxial stress applied to the material, ε is the strain under the applied stress, b is the bulking factor, and σ_c is strength of the rock pieces and ε_m can be determined as:

$$\varepsilon_m = \frac{b-1}{b} \tag{4}$$

2.3. Mechanical Behavior of Fracturing-Weakening in the Fractured Zone

The rock mass in the FZ was heavily fractured, which would obviously have led to weaker mechanical properties of the rock mass. Because of the stratification feature of coal mine strata, tensile failure and tension-induced fracturing play dominant roles in the FZ, since the strata bend like beams with different deflections. Because the roof beams bend and separate in tension, the tensile failure and tension-induced weakening on the rock mass should not be ignored in the analysis, where the behaviors of the strata play a significant role.

Numerical simulation is a powerful technique for studies on rock mechanics and engineering, but its accuracy and reliability lie on the used simulation approach, constitutive model, material properties, etc. With the numerical simulation method, many studies were conducted on the stress redistribution and strata movement induced by longwall mining, among which the inherent perfect elastoplastic and strain-softening models using Mohr–Coulomb failure criterion are most commonly used [4,19–24].

However, for both constitutive models embedded in FLAC3D, the Young's modulus is invariable regardless of the occurrence of failure. It is well accepted that Young's modulus of the rock mass and the characteristics of fractures and joints is directly related [14,24–27]. The variation of Young's modulus induced by tensile failure must be considered for numerical simulations related to the FZ.

Many numerical methods such as the discrete element and extended finite element method have strength in the simulation of fracture occurrence and propagation, but they require enormous calculation time and memory to run a mine-wide model. For the behavior and response of surrounding strata after vast coal extraction, the finite difference element method with the consideration of the tensile failure in the equivalent material method is more efficient and applicable. In this study, the tension-weakening model, developed by Jiang et al. [28] based on FLAC3D, was adaptable for the numerical analysis of fractured rock mass with the consideration of the fracturing intensity and its weakening effect. The tension-weakening model can monitor the failure state of each element in the entire 3D model during calculation. When shear failure is detected at a certain zone, the softened properties (cohesion, tensile strength, friction or dilation angle) are reassigned to this particular zone according to the law of the strain-softening model. When tensile failure is detected, the residual Young's modulus is reassigned to the zone; for zones that experience both shear and tensile failures, all aforementioned properties will be softened. The conventional FLAC3D simulation with the Morh–Coulomb Model and Strain-softening Model is widely accepted for stability analysis under a soft rock condition (e.g., underground coal mining), however, Young's modulus is invariable regardless of the occurrence of failure by employing these models. By applying the tension-weakening model, the calculation with the tension-weakening model will proceed during the entire numerical analysis, which overcomes the limitations of the conventional simulation techniques on the rock mass post-peak behavior, that is, Young's modulus is invariable regardless of the fracture characteristics within a rock mass. Proposed by Jiang et al. [28] based on the *GSI* (Geological Strength Index) system [25], a new indicator GSI_t was defined as the *GSI* value for the rock mass with fractures induced by tension. The *GSI* system developed by Hoek et al. [25] can provide a quantitative description of the structural characterization and intensity of discontinuities, and is widely used to determine the rock mass deformability and strength on the basis of intact rock properties and the quantitative index of *GSI*. By describing the intensity of the tensile fractures, $GSI_t = 90$ indicates that no tensile failure occurs, or no fractures are induced by the tensile failure. The range and unit of GSI_t follow the system of *GSI*. The residual elastic modulus E_r of the rock mass after tensile failure was described by the following expression:

$$E_r = \sqrt{\frac{\sigma_m}{100}} \times 10^{(\frac{GSI_t-10}{40})}$$

(5)

3. Numerical Modeling Procedure

3.1. Overview of the Study Site

The numerical 3D model of this study was built based on the geological and geotechnical information of an underground coal mine case study in China [28]. The mine used the retreat longwall mining method; the target seam was horizontally buried at −600 m depth. Instead of slicing mining, a full-seam mining technique was used to extract the 6 m-thick seam in one retreat. As shown in Figure 2, the target panel was 180 m along its dip and 2000 m along its strike. As illustrated in Figure 3, the roof structure was identified by borehole core loggings. A soft and thin layer of mudstone was located immediately above the coal seam as the immediate roof. The main roof mainly consisted of sandstone and sandy mudstone; no hard-to-cave strong stratum was found from the borehole loggings. The floor was mainly sandy mudstone.

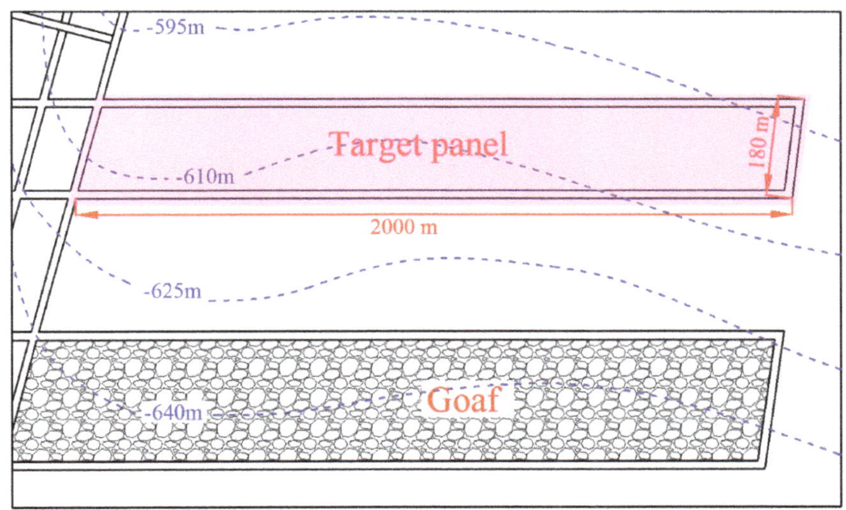

Figure 2. Plan view of the target panel.

Lithology	Thickness(m)
Sandy mudstone	10.53
Sandstone	21.92
Sandy mudstone	13.95
Mudstone	1.86
Coal	6.12
Sandy mudstone	13.88
Siltstone	2.00
Sandy mudstone	15.82
Siltstone	8.41

Figure 3. Roof and floor strata information.

Rock core samples were collected on site and laboratory tests were conducted to obtain the mechanical properties of the rocks, and coefficients c1~c4 in Equation (1) were obtained accordingly. Then, based on the estimation method described in Section 2.1, the height of the CZ and FZ of this panel were estimated to be 12.7 m and 45.5 m, respectively.

The double-yield model, an inherent constitutive model of FLAC3D, is capable of simulating material's hardening behaviors with increasing strain in compression, and a widely used approach to simulate the mechanical behavior of CZ consolidation [4,17,18,29]. The input properties of the double-yield model were back-analyzed to match the mechanical behavior of Equation (3) by Jiang et al. [30].

3.2. Model Description and Simulation Procedure

Considering that the longwall panel is symmetrical around its centerline, a 3D model is generated based on the study site, as shown in Figure 4. According to the geological column, this model is 190 m wide, 350 m long and 120 m high, and the model size and mesh density were determined based on a sensitivity analysis. The panel was 90 m wide and 250 m long, which left sufficient space to eliminate the boundary effect. It should be noted that, all parameters applied in the numerical simulation came from a previous case study presented in [28] and [30], the parameters for boundary conditions were based on in situ stress measurement, and the properties of the rock masses were based on the laboratory tests. To simulate the in situ stress state, a 15 MPa load was vertically applied to the top boundary; according to the in situ stress measurement and previous study [28], a 12 MPa horizontal load was applied perpendicular to the direction of retreat mining, and an 18 MPa horizontal load was applied along the direction of retreat mining. No displacement in the direction perpendicular to the model lateral boundaries was allowed. In situ stresses were applied as described in Section 3.1. The rock mass properties for the simulation were estimated from the intact rock properties, as listed in Table 1. A flowchart of the proposed numerical modeling approach for longwall mining is illustrated in Figure 5. The simulation process was as follows: (i) Generate the 3D model and assign the constitutive model and properties; (ii) simulate the in situ stress state; (iii) mine the coal seam in the retreat by 10 m for each cycle; (iv) fill the CZ with the double-yield model; (v) run the calculation with the tension-weakening model in the FZ to simulate the rock mass weakening induced by fracturing; (vi) after the calculation reaches equilibrium, repeat (iii) and (iv) until the entire panel is completely mined.

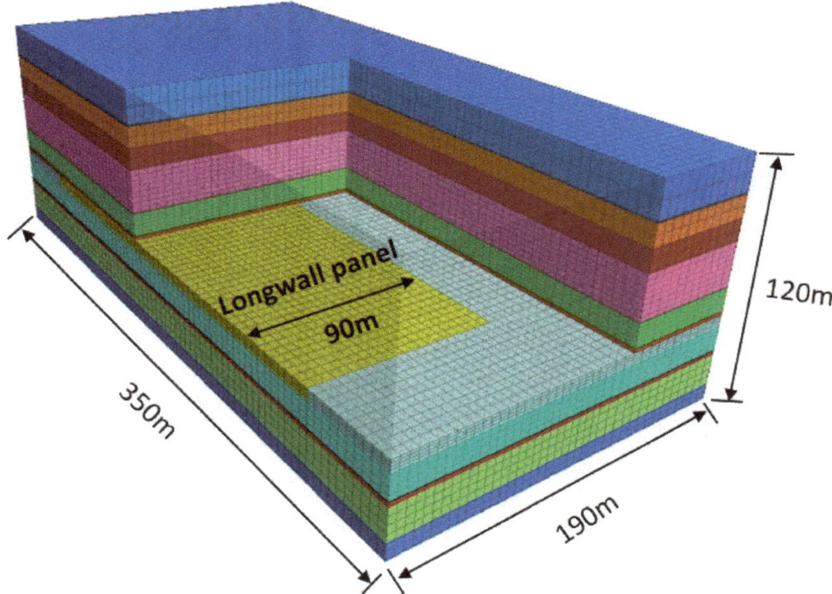

Figure 4. Layout of the three-dimensional panel-wide model.

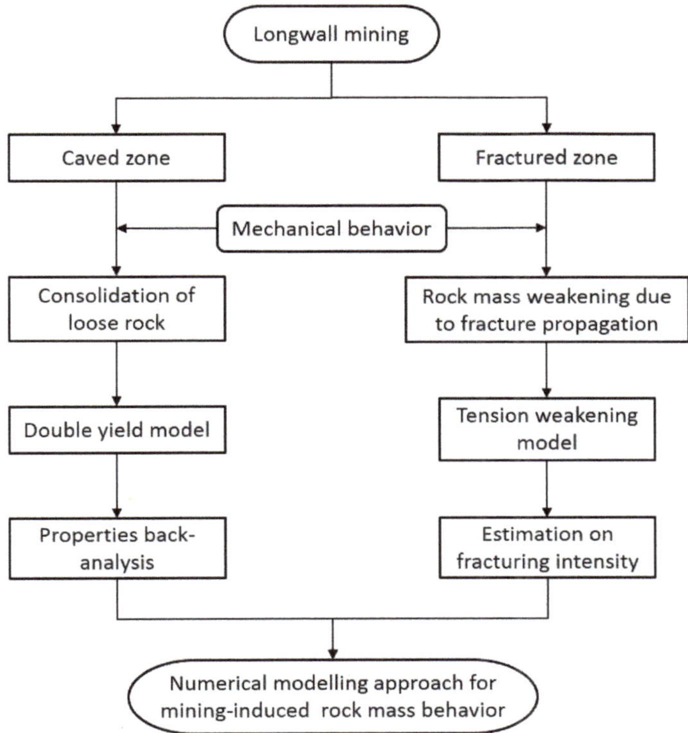

Figure 5. Flowchart of the numerical modeling approach for longwall mining.

Table 1. Rock mass mechanical properties [b].

Strata	Lithology	K (GPa)	G (GPa)	c (MPa)	σ_t (MPa)	ϕ (deg.)	ρ (N/m)
	Sandstone	9.1	5.9	3.9	2.3	45	28263
Roof	Sandy mudstone	5.2	3.1	3.2	1.8	40	25392
	Mudstone	2.4	1.1	2.1	0.8	35	25509
Coal seam	Coal	1.3	0.6	1.4	0.3	31	14063
Floor	Sandy mudstone	7.2	4.0	3.4	2.2	37	25764
	Siltstone	9.6	6.5	4.2	3.5	47	26989

[a] K is the bulk modulus; G is the shear modulus; c is the cohesion; σ_t is the tensile strength; ϕ is the friction angle.

3.3. Parametric Study on the Fracture Intensity of the Fractured Zone

To investigate the tensile failure development in the FZ after the longwall mining had been started, the zones with tensile failure states were identified and are marked in Figure 6. Many tensile failure zones occurred in the FZ, which was consistent with the description in Section 2.3. Such extensive rock mass failure in tension requires the consideration of the weakening effect on the rock mass behavior due to failure and fracturing in tension.

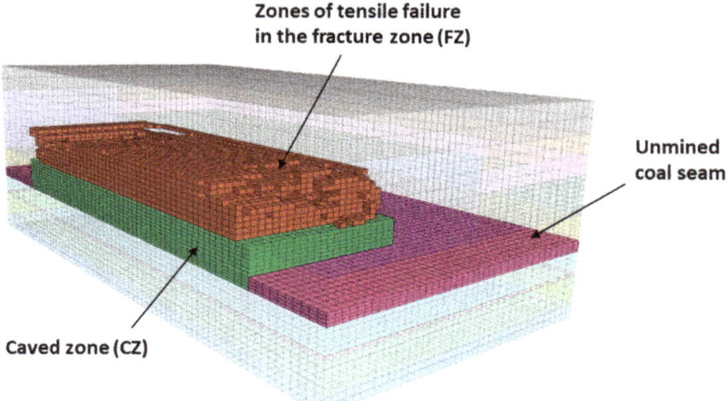

Figure 6. Distribution of tensile failure zones in the fractured zone.

Therefore, to investigate the stress distribution and strata behavior with respect to various tensile fracturing intensities, a parametric study of the tension weakening model was performed with four different GSI_t values (as shown in Table 2), where E_r was the residual Young's modulus of rock mass after tensile failure, E_m was the Young's modulus of rock mass before failure, and a high GSI_t indicated high fracture intensity and consequently a low residual Young's modulus of rock mass according to Equation (6).

Table 2. Parameters for the parametric study.

GSI_t	E_r/E_m
10	0.01
30	0.03
50	0.1
70	0.3

4. Fracturing Weakening Effect on the Mining-Induced Stress Redistribution and Strata Movement

The mining-induced stress redistribution and strata movement were three-dimensional. It is well accepted that analysis on these behaviors could be done in two-dimension analytical analysis, by using separate beam models along the strike direction and dip direction of the longwall panel.

Considering that the panel was symmetrical about its centerline, Figure 7 shows the mining-induced stress redistribution after the panel had been mined in the retreat by 200 m along its strike with respect to different GSI_t. The tensile fracturing intensity of the FZ significantly affected the stress redistribution due to mining. Ellipses were marked to outline the areas of the goaf where the recovered stress exceeded 12.5 MPa (76.9% of the overburden pressure). In Figure 7, we also marked the perpendicular distances from the peak front abutment stress (σ_f) to the working face (L_f), from the peak side abutment stress to the face (L_s), from the face to the edge of the stress-recovered area (L_y) and from the edge of the stress-recovered area to the lateral goaf edge (L_x).

As observed from Figure 7, when the fracture intensity increased (GSI_t decreases), the mining stress field responded in two aspects: (1) Increase in the front abutment stress; (2) change in patterns of stress recovery in the goaf. The simulation results of the three aspects and their behaviors are analyzed in the following text.

The induced stress concentration ahead of the face, which is known as the front abutment stress, is a paramount parameter for many design practices in longwall mining. As shown in Figure 7, in the case of $GSI_t = 10$, which indicates that the FZ had the highest fracture intensity, the peak abutment stress (33.0 MPa) due to mining occurred at 30.6 m ahead of the working face. With the increase in GSI_t, which indicated less tensile fracture in the rock mass, the abutment stress decreased to 28.4 MPa, but the distance to the peak abutment stress hardly moved toward the face. In addition, the area of highly concentrated stress ahead of the face dramatically increased with the decrease in GSI_t. Because the rock mass in the FZ was more fractured (decrease in GSI_t), the massive mining-induced stress tended to redistribute to surrounding rock mass, which resulted in more stress concentrated ahead of the face.

The stress data along the panel centerline in the goaf is shown in Figure 8 for the quantitative analysis. The stress in the goaf gradually increased because of the compression of the overlying strata and approached the cover pressure far behind the face. Within the first 80 m behind the face, the recovered stress presented a negative correlation with GSI_t, i.e., the cases with a high fracturing intensity in the roof would have more stress concentrated at a certain location in the goaf. However, this trend reversed at a certain point that was far behind the face. The aforementioned simulation results were consistent with one another and can be well explained with the strata structural behavior induced by mining. Numerous studies [31–33] have demonstrated that the overlying strata behave like beams along the strike direction of the panel, as illustrated in Figure 1; applying overburden stress to the goaf and unmined seam will break and cave with an increase in the suspending length because of face advancing, which is called the first and periodic roof weighting phenomenon [1,34]. With the increase in fracture intensity, the roof strata are more prone to break and cave, the suspending length of the roof beam will reduce, which makes the strata sufficient enough to break, cave and transfer the overburden load to loose rocks in the goaf (cases of $GSI_t = 10$ and 30 in Figures 5 and 6). When there are few tensile fractures in the strata, the roof beam structure has more stability. A hard-to-break beam with large suspending length over the goaf can only apply an overburden load to the CZ at its pivot far behind the face. This mechanism is well presented in the stress data of $GSI_t = 50$ and 70 in Figure 6.

An interesting phenomenon was noticed from the differences in shape of the high-stress-concentration areas in the goaf (marked with ellipses). Taking the axial length along the x-direction as a and the axial length along the y-direction as b, the stress ellipse had a shape of $a{:}b \approx 1.65$ when $GSI_t = 70$ and was located at 100.9 m behind the face and 36.7 m to the lateral goaf-edge; the high-stress area in the goaf extended more along the x-direction than along the y-direction, which indicated that the roof beams along the dip direction of the panel (x-direction) more sufficiently caved than those along the strike direction. When GSI_t decreased, the shape and relative location of the ellipse monotonically changed and eventually became an ellipse at 76.8 m behind the face and 63.5 m to the lateral goaf-edge with $a{:}b \approx 0.88$ when $GSI_t = 10$. These results indicated that when the fracturing intensity increased, the roof beams were prone to breaking and caving, which made the high-stress area in the goaf closer to the face (L_y reduces from 100.9 m to 76.8 m), and the caving along the strike direction of the panel became the dominant overlying strata movement. When GSI_t was less than 50, the dominant movement caved along the dip direction. This analysis result could also be obtained from the decrease in perpendicular distance from the peak side abutment stress to the face (from 104.8 m to 87.6 m) when GSI_t decreased.

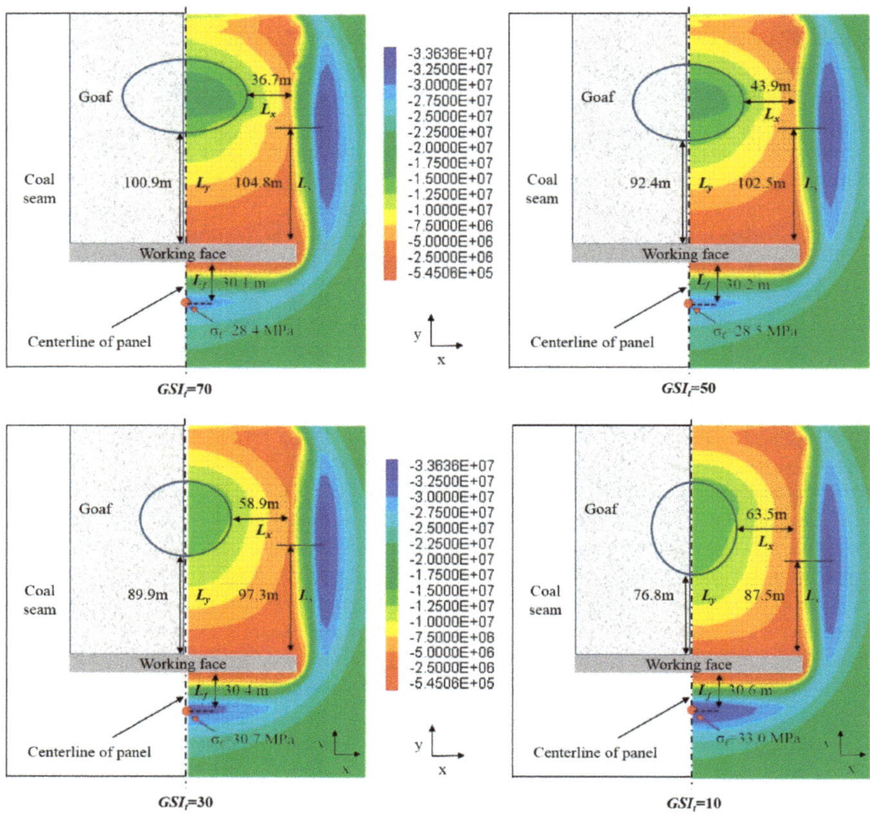

Figure 7. Distribution of the mining-induced vertical concentrated stress with respect to different GSI_t values.

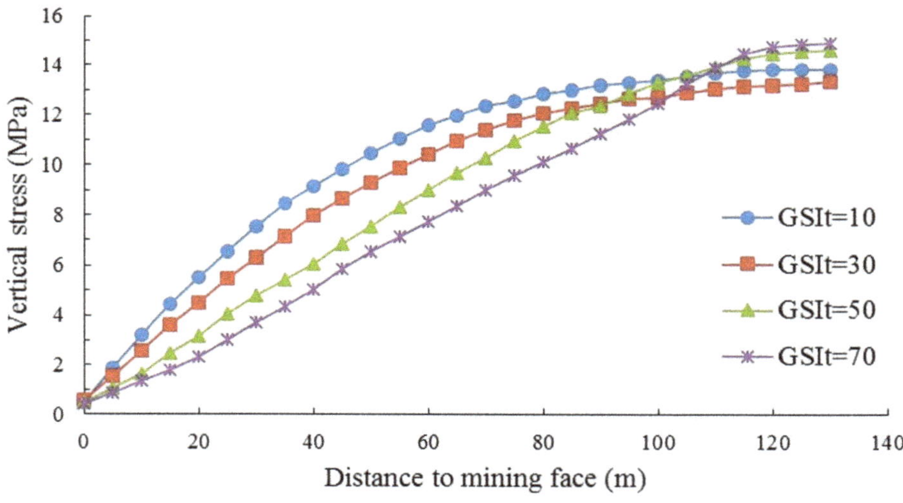

Figure 8. Stress data along the panel centerline in the goaf with respect to different GSI_t values.

5. Discussion and Validation

Using a numerical approach, which can consider the fracture-induced weakening, we investigated the fracture-weakening effect on the mining-induced stress redistribution through a parametric study with respect to the indicator of fracture intensity in tension, GSI_t. According to the analysis in Section 4, the tensile fracture intensity in the FZ strongly affected the mining-induced stress distribution and strata movement. Thus, this factor should not be ignored in studies related to the behaviors of the FZ, mining-induced stress redistribution and overlying strata movement.

The numerical results of the parametric study showed an interesting and rational changing law, which was consistent with the behaviors of the overlying strata. The changes in fracture intensity led to different abutment stress distributions and roof caving, which contributed to the design of the shield support in the working face.

The in situ stress measurement in longwall mining is generally difficult; there is only a small amount of measured data in this regard [35–37]. The extreme difficulties in stress measurement in the non-backfill goaf area [1] is due to the inaccessibility of the goaf. Even though the stress monitoring cells can be put in the goaf, but are easily damaged by the fallen rocks. Therefore, few attempts have successfully acquired the field stress data in the goaf and have used it to validate the numerical results.

Consequently, the studies on the mining-induced stress and its indicated overlying strata movement often rely on empirical or analytical methods, but only few analytical models are suitable for jointed strata. Singh and Dubey [38] proposed an expression to estimate the caving span of the jointed roof beam structure L as follows:

$$L = F\sqrt{\frac{2KRtT_r}{\lambda}} \tag{6}$$

where F is the orientation factor, t is the thickness of the bed, T_r is the rock tensile strength, and K and R are the jointing factor and stratification factor, respectively, which are determined by the core study. It was noted that K and R had a positive correlation with L. According to Singh and Dubey's study, a roof with more fractures would have a lower K and R. Therefore, highly jointed and fractured roof strata would have a small caving span, which is consistent with the numerical results presented in Section 4. Unfortunately, the study of Singh and Dubey [38] did not describe how to quantify these parameters. Hence, at this stage, Equation (6) could not be directly used to quantitatively validate this numerical study. However, this equation and its indicated relationship is consistent with the trend obtained from the numerical analysis (presented in Section 4) and hereby served to validate the present study.

According to the aforementioned analysis, the further studies on the quantitative analysis, by means of analytical or numerical analysis, in this regard is surely need to investigate the behaviors of mining-induced roof strata. In addition, the following instructions for applying the Tension weakening model can be drawn: The fracture intensity of the FZ should be case-dependent in light of geological and geotechnical conditions; differences in geological (strata lithology, thickness, properties, etc.) and geotechnical (mining height, length of each mining cycle) conditions will lead to different extents and intensities of the FZ. Hence, it should be estimated based on the core loggings with long roof boreholes or back-analyzed according to field measurements.

This study focused on investigating the mining-induced roof behaviors with a newly proposed simulation technique, and the study was carried out based on a case study with typical geological conditions of underground coal mines. The results of this study may vary with the geological structures and mechanical properties of strata, the sensitivity analysis of different structures and hence these properties will be investigated in future studies.

6. Conclusions

The mining-induced surrounding strata behaviors are characterized into caved zone consolidation and fractured zone development, and their extents and mechanical behaviors have been analyzed and quantified. Accordingly, a numerical modeling approach was proposed by considering the caved zone consolidation and tension-induced fracture-weakening in the fractured zone.

Based on the geological and geotechnical conditions of a study site, a 3D numerical model was built with FLAC3D, and the mechanical properties of the caved zone material were obtained from the back analysis. Extensive rock mass tensile failure occurred in the fractured zone after the retreat-longwall mining began.

To investigate the fracturing weakening effect on the mining-induced stress redistribution and strata movement, a parametric study with respect to various tensile fracturing intensity parameters was conducted. The numerical results showed that the tensile fracture intensity had a notable effect on the mining-induced stress distribution in two aspects: (1) Increase in peak and area of the front abutment stress; when the rock mass in the FZ became more fractured, the massive mining-induced stress tended to redistribute to the surrounding rock mass, which resulted in higher stress concentrated ahead of the face; (2) variation in the patterns of stress recovery in the goaf. The stress data obtained from the numerical simulation represent and help to back-analyze the behaviors (failure, movement) of the overlying strata. The high stress on the coal seam indicated that the strata lay on and transferred loads to the seam, whereas a low stress indicated the detachment between the seam and the suspending strata. With the increase in fracture intensity, the roof strata were more prone to break and cave, and the suspending length of roof beam would decrease, which made the strata sufficiently break, cave and transfer the overburden load to loose rocks in the goaf; the caving along the strike direction of the panel became the dominant overlying strata movement. In the case of strong and intact overlying strata with few tensile fractures, the dominant movement caved along the dip direction.

Considering the fracturing intensity in tension and its induced weakening effect on the rock mass, this study provides a novel simulation approach for studies on the behaviors of the fractured zone, mining-induced stress redistribution and overlying strata movement. According to the simulation results, the tensile fracture intensity in the FZ has a notable effect on the mining-induced stress distribution and strata structural movement. Thus, this factor should not be ignored in studies related to the behaviors of the FZ, mining-induced stress redistribution and overlying strata movement. Based on a proper estimation or back analysis on the roof fracturing state, this presented study and modeling approach can contribute to multiple research areas (such as the shield support design, panel layout and roof water-permeability estimation) when extensive tensile failure is expected in the fractured zone, such as the design of shield support in the working face and panel layout.

Author Contributions: Methodology, P.K.; software, P.W.; validation, J.S.; writing—original draft, L.J.; writing—review and editing, P.Z.

Acknowledgments: This research was supported by the National Key R&D Program of China (2018YFC0604703), National Natural Science Foundation of China (Grant No. 51704182, 51574155, 51704185).

Conflicts of Interest: The authors declare no conflicts of interest.

References

1. Peng, S.S. *Coal Mine Ground Control*, 3rd ed.; Woodhead Publishing: Morgantown, WV, USA, 2008; pp. 229–237, 260–267, 422–423.
2. Bai, Q.S.; Tu, S.H.; Zhang, X.G.; Zhang, C.; Yuan, Y. Numerical modeling on brittle failure of coal wall in longwall face—A case study. *Arab. J. Geosci.* **2014**, *7*, 5067–5080. [CrossRef]
3. Wang, M.; Bai, J.; Li, W.; Wang, X.; Cao, S. Failure mechanism and control of deep gob-side entry. *Arab. J. Geosci.* **2015**, *8*, 9117–9131. [CrossRef]
4. Li, W.; Bai, J.; Peng, S.; Wang, X.; Xu, Y. Numerical modeling for yield pillar design: a case study. *Rock Mech. Rock Eng.* **2015**, *48*, 305–318. [CrossRef]

5. Song, G.; Yang, S. Investigation into strata behaviour and fractured zone height in a high-seam longwall coal mine. *J. S. Afr. Inst. Min. Metall.* **2015**, *115*, 781–788. [CrossRef]
6. Wang, F.; Tu, S.; Zhang, C.; Zhang, Y.; Bai, Q. Evolution mechanism of water-flowing zones and control technology for longwall mining in shallow coal seams beneath gully topography. *Environ. Earth Sci.* **2016**, *75*, 1309. [CrossRef]
7. Adhikary, D.P.; Guo, H. Measurement of longwall mining induced strata permeability. *Geotech. Geol. Eng.* **2014**, *32*, 617–626. [CrossRef]
8. Palchik, V. Formation of fractured zones in overburden due to longwall mining. *Environ. Geol.* **2003**, *44*, 28–38. [CrossRef]
9. Booth, C.J. Groundwater as an environmental constraint of longwall coal mining. *Environ. Geol.* **2006**, *49*, 796–803. [CrossRef]
10. Zhang, D.; Fan, G.; Liu, Y.; Ma, L. Field trials of aquifer protection in longwall mining of shallow coal seams in china. *Int. J. Rock Mech. Min. Sci.* **2010**, *47*, 908–914. [CrossRef]
11. Miao, X.; Cui, X.; Wang, J.; Xu, J. The height of fractured water-conducting zone in undermined rock strata. *Eng. Geol.* **2011**, *120*, 32–39. [CrossRef]
12. Hoek, E.; Marinos, P.; Benissi, M. Applicability of the geological strength index (GSI) classification for very weak and sheared rock masses. The case of the Athens Schist Formation. *Bull. Eng. Geol. Environ.* **1998**, *57*, 151–160. [CrossRef]
13. Cai, M.; Kaiser, P.K. Numerical simulation of the Brazilian test and the tensile strength of anisotropic rocks and rocks with pre-existing cracks. *Int. J. Rock Mech. Min. Sci.* **2004**, *41*, 450–451. [CrossRef]
14. Mitri, H.S.; Edrissi, R.; Henning, J.G. Finite-element modeling of cable-bolted stopes in hard-rock underground mines. *Trans. Soc. Min. Metall. Explor. Inc.* **1995**, *298*, 1897–1902.
15. Bai, M.; Kendorski, F.S.; Van Roosendaal, D.J. Chinese and north american high-extraction underground coal mining strata behavior and water protection experience and guidelines. In Proceedings of the 14th International Conference on Ground Control in Mining, Morgantown, WV, USA, 1–3 August 1995.
16. Salamon, M. Mechanism of caving in longwall coal mining, Rock mechanics contribution and challenges. In Proceedings of the 31st US Symposium of Rock Mechanics, Golden, CO, USA, 18–20 June 1990; pp. 161–168.
17. Yavuz, H. An estimation method for cover pressure re-establishment distance and pressure distribution in the goaf of longwall coal mines. *Int. J. Rock Mech. Min. Sci.* **2004**, *41*, 193–205. [CrossRef]
18. Zhang, Z.Z.; Yu, X.Y.; Wu, H.; Deng, M. Stability Control for Gob-Side Entry Retaining with Supercritical Retained Entry Width in Thick Coal Seam Longwall Mining. *Energies* **2019**, *12*, 1375. [CrossRef]
19. Jiang, L.; Kong, P.; Shu, J.; Fan, K. Numerical analysis of support designs based on a case study of a longwall entry. *Rock Mech. Rock Eng.* **2019**. [CrossRef]
20. Yan, S.; Bai, J.; Wang, X.; Huo, L. An innovative approach for gateroad layout in highly gassy longwall top coal caving. *Int. J. Rock Mech. Min. Sci.* **2013**, *59*, 33–41. [CrossRef]
21. Wang, P.; Jiang, L.; Li, X.; Qin, G.; Wang, E. Physical simulation of mining effect caused by a fault tectonic. *Arab. J. Geosci.* **2018**, *11*. [CrossRef]
22. Sun, W.; Du, W.; Zhou, F.; Shao, J. Experimental study of crack propagation of rock-like specimens containing conjugate fractures. *Geomech. Eng.* **2019**, *17*, 323–333.
23. Jiang, L.; Wang, P.; Zheng, P.; Luan, H.; Zhang, C. Influence of Different Advancing Directions on Mining Effect Caused by a Fault. *Adv. Civ. Eng.* **2019**, *2019*, 1–10. [CrossRef]
24. Shu, J.; Jiang, L.; Kong, P.; Wang, Q. Numerical Analysis of the Mechanical Behaviors of Various Jointed Rocks under Uniaxial Tension Loading. *Appl. Sci.* **2019**, *9*, 1824. [CrossRef]
25. Hoek, E.; Carranza-Torres, C.T.; Corkum, B.; Hoek, E.; Carranza-Torres, C. Hoek-Brown failure criterion—2002 Edition. In Proceedings of the 17th Tunnelling Association of Canada Conference, Toronto, ON, Canada, 7–10 July 2002.
26. Cai, M.; Kaiser, P.K.; Martin, C.D. Quantification of rock mass damage in underground excavations from microseismic event monitoring. *Int. J. Rock Mech. Min. Sci.* **2001**, *38*, 35–1145. [CrossRef]
27. Golsanami, N.; Sun, J.; Liu, Y.; Yan, W.; Lianjun, C.; Jiang, L.; Dong, H.; Zong, C.; Wang, H. Distinguishing fractures from matrix pores based on the practical application of rock physics inversion and NMR data: A case study from an unconventional coal reservoir in China. *J. Nat. Gas Sci. Eng.* **2019**, *65*, 145–167. [CrossRef]
28. Jiang, L.; Sainoki, A.; Mitri, H.S.; Ma, N.; Liu, H.; Hao, Z. Influence of fracture-induced weakening on coal mine gateroad stability. *Int. J. Rock Mech. Min. Sci.* **2016**, *88*, 307–317. [CrossRef]

29. Li, Y.; Wu, W.; Li, B. An analytical model for two-order asperity degradation of rock joints under constant normal stiffness conditions. *Rock Mech. Rock Eng.* **2018**, *51*, 431–1445. [CrossRef]

30. Jiang, L.; Zhang, P.; Chen, L.; Hao, Z.; Sainoki, A.; Mitri, H.S.; Wang, Q. Numerical approach for goaf-side entry layout and yield pillar design in fractured ground conditions. *Rock Mech. Rock Eng.* **2017**, *50*, 3049–3071. [CrossRef]

31. Jiang, L.; Wu, Q.S.; Wu, Q.L.; Wang, P.; Xue, Y.; Kong, P.; Gong, B. Fracture failure analysis of hard and thick key layer and its dynamic response characteristics. *Eng. Fail. Anal.* **2019**, *98*, 118–130. [CrossRef]

32. Unver, B.; Yasitli, N.E. Modelling of strata movement with a special reference to caving mechanism in thick seam coal mining. *Int. J. Coal Geol.* **2006**, *66*, 227–252. [CrossRef]

33. Xu, N.; Zhang, J.; Tian, H.; Mei, G.; Ge, Q. Discrete element modeling of strata and surface movement induced by mining under open-pit final slope. *Int. J. Rock Mech. Min. Sci.* **2016**, *88*, 61–76. [CrossRef]

34. Shabanimashcool, M.; Li, C.C. Numerical modelling of longwall mining and stability analysis of the gates in a coal mine. *Int. J. Rock Mech. Min. Sci.* **2012**, *51*, 24–34. [CrossRef]

35. Rezaei, M.; Hossaini, M.F.; Majdi, A. Determination of longwall mining-induced stress using the strain energy method. *Rock Mech. Rock Eng.* **2015**, *48*, 2421–2433. [CrossRef]

36. Wang, P.; Jiang, L.; Jiang, J.; Zheng, P.; Li, W. Strata behaviors and rock burst–inducing mechanism under the coupling effect of a hard, thick stratum and a normal fault. *Int. J. Geomech.* **2018**, *18*, 04017135. [CrossRef]

37. Shen, W.L.; Bai, J.B.; Li, W.F.; Wang, X.Y. Prediction of relative displacement for entry roof with weak plane under the effect of mining abutment stress. *Tunn. Undergr. Space Technol.* **2018**, *71*, 309–317. [CrossRef]

38. Singh, T.N.; Dubey, B.K. Cavability study of a competent roof—A case study. *Int. J. Rock Mech. Min. Sci. Geomech. Abstr.* **1995**, *32*, 417A.

Article

Studies of Fracture Damage Caused by the Proppant Embedment Phenomenon in Shale Rock

Mateusz Masłowski *, Piotr Kasza, Marek Czupski, Klaudia Wilk and Rafał Moska

Oil and Gas Institute – National Research Institute, Lubicz 25A Str., 31-503 Krakow, Poland; piotr.kasza@inig.pl (P.K.); marek.czupski@inig.pl (M.C.); klaudia.wilk@inig.pl (K.W.); rafal.moska@inig.pl (R.M.)
* Correspondence: mateusz.maslowski@inig.pl

Received: 29 April 2019; Accepted: 24 May 2019; Published: 29 May 2019

Featured Application: The result of laboratory imaging of the embedment phenomenon may be one of the preliminary assessments of the effectiveness of hydraulic fracturing at the design stage.

Abstract: This paper concerns the effect of proppant embedment related to hydraulic fracturing treatment. This phenomenon occurs if the strength of a dry reservoir rock is lower than that of proppant grains. The aim of this research was the laboratory determination of the loss of width of the proppant pack built of light ceramic grains. A laboratory simulation of the embedment phenomenon was carried out for a shale rock on a hydraulic press in a heated embedment chamber specially prepared for this purpose. Tests were conducted at high temperature and axial compressive stress conditions. The surfaces of cylindrical core plugs (fracture faces) were imaged under an optical microscope equipped with 3D software. The fracture faces were examined and compared before and after the embedment phenomenon. Analysis of the obtained images of the fracture face was done, based on a research method of the embedment phenomenon developed at the Oil and Gas Institute—National Research Institute. On the basis of the laboratory tests, the parameters characterizing the embedment phenomenon were defined and discussed. In addition, the percentage reduction in the width of the proppant pack was determined.

Keywords: embedment; shale rock; proppant pack; fracture width

1. Introduction

Hydraulic fracturing is one of the main methods for stimulating unconventional hydrocarbon reservoirs. In the case of shale formations, which are characterized by increased content of clay minerals, for intensification treatments to be effective, numerous fractures and cracks should be created [1–7], as shown in Figure 1 [2,8–10].

The producing formation is fractured using hydraulic pressure, and then proppants are pumped into the fractures with a fracturing fluid [11]. The industry has been making use of slickwater, where the proppant transport is governed by the high velocity of the injected water, unlike polymer-based fluids for which the transporting mechanism is based on viscosity [12]. The literature [12–14] has reported a significant use of hybrid technologies that combine slickwater and polymer fluids. In hydraulic fracturing, energized fluids are also used (fluids with one compressible component such as nitrogen or carbon dioxide) [15]. The use of a gas component helps to reduce the hydrostatic pressure. It also supports wellbore and fracture clean up. Polymer-based fluids are still the most commonly used type of fracturing fluids [12]. The material used for proppants can range from natural sand grains called frac sand and resin-coated sand to high-strength ceramic materials and resin-coated ceramic materials [11]. The typical proppant sizes in shale reservoirs hydraulic fracturing are generally between 30 and 50 mesh (from 0.300 mm to 0.600 m) and between 40 and 70 mesh (from 0.212 mm to 0.420 mm).

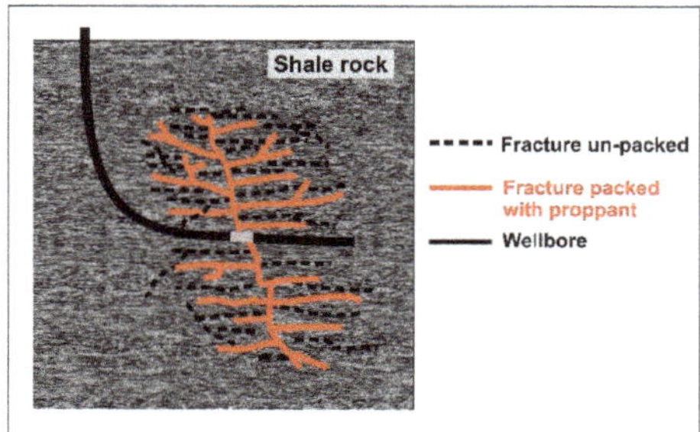

Figure 1. Visualization of numerous fractures and microcracks that allow an absorbed gas to be release from the shale rock.

Numerous fractures and cracks created in shale formations are characterized by low width and large values of the stimulated reservoir volume (SRV). It must also be pointed out that a proper selection of the proppant should ensure high conductivity of the whole generated system of fractures distributing the proppant to the furthest parts of the fractures [11,16,17]. Apart from the way by which the proppant is transported and placed in the fractures, the phenomena presented in Figure 2 have a significant influence on the effective packaging of fractures with the proppant [10,17]. This occurs after the treatment, when compressive stress closes the fracture on the proppant.

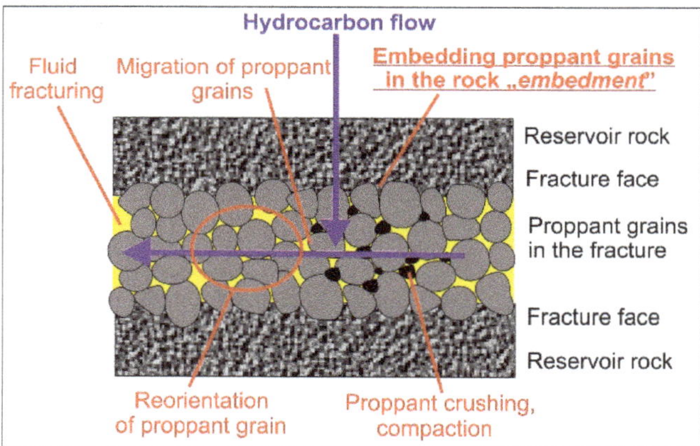

Figure 2. The phenomena influencing the effective packaging of a fracture and the achievement of high conductivity.

The proppant embedment phenomenon presented in this model causes the decrease of the width W_f of the created fracture (Figure 3) [7,10,18–21] and an increase in the damage of the fracture face, which results in a decrease of its permeability and conductivity [22–25].

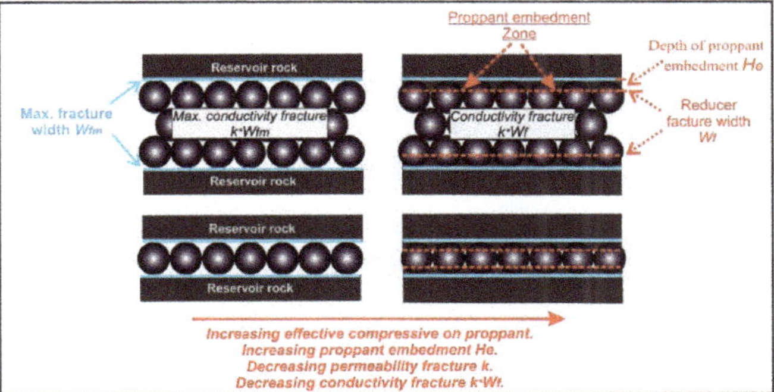

Figure 3. The effect of the embedment phenomenon on the width of a packed fracture for various proppant surface concentration.

For many years, a number of laboratory tests, imaging tests, and mathematical modeling of the embedment phenomenon have been performed. They are constantly being modernized as the capabilities of hardware and software increase.

In a research project [23], the fracture surface strength and fracture conductivity for American shale reservoirs (Barnett, Haynesville, Marcellus) were studied, and the kneading depth of small proppant grains into the fracture face was determined [23,26]. In the mentioned study, long-term conductivity of the dry and pre-saturated fracture was determined.

A study of proppant embedment in shales and its effect on hydraulic fracture conductivity is present also in the literature [4]. It described the relations between rock mineralogy, mechanical properties, fluid composition, and proppant embedment. Initial results showed a close correlation between the amount of proppant embedment at a given stress and the rock stiffness, which is affected by its mineral content, mainly, the amount and type of clay minerals. These correlations are used to predict the amount of conductivity loss due to proppant embedment in different unconventional reservoirs [4].

Another study [27] presents proppant embedment, which occurs in rock formations and can result in rapid decline of hydrocarbon production. These studies highlight the importance of the creep phenomenon (a function of confinement and temperature), of the percentage of clay content, and of the surface roughness in proppant embedment. Other parameters, such as time, temperature, and fracture fluid, can also impact the rate of proppant embedment. Also presented are numerical and analytical models representing proppant embedment [27].

The purpose of the laboratory tests presented in this paper was to determine the quantities characterizing the embedment phenomenon for dry polish shale rock (i.e., the total depth and width of the dents of proppant grains in the fracture face). The obtained values allowed to determine the percentage damage of the surface of the fracture face, the fracture width packed with proppant, and the percentage reduction in the fracture width, under given conditions at high temperature and axial compressive stress. The analyses of the obtained images of the fracture face were done on the basis of the research method of the embedment phenomenon developed at the Oil and Gas Institute-NRI.

2. Materials and Methods

2.1. Characteristics of the Reservoir Rock and Proppant Material Used for Testing

Shale rock (Figure 4a) containing 47.7% of clay minerals was used for testing. The content of quartz amounted to 24.4%, that of carbonates to 14.2%, and that of other components to 13.7%. A lightweight ceramic proppant 30/50 mesh (Figure 4b) with a grain size from 0.600 to 0.300 mm was used as the proppant material. The mean diameter of the proppant grains was 0.450 mm. The roundness and the sphericity of the grains was 0.9, the bulk density was 1.51 g/cm^3.

(a)   (b)

Figure 4. Materials used for testing: (**a**) Shale rock; (**b**) Proppant.

2.2. Methodology for Studying the Embedment Phenomenon

The research methodology developed in the Oil and Gas Institute – NRI was used [7,10,20,21,28]. It allowed an initial determination of the primary roughness of the fracture face. It was determined for several selected areas, and then the average roughness was calculated from the roughness profiles along the selected measurement sections. The method of determination of the surface roughness and the measurements are presented in Figure 5 [7,10,20,21]. Equation (1) was used [7,10,20,21,28].

$$R = \frac{\sum_{i=0}^{n} H_{p_i} + \sum_{i=0}^{n} H_{v_i}}{n_p + n_v} \tag{1}$$

where R is the roughness of the profile surface along the measurement section (mm), H_p is the peak height (mm), H_v is the valley depth (mm), n_p is the total number of peaks (-), n_v is the total number of valleys (-).

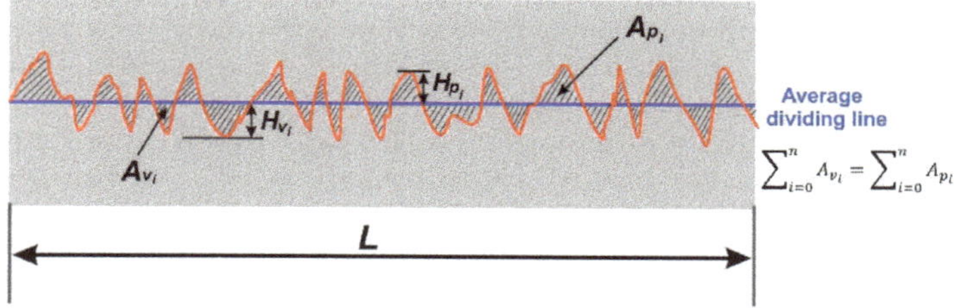

Figure 5. An example of the surface roughness profile along the measurement section for the selected area on the surface of the fracture face.

The average primary roughness R_a for the entire surface of the fracture face was determined as an arithmetic average of the roughness of the profiles determined for the individually selected areas.

Laboratory simulation of the embedment phenomenon consisted of placing a proppant between two cylindrical core plugs and then exposing it to the set axial compression stress, at the set temperature, for the set period of time (Figure 6a,b) [7,10,21,22].

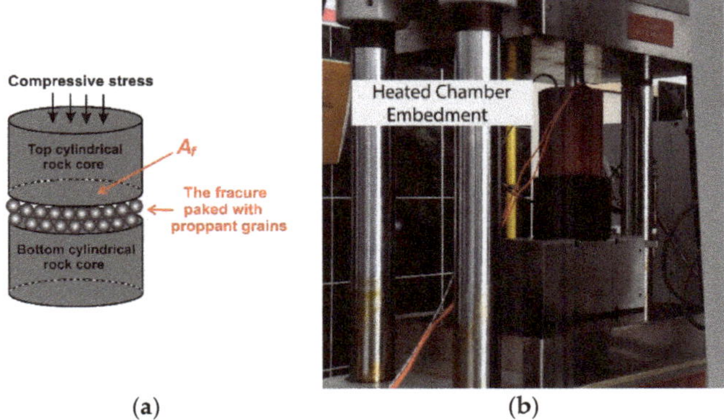

(a)   (b)

Figure 6. Test unit: (**a**) Arrangement of cores and proppant into the chamber; (**b**) Hydraulic press with heating chamber for the simulation of the embedment phenomenon.

The amount of the proppant material needed to pack the fracture and obtain the specified surface concentration was determined according to Equation (2) [10,28,29]:

$$m_p = A_f \cdot 10^{-1} \cdot C \tag{2}$$

where m_p is the weight of the proppant (g), C is the surface concentration of the proppant (kg/m^2), A_f is the surface area of the fracture face subjected to compression stress (cm^2).

The analysis of the fracture face after simulation of the embedment phenomenon consisted in the determination of the average depth of embedment of the proppant grains and of the damage of its surface. The method of determination of embedment depth and damage of the fracture face along the measurement section is presented in Figure 7 [7,10,20,21,26] and in Equation (3) [7,10,20,21,28].

$$H_e = \frac{\sum_{i=0}^{n} H_{e_i}}{n_e} \tag{3}$$

where He is the average depth of proppant embedment in the fracture face of the profile along the measurement section (mm), He_i is the valley depth (embedment of a proppant grain in the fracture face) (mm), n_e is the total number of valleys (embedment of proppant grains in the fracture face) (-).

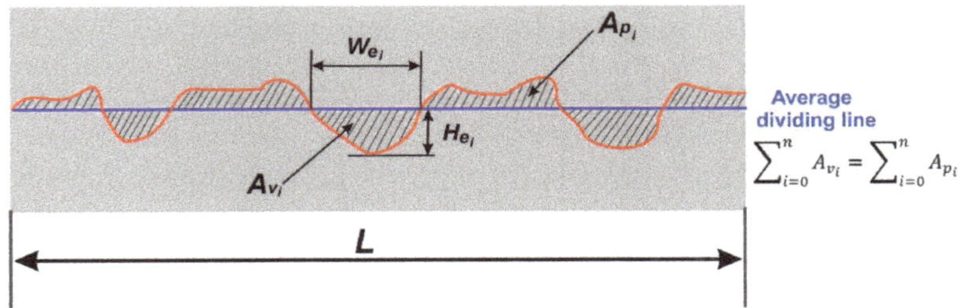

Figure 7. Sample profile of the depth and width of grain embedment (valleys) along the measurement section in the selected area, on the surface of the fracture face.

The total average depth He_t of proppant embedment in the fracture faces (rock), expressed in mm, was determined according to Equation (4) [7,10,20,21,28]:

$$H_{e_t} = H_{e_{T.a}} + H_{e_{B.a}} \tag{4}$$

where $He._{T.a}$ is the average depth of proppant embedment in the top fracture face, corresponding to the arithmetic average of the obtained values for individually specified areas (mm), $He._{B.a}$ is the average depth of proppant embedment in the bottom fracture face, corresponding to the arithmetic average of the obtained values for individually specified areas (mm).

The percentage damage of the fracture surface (PDW_e) for the profile, along the measurement section was determined according to Equation (5) [7,10,20,21], expressed in (%):

$$PDW_e = \frac{\sum_{i=0}^{n} W_{e_i}}{L} \cdot 100 \tag{5}$$

where W_{e_i} is the valley width, i.e., the embedment of a proppant grain in the fracture face (mm), and L is the length of the measurement section (mm).

The total percentage damage of the fracture surface PDW_{e_t} (embedment of the embedding proppant grains on the surface of the fracture faces) was determined according to Equation (6) [7,10,20,21], expressed in (%):

$$PDW_{e_t} = \frac{PDW_{e_{T.a}} + PDW_{e_{B.a}}}{2} \tag{6}$$

where $PDW_{e_{T.a}}$ is the average percentage damage of the surface of the top fracture face (rock), corresponding to the arithmetic average of the obtained values for individually specified areas (%), and $PDW_{e_{B.a}}$ is the average percentage damage of the surface of the bottom fracture face (rock), corresponding to the arithmetic average of the obtained values for individually specified areas (%).

The effect of the embedment phenomenon on the effective width of the fracture packed with proppant after exposure to axial compression stress was determined using Equations (7) and (8) [7,10, 20,21,28]:

$$W_f = W_{f_m} - H_{e_t} \tag{7}$$

where W_f is the fracture width packed with proppant, taking into account the embedment phenomenon (mm), and W_{f_m} is the maximum fracture width packed with proppant, without the occurrence of the embedment phenomenon (mm).

The percentage reduction of the fracture width (PRW_f) packed with proppant, taking into account the embedment phenomenon, was determined according to Equation (8) [7,10,20,21], expressed in (%):

$$PRW_f = \frac{H_{e_t}}{W_{f_m}} \cdot 100 \tag{8}$$

The maximum width W_{f_m} of the fracture packed with proppant, without the occurrence of the embedment phenomenon, was determined according to the research procedure previously mentioned in this paper. The only difference was the use of cylindrical steel plugs instead of cylindrical core plugs, which have a steel hardness of more than 43 on the Rockwell C scale (HRC). The maximum width W_{f_m} of the fracture packed with proppant was measured throughout the testing with the use of an LVDT (Linear Variable Differential Transformer) device. LVDT readings took into account the amount of deformation of the test unit (i.e., hydraulic press, measuring chamber, and steel plugs) under the specified conditions of axial compressive stress and temperature.

3. Execution of a Laboratory Simulation of the Embedment Phenomenon and Analysis of the Obtained Test Results

The tests were performed on cylindrical core plugs with a diameter of 2.54 cm. Firstly, the average primary roughness Ra of the entire surface of the core plug face (for the top and bottom fracture face), presented in Figure 7, was determined according to the test procedure described in the previous part of the paper. It was determined as an arithmetic average of two selected areas on the face of the tested core plug, from one profile running across the tested area. These tests were performed using an optical microscope (Figure 8), and the results are presented in Figures 9 and 10.

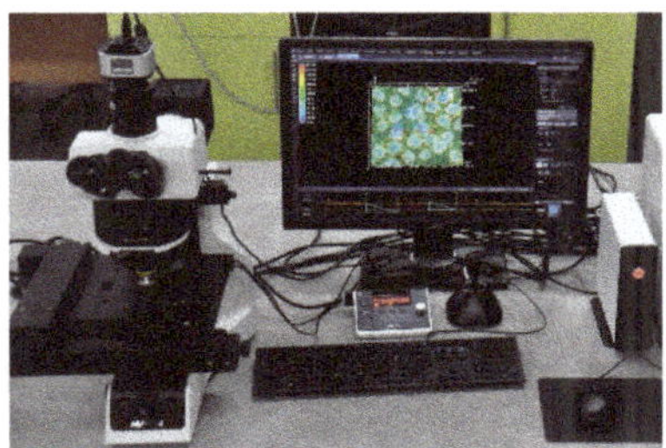

Figure 8. Optical microscope with 3D software.

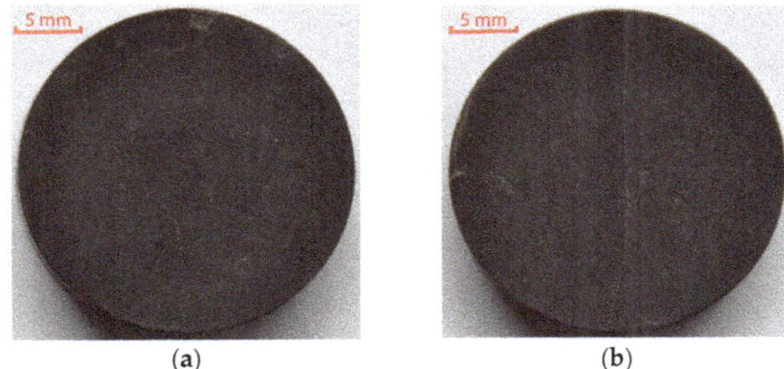

(**a**) (**b**)

Figure 9. Surface of the core plug sample with a diameter of 2.54 cm before the embedment test: (**a**) Top; (**b**) Bottom.

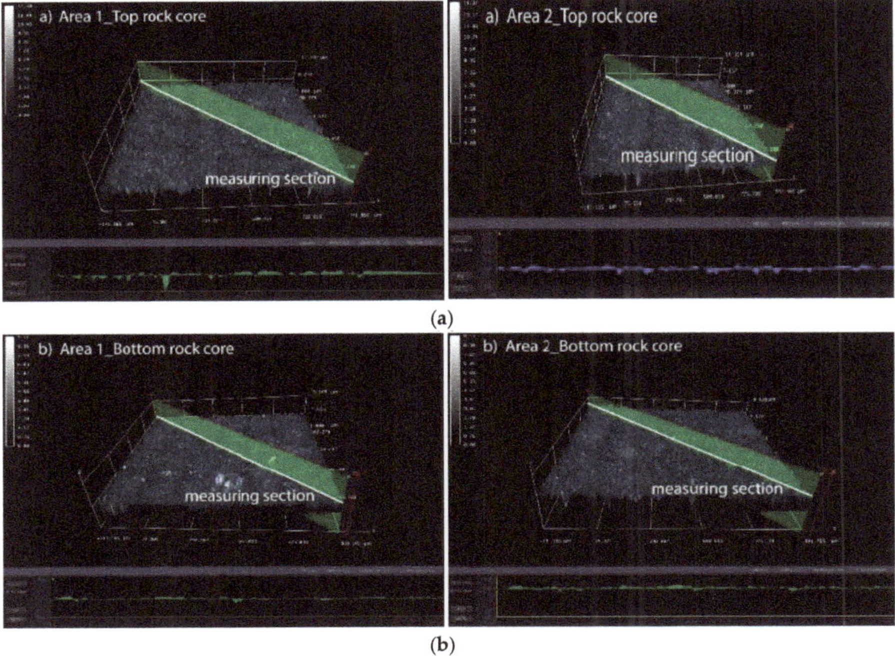

Figure 10. Determination of the primary roughness of the core plug face (fracture faces): (**a**) Top; (**b**) Bottom.

The average primary roughness R_a of the entire face of the top core plug amounted to 0.00066 mm +/− 0.00015 mm. For the bottom core plug, it amounted to 0.00033 mm +/− 0.00007 mm.

Next, a laboratory simulation of the phenomenon of proppant embedment in the fracture faces, on the test unit presented in Figure 6, was carried out.

The test conditions are presented in Table 1.

Table 1. Conditions for the tests no. 1 and no. 2.

Conditions for Test	
Temperature, (°C)	70.0
Surface concentration of proppant, (kg/m^2)	2.44
Compressive stress, (MPa)	48.3
Exposure to the defined compressive stress, (hours)	6

The result of test no. 1 is presented in Figures 11–13 and in Tables 2 and 3.

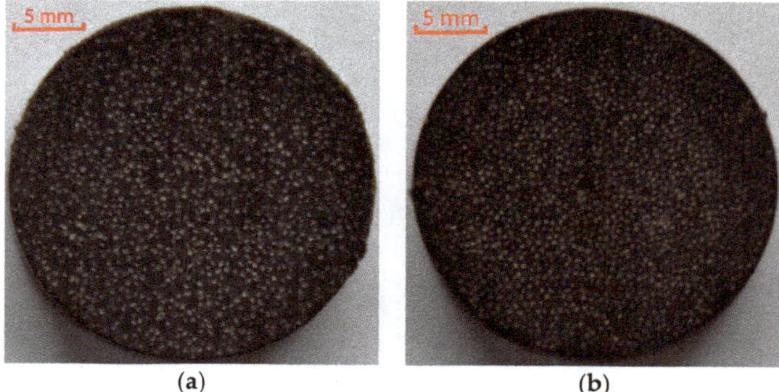

(a)   (b)

Figure 11. Surface of the core plug sample with a diameter of 2.54 cm after the embedment test: (a) Top; (b) Bottom.

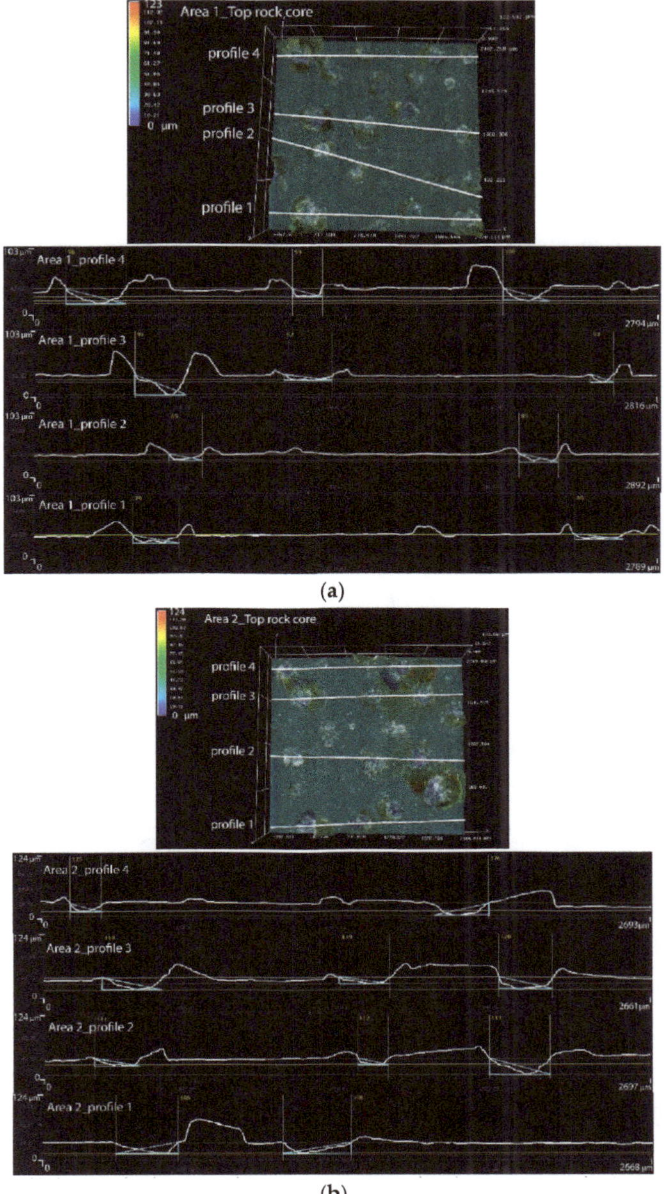

Figure 12. Average depth $H_{e_{T,a}}$ and average percentage surface damage $PDW_{e_{T,a}}$ for the top core plug (Test 1): (**a**) Area 1; (**b**) Area 2.

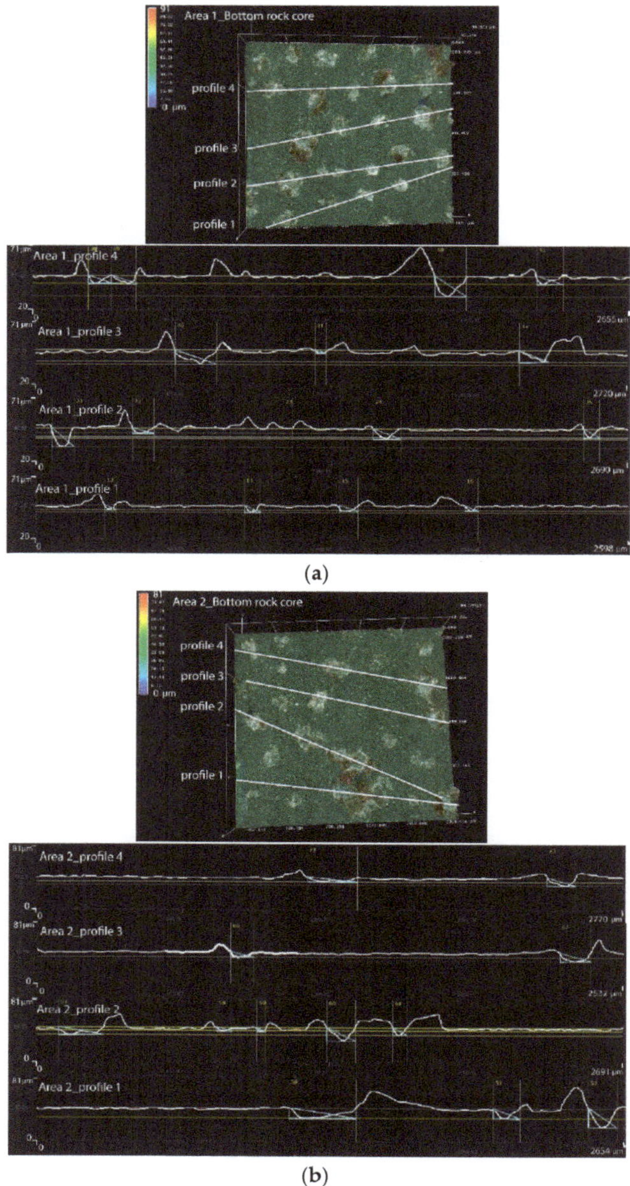

Figure 13. Average depth $H_{e_{B,a}}$ and average percentage surface damage $PDW_{e_{B,a}}$ for the bottom core plug (Test 1): (**a**) Area 1; (**b**) Area 2.

Table 2. Total average depth of proppant embedment in the fracture faces—Test no. 1.

Fracture Face	No. of the Tested Area	Surface Area (mm^2)	Average Measurement Section Length (mm)	Total Measurement Sections Length (mm)	H_e (mm)	He_a (mm)	He_t (mm)
Top	1	7.2188	2.8230	11.2919	0.0141	0.0170	
	2	6.8904	2.6717	10.6866	0.0200		0.0254
Bottom	1	6.8251	2.6658	10.6632	0.0065	0.0084	
	2	6.9765	2.6620	10.6481	0.0102		

Table 3. Average percentage damage of the fracture surface—Test no.1.

Fracture Face	No. of the Tested Area	W_e (mm)	PDW_e (%)	$PDWe_a$ (%)	$PDWe_t$ (%)
Top	1	1.9268	17.1	19.2	
	2	2.2712	21.2		17.1
Bottom	1	1.5250	14.3	14.9	
	2	1.6647	15.6		

Test no. 2 was performed in order to determine the maximum achievable width of the fracture packed with a light ceramic proppant without embedment. The test took into account the width reduction which can occur as a result of proppant crushing and proppant grains rearrangement within the fracture. The conditions of test no. 2 are presented in Table 1. In test no. 2, cylindrical core plugs were replaced with cylindrical steel plugs.

After 6 hours of exposure to the defined axial compressive stress, a maximum fracture width W_{fm} of 1.514 mm was obtained.

The uncertainty of the estimated width of the fracture packed with proppant was determined on the basis of the accuracy of the LVDT fracture gauge +/- 0.001 mm. The uncertainty of the estimated total average depth of proppant embedding in the fracture faces was determined on the basis of the standard deviation from the average value. The parameters of the fracture effectively packed with the proppant are presented in Figure 14.

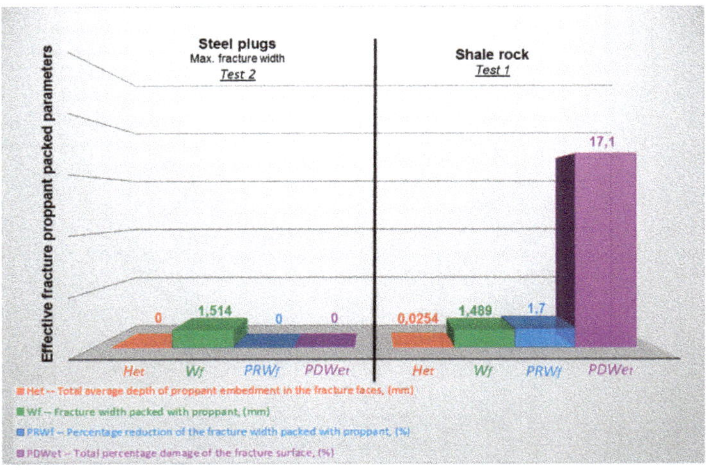

Figure 14. Parameters of the fracture effectively packed with the proppant for cylindrical steel plugs and shale rock.

In addition, an attempt was made to simulate the effect of embedment on an effectively packed fracture with only one proppant layer. The maximum fracture width, corresponding to the average diameter of the tested proppant grains, amounting to 0.450 mm, was used for the calculation. It was assumed that the value of the average depth of proppant embedment He_t and surface damage $PDWe_t$, were equal to the values obtained in test no. 1. The results of this analysis are presented in Figure 15.

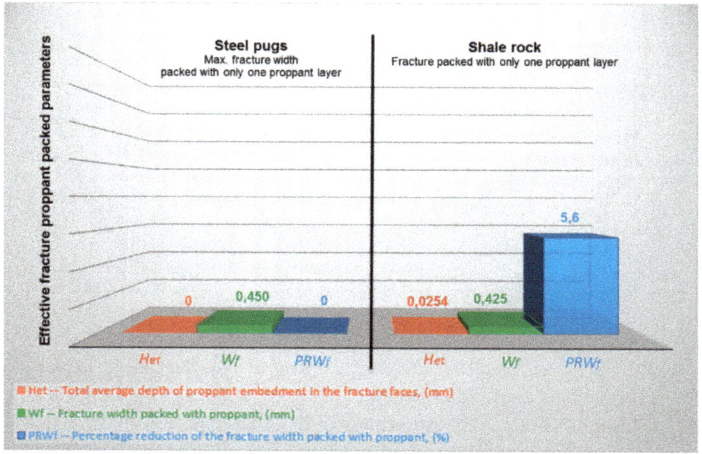

Figure 15. Parameters of the fracture effectively packed with the proppant for shale rock packed with one layer of proppant (grains size from 0.600 to 0.300 mm).

4. Discussion

Measurements were performed for a light ceramic proppant consisting of grains with size from 0.600 to 0.300 mm, with a low surface concentration of proppant of 2.44 kg/m^2 (several layers of proppant grains), and axial compression stress of 48.3 MPa for 6 hours at 70 °C. The average diameter of the proppant grains was 0.450 mm. The tested dry shale rock was characterized by:

1. On the basis of the analyzed embedment profiles, it was concluded that 17.1% of the total surface was damaged by the proppant grains.
2. The total depth of proppant embedment in the fracture faces was 0.0254 mm.
3. The obtained width of the fracture was 1.489 mm, therefore 1.7% less than the maximum achievable fracture width, which could be 1.514 mm, for the specified test conditions.

For the additionally simulated maximum fracture width (0.450 mm), corresponding to only one layer of the tested proppant, a decrease of the maximum fracture width by 5.6% was obtained. In this case, the depth of the proppant embedment of 0.0254 mm was used for the calculation. The final width of such packed fracture was 0.425 mm.

The size of the fracture width determines the flow of hydrocarbons through the fracture packed with proppant grains to the wellbore.

The tested dry shale rock allowed to maintain the width of the packed fracture in order for hydrocarbons to flow.

Test results indicate that the developed method of measurement may be used for preliminary assessments when choosing the proppant type and fracturing fluid for hydraulic fracturing of unconventional reservoirs, especially shale rocks.

Author Contributions: Conceptualization, M.M., P.K., M.C., K.W. and R.M.; Formal analysis, M.M., P.K., M.C., K.W. and R.M.; Funding acquisition, M.M.; Investigation, M.M. and P.K.; Methodology, M.M.; Software, M.M.;

Supervision, P.K., M.C. and K.W.; Validation, M.M. and R.M.; Visualization, M.M.; Writing—Original Draft, M.M.; Writing—Review & Editing, P.K., M.C. and K.W.

Funding: This paper is based on the results from statutory work. Archive no. DK-4100-56/17.

Conflicts of Interest: The authors declare no conflict of interest.

References

1. Economides, M.J.; Nolte, K.G. *Reservoir Stimulation*, 2nd ed.; Schlumberger Educational Services: Houston, TX, USA, 1989.
2. Chong, K.K.; Grieser, W.V.; Passman, A.; Tamayo, C.H.; Modeland, N.; Burke, B. A completions Guide Book to Shale-Play Development: A Review of Successful Approaches Towards Shale-Play Stimulation in the Last Two Decades. In Proceedings of the Canadian Unconventional Resources and International Petroleum Conference, SPE 133874-MS, Calgary, AB, Canada, 19–21 October 2010. [CrossRef]
3. Kasza, P. Stimulation treatments in unconventional hydrocarbon reservoirs. *Nafta-Gaz* **2011**, *10*, 697–701.
4. Alramahi, B.; Sundberg, M.I. Proppant embedment and conductivity of hydraulic fractures in Shales. *Am. Rock Mech. Assoc.* **2012**, *ARMA 12-291*, 1–6.
5. Kasza, P.; Wilk, K. Completion of shale gas formations by hydraulic fracturing. *Przemysł Chem.* **2012**, *91*, 608–612.
6. Masłowski, M. Studies of the conductivity of proppant material for the wet gas (nitrogen) after the hydraulic fracturing treatment of unconventional reservoirs. *Nafta-Gaz* **2016**, *3*, 177–185. [CrossRef]
7. Masłowski, M.; Biały, E. Studies of the embedment phenomenon in stimulation treatments. *Nafta-Gaz* **2016**, *12*, 1101–1106. [CrossRef]
8. King, G.E. Apache Corporation: Thirty Years of Gas Shale Fracturing: What Have We Learned? In Proceedings of the SPE Annual Technical Conference and Exhibition, SPE 133456-MS, Florence, Italy, 19–22 September 2010. [CrossRef]
9. Morales, H.; Tek, T. Sustaining Fracture Area and Conductivity of Gas Shale Reservoirs for Enhancing Long-Term Production and Recovery. In *RPSEA Unconventional Gas Conference 2012*; Geology, the Environment, Hydraulic Fracturing: Canonsburg, PA, USA, 2012.
10. Masłowski, M.; Kasza, P.; Wilk, K. Studies on the effect of the proppant embedment phenomenon on the effective packed fracture in shale rock. *Acta Geodyn. Geomater.* **2018**, *15*, 105–115. [CrossRef]
11. Beckmann, G.; Retsch Technology GmbH. Measuring the Size and Shape of Frac Sand and other Proppants. *Webinar Presentation* **2012**, *9*.
12. Ghaithan, A.; Al-Muntasheri; Aramco Research Centers-Houston & Saudi Aramco. A Critical Review of Hydraulic Fracturing Fluids over the Last Decade. In Proceedings of the SPE Western North American and Rocky Mountain Joint Regional Meeting, SPE 169552-MS, Denver, CO, USA, 16–18 April 2014. [CrossRef]
13. Coronado, J.A. Success of Hybrid Fracs in the Basin. In Proceedings of the Society of Petroleum Engineers. Production and Operations Symposium, SPE-106758-MS, Oklahoma City, OK, USA, 31 March–3 April 2007. [CrossRef]
14. Handren, P.; Palisch, T. Successful Hybrid Slickwater Fracture Design Evolution—An East Texas Cotton Valley Taylor Case History. In Proceedings of the SPE Annual Technical Conference and Exhibition, SPE 110 451, Anaheim, CA, USA, 11–14 November 2007. [CrossRef]
15. Cawiezel, K.E.; Gupta, D.V.S. Successful Optimization of Viscoelastic Foamed Fracturing Fluids with Ultra lightweight Proppants for Ultralow-Permeability Reservoirs. *J. SPE Prod. Facil.* **2010**, *25*, 80–88. [CrossRef]
16. Masłowski, M. Proppant material for hydraulic fracturing in unconventional reservoirs. *Nafta-Gaz* **2014**, *2*, 75–86.
17. Mueller, M.; Amro, M. Indentaion Hardness for Improved Proppant Embedment Prediction in Shale Formation. In Proceedings of the SPE European Formation Damage Conference and Exhibition, SPE 174227-MS, Budapest, Hungary, 3–5 June 2015.
18. Sato, K.; Wright, C.; Ichikawa, M. Post-Frac analysis indicating multiple fractures created in a volcanic formation. In Proceedings of the SPE India Oil and Gas Conference and Exhibition, SPE 39513-MS, New Delhi, India, 17–19 February 1998.
19. Legarth, B.; Huenges, E.; Zimmermann, G. Hydraulic fracturing in a sedimentary geothermal reservoir: Results and implications. *Int. J. Rock Mech. Mining Sci.* **2005**, *42*, 1028–1041. [CrossRef]

20. Masłowski, M. Studies of the embedment phenomenon after the hydraulic fracturing treatment of unconventional reservoirs. *Nafta-Gaz* **2015**, *7*, 461–471.
21. Masłowski, M.; Kasza, P.; Czupski, M. Studies of the susceptibility of the tight gas rock to the phenomenon of embedment, limiting the effectiveness of hydraulic fracturing. *Nafta-Gaz* **2016**, *10*, 822–832. [CrossRef]
22. Zhang, J.; Ouyang, L.; Hill, A.D.; Zhu, D. Experimental and Numerical Studies of Reduced Fracture Conductivity due to Proppant Embedment in Shale Reservoirs. In Proceedings of the PE Annual Technical Conference and Exhibition, SPE 170775-MS, Amsterdam, The Netherlands, 27–29 October 2014; pp. 1–15. [CrossRef]
23. Ghassemi, A.; Suarez-Rivera, R. Sustaining Fracture Area and Conductivity of Gas Shale Reservoirs for Enhancing Long-Term Production and Recovery. Research Partnership to Secure Energy for America; Final Report of Project, no. 08122-48; 2012. Available online: https://edx.netl.doe.gov/dataset/sustaining-fracture-area-and-conductivity-of-gas-shale-reservoirs-for-enhancing-long-term-productio/revision_resource/32b62b9d-1e36-0db5-28a6-a447e46e02fe (accessed on 29 May 2019).
24. Guo, J.; Liu, Y. Modeling of Proppant Embedment: Elastic Deformation and Creep Deformation. In Proceedings of the SPE International Production and Operations Conference and Exhibition, SPE 157449-MS, Doha, Qatar, 14–16 May 2012.
25. Zhang, F.; Zhu, H.; Zhou, H.; Guo, J.; Bo, H. Discrete-Element-Method/Computational-Fluid-Dynamics Coupling Simulation od Proppant Embedment and Fracture Conductivity after Hydraulic Fracturing. *SPE J.* **2017**, *22*, 632–644. [CrossRef]
26. Ghassemi, A.; Suarez-Rivera, R. Sustaining Fracture Area and Conductivity of Gas Shale Reservoirs for Enhancing Long-Term Production and Recovery. Appendix 5—Proppant Embedment Standard Testing Procedure. 2012. Available online: http://rpsea.org/sites/default/files/2018-04/08122-48-FR-Appendix_5_Proppant_Embedment_Testing_Procedure-05-15-13_P.pdf (accessed on 29 May 2019).
27. Bandara, K.M.A.S.; Ranjith, P.G.; Rathnaweera, T.D. Improved understanding of proppant embedment behavior under reservoir conditions: A review study. *Powder Technol.* **2019**, *352*, 170–192. [CrossRef]
28. Masłowski, M.; Kasza, P.; Czupski, M.; Wilk, K. Sposób Wyznaczania Zmniejszenia Rozwartości Podsadzonej Szczeliny. Patent no. 228609, 10 April 2018.
29. *International Standard ISO 13503-5:2006: Petroleum and Natural Gas Industries—Completion Fluids and Materials—Part 5: Procedures for Measuring the Long-Term Conductivity of Proppants*, 1st ed.; International Organization for Standardization: Geneva, Switzerland, 2006.

Article

Experimental Investigation on Chemical Grouting in a Permeated Fracture Replica with Different Roughness

Lichuang Jin [1], Wanghua Sui [1,*] and Jialu Xiong [2]

[1] School of Resources and Geosciences, China University of Mining and Technology, Xuzhou 221116, China
[2] Jiangsu Transportation Institute, Nanjing 210019, China
*   Correspondence: suiwanghua@cumt.edu.cn; Tel.: +86-0516-8359-1015

Received: 9 May 2019; Accepted: 4 July 2019; Published: 9 July 2019

Abstract: This paper presents an experimental investigation on chemical grouting in a permeated fracture replica considering its roughness. Tests of grouting with flowing water in the fracture replica were carried out under different Bardon's standard roughness profiles. The interactions between influential factors were considered and an experimental platform for grouting in rough fractures with flowing water was established. The effect of chemical grouting in fractures with flowing water was investigated using orthogonal experiment. The joint roughness coefficient (*JRC*), the initial water flow rate, the gel time, and the fracture opening were selected as factors in the orthogonal experiment. The results show that there is a positive correlation between the water plugging rate and *JRC*, and negative correlations between the water plugging rate and the initial water flow rate, gel time, and fracture opening. The change curve of the water flow rate is divided into three categories: Single platform decreasing type, double platform decreasing type, and multi-peak fluctuating type. The curve of seepage pressure contains three categories: Single peak type, multi-peak type and platform type. The results provide a reference for grouting in rock fractures.

Keywords: chemical grouting; fracture; flowing water; water plugging rate; joint roughness coefficient

1. Introduction

The problem of water inrush with fracture rock mass seriously affects underground engineering. Grouting technology has become an important method for the prevention and governance of water inrush disasters. It is very necessary to enhance the research on the effects of grouting and water plugging with flowing water.

Abundant achievements have been acquired in grouting technology. Grouting technology is widely used to control water inrush during the construction of tunnels and mines. Grouting plays an important role in mining because fractures are often caused by water inrushes [1], but grouting in rock fractures with dynamic water is still a challenge for engineers [2].

The geological conditions of grouting in rock mass with fractures are complex, therefore, the factors that influence the grouting sealing effect in fractures are varied, for example, the density of joints, fracture opening, grouting materials, flowing water, etc. Many grouting experiments have been carried out in the field and laboratory, and many factors should be considered during grouting. Grouting materials can be divided into multiple categories, and cement grout and chemical grout are the mainly grouting materials used in grouting engineering.

The grouting areas often fracture with flowing water. Meanwhile, the pressure of underground water will increase with the difficulty involved in grouting [3]. The flowing water in fractures also affects the grout diffusion, and the flowing water also reduces the strength of adjacent clay [4]. Visualization of grout propagation is achieved through laboratory experiments, using the grout diffusion distance and the shape of grout diffusion as the main research objects. The propagation length of cement grout has positive correlations with the water flow rate and grouting pressure [1], and methods for calculating the diffusion distance of slurry have been proposed [5,6]. To study the shape of grout diffusion, Li proposed the U-shaped diffusion theory of cement grouting in a single plate fracture with flowing water [7]. Zhang compared the diffusion law of chemical grouting under static water and flowing water and found that the diffusion shape of grout is nearly round under static water conditions [8].

The goal of grouting is to seal the fracture and block water, and it can be used to improve the shear strength of rock joints [9]. Kohkichi discussed the grouting effects of rock masses by situ experiments and determined that grouting can improve the deformability of rock mass [10]. Chemical grouting was simulated in a model of an indoor coal mine shaft by Wang [11], and grout propagation in soils was determined. Sui investigated the changes in seepage pressure and the grout diffusion law in a plate fracture with flowing water, and determined the relationships among the water flow velocity, gel time, aperture width, and sealing effect of grout [12]. Liang investigated the sealing effect of chemical grouting in an inclined fracture with water and sand, the results showed that grouting plays an important role in controlling the sand and water [13]. In addition, grouting pressure plays an important role in grouting and is a crucial factor for grouting results [14].

However, the natural fracture surfaces are rough in actual grouting projects. The roughness is one of the basic characteristics of rock masses. The concept of the roughness coefficient of structural surfaces was presented by Bardon [15], and the roughness of fractures was divided into 10 grades by Bardon [16] according to the degree of fluctuation, where the joint roughness coefficient (*JRC*) values were from 0 to 20. Therefore, enhancing research on grouting in rough fractures is necessary. This paper investigated chemical grouting in rough fractures with flowing water, which has a good correlation between theory and actual engineering.

2. Materials and Methods

2.1. Grouting Materials

Experiments were conducted to study flowing water grouting in fractures with different roughness. Modified urea formaldehyde resin (liquid A) and oxalate acid (liquid B) were chosen as grouting materials. These materials are widely used in underground engineering in China. This study mainly investigated the rheological properties of grout. The gel time was used as the chief index of the rheological properties of grout. It was adjusted by adjusting the ratio of liquid A to liquid B. The gel time of the grout was defined as when the fluidity of the liquid A to liquid B was lost. The ratio of liquid A to liquid B was 1:1. The gel time of the grout was controlled by changing the concentration of liquid B. Figure 1 shows the relationship between the gel time and the concentration of oxalic acid in liquid B.

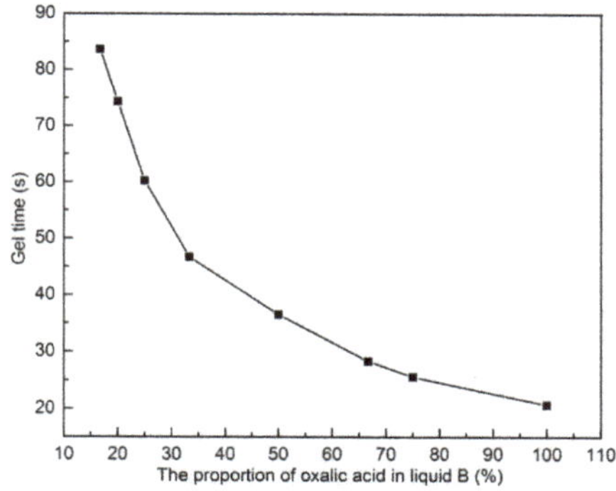

Figure 1. The relationship between gel time and the concentration of oxalic acid in liquid B.

The result shows that the gel time decreased as the oxalic acid's concentration increased. The gel time ranged from 20.7 to 83.6 s. Figure 2 shows the relationship between the viscosity of chemical grout and time measured by viscometer with different proportions of oxalic acid in liquid. The viscosity of the chemical grout increases over time, and the higher the proportion of oxalic acid in liquid B was, the faster the viscosity increased. The viscosity changed slightly before the chemical grout solidifies. The viscosity increased rapidly after solidification of the chemical grout.

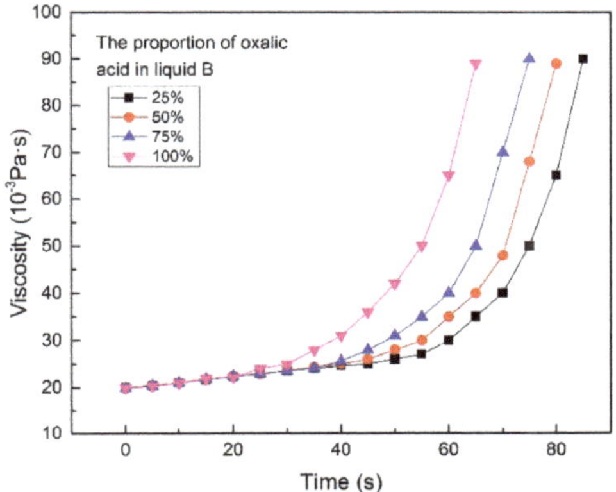

Figure 2. Curve of viscosity versus time under different proportions of oxalic acid in liquid B.

Figure 3 shows the effect of different proportions of oxalic acid in liquid B on the gel shear resistance. The gel shear resistance increased with time, and the higher the concentration of the liquid was, the faster the gel shear resistance increased.

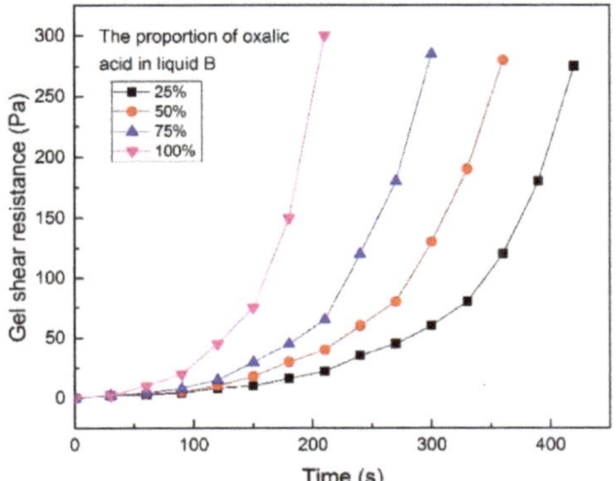

Figure 3. Curve of gel shear resistance versus time under different proportions of oxalic acid in liquid B.

2.2. Method of Generating the Fracture Replica

There are two representative methods to generate fracture replica: Numerical and physical. One of the numerical methods is the exponential variogram, which can be used to characterize the heterogeneity and anisotropy of fractures media as an effective method for studying permeability. However, normal transformation increases the degree of heterogeneity of fracture media. Another method is the self-affine surface, which can be used to simulate the form of a fracture. Three fractures were numerically simulated and the power-law fluid flow in rough fracture was researched by Lavrov [17] using this method. An isotropic self-affine fractal method was used to simulate the aperture fluctuations, and a numerical model of rough fractures was established to describe a single-phase viscous flow in rough fractures in the work of Meheust [18]. In terms of physical methods, Scesi and Gattinoni used a cement layer and Bardon's roughness profiles to prepare sample for permeability testing [19]. This method was adopted to make fracture replicas with different roughness in this paper. The system used on simulated rough fractures was composed of a PC board, a model box, and a rough fracture, which was made from cement mortar. Grade C25 strength Portland cement with river sand with a particle size of less than 0.1 mm were used in a ratio is 1:3 as the materials to simulate rough fracture. The ratio of water to cement was 1:0.5. This was mixed with river sand form cement mortar. The cement mortar was poured into the model box and the surface was pave. This was maintained for two hours. The steel plates were cut into the shapes of Bardon's standard roughness profiles using laser cutting technology. The surface of the cement mortar was scraped with the steel plate to form a standard roughness fracture surface. Figure 4 shows the steel sheets with the Bardon's standard roughness profiles. Figure 5 shows the rough fracture simulation model.

Figure 4. Steel sheets with Bardon's roughness profiles.

Figure 5. The roughness fracture simulation model.

2.3. Experimental Device

The platform used for chemical grouting in the rough fracture with flowing water was assembled in the laboratory. Figure 6 shows the experimental device, which consisted of a stable water head system (labeled 1 in Figure 6), the simulated fracture (2), a grouting pump (3), pore pressure transducers (4), a data acquisition instrument (5), a weighing system (6), an image acquisition system (7), and a computer (8).

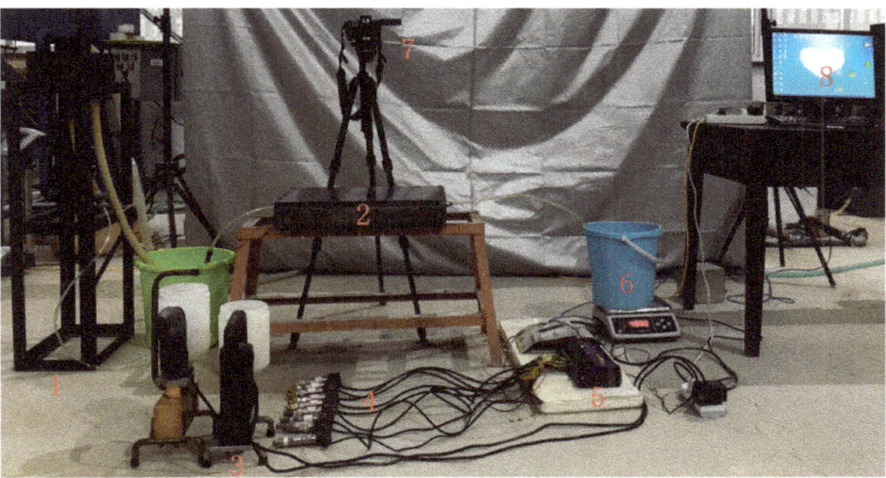

Figure 6. Picture of the experimental device. 1—Stable water head system; 2—the simulated fracture; 3—grouting pump; 4—pore pressure transducers; 5—data acquisition instrument; 6—weighing system; 7—image acquisition system; 8—computer.

The stable water head system was used to provide a constant flow of water by adjusting the height of the head to control the water velocity. The height of the water head remained constant during the grouting process. The same volumes of liquid A and liquid B were poured into the grouting pumps. Next, the grouting pumps were connected with the grouting hole through the PU tube. There were seven holes on the bottom plate of the model, including one grouting hole and six transducer holes. Figure 7 shows the layout chart and physical map of each transducer. A rectangular coordinate system was established, and the position of the grouting hole was defined as the origin. The flow direction was defined as the X forward direction, and four pressure measuring holes were arranged in the direction of the X axis, which were located at −10 cm, +10 cm, +20 cm, and +30 cm away from the grouting hole, respectively. There were two pressure measuring holes along the Y axis that were located at a distance from grouting hole of 7.5 cm in each direction. An inlet hole and an outlet hole with an inner diameter of 10 mm were put onto the side wall of the model. The inlet hole was connected with the stable water head system. The data acquisition instrument collected the seepage pressure data in real time. The internal size of the model box used to simulate the rough rock fracture was 50 cm × 30 cm × 10 cm.

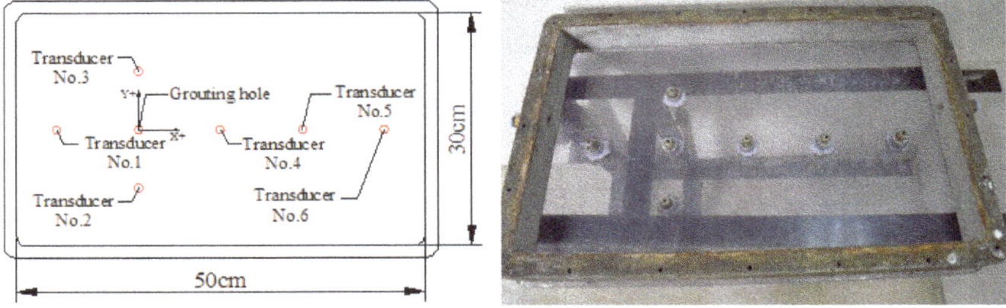

Figure 7. The layout chart and physical map of each transducer.

2.4. Similarity Theory and Criterion

In the process of grouting with flowing water in rock mass fractures, the movement of grout and water is an interaction among the rock mass, groundwater and grouts. The viscous and rheological properties of grouts are different from those of water. The grouting process in rock mass fractures is very complicated, and similarity criteria for fluid mechanical models should be considered.

Inertia force affects the flow directly, while the pressure, viscous force and gravity affected the fluid as external forces. So, the proportional relationship of forces should be compared with the inertia force [20,21]. According to dynamic similarity, it can be obtained that:

$$F = ma = \rho Va = \rho l^3 \times \frac{l}{T^2} = \rho l^4 T^{-2} = \rho l^2 u^2 \tag{1}$$

$$K = \frac{\rho_p l_p u_p}{\rho_m l_m u_m} = \lambda_\rho \lambda_l^2 \lambda_u^2 \tag{2}$$

$$K = 1 \tag{3}$$

$$(Ne)_p = (Ne)_m \tag{4}$$

where p is the prototype; m is the model; l is the length; ρ is the density; u is the velocity; K is the similarity index; Ne is the Newton number.

The Newton number of the model was equal to that of prototype, which ensured power similarity.

The fracture can be classified as follows; widening fractures (fracture width > 5 mm), opening fractures (3–5 mm), micro-tensioning fractures (1–3 mm) and closing fractures (<1 mm). The aperture widths simulated in this paper was 2, 3, 4 and 5 mm. This means that the similarity of the fracture was 1:1, which is a full-scale model.

2.5. Design of Experiment

Considering the diversity of the factors that affect the grouting effect and based on our previous research, the roughness of fracture (*JRC*) (*A*), initial water flow rate (*B*), gel time (*C*), and fracture opening (*D*) were selected as the studied factors in the experiment. An orthogonal experiment with four factors and four levels was used.

The length of the Bardon's standard roughness profiles was 10 cm in the horizontal direction, and each profile was copied four times in the direction of the length. Then, standard roughness profiles of 50 cm length were formed. Steel plates were used to simulate 10 Bardon's standard roughness profiles, and four steels were selected as the research objects based on the degree of fluctuation, which included the *JRC* values of 2–4, 6–8, 12–14, and 18–20.

By adjusting the height of the water head to control the initial water flow rate, it was ensured that the tank with the stable water head device was always full of water during the experiment. Water was collected from outlet into the cylinder and the time and volume were recorded. The velocity of water flow was determined by calculating the volume of water across the fracture over time. Initial water flow speeds of 0.4, 0.8, 1.2, and 1.6 cm/s were selected as the four levels.

In this paper, gel times of 36.5, 46.7, 60.2, and 83.6 s were chosen as the different levels.

The concept of equivalent opening was introduced to maintain the consistency of fracture opening in the tests. Zheng put forward the concept of "equivalent hydraulic opening" based on the fracture dip angle and azimuth angle [22]. The equivalent fracture opening was determined by measuring the excess water flow across the rough fracture surface. The equivalent fracture opening levels selected were 2, 3, 4, and 5 mm.

The four factors and four levels used in the experiment are shown in Table 1.

Table 1. The four factors and four levels used in the experiment.

Level	The Roughness of Fracture (JRC) A	Initial Water Flow Rate (cm/s) B	Gel Time (s) C	Fracture Opening (mm) D
1	2–4 (A1)	0.4 (B1)	36.5 (C1)	2 (D1)
2	6–8 (A2)	0.8 (B2)	46.7 (C2)	3 (D2)
3	12–14 (A3)	1.2 (B3)	60.2 (C3)	4 (D3)
4	18–20 (A4)	1.6 (B4)	83.6 (C4)	5 (D4)

The orthogonal array for chemical grouting in a rough fracture with flowing water is shown in Table 2.

Table 2. Orthogonal array for chemical grouting in a rough fracture with flowing water.

Trial No.	Symbol for Trial	Factors and Levels			
		The Roughness of Fracture (JRC) A	Initial Water Flow Rate (cm/s) B	Gel Time (s) C	Fracture Opening (mm) D
1	A1B1C1D1	2–4	0.4	36.5	2
2	A1B2C2D2	2–4	0.8	46.7	3
3	A1B3C3D3	2–4	1.2	60.2	4
4	A1B4C4D4	2–4	1.6	83.6	5
5	A2B1C2D3	6–8	0.4	46.7	4
6	A2B2C1D4	6–8	0.8	36.5	5
7	A2B3C4D1	6–8	1.2	83.6	2
8	A2B4C3D2	6–8	1.6	60.2	3
9	A3B1C3D4	12–14	0.4	60.2	5
10	A3B2C4D3	12–14	0.8	83.6	4
11	A3B3C1D2	12–14	1.2	36.5	3
12	A3B4C2D1	12–14	1.6	46.7	2
13	A4B1C4D2	18–20	0.4	83.6	3
14	A4B2C3D1	18–20	0.8	60.2	2
15	A4B3C2D4	18–20	1.2	46.7	5
16	A4B4C1D3	18–20	1.6	36.5	4

3. Tests Results and Analyses

3.1. Grout Propagation Process

Sixteen groups of experiments were carried out at room temperature (22 degrees), and the whole process of grout diffusion was recorded. Five sets of test images were selected as the representatives for the analysis. Figure 8a presents the grout diffusion process for the condition *JRC* = 2–4. In this case, the fluctuation of the simulated rough fracture was smaller. The diffusion process could be divided into two stages: Near circular diffusion and the whole passage near rectangular stage. The first stage lasted for 100 s, and the same slurry diffusion distance was formed in each direction. The next stage was sustained from 100 to 280 s, after grouting for 100 s, the grout began to spread to the boundary. Afterwards, in the direction of the X axis, with the scouring effect of water flow, the velocity of grout diffusion in the direction of flow became faster than that in the direction of the reverse flow. In the direction of the Y axis, the diffusion distance was basically the same in the direction of the vertical flow. Figure 8b shows pictures of the grout diffusion in trial No.6, where the grout diffusion was similar to that in trial No.3. The gel time was the shortest as a result of the grout solidifying quickly. At 120 s, a water gushing channel was formed under the inrush of water. A compete grout diffusion process was reached with continuous grouting. Figure 8c presents pictures showing the fair effect of water plugging. The grout was washed away easily since the rate of the initial water flow was the fastest, and the grout was taken away by water to the outlet. The grout also did not spread to the boundary.

Figure 8d shows images of the grout diffusion in test No.13 with simulation of a rough fracture of *JRC* = 18–20. The maximum fluctuation degree increased the resistance of grout diffusion. The grout spread along the direction of the Y axis first. Then, it began to expand across the rough fracture with the increase of grouting pressure and finally reached the boundary of the model. Figure 8e presents the grout diffusion in trial No.14, which had the best sealing effect in orthogonal experiment, and the whole fracture was filled with grout. The process of grout diffusion was similar to that trial No.13.

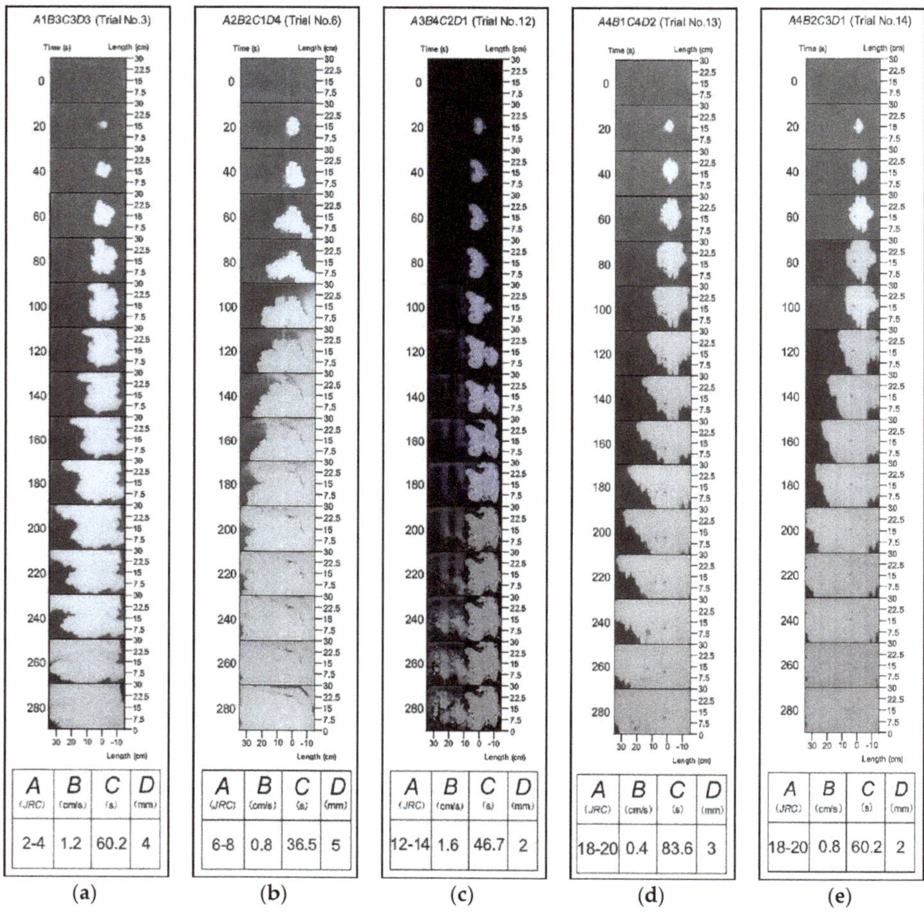

Figure 8. Images of grout diffusion along rough fracture. (**a**) Trial No.3. (**b**) Trial No.6. (**c**) Trial No.12. (**d**) Trial No.13. (**e**) Trial No.14.

3.2. Analysis of the Grouting Seepage Pressure in the Rough Fracture

A series of curves was acquired on the basis of the data from the pressure sensors in the orthogonal experiment. The curves of seepage pressure had good regularity, the seepage pressure change curve in the orthogonal experiment on grouting in a rough fracture with flowing water were divided into "single peak types", "multi-peak types", and "platform types". The change curves of seepage pressure were divided into three types, which are shown in Figure 9.

(1) "Single peak type" change curve of seepage pressure

The basic characteristics of the "single peak type" are shown in Figure 9a. By regulating the velocity of the flowing water to the level of the test, the model of the simulated rough fracture became full of water. After starting grouting, the grout diffused to the measuring point during the prophase of grouting, and the seepage pressure had little change in this process. The grouting pressure increased with the consolidation of grout, and the seepage pressure reached the peak value while the grout filled the whole fracture. Then, the seepage pressure began to decrease after the grouting ended.

(2) "Multi-peak type" change curve of seepage pressure

The seepage pressure field of trials No.2, No.6, No.7, and No.13 in the orthogonal experiment belonged to the "multi-peak type" change curve of seepage pressure. Representatives of the "multi-peak type" are shown in Figure 9b,c. The basic features of the "multi-peak type" were as follows: The grout was diffused under grouting pressure and flowing water. The seepage pressure increased to the peak value as time went by, and the fracture began to be blocked by the grout at the same time. Then, the seepage fracture declined since the grout had not been condensed completely, and this was washed away by flowing water. Afterwards, this process repeated until the grout plugged up the fracture completely, and this led to a multi peak value.

(3) "Platform type" of the seepage pressure change curve.

There were five groups with the "platform type" change curve of seepage pressure in the orthogonal experiment: Trials No.4, No.8, No.9, No.15, and No.16. A typical "platform type" change curve of the seepage pressure is shown in Figure 9d. The seepage pressure increased to the first peak value in early grouting. Then, the value had little change for a period of time. The reason for this phenomenon was that the grout did not spread to the whole fracture. Chemical grout spread to each measurement point first. After that, the grout diffused to the boundary of the fracture and the seepage pressure showed little change in the meantime. After the grout spread to the boundary, the passage of water was blocked which caused the seepage pressure to rise. With the end of grouting, the value of seepage pressure started to fall.

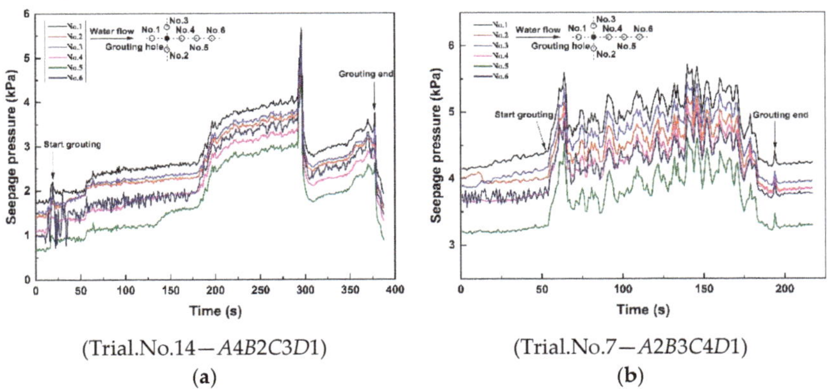

(Trial.No.14—*A4B2C3D1*)

(**a**)

(Trial.No.7—*A2B3C4D1*)

(**b**)

Figure 9. *Cont.*

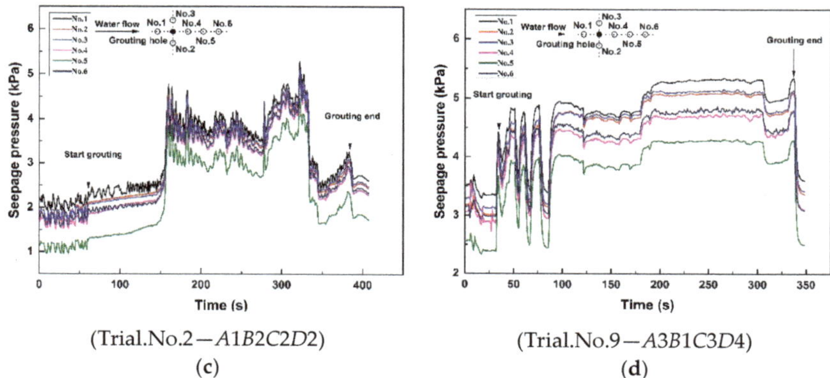

(Trial.No.2 — A1B2C2D2)

(c)

(Trial.No.9 — A3B1C3D4)

(d)

Figure 9. The types of seepage pressure change curves. (**a**) "Single peak type" of seepage pressure. (**b**) "Multi-peak type" of seepage pressure. (**c**) "Multi-peak type" of seepage pressure. (**d**) "Platform type" of seepage pressure.

3.3. Analysis of the Change in the Water Flow Rate during the Grouting

Grout and water flowed out through the outlet to the weighing platform during the grouting and flowing water rate changed constantly. The change in the water flow rate was monitored at all times by using a camera. The flow change was expressed by the variation in the platform scale number. The change curve of the flowing water rate in the orthogonal experiment can be divided into three categories: "Single platform decreasing type", "double platform decreasing type", and "multi-peak fluctuating type". The change curve of the flowing water is shown in Figure 10.

(1) "Single platform decreasing type" change curve of the water flow rate

The single platform decreasing type was split into three stages, as shown in Figure 10a. The first stage was the initial rising stage, which indicated the start of grouting. The flow rate rose to the peak value as the grout began to enter the fracture. The reason for this is that the grout washed the original water in the fracture away and caused the water flow rate to reach its peak value. The second stage was the stable platform stage. As grouting continued, chemical grout was injected into the fracture constantly, and the common movement of two kinds of fluid with chemical grout and flowing water was maintained. The flow of water at the end of the fracture was kept steady. In the third stage, the flow declined rapidly and eventually became steady. The chemical grout gradually spread to the edge of the fracture until it filled the fracture and caused concretion. An effective barrier to flowing water flow was realized, which caused the flow to decline and become steady.

(2) "Double platform decreasing type" change curve of the water flow rate

The "double platform decreasing type" change curve of the water flow rate is shown in Figure 10b. It can be divided into three stages. In the first stage, from 0 to 275 s, the grout diffused slowly because of the large fluctuation (*JRC* of 18–20). The grout primarily spread into the grooves and had little effect on the water flow rate. After the grout had filled the grooves in the direction of the *Y* axis, the grout spread all the around and the water flow rate declined rapidly. In the second stage, from 275 to 350 s, the chemical grout was injected into the fracture and a stable grouting stage formed. After 350 s, the water flow rate reduced obviously while the grout filled the fracture. The final change in the water flow rate was close to 0 and a good water plugging efficiency was achieved.

(3) "Multi-peak fluctuating type" change curve of the water flow rate

The "multi peak fluctuating type" change curve with flowing water is shown in Figure 10c,d. Figure 10c shows a type of water plugging failure. The initial water flow rate is maximum when *JR C*= 2–4, where the surface of the rough fracture is near the horizontal. The grout was taken away under the flowing water, and the blocking channel did not form. Thus, the grouting effect failed completely. Another situation of the multi peak fluctuating type is shown in Figure 10d, with the first phase from 0 to 67 s. There are many peak values owing to the large changes in the water flow. Chemical grout was injected into the fracture and the balance of the single flowing fluid was broken. From 67 to 73 s, temporary plugging formed, and the grout that had not completely solidified was washed away under the conditions of flowing water. The water flow fluctuated repeatedly with the continuous grouting. At 125 s, the water gushing channel was blocked by chemical grout, and the water flow showed an effective decline.

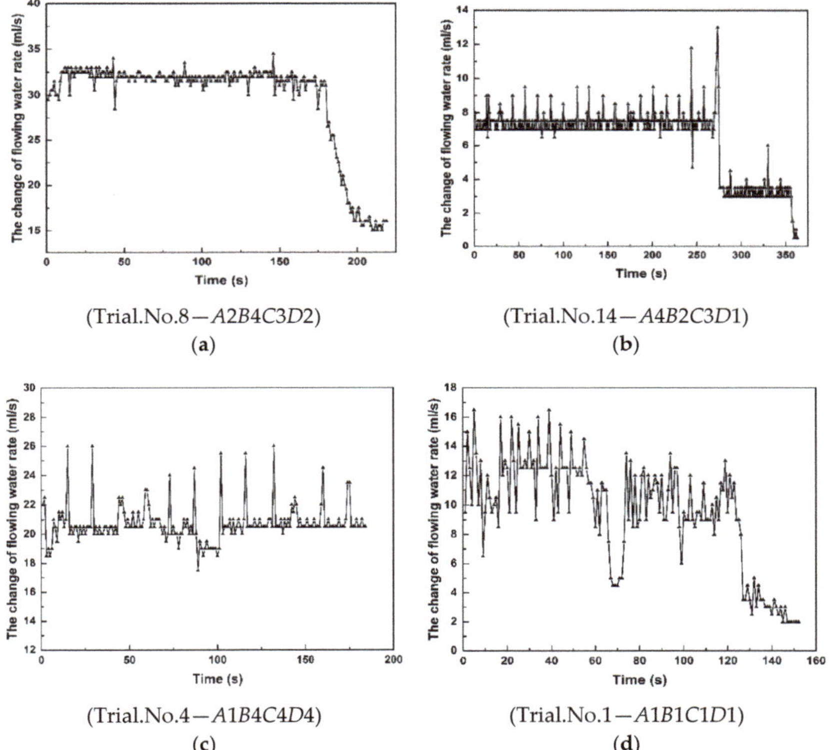

(Trial.No.8 — *A2B4C3D2*)

(**a**)

(Trial.No.14 — *A4B2C3D1*)

(**b**)

(Trial.No.4 — *A1B4C4D4*)

(**c**)

(Trial.No.1 — *A1B1C1D1*)

(**d**)

Figure 10. The change curves of flowing water. (**a**) Single platform decreasing type. (**b**) Double platform decreasing type. (**c**) Multi-peak fluctuating type. (**d**) Multi-peak fluctuating type.

3.4. Sealing Effect

The whole process of water flow change was detected during the simulated rough fracture grouting. The sealing effect can be used as the evaluation criterion of grouting plugging. After the grouting finished and stabilized, the ratio of the reduction of water flow to the initial water flow was defined as the water sealing effect (*SE*). The formula of *SE* was put forward by Eriksson [23], as follows:

$$SE(\%) = \frac{Q_0 - Q_{grout}}{Q_0} \cdot 100 \tag{5}$$

where Q_0 is the initial water flow before grouting and Q_{grout} is the water flow after grouting.

A classification system for the grouting effect was proposed by Sui [12], as shown in Table 3.

Table 3. The classification system for the grouting effect proposed in this study.

Grade	Sealing Effect (SE) (%)	Grouting Quality
1	$90 \leq SE$	Excellent
2	$80 \leq SE < 90$	Good
3	$50 \leq SE < 80$	Fair
4	$30 \leq SE < 50$	Poor
5	$10 \leq SE < 30$	Very poor
6	$10 > SE$	Fail

Table 4 showed the results of the orthogonal test of chemical grouting in a rough fracture with flowing water. Trial number 14 which had a *JRC* of 18–20, an initial water flow speed of 0.8 cm/s, a gel time of 60.2 s, and a fracture opening of 2 mm, had the largest water plugging rate and the best plugging effect. Trial number 4 which had a *JRC* of 2–4, an initial water flow speed of 1.6 cm/s, a gel time of 83.6 s, and a fracture opening of 5 mm, had the smallest water plugging rate and the worst plugging effect. The value of sealing effect ranged from 6.82% to 93.75% across the whole test.

Table 4. Results of the orthogonal test of chemical grouting in a rough fracture with flowing water.

Trial No.	Symbol for Trial	The Joint Roughness of Fracture (JRC)A	Initial Water Flow Speed (cm/s) B	Gel Time (s) C	Fracture Opening (mm) D	Sealing Effect SE (%)	Grouting Quality
1	A1B1C1D1	2–4	0.4	36.5	2	78.95	Fair
2	A1B2C2D2	2–4	0.8	46.7	3	64.71	Fair
3	A1B3C3D3	2–4	1.2	60.2	4	25	Very poor
4	A1B4C4D4	2–4	1.6	83.6	5	6.82	Fail
5	A2B1C2D3	6–8	0.4	46.7	4	76	Fair
6	A2B2C1D4	6–8	0.8	36.5	5	48.15	Poor
7	A2B3C4D1	6–8	1.2	83.6	2	40.63	Poor
8	A2B4C3D2	6–8	1.6	60.2	3	45.76	Poor
9	A3B1C3D4	12–14	0.4	60.2	5	35.72	Poor
10	A3B2C4D3	12–14	0.8	83.6	4	55	Fair
11	A3B3C1D2	12–14	1.2	36.5	3	78.26	Fair
12	A3B4C2D1	12–14	1.6	46.7	2	56.52	Fair
13	A4B1C4D2	18–20	0.4	83.6	3	80	Fair
14	A4B2C3D1	18–20	0.8	60.2	2	93.75	Excellent
15	A4B3C2D4	18–20	1.2	46.7	5	67.44	Fair
16	A4B4C1D3	18–20	1.6	36.5	4	60	Fair

Table 5 lists the range of influences on the sealing effect for different variables, SE_i ($i = 1, 2, 3,$ 4) indicates the average values for factors with the same level. The factors and levels used in the experimental are shown in Table 1. SE_1 represents the average value of sealing effect with the first level of each factor. SE_2 represents that the average value of sealing effect with the second level of each factor. For example, when $JRC = 2$–4 was the first level of fracture roughness, the SE_1 of the fracture roughness was obtained by calculating the average of trials 1, 2, 3 and 4, as shown in Table 5. The value of 0.4 cm/s was the first level of the initial water flow speed, so the SE_1 of the initial water flow speed was obtained by calculating the average of trial 1, 5, 9, and 13 as shown in Table 5. The difference between the maximum and minimum values of diverse levels with same factors was regarded as an important indicator to reflect the date fluctuation. The larger the range of values was, the greater influence of water plugging rate was. The relationship of various factors from Table 5 was in the order $R_A > R_D > R_B > R_C$, and the factors of affecting grouting plugging were followed by A, D, B, and then C.

Table 5. Range analysis of the influence of different levels on the sealing effect.

For Levels	Average *SE* (%) for Factors			
	The Roughness of Fracture (*JRC*) *A*	Initial Water Flow Rate (cm/s) *B*	Gel Time (s) *C*	Fracture Opening (mm) *D*
SE_1	43.870	67.668	66.340	67.463
SE_2	52.635	65.403	66.168	67.183
SE_3	56.375	52.833	50.058	54.000
SE_4	75.298	42.275	45.613	39.533
Range (A,B,C,D)	31.428	25.393	20.727	27.930

A visual analysis chart was drawn according to the Table 5. Four factors showed a good correlation with the sealing effect. The sealing effect increased as the fracture roughness increased, while the sealing effect decreased as the initial water flow speed, gel time, and fracture opening increased. The best orthogonal results are shown in Figure 11, where a *JRC* of 18–20 is shown to be the best level of fracture roughness, the velocity of 0.4 cm/s is the best initial water flow speed, 36.5 s is the best gel time and 2 mm is the best fracture opening. Therefore, the best combination of test levels is *A4B1C1D1* in theory. The best combination of test levels was shown to be *A4B2C3D1* in the orthogonal test. Therefore, *A4B1C1D1* was test under in the same experimental conditions, and the sealing effect reached 100%. The sealing effect was better than that of *A4B2C3D1*, which verified the correctness of orthogonal experiment. The worst combination of test levels in the test was *A1B4C4D4*, which is in agreement with the theory. These results confirm the accuracy of the test results.

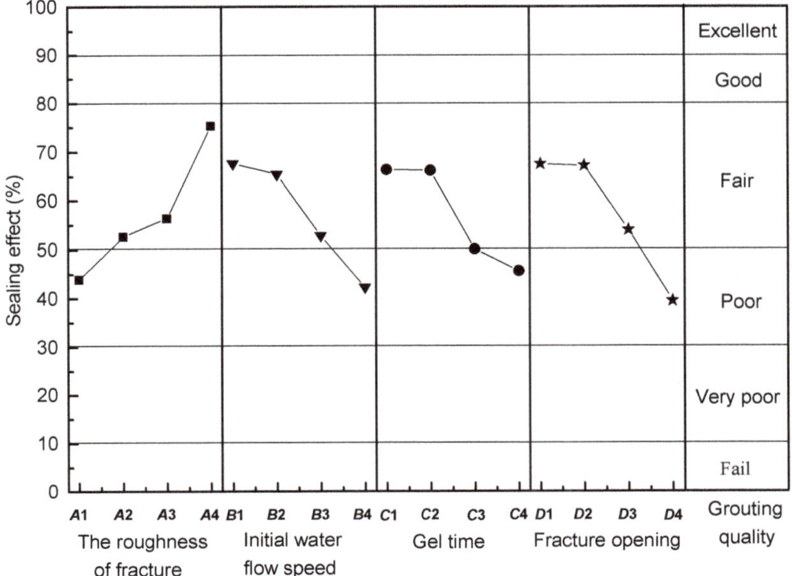

Figure 11. The analysis chart of *SE* for four factors.

4. Discussion and Limitations

Previous studied on grouting were based on the simulated single plate device. The experimental device used in this paper realized the visualization of the grouting process, and allowed the value of grout diffusion to be observed. The rock fracture in Bardon's standard roughness coefficient curves were used as a simulation model, as rough fractures are more similar to real fractures. The sealing effect of grouting in orthogonal experiments was acquired by monitoring the changes in the initial water flow rate, and the sequence of the influential factors in the sealing effect was obtained to provide the basis for choosing factors during actual grouting engineering. However, the scale model of the experiment was simplified and idealized and only four factors that affect the sealing effect were considered. In further studies, experiments on grouting in rough fracture networks with flowing water should be carried out, and 3D printing technology could be used to simulate rock masses. Moreover, an experimental scheme should be combined with an actual engineering process.

5. Conclusions

In this paper, a set of chemical grouting tests were carried out to simulate rough fractures using Bardon's standard roughness curves. The fracture roughness, initial water flow speed, gel time, and fracture opening were selected as the factors for the orthogonal test. The changes in seepage pressure, grout diffusion, and water flow were monitored in real time during the grouting process. The optimal combination of grouting plugging was determined according to the analysis of the sealing effect of each test. The main achievements and conclusions are as follows:

(1) The factors affecting the grouting effect in the chemical grouting experiment of a rough fracture with flowing water were as follows: The fracture roughness, initial water flow speed, fracture opening, and gel time. The optimal combination of levels in the orthogonal test is $A4B1C1D1$. The fracture joint roughness coefficient was shown to have the greatest influence on the grouting effect.

(2) The grout diffusion pattern of grout is related to the fracture roughness. The smaller the value of JRC is, the easier the grout diffusion is. The larger the initial water flow speed is, the less complete the grout shape is.

(3) The seepage pressure change curves of grouting plugging tests with the fracture roughness coefficient can be divided into three types: Single peak type, multi-peak type, and platform type.

(4) The curves of the water flow across the rough fracture can be divided into three categories: Single platform decreasing type, double platform decreasing type, and multi-peak type.

Author Contributions: Conceptualization, W.S., J.X. and L.J.; data curation, L.J.; formal analysis, L.J. and J.X.; investigation, L.J.; methodology, J.X., W.S., and L.J.; validation, L.J. and J.X.; writing—original draft, L.J.; writing—review and editing, W.S. and L.J.

Acknowledgments: The authors thank the support of the National Key R & D Project under Grant No. 2017YFC1501303 and the National Natural Science Foundation of China under Grant No.-41472268.

Conflicts of Interest: The authors declare no conflict of interest.

References

1. Zhang, J. Investigations of water inrushes from aquifers under coal seams. *Int. J. Rock Mech. Min. Sci.* **2005**, *42*, 350–360. [CrossRef]
2. Yang, P.; Li, T.; Song, L.; Deng, T.; Xue, S. Effect of different factors on propagation of carbon fiber composite cement grout in a fracture with flowing water. *Constr. Build. Mater.* **2016**, *121*, 501–506. [CrossRef]
3. Mao, D.; Lu, M.; Zhao, Z.; Ng, M. Effects of water related factors on pre-grouting in hard rock tunneling. *Procedia Eng.* **2016**, *165*, 300–307. [CrossRef]
4. Mohammed, M.H.; Pusch, R.; Knutsson, S. Study of cement-grout penetration into fractures under static and oscillatory conditions. *Tunn. Undergr. Space Technol.* **2015**, *45*, 10–19. [CrossRef]
5. Funehag, J.; Fransson, A. Sealing narrow fractures with a Newtonian fluid model prediction for grouting verified by field study. *Tunn. Undergr. Space Technol.* **2006**, *21*, 492–498. [CrossRef]

6. Yang, M.; Yue, Z.; Lee, P.K.; Su, B.; Tham, L.G. Prediction of grout penetration in fractured rocks by numerical simulation. *Can. Geotech. J.* **2011**, *39*, 1384–1394. [CrossRef]

7. Li, S.; Zhang, X.; Zhang, Q.; Sun, K.; Xu, Y.; Zhang, W.; Li, H.; Liu, R.; Li, P. Research on mechanism of grout diffusion of dynamic grouting and plugging method in water inrush of underground engineering. *Chin. J. Rock Mech. Eng.* **2011**, *30*, 2377–2396.

8. Zhang, G.; Zhan, K.; Gao, Y.; Wang, W. Comparative experimental investigation of chemical grouting into a fracture with flowing and static water. *Min. Sci. Technol.* **2011**, *21*, 201–205.

9. Salimian, M.H.; Baghbanan, A.; Hashemolhosseini, H.; Dehghanipoodeh, M.; Norouzi, S. Effect of grouting on shear behavior of rock joint. *Int. J. Rock Mech. Min. Sci.* **2017**, *98*, 159–166. [CrossRef]

10. Kikuchi, K.; Igari, T.; Mito, Y.; Utsuki, S. In situ experimental studies on improvement of rock massed by grouting treatment. *Int. J. Rock Mech. Min. Sci.* **1997**, *34*, 138. [CrossRef]

11. Wang, D.; Sui, W. Grout diffusion characteristics during chemical grouting in a deep water-bearing sand layer. *Int. J. Min. Sci. Technol.* **2012**, *22*, 589–593. [CrossRef]

12. Sui, W.; Liu, J.; Hu, W.; Qi, J.; Zhan, K. Experimental investigation on sealing efficiency of chemical grouting in rock fracture with flowing water. *Tunn. Undergr. Space Technol.* **2015**, *50*, 239–249. [CrossRef]

13. Liang, Y.; Sui, W.; Qi, J. Experimental investigation on chemical grouting of inclined fracture to control sand and water flow. *Tunn. Undergr. Space Technol.* **2019**, *83*, 82–90. [CrossRef]

14. Rafi, J.Y.; Stille, H. Control of rock jacking considering spread of grout and grouting pressure. *Tunn. Undergr. Space Technol.* **2014**, *40*, 1–15. [CrossRef]

15. Bardon, N. Review of a new shear-strength criterion for rock joints. *Eng. Geol.* **1973**, *7*, 287–332. [CrossRef]

16. Bardon, N.; Choubey, V. The shear strength of rock joints in theory and practice. *Rock Mech.* **1977**, *10*, 1–54. [CrossRef]

17. Lavrov, A. Redirection and channelization of power-law fluid flow in a rough-walled fracture. *Chem. Eng. Sci.* **2013**, *99*, 81–88. [CrossRef]

18. Meheust, Y.; Schmittbuhl, J. Geometrical heterogeneities and permeability anisotropy of rough fractures. *J. Geophys. Res.* **2001**, *106*, 2089–2102. [CrossRef]

19. Sces, L.; Gattinoni, P. Roughness control on hydraulic conductivity in fractured rocks. *Hydrogeol. J.* **2007**, *15*, 201–211. [CrossRef]

20. Wu, Y. *Engineering Fluid Mechanics*; China Architecture & Building Press: Beijing, China, 2006.

21. Zhang, J. *Fluid Dynamics*; China University of Mining and Technology Press: Jiangsu, China, 2001.

22. Zheng, C. Simulation Research of Grouting in Fractured Rock Mass. Ph.D. Thesis, Central South University, Changsha, China, 1999.

23. Eriksson, M. Prediction of Grout Spread and Sealing Effect: A Probabilistic Approach. Ph.D. Thesis, Royal Institute of Technology, Stockholm, Sweden, 2002.

Article

Numerical Investigation of Mineral Grain Shape Effects on Strength and Fracture Behaviors of Rock Material

Zhenhua Han [1,2,3,*], Luqing Zhang [1,2] and Jian Zhou [1,2]

[1] Key Laboratory of Shale Gas and Geoengineering, Institute of Geology and Geophysics, Chinese Academy of Science, Beijing 100029, China

[2] Institutions of Earth Science, Chinese Academy of Sciences, Beijing 100029, China

[3] College of Earth Science, University of Chinese Academy of Sciences, Beijing 100049, China

* Correspondence: hanzhenhua@mail.iggcas.ac.cn; Tel.: +8610-8299-8644; Fax: +86-10-6201-0846

Received: 13 June 2019; Accepted: 15 July 2019; Published: 17 July 2019

Abstract: Rock is an aggregate of mineral grains, and the grain shape has an obvious influence on rock mechanical behaviors. Current research on grain shape mostly focuses on loose granular materials and lacks standardized quantitative methods. Based on the CLUMP method in the two-dimensional particle flow code (PFC^{2D}), three different grain groups were generated: strip, triangle, and square. Flatness and roughness were adopted to describe the overall contour and the surface morphology of the mineral grains, respectively. Simulated results showed that the grain shape significantly affected rock porosity and further influenced the peak strength and elastic modulus. The peak strength and elastic modulus of the model with strip-shaped grains were the highest, followed by the models with triangular and square grains. The effects of flatness and roughness on rock peak strength were the opposite, and the peak strength had a significant, positive correlation with cohesion. Tensile cracking was dominant among the generated microcracks, and the percentage of tensile cracking was maximal in the model with square grains. At the postpeak stage, the interlocking between grains was enhanced along with the increased surface roughness, which led to a slower stress drop.

Keywords: mineral grain shape; particle flow code; uniaxial compression simulation; rock mechanical property

---

1. Introduction

Rock is a natural heterogeneous material composed of a variety of minerals of different geometries, strengths, and deformation characteristics. Rock heterogeneity can be defined as the uneven changes of the mineral composition and microstructure in the spatial distribution influenced by the diagenesis and tectonics [1]. The macroscopic failure of rock is the gradual evolution of internal microcracks [2–4], so microheterogeneity significantly affects the macroscopic mechanical properties of rock [5–8]. Rock microheterogeneity mainly includes microstructure heterogeneity due to different grain shapes, sizes, and arrangements; elastic heterogeneity; and microcontact heterogeneity [9]. Heterogeneity is one of the fundamental reasons for rock strength differences and has always been an important research topic [10,11]. Affected by a diagenetic environment, the mineral grain shapes are often diverse and irregular, which has a great impact on rock mechanical properties [7]. Previous studies have explored the influences of mineral grain shape on rock macroscopic properties from a microscale perspective based on experiments and simulations.

Due to the complexity of the mineral grain shape, it is difficult to analyze the influence of mineral grain shape through rock prototype experiments. Therefore, self-made samples have often been used to study the grain shape effect. Shinohara et al. [12] found that the intergranular interlocking effect and the internal friction angle of stainless-steel powder were increased with the increase of the particle edge angle by a triaxial compression test. Härtl et al. [13] studied the effect of grain shape on the shear behavior of a model using glass beads and found that the interlocking between different grain shapes could significantly increase the internal friction angle of the model. Liu et al. [14] quantified the grain shape of four different sand particles and glass spheres with the help of ImageJ software, which can easily achieve binary conversion of images. The results showed that the grain shape parameters (integral contour coefficient, sphericity, and angular angle) were well correlated with the friction angle and the dilatancy angle of the sand material. Johanson et al. [15] made samples from plastic pellets with different shapes (circular, heart shaped, and star shaped) and studied the effect of grain shape on the unconfined yield strength of the specimens by a shear test. The results showed that the number and direction of the contacts were key factors influencing the strength of the sample.

Because the information obtained from experiments is limited and it is difficult to make a sample composed of a complex grain shape, numerical simulations have often been used to study the effects of grain shape, as they can easily realize the formation of complex shapes by the combination of elements. Due to the heterogeneity, discontinuity, and anisotropy of rock materials, the theory of continuum mechanics has great limitations in simulating rock damage. As an important numerical simulation method, the discrete element method can effectively solve the mechanical problems of discontinuous material. Hence, the particle flow code (PFC) has also become an important tool for studying rock materials [16–20]. The traditional granule model in PFC is constructed from circular or spherical particles, but it can construct a cluster of particles of any shape using the CLUMP method [21]. In this method, two or more circular particles are joined together to form a complex shape, which can be regarded as a mineral grain. Based on this method, Santamarina and Cho [22] studied the effects of grain shape on the inherent anisotropy and the stress-induced anisotropy of sand materials. Shi et al. [23,24] studied the shear mechanical properties of nonrounded granular sands and found that the grain shape affected the peak strength, deformation characteristics, and the shear zone thickness of the specimen. Kong et al. [25] defined the shape coefficient using roundness and concavity to describe sand-like grains. Using biaxial and direct shear tests, negative linear correlations were found between the shape coefficient and the peak strength, internal friction angle, and shear strength of the sand material. Kerimov [26] investigated the effects of irregularly shaped grains on the porosity, permeability, and elastic bulk modulus of granular porous media, and they found that the grain shape had the greatest effect on permeability.

The above studies have revealed the relationship between the grain shape and the mechanical properties of loose granular material such as sand and soil. However, few papers have discussed the influence of mineral grain shape on rock mechanical behavior. Unlike loose material, the mineral grains of rock are bonded together, hence the grain shape effects are different with loose material. In the study of rock mineral shape effects, quantitative descriptions of grain shape characteristics are essential. Because of the irregularity of the mineral grain shape, studies on the relationship between the shape parameters and rock macroscopic mechanical properties are lacking. Based on PFC, Cho and Martin [21] concluded that the models with complex grain shapes have higher tensile strength to compressive strength ratio compared with the circular grain model. Kem et al. [27] thought that the oblate plate grain shape and corresponding shape orientation distributions would lead to an increase in the calculated elastic anisotropy of rocks. Rong et al. [7] investigated the effects of grain shape on rock mechanical properties using three-dimensional PFC (PFC^{3D}). He used the sphericity index to quantitatively characterize the grain shape and found that the initiation, damage, and peak strength all decreased with an increasing sphericity index. However, it is not comprehensive to quantify the effect of grain shape on rock mechanical properties only using the indicator of sphericity because sphericity only reflects the proximity of the grain to the sphere and it cannot reflect the roughness of the grain surface. Thus, there is a need for further investigation of the mineral grain shape effects on rock mechanical properties.

The specific problem we addressed in the present work is a preliminary investigation of the irregular grain shape effects on the rock strength and fracture behaviors. Lac du Bonnet (LDB) granite was used for model calibration, which is a candidate host rock for the repository of high-level radioactive waste. Previous studies on grain shape are limited to a specific set of four or five irregularly shaped grains. In the present work, three different grain groups were constructed based on the CLUMP method in two-dimensional PFC (PFC^{2D}), respectively strip, triangle, and square. In each shape group, a series of six grains was included for systematic investigation. Combined with the grain characteristics, the shape parameters, which can describe the overall contour and surface morphology of the mineral grain, were defined. The influences of mineral grain shape on rock mechanical properties under uniaxial compressive conditions were further analyzed.

2. Modeling Methodology

PFC^{2D} is based on the distinct element theory, which was first proposed by Cundall [16]. It is suitable for analyzing material mechanics problems under quasi-static and dynamic conditions. In this method [17], the material is represented as an aggregate of numerous rigid circular particles. Noncontinuous mechanical behaviors of rock material can be observed as a result of the interactions and movements of the rigid particles. The interaction of the particles is treated as a dynamic process with states of equilibrium developing whenever the internal forces balance. The calculation process can be seen from Figure 1. Newton's second law was used to calculate the translational and rotational motion of each particle arising from the forces acting upon it, and the force–displacement law was used to update the contact forces arising from the relative motion at each contact. A time-step algorithm was used repeatedly on each particle. For further details, please refer to Potyondy and Cundall [17].

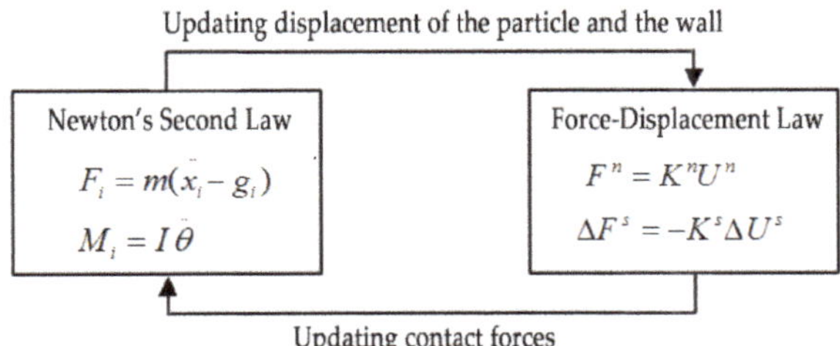

Figure 1. Calculation principles of the particle discrete element method [1].

PFC provides three basic contact models: parallel bond model, contact bond model, and sliding model. The parallel bond model is among the most frequently used models in rock mechanics research. The parallel bond model can be assumed to be a cemented polymer (Figure 2), where particles are cemented together at the point of contact so as to provide certain resistance to the exterior load. The contact force between particles is represented by F_i. The force and moment carried by the parallel bond are represented by $\overline{F}_i$ and $\overline{M}_i$, denoted in Figure 2. When an external force acting on each parallel bond exceeds its limit shear or tensile strength, the bond is damaged, and a shear or tensile microcrack develops at the corresponding position. With the ongoing generation of microcracks, a macrofracture can be formed by the linking of these individual microcracks.

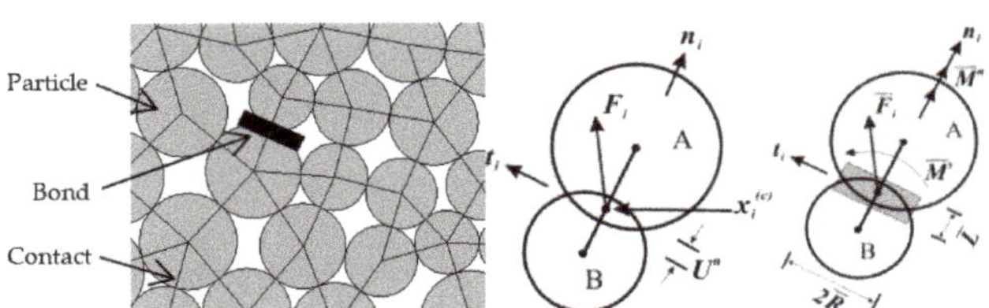

Figure 2. Force–displacement behavior of the grain-bonding system after [17].

3. Model Descriptions

3.1. Construction of Mineral Grains of Different Shapes

The basic particle shape is a circle, but noncircular grains can be achieved by the combination of two or more circular particles using the CLUMP method. In this method, grains with irregular shapes can be represented as single objects, which means the particles within a clump may overlap to any extent, but contact forces are not generated between these particles. The mineral grain is internally equivalent to a rigid body, and there is only a slight deformation at the contact boundary between the mineral grains. Considering the complexity of real mineral grains, it is difficult to simulate each mineral grain shape. Therefore, some simple grain shapes were adopted in this study. As shown in Figure 3, three mineral grain groups of different shapes were constructed: strip, triangle, and square. There were six grains of similar shapes in each group.

1. The strip-shaped grain was composed of two circular particles of the same radius R. The six grains in this group were represented as S1–S6. From S1 to S6, the distance d between the centers of the two circular particles gradually increased, and the ratios of d to R were 0, 0.25, 0.5, 1.0, 1.25, and 1.5, respectively.
2. The triangular grain was composed of three circular particles of the same radius R. The line connecting the centers of the three circular particles was an equilateral triangle with a side length of d. The six grains in this group were represented as T1–T6. From T1 to T6, the value of d was gradually increased. The variation of d/R was consistent with the strip-shaped grains.
3. The square grain was composed of four circular particles of the same radius R. The line connecting the centers of the four circular particles was a square with a side length of d. The six grains in this group were represented as R1–R6. From R1 to R6, the value of d was gradually increased. The variation of d/R was also consistent with the strip-shaped grains.

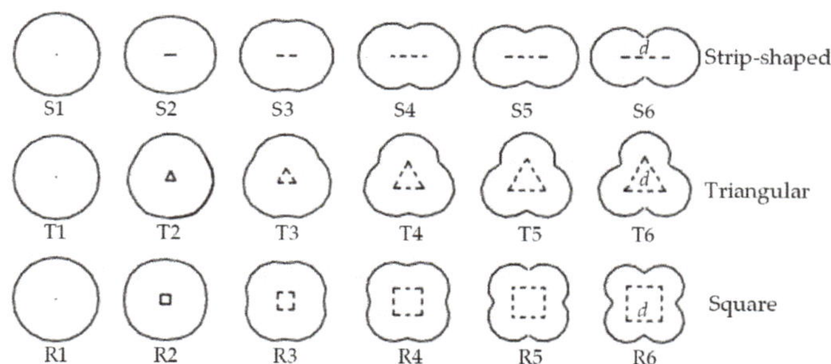

Figure 3. The construction of three particle shapes using the CLUMP method.

Taking the strip-shaped grain as an example, the circular particle S1 could be directly generated, but grains S2–S6 needed to be generated by the CLUMP method of PFC^{2D}. In the CLUMP method, the circular particles were first created, then the circular particles were replaced with grains S2–S6. The replacements followed three principles: "area equivalence", "area center equivalent", and "directional randomization". This means that the area and centroid of grains S2–S6 were the same as grain S1, but their orientations were random.

3.2. Quantitative Description of Grain Shape

For an irregular grain, the shape can be described from two aspects [14]. First is the overall contour, which can be simply described as circular, square, plate, or column. The second aspect is the surface morphology, which refines the grain outline based on the first aspect. It focuses on the irregularity of the grain surface or the fluctuation of the grain boundary. Flatness was adopted in this study to quantify the overall contour of the grain, and the surface morphology of the grain was described by roughness [28,29]. As shown in Equation (1), flatness is the ratio of the maximum Feret's diameter of the grain to the minimum Feret's diameter. Feret's diameter means the vertical distance between two parallel lines along the grain boundaries, as shown in Figure 4. Flatness characterizes the elongation property of the grain with a minimum value of 1. The grain approaches are spherical or circular when this value is closer to 1. On the contrary, the larger the flatness value, the flatter and narrower the grain. As shown in Equation (2), roughness is defined as the square of the ratio of two perimeters, which are the perimeter of the grain itself and the perimeter of the smallest circumscribing polygon of the grain (Figure 4). Roughness characterizes the fluctuation of the grain boundary curve, and greater roughness means a more irregular surface morphology of the grain.

$$e = L/B; \tag{1}$$

$$r = (P/PC)^2; \tag{2}$$

where e is the flatness, and L and B are the maximum and minimum Feret's diameters, respectively. Roughness is expressed by r, P is the perimeter of the grain, and Pc is the perimeter of the smallest circumscribing polygon of the grain.

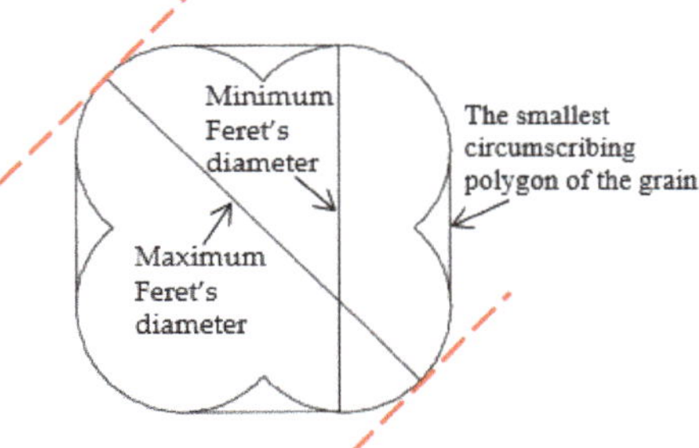

Figure 4. Basic dimension parameters of irregular grains used for calculating flatness and roughness.

According to Equations (1) and (2), the flatness and roughness of the three mineral grain groups mentioned above were calculated, and the results are shown in Figure 5. As can be seen from Figure 5a, the flatness of the strip-shaped grain group was significantly higher than that of the triangular and square grain groups. In the strip-shaped grain group, from S1 to S6, the flatness increased rapidly with the increase of d/R. The flatness of the square grain group was slightly higher than that of the triangular grain group, and the flatness increased slightly from T1 to T6 and R1 to R6. It can be seen from Figure 5b that the square grain group had the highest surface roughness, followed sequentially by the triangular and the strip-shaped grain group. With the increase of d/R, the roughness of the three grain groups all increased. When the d/R value was low (shape numbers 1–3), an increase in the roughness was not obvious; however, it increased faster when the d/R value was higher (shape numbers 4–6).

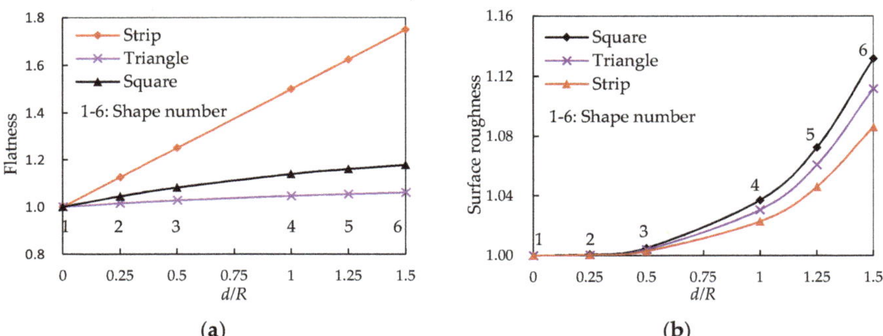

Figure 5. Shape factor variation of three representative grain groups: (**a**) flatness; (**b**) roughness.

3.3. Calibration of the Microparameters

The microparameters used in PFC^{2D} cannot be directly acquired from experimental laboratory tests, which need to be calibrated using a trial-and-error procedure. In this procedure, the microparameters needed to be optimized continuously until the simulated macro parameters matched the experimental results. The macro parameters used for calibration generally include the elastic modulus, Poisson's ratio, uniaxial compressive strength (UCS), and tensile strength. As shown in Figure 6, the circular-shaped grain was used for microparameter calibration, and the model was 50 mm in width and 100 mm in height, with the particle radii ranging from 0.5 to 0.83 mm following a normal distribution. The walls were used to apply a stress-boundary condition, and the wall velocity was controlled by a numerical servo-control mechanism in order to maintain a specified wall stress.

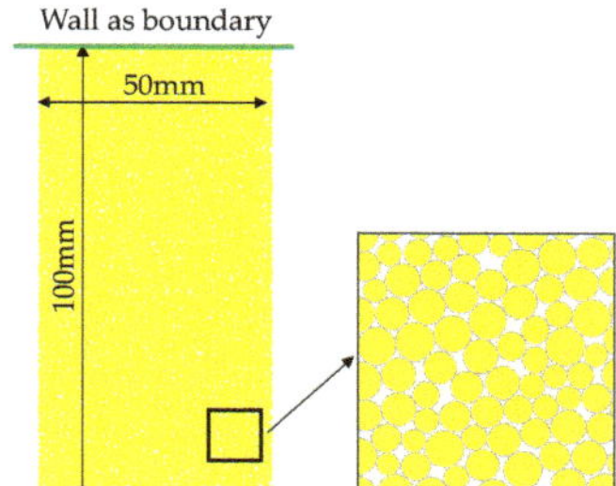

Figure 6. Numerical model for the uniaxial compression simulation.

In this study, the microparameters were calibrated to match the macroproperties of the uniaxial compression test of Lac du Bonnet (LDB) granite, including the elastic modulus, peak strength, and Poisson's ratio, which were 70 GPa, 224 MPa, and 0.26, respectively [9]. LDB granite is a representative brittle rock. Because of its comprehensive rock mechanics parameters, it is often used for model validation [9,18,30]. After a trial-and-error process, the elastic modulus, peak strength, and Poisson's ratio of the calibration model were determined to be 70 GPa, 207 MPa, and 0.26, respectively, which were consistent with the corresponding experimental results. The mesoscopic physical and mechanical parameters of the model are shown in Table 1. As can be seen, it includes the micromechanical parameters of the particles and the parallel bond between them. According to the existing literature [31,32] and our calibration process, the macro elastic modulus of the model was controlled by the contact modulus, and Poisson's ratio was affected by the ratio of normal stiffness to shear stiffness. The strength of the model was mainly decided by the tensile and shear strengths of the bond.

Table 1. Micro-physico-mechanical parameters of the particle and parallel bond.

Classification	Microparameters	Notations	Values
Particle	Density (kg/m^3)		2630
	Contact modulus (GPa)	E	62
	Normal-to-shear stiffness ratio	k_n/k_s	2.5
	Friction coefficient	u	0.5
Parallel bond between particles	Tensile strength (MPa)	σ_c	157
	Standard deviation /MPa	σ_{cs}	36
	Normal-to-shear stiffness ratio	$\overline{k_n}/\overline{k_s}$	2.5
	Shear strength (MPa)	τ_c	175
	Standard deviation /MPa	τ_{cs}	40
	Modulus (GPa)	E_c	62
	Radius multiplier	λ	1

4. Analysis of Grain Shape Effects

A series of numerical models were established based on the grain shapes shown in Figure 3. The microparameters shown in Table 1 were adopted to conduct uniaxial compression simulations. According to the results, the influences of grain shape on the macroscopic parameters and failure mode of rock were analyzed.

4.1. Effects of Grain Shape on Rock Macroscopic Parameters

Figure 7 shows the effects of grain shape on rock peak strength and elastic modulus. It can be seen that the peak strength and elastic modulus of the model with the strip-shaped grain was the largest, followed by the triangular grain and square grain models. Compared with the circular grain (shape number 1), the elastic modulus of the models with clumped grains was significantly increased (shape numbers 2–6). According to our simulated results, the effects of grain shape on rock peak strength and elastic modulus were closely related to the model's porosity. Figure 8 shows the effects of grain shape on porosity. It can be seen that the porosity of the square grain model was the largest, followed by the triangular grain and strip-shaped grain models, contrary to the variation of peak strength and elastic modulus.

Numerous experiments have shown that increased porosity can reduce rock strength and elastic modulus [33–35]. Hudyma et al. [36] studied the mechanical properties of porous tuff and found that the compressive strength and elastic modulus of tuff decreased with increasing porosity. Al-Harthi et al. [37] studied the effect of pores on the mechanical properties of basalt based on image analysis techniques and came to a similar conclusion as Hudyma et al. Using PFC^{3D} simulations, Schopfer et al. [38] found that rock strength and elastic modulus decreased with increasing porosity. Therefore, the macroscopic mechanical parameters decreased with the increase of the porosity in this study. It should be noted that the porosity of a model in PFC is a set value, which was set to 0.16 in our simulation. However, in the process of adjusting the model's internal stress, the porosity of the model will change until the model reaches an isotropic stress state, especially when the grain shape is complicated. It can be seen from Figure 8 that the actual porosity of the model was approximately 0.16 when the grain shape was circular (shape number 1), which was in accordance with the set value. However, when clumped grains were used, the actual porosity of the model varied greatly from the set value, which indicated that the grain shape affected the model porosity and further affected rock elastic modulus and peak strength.

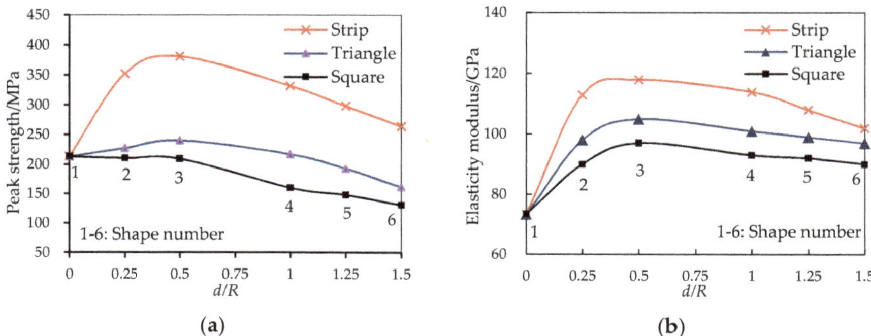

Figure 7. Effects of grain shape on rock (**a**) peak strength and (**b**) elastic modulus.

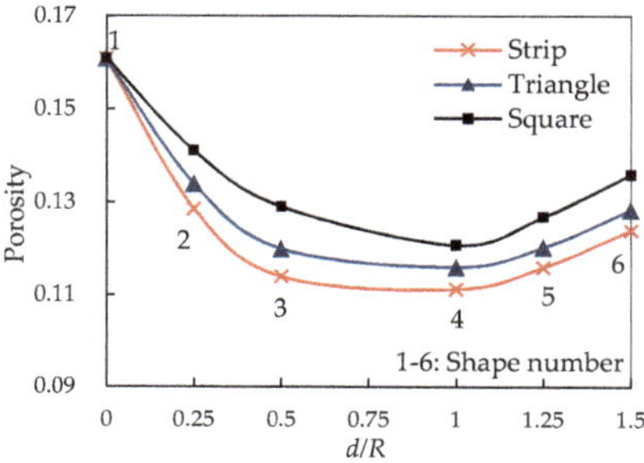

Figure 8. Effects of grain shape on rock porosity.

It can also be seen from Figure 7a that with the increase of the d/R value, the peak strength of the models with strip-shaped and triangular grains first increased and then decreased, but it only decreased when the grain shape was square. This variation was closely related to the flatness and roughness of the mineral grain. Firstly, as can be seen in Figure 5a, the flatness of the strip-shaped grain was obviously higher than that of the triangular and square grains. Also, the peak strength of model with strip-shaped grains was significantly higher, indicating that an increase in flatness can enhance rock peak strength, which was mainly because the strip-shaped grains were difficult to rotate and displace when they were assembled together. Zhang et al. [39] also thought that a model with strip-shaped grains would have higher strength. Secondly, because the flatness of the square and triangular grains was significantly lower (Figure 5a), the overall contour was close to circular. Therefore, the peak strength was slightly affected by flatness in the models with square and triangular grains. However, the grain surface roughness was relatively higher, and the main influencing factor affecting the peak strength was roughness. As can be seen from Figure 7a, the peak strength decreased with increasing roughness. Therefore, the variations of rock peak strength shown in Figure 7a were due to the combined effects of mineral flatness and roughness. The effects of shape factors on the peak strength of models with different grain shapes are discussed in detail as follows:

1. Strip-shaped grain

It can be seen from Figure 5 that when the d/R value was low, the surface roughness of the strip-shaped grain increased slowly, but the flatness increased rapidly. Therefore, the obvious increase of peak strength from models S1 to S3 was mainly controlled by the flatness. When the d/R value became higher, the roughness increased rapidly, and the peak strength decreased significantly, indicating that the peak strength was mainly affected by the roughness from models S4 to S6. It can be concluded that when the roughness exceeds a certain limit, it will be the main factor of strength change.

2. Triangular grain

When the *d/R* value was low (shape numbers T1–T3), similar to the model with the strip-shaped grain, the increase of peak strength of models with triangular grains was also controlled by the flatness. However, because the increase of the flatness of triangular grain was much lower than that of the strip-shaped grain, the increase of peak strength of models with triangular grains was also obviously lower than that of the models with strip-shaped grains. When the *d/R* value was higher (shape numbers T4–T6), the decrease of the peak strength was affected by the increasing roughness.

3. Square grain

For the models with square grains, the decrease of the peak strength was controlled by the roughness. The roughness increased slowly from models R1 to R3, which led to the slow decrease of the peak strength. The increase of the roughness was accelerated from models R4 to R6; thus, the decrease of the peak strength was also significantly reduced.

In addition, it was found that the grain shape also had effects on the cohesion and internal friction angle of the model. These are two important parameters for characterizing the mechanical properties of mineral grains. Generally, the greater grain surface roughness means a higher internal friction angle. In bulk materials such as sand, samples with complex grain shapes generally have higher strength, which is mainly because the interlocking between mineral grains can increase the internal friction angle [12,13,40]. In other words, the strength of the bulk material is positively correlated with the internal friction angle. However, in rock materials, the strength of rock was found to be positively correlated with cohesion. Taking the model with circular grains (shape number 1) as an example, as shown in Figure 9, the peak compressive strengths of the model under three confining pressures (0, 50, and 200 MPa) were obtained. Three ultimate stress circles and their common tangent line were further obtained according to the confining pressure and the corresponding peak compressive strength. The intersection of the common tangent line and the τ axis was the cohesion of the model. Figure 10 shows the effects of grain shape on rock cohesion. It can be seen that the effects of grain shape on rock cohesion were similar to the effects on rock uniaxial peak strength. In PFC models, the mineral grains are bonded together, and from a macroscopic view, the cohesion controls the strength of the model. Here, the peak strength of the rock increased with the increase of the cohesion.

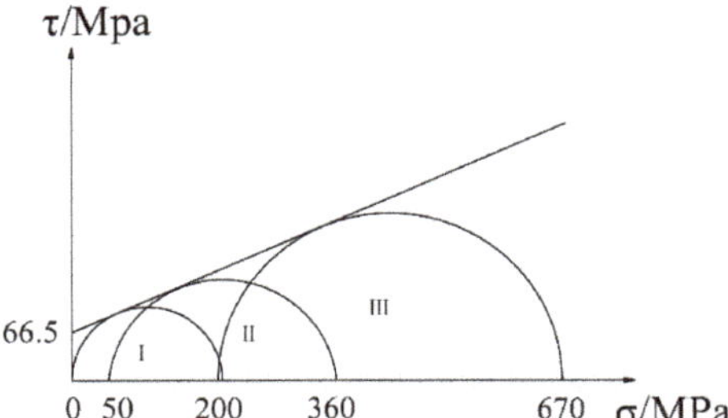

Figure 9. Cohesion of the model with circular grains (shape number 1).

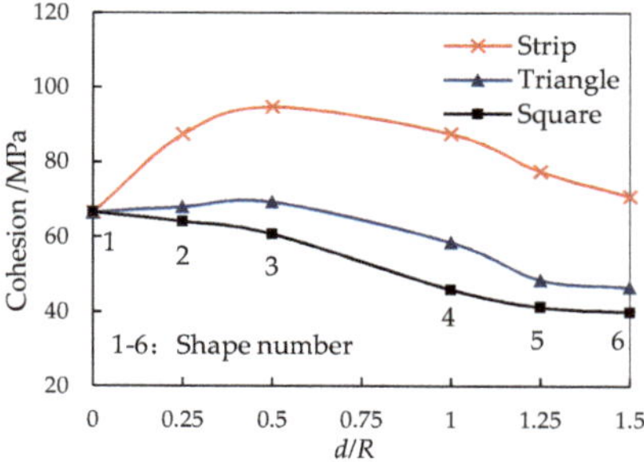

Figure 10. Effects of grain shape on rock cohesion.

4.2. Effects of Grain Shape on Rock Failure Mode

The macroscopic failure of rock is a gradual evolutionary process of internal microcracks, and the microcracking behavior has a great influence on the ultimate failure mode of rock. Figure 11 presents the spatial distributions of microcracks for numerical models with different grain shapes when the models reached the postpeak stage and the axial stress was 80% of the peak strength. The generated microcracks can be divided into two types—tensile and shear cracks—represented by black and red segments, respectively. A large number of previous studies [41–43] have shown that tensile failure plays a dominant role in rock failure mechanisms. In this study, the microcracks were dominated by tensile cracks (Table 2), accounting for more than 80% of the total number of cracks. It can also be seen from Table 2 that the percentage of tensile cracks was maximal in the model with square grains, followed by the triangular grain and strip-shaped grain models.

Table 2. Percentage of tensile cracks of rock with three particle shapes in postpeak stage $\sigma = 0.8\ \sigma_c$.

Ratio of Tensile Crack (%)	Shape Number					
	1	**2**	**3**	**4**	**5**	**6**
Strip-shaped grain	88	87	83	85	79	83
Triangular grain	88	92	90	92	92	90
Square grain	88	92	95	93	93	91

Figure 12 shows the simulated stress–strain curves of the models with triangular grains (taking the models with shape numbers T3, T4, and T6 as examples). It can be seen that the stress dropped after the peak strength gradually became slower from T3 to T6. This can be explained by the fact that from T3 to T6, the surface roughness of the grain increased, and the irregularity of the grain shape increased. In the pre-peak stage, the mineral grains were cemented together. The bond damage increased after the rock reached the peak strength, then the number of microcracks increased rapidly and formed macrocracks. The mineral grains on both sides of the crack began to move around each other along the grain surface and produced significant displacement. In mineral grains with a high surface roughness, the interlocking effect and friction between grains were stronger, which made samples have a higher residual strength and resulted in slower stress drop. Figure 13 shows the stress–strain curves of the models with three grain shapes (taking the models with shape numbers S6, T6, and R6 as examples). As can be seen, from the strip-shaped grain to the square grain, the surface roughness increased, and the postpeak stress drop of the three models also became slower, which further confirmed the influence of grain shape on rock characteristics after peak strength.

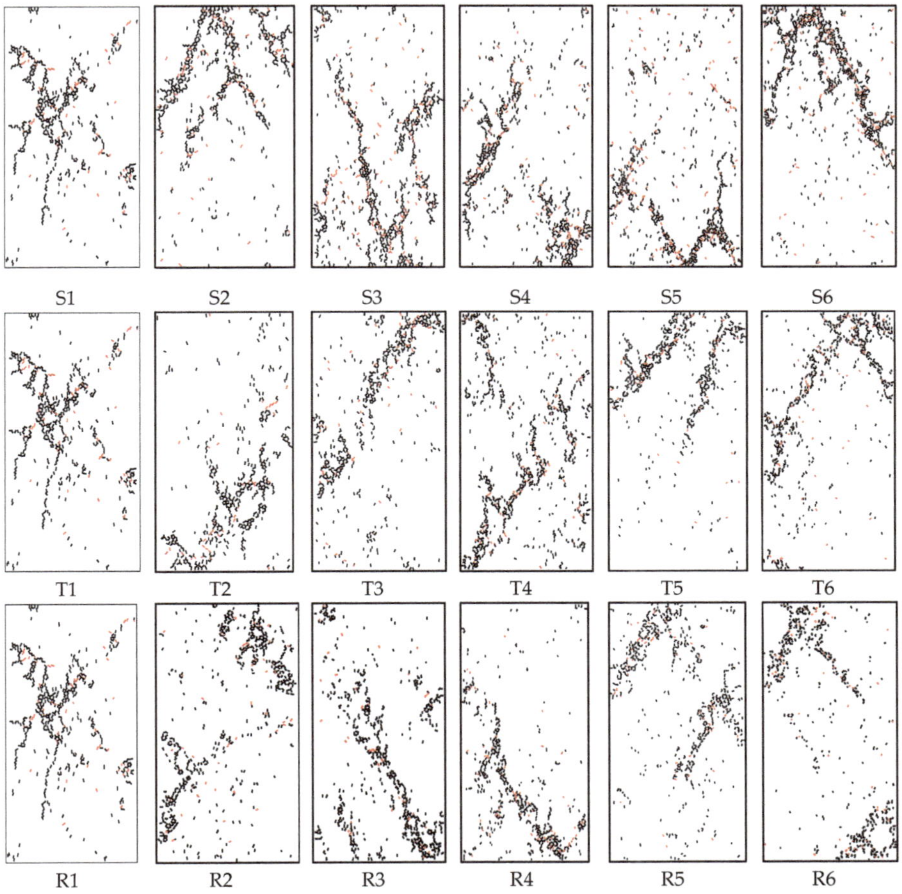

Figure 11. Failure mode of rock with three grain shapes in the postpeak stage when $\sigma = 0.8\ \sigma_c$ (black segments represent tensile cracks and red segments represent shear cracks).

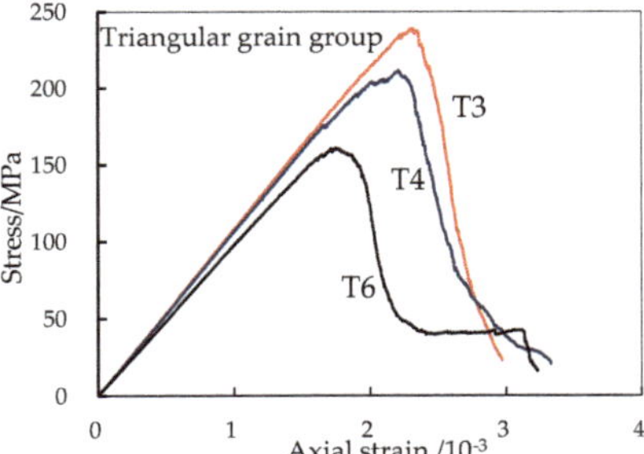

Figure 12. Stress–strain curves of different triangular grain models (case study of models T3, T4, and T6).

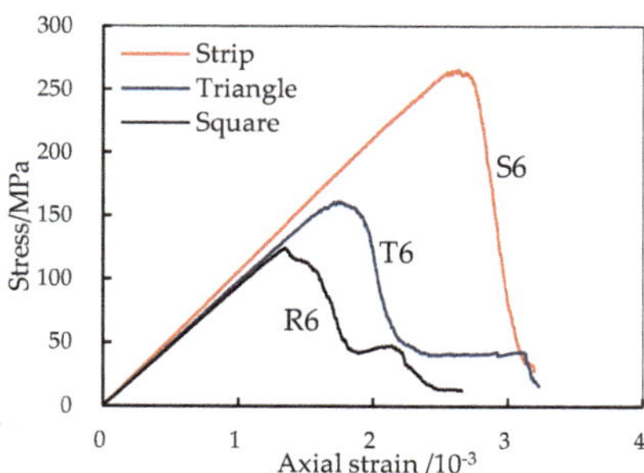

Figure 13. Stress–strain curves of three different grain shape models (case study of model with shape number 6).

According to the above analysis, the PFC model was composed of grains by cementation, and the failure of the model was essentially the break of the intergrain bond. Hence, the peak strength of the model was macroscopically controlled by cohesion. After the axial stress reached the peak strength of the model, obvious displacement started to appear along the failure surface. The interlocking between grains played a major role, the relative sliding between the grains became difficult with the increasing surface roughness, and the macroresponse was that the stress drop became slower.

5. Conclusions

In this study, the effects of mineral grain shape on the mechanical properties of LDB granite were investigated by means of uniaxial compression simulations. Based on the CLUMP method in PFC^{2D}, three different grain shapes were constructed: strip, triangle, and square. The relationship between the grain shapes and rock mechanical properties was analyzed, and the main conclusions are as follows:

1. The elastic modulus and peak strength of rock are affected by the mineral grain shape. In our study, the model with strip-shaped grains had the largest elastic modulus and peak strength, followed by the triangular grain and square grain models. The mechanism of grain shape effect on the peak strength and elastic modulus was mainly achieved by affecting the porosity of the model. The increasing porosity led to a smaller peak strength and elastic modulus.

2. Flatness and roughness can describe the overall contour and surface morphology of mineral grains well. The effects of grain flatness and surface roughness on rock peak strength are the opposite. A flat grain helps to increase rock strength, while rock strength is reduced due to increased porosity if the grain has higher roughness. Further, the relationship between the peak strength and the cohesion of rock is positively correlated.

3. Here, the generated microcracks were mainly of the tension type, and the model with square grains had the maximum ratio of tensile cracking, followed by the triangular grain and strip-shaped grain models. The shape of the mineral grain also had a significant effect on rock mechanical behaviors at the postpeak stage. The stress drop became slower with an increasing surface roughness because the interlocking restrained the slip and rotation of grains on the fracture surface.

However, there were some defects in the modeling. The grain shapes adopted in this study were simple and single, while real mineral grain shapes are very complex. Since the CLUMP method can realize the construction of complex grain shapes, future works should connect this study to real rocks. In addition, three-dimensional grain shape effects should also be investigated in future studies.

Author Contributions: Z.H. and L.Z. designed the theoretical framework; Z.H. wrote the basic code of the program and designed the numerical simulation; J.Z. checked the simulation results.

Funding: This research was funded by the National Natural Science Foundation of China (Grant Nos. 41572312, 41672321), China Postdoctoral Science Foundation (Grant Nos. 2018M630204, 2019T120133), and China Scholarship Council (Grant No. 201804910293) to the third author.

Conflicts of Interest: The authors declare no conflict of interest.

References

1. Han, Z.H.; Zhou, J.; Zhang, L.Q. Influence of Grain Size Heterogeneity and In-Situ Stress on the Hydraulic Fracturing Process by PFC 2D Modeling. *Energies* **2018**, *11*, 1413. [CrossRef]
2. Lajtai, E.Z. Brittle fracture in compression. *Int. J. Fra.* **1974**, *10*, 525–536. [CrossRef]
3. Nicksiar, M.; Martin, C.D. Evaluation of Methods for Determining Crack Initiation in Compression Tests on Low-Porosity Rocks. *Rock Mech. Rock Eng.* **2012**, *45*, 607–617. [CrossRef]
4. Basu, A.; Mishra, D.A.; Roychowdhury, K. Rock failure modes under uniaxial compression, Brazilian, and point load tests. *Bull. Eng. Geol. Environ.* **2013**, *72*, 457–475. [CrossRef]
5. Koyama, T.; Jing, L. Effects of model scale and particle size on micro-mechanical properties and failure processes of rocks-A particle mechanics approach. *En. Anal. Bound. Elem.* **2007**, *31*, 458–472. [CrossRef]
6. Hosseininia, E.S. Erratum to: Investigating the micromechanical evolutions within inherently anisotropic granular materials using discrete element method. *Granular Matter.* **2012**, *14*, 483–503. [CrossRef]
7. Rong, G.; Liu, G.; Hou, D.; Zhou, C.B. Effect of Particle Shape on Mechanical Behaviors of Rocks: A Numerical Study Using Clumped Particle Model. *Sci. World J.* **2013**, *2013*, 589215. [CrossRef]
8. Wiącek, J.; Molenda, M.; Horabik, J.; Ooi, J.Y. Influence of grain shape and intergranular friction on material behavior in uniaxial compression: Experimental and DEM modeling. *Powder Technol.* **2012**, *217*, 435–442. [CrossRef]
9. Lan, H.X.; Martin, C.D.; Hu, B. Effect of heterogeneity of brittle rock on micromechanical extensile behavior during compression loading. *J. Geophys. Res-Sol. Ea.* **2010**, *115*, 414–431. [CrossRef]
10. Manouchehrian, A.; Cai, M. Influence of material heterogeneity on failure intensity in unstable rock failure. *Comput. Geotech.* **2016**, *71*, 237–246. [CrossRef]
11. Peng, J.; Wong, L.N.Y.; The, C.I. Influence of grain size heterogeneity on strength and microcracking behavior of crystalline rocks. *J. Geophys. Res-Sol. Ea.* **2017**, *122*, 1054–1073. [CrossRef]
12. Shinohara, K.; Oida, M.; Golman, B. Effect of particle shape on angle of internal friction by triaxial compression test. *Powder Technol.* **2000**, *107*, 131–136. [CrossRef]
13. Härtl, J.; Jin, Y.O. Numerical investigation of particle shape and particle friction on limiting bulk friction in direct shear tests and comparison with experiments. *Powder Technol.* **2011**, *212*, 231–239. [CrossRef]
14. Liu, Q.B.; Xiang, W.; Budhu, M.; Cui, D.-S. Study of particle shape quantification and effect on mechanical property of sand. *Rock Soil Mech.* **2011**, *32*, 190–197.
15. Johanson, K. Effect of particle shape on unconfined yield strength. *Powder Technol.* **2009**, *194*, 246–251. [CrossRef]
16. Cundall, P.A. A computer model for simulating progressive, large-scale movements in blocky rock systems. *Proc. Int. symp. on Rock Fracture* **1971**, *2*, 2–8.
17. Potyondy, D.O.; Cundall, P.A. A Bonded-Particle Model for Rock. *Int. J. Rock Mech. Min.* **2004**, *41*, 1329–1364. [CrossRef]
18. Potyondy, D.O. The bonded-particle model as a tool for rock mechanics research and application: Current trends and future directions. *Geo. Eng.* **2015**, *18*, 1–28. [CrossRef]
19. Zhou, J.; Lan, H.X.; Zhang, L.Q.; Yang, D.X.; Song, J.; Wang, S. Novel Grains Bonded Model for Simulation of Brittle Failure of Alxa Porphyritic Granite. *Eng. Geo.* **2019**, *251*, 100–114. [CrossRef]
20. Zhou, J.; Zhang, L.Q.; Yang, D.X.; Braun, A.; Zhenhua, H. Investigation of the Quasi-Brittle Failure of Alashan Granite Viewed from Laboratory Experiments and Grain-Based Discrete Element Modeling. *Materials* **2017**, *10*, 835. [CrossRef]
21. Cho, N.; Martin, C.D.; Sego, D.C. A clumped particle model for rock. *Int. J. Rock Mech. Min.* **2007**, *44*, 997–1010. [CrossRef]
22. Santamarina, J.C.; Cho, G.C. Soil behaviour: The role of particle shape. In Proceedings of the Skempton Conference, London, UK, 29–31 March 2004.
23. Shi, D.D.; Zhou, J.; Liu, W.B.; Jia, M.C. Numerical simulation for behaviors of sand with non-circular particles under monotonic shear loading. *Chinese J. Geotech. Eng.* **2008**, *30*, 1361–1366.
24. Shi, D.D.; Zhou, J.; Liu, W.B.; Deng, Y.B. Exploring macro- and micro-scale responses of sand in direct shear tests by numerical simulations using non-circular particles. *Chinese J. Geotech. Eng.* **2010**, *32*, 1557–1565.

25. Kong, L.; Peng, R. Particle flow simulation of influence of particle shape on mechanical properties of quasi-sands. *Chinese J. Rock Mech. Eng.* **2011**, *30*, 2112–2118.

26. Kerimov, A.; Mavko, G.; Mukerji, T.; Dvorkin, J.; Al Ibrahim, M.A. The influence of convex particles irregular shape and varying size on porosity, permeability, and elastic bulk modulus of granular porous media: Insights from numerical simulations. *J. Geophys. Res-Sol. Ea.* **2018**, *123*, 10563–10582. [CrossRef]

27. Kern, H.; Ivankina, T.I.; Nikitin, A.N.; Lokajicek, T.; Pros, Z. The effect of oriented microcracks and crystallographic and shape preferred orientation on bulk elastic anisotropy of a foliated biotite gneiss from Outokumpu. *Tectonophysics.* **2008**, *457*, 143–149. [CrossRef]

28. Mark, L.H.; Neil, W.P. Selection of Descriptors for Particle Shape Characterization. *Part. Part. Sys. Char.* **2010**, *20*, 25–38.

29. Mora, C.; Kwan, A. Sphericity, shape factor, and convexity measurement of coarse aggregate for concrete using digital image processing. *Cement Concrete Res.* **2000**, *30*, 351–358. [CrossRef]

30. Hazzard, J.F.; Young, R.P.; Maxwell, S.C. Micromechanical modeling of cracking and failure in brittle rocks. *J. Geophys. Res-Sol. Ea.* **2000**, *105*, 16683–16697. [CrossRef]

31. Weng, M.C.; Li, H.H. Relationship between the deformation characteristics and microscopic properties of sandstone explored by the bonded-particle model. *Int. J. Rock Mech. Min. Sci.* **2012**, *56*, 34–43. [CrossRef]

32. Yoon, J. Application of experimental design and optimization to PFC model calibration in uniaxial compression simulation. *Int. J. Rock Mech. Min. Sci.* **2007**, *44*, 871–889. [CrossRef]

33. Vernik, L.; Bruno, M.; Bovberg, C. Empirical relations between compressive strength and porosity of siliciclastic rocks. *Int. J. Rock Mech. Min.* **1993**, *30*, 677–680. [CrossRef]

34. Palchik, V. Influence of porosity and elastic modulus on uniaxial compressive strength in soft brittle porous sandstones. *Rock Mech. Rock Eng.* **1999**, *32*, 303–309. [CrossRef]

35. Bell, F.G.; Lindsay, P. The petrographic and geomechanical properties of some sandstones from the Newspaper Member of the Natal Group near Durban, South Africa. *Eng. Geo.* **1999**, *53*, 57–81. [CrossRef]

36. Hudyma, N.; Avar, B.B.; Karakouzian, M. Compressive strength and failure modes of lithophysae-rich Topopah Spring Tuff specimens and analog models containing cavities. *Eng. Geo.* **2004**, *73*, 179–190. [CrossRef]

37. Al-Harthi, A.A.; Al-Amri, R.M.; Shehata, W.M. The porosity and engineering properties of vesicular basalt in Saudi Arabia. *Eng. Geo.* **1999**, *54*, 313–320. [CrossRef]

38. Schopfer, M.P.J.; Abe, S.; Childs, C.; Walsh, J.J. The impact of porosity and crack density on the elasticity, strength and friction of cohesive granular materials: Insights from DEM modelling. *Int. J. Rock Mech. Min.* **2009**, *46*, 250–261. [CrossRef]

39. Zhang, Z.; Shu, G.P. Effect of particle shape on biaxial tests simulated by particle flow code. *Chinese J. Geotech. Eng.* **2009**, *31*, 1281–1286.

40. Cho, G.C.; Dodds, J.; Santamarina, J.C. Particle Shape Effects on Packing Density, Stiffness, and Strength: Natural and Crushed Sands. *J. Geotech. Geo. Eng.* **2006**, *133*, 591–602. [CrossRef]

41. Backers, T.; Stanchits, S.; Dresen, G. Tensile fracture propagation and acoustic emission activity in sandstone: The effect of loading rate. *Int. J. Rock Mech. Min.* **2005**, *42*, 1094–1101. [CrossRef]

42. Brace, W.F.; Bombolakis, E.G. A note on brittle crack growth in compression. *J. Geophys. Res.* **1963**, *68*, 3709–3713. [CrossRef]

43. Diederichs, M.S.; Kaiser, P.K.; Eberhardt, E. Damage initiation and propagation in hard rock during tunneling and the influence of near-face stress rotation. *Int. J. Rock Mech. Min.* **2004**, *41*, 785–812. [CrossRef]

Article

A Numerical Evaluation of SIFs of 2-D Functionally Graded Materials by Enriched Natural Element Method

Jin-Rae Cho

Department of Naval Architecture and Ocean Engineering, Hongik University, Jochiwon, Sejong 30016, Korea; jrcho@hongik.ac.kr; Tel.: +82-44-860-2546

Received: 29 July 2019; Accepted: 25 August 2019; Published: 1 September 2019

Featured Application: Prediction of crack propagation of functionally graded materials (FGMs).

Abstract: This paper presents the numerical prediction of stress intensity factors (SIFs) of 2-D inhomogeneous functionally graded materials (FGMs) by an enriched Petrov-Galerkin natural element method (PG-NEM). The overall trial displacement field was approximated in terms of Laplace interpolation functions, and the crack tip one was enhanced by the crack-tip singular displacement field. The overall stress and strain distributions, which were obtained by PG-NEM, were smoothened and improved by the stress recovery. The modified interaction integral $\widetilde{M}^{(1,2)}$ was employed to evaluate the stress intensity factors of FGMs with spatially varying elastic moduli. The proposed method was validated through the representative numerical examples and the effectiveness was justified by comparing the numerical results with the reference solutions.

Keywords: enhanced PG-NEM; functionally graded material (FGM); stress intensity factor (SIF); modified interaction integral

1. Introduction

In the late 1980s, a new material concept called functionally graded material (FGM) was proposed to resolve the inherent problem of traditional lamination-type composites [1]. The sharp material discontinuity across the layer interface causes the stress concentration, which may trigger the layer delamination. This crucial stress concentration can be significantly minimized by inserting a graded layer between two distinct homogeneous material layers [2,3]. This is because the material discontinuity completely disappears according to the material composition gradation, where the constituent particles of two base materials are mixed up in a random microstructure within a graded layer to maximize the desired performance [4–6]. Naturally, a functionally graded material is an inhomogeneous material, with spatially non-uniform material properties characterized by continuity and functionality. In addition to the suppression of stress concentration, the material concept of FGM rapidly and continuously spread throughout engineering and fields [7–10].

Early research efforts were concentrated on material characterization, fabrication, modeling, and analysis [1,11,12]. This was because the mechanical behaviors of FGMs are governed by the geometric dimensions and orientation, microstructure, and volume fractions of constituent particles. Recently, however, the concern toward the crack problems has increased because the structural failure of FGMs is dominated by micro-cracking [7,13,14]. In this regard, an accurate numerical prediction of stress intensity factors and the crack propagation has been an essential research subject [15,16]. For these subjects, an accurate reproduction of the $\frac{1}{\sqrt{r}}$ singularity in the near-tip stress field in highly heterogeneous media becomes a key task [17–20].

To evaluate the stress intensity factors of FGMs with cracks, one can consider the use of the well-known $J-$ or $M-$integral methods. However, these conventional indirect integral methods cannot reflect the spatially varying material properties of FGMs. The studies on the fracture mechanics of inhomogeneous bodies were initiated in the 1970~80s by assuming the spatially varying elastic modulus as an exponential function [17,21]. The standard integral methods were modified and/or refined to reflect the spatial non-uniformity of material properties by subsequent investigators. The most works were made by utilizing the finite element method [15,22–25]. However, since late 1990s, the employment of meshless methods to crack problems of inhomogeneous bodies has been actively advanced, particularly for the computation of SIFs by the modified $M-$integral method [26,27]. Here, the extension of the natural element method (NEM) is worth noting, even though it was restricted to 2-D homogenous material [28].

The natural element method was introduced to overcome the demerits of conventional meshless methods [29], the difficulty in enforcing the displacement constraint and the numerical integration. The Laplace interpolation functions in NEM strictly obey the Kronecker delta property so that the imposition of displacement constraint becomes easy. In addition, Delaunay triangles, which were produced during the process for introducing Laplace interpolation functions, also serve as a background mesh for the numerical integration. In particular, PG-NEM can further improve the numerical integration accuracy by maintaining the consistency between the Delaunay triangle and the integration region [30]. Although Laplace interpolation functions provide the high smoothness of C^1-continuity, there is still room for further improvement in capturing the high stress singularity at the crack tip.

In this context, this paper introduces an enriched PG-NEM to explore whether and how much the enrichment of interpolation function increases the prediction reliability of stress intensity factors for FGMs. The validity of enrichment was reported for homogeneous materials [31,32], but it was rarely reported for inhomogeneous materials. The trial function is enriched by the asymptotic displacement fields of mode I and II, and the approximated overall stress field is enhanced by the patch recovery technique. The proposed method was validated through the illustrative numerical examples and its effectiveness was quantitatively evaluated.

2. 2-D Inhomogeneous Cracked Bodies

2.1. Linear Elasticity of 2-D Cracked Bodies

Figure 1 represents a 2-D linearly elastic isotropic inhomogeneous material with an edge crack which is contained within a bounded domain $\Omega \in \mathfrak{R}^2$ with the boundary $\partial\Omega = \overline{\Gamma_D \cup \Gamma_N \cup \Gamma_c}$. Here, Γ_D and Γ_N indicate the displacement and force boundary regions, and $\Gamma_c = \overline{\Gamma_c^+ \cup \Gamma_c^-}$ denotes the traction-free crack surface. As a representative non-homogeneous material, FGMs are characterized by the spatially varying elastic modulus E and Poisson's ratio v over the bounded domain Ω. For the mathematical description purpose, two Cartesian co-ordinate systems are employed, $\{x, y\}$ for the 2-D linear elasticity problem and $\{x', y'\}$ for the SIF evaluation of crack respectively. Assuming the crack surface is traction-free and neglecting the body force b, then the displacement field $u(x)$ in the Cartesian coordinate system $\{x, y\}$ is subjected to the equilibrium equations

$$\nabla \cdot \sigma = 0 \quad in \quad \Omega \tag{1}$$

with the displacement constraint

$$u = \hat{u} \quad on \quad \Gamma_D \tag{2}$$

and the force boundary condition

$$\sigma \cdot n = \begin{cases} \hat{t} & on \quad \Gamma_N \\ 0 & on \quad \Gamma_c^\pm \end{cases} \tag{3}$$

Here, σ are the Cauchy stresses, n the outward unit vector normal to $\partial\Omega$, and $\hat{t}$ the contour traction. When the displacement and strains are assumed to be small, the Cauchy strain ε is constituted to the Cauchy stress σ via the (3×2) gradient-like operator L such that

$$\varepsilon = \varepsilon(u) = Lu \tag{4}$$

Letting D be the constitutive tensor, the stresses and strains are constituted by

$$\sigma = D : \varepsilon \tag{5}$$

Note that the displacement, strains, and stresses are calculated based on the co-ordinate system $\{x, y\}$ and transformed into the co-ordinate system $\{x', y'\}$.

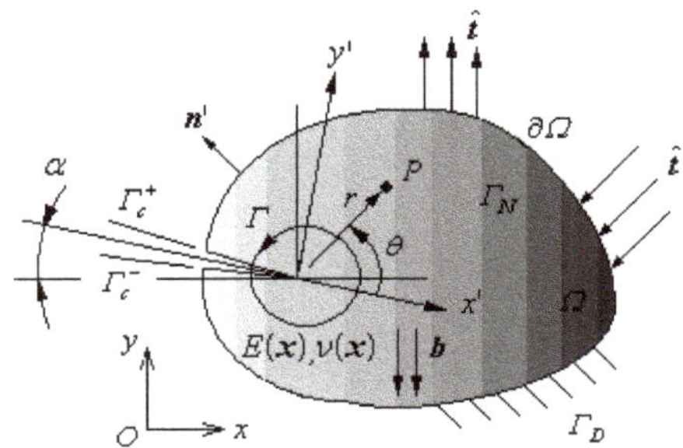

Figure 1. An inhomogeneous isotropic body with an edge crack.

For a homogeneous cracked body, the energy release rate per unit crack propagation along the x'−axis can be estimated by the path-independent J−integral defined by

$$J = \int_{\Gamma}\left(W\delta_{1j} - \sigma_{ij}\frac{\partial u_i}{\partial x'_1}\right)n'_j ds \tag{6}$$

using the indicial notation (i.e., $x'_1 = x'$ and $x'_2 = y'$), the Dirac delta function δ_{1j}, and the strain-energy density $W = \sigma \cdot \varepsilon/2 = \varepsilon_{ij}D_{ijkl}\varepsilon_{kl}/2$. Here, Γ indicates an arbitrary closed path, which surrounds the crack tip counterclockwise. As shown in Figure 2, it is expanded to $C = \Gamma + \Gamma_c^- + \Gamma_c^+ + \Gamma_o$ in order to generate a grayed donut-type region A, in which a smooth function $q(x)\,(0 \leq q \leq 1)$ is introduced to recast the integral into an equivalent domain form [33]. The function q, called by weighting function, becomes unity on Γ, zero on Γ_o, and arbitrary value between 0 and 1 within the grayed donut region A. Then, the above Equation (6) can be expanded as following

$$J = \int_A\left(\sigma_{ij}\frac{\partial u_i}{\partial x'_1} - W\delta_{1j}\right)\frac{\partial q}{\partial x'_j}dA + \int_A \frac{\partial}{\partial x'_j}\left(\sigma_{ij}\frac{\partial u_i}{\partial x'_1} - W\delta_{1j}\right)qdA \tag{7}$$

according to the divergence theorem, together with $n' = -n'_i$ on Γ in C. By further expanding the second term on the right hand side, Equation (7) becomes

$$J = \int_A\left(\sigma_{ij}\frac{\partial u_i}{\partial x'_1} - W\delta_{1j}\right)\frac{\partial q}{\partial x'_j}dA + \int_A\left(\frac{\partial\sigma_{ij}}{\partial x'_j}\frac{\partial u_i}{\partial x'_1} + \sigma_{ij}\frac{\partial^2 u_i}{\partial x'_j\partial x'_1} - \sigma_{ij}\frac{\partial\varepsilon_{ij}}{\partial x'_1} - \frac{1}{2}\varepsilon_{ij}\frac{\partial D_{ijkl}}{\partial x'_1}\varepsilon_{kl}\right)qdA \tag{8}$$

But, the second integral on the right-hand side of Equation (8) becomes zero according to the equilibrium (1), compatibility (4), and the material uniformity. Therefore, the J−integral for homogeneous materials becomes the area integral defined by

$$J = \int_A \left(\sigma_{ij} \frac{\partial u_i}{\partial x'_1} - W\delta_{1j} \right) \frac{\partial q}{\partial x'_j} dA \tag{9}$$

But, for non-homogeneous materials, the last material derivation term within the second integrand of Equation (8) does not become zero. Therefore, Equation (8) ends up with a more general $\widetilde{J}$−integral [34], which is given by

$$\widetilde{J} = \int_A \left(\sigma_{ij} \frac{\partial u_i}{\partial x'_1} - W\delta_{1j} \right) \frac{\partial q}{\partial x'_j} dA - \int_A \frac{1}{2} \varepsilon_{ij} \frac{\partial D_{ijkl}}{\partial x'_1} \varepsilon_{kl} q \, dA \tag{10}$$

The last term in Equation (10) becomes extremely small as a contour Γ_0 shrinks to the crack tip, but its contribution is not negligible when the domain of integration A is reasonably large.

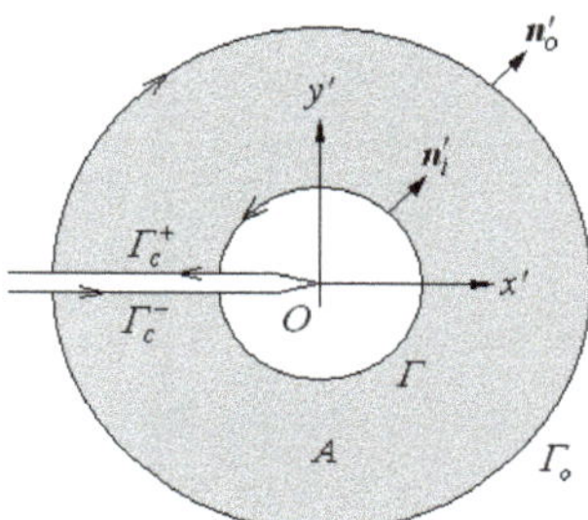

Figure 2. An extension of contour Γ and a donut-type domain of integration A.

2.2. Modified Interaction Integral $\widetilde{M}^{(1,2)}$

In order to extract K_I and K_{II} from J−integral, one needs to employ the interaction integral, in which two equilibrium states of a cracked body are considered. State 1 is the real equilibrium state of a body that is subjected to the prescribed boundary conditions, while state 2 stands for an auxiliary equilibrium state which would be the asymptotic near-tip fields for modes I or II. Another equilibrium state, called state S, could be established by combining these two states, for which the $\widetilde{J}$−integral in Equation (10) is rewritten as [21]

$$\widetilde{J}^{(S)} = \int_A \left(\left(\sigma_{ij}^{(1)} + \sigma_{ij}^{(2)} \right) \frac{\partial \left(u_i^{(1)} + u_i^{(2)} \right)}{\partial x'_1} - W^{(S)}\delta_{1j} \right) \frac{\partial q}{\partial x'_j} dA + \int_A \frac{\partial}{\partial x'_j} \left(\left(\sigma_{ij}^{(1)} + \sigma_{ij}^{(2)} \right) \frac{\partial \left(u_i^{(1)} + u_i^{(2)} \right)}{\partial x'_1} - W^{(S)}\delta_{1j} \right) q \, dA \tag{11}$$

with three different strain energy densities, $W^{(1)}$, $W^{(2)}$ and $W^{(1,2)}$, defined by

$$\begin{aligned} W^{(S)} &= \tfrac{1}{2}\left(\sigma_{ij}^{(1)} + \sigma_{ij}^{(2)} \right)\left(\varepsilon_{ij}^{(1)} + \varepsilon_{ij}^{(2)} \right) = \tfrac{1}{2}\sigma_{ij}^{(1)}\varepsilon_{ij}^{(1)} + \tfrac{1}{2}\sigma_{ij}^{(2)}\varepsilon_{ij}^{(2)} + \tfrac{1}{2}\left(\sigma_{ij}^{(1)}\varepsilon_{ij}^{(2)} + \sigma_{ij}^{(2)}\varepsilon_{ij}^{(1)} \right) \\ &= W^{(1)} + W^{(2)} + W^{(1,2)} \end{aligned} \tag{12}$$

By utilizing the equilibrium (1) (i.e., $\partial\sigma_{ij}^{(\cdot)}/\partial x_j = 0$) and the compatibility (4) (i.e., $\varepsilon_{ij}^{(\cdot)} = \partial u_i^{(\cdot)}/\partial x_j$), Equation (11) can be further simplified as

$$
\begin{aligned}
\widetilde{J}^{(S)} &= \int_A \left(\left(\sigma_{ij}^{(1)} + \sigma_{ij}^{(2)} \right) \frac{\partial\left(u_i^{(1)} + u_i^{(2)} \right)}{\partial x'_1} - \left(W^{(1)} + W^{(2)} + W^{(1,2)} \right)\delta_{1j} \right) \frac{\partial q}{\partial x'_j} dA \\
&\quad + \int_A \frac{1}{2}\left(-\varepsilon_{ij}^{(1)} \frac{\partial D_{ijkl}}{\partial x'_1}\varepsilon_{kl}^{(1)} + \sigma_{ij}^{(1)}\frac{\partial\varepsilon_{ij}^{(2)}}{\partial x'_1} - \frac{\partial\sigma_{ij}^{(2)}}{\partial x'_1}\varepsilon_{ij}^{(1)} + \sigma_{ij}^{(2)}\frac{\partial\varepsilon_{ij}^{(1)}}{\partial x'_1} - \frac{\partial\sigma_{ij}^{(1)}}{\partial x'_1}\varepsilon_{ij}^{(2)} \right) qdA \\
&= \widetilde{J}^{(1)} + \widetilde{J}^{(2)} + \widetilde{M}^{(1,2)}
\end{aligned}
\tag{13}
$$

Here,

$$
\widetilde{J}^{(1)} = \int_A \left(\sigma_{ij}^{(1)}\frac{\partial u_i^{(1)}}{\partial x'_1} - W^{(1)}\delta_{1j} \right) \frac{\partial q}{\partial x'_j}dA - \frac{1}{2}\int_A \varepsilon_{ij}^{(1)}\frac{\partial D_{ijkl}}{\partial x'_1}\varepsilon_{kl}^{(1)} qdA
\tag{14}
$$

$$
\widetilde{J}^{(2)} = \int_A \left(\sigma_{ij}^{(2)}\frac{\partial u_i^{(2)}}{\partial x'_1} - W^{(2)}\delta_{1j} \right) \frac{\partial q}{\partial x'_j}dA
\tag{15}
$$

denote the $\widetilde{J}$−integrals for state 1 and state 2 respectively, and

$$
\widetilde{M}^{(1,2)} = \int_A \left(\sigma_{ij}^{(1)}\frac{\partial u_i^{(2)}}{\partial x'_1} + \sigma_{ij}^{(2)}\frac{\partial u_i^{(1)}}{\partial x'_1} - W^{(1,2)}\delta_{1j} \right) \frac{\partial q}{\partial x'_j}dA + \int_A \frac{1}{2}\left(\sigma_{ij}^{(1)}\frac{\partial\varepsilon_{ij}^{(2)}}{\partial x'_1} - \frac{\partial\sigma_{ij}^{(2)}}{\partial x'_1}\varepsilon_{ij}^{(1)} + \sigma_{ij}^{(2)}\frac{\partial\varepsilon_{ij}^{(1)}}{\partial x'_1} - \frac{\partial\sigma_{ij}^{(1)}}{\partial x'_1}\varepsilon_{ij}^{(2)} \right) qdA
\tag{16}
$$

indicates the modified interaction integral. All the quantities in Equation (16) are evaluated based on a coordinate system aligned to the crack tip, and the identification of domain of integration A and the weighting function $q(x)$ will be described in details in Section 3.2.

Since two $\widetilde{J}$−integrals for inhomogeneous cracked bodies which are subject to mixed-mode loading also represent the energy release rates, one can obtain

$$
\widetilde{J}^{(1)} = \frac{1}{\overline{E}_{tip}}\left(K_I^{(1)^2} + K_{II}^{(1)^2} \right)
\tag{17}
$$

$$
\widetilde{J}^{(2)} = \frac{1}{\overline{E}_{tip}}\left(K_I^{(2)^2} + K_{II}^{(2)^2} \right)
\tag{18}
$$

And

$$
\widetilde{J}^{(S)} = \widetilde{J}^{(1)} + \widetilde{J}^{(2)} + \frac{2}{\overline{E}_{tip}}\left(K_I^{(1)}K_I^{(2)} + K_{II}^{(1)}K_{II}^{(2)} \right)
\tag{19}
$$

where $\overline{E}_{tip}$ at the crack tip is E_{tip} for plane stress, while it is $E_{tip}/\left(1-v^2\right)$ for plane strain. By equating Equation (13) with Equation (19), one can obtain

$$
\widetilde{M}^{(1,2)} = \frac{2}{\overline{E}_{tip}}\left(K_I^{(1)}K_I^{(2)} + K_{II}^{(1)}K_{II}^{(2)} \right)
\tag{20}
$$

In 2-D linear fracture mechanics, the closed-form near-tip displacement fields are available for mode I and mode II [35]. The stress intensity factor $K_I^{(1)}$ for mode I can be obtained by letting state 2 be the pure mode-I asymptotic field (i.e., $K_I^{(2)} = 1$ and $K_{II}^{(2)} = 0$):

$$
M^{(1,\,\text{ModeI})} = \frac{2}{\overline{E}_{tip}}K_I^{(1)}
\tag{21}
$$

In a similar manner, the stress intensity factor K_{II} for mode-II can be also determined.

3. Enriched Petrov-Galerkin Natural Element Method

3.1. Enriched NEM Approximation

The boundary value problem expressed by Equations (1)–(3) is converted to the weak form according to the virtual work principle: Find $u(x)$ such that

$$\int_{\Omega} \varepsilon(v) : \sigma(u)\, d\Omega = \int_{\Gamma_N} \hat{t} \cdot v\, ds \tag{22}$$

for all the test displacements $v(x)$ in the Cartesian coordinate system $\{x, y\}$. Next, a non-convex NEM grid $\mathfrak{I}_{NEM}$, shown in Figure 3a, is constructed using N nodes and a number of Delaunay triangles. Then, for a given NEM grid, trial and test displacement fields $u(x)$ and $v(x)$ for PG-NEM approximation are interpolated as

$$u_h(x) = \sum_{J=1}^{N} \bar{u}_J \phi_J(x) + k_1 \begin{bmatrix} Q_{11}(x) \\ Q_{12}(x) \end{bmatrix} + k_2 \begin{bmatrix} Q_{21}(x) \\ Q_{22}(x) \end{bmatrix} \tag{23}$$

$$v_h(x) = \sum_{I=1}^{N} \bar{v}_I \Psi_I(x) + \lambda_1 \begin{bmatrix} Q_{11}(x) \\ Q_{12}(x) \end{bmatrix} + \lambda_2 \begin{bmatrix} Q_{21}(x) \\ Q_{22}(x) \end{bmatrix} \tag{24}$$

with Laplace interpolation functions $\phi_J(x)$ represented in Figure 3b. Here, $\psi_I(x)$ are constant-strain FE basis functions, which are defined over Delaunay triangles.

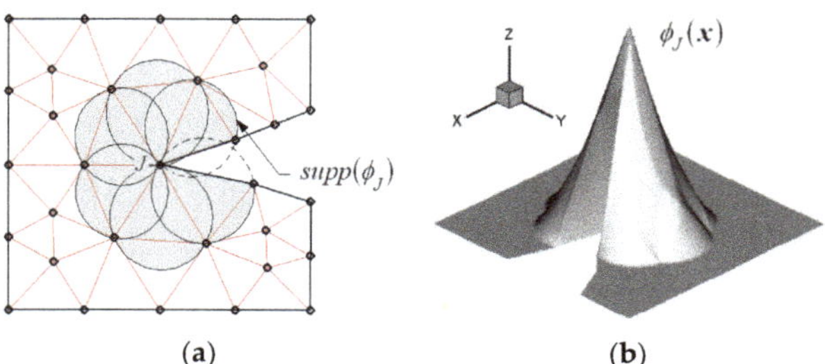

(a) (b)

Figure 3. (a) Non-convex NEM grid $\mathfrak{I}_{NEM}$, (b) Laplace interpolation function $\phi_J(x)$ at the crack tip.

In addition, k_1 and k_2 are two unknown constants associated with a crack, and $Q_{1\alpha}(x)$ and $Q_{2\alpha}(x)$ represent the near-tip singular displacement fields given by

$$Q_{11}(x) = \frac{1}{2\mu} \sqrt{\frac{r}{2\pi}} \cos\left(\frac{\theta}{2}\right)\left[\kappa - 1 + 2\sin^2\left(\frac{\theta}{2}\right)\right] \tag{25}$$

$$Q_{12}(x) = \frac{1}{2\mu} \sqrt{\frac{r}{2\pi}} \sin\left(\frac{\theta}{2}\right)\left[\kappa + 1 - 2\cos^2\left(\frac{\theta}{2}\right)\right] \tag{26}$$

$$Q_{21}(x) = \frac{1}{2\mu} \sqrt{\frac{r}{2\pi}} \sin\left(\frac{\theta}{2}\right)\left[\kappa + 1 + 2\cos^2\left(\frac{\theta}{2}\right)\right] \tag{27}$$

$$Q_{22}(x) = \frac{-1}{2\mu} \sqrt{\frac{r}{2\pi}} \cos\left(\frac{\theta}{2}\right)\left[\kappa - 1 - 2\sin^2\left(\frac{\theta}{2}\right)\right] \tag{28}$$

Referring to Figure 1, r is the radial distance from the crack tip, while θ is the angle from the x'−axis. Meanwhile, μ indicates the shear modulus, and κ is the Kolosov constant given by

$$\kappa = \begin{cases} 3 - 4v & \text{for plane strain} \\ (3 - v)/(1 + v) & \text{for plane stress} \end{cases} \tag{29}$$

with v being the Poisson's ratio. From Equation (23), the overall nodal coefficients $\bar{u}_J$ are defined by

$$\bar{u}_J(x) = u_h(x_J) - k_1 \begin{bmatrix} Q_{11}(x_J) \\ Q_{12}(x_J) \end{bmatrix} - k_2 \begin{bmatrix} Q_{21}(x_J) \\ Q_{22}(x_J) \end{bmatrix} \tag{30}$$

and the overall stress and strain fields corresponding to $\bar{u}_J$ are recovered by the global recovery technique. Meanwhile, the original essential boundary condition (2) is modified as

$$\hat{u} = \bar{\hat{u}} - k_1 \begin{bmatrix} Q_{11}(\hat{x}) \\ Q_{12}(\hat{x}) \end{bmatrix} - k_2 \begin{bmatrix} Q_{21}(\hat{x}) \\ Q_{22}(\hat{x}) \end{bmatrix}, \quad \hat{x} \quad on \quad \Gamma_D \tag{31}$$

Substituting Equations (23) and (24) into Equation (22), together with Equations (4) and (5), ends up with

$$\begin{bmatrix} K_{IJ} & K_{I1} & K_{I2} \\ K_{I1}^T & K_{11} & K_{12} \\ K_{I2}^T & K_{12}^T & K_{22} \end{bmatrix} \begin{Bmatrix} \bar{u}_J \\ k_1 \\ k_2 \end{Bmatrix} = \begin{Bmatrix} F_I \\ f_1 \\ f_2 \end{Bmatrix} \tag{32}$$

In which the global stiffness matrices K_{IJ} and load vectors F_I are defined by

$$K_{IJ} = \int_\Omega B_I^T DB_J d\Omega, \quad F_I = \int_{\Gamma_N} \Phi_I^T \bar{t} ds, \quad I, J = 1, 2, \cdots, N \tag{33}$$

where

$$B_I^T = \begin{bmatrix} \phi_{I.x} & 0 & \phi_{I,y} \\ 0 & \phi_{I,y} & \phi_{I,x} \end{bmatrix} \tag{34}$$

$$D = \frac{E}{(1 + v)(1 - 2v)} \begin{bmatrix} 1 - v & v & 0 \\ v & 1 - v & 0 \\ 0 & 0 & \frac{1-2v}{2} \end{bmatrix} \text{ for plane strain} \tag{35}$$

Meanwhile, the enriched stiffness matrices $K_{\alpha\beta}$, load vectors f_α, and the interface matrices $K_{I\alpha}$ are defined by, respectively

$$K_{\alpha\beta} = \int_\Omega \hat{B}_\alpha^T D\hat{B}_\beta d\Omega, \quad f_\alpha = \int_{\Gamma_N} \hat{\phi}_\alpha^T \bar{t} ds, \quad \alpha, \beta = 1, 2 \tag{36}$$

$$K_{I\alpha} = \int_\Omega B_I^T D\hat{B}_\alpha d\Omega \tag{37}$$

with

$$\hat{\phi}_\alpha^T = [Q_{\alpha 1}(x), Q_{\alpha 2}(x)] \tag{38}$$

$$\hat{B}_\alpha = \begin{bmatrix} Q_{\alpha\alpha,x}(x) \\ Q_{\alpha\beta,y}(x) \\ Q_{\alpha\alpha,y}(x) + Q_{\alpha\beta,y}(x) \end{bmatrix}, \quad \alpha\alpha = \text{no sum} \tag{39}$$

3.2. Numerical Implementation of $\widetilde{M}^{(1,2)}$

Figure 4 schematically represents the identification of domain of integration A and the definition of weighting function $q(x)$, which are prerequisite for the numerical implementation of $\widetilde{M}^{(1,2)}$ in Eq. (6). For these two things, a circle of the radius r_{int}, which is originated at the crack tip, is first imaginarily drawn on the NEM grid. Then, the nodal values of $q(x)$ are assigned based on this circle, such that unity is specified to all the interior nodes while zero to the outer remaining nodes. Then, referring to Figure 2, the boundary of a square composed of eight darkened Delaunay triangles serves as an internal path Γ. On the other hand, the grayed region outside the internal path Γ automatically becomes a domain of integration A. According to the properties of Laplace interpolation functions [29], the defined weighting function $q(x)$ becomes unity within a set of darkened Delaunay triangles while it has the values between zero and unity within the grayed domain of integration A.

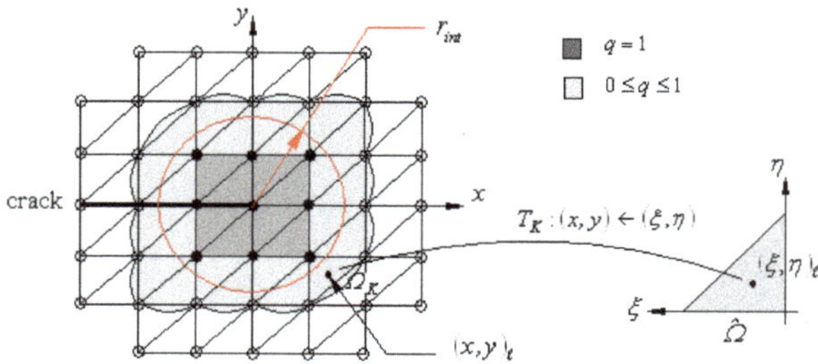

Figure 4. A crack tip-originated circle and a graded domain of integration A.

Both the weighting function $q(x)$ and the derivative $\partial q / \partial x_j$ become zero outside the domain of integration A. Furthermore, the derivative $\partial q / \partial x_j$ vanishes within the darkened squares. So, only the grayed Delaunay triangles are taken for the modified interaction integral $\widetilde{M}^{(1,2)}$ such that

$$\widetilde{M}^{(1,2)} = \sum_{K=1}^{M_G} \widetilde{M}_K^{(1,2)} \tag{40}$$

Here, M_G indicates the total number of grayed Delaunay triangles, while $\widetilde{M}_K^{(1,2)}$ stand for the triangle-wise modified interaction integrals, which are defined by

$$\widetilde{M}_K^{(1,2)} = \sum_{\ell=1}^{INT} \left\{ \left[\left[\sigma_{ij}^{(1)} \frac{\partial u_i^{(2)}}{\partial x_1} + \sigma_{ij}^{(2)} \frac{\partial u_i^{(1)}}{\partial x_1} - W^{(1,2)} \delta_{1j} \right] \frac{\partial q}{\partial x_j} \Big|_{x_\ell} x_\ell w_\ell |J|_{x_\ell} \right. \\ \left. + \frac{1}{2} \left[\sigma_{ij}^{(1)} \frac{\partial \varepsilon_{ij}^{(2)}}{\partial x_1} - \frac{\partial \sigma_{ij}^{(2)}}{\partial x'_1} \varepsilon_{ij}^{(1)} + \sigma_{ij}^{(2)} \frac{\partial \varepsilon_{ij}^{(1)}}{\partial x_1} - \frac{\partial \sigma_{ij}^{(1)}}{\partial x_1} \varepsilon_{ij}^{(2)} \right] q(x_\ell) w_\ell |J|_{x_\ell} \right] \right\} \tag{41}$$

with INT, x_ℓ, and w_ℓ being the total number, co-ordinates, and weights of sampling points for Gauss numerical integration for triangles. Meanwhile, T_K in Figure 4 indicates the geometry transformation between the master element $\hat{\Omega}$ and the grayed triangle Ω_K and the master element $\hat{\Omega}$, which is defined by

$$T_K : x_\ell = \sum_{i=1}^{3} x_i \psi_i(\xi, \eta)_\ell, \quad y_\ell = \sum_{i=1}^{3} y_i \psi_i(\xi, \eta)_\ell \tag{42}$$

where ψ_i stand for the constant-strain FE basis functions, which are employed for the test displacement field $v(x)$.

4. Numerical Experiments

The enriched NEM module was developed in Fortran and combined into the previously developed PG-NEM code [30] having the stress recovery module [36]. To demonstrate and validate the present method, the crack problem of an infinite plate with collinear cracks depicted in Figure 5a is considered. Differing from the cracks in plates of finite size which are of great practical interest, the present infinite plate with periodic cracks provides the closed form solution. Meanwhile, for the numerical study, one may take two kinds of periodic finite crack problems. One is obtained when the plate is cut along EF and GH, while the other is obtained if the plate is cut along AB and CD. Since the first one has been widely taken for the numerical studies (i.e., Ryicki and Kanninen [37] and Chow and Atluri [38]), the second periodic analysis model is taken for the current numerical experiments. Figure 5b represents the detailed second periodic model, as well as a NEM grid having higher grid density toward the crack tip for an upper left darkened quarter. The SIF for this infinite plate with collinear cracks is expressed in the following closed form (the correction factor C of 1.02) [39]:

$$K_I = \sigma_\infty \sqrt{\pi a}\left(\frac{2b}{\pi a}\tan\frac{\pi a}{2b}\right)\cdot C \tag{43}$$

with a being the half crack length, where a circular path Γ around the left crack tip has the relative radius $r/a = 0.0139$, and the near-tip stress distributions are to be evaluated along this circular path. The Young's modulus E is 200 GPa and Poisson's ratio ν is 0.3.

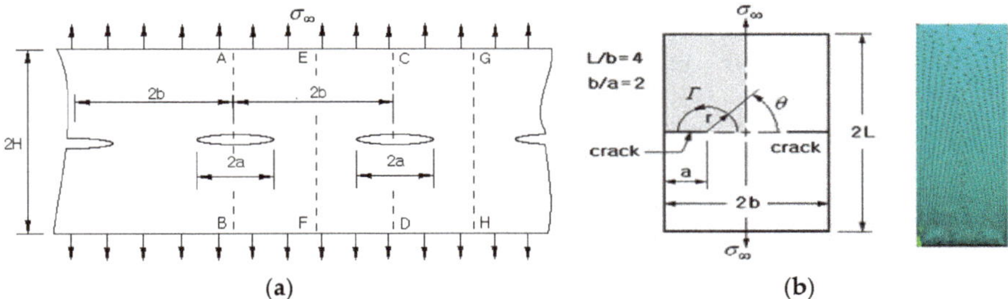

Figure 5. (a) Infinite plate with collinear cracks in plane strain state [2], (b) periodic model with double edge cracks and a gradient natural element method (NEM) grid (unit: m, $N = 2516$).

The problem was solved with 13 Gauss integration points, and the stress field approximated by PG-NEM was smoothened by the stress recovery technique [36] with the Laplace interpolation functions. The stress distributions were evaluated along the circular path Γ and are compared in Figure 6. It can be observed that the case without enrichment shows similar stress distributions to the exact ones, but it provides a significantly low stress level. Thus, one can realize that only the fine gradient grid cannot sufficiently capture the near-tip stress distribution. Meanwhile, it is clear that the case with enrichment remarkably improves the stress level. Therefore, it has been justified that the enrichment of interpolation functions is effective in enhancing the near-tip stress approximation.

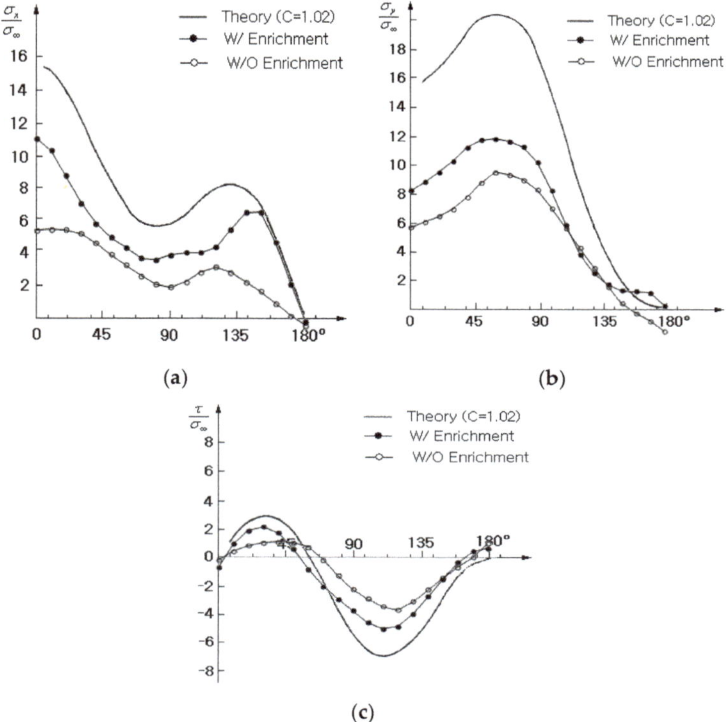

Figure 6. Comparison of the stress distributions along the path Γ ($r/a = 0.0139$): (a) σ_{xx}, (b) σ_{yy}, and (c) τ_{xy}.

Figure 7a shows a rectangular FGM plate with an edge crack which is in the plane strain condition, where W and H/W are 1.0 and 8.0 units, while a/W is set variable for the parametric investigation as follows: 0.2, 0.3, 0.4, 0.5, and 0.6. The Young's modulus E varies along the x−axis in a form of

$$E(x) = E_1 exp(\beta x), \quad 0 \le x \le W \tag{44}$$

but Poisson's ratio is uniform as 0.3. When E_1 and E_2 are specified to the elastic moduli at $x = 0$ and $x = L$, respectively, the parameter β is calculated according to $\beta = ln(E_2/E_1)$, with E_1 being 1.0 unit. Two material variation cases of E_2/E_1 are taken as follows: $E_2/E_1 = 0.1$ and 10.0, and two loading cases of tension in Figure 7b and bending in Figure 7c are considered. This example was initially studied by Erdogan and Wu [40], who also provided an analytical solution. The external loading is set by as follows: for tension and $\sigma_{yy}(x, \pm 4) = \pm 1.0$ for tension and $\sigma_{yy}(x, \pm 4) = \pm(-2x + 1)$ for bending.

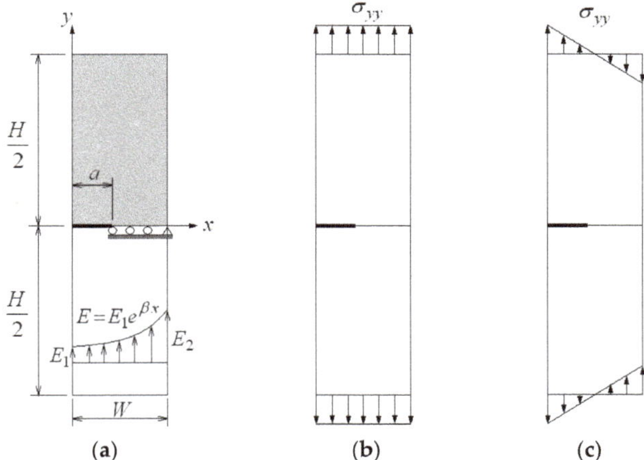

Figure 7. A cracked rectangular functionally graded material (FGM) plate with a varying Young's modulus: (**a**) Geometry dimensions, (**b**) tension loading, (**c**) bending.

An upper grayed half is taken for the crack analysis from the problem symmetry, and the following symmetric constraint is specified to the symmetric line. The non-cracked symmetric line is subjected to the vertical displacement constraint of $u_y = 0$, and the right end point is enforced by the lateral displacement constrain of $u_x = 0$. A 11×41 uniform NEM grid is used, and all the NEM analyses, the stress recovery, and the modified interaction integrals $\widetilde{M}^{(1,2)}$ were carried out with 13 Gauss points. The radius r_{int} for determining the domain of integration A was chosen using twice the side length of a square consisting of two Delaunay triangles [41].

Table 1 comparatively represents the normalized mode-I SIFs for uniform tensile loading and bending loading, respectively, where the values in parenthesis indicate the relative differences (%). It is confirmed that the proposed enriched PG-NEM provides the SIFs, which are consistent with the analytical solutions [40], for all the combinations of E_2/E_1 and a/W of two types of loading, except for the relatively remarkable relative difference 6.807% in the tension case with $E_2/E_1 = 10$ and $a/W = 0.3$. For the further comparison, the numerical results by Kim and Paulino [15] and by Rao and Rahman [34] are also given in Table 1. It is observed that the SIFs of proposed method are correlated satisfactorily with the FEM results by the J_k^*-integral and the EFGM results by the modified $\widetilde{M}^{(1,2)}$-interaction integral. Table 2 represents the comparison with the results obtained without using the enrichment. In the uniform tensile loading, the case without enrichment provides the SIFs, which are lower than those with enrichment, such that the range of relative differences is from 27.59% and 51.04%. Meanwhile, in the bending, it is observed that the difference between with and without enrichment is not significant. Furthermore, the case without using enrichment provides the SIFs which are higher than those obtained by the enrichment. It is inferred that the near-tip stress field produced by bending is totally different from that by uniform tensile loading. The relative difference becomes smaller as the relative crack length a/W increases. It is observed from the detailed numerical values that the relative difference ranges from 2.41% to 24.68%. Thus, it has been found that the proposed enrichment method is influenced by the loading type.

Table 1. Normalized stress intensity factors $K_I / \sigma_{yy} \sqrt{\pi a}$ for an edged cracked plate.

	Method	E_2/E_1	Relative Crack Length a/W			
			0.2	0.3	0.4	0.5
Uniform tension	Proposed Method	0.1	1.302	1.843	2.557	3.504
		10	0.960	1.150	1.581	2.176
	Erdogan and Wu [40]	0.1	1.297 (0.386)	1.858 (−0.807)	2.570 (−0.506)	3.570 (−1.849)
		10	1.002 (−4.192)	1.229 (−6.428)	1.588 (−0.441)	2.176 (0.000)
	Kim and Paulino $[J_k^\star]$ [15]	0.1	1.284 (1.402)	1.846 (−0.163)	2.544 (0.118)	3.496 (0.229)
		10	1.003 (−4.287)	1.228 (−6.352)	1.588 (−0.441)	2.175 (0.046)
	Rao & Rahman $[\widetilde{M}^{(1,2)}]$ [34]	0.1	1.337 (−2.618)	1.898 (−2.898)	2.594 (−1.426)	3.547 (−1.212)
		10	0.996 (−3.614)	1.234 (−6.807)	1.598 (−1.064)	2.189 (−0.594)
Bending	Proposed Method	0.1	1.885	1.872	1.908	2.141
		10	0.583	0.636	0.812	1.069
	Erdogan and Wu [40]	0.1	1.904 (−0.472)	1.886 (−0.742)	1.978 (−3.539)	2.215 (−3.341)
		10	0.565 (3.186)	0.659 (−3.490)	0.804 (0.995)	1.035 (3.285)
	Kim and Paulino $[J_k^\star]$ [15]	0.1	1.888 (−0.159)	1.864 (0.429)	1.943 (−0.035)	2.145 (−0.140)
		10	0.565 (3.186)	0.659 (−3.490)	0.804 (1.801)	1.035 (3.285)
	Rao & Rahman $[\widetilde{M}^{(1,2)}]$ [34]	0.1	1.903 (−0.420)	1.875 (−0.160)	1.954 (−2.354)	2.155 (−0.650)
		10	0.564 (3.369)	0.664 (−4.217)	0.812 (0.000)	1.045 (2.297)

Table 2. Comparison of $K_I / \sigma_{yy} \sqrt{\pi a}$ between with and without enrichment.

	Method	E_2/E_1	Relative Crack Length a/W			
			0.2	0.3	0.4	0.5
Uniform tension	With enrichment	0.1	1.3023 (46.902)	1.8429 (29.041)	2.5722 (37.785)	3.5643 (27.587)
		10	0.9960 (45.050)	1.2225 (47.706)	1.5210 (47.843)	2.1473 (51.041)
	Without enrichment	0.1	0.6915	1.3077	1.6003	2.5810
		10	0.5473	0.6393	0.7933	1.0513
Bending	With enrichment	0.1	1.8847 (−24.683)	1.8719 (−12.132)	1.9083 (−4.910)	2.1406 (−2.406)
		10	0.5826 (18.641)	0.6361 (7.436)	0.8122 (9.554)	1.0686 (15.319)
	Without enrichment	0.1	2.3499	2.0990	2.0020	2.1921
		10	0.4740	0.5888	0.7346	0.9049

The third example is a slant edge crack in a 2-D FGM plate in plane stress condition with the height $H = 2$ units and the width $W = 1$ unit. Referring to Figure 8a, the crack angle α is $45°$ and the relative crack length is a/W is $0.4 \sqrt{2}$, where the enrichment and the crack evaluation are performed within the inclined Cartesian co-ordinate system $\{x', y'\}$ originated at the crack tip O'. The Poison's ratio is kept uniform as $v = 0.3$, but the elastic modulus varies in an exponential function along the x–direction

$$E(x) = \overline{E}exp[\eta(x-0.5)], \quad 0 \leq x \leq W \tag{45}$$

Here, $\overline{E} = 1$ unit and η is variable for the parametric study such as $\eta a = 0, 0.1, 0.25, 0.5, 0.75,$ and 1.0. As external load, an exponentially varying distributed load is applied to the upper edge with $\sigma_{yy}(x,1) = \overline{\varepsilon}\overline{E}exp[\eta(x-0.5)]$, with $\overline{\varepsilon}$ being 1. The whole bottom edge is constrained in the vertical direction (i.e., $u_y = 0$). At the same time, the right end is constrained in the horizontal direction. A uniform NEM grid, given in Figure 8b, is employed which is discretized by 11×21, where a darkened area indicates the domain A for the interaction integral. The total number of grid points is 235 because four additional grid points are needed along the crack to split a crack line into upper and lower faces. All the NEM analyses, patch recovery, and the modified interaction integrals were performed with 13 Gauss points.

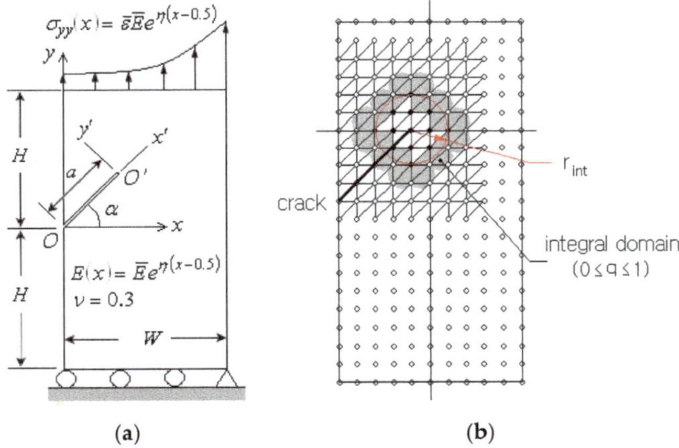

(a)   (b)

Figure 8. (a) An FGM plate with an inclined edge crack under the exponential distribution load, (b) a uniform NEM grid and the domain of integration (235 nodes).

Table 3 comparatively represents the predicted normalized mixed-mode SIFs $K_I/\overline{\varepsilon}\overline{E}\sqrt{\pi a}$ and $K_{II}/\overline{\varepsilon}\overline{E}\sqrt{\pi a}$ evaluated by three different methods, the present PG-NEM, J_k^*–integral using FEM, and EFGM for seven different ηa values. It is confirmed that the present method is in good agreement with FEM and EFGM, such that the biggest relative difference is 4.37% at $K_{II}/\overline{\varepsilon}\overline{E}\sqrt{\pi a}$ for $\eta a = 0.25$. The values in [*] are the normalized mixed-mode SIFs obtained by PG-NEM without using enrichment. It is observed that the values are significantly smaller than those obtained using enrichment, such that the relative difference ranges from 23.32% to 65.63%. The relative difference increases in proportional to the exponential index ηa. Hence, it was confirmed that the enrichment successfully and significantly improves the prediction accuracy of mixed-mode SIFs of the highly heterogeneous FGM, even for the coarse NEM grid.

Table 3. Normalized stress intensity factors (SIFs) for an inclined crack in a functionally graded plate.

ηa	Present Method		Kim and Paulino [J_k^*] [15]		Rao and Rahman [$\widetilde{M}^{(1,2)}$] [34]	
	$\dfrac{K_I}{\bar{\varepsilon}\bar{E}\sqrt{\pi a}}$	$\dfrac{K_{II}}{\bar{\varepsilon}\bar{E}\sqrt{\pi a}}$	$\dfrac{K_I}{\bar{\varepsilon}\bar{E}\sqrt{\pi a}}$	$\dfrac{K_{II}}{\bar{\varepsilon}\bar{E}\sqrt{\pi a}}$	$\dfrac{K_I}{\bar{\varepsilon}\bar{E}\sqrt{\pi a}}$	$\dfrac{K_{II}}{\bar{\varepsilon}\bar{E}\sqrt{\pi a}}$
0.0	1.449 [0.951]	0.606 [0.354]	1.451 (−0.138)	0.604 (0.331)	1.448 (0.069)	0.610 (−0.656)
0.1	1.377 [0.831]	0.589 [0.335]	1.396 (−1.361)	0.579 (1.727)	1.391 (−1.006)	0.585 (0.684)
0.25	1.277 [0.673]	0.525 [0.266]	1.316 (−2.964)	0.544 (−3.493)	1.312 (−2.667)	0.549 (−4.372)
0.5	1.163 [0.460]	0.488 [0.241]	1.196 (−2.759)	0.491 (−0.611)	1.190 (−2.269)	0.495 (−1.414)
0.75	1.087 [0.357]	0.434 [0.159]	1.089 (−0.184)	0.443 (−2.032)	1.082 (0.462)	0.446 (−2.691)
1.0	1.029 [0.240]	0.411 [0.147]	0.993 (3.625)	0.402 (2.239)	0.986 (4.361)	0.404 (1.733)

[*] indicate the SIFs obtained without using the enrichment, while (*) indicate the relative differences (%).

5. Conclusions

In this paper, an enriched PG-NEM was introduced for the reliable crack analysis of inhomogeneous functionally graded materials (FGMs). The Laplace interpolation function was enhanced by the crack-tip singular displacement field and the essential boundary condition was modified accordingly. The unreached global displacement field was extracted, and its stress field was smoothened by the stress recovery. The validity and effectiveness of proposed method was illustrated and justified through three representative examples.

Through the numerical experiments, it was found that enriched PG-NEM is more effective to represent the near-tip stress singularity. The enrichment provides the stress level, which is more close to the exact solution when compared to the use of locally refined NEM grid. In addition, it was observed that enriched PG-NEM accurately predicts the mixed-mode SIFs of inhomogeneous functionally graded materials with exponentially varying elastic modulus. When compared to the case without using enrichment, the prediction accuracy was improved up to four times. Meanwhile, with respect to the analytical solution, the maximum relative difference was found to be less than 6.5%.

Funding: This research was supported by Basic Science Research Program through the National Research Foundation of Korea (NRF) funded by the Ministry of Education (Grant No. NRF-2017R1D1A1B03028879).

Conflicts of Interest: The author declares no conflicts of interest.

References

1. Miyamoto, Y.; Kaysser, W.W.; Rabin, B.H.; Kawasaki, A.; Ford, R.G. *Functionally Graded Materials: Design, Processing and Applications*; Springer Science Business Media: New York, NY, USA, 1999.
2. Giannakopolus, A.E.; Suresh, S.; Olsson, M. Elastoplastic analysis of thermal cycling: Layered materials with compositional gradients. *Acta Metall. Mater.* **1995**, *43*, 1335–1354. [CrossRef]
3. Cho, J.R.; Oden, J.T. Functionally graded material: A Parametric study on thermal stress characteristics using the Crank-Nicolson-Galerkin scheme. *Comput. Method. Appl. Mech. Eng.* **2000**, *188*, 17–38. [CrossRef]
4. Reiter, T.; Dvorak, G.J. Micromechanical models for graded composite materials: II. Thermomechanical loading. *J. Phys. Solids* **1998**, *46*, 1655–1673. [CrossRef]
5. Cho, J.R.; Ha, D.Y. Volume fraction optimization for minimizing thermal stress in Ni-Al2O3 functionally graded materials. *Mater. Sci. Eng. A* **2002**, *334*, 147–155. [CrossRef]
6. Apalak, M.K. Functionally graded adhesively bonded joints. *Rev. Adhesion Adhesives* **2014**, *1*, 56–84. [CrossRef]
7. Zhang, C.; Sladek, J.; Sladek, V. Crack analysis in unidirectionally and bidirectionally functionally graded materials. *Int. J. Fract.* **2004**, *129*, 385–406. [CrossRef]

8. Yahia, S.A.; Atmane, H.A.; Houari, M.S.A.; Tounsi, T. Wave propagation in functionally graded plates with porosities using various higher-order shear deformation plate theories. *Struct. Eng. Mech.* **2015**, *53*, 1143–1165. [CrossRef]

9. Abdelaziz, H.H.; Meziane, M.A.A.; Bousahla, A.A.; Tounsi, A.; Mahmoud, S.R.; Alwabli, A.S. An efficient hyperbolic shear deformation theory for bending, buckling and free vibration of FGM sandwich plates with various boundary conditions. *Steel Comp. Struct.* **2017**, *25*, 693–704.

10. Bourada, F.; Bousahla, A.A.; Bourada, M.; Azzaz, A.; Zinata, A.; Tounsi, A. Dynamic investigation of porous functionally graded beam using a sinusoidal shear deformation theory. *Wind Struct.* **2019**, *28*, 19–30.

11. Cho, J.R.; Ha, D.Y. Averaging and finite-element discretization approaches in the numerical analysis of functionally graded materials. *Mater. Sci. Eng. A* **2001**, *302*, 187–196. [CrossRef]

12. Birman, V.; Byrd, L.W. Modeling and analysis of functionally graded materials and structures. *Appl. Mech. Rev.* **2007**, *60*, 195–216. [CrossRef]

13. Mahamood, R.M.; Akinlabi, E.T.; Shukla, M.; Pityana, S. Functionally graded material: An overview. In Proceedings of the World Congress on Engineering 2012 (WCE 2012), London, UK, 4–6 July 2012.

14. Ivanov, I.V.; Sadowski, T.; Pietras, D. Crack propagation in functionally raded strip under thermal shock. *Eur. Phys. J. Spec. Top.* **2013**, *222*, 1587–1595. [CrossRef]

15. Kim, J.H.; Paulino, G.J. Finite element evaluation of mixed mode stress intensity factors in functionally graded materials. *Int. J. Numer. Methods Eng.* **2002**, *53*, 1903–1935. [CrossRef]

16. Tilbrook, M.T.; Moon, R.J.; Hoffman, M. Crack propagation in graded composites. *Compos. Sci. Technol.* **2005**, *65*, 201–220. [CrossRef]

17. Delale, F.; Erdogan, F. The crack problem for a nonhomogeneous plane. *J. Appl. Mech.* **1983**, *50*, 609–614. [CrossRef]

18. Eischen, J.W. Fracture of nonhomogeneous materials. *Int. J. Fract.* **1987**, *34*, 3–22.

19. Anlas, G.; Lambros, J.; Santare, M.H. Dominance of asymptotic crack tip fields in elastic functionally graded materials. *Int. J. Fract.* **2002**, *115*(2), 193–204. [CrossRef]

20. Ayhan, A.O. Stress intensity factors for three-dimensional cracks in functionally graded materials using enriched finite elements. *Int. J. Solids Struct.* **2007**, *44*, 8579–8599. [CrossRef]

21. Atkinson, C.; List, R.D. Steady state crack propagation into media with spatially varying elastic properties. *Int. J. Eng. Sci.* **1978**, *16*, 717–730. [CrossRef]

22. Gu, P.; Dao, M.; Asaro, R.J. A simplified method for calculating the crack tip field of functionally graded materials using the domain integral. *J. Appl. Mech.* **1999**, *66*, 101–108. [CrossRef]

23. Anlas, G.; Santare, M.H.; Lambros, J. Numerical calculation of stress intensity factors in functionally graded materials. *Int. J. Fract.* **2000**, *104*, 131–143. [CrossRef]

24. Dolbow, J.E.; Gosz, M. On the computation of mixed-mode stress intensity factors in functionally graded materials. *Int. J Solids Struct.* **2002**, *39*, 2557–2574. [CrossRef]

25. Kaczmarczyk, J.; Grajcar, A. Numerical simulation and experimental investigation of cold-rolled steel cutting. *Materials* **2018**, *11*, 1263. [CrossRef] [PubMed]

26. Rao, B.N.; Rahman, S. Mesh-free analysis of cracks in isotropic functionally graded materials. *Eng. Fract. Mech.* **2003**, *70*, 1–27. [CrossRef]

27. Liu, K.Y.; Long, S.Y.; Li, G.Y. A meshless local Petrov-Galerkin method for the analysis of cracks in the isotropic functionally graded material. *Comp. Model. Eng. Sci.* **2008**, *7*, 43–57.

28. Cho, J.R.; Lee, H.W. Calculation of stress intensity factors in 2-D linear fracture mechanics by Petrov–Galerkin natural element method. *Int. J. Numer. Methods Eng.* **2014**, *98*, 819–839. [CrossRef]

29. Sukumar, N.; Moran, A.; Belytschko, T. The natural element method in solid mechanics. *Int. J. Numer. Methods Eng.* **1998**, *43*, 839–887. [CrossRef]

30. Cho, J.R.; Lee, H.W. A Petrov–Galerkin natural element method securing the numerical integration accuracy. *J. Mech. Sci. Technol.* **2006**, *20*, 94–109. [CrossRef]

31. Fleming, M.; Chu, Y.A.; Moran, B.; Belytschko, T. Enriched element-free Galerkin methods for crack tip fields. *Int. J. Numer. Methods Eng.* **1997**, *40*, 1483–1504. [CrossRef]

32. Pant, M.; Singh, I.V.; Mishra, B.K. A novel enrichment criterion for modeling kinked cracks using element free Galerkin method. *Int. J. Mech. Sci.* **2013**, *68*, 140–149. [CrossRef]

33. Moran, B.; Shih, F. Crack tip and associated domain integrals form momentum and energy balance. *Eng. Fract. Mech.* **1987**, *27*, 615–642. [CrossRef]

34. Rao, B.N.; Rahman, S. An efficient meshless method for fracture analysis of cracks. *Comput. Mech.* **2000**, *26*, 398–408. [CrossRef]

35. Anderson, T.L. *Fracture Mechanics: Fundamentals and Applications*, 1st ed.; CRC Press: Boca Raton, FL, USA, 1991.

36. Cho, J.R. Stress recovery techniques for natural element method in 2-D solid mechanics. *J. Mech. Sci. Technol.* **2016**, *30*, 5083–5091. [CrossRef]

37. Ryicki, E.F.; Kanninen, M.F. A finite element calculation of stress intensity factors by a modified crack closure integral. *Eng. Fract. Mech.* **1977**, *9*, 931–938. [CrossRef]

38. Chow, W.T.; Atluri, S.N. Finite element calculation of stress intensity factors for interfacial crack using virtual crack closure integral. *Comput. Mech.* **1995**, *16*, 417–425. [CrossRef]

39. ASTM. *Fracture Toughness Testing and Its Applications*; ASTM International: West Conshohocken, PA, USA, 1965; Volume 381, pp. 43–51.

40. Erdogan, F.; Wu, B.T. The surface crack problem for a plate with functionally graded properties. *ASME J. Appl. Mech.* **1997**, *64*, 449–456. [CrossRef]

41. Moës, N.; Dolbow, J.; Belytschko, T. A finite element method for crack growth without remeshing. *Int. J. Numer. Meth. Eng.* **1999**, *46*, 131–150. [CrossRef]

Review

Understanding the Fracture Behaviors of Metallic Glasses—An Overview

Guan-Nan Yang [1,2,3,*], Yang Shao [2,3] and Ke-Fu Yao [2,3,*]

[1] School of Electromechanical Engineering, Guangdong University of Technology, Guangzhou 510006, China
[2] School of Material Science and Engineering, Tsinghua University, Beijing 100084, China; shaoyang@tsinghua.edu.cn
[3] Key Laboratory for Advanced Materials Processing Technology, Ministry of Education, Beijing 100084, China
* Correspondence: ygn@gdut.edu.cn (G.-N.Y.); kfyao@tsinghua.edu.cn (K.-F.Y.)

Received: 17 September 2019; Accepted: 3 October 2019; Published: 12 October 2019

Abstract: Fracture properties are crucial for the applications of structural materials. The fracture behaviors of crystalline alloys have been systematically investigated and well understood. The fracture behaviors of metallic glasses (MGs) are quite different from that of conventional crystalline alloys and have drawn wide interests. Although a few reviews on the fracture and mechanical properties of metallic glasses have been published, an overview on how and why metallic glasses fall out of the scope of the conventional fracture mechanics is still needed. This article attempts to clarify the up-to-date understanding of the question. We review the fracture behaviors of metallic glasses with the related scientific issues including the mode I fracture, brittle fracture, super ductile fracture, impact toughness, and fatigue fracture behaviors. The complex fracture mechanism of MGs is further discussed from the perspectives of discontinuous stress/strain field, plastic zone, and fracture resistance, which deviate from the classic fracture mechanics in polycrystalline alloys. Due to the special deformation mechanism, metallic glasses show a high variability in fracture toughness and other mechanical properties. The outlook presented by this review could help the further studies of metallic glasses. The review also identifies some key questions to be answered.

Keywords: metallic glasses; fracture; shear bands; mechanical properties; fracture mechanism

1. Introduction

Since the discovery of metallic glasses (MGs), especially bulk metallic glasses (BMGs), the mechanical behavior of MGs is attracting increasing attention for both potential structural applications and scientific interests [1]. A series of distinguished mechanical properties, including high compressive plasticity, hardness, ultimate strength, and fracture toughness have been reported [1–8]. These excellent properties and the unusual deformation mechanism of viscous flow and shear band motion make MGs a special member in the family of structural materials.

At room temperature, the deformation behaviors of MGs are controlled by shear bands, which are kinds of localized viscous flow [9]. The shear bands are very different from the slip bands formed via dislocations in crystalline alloys, and result in the special mechanical behaviors of MGs. The information about the topics of shear bands, mechanical properties, and fracture behaviors of MGs can be found in some review papers [9–12]. However, a review on the topic of how and why MGs fall out of the scope of conventional fracture mechanics is still lacking. In this article, we attempt to summarize the up-to-date understanding on this issue from several aspects about the fracture behaviors and fracture mechanism. We focus on the main fracture behaviors of metallic glasses, including the mode I fracture, brittle fracture, super ductile fracture, impact toughness, and fatigue fracture behaviors. The complex fracture mechanism of metallic glasses are discussed from the perspectives of discontinuous stress/strain field, plastic zone, and fracture resistance, which deviate from the classic fracture mechanics in polycrystalline alloys. This work would help researchers and engineers in their scientific and engineering studies of metallic glasses.

2. Fracture Behaviors of MGs

The fracture criterion is an important parameter to understand the deformation and fracture mechanism of a material. For MGs, due to the lack of work-hardening effect, the fracture strength in uniaxial compression tests generally will be equal to or close to the yielding strength [13]. The intensity and direction of stress at the moment of yielding or fracture reflect the critical stress condition to trigger the shear band avalanche [14,15]. From the microcosmic aspect, the yielding criterion reflects the critical stress condition to trigger the localization of shear transformation of atomic groups [16,17]. Previous studies on this issue were mostly based on the uniaxial tensile and compression experiments or simulations [18–20]. It has been discovered that MGs generally show a shear mode failure along a shear/fracture angle (the angle between the shear band/fracture surface and the load axis) near 45° [18–20]. The angle will be slightly larger than 45° under tension, and will be slightly smaller than 45° under compression. Some brittle MG systems can show cleavage or split mode failure, which will be discussed later. To understand the asymmetric compression and tension behaviors, Schuh et al. indicates that the microstructure of MGs is analogous to that of randomly packed particles in a granular solid [18]. Therefore, the yielding criterion of MGs could be described by the Mohr-Coulomb criterion:

$$\tau_y = \tau_0 - \alpha\sigma_n \tag{1}$$

where τ_y stands for the shear yield stress, τ_y is a constant, σ_n is the normal stress on the shear plane, α is a coefficient that reflects the degree of internal friction in the system. This criterion explains the asymmetric compression and tension strength of MGs, but show deviations in the estimation of shear angle. On this basis, Z.F. Zhang et al. further proposed the elliptical criterion [19,20]:

$$\frac{\sigma^2}{\sigma_0^2} + \frac{\tau^2}{\tau_0^2} \geq 1 \tag{2}$$

where τ and σ stand for the shear stress and normal stress on a shear plane, τ_0 and σ_0 are material dependent constants. The elliptical criterion provides a applicable model to comprehensively explain the strength and shear angle of MGs under the simple loading condition of uniaxial compression/tension. However, the fracture behaviors of MGs under bending, fast loading, fatigue loading and other loading conditions are still complicated to be understood.

2.1. Typical Mode I Fracture of MGs

The fracture properties are very important for the engineering applications of structural materials. The mode I fracture properties of metallic glasses is widely studied by 3-point/4-point bending and compact tension tests [21,22]. The measured fracture toughness of most MGs falls in the range of a few MPa·m$^{1/2}$ to more than one hundred MPa·m$^{1/2}$ [11,12]. This value is much lower than that of steel and other commonly used engineering alloys [23]. The fracture brittleness is considered as one of the main problems for the engineering application of MGs.

The low fracture toughness of MGs can be understood by their special fracture mechanism. According to the experiments and literature, a typical mode I fracture process of MGs can roughly be divided into three stages:

2.1.1. Multiple Shear Bands Formation and Sliding

With the increasing notch-tip stress intensity during loading, shear bands are formed. Different from the plane shear bands formed in uniaxial compression/tension tests, the shear bands formed during mode I fracture are curved. This is because the propagation of shear bands in MGs follows the direction of the local maximum shear stress [24]. Therefore, the shear bands will reproduce the shear stress direction in the notch-tip stress field. After formation, the shear bands can slide for a distance, which ranges from less than one micron to tens of microns [25,26]. Figure 1 shows the evolution of shear-offsets and load–displacement curve of a typical ductile $Pd_{77.5}Cu_6Si_{16.5}$ metallic glass during 3-point bending test, with a sample size of $3 \times 6 \times 30$ mm^3 [27]. Curved shear offsets could be observed on the sample's side surface (Figure 1a–e, taken by an optical camera, Nikon D810, Japan), and some small stress-drops/serrations could be noticed on the loading curve (Figure 1g), representing the activation and sliding of these shear bands. Enlarged SEM (taken by a LEO-1530 field emission scanning electron microscope, German) image shows that the shear direction of the shear offsets points into the sample (Figure 1f). Due to the randomness in the shear band motions of MGs, it is almost impossible to predict the actual formation moment, sequence, position and sliding distance for each shear band in the real case.

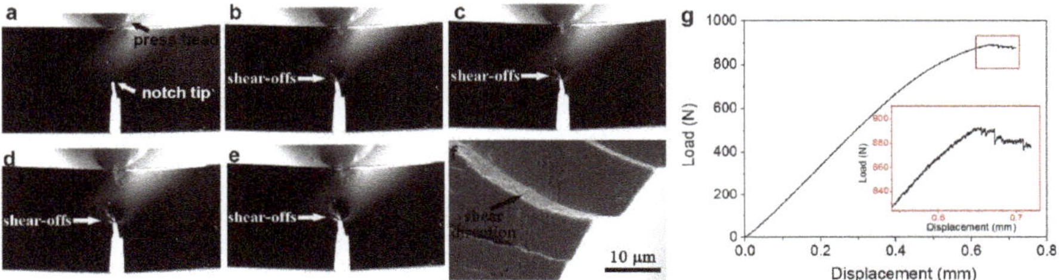

Figure 1. (**a–f**) The side view of a $Pd_{77.5}Cu_6Si_{16.5}$ metallic glass (MG) sample with a size of $3 \times 6 \times 30$ mm^3 during 3-point bending test. (**g**) The load-displacement curve.

2.1.2. Shear Band Delamination

After the multiple shear bands formation and sliding, the material could reach the stage of shear band delamination. At present, the critical condition for shear band delamination is still complicated to be quantitatively predicted. In experiments, the shear bands in MGs will develop into fracture after sliding for a distance of a few tens of microns or less [6,26,28–30]. The physics behind the shear band delamination can be understood from different perspectives. As the model of Taylor's fluid meniscus instability [31,32] suggested, the very thin shear band could be considered as a viscous layer, which would delaminate when a critical stress condition (a critical suction stress gradient along the shear band) is met. On the other hand, the shear band will gradually become looser and softer during sliding, due to the shear dilatation effect [33–35]. Irreversible microstructure change (such as nano cavitation [36] and nanocrystallization [37]), micrometer-scale softening and cavitation [38] can also occur during this process and weaken the shear band. Finally, the shear band reaches the critical condition of delamination. When the delamination occurs, the crack tip will propagate rapidly along the shear band path, and the fractograph will reproduce the shape of the shear band, denominated as "flat zone" or "planar zone" in some reports [26,28]. Figure 2 shows the fractograph of a $Pd_{77.5}Cu_6Si_{16.5}$ metallic glass, which reveals that the "flat zone" on the fractograph is actually not straightforward but will deviate from the original notch tip direction [27]. The morphologies of the shear bands and the fractograph formed by shear band delamination depend on the local stress condition. Near the side surface of the sample, the local maximum shear stress points into the sample, and cone shape shear bands will form (Figure 2b,c). Near the notch-tip inside the sample, the local maximum shear stress deviates from the notch-tip direction, and curved shape shear bands will form (Figure 2c,d). The shear direction and sliding distance of the shear band before delamination can be determined by the smooth shear-offsets at the edge of the fractograph [26,28,39].

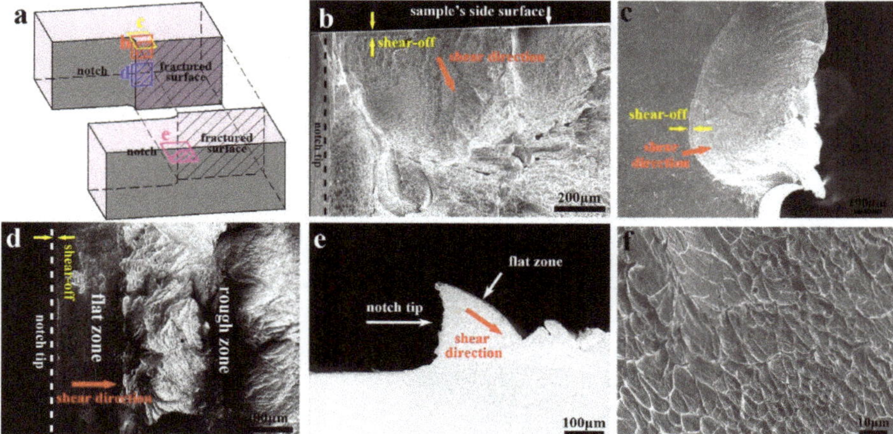

Figure 2. Fractograph of a $Pd_{77.5}Cu_6Si_{16.5}$ metallic glass. (**a**) Sketch of the fractured sample with four marked regions. (**b**,**c**) The fractograph near the side surface of the sample. (**d**) The fractograph inside the sample. (**e**) The cross section profile inside the sample. (**f**) Enlarged image of the viscous features on the fractograph.

2.1.3. Rapid Fracture

After the shear band delamination, the crack can continue propagating rapidly through the sample. As there is no existing shear band and preferred crack propagation direction in this area, irregular rough fractograph is formed (Figure 2d), denominated as "rough zone" in some studies [26,28]. As the external stress is higher than the required stress for crack propagation at this stage, this part of fractograph is also called "over load region" [39]. Due to the lack of measuring method, there still lacks systematic researches on the fracture mechanism of this rapid fracture process.

Through the above three stages of fracture, a typical mode I fractograph of MG consisting of smooth region (shear band formation and sliding), flat zone (shear band delamination) and rough zone (rapid fracture) is formed. Plentiful viscous fracture features including dimples, veins and droplet patterns can be observed on the fractographs [10,39,40] (Figure 2f). These kinds of features are typical in MGs. As MGs are a kind of shear softened material, the deformed position (i.e. shear bands) will go through significant softening and viscosity drop [33–35].

When the fracture process develops into the stage of shear band delamination, the stress required for further crack extension will drop dramatically. For most MG samples, the stage of multiple shear bands formation and sliding in the beginning of the fracture process is short and unsustainable. Some brittle MGs can hardly form multiple shear bands at the notch-tip [41]. A large sample size of MGs is also against the multiple shear banding [42,43]. As a result, these MG samples will experience shear band delamination and generally show relatively poor fracture toughness/fracture resistance in the fracture tests.

In the widely used uniaxial compress and tension tests, MGs generally show a shear mode of fracture. In this case, the shear bands can propagate through the sample and are approximately plane. The fracture plane and the applied load direction show an angle near 45° [18–20]. Viscous features and smooth shear-offset can be observed on the fractograph [6,44], indicating that the fracture process also consists of shear band formation, sliding, and delamination

2.2. Brittle Fracture of MGs

Although the fracture behaviors of most BMGs are controlled by the shear band deformation mechanism, there are exceptions, of course. In some Fe-based, Ca-based, Mg-based, Rare earth-based and other brittle MGs, the fracture mode under compression/tension loading can change from shear mode (fracture along the shear band path, which is ~45°from the loading axis) to cleavage mode (fracture along the plane perpendicular to the applied stress), split mode [45] (fracture along the plane parallel to the applied stress) or fragmentation mode. The cleavage or split surface of these brittle MGs can be divided into mirror, mist, and hackle regions [41,46]. The cleavage surface of MGs is smooth and continuous, with many radial ridge patterns along the crack propagation direction [47]. The split or fragment surface of MGs consists of many smooth conchoidal morphologies [46]. Plentiful periodic nanoscale wavy corrugations can be observed on the enlarged fracture surface (Figure 3). The formation mechanism of such features has been explained within the framework of the meniscus instability, plastic zone theory and local softening mechanism [41,48]. To explain the difference between the ductile and brittle fractographs, some studies point out that there is a positive correlation between the fracture toughness and the dimple size on the fractographs of MGs [49]. Those brittle MGs tends to form small dimples during fracture, and the most brittle MGs will form nanoscale wavy corrugations.

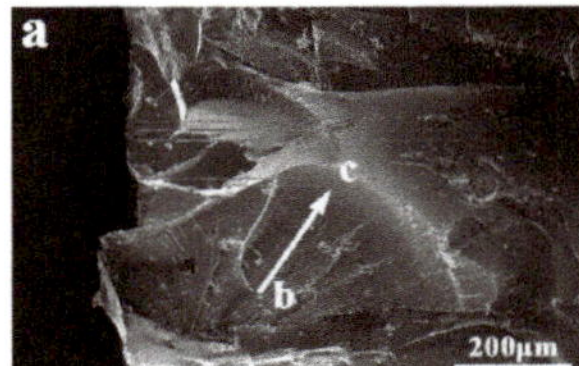
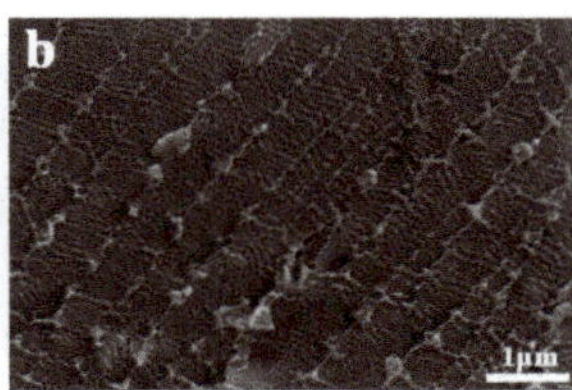
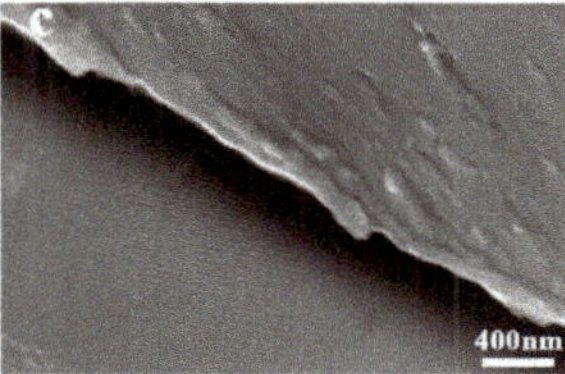

Figure 3. Fractograph of a brittle $Mg_{54}Cu_{28}Ag_7Y_{11}$ MG with many periodic nanoscale wavy corrugations. (**a**) The overall appearance of the conchoidal fractograph. (**b**) Enlarged image of wavy corrugations on the conchoidal fractograph. (**c**) Enlarged image of nanoscale wavy corrugations at the edge of the conchoidal fractograph.

2.3. "Super Ductile" Fracture of MGs

In the last decade, some Zr-based and Pd-based MGs with extremely large fracture toughness have been reported. The measured J-integral toughness of the Zr-based and Pd-based MGs can reach 150 MPa·m$^{1/2}$ and more than 220 MPa·m$^{1/2}$, respectively [6,50]. The Zr-based MG can form plentiful multiple shear bands during fracture. The Pd-based MGs can sustain multiple shear bands formation and sliding, without shear band delamination. Figure 4 shows the appearance of another $Pd_{77.5}Cu_6Si_{16.5}$ MG sample with a size of size is 2.2 × 3.8 × 25 mm^3 during 3-point bending [51]. The sample can be severely deformed without fracture (Figure 4a). Plentiful shear offsets can be observed on the side surface of the sample (Figure 4b). The fractograph shows abundant strip features perpendicular to the crack extension direction (Figure 4d,f), which are formed through multiple shear banding. Such a "Super ductile" fracture behavior is inspiring, however, it is only observed in small size samples. The reasons for this fracture behavior and high toughness can be ascribed to:

1. The Pd-based MGs have better deformability than other MG systems and can more easily form multiple shear bands in different loading conditions.
2. The size of the samples used in these studies is small (the thickness and ligament width are not more than 3 mm). Since a smaller sliding distance is needed to release the same stress in a smaller sample, it will be less easy for small samples to reach the critical condition of shear band delamination, i.e. small sample show higher shear band stability. Therefore, sustainable multiple shear banding can be achieved.
3. The value of J-integral toughness depends on the energy absorption during crack extension, rather than the crack-tip stress intensity [22,52]. Due to the massive energy dissipation via the multiple shear band movements, a high J-integral toughness can be obtained.

Besides this review, another recent report also indicates that with decreasing sample size, the fracture mode of the $Pd_{77.5}Cu_6Si_{16.5}$ MG transits from multiple shear banding-shear band delamination to sustainable multiple shear banding [43]. The fracture toughness increases significantly when the transition occurs. Similar transition can also be observed in other MG systems. Figure 5 shows the fractograph of a pre-notched $Zr_{41.2}Ti_{13.8}Cu_{12.5}Ni_{10}Be_{22.5}$ (Vitreloy-1) MG rods after 3-point bending. A long sustainable multiple shear banding can be achieved when the sample size decreases down to φ2 mm. A large region of smooth stripes (shear offsets) can be observed on the fractograph (Figure 5a–e). A long serration process can be observed on the loading curve (Figure 5f). In the previous mechanical studies of MGs, a correlation of better properties and smaller sample size is widely observed [42]. This law is also applicable in the fracture performance.

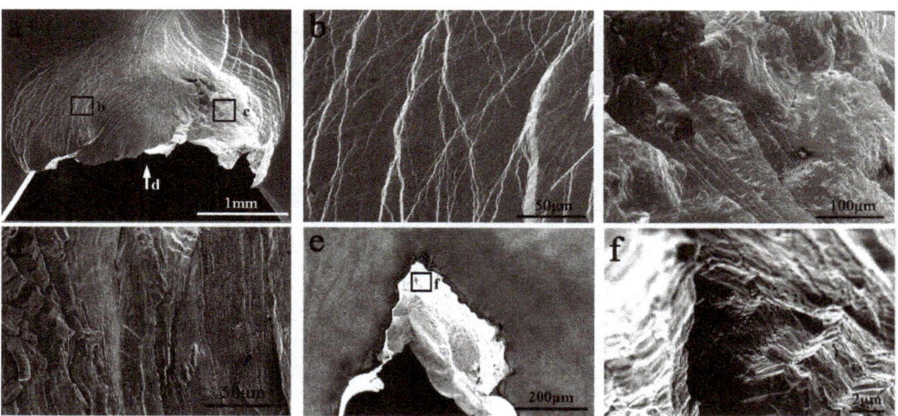

Figure 4. Appearance of a $Pd_{77.5}Cu_6Si_{16.5}$ MG sample with a size of $2.2 \times 3.8 \times 25$ mm^3 after 3-point bending. (**a**–**c**) Side surface appearance. (**d**–**f**) Fracture surface appearance.

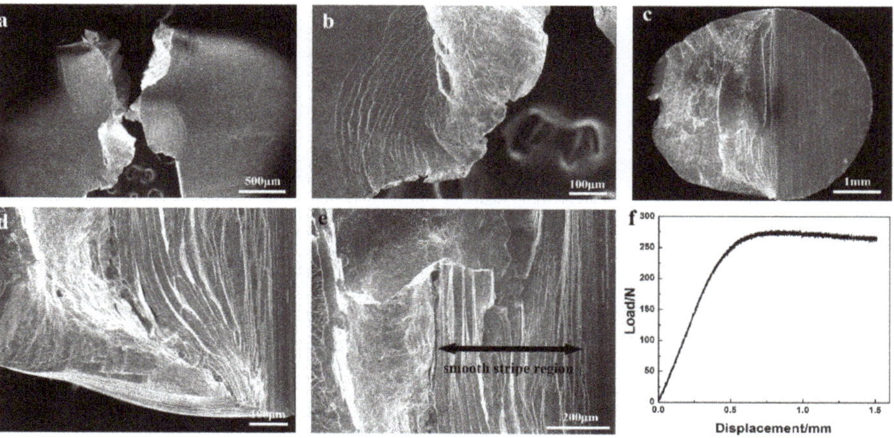

Figure 5. Appearance of a Vitreloy-1 MG rod with a size of φ2 after 3-point bending. (**a**,**b**) Side surface appearance. (**c**–**e**) Fracture surface appearance. (**f**) Loading curve.

2.4. Impact Toughness of MGs

The impact toughness reflects the damage resistance of a material under impact loading. At present, the impact toughness data of MGs are still few. As the size requirement of the standard test ($10 \times 10 \times 55$ mm) can hardly be reached by most MGs. Only a small number of MGs can achieve a critical size over 10 mm [53–58]. Many impact toughness data are measured through nonstandard small samples. Table 1 shows a collection of reported impact toughness data of several MGs. Although there is doubt about the reliability of the data measured in nonstandard samples, the highest measured impact toughness in MGs is not higher than 160 kJ/m^2. This value is relatively smaller than that of AISI-1018 steel [59].

Table 1. Impact toughness data of different metallic glasses.

Composition	Sample Size (mm × mm × mm)	Deepness and Shape of Notch	Impact Energy (kJ/m^2)
$(Ti_{36.1}Zr_{33.2}Ni_{5.8}Be_{24.9})_{95}Cu_5$	10 × 10 × 55	U-shape	50 [60]
$Zr_{55}Al_{10}Cu_{30}Ni_5$	2.6 × 10 × 55	U-shape	63 [61]
$Zr_{50}Cu_{40}Al_{10}$	5 × 10 × 55	U-shape	100 [62]
$Zr_{41.2}Ti_{13.8}Cu_{12.5}Ni_{10}Be_{22.5}$	3 × 3 × 30	1 mm V-shape	80 ~ 160 [63]
$Zr_{41.2}Ti_{13.8}Cu_{12.5}Ni_{10}Be_{22.5}$	3 × 6 × 30	3 mm V-shape	133 [64]
$La_{55}Al_{25}Cu_{10}Ni_5Co_5$	2.6 × 5 × 22	1 mm	77 [65]
$Zr_{41.2}Ti_{13.8}Cu_{12.5}Ni_{10}Be_{22.5}$	3 × 6 × 30	3 mm	90 ~ 133 [66]
$Pd_{40}Cu_{30}Ni_{10}P_{20}$	2.5 × 10 × 55	2 mm U-shape	70 [67]
$Ti_{32.8}Zr_{30.2}Be_{26.6}Ni_{5.3}Cu_9$	10 × 10× 55	2 mm U-shape	50 [60]
$(Ti_{41}Zr_{25}Be_{28}Fe_6)_{91}Cu_9$	10 × 10 × 55	2 mm U-shape	60
AISI-1018 steel	10 × 10 × 55	2 mm V-shape	~500 [59]

Figure 6 shows a typical impact fracture morphology of a $(Ti_{41}Zr_{25}Be_{28}Fe_6)_{91}Cu_9$ MG, which is very similar to that of mode I fracture. The fractograph starts with a smooth region and flat zone (Figure 6b), which are formed during shear band sliding and delamination. The other positions of the fractograph are rough, with many ridge patterns. Plentiful vein patterns and viscous features can be observed on the enlarged images of the fractograph (Figure 5d–f). The fracture direction changes obviously near the end of the sample (Figure 6c), which results from the changed stress condition at this position and the existence of shear bands generated by the impact hammer.

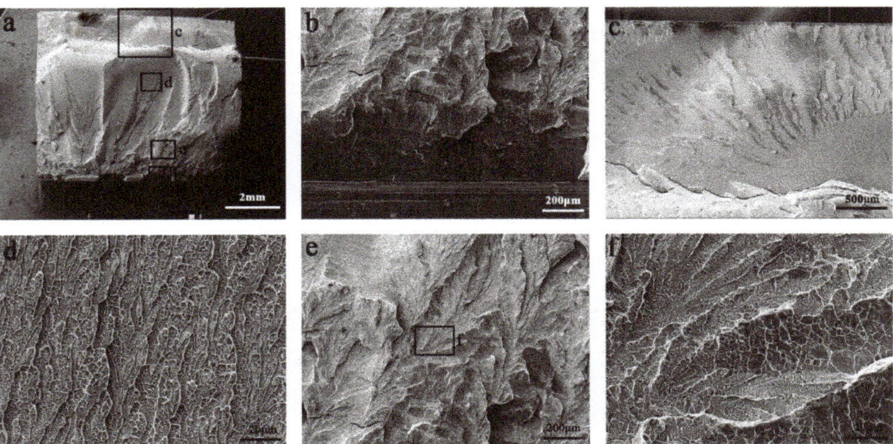

Figure 6. Fractograph morphology of a $(Ti_{41}Zr_{25}Be_{28}Fe_6)_{91}Cu_9$ MG after impact toughness test. (**a**) The overall appearance of the fractograph. (**b**) The smooth region and flat zone on the fractograph near the initial notch tip. (**c**) Changed fracture direction and fractograph morphology near the end of the sample. (**d–f**) The vein patterns and viscous features at different positions of the fractograph.

The loading rate of impact toughness test is much faster than that of conventional static tests, and is also much faster than that of shear band sliding. Generally, the shear band sliding speed in MGs is at the magnitude of 1 mm/s or less [68,69]. The speed of pendulum hammer in the impact toughness test is at the magnitude of 4 m/s. Therefore, the fractograph of MGs after impact tests is similar to the over load region formed during rapid fracture. It has been reported that the mechanical behaviors of MGs are sensitive to the strain rate [70]. However, systematic researches on the dynamic fracture mechanism and influence of high stain rate in MGs during impact toughness test are still rare.

2.5. Fatigue Performance of MGs

Fatigue rupture is one of the main failure modes in engineering material [71]. The fatigue study of MGs can be traced back to the mid-1970 [72,73]. Due to the size limit, the tests of the early discovered MGs were carried out with ribbon samples. The Pd-Si MG ribbons show a fatigue crack growth threshold as high as $\Delta K_{th} = 9$ MPa·m$^{1/2}$ [74], which is even higher than that of high-strength steel [75]. With the development of bulk metallic glasses, the fatigue properties can be measured by bulk samples under tension, compression, and bending tests. Although the fatigue ratio (the ratio of the fatigue endurance limit to the yielding stress) of most BMGs is near 0.3 [76], the fatigue endurance limit (the maximum stress amplitude to which the material is subjected for 10^7 cycles without failure) is still high, due to their high yielding strength. Some Co-based and Fe-based MGs can reach a fatigue limit as high as 2 GPa [77]. Table 2 shows the data of the fatigue endurance limit and the fatigue ratio of some reported MGs with a size larger than 2 mm and other commonly used metal materials including steel and Titanium alloy.

Table 2. Fatigue properties of some MGs and some commonly used metal materials.

Constituent	Fatigue Endurance Limit σ_L (MPa)	Sample Size/mm	Fatigue Ratio	Test Mode	Ref.
$Pd_{40}Cu_{30}Ni_{10}P_{20}$	340	φ5 × 10	0.2	bending	[67]
$Fe_{48}Cr_{15}Mo_{14}Er_2C_{15}B_6$	682	3 × 3 × 25	0.17	bending	[78,79]
Vit 1	152	3 × 3 × 50	0.08	bending	[80]
Vit 1	703	φ2.98	0.38	tension	[81]
Vit 1	615	φ2.98	0.33	tension	[81]
Ti-6Al-4V	515		0.50		[82]
300 M steel	800		0.4		[82]
$Cu_{47.5}Zr_{47.5}Al_5$	224	3 × 3 × 25	0.12	bending	[83]
2090-T18 Al-Li alloy	250		0.48		[82]
$Zr_{52.5}Ti_5Cu_{17.9}Ni_{14.6}Al_{10}$	907	φ2.98	0.53	tension	[84]
$Zr_{52.5}Ti_5Cu_{17.9}Ni_{14.6}Al_{10}$	850	3.5 × 3.5 × 30	0.5	bending	[85]
$Zr_{50}Cu_{40}Al_{10}$	752	φ2.98	0.41	tension	[86]
$Zr_{50}Cu_{30}Al_{10}Ni_{10}$	865	φ2.98	0.45	tension	[86]
$Zr_{50}Cu_{37}Al_{10}Pd_3$	983	φ2.98	0.52	tension	[87]
$Ca_{65}Mg_{15}Zn_{20}$	140	4 × 4 × 4	0.38	compression	[88]

It is noticed that the measured fatigue properties of MGs are sensitive to the test method, test environment, sample quality, sample size, surface condition, and etc. Different results have been reported by different studies [81,89,90]. For example, the pour casting fabricated $Zr_{55}Cu_{30}Ni_5Al_{10}$ MG shows better fatigue properties, as the size and distribution of flaws could be tuned by the fabrication method. Thermal relaxed MGs could show higher fatigue limit, due to the relatively low free volume content [91–93]. The Poisson's ratio, sample size [94] and the content of some alloying elements [87] also have significant influences on the fatigue behavior.

Figure 7 shows a typical fractograph of a brittle Mg-Cu-Ag-Y MG after tension fatigue fracture. The fractograph of the fatigue cracked MGs can be divided into the crack initiation site, the crack growth region and the fast fracture region [86,95]. There can be more than one crack initiation site on one sample. The crack growth region is relative smooth, with many regular fatigue striations perpendicular to the crack growth direction (Figure 7c). The spacing between striations is at micron order or submicron order. The fast fracture region is rough (Figure 7c). Vein patterns and melting features can be observed on the fractograph (Figure 7d), which result from the local heating and viscosity drop during the rapid fracture.

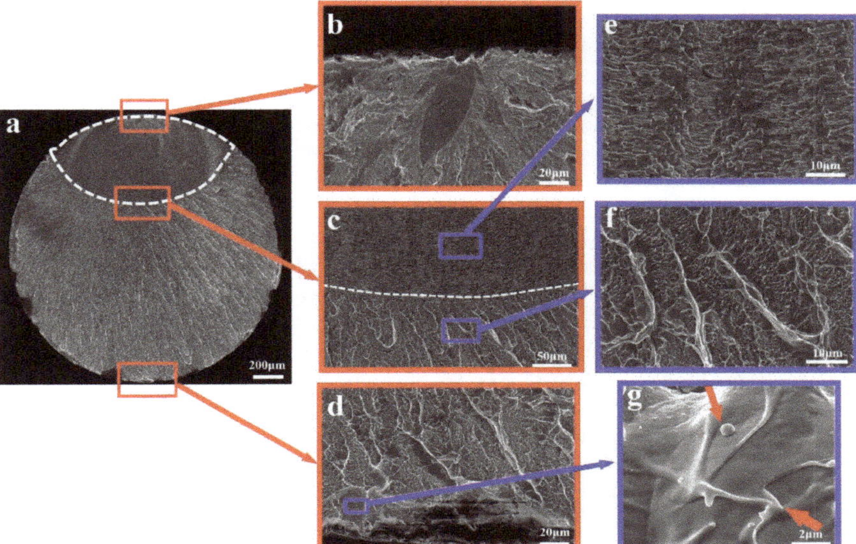

Figure 7. Fractograph morphology of a brittle Mg-Cu-Ag-Y MG after tension fatigue fracture. (**a**) The overall appearance of the fractograph. (**b**) Enlarged image of the crack initiation site. (**c**) The boundary between the regular fatigue striations and the fast fracture region. (**d**) The morphology at the end of the fractograph. (**e**) Enlarge image of the regular fatigue striations. (**f**) Enlarged image of the fast fracture region (**g**) Enlarge image of the droplet feature at the end of the fractograph.

3. Fracture Mechanism of MGs

The fracture process in MGs discussed above is very different from that in conventional polycrystalline alloys. With work-hardening effect and the deformation mechanism of dislocations, the deformation of many polycrystalline alloys will experience an elastic, yielding, work-hardening, and fracture process. The mode I fracture process in these materials can be described by the classic linear elastic fracture mechanics (LEFM), which depicts the plastic zone by the contour line of yielding stress [52]. The material inside the plastic zone could deform plastically, and the material outside the zone will be still elastic. The plastic zone moves accordingly with the extension of crack-tip. The stress intensity for crack extension (fracture toughness) and energy absorption during the crack extension (fracture resistance) can be considered as material dependent constants. Compared to such a fracture process in polycrystalline alloys, the characteristics of fracture behaviors in MGs are summarized as follows:

3.1. Discontinuous Stress/Strain Field

For polycrystalline alloys with work hardening effect, the deformation and stress/strain field can be approximately continuous. For MGs, however, due to the shear softening effect [33,34,96], the viscous flow in MGs at room temperature will localize and form shear bands [16,34]. Since the shear transformations and shear band motions can be considered as thermal activated events [97], the position and the moment of shear band formation are random. After formation, the shear band can slide at a relatively lower stress [16,27,34], with a strain rate much faster than the external load [69,98]. During further loading, the shear bands delaminate and cause rapid fracture of the material [38]. As a result, the spatial and temporal distribution of strain/stress field in MGs during the fracture process will be discontinuous. The global stress/strain will also change discontinuously, observed as the serrations on the loading curves [13,25,99]. The strain in the shear bands is highly localized, while the other places are still elastic.

3.2. Fracture Resistance

Since the fracture process of MGs can be divided into different stages, the fracture resistance and plastic zone size might not be constant during the crack extension. In the stage of multiple shear banding, the stress intensity should be large enough that the shear bands can nucleation from the notch-tip. Part of mechanical energy will be released by plastic deformation in the manner of shear band sliding. It corresponds to a ductile fracture behavior. When the shear band delamination begins, the stress intensity required for further delamination will drop suddenly (due to the stress concentration at the sharp tip of the delaminated shear band and the relatively low viscosity inside the shear band). At the moment, the stress intensity in the sample is higher than the stress required for crack extension. The over load makes the shear band delamination a transient non-static process. Later, the crack extends rapidly and forms the rough zone on the fractograph. In experiments, the rapid fracture of MGs usually takes a very short moment (<<1 s), and sometimes is accompanied by sparking [100]. Since the crack extension rate is much faster than the shear band sliding rate, there will be not enough time for the material to deform by shear band sliding. The plastic deformation and energy absorption during the shear band delamination and rapid fracture will be small. It corresponds to a brittle fracture behavior.

3.3. Plastic Zone

The crack-tip plastic zone of MGs also differs from that of polycrystalline alloys. In the stage of multiple shear banding, the plastic zone depends on the distribution of shear bands. It has been noticed that the notch-tip shear bands distribution on the side surface of MGs during mode I fracture deviates from the local stress distribution [6,27,50] (Figure 8). The reason for such deviation is that the shear bands in MGs propagates along the local maximum shear stress direction (or the direction of the local maximum shear stress gradient) [101]. Due to the stress concentration at shear band tip, the shear band is able to propagate into the region with relatively low stress. Therefore, the actual plastic deformed region of MGs depends on the distribution of shear stress direction, instead of the distribution of stress intensity. With the increasing stress intensity during loading, new shear bands are formed from the notch-tip, and the plastic deformed region gradually extends. Later, the shear band delamination is triggered when the critical condition is reached. At the moment, the sharp crack tip extends rapidly along the shear band path. The plastic zone depends on the softened region near the crack tip. Since the opposite sides of the delaminated shear bands could match well with each other, the plastic zone size might be very small. If the plastic region has a similar size to the dimples on the fractograph (a few microns to tens of microns), it will be much smaller than the shear band itself (the shear bands in BMGs can be millimeter-scale). The decreased plastic zone size explains the reduced fracture resistance during the shear band delamination.

Based on the above discussions, the shear band controlled fracture process of MGs can be considered as a dynamic process, in which the plastic deformation is achieved through individual shear band formation and sliding instead of continuous deformation. With different sample size, sample shape and loading condition, the stability, direction, formation stress, morphology and sliding distance of the shear bands in MGs will change accordingly, which will result in different deformation behaviors. Therefore, those mechanical properties including plasticity and fracture toughness will strongly depend on both the material and the external loading condition. For instance, straightforward shear bands are formed under uniaxial loading, and the material will show a shear mode fracture [18–20,44]. Due to the normal stress sensitivity of the shear band nucleation, MGs show different yielding stress and shear directions in compression and tension [18–20]. Under different constrained loading conditions, different failure mode, plasticity, and serration behavior can be obtained [102,103], etc. In the case of mode I fracture, the energy absorption is mainly attributed to the stage of multiple shear bands formation and sliding. In small size samples, sustainable multiple shear banding can be more easily achieved and better fracture toughness/fracture resistance can be obtained. Under different loading mode, the notch tip shear band distribution (plastic zone) will change accordingly, which will result in different fracture properties [104]. The ductility and toughness of the material depend on the crucial moment of shear band delamination, which is sensitive to the local structure and stress condition. In different tests, different samples, and different positions of a sample, the critical condition of shear band delamination might show a high degree of variability and randomness. With this understanding, it becomes doubtful to treat those mechanical properties in MGs as intrinsic material properties.

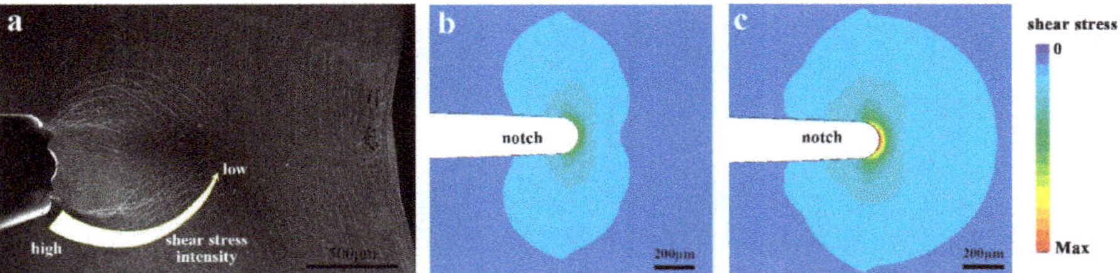

Figure 8. A comparison of shear band distribution and shear stress distribution near the notch tip of a $Pd_{77.5}Cu_6Si_{16.5}$ MG during 3-point bending. (**a**) The shear band distribution near the notch tip of the sample. (**b**) Simulated shear stress distribution on the side surface of the sample. (**c**) Simulated shear stress distribution in the middle of the sample.

4. Characterizing the Fracture Properties of Metallic Glasses

At present, how to characterize the fracture property of MGs is still under debate. The main points in controversy are summarized as below:

4.1. Size Limit of MGs

According to the standard test method of mode I fracture toughness [21,22], a large sample size is required to ensure the validity of the measured results. For a sample with a fracture toughness of K_{IC}, the required thickness of the sample can be expressed as:

$$B, a \geq 2.5\left(\frac{K_Q}{\sigma_Y}\right)^2 \tag{3}$$

where B represents the sample's thickness, a represents the crack length, K_Q represents the toughness value, and σ_Y represents the yielding strength. As the mode I fracture toughness value of MGs falls in the range of a few to more than one hundred $MPa \cdot m^{0.5}$ [21,22], the required sample size will be a few millimeters to more than one centimeter. However, such a size requirement can hardly be reached by most MGs, due to their poor glass-forming ability. Until now, the BMGs with a critical size of more than one centimeter are still rare. Most fracture tests of MGs are carried out with a sample size of a few millimeters, which can hardly reach the size requirement of standard test [21,22]. Besides, crack opening displacement should be measured during the fracture process as required by the test standard. However, it will be difficult to install a very small extensometer onto the small MG samples.

4.2. Fatigue Pre-Crack

According to the test standard [21,22], a fatigue pre-crack with enough length should be prepared to ensure the sharpness of the initial crack tip. However, this requirement can hardly be achieved by MGs with small size. As the ligament length of most MG samples is small, the fatigue load should decrease according with the extension of the fatigue crack. Moreover, it is difficult to prepare a straight fatigue crack in MGs. As the shear bands near the notch tip are curved and deviated from the original notch direction, the fatigue crack also will easily deviate from the original direction. A recent report discovered that straightforward fatigue crack can be prepared under the pure shear stress condition achieved by asymmetric four-point bending [105], which provides a feasible solution for this question.

4.3. Variability of Fracture Toughness

As MG samples can hardly meet the requirement of size and fatigue pre-crack in the test standard, some nonstandard measurements were used for substitution. Some studies used the notch toughness (without fatigue pre-crack) as an approximation to characterize the fracture toughness of MGs [30,106]. In some studies, the fatigue pre-cracking process was replaced by machining a notch tip with a very small radius [6]. J-integral toughness tests were also used, as the size requirement of J-integral toughness test is smaller than that of standard measurement [22,52]. In these studies under different loading conditions and different test methods, a high variability in the measured fracture toughness is observed.

Experiments of notch toughness show that with increasing notch-tip radius, the degree of stress concentration at the notch-tip decreases, and a greater load is needed to form shear bands from the notch-tip. A larger fracture toughness can be measured [107]. In small size samples, a shorter shear band sliding distance is required to release the same stress. The material prefers to form multiple shear bands rather than shear band delamination, and a higher fracture resistance can be achieved [42,43,51]. The measured fracture toughness is also sensitive to the temperature [108], loading rate, loading mode [107], heat history [109], residual stress [110], geometry [111] and the distribution of impurities/flaws [112]. For example, the cooling rate directly affects the fracture energy of a Ti-Zr-Cu-Ni-Be MG, which can range from 148.9 to only 0.2 kJ m^{-2} [26]. The measured fracture toughness of Vitreloy-1 MG can vary widely from ~18 to ~130 MPa·m$^{1/2}$ in different reports with different test methods [39,75,113].

To understand the variability of fracture toughness in MGs. Some studies suggest that the properties of MGs are "intrinsically variable" [114], as the shear band behaviors of MGs depends on both the material and the loading condition [42,107,115]. Some studies suggest that the sample geometry to measure the fracture toughness of BMGs should be much smaller than suggested by the standards for crystalline alloys [111]. According to the discussions of this review, the variability of mechanical properties of MGs actually reflects the sensibility of the shear band behaviors to the external loading condition and the randomness in the shear band motions. We have seen the tendency that new experimental methods, theoretical criteria, or definitions will be proposed to further study these shear band controlled materials, especially for the engineering important metallic glasses.

5. Prospects

Here a brief review has been provided to understand the fracture behavior and fracture mechanism of MGs. As present, the study of fracture in MGs is still preliminary, due to the complex fracture mechanism of MGs compared to the classic fracture mechanism in polycrystalline alloys. Here we would like to raise some representative unsolved issues in this field, which to our knowledge have relatively high interest and importance.

5.1. Characterizing the Dynamic Fracture Properties of MGs

As discussed above, the different fracture modes, dynamically changing fracture resistance during different stages of fracture, and high variability in the measured fracture toughness of MGs result from their complex fracture mechanism. Due to the limitation of glass-forming ability, standard test methods of fracture properties might not be applicable in those MGs with small critical size. Appropriate experimental methods and theoretical tools are needed to characterize and further study the dynamic fracture behavior of MGs and other shear band controlled materials. For those small size MGs that cannot meet the requirement of test standard, new methods are needed to be developed to characterize their toughness and fracture resistance.

5.2. Toughening Method of MGs

Although some MGs show high damage tolerance, the fracture toughness of the most widely used Fe-based magnetic MGs in industry is still poor. The poor tensile ductility and low impact toughness of MGs also limit their further applications in many cases. How to enhance the mechanical performance of Fe-based MG and their composites without negative effects on their magnetic properties, and how to achieve tensile ductility and toughness in large MG samples are of great significance.

5.3. Performance Under Special Conditions

As the shear band behaviors can be tuned by external loading condition, MGs can possibly show better mechanical performance under special conditions, such as high temperature, cryogenic temperature, small sample size, irregular sample shape, irradiation environment, high strain rate, constrained loading, etc. These factors could provide potential applications for MGs.

5.4. Rapid Fracture Mechanism of MGs

The formation process of the rough zone on the fractograph and the rapid fracture mechanism under impact loading and other rapid loading conditions are still unclear. To optimize the fracture toughness and impact toughness of MGs, the physics behind the rapid fracture of MGs should be studied and clarified.

5.5. Mode II and Mode III Fracture Mechanism of MGs

At present, the studies of fracture mechanism of MGs under Mode II and mode III fracture are still rare. The different fracture properties of MGs under different load mode might provide an approach to understand the correlation between the mechanical behaviors and the loading conditions of MGs.

Author Contributions: Conceptualization, G.-N.Y., Y.S. and K.-F.Y.; methodology, G.-N.Y., Y.S. and K.-F.Y.; experiment and simulation, G.-N.Y.; data analysis, G.-N.Y., Y.S. and K.-F.Y.; writing—original draft preparation, G.-N.Y.; writing—review and editing, Y.S. and K.-F.Y.; visualization, G.-N.Y.; supervision, K.-F.Y.; funding acquisition, K.-F.Y. and G.-N.Y.

Funding: This work is supported by the National Natural Science Foundation of China (Grand Nos. 51771096 and 51571127) and the Fostering Talents Foundation of Guangdong University of Technology (Grand No. 262511006).

Acknowledgments: We thank Pan Gong and Xin Wang for sharing the precious data and helpful discussion.

Conflicts of Interest: The authors declare no conflict of interest.

References

1. Inoue, A. Stabilization of metallic supercooled liquid and bulk amorphous alloys. *Acta Mater.* **2000**, *48*, 279–306. [CrossRef]
2. Wang, W.H.; Dong, C.; Shek, C.H. Bulk metallic glasses. *Mater. Sci. Eng. R* **2004**, *44*, 45–89. [CrossRef]
3. Jan, S. Processing of bulk metallic glass. *Adv. Mater.* **2010**, *22*, 1566–1597.
4. Greer, A.L.; Ma, E. Bulk metallic glasses: At the cutting edge of metals research. *MRS Bull.* **2007**, *32*, 611–615. [CrossRef]
5. Akihisa, I.; Baolong, S.; Hisato, K.; Hidemi, K.; Yavari, A.R. Cobalt-based bulk glassy alloy with ultrahigh strength and soft magnetic properties. *Nat. Mater.* **2003**, *2*, 661–663.
6. Demetriou, M.D.; Launey, M.E.; Garrett, G.; Schramm, J.P.; Hofmann, D.C.; Johnson, W.L.; Ritchie, R.O. A damage-tolerant glass. *Nat. Mater.* **2011**, *10*, 123–128. [CrossRef]
7. Yao, K.F.; Ruan, F.; Yang, Y.Q.; Chen, N. Superductile bulk metallic glass. *Appl. Phys. Lett.* **2006**, *88*, 122106. [CrossRef]
8. Yao, K.F.; Zhang, C.Q. Fe-based bulk metallic glass with high plasticity. *Appl. Phys. Lett.* **2007**, *90*, 061901. [CrossRef]
9. Greer, A.L.; Cheng, Y.Q.; Ma, E. Shear bands in metallic glasses. *Mater. Sci. Eng. R* **2013**, *74*, 71–132. [CrossRef]

10. Schuh, C.A.; Hufnagel, T.C.; Ramamurty, U. Overview No.144—Mechanical behavior of amorphous alloys. *Acta Mater.* **2007**, *55*, 4067–4109. [CrossRef]

11. Sun, B.A.; Wang, W.H. The fracture of bulk metallic glasses. *Prog. Mater Sci.* **2015**, *74*, 211–307. [CrossRef]

12. Xu, J.; Ramamurty, U.; Ma, E. The fracture toughness of bulk metallic glasses. *JOM* **2010**, *62*, 10–18. [CrossRef]

13. Yang, G.N.; Chen, S.Q.; Gu, J.L.; Zhao, S.F.; Li, J.F.; Shao, Y.; Wang, H.; Yao, K.F. Serration behaviours in metallic glasses with different plasticity. *Philos. Mag.* **2016**, *96*, 2243–2255. [CrossRef]

14. Wu, Y.; Bei, H.; Wang, Y.L.; Lu, Z.P.; George, E.P.; Gao, Y.F. Deformation-induced spatiotemporal fluctuation, evolution and localization of strain fields in a bulk metallic glass. *Int. J. Plast.* **2015**, *71*, 136–145. [CrossRef]

15. Lan, S.; Wu, Z.; Wei, X.; Zhou, J.; Lu, Z.; Neuefeind, J.; Wang, X.-L. Structure origin of a transition of classic-to-avalanche nucleation in Zr-Cu-Al bulk metallic glasses. *Acta Mater.* **2018**, *149*, 108–118. [CrossRef]

16. Steif, P.S.; Spaepen, F.; Hutchinson, J.W. Strain localization in amorphous metals. *Acta Metall.* **1982**, *30*, 447–455. [CrossRef]

17. Cao, A.J.; Cheng, Y.Q.; Ma, E. Structural processes that initiate shear localization in metallic glass. *Acta Mater.* **2009**, *57*, 5146–5155. [CrossRef]

18. Schuh, C.A.; Lund, A.C. Atomistic basis for the plastic yield criterion of metallic glass. *Nat. Mater.* **2003**, *2*, 449–452. [CrossRef]

19. Zhang, Z.F.; He, G.; Eckert, J.; Schultz, L. Fracture mechanisms in bulk metallic glassy materials. *Phys. Rev. Lett.* **2003**, *91*, 045505. [CrossRef]

20. Zhang, Z.F.; Eckert, J. Unified tensile fracture criterion. *Phys. Rev. Lett.* **2005**, *94*, 094301. [CrossRef]

21. ASTM International. *ASTM E399, Standard Test Method for Linear-Elastic Plane-Strain Fracture Toughness KIc of Metallic Materials*; ASTM International: West Conshohocken, PA, USA, 2009.

22. ASTM International. *ASTM E1820, Standard Test Method for Measurement of Fracture Toughness*; ASTM International: West Conshohocken, PA, USA, 2008.

23. Yu, M.; Zhen, L.; Chao, Y.J. An Assessment of Mechanical Properties of A508-3 Steel Used in Chinese Nuclear Reactor Vessels. *J. Press. Vess.Technol. ASME* **2015**, *137*, 031402. [CrossRef]

24. Klaumünzer, D.; Maaß, R.; Löffler, J.F. Stick-slip dynamics and recent insights into shear banding in metallic glasses. *J. Mater. Res.* **2011**, *26*, 1453–1463. [CrossRef]

25. Sun, B.A.; Pauly, S.; Tan, J.; Stoica, M.; Wang, W.H.; Kühn, U.; Eckert, J. Serrated flow and stick–slip deformation dynamics in the presence of shear-band interactions for a Zr-based metallic glass. *Acta Mater.* **2012**, *60*, 4160–4171. [CrossRef]

26. Gu, X.J.; Poon, S.J.; Shiflet, G.J.; Lewandowski, J.J. Ductile-to-brittle transition in a Ti-based bulk metallic glass. *Scr. Mater.* **2009**, *60*, 1027–1030. [CrossRef]

27. Yang, G.N.; Shao, Y.; Yao, K.F. The shear band controlled deformation in metallic glass: A perspective from fracture. *Sci. Rep.* **2016**, *6*, 21852. [CrossRef] [PubMed]

28. Tandaiya, P.; Narasimhan, R.; Ramamurty, U. On the mechanism and the length scales involved in the ductile fracture of a bulk metallic glass. *Acta Mater.* **2013**, *61*, 1558–1570. [CrossRef]

29. Lewandowski, J.J.; Gu, X.J.; Shamimi Nouri, A.; Poon, S.J.; Shiflet, G.J. Tough Fe-based bulk metallic glasses. *Appl. Phys. Lett.* **2008**, *92*, 091918. [CrossRef]

30. Jia, P.; Zhu, Z.D.; Ma, E.; Xu, J. Notch toughness of Cu-based bulk metallic glasses. *Scr. Mater.* **2009**, *61*, 137–140. [CrossRef]

31. Argon, A.S.; Salama, M. Mechanism of fracture in glassy materials capable of some inelastic deformation. *Mat. Sci. Eng.* **1976**, *23*, 219–230. [CrossRef]

32. Saffman, P.G.; Taylor, G. The penetration of a fluid into a porous medium or hele-shaw cell containing a more viscous liquid. *Proc. R. Soc. London, Ser. A* **1958**, *245*, 312–329.

33. Spaepen, F. A microscopic mechanism for steady state inhomogeneous flow in metallic glasses. *Acta Metall.* **1977**, *25*, 407–415. [CrossRef]

34. Argon, A.S. Plastic deformation in metallic glasses. *Acta Metall.* **1979**, *27*, 47–58. [CrossRef]

35. Pan, J.; Chen, Q.; Liu, L.; Li, Y. Softening and dilatation in a single shear band. *Acta Mater.* **2011**, *59*, 5146–5158. [CrossRef]

36. Shao, Y.; Yang, G.N.; Yao, K.F.; Liu, X. Direct experimental evidence of nano-voids formation and coalescence within shear bands. *Appl. Phys. Lett.* **2014**, *105*, 181909. [CrossRef]

37. Wang, K.; Fujita, T.; Zeng, Y.Q.; Nishiyama, N.; Inoue, A.; Chen, M.W. Micromechanisms of serrated flow in a Ni50Pd30P20 bulk metallic glass with a large compression plasticity. *Acta Mater.* **2008**, *56*, 2834–2842. [CrossRef]

38. Maaß, R.; Birckigt, P.; Borchers, C.; Samwer, K.; Volkert, C.A. Long range stress fields and cavitation along a shear band in a metallic glass: The local origin of fracture. *Acta Mater.* **2015**, *98*, 94–102. [CrossRef]

39. Lowhaphandu, P.; Ludrosky, L.A.; Montgomery, S.L.; Lewandowski, J.J. Deformation and fracture toughness of a bulk amorphous Zr–Ti–Ni–Cu–Be alloy. *Intermetallics* **2000**, *8*, 487–492. [CrossRef]

40. Raghavan, R.; Murali, P.; Ramamurty, U. On factors influencing the ductile-to-brittle transition in a bulk metallic glass. *Acta Mater.* **2009**, *57*, 3332–3340. [CrossRef]

41. Wang, G.; Zhao, D.Q.; Bai, H.Y.; Pan, M.X.; Xia, A.L.; Han, B.S.; Xi, X.K.; Wu, Y.; Wang, W.H. Nanoscale periodic morphologies on the fracture surface of brittle metallic glasses. *Phys. Rev. Lett.* **2007**, *98*, 235501. [CrossRef]

42. Yang, Y.; Liu, C.T. Size effect on stability of shear-band propagation in bulk metallic glasses: An overview. *J. Mater. Sci.* **2012**, *47*, 55–67. [CrossRef]

43. Gludovatz, B.; Granata, D.; Thurston, K.V.S.; Löffler, J.F.; Ritchie, R.O. On the understanding of the effects of sample size on the variability in fracture toughness of bulk metallic glasses. *Acta Mater.* **2017**, *126*, 494–506. [CrossRef]

44. Jun, H.J.; Lee, K.S.; Kim, C.P.; Chang, Y.W. Ductility enhancement of a Ti-based bulk metallic glass through annealing treatment below the glass transition temperature. *Intermetallics* **2012**, *20*, 47–54. [CrossRef]

45. Raphael, J.; Wang, G.Y.; Liaw, P.K.; Senkov, O.N.; Miracle, D.B. Fatigue and Fracture Behavior of a Ca-Based Bulk-Metallic Glass. *Metall. Mater. Trans. A* **2010**, *41*, 1775–1779. [CrossRef]

46. Wang, G.Y.; Liaw, P.K.; Senkov, O.N.; Miracle, D.B. The Duality of Fracture Behavior in a Ca-based Bulk-Metallic Glass. *Metall. Mater. Trans. A* **2011**, *42*, 1499–1503. [CrossRef]

47. Zhang, Z.F.; Wu, F.F.; Gao, W.; Tan, J.; Wang, Z.G.; Stoica, M.; Das, J.; Eckert, J.; Shen, B.L.; Inoue, A. Wavy cleavage fracture of bulk metallic glass. *Appl. Phys. Lett.* **2006**, *89*, 2273. [CrossRef]

48. Wang, G.; Han, Y.N.; Xu, X.H.; Ke, F.J.; Han, B.S.; Wang, W.H. Ductile to brittle transition in dynamic fracture of brittle bulk metallic glass. *J. Appl. Phys.* **2008**, *103*, 093520. [CrossRef]

49. Xi, X.K.; Zhao, D.Q.; Pan, M.X.; Wang, W.H.; Wu, Y.; Lewandowski, J.J. Fracture of brittle metallic glasses: Brittleness or plasticity. *Phys. Rev. Lett.* **2005**, *94*, 125510. [CrossRef]

50. He, Q.; Cheng, Y.Q.; Ma, E.; Xu, J. Locating bulk metallic glasses with high fracture toughness: Chemical effects and composition optimization. *Acta Mater.* **2011**, *59*, 202–215. [CrossRef]

51. Yang, G.N.; Shao, Y.; Yao, K.F. A non-viscous-featured fractograph in metallic glasses. *Philos. Mag.* **2016**, *96*, 542–550. [CrossRef]

52. Anderson, T.L. *Fracture Mechanics: Fundamentals and Applications*, 3rd ed.; Taylor & Francis: Abingdon, UK, 2011.

53. Gu, J.L.; Yang, X.L.; Zhang, A.L.; Shao, Y.; Zhao, S.F.; Yao, K.F. Centimeter-sized Ti-rich bulk metallic glasses with superior specific strength and corrosion resistance. *J. Non-Cryst. Solids* **2019**, *512*, 206–210. [CrossRef]

54. Ding, H.Y.; Shao, Y.; Gong, P.; Li, J.F.; Yao, K.F. A senary TiZrHfCuNiBe high entropy bulk metallic glass with large glass-forming ability. *Mater. Lett.* **2014**, *125*, 151–153. [CrossRef]

55. Zhao, S.F.; Yang, G.N.; Ding, H.Y.; Yao, K.F. A quinary Ti-Zr-Hf-Be-Cu high entropy bulk metallic glass with a critical size of 12 mm. *Intermetallics* **2015**, *61*, 47–50. [CrossRef]

56. Gong, P.; Wang, X.; Shao, Y.; Chen, N.; Liu, X.; Yao, K.F. A Ti-Zr-Be-Fe-Cu bulk metallic glass with superior glass-forming ability and high specific strength. *Intermetallics* **2013**, *43*, 177–181. [CrossRef]

57. Zhao, S.F.; Shao, Y.; Liu, X.; Chen, N.; Ding, H.Y.; Yao, K.F. Pseudo-quinary Ti20Zr20Hf20Be20(Cu20-xNix) high entropy bulk metallic glasses with large glass forming ability. *Mater. Des.* **2015**, *87*, 625–631. [CrossRef]

58. Zhao, S.F.; Gong, P.; Li, J.F.; Chen, N.; Yao, K.F. Quaternary Ti-Zr-Be-Ni bulk metallic glasses with large glass-forming ability. *Mater. Des.* **2015**, *85*, 564–573. [CrossRef]

59. Chao, Y.J.; Ward, J.D., Jr.; Sands, R.G. Charpy impact energy, fracture toughness and ductile–brittle transition temperature of dual-phase 590 Steel. *Mater. Des.* **2007**, *28*, 551–557. [CrossRef]

60. Tang, M.Q.; Zhang, H.F.; Zhu, Z.W.; Fu, H.M.; Wang, A.M.; Li, H.; Hu, Z.Q. TiZr-base Bulk Metallic Glass with over 50 mm in Diameter. *J. Mater. Sci. Technol.* **2010**, *26*, 481–486. [CrossRef]

61. Inoue, A.; Zhang, T. Impact Fracture Energy of Bulk Amorphous Zr55Al10Cu30Ni5 Alloy. *Mater. Trans.* **1996**, *37*, 1726–1729. [CrossRef]

62. Yokoyama, Y.; Akeno, Y.; Yamasaki, T.; Liaw, P.K.; Buchanan, R.A.; Inoue, A. Evolution of Mechanical Properties of Cast Zr50Cu40Al10 Glassy Alloys by Structural Relaxation. *Mater. Trans.* **2005**, *46*, 2755–2761. [CrossRef]

63. Roberts, S.; Zachrisson, C.; Kozachkov, H.; Ullah, A.; Shapiro, A.A.; Johnson, W.L.; Hofmann, D.C. Cryogenic Charpy impact testing of metallic glass matrix composites. *Scr. Mater.* **2012**, *66*, 284–287. [CrossRef]

64. Raghavan, R.; Murali, P.; Ramamurty, U. Ductile to brittle transition in the Zr 41.2Ti 13.75Cu 12.5Ni 10Be 22.5 bulk metallic glass. *Intermetallics* **2006**, *14*, 1051–1054. [CrossRef]

65. Ramamurty, U.; Lee, M.L.; Basu, J.; Li, Y. Embrittlement of a bulk metallic glass due to low-temperature annealing. *Scr. Mater.* **2002**, *47*, 107–111. [CrossRef]

66. Szuecs, F.; Kim, C.P.; Johnson, W.L. Mechanical properties of Zr56.2Ti13.8Nb5.0Cu6.9Ni5.6Be12.5 ductile phase reinforced bulk metallic glass composite. *Acta Mater.* **2001**, *49*, 1507–1513. [CrossRef]

67. Yokoyama, Y.; Nishiyama, N.; Fukaura, K.; Sunada, H.; Inoue, A. Rotating-Beam Fatigue Strength of Pd40Cu30Ni10P20 Bulk Amorphous Alloy. *Mater. Trans.* **1999**, *40*, 696–699. [CrossRef]

68. Song, S.X.; Nieh, T.G. Flow serration and shear-band viscosity during inhomogeneous deformation of a Zr-based bulk metallic glass. *Intermetallics* **2009**, *17*, 762–767. [CrossRef]

69. Maaß, R.; Löffler, J.F. Shear-Band Dynamics in Metallic Glasses. *Adv. Funct. Mater.* **2015**, *25*, 2353–2368. [CrossRef]

70. Ma, W.F.; Kou, H.C.; Li, J.S.; Chang, H.; Zhou, L. Effect of strain rate on compressive behavior of Ti-based bulk metallic glass at room temperature. *J. Alloys Compd.* **2009**, *472*, 214–218. [CrossRef]

71. Mann, J.Y. *Fatigue Materials*; Melbourne University Press: Melbourne, Australia, 1967.

72. Davis, L.A. Fracture of Ni-Fe base metallic glasses. *J. Mater. Sci.* **1975**, *10*, 1557–1564. [CrossRef]

73. Ogura, T.; Masumoto, T.; Fukushima, K. Fatigue fracture of amorphous Pd-20at-percent Si alloy. *Scr. Metall.* **1975**, *9*, 109–113. [CrossRef]

74. Ogura, T.; Fukushima, K.; Masumoto, T. Propagation of fatigue cracks in amorphous metals. *Mater. Sci. Eng.* **1976**, *23*, 231–235. [CrossRef]

75. Gilbert, C.J.; Schroeder, V.; Ritchie, R.O. Mechanisms for fracture and fatigue-crack propagation in a bulk metallic glass. *Metall. Mater. Trans. A* **1999**, *30*, 1739–1753. [CrossRef]

76. Chen, H.S. Glassy metals. *Rep. Prog. Phys.* **1980**, *43*, 353. [CrossRef]

77. Fujita, K.; Zhang, W.; Shen, B.; Amiya, K.; Ma, C.L.; Nishiyama, N. Fatigue properties in high strength bulk metallic glasses. *Intermetallics* **2012**, *30*, 12–18. [CrossRef]

78. Ponnambalam, V.; Poon, S.J.; Shiflet, G.J. Fe-based bulk metallic glasses with diameter thickness larger than one centimeter. *J. Mater. Res.* **2004**, *19*, 1320–1323. [CrossRef]

79. Qiao, D.C.; Wang, G.Y.; Liaw, P.K.; Ponnambalam, V.; Poon, S.J.; Shiflet, G.J. Fatigue behavior of an Fe48Cr15Mo14Er2C15B6 amorphous steel. *J. Mater. Res.* **2007**, *22*, 544–550. [CrossRef]

80. Gilbert, C.J.; Lippmann, J.M.; Ritchie, R.O. Fatigue of a Zr-Ti-Cu-Ni-Be bulk amorphous metal: Stress/life and crack-growth behavior. *Scr. Mater.* **1998**, *38*, 537–542. [CrossRef]

81. Wang, G.Y.; Liaw, P.K.; Peker, A.; Yang, B.; Benson, M.L.; Yuan, W.; Peter, W.H.; Huang, L.; Freels, A.; Buchanan, R.A.; et al. Fatigue behavior of Zr-Ti-Ni-Cu-Be bulk-metallic glasses. *Intermetallics* **2005**, *13*, 429–435. [CrossRef]

82. Wang, G.Y.; Liaw, P.K.; Peter, W.H.; Yang, B.; Yokoyama, Y.; Benson, M.L.; Green, B.A.; Kirkham, M.J.; White, S.A.; Saleh, T.A.; et al. Fatigue behavior of bulk-metallic glasses. *Intermetallics* **2004**, *12*, 885–892. [CrossRef]

83. Qiao, D.C.; Fan, G.J.; Liaw, P.K.; Choo, H. Fatigue behaviors of the Cu47.5Zr47.5Al5 bulk-metallic glass (BMG) and Cu(47.5)Zr(38)Hf(9.5)Al(5)BMG composite. *Int. J. Fatigue* **2007**, *29*, 2149–2154. [CrossRef]

84. Peter, W.H.; Liaw, P.K.; Buchanan, R.A.; Liu, C.T.; Brooks, C.R.; Horton, J.A.; Carmichael, C.A.; Wright, J.L. Fatigue behavior of Zr52.5Al10Ti5Cu17.9Ni14.6 bulk metallic glass. *Intermetallics* **2002**, *10*, 1125–1129. [CrossRef]

85. Morrison, M.L.; Buchanan, R.A.; Liaw, P.K.; Green, B.A.; Wang, G.Y.; Liu, C.; Horton, J.A. Four-point-bending-fatigue behavior of the Zr-based Vitreloy 105 bulk metallic glass. *Mater. Sci. Eng. A Struct. Mater. Prop. Microstruct. Process.* **2007**, *467*, 190–197. [CrossRef]

86. Wang, G.Y.; Liaw, P.K.; Peter, W.H.; Yang, B.; Freels, M.; Yokoyama, Y.; Benson, M.L.; Green, B.A.; Saleh, T.A.; McDaniels, R.L.; et al. Fatigue behavior and fracture morphology of Zr50Al10Cu40 and Zr50Al10Cu30Ni10 bulk-metallic glasses. *Intermetallics* **2004**, *12*, 1219–1227. [CrossRef]

87. Wang, G.Y.; Liaw, P.K.; Yokoyama, Y.; Freels, M.; Inoue, A. The influence of Pd on tension-tension fatigue behavior of Zr-based bulk-metallic glasses. *Int. J. Fatigue* **2010**, *32*, 599–604. [CrossRef]

88. Wang, G.; Liaw, P.K.; Senkov, O.N.; Miracle, D.B.; Morrison, M.L. Mechanical and Fatigue Behavior of Ca65Mg15Zn20 Bulk-Metallic Glass. *Adv. Eng. Mater.* **2009**, *11*, 27–34. [CrossRef]

89. Gilbert, C.J.; Ritchie, R.O.; Johnson, W.L. Fracture toughness and fatigue-crack propagation in a Zr-Ti-Ni-Cu-Be bulk metallic glass. *Appl. Phys. Lett.* **1997**, *71*, 476–478. [CrossRef]

90. Menzel, B.C.; Dauskardt, R.H. The fatigue endurance limit of a Zr-based bulk metallic glass. *Scr. Mater.* **2006**, *55*, 601–604. [CrossRef]

91. Launey, M.E.; Busch, R.; Kruzic, J.J. Influence of structural relaxation on the fatigue behavior of a Zr41.25Ti13.75Ni10CU12.5Be22.5 bulk amorphous alloy. *Scr. Mater.* **2006**, *54*, 483–487.

92. Launey, M.E.; Busch, R.; Kruzic, J.J. Effects of free volume changes and residual stresses on the fatigue and fracture behavior of a Zr-Ti-Ni-Cu-Be bulk metallic glass. *Acta Mater.* **2008**, *56*, 500–510. [CrossRef]

93. Launey, M.E.; Hofmann, D.C.; Johnson, W.L.; Ritchie, R.O. Solution to the problem of the poor cyclic fatigue resistance of bulk metallic glasses. *Proc. Natl. Acad. Sci. USA* **2009**, *106*, 4986–4991. [CrossRef]

94. Wang, G.Y.; Liaw, P.K.; Yokoyama, Y.; Inoue, A. Size effects on the fatigue behavior of bulk metallic glasses. *J. Appl. Phys.* **2011**, *110*, 42. [CrossRef]

95. Song, Z.Q.; He, Q.; Ma, E.; Xu, J. Fatigue endurance limit and crack growth behavior of a high-toughness Zr61Ti2Cu25Al12 bulk metallic glass. *Acta Mater.* **2015**, *99*, 165–175. [CrossRef]

96. Spaepen, F. Metallic glasses: Must shear bands be hot? *Nat. Mater.* **2006**, *5*, 7–8. [CrossRef]

97. Mayr, S.G. Activation energy of shear transformation zones: A key for understanding rheology of glasses and liquids. *Phys. Rev. Lett.* **2006**, *97*, 195501. [CrossRef] [PubMed]

98. Maaß, R.; Klaumünzer, D.; Löffler, J.F. Propagation dynamics of individual shear bands during inhomogeneous flow in a Zr-based bulk metallic glass. *Acta Mater.* **2011**, *59*, 3205–3213. [CrossRef]

99. Sun, B.A.; Yu, H.B.; Jiao, W.; Bai, H.Y.; Zhao, D.Q.; Wang, W.H. Plasticity of Ductile Metallic Glasses: A Self-Organized Critical State. *Phys. Rev. Lett.* **2010**, *105*, 035501. [CrossRef] [PubMed]

100. Liu, C.T.; Heatherly, L.; Easton, D.S.; Carmichael, C.A.; Schneibel, J.H.; Chen, C.H.; Wright, J.L.; Yoo, M.H.; Horton, J.A.; Inoue, A. Test environments and mechanical properties of Zr-base bulk amorphous alloys. *Metall. Mater. Trans. A* **1998**, *29*, 1811–1820. [CrossRef]

101. Yang, G.N.; Shao, Y.; Yao, K.F. The material-dependence of plasticity in metallic glasses: An origin from shear band thermology. *Mater. Des.* **2016**, *96*, 189–194. [CrossRef]

102. Yang, G.N.; Gu, J.L.; Chen, S.Q.; Shao, Y.; Wang, H.; Yao, K.F. Serration Behavior of a Zr-Based Metallic Glass Under Different Constrained Loading Conditions. *Metall. Mater. Trans. A* **2016**, *47*, 5395–5400. [CrossRef]

103. Pan, J.; Zhou, H.F.; Wang, Z.T.; Li, Y.; Gao, H.J. Origin of anomalous inverse notch effect in bulk metallic glasses. *J. Mech. Phys. Solids.* **2015**, *84*, 85–94. [CrossRef]

104. Tandaiya, P.; Ramamurty, U.; Narasimhan, R. Mixed mode (I and II) crack tip fields in bulk metallic glasses. *J. Mech. Phys. Solids.* **2009**, *57*, 1880–1897. [CrossRef]

105. Bernard, C.; Keryvin, V.; Doquet, V.; Hin, S.; Yokoyama, Y. A sequential pre-cracking procedure to measure the mode-I fracture toughness of ultra pure bulk metallic glasses. *Scr. Mater.* **2017**, *141*, 58–61. [CrossRef]

106. Zhu, Z.D.; Jia, P.; Xu, J. Optimization for toughness in metalloid-free Ni-based bulk metallic glasses. *Scr. Mater.* **2011**, *64*, 785–788. [CrossRef]

107. Hassan, H.A.; Kecskes, L.; Lewandowski, J.J. Effects of changes in test temperature and loading conditions on fracture toughness of a Zr-based bulk metallic glass. *Metall. Mater. Trans. A* **2008**, *39*, 2077–2085. [CrossRef]

108. Raut, D.; Narayan, R.L.; Tandaiya, P.; Ramamurty, U. Temperature-dependence of mode I fracture toughness of a bulk metallic glass. *Acta Mater.* **2018**, *144*, 325–336. [CrossRef]

109. Chen, W.; Zhou, H.F.; Liu, Z.; Ketkaew, J.; Li, N.; Yurko, J.; Hutchinson, N.; Gao, H.J.; Schroers, J. Processing effects on fracture toughness of metallic glasses. *Scr. Mater.* **2017**, *130*, 152–156. [CrossRef]

110. Zhang, Y.; Wang, W.H.; Greer, A.L. Making metallic glasses plastic by control of residual stress. *Nat. Mater.* **2006**, *5*, 857–860. [CrossRef] [PubMed]

111. Chen, W.; Zhou, H.F.; Liu, Z.; Ketkaew, J.; Shao, L.; Li, N.; Gong, P.; Samela, W.; Gao, H.J.; Schroers, J. Test sample geometry for fracture toughness measurements of bulk metallic glasses. *Acta Mater.* **2018**, *145*, 477–487. [CrossRef]

112. Zhao, Y.Y.; Ma, E.; Xu, J. Reliability of compressive fracture strength of Mg-Zn-Ca bulk metallic glasses: Flaw sensitivity and Weibull statistics. *Scr. Mater.* **2008**, *58*, 496–499. [CrossRef]

113. Conner, R.D.; Rosakis, A.J.; Johnson, W.L.; Owen, D.M. Fracture toughness determination for a beryllium-bearing bulk metallic glass. *Scr. Mater.* **1997**, *37*, 1373–1378. [CrossRef]
114. Narayan, R.L.; Tandaiya, P.; Garrett, G.R.; Demetriou, M.D.; Ramamurty, U. On the variability in fracture toughness of 'ductile' bulk metallic glasses. *Scr. Mater.* **2015**, *102*, 75–78. [CrossRef]
115. Kumar, G.; Desai, A.; Schroers, J. Bulk metallic glass: The smaller the better. *Adv. Mater.* **2011**, *23*, 461–476. [CrossRef] [PubMed]

Article

An Elastic Interface Model for the Delamination of Bending-Extension Coupled Laminates

Stefano Bennati, Paolo Fisicaro, Luca Taglialegne and Paolo S. Valvo *

Department of Civil and Industrial Engineering, University of Pisa, Largo Lucio Lazzarino, I-56122 Pisa, Italy
* Correspondence: p.valvo@ing.unipi.it; Tel.: +39-050-2218223

Received: 23 July 2019; Accepted: 24 August 2019; Published: 30 August 2019

Abstract: The paper addresses the problem of an interfacial crack in a multi-directional laminated beam with possible bending-extension coupling. A crack-tip element is considered as an assemblage of two sublaminates connected by an elastic-brittle interface of negligible thickness. Each sublaminate is modeled as an extensible, flexible, and shear-deformable laminated beam. The mathematical problem is reduced to a set of two differential equations in the interfacial stresses. Explicit expressions are derived for the internal forces, strain measures, and generalized displacements in the sublaminates. Then, the energy release rate and its Mode I and Mode II contributions are evaluated. As an example, the model is applied to the analysis of the double cantilever beam test with both symmetric and asymmetric laminated specimens.

Keywords: fiber-reinforced composite laminate; multi-directional laminate; delamination; elastic interface; energy release rate; mixed-mode fracture

---

1. Introduction

Delamination, or interlaminar fracture, is the main life-limiting failure mode for fiber-reinforced composite laminates [1]. Delamination cracks may originate from localized defects and propagate due to peak interlaminar stresses. A huge number of analytical, numerical, and experimental studies have been devoted to this problem during the last few decades [2–6]. The phenomenon is also relevant for many similar layered structures, such as glued laminated timber beams [7] and multilayered ceramic composites [8].

Within the framework of linear elastic fracture mechanics (LEFM), a delamination crack is expected to propagate when the energy release rate, $\mathcal{G}$, attains a critical value, or fracture toughness, $\mathcal{G}_\mathrm{c}$. Since delamination cracks preferentially propagate along the weak interfaces between adjoining plies (laminae), propagation generally involves a mix of the three basic fracture modes (Mode I or opening, Mode II or sliding, and Mode III or tearing). Each fracture mode furnishes an additive contribution to the total energy release rate, $\mathcal{G}_\mathrm{I}$, $\mathcal{G}_\mathrm{II}$, and $\mathcal{G}_\mathrm{III}$ [9], and corresponds to a different value of interlaminar fracture toughness, $\mathcal{G}_\mathrm{Ic}$, $\mathcal{G}_\mathrm{IIc}$, and $\mathcal{G}_\mathrm{IIIc}$ [10]. Thus, to predict delamination crack propagation, it is necessary (i) to assess by experiments the interlaminar fracture toughness (in pure and mixed fracture modes), (ii) to define a suitable mixed-mode fracture criterion, and (iii) to use a theoretical model to evaluate the energy release rate and its modal contributions [11].

An effective modeling approach to evaluate the energy release rate is to consider a delaminated laminate as composed of sublaminates connected by an elastic interface, i.e., a continuous distribution of linearly elastic–brittle springs. Such an elastic interface model was developed first by Kanninen for the double cantilever beam (DCB) test specimen [12] and later adopted by many authors for the analysis of the delamination of composite laminates [13–26]. Elastic interface models can be regarded as particular cases of the more general cohesive-zone models [27,28]. The latter are increasingly used to model delamination associated with large-scale bridging and other non-linear damage phenomena [29–38]. A parallel line of research concerns adhesively-bonded joints, for which models of growing complexity have been proposed in the literature [39]: from elastic interface models [40–46] to non-linear cohesive-zone models [47,48].

Bending-extension coupling and other elastic couplings are a typical feature of composite laminated beams and plates [49]. Nevertheless, only a few theoretical models for the study of delamination take into account elastic couplings. Among these, it is worth citing the pioneering work by Schapery and Davidson on the prediction of the energy release rate in mixed-mode fracture conditions [50]. In recent years, Xie et al. considered bending-extension coupling in the analysis of delamination toughness specimens [36] and composite laminated plates subjected to flexural loading [25]. Dimitri et al. presented a general formulation of the elastic interface model including bending-extension and shear deformability, but limited their analytical solution to the case with no elastic coupling [26]. Valvo analyzed the delamination of shear-deformable laminated beams with bending-extension coupling based on a rigid interface model [51]. Tsokanas and Loutas extended the above-mentioned analysis to include the effects of crack tip rotations and hygrothermal stresses [52]. To the best of our knowledge, a complete analytical solution for the elastic interface model of delaminated beams with bending-extension coupling and shear deformability has not yet been presented in the literature.

This paper analyses the problem of an interfacial crack in a multi-directional laminated beam with possible bending-extension coupling [53]. A crack-tip element is defined as a laminate segment extending ahead and behind the crack tip cross-section, such that the fracture process zone is fully included within [54,55]. In line with the elastic interface modeling approach, the crack-tip element is considered as an assemblage of two sublaminates connected by an elastic-brittle interface of negligible thickness. Each sublaminate is modeled as an extensible, flexible, and shear-deformable laminated beam [56]. The mathematical problem is reduced to a set of two differential equations in the interfacial stresses. A complete analytical solution is derived including explicit expressions for the internal forces, strain measures, and generalized displacements in the sublaminates. Then, the energy release rate and its Mode I and Mode II contributions are evaluated based on Rice's $\mathcal{J}$-integral [57]. By way of illustration, the model is applied to the analysis of the DCB test. Three laminated specimens are analyzed, made up of uni-directional, cross-ply, and multi-directional sublaminates.

2. Crack-Tip Element

2.1. Mechanical Model

Let us consider a composite laminated beam of length L, width B, and thickness $H = 2h$, with a through-the-width delamination crack of length a. The delamination plane ideally subdivides the laminate into two sublaminates (labeled 1 and 2) of thicknesses $H_1 = 2h_1$ and $H_2 = 2h_2$. We fix a right-handed global Cartesian reference system, $Oxyz$, with the origin O at the center of the crack tip cross-section and the x-, y-, and z-axes aligned with the laminate longitudinal, width, and thickness directions, respectively (Figure 1).

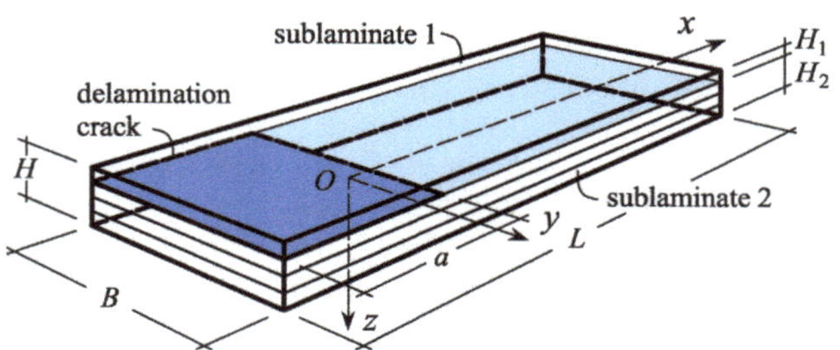

Figure 1. Laminated beam with through-the-width delamination.

The present mechanical model focuses on a *crack-tip element (CTE)*, here defined as a laminate segment of length ℓ, extending ahead and behind the crack tip cross-section, which fully includes the fracture process zone [54,55]. In practice, the end sections of the CTE, A and B, should be chosen such that A is immediately ahead of the delamination front and B is sufficiently far from it to make sure that the laminate behind behaves as a monolithic beam (Figure 2a). We model the sublaminates as two plane beams connected by an elastic-brittle interface of negligible thickness. The *generalized displacements* for the sublaminates are defined as follows: u_α and w_α denote the sublaminate mid-plane displacements along the x- and z-directions, respectively; φ_α denote their cross-section rotations about the y-axis (positive if counter-clockwise). Henceforth, the subscript α is used to refer to the upper ($\alpha = 1$) and lower ($\alpha = 2$) sublaminates (Figure 2b).

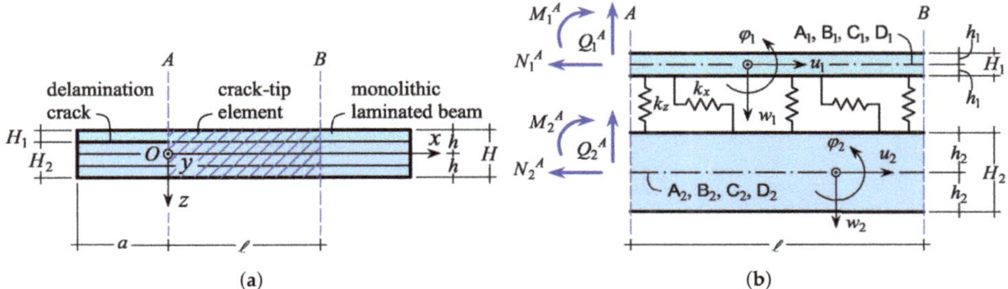

Figure 2. Crack-tip element: (**a**) definition; (**b**) model.

In line with the kinematics of Timoshenko's beam theory [56], we define the *strain measures* for the sublaminates as the *mid-plane axial strain*, ϵ_α, *average shear strain*, γ_α, and *pseudo-curvature*, κ_α. These are related to the generalized displacements as follows:

$$\epsilon_\alpha = \frac{du_\alpha}{dx}, \quad \gamma_\alpha = \frac{dw_\alpha}{dx} + \varphi_\alpha, \quad \text{and} \quad \kappa_\alpha = \frac{d\varphi_\alpha}{dx}. \tag{1}$$

Next, we introduce the conjugated *internal forces*, namely the *axial force*, N_α, *shear force*, Q_α, and *bending moment*, M_α. For a laminated beam with bending-extension coupling, the internal forces are related to the strain measures as follows:

$$N_\alpha = A_\alpha \epsilon_\alpha + B_\alpha \kappa_\alpha, \quad Q_\alpha = C_\alpha \gamma_\alpha, \quad \text{and} \quad M_\alpha = B_\alpha \epsilon_\alpha + D_\alpha \kappa_\alpha, \tag{2}$$

where A_α, B_α, C_α, and D_α respectively denote the sublaminate *extensional stiffness, bending-extension coupling stiffness, shear stiffness,* and *bending stiffness*. In general, such stiffnesses should be calculated according to classical laminated plate theory [49], as detailed in Appendix A.

For what follows, it is convenient to introduce also the sublaminate compliances,

$$a_\alpha = \frac{D_\alpha}{A_\alpha D_\alpha - B_\alpha^2}, \quad b_\alpha = -\frac{B_\alpha}{A_\alpha D_\alpha - B_\alpha^2}, \quad c_\alpha = \frac{1}{C_\alpha}, \quad \text{and} \quad d_\alpha = \frac{A_\alpha}{A_\alpha D_\alpha - B_\alpha^2}, \tag{3}$$

and write the constitutive relations, Equation (2), in inverse form:

$$\epsilon_\alpha = a_\alpha N_\alpha + b_\alpha M_\alpha, \quad \gamma_\alpha = c_\alpha Q_\alpha, \quad \text{and} \quad \kappa_\alpha = b_\alpha N_\alpha + d_\alpha M_\alpha. \tag{4}$$

The elastic interface between the upper and lower sublaminates transfers both normal and shear *interfacial stresses*, respectively given by:

$$\sigma = k_z \Delta w \quad \text{and} \quad \tau = k_x \Delta u, \tag{5}$$

where k_z and k_x are the *interface elastic constants* and:

$$\Delta w = w_2 - w_1 \quad \text{and} \quad \Delta u = u_2 - u_1 - \varphi_2 h_2 - \varphi_1 h_1 \tag{6}$$

are the transverse and longitudinal *relative displacements* at the interface, respectively. By substituting Equation (6) into (5), we obtain:

$$\sigma = k_z (w_2 - w_1) \quad \text{and} \quad \tau = k_x (u_2 - u_1 - \varphi_2 h_2 - \varphi_1 h_1). \tag{7}$$

2.2. Differential Problem

The equilibrium equations for the upper and lower sublaminates are:

$$\frac{dN_\alpha}{dx} + n_\alpha = 0, \quad \frac{dQ_\alpha}{dx} + q_\alpha = 0, \quad \text{and} \quad \frac{dM_\alpha}{dx} + m_\alpha - Q_\alpha = 0, \tag{8}$$

where n_α, q_α, and m_α respectively are the *distributed axial load, transverse load,* and *bending couple,* due to the normal and shear stresses transferred by the interface:

$$n_1 = -n_2 = B\tau, \quad q_1 = -q_2 = B\sigma, \quad \text{and} \quad m_\alpha = B\tau h_\alpha. \tag{9}$$

We substitute Equation (9) into (8) and differentiate the latter with respect to x. Then, by using Equations (1)–(4), we obtain the set of governing differential equations for the upper sublaminate,

$$\begin{aligned}
\frac{1}{B} \frac{d^3 u_1}{dx^3} &= -(a_1 + b_1 h_1) \frac{d\tau}{dx} - b_1 \sigma, \\
\frac{1}{B} \frac{d^4 w_1}{dx^4} &= (b_1 + d_1 h_1) \frac{d\tau}{dx} - c_1 \frac{d^2 \sigma}{dx^2} + d_1 \sigma, \quad \text{and} \\
\frac{1}{B} \frac{d^3 \varphi_1}{dx^3} &= -(b_1 + d_1 h_1) \frac{d\tau}{dx} - d_1 \sigma,
\end{aligned} \tag{10}$$

and for the lower sublaminate,

$$\begin{aligned}
\frac{1}{B} \frac{d^3 u_2}{dx^3} &= (a_2 - b_2 h_2) \frac{d\tau}{dx} + b_2 \sigma, \\
\frac{1}{B} \frac{d^4 w_2}{dx^4} &= -(b_2 - d_2 h_2) \frac{d\tau}{dx} + c_2 \frac{d^2 \sigma}{dx^2} - d_2 \sigma, \quad \text{and} \\
\frac{1}{B} \frac{d^3 \varphi_2}{dx^3} &= (b_2 - d_2 h_2) \frac{d\tau}{dx} + d_2 \sigma.
\end{aligned} \tag{11}$$

2.3. Boundary Conditions

The crack-tip element is subjected to known internal forces on the upper and lower sublaminates at cross-section A. Hence, the following six static boundary conditions can be written:

$$\begin{aligned}
N_1|_{x=0} &= N_1^A, \quad Q_1|_{x=0} = Q_1^A, \quad M_1|_{x=0} = M_1^A; \\
N_2|_{x=0} &= N_2^A, \quad Q_2|_{x=0} = Q_2^A, \quad M_2|_{x=0} = M_2^A.
\end{aligned} \tag{12}$$

At cross-section B, three kinematic boundary conditions can be obtained from the assumption that the laminate behind behaves as a monolithic beam:

$$\begin{aligned}
\epsilon_1|_{x=\ell} + \kappa_1|_{x=\ell} h_1 &= \epsilon_2|_{x=\ell} - \kappa_2|_{x=\ell} h_2, \\
\gamma_1|_{x=\ell} &= \gamma_2|_{x=\ell}, \quad \text{and} \quad \kappa_1|_{x=\ell} = \kappa_2|_{x=\ell}.
\end{aligned} \tag{13}$$

To sum up, we have nine boundary conditions to complete the formulation of the differential problem. We note that, in general, no restraints are given to prevent rigid-body motions of the CTE.

3. Solution of the Differential Problem

3.1. Solution Strategy

Following a solution strategy already adopted to solve some specific problems [22,43,44,53], the interfacial stresses are here assumed as the main unknowns. To this aim, we differentiate the normal and shear interfacial stresses, Equation (7), four and three times, respectively, with respect to x. Then, we substitute Equations (10) and (11) into the result. After some simplifications, we obtain the following differential equation set:

$$\frac{d^4\sigma}{dx^4} - \alpha_1 \frac{d^2\sigma}{dx^2} + \alpha_2\sigma + k_z\beta_0\frac{d\tau}{dx} = 0,$$
$$\frac{d^3\tau}{dx^3} - \alpha_3\frac{d\tau}{dx} - k_x\beta_0\sigma = 0,$$

(14)

where:

$$\alpha_1 = k_z B\left(c_1 + c_2\right), \quad \alpha_2 = k_z B\left(d_1 + d_2\right), \quad \text{and}$$
$$\alpha_3 = k_x B\left(a_1 + a_2 + 2b_1 h_1 - 2b_2 h_2 + d_1 {h_1}^2 + d_2 {h_2}^2\right)$$

(15)

are constant coefficients and:

$$\beta_0 = B\left(b_1 + b_2 + d_1 h_1 - d_2 h_2\right)$$

(16)

is a coupling parameter, here called the *unbalance parameter* by analogy with the literature on adhesive joints [39].

Two cases have to be considered in solving the differential problem:

(1) the *balanced case*, when $\beta_0 = 0$, i.e., $d_1 h_1 + b_1 = d_2 h_2 - b_2$. This condition is trivially fulfilled, for instance, by homogeneous laminates with mid-plane delaminations (for which $b_1 = b_2 = 0$, $d_1 = d_2$, and $h_1 = h_2$) or symmetrically-stacked laminates with mid-plane delaminations (for which $b_1 = -b_2$, $d_1 = d_2$, and $h_1 = h_2$). Besides, the balance condition is satisfied by homogeneous sublaminates with bending stiffness ratio $\frac{D_1}{D_2} = \frac{h_1}{h_2}$ [38,58]. More generally, it is possible to conceive of non-trivial stacking sequences, where the upper and lower sublaminates have unequal thicknesses, but fulfil the balance condition [59,60]. In the balanced case, Equations (14) are uncoupled and can be solved separately to obtain the normal and shear interfacial stresses;

(2) the *unbalanced case*, when $\beta_0 \neq 0$, i.e., $d_1 h_1 + b_1 \neq d_2 h_2 - b_2$. This condition corresponds to laminates with generic stacking sequences and arbitrarily-located delaminations (with respect to the mid-plane). In the unbalanced case, Equations (14) are coupled and must be solved simultaneously.

In the following subsections, we will show how to solve the stated differential problem and obtain the expressions for the interfacial stresses. Then, by using Equations (1), (4) and (8), it is straightforward to deduce also the expressions for the internal forces, strain measures, and generalized displacements.

3.2. Balanced Case

3.2.1. Interfacial Stresses

In the balanced case, the differential problem, Equation (14), reduces to the following two uncoupled differential equations:

$$\frac{d^4\sigma}{dx^4} - \alpha_1 \frac{d^2\sigma}{dx^2} + \alpha_2\sigma = 0,$$
$$\frac{d^3\tau}{dx^3} - \alpha_3 \frac{d\tau}{dx} = 0, \tag{17}$$

whose coefficients are defined in Equation (15).

The general solution for the interfacial stresses in the balanced case can be written as follows:

$$\sigma(x) = \frac{1}{B} \sum_{i=1}^{4} f_i \exp(-\lambda_i x), \tag{18}$$
$$\tau(x) = \frac{1}{B} \left[f_5 \exp(-\lambda_5 x) + f_6 \exp(-\lambda_6 x) + f_7 \right],$$

where $f_1, f_2, ..., f_7$ are integration constants and:

$$\lambda_1 = \frac{1}{\sqrt{2}} \sqrt{\alpha_1 - \sqrt{\alpha_1^2 - 4\alpha_2}}, \quad \lambda_2 = -\lambda_1,$$
$$\lambda_3 = \frac{1}{\sqrt{2}} \sqrt{\alpha_1 + \sqrt{\alpha_1^2 - 4\alpha_2}}, \quad \lambda_4 = -\lambda_3, \tag{19}$$
$$\lambda_5 = \sqrt{\alpha_3}, \quad \text{and} \quad \lambda_6 = -\lambda_5$$

are the non-zero roots of the characteristic equation of the differential problem (17):

$$\left(\lambda^4 - \alpha_1\lambda^2 + \alpha_2\right)\left(\lambda^2 - \alpha_3\right) = 0. \tag{20}$$

Strictly speaking, the solution expressed by Equation (18) holds provided that the interface constant k_z is not equal to the following "critical" value:

$$\hat{k}_z = \frac{4}{B} \frac{d_1 + d_2}{(c_1 + c_2)^2}, \tag{21}$$

for which Equation (20) has coincident roots and the solution requires a different expression. However, for material properties corresponding to common fiber-reinforced laminates, the numerical values of k_z are usually greater than $\hat{k}_z$, and the roots of the characteristic equation are real-valued (see Appendix B.1). For very compliant interfaces, e.g., if the present model is applied to the analysis of adhesively-bonded joints with weak adhesive, it may happen that $k_z < \hat{k}_z$, and the characteristic equation has complex roots. In such cases, the present solution is still valid, provided that the complex exponential functions are considered in Equation (18) and the following. Similar considerations apply also to the unbalanced case.

3.2.2. Internal Forces

By substituting Equation (18) into (9), the latter into (8), and integrating with respect to x, we obtain the expressions for the internal forces in the upper sublaminate,

$$N_1(x) = \frac{f_5}{\lambda_5} \exp(-\lambda_5 x) + \frac{f_6}{\lambda_6} \exp(-\lambda_6 x) - f_7 x - f_8,$$
$$Q_1(x) = \sum_{i=1}^{4} \frac{f_i}{\lambda_i} \exp(-\lambda_i x) - f_{10}, \quad \text{and} \tag{22}$$
$$M_1(x) = -\sum_{i=1}^{4} \frac{f_i}{\lambda_i^2} \exp(-\lambda_i x) + \frac{h_1 f_5}{\lambda_5} \exp(-\lambda_5 x) + \frac{h_1 f_6}{\lambda_6} \exp(-\lambda_6 x) - (f_{10} + h_1 f_7) x - f_{12},$$

and in the lower sublaminate,

$$N_2(x) = -\frac{f_5}{\lambda_5}\exp(-\lambda_5 x) - \frac{f_6}{\lambda_6}\exp(-\lambda_6 x) + f_7 x + f_9,$$

$$Q_2(x) = -\sum_{i=1}^{4}\frac{f_i}{\lambda_i}\exp(-\lambda_i x) + f_{11}, \quad \text{and} \tag{23}$$

$$M_2(x) = \sum_{i=1}^{4}\frac{f_i}{\lambda_i^2}\exp(-\lambda_i x) + \frac{h_2 f_5}{\lambda_5}\exp(-\lambda_5 x) + \frac{h_2 f_6}{\lambda_6}\exp(-\lambda_6 x) + (f_{11} - h_2 f_7)x + f_{13},$$

where $f_8, f_9, \ldots, f_{13}$ are integration constants.

3.2.3. Strain Measures

By substituting Equations (22) and (23) into (4), we obtain the expressions for the strain measures in the upper sublaminate,

$$\epsilon_1(x) = (a_1 + b_1 h_1)\left[\frac{f_5}{\lambda_5}\exp(-\lambda_5 x) + \frac{f_6}{\lambda_6}\exp(-\lambda_6 x)\right] - b_1\sum_{i=1}^{4}\frac{f_i}{\lambda_i^2}\exp(-\lambda_i x) +$$

$$- [(a_1 + b_1 h_1)f_7 + b_1 f_{10}]x - a_1 f_8 - b_1 f_{12},$$

$$\gamma_1(x) = c_1\sum_{i=1}^{4}\frac{f_i}{\lambda_i}\exp(-\lambda_i x) - c_1 f_{10}, \quad \text{and} \tag{24}$$

$$\kappa_1(x) = -d_1\sum_{i=1}^{4}\frac{f_i}{\lambda_i^2}\exp(-\lambda_i x) + (d_1 h_1 + b_1)\left[\frac{f_5}{\lambda_5}\exp(-\lambda_5 x) + \frac{f_6}{\lambda_6}\exp(-\lambda_6 x)\right] +$$

$$- [(d_1 h_1 + b_1)f_7 + d_1 f_{10}]x - b_1 f_8 - d_1 f_{12},$$

and in the lower sublaminate,

$$\epsilon_2(x) = -(a_2 - b_2 h_2)\left[\frac{f_5}{\lambda_5}\exp(-\lambda_5 x) + \frac{f_6}{\lambda_6}\exp(-\lambda_6 x)\right] + b_2\sum_{i=1}^{4}\frac{f_i}{\lambda_i^2}\exp(-\lambda_i x) +$$

$$+ [(a_2 - b_2 h_2)f_7 + b_2 f_{11}]x + a_2 f_9 + b_2 f_{13},$$

$$\gamma_2(x) = -c_2\sum_{i=1}^{4}\frac{f_i}{\lambda_i}\exp(-\lambda_i x) + c_2 f_{11}, \quad \text{and} \tag{25}$$

$$\kappa_2(x) = d_2\sum_{i=1}^{4}\frac{f_i}{\lambda_i^2}\exp(-\lambda_i x) + (d_2 h_2 - b_2)\left[\frac{f_5}{\lambda_5}\exp(-\lambda_5 x) + \frac{f_6}{\lambda_6}\exp(-\lambda_6 x)\right] +$$

$$- [(d_2 h_2 - b_2)f_7 - d_2 f_{11}]x + b_2 f_9 + d_2 f_{13}.$$

3.2.4. Generalized Displacements

Lastly, by substituting Equations (24) and (25) into (1) and integrating the latter with respect to x, we obtain the expressions for the generalized displacements in the upper sublaminate,

$$u_1(x) = -(a_1 + b_1 h_1)\left[\frac{f_5}{\lambda_5^2}\exp(-\lambda_5 x) + \frac{f_6}{\lambda_6^2}\exp(-\lambda_6 x)\right] + b_1\sum_{i=1}^{4}\frac{f_i}{\lambda_i^3}\exp(-\lambda_i x) +$$

$$-\left[(a_1 + b_1 h_1)f_7 + b_1 f_{10}\right]\frac{x^2}{2} - (a_1 f_8 + b_1 f_{12})x + f_{14},$$

$$w_1(x) = \sum_{i=1}^{4}\left(\frac{d_1}{\lambda_i^2} - c_1\right)\frac{f_i}{\lambda_i^2}\exp(-\lambda_i x) - (d_1 h_1 + b_1)\left[\frac{f_5}{\lambda_5^3}\exp(-\lambda_5 x) + \frac{f_6}{\lambda_6^3}\exp(-\lambda_6 x)\right] +$$

$$+\left[(d_1 h_1 + b_1)f_7 + d_1 f_{10}\right]\frac{x^3}{6} + (b_1 f_8 + d_1 f_{12})\frac{x^2}{2} - (f_{16} + c_1 f_{10})x + f_{15},\quad \text{and} \tag{26}$$

$$\varphi_1(x) = d_1\sum_{i=1}^{4}\frac{f_i}{\lambda_i^3}\exp(-\lambda_i x) - (d_1 h_1 + b_1)\left[\frac{f_5}{\lambda_5^2}\exp(-\lambda_5 x) + \frac{f_6}{\lambda_6^2}\exp(-\lambda_6 x)\right] +$$

$$-\left[(d_1 h_1 + b_1)f_7 + d_1 f_{10}\right]\frac{x^2}{2} - (b_1 f_8 + d_1 f_{12})x + f_{16},$$

and in the lower sublaminate,

$$u_2(x) = (a_2 - b_2 h_2)\left[\frac{f_5}{\lambda_5^2}\exp(-\lambda_5 x) + \frac{f_6}{\lambda_6^2}\exp(-\lambda_6 x)\right] - b_2\sum_{i=1}^{4}\frac{f_i}{\lambda_i^3}\exp(-\lambda_i x) +$$

$$+\left[(a_2 - b_2 h_2)f_7 + b_2 f_{11}\right]\frac{x^2}{2} + (a_2 f_9 + b_2 f_{13})x + f_{17},$$

$$w_2(x) = -\sum_{i=1}^{4}\left(\frac{d_2}{\lambda_i^2} - c_2\right)\frac{f_i}{\lambda_i^2}\exp(-\lambda_i x) - (d_2 h_2 - b_2)\left[\frac{f_5}{\lambda_5^3}\exp(-\lambda_5 x) + \frac{f_6}{\lambda_6^3}\exp(-\lambda_6 x)\right] +$$

$$+\left[(d_2 h_2 - b_2)f_7 - d_2 f_{11}\right]\frac{x^3}{6} - (b_2 f_9 + d_2 f_{13})\frac{x^2}{2} - (f_{19} - c_2 f_{11})x + f_{18},\quad \text{and} \tag{27}$$

$$\varphi_2(x) = -d_2\sum_{i=1}^{4}\frac{f_i}{\lambda_i^3}\exp(-\lambda_i x) - (d_2 h_2 - b_2)\left[\frac{f_5}{\lambda_5^2}\exp(-\lambda_5 x) + \frac{f_6}{\lambda_6^2}\exp(-\lambda_6 x)\right] +$$

$$-\left[(d_2 h_2 - b_2)f_7 - d_2 f_{11}\right]\frac{x^2}{2} + (b_2 f_9 + d_2 f_{13})x + f_{19},$$

where $f_{14}, f_{15}, ..., f_{19}$ are further integration constants.

3.2.5. Supernumerary Integration Constants

In the obtained analytical solution, nineteen integration constants appear. However, seven of them are supernumerary as a consequence of the adopted solution strategy (see Section 3.1). To determine the supernumerary integration constants, we substitute the solution for the interfacial stresses, Equation (18), and generalized displacements, Equations (26) and (27), into Equation (7). In this way, two identities are obtained, which need to be fulfilled for all values of x. The following seven relations among the integration constants are determined:

$$f_{10} = -\frac{\alpha_3}{B k_x \alpha_4}d_2 f_7, \quad f_{11} = \frac{\alpha_3}{B k_x \alpha_4}d_1 f_7,$$

$$f_{12} = -\frac{(a_1 + b_1 h_1)d_2 - (b_2 - d_2 h_2)b_1}{\alpha_4}f_8 - \frac{a_2 d_2 - b_2^2}{\alpha_4}f_9,$$

$$f_{13} = \frac{a_1 d_1 - b_1^2}{\alpha_4}f_8 + \frac{(a_2 - b_2 h_2)d_1 - (b_1 + d_1 h_1)b_2}{\alpha_4}f_9, \tag{28}$$

$$f_{14} = f_{17} - (h_1 + h_2)f_{19} - \frac{1}{k_x}\left[\frac{1}{B} + \frac{\alpha_3}{B\alpha_4}(c_1 d_2 - c_2 d_1)h_1\right]f_7,$$

$$f_{15} = f_{18}, \quad \text{and} \quad f_{16} = f_{19} + \frac{\alpha_3}{B k_x \alpha_4}(c_1 d_2 - c_2 d_1)f_7,$$

where:

$$\alpha_4 = b_1 d_2 - b_2 d_1 + d_1 d_2 (h_1 + h_2). \tag{29}$$

The values of the remaining twelve independent integration constants (f_1, f_2, ..., f_9 and f_{17}, f_{18}, and f_{19}) have to be determined by imposing the boundary conditions of the particular problem being solved. It can be noted that the last three independent integration constants describe a rigid motion of the specimen and can be assessed only when appropriate kinematic restraints are prescribed. The adopted solution strategy enables the determination of the internal forces, strain measures, and relative displacements, even though the values of the last three constants remain undetermined.

3.3. Unbalanced Case

3.3.1. Interfacial Stresses

In the unbalanced case, Equations (14) are coupled. To separate the unknowns, we solve the first of such equations w.r.t. to $d\tau/dx$ and substitute the result into the second equation. After some simplifications, we obtain:

$$\begin{aligned}
&\frac{d^6\sigma}{dx^6} - \beta_1 \frac{d^4\sigma}{dx^4} + \beta_2 \frac{d^2\sigma}{dx^2} - \beta_3\sigma = 0, \\
&\frac{d\tau}{dx} = -\frac{1}{k_z\beta_0}\left(\frac{d^4\sigma}{dx^4} - \alpha_1 \frac{d^2\sigma}{dx^2} + \alpha_2\sigma\right),
\end{aligned} \tag{30}$$

where:

$$\beta_1 = \alpha_1 + \alpha_3, \quad \beta_2 = \alpha_2 + \alpha_1\alpha_3, \quad \text{and} \quad \beta_3 = \alpha_2\alpha_3 - k_x k_z \beta_0^2, \tag{31}$$

with α_1, α_2, α_3, and β_0 already defined in Equations (15) and (16), respectively.

The general solution for the interfacial stresses in the unbalanced case can be written as follows:

$$\begin{aligned}
&\sigma(x) = \frac{1}{B}\sum_{i=1}^{6} g_i \exp(-\mu_i x), \\
&\tau(x) = -\frac{k_x\beta_0}{B}\sum_{i=1}^{6} \frac{g_i \exp(-\mu_i x)}{\mu_i(\mu_i^2 - \alpha_3)} + \frac{g_7}{B},
\end{aligned} \tag{32}$$

where g_1, g_2, ..., g_7 are integration constants and μ_1, μ_2, ..., μ_6 are the non-zero roots of the characteristic equation of the differential problem (30):

$$\mu^6 - \beta_1\mu^4 + \beta_2\mu^2 - \beta_3 = 0. \tag{33}$$

By substituting Equation (31) into (33), the latter can be rearranged as:

$$\left(\mu^2 - \lambda_1^2\right)\left(\mu^2 - \lambda_3^2\right)\left(\mu^2 - \lambda_5^2\right) + k_x k_z \beta_0^2 = 0, \tag{34}$$

where λ_1, λ_3, and λ_5 are given by Equation (19). From Equation (34), it can be easily concluded that the roots of the characteristic equation for the unbalanced case are different from those for the balanced case. Conversely, by assuming $\beta_0 = 0$, Equation (34) turns out to be equivalent to Equation (20).

In general, Equation (33) can be solved numerically. In Appendix B.2, the conditions are given for which μ_1, μ_2, ..., μ_6 are real, along with their analytical expressions in this case. It should be noted that, strictly speaking, the solution expressed by Equation (32) is valid only when the characteristic equation has no coincident roots, which is the case for common geometric and material properties.

3.3.2. Internal Forces

By substituting Equation (32) into (9), the latter into (8), and integrating with respect to x, we obtain the expressions for the internal forces in the upper sublaminate,

$$N_1(x) = -k_x\beta_0 \sum_{i=1}^{6} \frac{g_i}{\mu_i^2(\mu_i^2 - \alpha_3)} \exp(-\mu_i x) - g_7 x - g_8,$$

$$Q_1(x) = \sum_{i=1}^{6} \frac{g_i}{\mu_i} \exp(-\mu_i x) - g_{10}, \quad \text{and} \tag{35}$$

$$M_1(x) = -\sum_{i=1}^{6} \frac{g_i}{\mu_i^2}\left(1 + \frac{h_1 k_x \beta_0}{\mu_i^2 - \alpha_3}\right) \exp(-\mu_i x) - (h_1 g_7 + g_{10})x - g_{12},$$

and in the lower sublaminate,

$$N_2(x) = k_x\beta_0 \sum_{i=1}^{6} \frac{g_i}{\mu_i^2(\mu_i^2 - \alpha_3)} \exp(-\mu_i x) + g_7 x + g_9,$$

$$Q_2(x) = -\sum_{i=1}^{6} \frac{g_i}{\mu_i} \exp(-\mu_i x) + g_{11}, \quad \text{and} \tag{36}$$

$$M_2(x) = \sum_{i=1}^{6} \frac{g_i}{\mu_i^2}\left(1 - \frac{h_2 k_x \beta_0}{\mu_i^2 - \alpha_3}\right) \exp(-\mu_i x) - (h_2 g_7 - g_{11})x + g_{13},$$

where $g_8, g_9, \ldots, g_{13}$ are integration constants.

3.3.3. Strain Measures

By substituting Equations (35) and (36) into (4), we obtain the expressions for the strain measures in the upper sublaminate,

$$\epsilon_1(x) = -\sum_{i=1}^{6} \frac{g_i}{\mu_i^2}\left[\frac{k_x\beta_0}{\mu_i^2 - \alpha_3}(a_1 + b_1 h_1) + b_1\right]\exp(-\mu_i x) +$$
$$- [(a_1 + b_1 h_1)g_7 + b_1 g_{10}]x - a_1 g_8 - b_1 g_{12},$$

$$\gamma_1(x) = c_1 \sum_{i=1}^{6} \frac{g_i}{\mu_i}\exp(-\mu_i x) - c_1 g_{10}, \quad \text{and} \tag{37}$$

$$\kappa_1(x) = -\sum_{i=1}^{6} \frac{g_i}{\mu_i^2}\left[d_1 + \frac{k_x\beta_0}{\mu_i^2 - \alpha_3}(d_1 h_1 + b_1)\right]\exp(-\mu_i x) +$$
$$- [(d_1 h_1 + b_1)g_7 + d_1 g_{10}]x - b_1 g_8 - d_1 g_{12},$$

and in the lower sublaminate,

$$\epsilon_2(x) = \sum_{i=1}^{6} \frac{g_i}{\mu_i^2}\left[\frac{k_x\beta_0}{\mu_i^2 - \alpha_3}(a_2 - b_2 h_2) + b_2\right]\exp(-\mu_i x) +$$
$$+ [(a_2 - b_2 h_2)g_7 + b_2 g_{11}]x + a_2 g_9 + b_2 g_{13},$$

$$\gamma_2(x) = -c_2 \sum_{i=1}^{6} \frac{g_i}{\mu_i}\exp(-\mu_i x) + c_2 g_{11}, \quad \text{and} \tag{38}$$

$$\kappa_2(x) = \sum_{i=1}^{6} \frac{g_i}{\mu_i^2}\left[d_2 - \frac{k_x\beta_0}{\mu_i^2 - \alpha_3}(d_2 h_2 - b_2)\right]\exp(-\mu_i x) +$$
$$- [(d_2 h_2 - b_2)g_7 - d_2 g_{11}]x + b_2 g_9 + d_2 g_{13}.$$

3.3.4. Generalized Displacements

Lastly, by substituting Equations (37) and (38) into (1) and integrating the latter with respect to x, we obtain the expressions for the generalized displacements in the upper sublaminate,

$$u_1(x) = \sum_{i=1}^{6} \frac{g_i}{\mu_i^3} \left[\frac{k_x \beta_0}{\mu_i^2 - \alpha_3} (a_1 + b_1 h_1) + b_1 \right] \exp(-\mu_i x) +$$

$$- \left[(a_1 + b_1 h_1) g_7 + b_1 g_{10} \right] \frac{x^2}{2} - (a_1 g_8 + b_1 g_{12}) x + g_{14},$$

$$w_1(x) = \sum_{i=1}^{6} \frac{g_i}{\mu_i^2} \left[\frac{d_1}{\mu_i^2} - c_1 + \frac{k_x \beta_0 (d_1 h_1 + b_1)}{\mu_i^2 (\mu_i^2 - \alpha_3)} \right] \exp(-\mu_i x) +$$

$$+ \left[(d_1 h_1 + b_1) g_7 + d_1 g_{10} \right] \frac{x^3}{6} + (b_1 g_8 + d_1 g_{12}) \frac{x^2}{2} - (c_1 g_{10} + g_{16}) x + g_{15}, \quad \text{and}$$

$$\varphi_1(x) = \sum_{i=1}^{6} \frac{g_i}{\mu_i^3} \left[\frac{k_x \beta_0}{\mu_i^2 - \alpha_3} (d_1 h_1 + b_1) + d_1 \right] \exp(-\mu_i x) +$$

$$- \left[(d_1 h_1 + b_1) g_7 + d_1 g_{10} \right] \frac{x^2}{2} - (b_1 g_8 + d_1 g_{12}) x + g_{16},$$

and in the lower sublaminate,

$$u_2(x) = -\sum_{i=1}^{6} \frac{g_i}{\mu_i^3} \left[\frac{k_x \beta_0}{\mu_i^2 - \alpha_3} (a_2 - b_2 h_2) + b_2 \right] \exp(-\mu_i x) +$$

$$+ \left[(a_2 - b_2 h_2) g_7 + b_2 g_{11} \right] \frac{x^2}{2} + (a_2 g_9 + b_2 g_{13}) x + g_{17},$$

$$w_2(x) = -\sum_{i=1}^{6} \frac{g_i}{\mu_i^2} \left[\frac{d_2}{\mu_i^2} - c_2 - \frac{k_x \beta_0 (d_2 h_2 - b_2)}{\mu_i^2 (\mu_i^2 - \alpha_3)} \right] \exp(-\mu_i x) +$$

$$+ \left[(d_2 h_2 - b_2) g_7 - d_2 g_{11} \right] \frac{x^3}{6} - (b_2 g_9 + d_2 g_{13}) \frac{x^2}{2} + (c_2 g_{11} - g_{19}) x + g_{18}, \quad \text{and}$$

$$\varphi_2(x) = \sum_{i=1}^{6} \frac{g_i}{\mu_i^3} \left[\frac{k_x \beta_0}{\mu_i^2 - \alpha_3} (d_2 h_2 - b_2) - d_2 \right] \exp(-\mu_i x) +$$

$$- \left[(d_2 h_2 - b_2) g_7 - d_2 g_{11} \right] \frac{x^2}{2} + (b_2 g_9 + d_2 g_{13}) x + g_{19},$$

where $g_{14}, g_{15}, \ldots, g_{19}$ are further integration constants.

3.3.5. Supernumerary Integration Constants

As for the balanced case (see Section 3.2.5), nineteen integration constants appear in the analytical solution for the unbalanced case. Again, seven of them are supernumerary and can be determined by substituting the solution for the interfacial stresses, Equation (32), and generalized displacements, Equations (39) and (40), into Equation (7). In this way, two identities are obtained, which need to be fulfilled for all values of x. Consequently, the following seven relations among the integration constants are determined:

$$g_{10} = \left(-\frac{\alpha_3}{Bk_x \alpha_4} d_2 + \beta_0 \frac{b_2 - d_2 h_2}{B\alpha_4} \right) g_7, \quad g_{11} = \left(\frac{\alpha_3}{Bk_x \alpha_4} d_1 - \beta_0 \frac{b_1 + d_1 h_1}{B\alpha_4} \right) g_7,$$

$$g_{12} = -\frac{(a_1 + b_1 h_1) d_2 - (b_2 - d_2 h_2) b_1}{\alpha_4} g_8 - \frac{a_2 d_2 - b_2^2}{\alpha_4} g_9,$$

$$g_{13} = \frac{a_1 d_1 - b_1^2}{\alpha_4} g_8 + \frac{(a_2 - b_2 h_2) d_1 - (b_1 + d_1 h_1) b_2}{\alpha_4} g_9,$$

$$g_{14} = g_{17} - (h_1 + h_2) g_{19} - \left[\frac{1}{Bk_x} + \frac{\alpha_3}{Bk_x \alpha_4} (c_1 d_2 - c_2 d_1) h_1 + \beta_0 \frac{b_1 c_2 - c_1 b_2 + c_1 d_2 h_2 + c_2 d_1 h_1}{B\alpha_4} h_1 \right] g_7,$$

$$g_{15} = g_{18}, \quad \text{and} \quad g_{16} = g_{19} + \left[\frac{\alpha_3}{Bk_x \alpha_4} (c_1 d_2 - c_2 d_1) + \beta_0 \frac{b_1 c_2 - c_1 b_2 + c_1 d_2 h_2 + c_2 d_1 h_1}{B\alpha_4} \right] g_7,$$

(39)

(40)

(41)

where α_4 is furnished by Equation (29).

The values of the remaining twelve independent integration constants (g_1, g_2, ..., g_9 and g_{17}, g_{18}, and g_{19}) have to be determined by imposing the boundary conditions of the particular problem being solved. Again, the last three independent integration constants describe a rigid motion of the specimen and can be assessed only when appropriate kinematic restraints are prescribed. The adopted solution strategy enables the determination of internal forces, strain measures, and relative displacements, even though the values of the last three constants remain undetermined.

4. Delamination Crack Analysis

4.1. Energy Release Rate

Under I/II mixed-mode fracture conditions, the *energy release rate* can be written as $\mathcal{G} = \mathcal{G}_I + \mathcal{G}_{II}$, where $\mathcal{G}_I$ and $\mathcal{G}_{II}$ are the contributions related to Fracture Modes I and II, respectively. In LEFM, the energy release rate identifies the path-independent $\mathcal{J}$-integral introduced by Rice [57]. Choosing an integration path encircling the interface between sublaminates, as detailed in Appendix C, we obtain:

$$\mathcal{J} = \frac{1}{2}k_z \Delta w^2(0) + \frac{1}{2}k_x \Delta u^2(0) \tag{42}$$

or, by recalling Equation (5),

$$\mathcal{J} = \frac{\sigma_0^2}{2k_z} + \frac{\tau_0^2}{2k_x}, \tag{43}$$

where $\sigma_0 = \sigma(0)$ and $\tau_0 = \tau(0)$ are the peak values of the interfacial stresses at the delamination front. The two addends in Equation (43) correspond to the Mode I and Mode II contributions to the energy release rate, respectively [41]. Since compressive normal stresses at the crack tip ($\sigma_0 < 0$) do not promote crack opening, the modal contributions to the energy release rate can be written as

$$\mathcal{G}_I = \mathcal{H}(\sigma_0)\frac{\sigma_0^2}{2k_z} \quad \text{and} \quad \mathcal{G}_{II} = \frac{\tau_0^2}{2k_x}, \tag{44}$$

where $\mathcal{H}(\cdot)$ denotes the Heaviside step function. By substituting Equation (18) into (44), we obtain the following expressions for the balanced case:

$$\mathcal{G}_I = \frac{\mathcal{H}(\sigma_0)}{2k_z B^2}\left(\sum_{i=1}^{4} f_i\right)^2 \quad \text{and} \quad \mathcal{G}_{II} = \frac{1}{2k_x B^2}(f_5 + f_6)^2. \tag{45}$$

Likewise, by substituting Equation (32) into (44), we obtain the following expressions for the unbalanced case:

$$\mathcal{G}_I = \frac{\mathcal{H}(\sigma_0)}{2k_z B^2}\left(\sum_{i=1}^{6} g_i\right)^2 \quad \text{and} \quad \mathcal{G}_{II} = \frac{1}{2k_x B^2}\left(k_x \beta_0 \sum_{i=1}^{6} \frac{g_i}{\mu_i(\mu_i^2 - \alpha_3)}\right)^2. \tag{46}$$

Lastly, to characterize the ratio between the modal contributions to the energy release rate, we introduce the *mode-mixity angle* [11],

$$\psi = \text{sign}(\tau_0)\arctan\sqrt{\frac{\mathcal{G}_{II}}{\mathcal{G}_I}}, \tag{47}$$

ranging from $0°$ (pure Mode I) to $90°$ (pure Mode II).

5. Results

5.1. Examples

By way of illustration, we applied the developed solution to the analysis of the DCB test (Figure 3). Laminated specimens with the following geometric sizes were considered: $L = 100$ mm, $B = 25$ mm, and $H = 3.2$ mm. A delamination length $a = 25$ mm and a reference load $P = 100$ N were used in all of the following calculations. Boundary conditions (12) and (13) were specialized to the present case by taking the crack tip internal forces as $N_1^A = N_2^A = 0$, $Q_1^A = -Q_2^A = P$, and $M_1^A = -M_2^A = Pa$ and assuming a CTE length $\ell = L - a$.

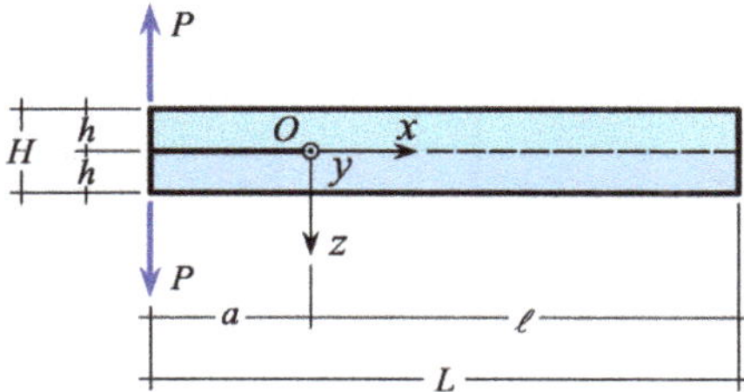

Figure 3. Double cantilever beam test.

Three laminated specimens with different stacking sequences were ideally built as follows. We started from a typical carbon-epoxy fiber-reinforced lamina of thickness $t_p = 0.2$ mm and elastic moduli shown in Table 1 [61]. Based on this, three stacking sequences were designed corresponding to uni-directional (UD), cross-ply (CP), and multi-directional (MD) sublaminates (Table 2). Lastly, the DCB test specimens were obtained by variously assembling the sublaminates one over the other:

- UD//UD
- UD//CP
- UD//MD

where the double slash (//) represents the position of the delamination crack.

Table 1. Lamina elastic moduli.

E_1 (MPa)	$E_2 = E_3$ (MPa)	$\nu_{12} = \nu_{13}$ –	ν_{23} –	$G_{12} = G_{13}$ (MPa)	G_{23} (MPa)
109,000	8819	0.342	0.380	4315	3200

Table 2. Sublaminate stacking sequences.

Sublaminate	Stacking Sequence
UD	$[0°_8]$
CP	$[0°/90°/0°_2/90°_3/0°]$
MD	$[45°/-45°_2/90°/45°_2/-45°/0°]$

The sublaminate plate stiffness matrices, calculated as explained in Appendix A, are reported in Table 3. It can be noted that no bending-extension coupling was exhibited by the UD sublaminate (all terms $B_{ij} = 0$), while bending-extension coupling was shown by both the CP (terms $B_{xx} = -B_{yy} \neq 0$) and MD sublaminates (all terms $B_{ij} \neq 0$, except $B_{xs} = B_{ys} = 0$).

Table 3. Sublaminate plate stiffnesses.

Sublaminate	$[A_{ij}]$ (N/mm)			$[B_{ij}]$ (N)			$[D_{ij}]$ (N mm)		
UD	176,066	4872	0	0	0	0	37,561	1039	0
	4872	14,245	0	0	0	0	1039	3039	0
	0	0	6904	0	0	0	0	0	1473
CP	95,156	4872	0	8091	0	0	25,155	1039	0
	4872	95,156	0	0	−8091	0	1039	15,445	0
	0	0	6904	0	0	0	0	0	1473
MD	66,477	33,550	0	10,959	−2868	0	19,419	6775	0
	33,550	66,477	0	−2868	−5223	0	6775	9710	0
	0	0	35,582	0	0	−2868	0	0	7209

The sublaminate equivalent beam stiffnesses, calculated via Equations (A10) and (A19), are given in Table 4. It can be noted that the UD sublaminate had the highest extensional and bending stiffnesses, A and D, while the MD sublaminate had the highest bending-extension coupling stiffness, B. The shear stiffness, C, showed little variations among the chosen stacking sequences.

Table 4. Sublaminate equivalent beam stiffnesses.

Sublaminate	A (kN)	B (kN mm)	C (kN)	D (kN mm^2)
UD	4360.0	0	143.8	930.1
CP	2372.4	201.5	129.0	627.0
MD	1238.5	314.2	123.8	367.0

A crucial point for the successful implementation of the present model is the appropriate choice of the values of the elastic interface constants. These can be assigned by conducting a compliance calibration with experimental tests or numerical simulations, if available [62]. In a previous study on asymmetric DCB test specimens [22], compliance calibration with finite element simulations gave:

$$k_x \approx 16.2\frac{G_{13}}{H} \quad \text{and} \quad k_z \approx 6.9\frac{E_3}{H}, \tag{48}$$

which, in the present case, yielded $k_x = 21{,}845 \text{ N/mm}^3$ and $k_z = 19{,}016 \text{ N/mm}^3$. Such values were adopted for the present examples.

Figure 4 shows the plots of the interfacial stresses and internal forces as functions of the x-coordinate in the neighborhood of the crack tip for the UD//UD specimen. This specimen was symmetric; hence, the unbalance parameter was $\beta_0 = 0$, and the analytical solution for the balanced case applied (see Section 3.2). For the DCB test, the specimen response also corresponded to pure Mode I fracture conditions: actually, Figure 4a shows that the shear stresses were identically null, while the normal stresses attained a peak value at the crack tip cross-section. Figure 4b shows that also the sublaminate axial forces were null. Instead, Figure 4c,d shows the trends of the sublaminate shear forces and bending moments, which started from the assigned values at the crack tip cross-section and then approached zero asymptotically. Shear forces and bending moments had symmetric trends in the upper and lower sublaminates because of the specimen symmetry.

Figures 5 and 6 show the plots of the interfacial stresses and internal forces as functions of the x-coordinate in the neighborhood of the crack tip for the UD//CP and UD//MD specimens, respectively. Both specimens were asymmetric with unbalance parameter $\beta_0 \neq 0$. Hence, the analytical solution for the unbalanced case applied (see Section 3.3). For the DCB test, the specimen response corresponded to I/II mixed-mode fracture conditions: actually, both Figures 5a and 6a show non-zero normal and shear stresses, which attained peak values at the crack tip cross-section. All of the internal forces started from the assigned values at the crack tip cross-section and then approached zero asymptotically.

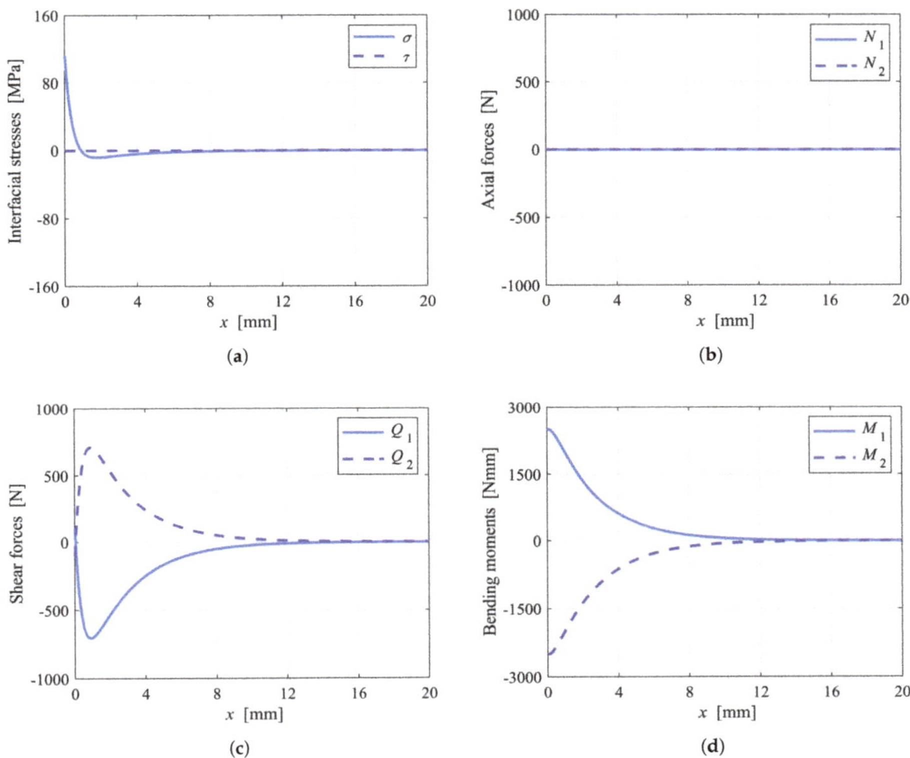

Figure 4. UD//UD specimen: (**a**) interfacial stresses, (**b**) axial forces, (**c**) shear forces, and (**d**) bending moments in the crack tip neighborhood.

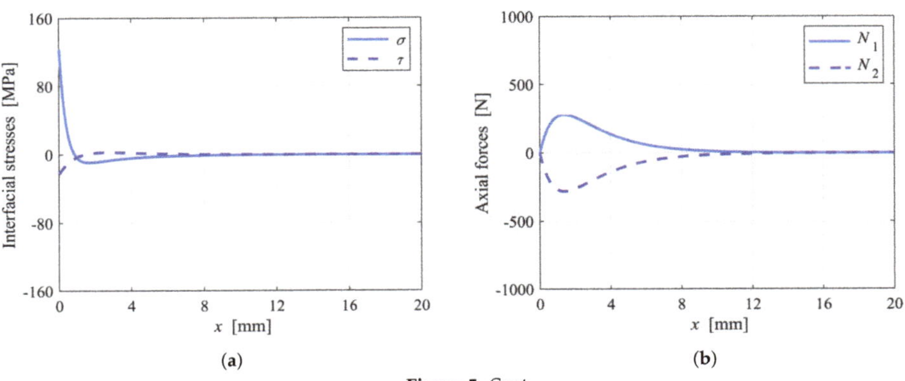

Figure 5. *Cont.*

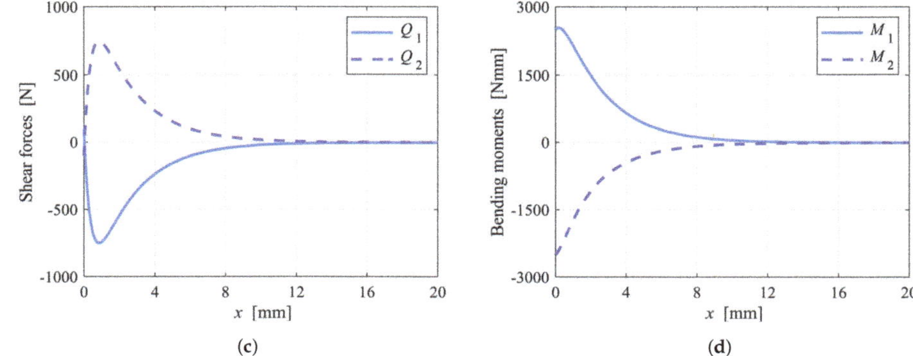

(c) **(d)**

Figure 5. UD//CP specimen: (**a**) interfacial stresses, (**b**) axial forces, (**c**) shear forces, and (**d**) bending moments in the crack tip neighborhood.

Table 5 reports the Mode I and II contributions to the energy release rate computed for the three analyzed specimens, along with the total energy release rate and mode-mixity angle. It is apparent that the relative amount of the Mode II contribution increased with the specimen asymmetry.

Table 5. Energy release rate and mode mixity.

Specimen	$\mathcal{G}_{\mathrm{I}}$ (J/m^2)	$\mathcal{G}_{\mathrm{II}}$ (J/m^2)	$\mathcal{G}$ (J/m^2)	ψ (deg)
UD//UD	339.7	0	339.7	0
UD//CP	403.4	11.4	414.8	-9.5
UD//MD	524.2	135.9	660.1	-27.0

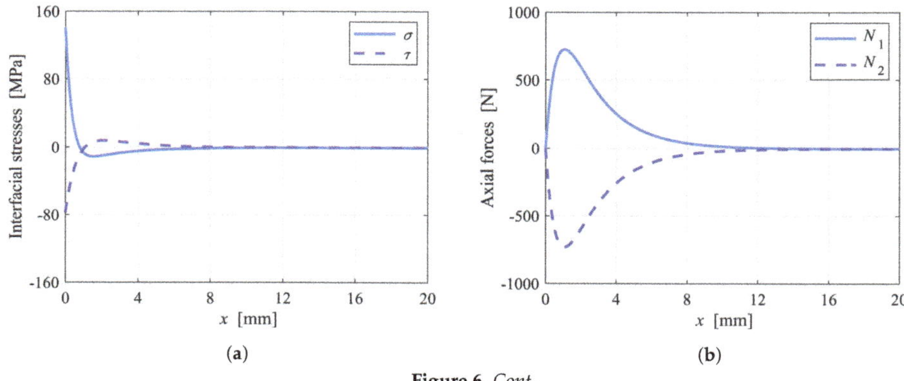

(a) **(b)**

Figure 6. *Cont.*

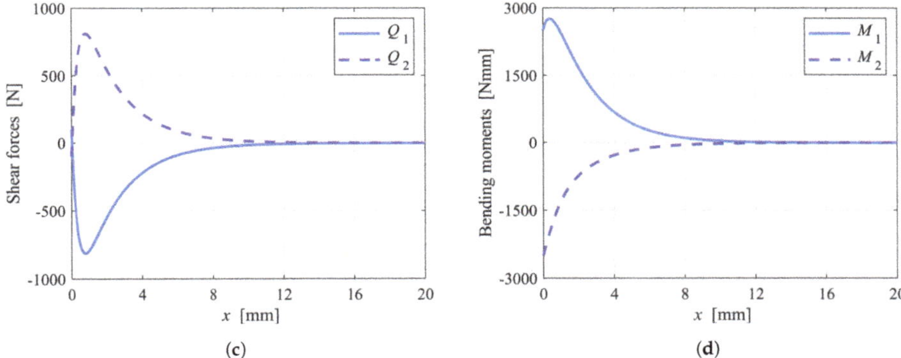

(c)   **(d)**

Figure 6. UD//MD specimen: (**a**) interfacial stresses, (**b**) axial forces, (**c**) shear forces, and (**d**) bending moments in the crack tip neighborhood.

5.2. Effects of Interface Stiffness

Figure 7 shows a schematic of the double cantilever beam specimen loaded by uneven bending moments (DCB-UBM) test [63]. Here, this particular test configuration was used to illustrate the effects of interface stiffness, i.e., of the values of the interface elastic constants, k_x and k_z, on the solution according to the proposed model.

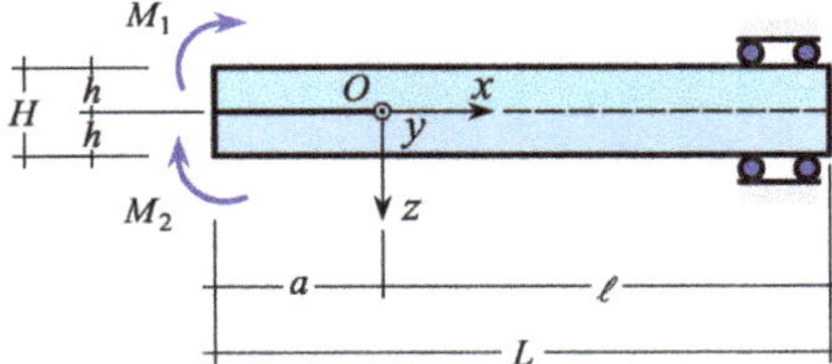

Figure 7. Double cantilever beam specimen with uneven bending moments.

The same laminated specimens considered in the examples of Section 5.1 were again analyzed, but subjected to different loading conditions. The upper sublaminate was loaded by a unit bending couple, $M_1 = 1$ N m, while the lower sublaminate was loaded by a variable bending couple, M_2. Figure 8 shows the dimensionless Mode I and Mode II contributions to the energy release rate, $\gamma_{\mathrm{I}} = \mathcal{G}_{\mathrm{I}}/\mathcal{G}$ and $\gamma_{\mathrm{II}} = \mathcal{G}_{\mathrm{II}}/\mathcal{G}$, as functions of the bending moment ratio, M_2/M_1.

For the UD//UD specimen (Figure 8a), the modal contributions to $\mathcal{G}$ turned out to be independent of the interface stiffness. This result was due to the mid-plane symmetry of the specimen and the independence of the crack tip bending moments on the delamination length. In this case, also the mode mixity depended uniquely on the bending moment ratio, M_2/M_1. In particular, it can be noted how pure Mode I fracture conditions ($\gamma_{\mathrm{II}} = 0$) were obtained for $M_2 = -M_1$, while pure Mode II fracture conditions ($\gamma_{\mathrm{I}} = 0$) were obtained for $M_2 = M_1$. Values of M_2/M_1 greater than one also corresponded to pure Mode II because of the incompressibility constraint introduced for $\mathcal{G}_{\mathrm{I}}$ in Equation (44).

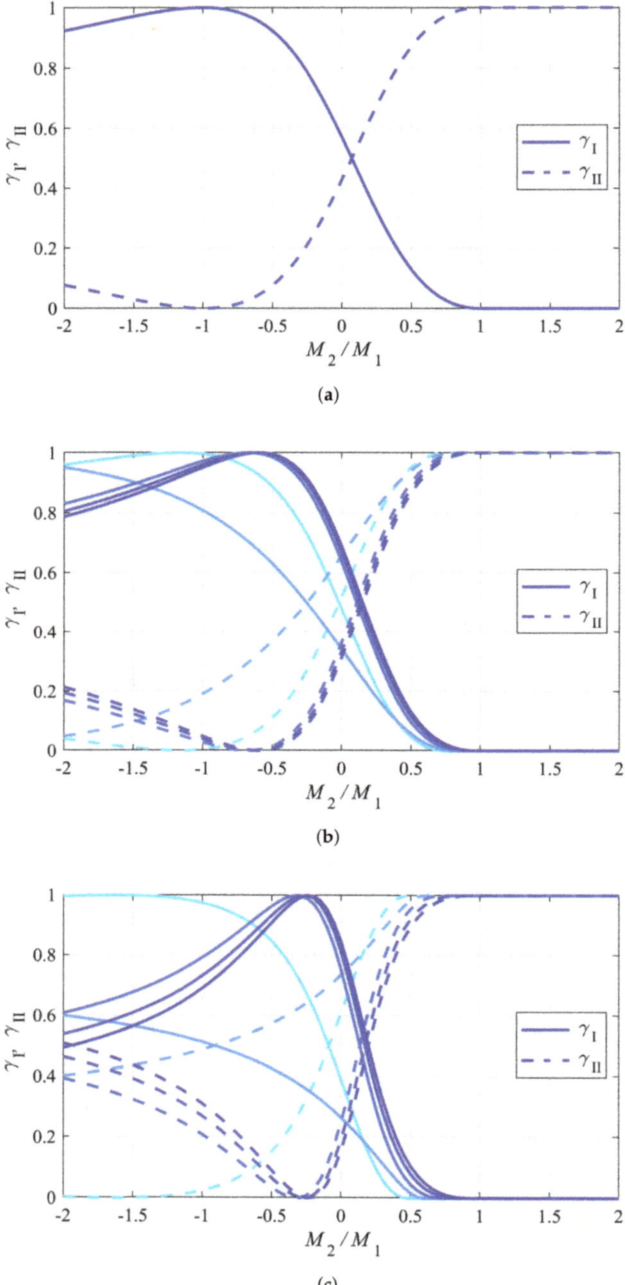

Figure 8. Dimensionless modal contributions to energy release rate vs. sublaminate bending moment ratio: (**a**) UD//UD specimen, (**b**) UD//CP specimen, and (**c**) UD//MD specimen.

Instead, for both the UD//CP (Figure 8b) and UD//MD specimens (Figure 8c), the modal contributions to $\mathcal{G}$ depended on the values of the interface elastic constants, k_x and k_z. For each modal contribution, the plots show five curves with different shades of blue: from lighter to darker, and these correspond to $k_x = k_z = 10^2, 10^3, 10^4$, and 10^5 N/mm^3; the darkest curves represent the limit case $k_x = k_z \rightarrow +\infty$, which corresponds to a rigid interface model [51]. For the specimens considered, also the mode mixity and the conditions for pure fracture modes depended on the interface stiffness.

The range of values adopted for k_x and k_z in the previous plots may look quite wide. Indeed, this range corresponds to real values that may be obtained when performing a compliance calibration with experimental results or numerical simulations [62]. The results obtained suggest that the elastic interface model is potentially capable of highly-accurate prediction and interpretation of experimental results, provided that the appropriate values of the interface elastic constants are selected. To this aim, it would be useful to validate the theoretical predictions obtained with experimental results, such as those obtained by Davidson et al. [64,65]. Such a comparison, similar to the one carried out by Harvey et al. to validate different mixed-mode partition theories [66,67], will be the subject of future work.

6. Conclusions

An elastic interface model was introduced for the analysis of delamination in bending-extension coupled laminated beams with through-the-width delamination cracks. The analytical solution obtained applies to a wide class of multi-directional fiber-reinforced composite laminates, including cross-ply laminates, provided that their stacking sequences allow them to be modeled as plane beams. Examples illustrated the application of the model to the analysis of the double cantilever beam test in I/II mixed-mode fracture conditions. Furthermore, the effects of the interface stiffness on the model predictions were discussed with reference to the DCB specimen loaded with uneven bending moments.

By suitably adapting the boundary conditions, the outlined solution can be easily used to analyze different delamination toughness test configurations, as well as delamination behavior in real structural elements and debonding of adhesively-bonded joints.

The proposed model is potentially capable of highly-accurate prediction and interpretation of experimental results, provided that the appropriate interface stiffness is defined. In this respect, it will be useful to compare the model predictions with the results of experimental tests available in the literature. Future developments also include the extension of the model to fully-coupled laminates, for which plate theory must be used, and I/II/III mixed-mode fracture conditions are expected.

Author Contributions: Conceptualization, S.B., P.F., L.T., and P.S.V.; methodology, S.B., P.F., L.T., and P.S.V.; software, P.F. and P.S.V.; validation, P.S.V.; formal analysis, P.F. and L.T.; investigation, P.F. and L.T.; resources, S.B. and P.S.V.; data curation, P.F. and P.S.V.; writing, original draft preparation, P.F., L.T. and P.S.V.; writing, review and editing, S.B.; visualization, P.F. and P.S.V.; supervision, S.B.; project administration, P.S.V.; funding acquisition, S.B. and P.S.V.

Funding: This research received no external funding.

Conflicts of Interest: The authors declare no conflict of interest.

Abbreviations

The following abbreviations are used in this manuscript:

CTE	crack-tip element
CP	cross-ply
DCB	double cantilever beam
DCB-UBM	double cantilever beam with uneven bending moments
LEFM	linear elastic fracture mechanics
MD	multi-directional
UD	uni-directional

Appendix A. Laminated Beam Stiffnesses

For a homogenous and special orthotropic beam of width B and thickness H, the extensional, bending-coupling, shear, and bending stiffnesses can be calculated respectively as:

$$A = E_x BH, \quad B = 0, \quad C = \frac{5}{6} G_{zx} BH, \quad \text{and} \quad D = \frac{1}{12} E_x BH^3, \tag{A1}$$

where E_x and G_{zx} respectively are the longitudinal Young's modulus and the transverse shear modulus of the material.

For laminated beams with generic stacking sequences, the above quantities have to be calculated as explained in the following sections.

Appendix A.1. Extension and Bending Stiffnesses

For a laminated plate with a generic stacking sequence, the constitutive laws are:

$$\begin{Bmatrix} N_x \\ N_y \\ N_{xy} \end{Bmatrix} = \begin{bmatrix} A_{xx} & A_{xy} & A_{xs} \\ A_{xy} & A_{yy} & A_{ys} \\ A_{xs} & A_{ys} & A_{ss} \end{bmatrix} \begin{Bmatrix} \epsilon_x \\ \epsilon_y \\ \gamma_{xy} \end{Bmatrix} + \begin{bmatrix} B_{xx} & B_{xy} & B_{xs} \\ B_{xy} & B_{yy} & B_{ys} \\ B_{xs} & B_{ys} & B_{ss} \end{bmatrix} \begin{Bmatrix} \kappa_x \\ \kappa_y \\ \kappa_{xy} \end{Bmatrix} \tag{A2}$$

and:

$$\begin{Bmatrix} M_x \\ M_y \\ M_{xy} \end{Bmatrix} = \begin{bmatrix} B_{xx} & B_{xy} & B_{xs} \\ B_{xy} & B_{yy} & B_{ys} \\ B_{xs} & B_{ys} & B_{ss} \end{bmatrix} \begin{Bmatrix} \epsilon_x \\ \epsilon_y \\ \gamma_{xy} \end{Bmatrix} + \begin{bmatrix} D_{xx} & D_{xy} & D_{xs} \\ D_{xy} & D_{yy} & D_{ys} \\ D_{xs} & D_{ys} & D_{ss} \end{bmatrix} \begin{Bmatrix} \kappa_x \\ \kappa_y \\ \kappa_{xy} \end{Bmatrix}, \tag{A3}$$

where N_x, N_y, and N_{xy} are the in-plane forces and M_x, M_y, and M_{xy} are the bending moments (per unit plate width); correspondingly, ϵ_x, ϵ_y, and γ_{xy} are the in-plane strains and κ_x, κ_y, and κ_{xy} are the bending curvatures; A_{ij}, B_{ij}, and D_{ij}, with $i, j \in \{x, y, s\}$, are the coefficients of the extensional, bending-extension coupling, and bending stiffness matrices, respectively, to be computed following classical laminated plate theory [49]:

$$\begin{Bmatrix} A_{ij} \\ B_{ij} \\ D_{ij} \end{Bmatrix} = \sum_{k=1}^{n} Q_{ij}^{(k)} \begin{Bmatrix} z_k - z_{k-1} \\ \frac{1}{2}(z_k^2 - z_{k-1}^2) \\ \frac{1}{3}(z_k^3 - z_{k-1}^3) \end{Bmatrix}, \tag{A4}$$

where $Q_{ij}^{(k)}$, with $i, j \in \{x, y, s\}$, are the in-plane elastic moduli of the kth lamina, included between the ordinates $z = z_{k-1}$ and $z = z_k$, in the global reference system, and n is the total number of laminae. For an orthotropic lamina, whose principal material axes, 1 and 2, are rotated by an angle θ_k with respect to the global axes x and y,

$$\begin{aligned}
Q_{xx}^{(k)} &= Q_{11}^{(k)} \cos^4 \theta_k + 2(Q_{12}^{(k)} + 2Q_{66}^{(k)}) \cos^2 \theta_k \sin^2 \theta_k + Q_{22}^{(k)} \sin^4 \theta_k, \\
Q_{xy}^{(k)} &= Q_{yx}^{(k)} = (Q_{11}^{(k)} + Q_{22}^{(k)} - 4Q_{66}^{(k)}) \cos^2 \theta_k \sin^2 \theta_k + Q_{12}^{(k)} (\cos^4 \theta_k + \sin^4 \theta_k), \\
Q_{xs}^{(k)} &= Q_{sx}^{(k)} = (Q_{11}^{(k)} - Q_{12}^{(k)} - 2Q_{66}^{(k)}) \cos^3 \theta_k \sin \theta_k + (Q_{12}^{(k)} - Q_{22}^{(k)} + 2Q_{66}^{(k)}) \cos \theta_k \sin^3 \theta_k, \\
Q_{yy}^{(k)} &= Q_{11}^{(k)} \sin^4 \theta_k + 2(Q_{12}^{(k)} + 2Q_{66}^{(k)}) \cos^2 \theta_k \sin^2 \theta_k + Q_{22}^{(k)} \cos^4 \theta_k, \\
Q_{ys}^{(k)} &= Q_{sy}^{(k)} = (Q_{11}^{(k)} - Q_{12}^{(k)} - 2Q_{66}^{(k)}) \cos \theta_k \sin^3 \theta_k + (Q_{12}^{(k)} - Q_{22}^{(k)} + 2Q_{66}^{(k)}) \cos^3 \theta_k \sin \theta_k, \quad \text{and} \\
Q_{ss}^{(k)} &= (Q_{11}^{(k)} + Q_{22}^{(k)} - 2Q_{12}^{(k)} - 2Q_{66}^{(k)}) \cos^2 \theta_k \sin^2 \theta_k + Q_{66}^{(k)} (\cos^4 \theta_k + \sin^4 \theta_k),
\end{aligned} \tag{A5}$$

where the in-plane elastic moduli of the lamina in its principal material reference system are:

$$Q_{11}^{(k)} = \frac{E_1^{(k)}}{1 - v_{12}^{(k)} v_{21}^{(k)}}, \quad Q_{12}^{(k)} = Q_{21}^{(k)} = \frac{v_{12}^{(k)} E_2^{(k)}}{1 - v_{12}^{(k)} v_{21}^{(k)}},$$

$$Q_{22}^{(k)} = \frac{E_2^{(k)}}{1 - v_{12}^{(k)} v_{21}^{(k)}}, \quad \text{and} \quad Q_{66}^{(k)} = G_{12}^{(k)} \tag{A6}$$

with $E_1^{(k)}$, $E_2^{(k)}$, $v_{12}^{(k)}$, $v_{21}^{(k)}$, and $G_{12}^{(k)}$ being the in-plane elastic moduli of the lamina in engineering notation.

For a plate to be modeled as a plane beam in the zx-plane, the out-of-plane internal forces and moments must be $N_y = N_{xy} = 0$ and $M_y = M_{xy} = 0$. Besides, the shear-extension and bend-twist coupling terms must vanish: $A_{xs} = A_{ys} = 0$ and $D_{xs} = D_{ys} = 0$. The bending-shear coupling terms must also be null: $B_{xs} = B_{ys} = 0$. Such conditions are fulfilled by laminates with single or multiple special orthotropic layers and antisymmetric cross-ply laminates. Furthermore, antisymmetric angle-ply and generic multi-directional laminates may satisfy the above-mentioned conditions for particular stacking sequences. As a consequence, Equations (A2) and (A3) reduce to:

$$\begin{Bmatrix} N_x \\ 0 \\ 0 \end{Bmatrix} = \begin{bmatrix} A_{xx} & A_{xy} & 0 \\ A_{xy} & A_{yy} & 0 \\ 0 & 0 & A_{ss} \end{bmatrix} \begin{Bmatrix} \epsilon_x \\ \epsilon_y \\ \gamma_{xy} \end{Bmatrix} + \begin{bmatrix} B_{xx} & B_{xy} & 0 \\ B_{xy} & B_{yy} & 0 \\ 0 & 0 & B_{ss} \end{bmatrix} \begin{Bmatrix} \kappa_x \\ \kappa_y \\ \kappa_{xy} \end{Bmatrix} \tag{A7}$$

and:

$$\begin{Bmatrix} M_x \\ 0 \\ 0 \end{Bmatrix} = \begin{bmatrix} B_{xx} & B_{xy} & 0 \\ B_{xy} & B_{yy} & 0 \\ 0 & 0 & B_{ss} \end{bmatrix} \begin{Bmatrix} \epsilon_x \\ \epsilon_y \\ \gamma_{xy} \end{Bmatrix} + \begin{bmatrix} D_{xx} & D_{xy} & 0 \\ D_{xy} & D_{yy} & 0 \\ 0 & 0 & D_{ss} \end{bmatrix} \begin{Bmatrix} \kappa_x \\ \kappa_y \\ \kappa_{xy} \end{Bmatrix}. \tag{A8}$$

For a laminated beam of width B, the axial force and bending moment respectively are $N = BN_x$ and $M = BM_x$; the axial strain and pseudo-curvature are $\epsilon = \epsilon_x$ and $\kappa = \kappa_x$. Thus, from Equations (A7) and (A8), the constitutive laws are derived as follows:

$$N = A\epsilon + B\kappa \quad \text{and} \quad M = B\epsilon + D\kappa, \tag{A9}$$

where the laminated beam stiffnesses are evaluated as:

$$A = B \left(A_{xx} - \frac{A_{xy}^2 D_{yy} - 2 A_{xy} B_{xy} B_{yy} + A_{yy} B_{xy}^2}{A_{yy} D_{yy} - B_{yy}^2} \right),$$

$$B = B \left[B_{xx} + \frac{A_{xy} (B_{yy} D_{xy} - B_{xy} D_{yy}) - A_{yy} B_{xy} D_{xy} + B_{xy}^2 B_{yy}}{A_{yy} D_{yy} - B_{yy}^2} \right], \quad \text{and} \tag{A10}$$

$$D = B \left(D_{xx} - \frac{A_{yy} D_{xy}^2 - 2 B_{xy} B_{yy} D_{xy} + B_{xy}^2 D_{yy}}{A_{yy} D_{yy} - B_{yy}^2} \right),$$

and the remaining plate strain measures turn out to be:

$$\epsilon_y = -\frac{A_{xy} D_{yy} - B_{xy} B_{yy}}{A_{yy} D_{yy} - B_{yy}^2} \epsilon - \frac{B_{xy} D_{yy} - B_{yy} D_{xy}}{A_{yy} D_{yy} - B_{yy}^2} \kappa,$$

$$\gamma_{xy} = 0,$$

$$\kappa_y = -\frac{A_{yy} B_{xy} - A_{xy} B_{yy}}{A_{yy} D_{yy} - B_{yy}^2} \epsilon - \frac{A_{yy} D_{xy} - B_{xy} B_{yy}}{A_{yy} D_{yy} - B_{yy}^2} \kappa, \quad \text{and} \tag{A11}$$

$$\kappa_{xy} = 0.$$

Appendix A.2. Shear Stiffness

The shear stiffness of a laminated plate can be evaluated by imposing a strain energy equivalence, as usually done for homogeneous beams [56] and recently extended to laminated plates by Xie et al. [25]. Here, we detail such a procedure for a laminated beam [68–70]. To this aim, first we recall the expression of the normal stress in the x-direction within the kth lamina [49]:

$$\sigma_x^{(k)}(z) = Q_{xx}^{(k)}\left(\epsilon_x + z\kappa_x\right) + Q_{xy}^{(k)}\left(\epsilon_y + z\kappa_y\right) + Q_{xs}^{(k)}\left(\gamma_{xy} + z\kappa_{xy}\right). \tag{A12}$$

Next, we substitute the expressions of the plate strain measures obtained in the previous section into Equation (A12). Thus, we obtain:

$$\sigma_x^{(k)}(z) = Q_{\epsilon 0}^{(k)}\epsilon + Q_{\kappa 0}^{(k)}\kappa + z\left(Q_{\epsilon 1}^{(k)}\epsilon + Q_{\kappa 1}^{(k)}\kappa\right), \tag{A13}$$

where:

$$Q_{\epsilon 0}^{(k)} = Q_{xx}^{(k)} - \frac{A_{xy}D_{yy} - B_{xy}B_{yy}}{A_{yy}D_{yy} - B_{yy}^2}Q_{xy}^{(k)}, \quad Q_{\kappa 0}^{(k)} = -\frac{B_{xy}D_{yy} - B_{yy}D_{xy}}{A_{yy}D_{yy} - B_{yy}^2}Q_{xy}^{(k)},$$

$$Q_{\epsilon 1}^{(k)} = -\frac{A_{yy}B_{xy} - A_{xy}B_{yy}}{A_{yy}D_{yy} - B_{yy}^2}Q_{xy}^{(k)}, \quad \text{and} \quad Q_{\kappa 1}^{(k)} = Q_{xx}^{(k)} - \frac{A_{yy}D_{xy} - B_{xy}B_{yy}}{A_{yy}D_{yy} - B_{yy}^2}Q_{xy}^{(k)}. \tag{A14}$$

Following the classical derivation of Jourawski's shear stress formula [56], we impose the static equilibrium in the x-direction of a short portion of beam defined by a given value of the z-coordinate. Thus, by recalling the inverse constitutive Equation (4) and static equilibrium Equation (8), with no distributed loads, the following expression is obtained for the shear stress in the kth lamina:

$$\tau_{zx}^{(k)}(z) = Q\left(R_k - S_k z - T_k z^2\right), \tag{A15}$$

where Q is the transverse shear force (acting in the z-direction) and:

$$R_k = S_k z_k + T_k z_k^2 + \sum_{j=k+1}^{n}\left[S_j\left(z_j - z_{j-1}\right) + T_j\left(z_j^2 - z_{j-1}^2\right)\right],$$

$$S_k = bQ_{\epsilon 0}^{(k)} + dQ_{\kappa 0}^{(k)}, \quad \text{and} \quad T_k = \frac{1}{2}\left(bQ_{\epsilon 1}^{(k)} + dQ_{\kappa 1}^{(k)}\right), \tag{A16}$$

where b and d respectively are the bending-extension and bending compliances of the laminated beam, as given by Equation (3).

The shear compliance, c, can now be calculated by imposing the following energy equivalence for a short beam segment:

$$\frac{1}{2}cQ^2 = \frac{1}{2}B\int_{-h}^{h}\frac{1}{G_{zx}^{(k)}}\left[\tau_{zx}^{(k)}(z)\right]^2 dz, \tag{A17}$$

where:

$$G_{zx}^{(k)} = \frac{1}{\dfrac{\sin^2\theta_k}{G_{23}^{(k)}} + \dfrac{\cos^2\theta_k}{G_{31}^{(k)}}} \tag{A18}$$

is the transverse shear modulus of the kth lamina in the laminate reference system, depending on its out-of-plane shear moduli, $G_{23}^{(k)}$ and $G_{31}^{(k)}$. By substituting Equation (A15) into (A17), the shear compliance is obtained:

$$c = B \sum_{k=1}^{n} \frac{1}{G_{zx}^{(k)}} \left[R_k^2 \left(z_k - z_{k-1} \right) - R_k S_k \left(z_k^2 - z_{k-1}^2 \right) + \right.$$
$$+ \frac{1}{3} \left(S_k^2 - 2R_k T_k \right) \left(z_k^3 - z_{k-1}^3 \right) +$$
$$\left. + \frac{1}{2} S_k T_k \left(z_k^4 - z_{k-1}^4 \right) + \frac{1}{5} T_k^2 \left(z_k^5 - z_{k-1}^5 \right) \right]. \tag{A19}$$

Lastly, the laminated beam shear stiffness is $C = 1/c$.

Appendix B. Properties of Characteristic Equation Roots

Appendix B.1. Balanced Case

The roots, $\lambda_1, \lambda_2, ..., \lambda_6$, of the characteristic Equation (20) for the uncoupled differential problem are furnished by Equation (19). They attain real values if:

$$\alpha_1 > 2\sqrt{\alpha_2} \quad \text{and} \quad \alpha_3 > 0, \tag{A20}$$

that is, recalling the definitions of Equation (15),

$$k_x > 0, \quad k_z > \hat{k}_z, \quad \text{and} \quad a_1 + d_1 h_1^2 + a_2 + d_2 h_2^2 > 2b_2 h_2 - 2b_1 h_1, \tag{A21}$$

with $\hat{k}_z$ introduced by Equation (21). The above conditions are usually fulfilled for geometric and material properties corresponding to common fiber-reinforced laminates.

Appendix B.2. Unbalanced Case

The roots, $\mu_1, \mu_2, ..., \mu_6$, of the characteristic Equation (33) for the coupled differential problem can be computed numerically. They attain distinct real values if the associated cubic equation has positive discriminant,

$$D = \beta_1^2 \beta_2^2 - 4\beta_2^3 - 4\beta_1^3 \beta_3 + 18\beta_1 \beta_2 \beta_3 - 27\beta_3^2 > 0, \tag{A22}$$

and positive roots,

$$\zeta_1 = \frac{\beta_1}{3} + \frac{2}{3}\sqrt{\beta_1^2 - 3\beta_2} \cos \frac{\vartheta}{3} > 0,$$
$$\zeta_2 = \frac{\beta_1}{3} + \frac{2}{3}\sqrt{\beta_1^2 - 3\beta_2} \cos \left(\frac{\vartheta}{3} + \frac{2\pi}{3} \right) > 0, \quad \text{and} \tag{A23}$$
$$\zeta_3 = \frac{\beta_1}{3} + \frac{2}{3}\sqrt{\beta_1^2 - 3\beta_2} \cos \left(\frac{\vartheta}{3} + \frac{4\pi}{3} \right) > 0,$$

where:

$$\vartheta = \cos^{-1} \frac{2\beta_1^3 - 9\beta_1\beta_2 + 27\beta_3}{2\sqrt{\left(\beta_1^2 - 3\beta_2 \right)^3}}. \tag{A24}$$

In this case, the characteristic equation roots are:

$$\mu_1 = -\mu_2 = \sqrt{\zeta_1}, \quad \mu_3 = -\mu_4 = \sqrt{\zeta_2}, \quad \text{and} \quad \mu_5 = -\mu_6 = \sqrt{\zeta_3}. \tag{A25}$$

Appendix C. $\mathcal{J}$-Integral Calculation for the Crack-Tip Element

The path-independent $\mathcal{J}$-integral was introduced by Rice [57] as:

$$\mathcal{J} = \int_{\Gamma} \left[\omega \, dz - \left(t_x \frac{\partial u}{\partial x} + t_z \frac{\partial w}{\partial x} \right) d\Gamma \right], \tag{A26}$$

where Γ is an arbitrary integration path surrounding the crack tip, ω is the *strain energy density* (per unit volume), t_x and t_z are the *stress vector components* (referred to the outer normal to Γ), and u and w are the *displacement vector components* along the coordinates x and z.

For the CTE, we choose an integration path, Γ, encircling the interface between sublaminates as depicted in Figure A1. The path is subdivided into three segments, Γ_{12}, Γ_{23}, and Γ_{34}, along which quantities entering the $\mathcal{J}$-integral have the expressions listed in Table A1.

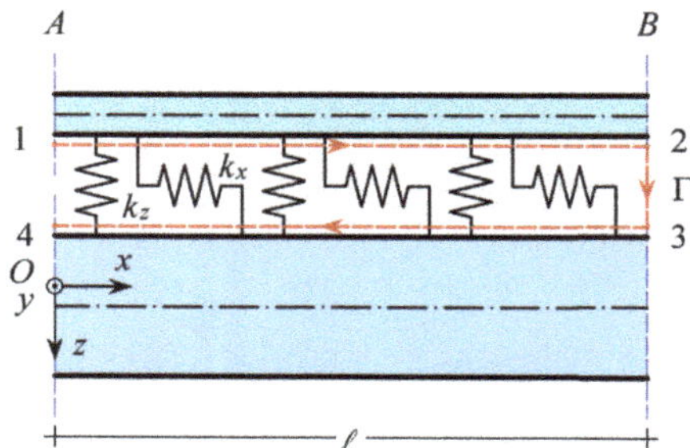

Figure A1. Integration path Γ chosen for the evaluation of the $\mathcal{J}$-integral for the CTE.

Table A1. Calculation of $\mathcal{J}$-integral.

Path Segment	dz	dΓ	ω	t_x	t_z	u	w
Γ_{12}	0	dx	-	$-\tau$	$-\sigma$	$u_1 + \varphi_1 h_1$	w_1
Γ_{23}	dz	dz	Φ/t	t_x	t_z	u	w
Γ_{34}	0	$-$dx	-	τ	σ	$u_2 - \varphi_2 h_2$	w_2

Accordingly, the following expression results:

$$\mathcal{J} = \int_0^\ell -\left[-\tau\frac{\mathrm{d}}{\mathrm{d}x}\left(u_1 + h_1\varphi_1\right) - \sigma\frac{\mathrm{d}w_1}{\mathrm{d}x}\right]\mathrm{d}x +$$
$$+ \int_0^t \left[\frac{\Phi}{t} - \left(t_x\frac{\mathrm{d}u}{\mathrm{d}x} + t_z\frac{\mathrm{d}w}{\mathrm{d}x}\right)\right]\mathrm{d}z + \tag{A27}$$
$$+ \int_\ell^0 -\left[\tau\frac{\mathrm{d}}{\mathrm{d}x}\left(u_2 - h_2\varphi_2\right) + \sigma\frac{\mathrm{d}w_2}{\mathrm{d}x}\right]\left(-\mathrm{d}x\right),$$

where t is the interface thickness and $\Phi = \Phi(\Delta w, \Delta u)$ is the *interface potential energy* [27], such that:

$$\sigma = \frac{\partial\Phi}{\partial\Delta w} \quad \text{and} \quad \tau = \frac{\partial\Phi}{\partial\Delta u}. \tag{A28}$$

By recalling the definition of relative displacements, Equation (6), and assuming a vanishing interface thickness ($t \to 0$), we obtain:

$$\mathcal{J} = \Phi\left[\Delta w(\ell), \Delta u\left(\ell\right)\right] - \int_0^\ell \left[\sigma(x)\frac{\mathrm{d}\Delta w(x)}{\mathrm{d}x} + \tau(x)\frac{\mathrm{d}\Delta u(x)}{\mathrm{d}x}\right]\mathrm{d}x. \tag{A29}$$

Next, by substituting Equation (A28) into (A29) and recalling that $\mathrm{d}\Phi$ is an exact differential, we deduce:

$$\mathcal{J} = \Phi\left[\Delta w(\ell), \Delta u\,(\ell)\right] - \int_0^\ell \left[\frac{\partial \Phi}{\partial \Delta w}\frac{\mathrm{d}\Delta w(x)}{\mathrm{d}x} + \frac{\partial \Phi}{\partial \Delta u}\frac{\mathrm{d}\Delta u(x)}{\mathrm{d}x}\right]\mathrm{d}x =$$

$$= \Phi\left[\Delta w(\ell), \Delta u\,(\ell)\right] - \int_{[\Delta w(0),\Delta u(0)]}^{[\Delta w(\ell),\Delta u(\ell)]} \frac{\partial \Phi}{\partial \Delta w}\,\mathrm{d}\Delta w + \frac{\partial \Phi}{\partial \Delta u}\,\mathrm{d}\Delta u = \tag{A30}$$

$$= \Phi\left[\Delta w(\ell), \Delta u\,(\ell)\right] - \int_{[\Delta w(0),\Delta u(0)]}^{[\Delta w(\ell),\Delta u(\ell)]}\mathrm{d}\Phi.$$

Hence,

$$\mathcal{J} = \Phi\left[\Delta w(0), \Delta u(0)\right], \tag{A31}$$

which holds for a general cohesive interface with potential-based traction-separation laws. For a brittle-elastic interface, the potential energy function specializes to:

$$\Phi(\Delta w, \Delta u) = \frac{1}{2}k_z\Delta w^2 + \frac{1}{2}k_x\Delta u^2. \tag{A32}$$

By substituting Equation (A32) into (A31), we obtain:

$$\mathcal{J} = \frac{1}{2}k_z\Delta w^2(0) + \frac{1}{2}k_x\Delta u^2(0), \tag{A33}$$

which is identical to Equation (42).

For the present elastic interface model, the two addends in Equation (A33) naturally identify the Mode I and II contributions to the energy release rate (see Section 4.1). This result holds for both the balanced and unbalanced cases, for which the analytical solution has been derived in Section 3. For a general cohesive interface model, Equation (A31) yields the value of the $\mathcal{J}$-integral, but its decomposition into Mode I and II contributions is not straightforward. The interested reader can find further hints about this issue in the papers by Wu et al. [37,38] and in the references recalled therein.

References

1. Talreja, R.; Singh, C.V. *Damage and Failure of Composite Materials*; Cambridge University Press: Cambridge, UK, 2012; ISBN 978-0-521-81942-8.
2. Garg, A.C. Delamination—A damage mode in composite structures. *Eng. Fract. Mech.* **1988**, *29*, 557–584. [CrossRef]
3. Bolotin, V.V. Delaminations in composite structures: its origin, buckling, growth and stability. *Compos. Part B Eng.* **1996**, *27*, 129–145. [CrossRef]
4. Tay, T.E. Characterization and analysis of delamination fracture in composites: An overview of developments from 1990 to 2001. *Appl. Mech. Rev.* **2003**, *56*, 1–31. [CrossRef]
5. Senthil, K.; Arockiarajan, A.; Palaninathan, R.; Santhosh, B.; Usha, K.M. Defects in composite structures: Its effects and prediction methods—A comprehensive review. *Compos. Struct.* **2013**, *106*, 139–149. [CrossRef]
6. Tabiei, A.; Zhang, W. Composite Laminate Delamination Simulation and Experiment: A Review of Recent Development. *Appl. Mech. Rev.* **2018**, *70*, 030801. [CrossRef]
7. Bucur, V. (Ed.) *Delamination in Wood, Wood Products and Wood-Based Composites*; Springer: Dordrecht, The Netherlands, 2011; ISBN 978-90-481-9549-7. [CrossRef]
8. Minatto, F.D.; Milak, P.; De Noni, A.; Hotza, D.; Montedo, O.R.K. Multilayered ceramic composites—A review. *Adv. Appl. Ceram.* **2015**, *114*, 127–138. [CrossRef]
9. Friedrich, K. (Ed.) *Application of Fracture Mechanics to Composite Materials*; Elsevier: Amsterdam, The Netherlands, 1989; ISBN 978-0-444-87286-9.
10. Carlsson, L.A.; Adams, D.F.; Pipes, R.B. *Experimental Characterization of Advanced Composite Materials*, 4th ed.; CRC Press: Boca Raton, FL, USA, 2014; ISBN 978-1-4398-4858-6.
11. Hutchinson, J.W.; Suo, Z. Mixed mode cracking in layered materials. *Adv. Appl. Mech.* **1991**, *29*, 63–191. [CrossRef]
12. Kanninen, M.F. An augmented double cantilever beam model for studying crack propagation and arrest. *Int. J. Fract.* **1973**, *9*, 83–92. [CrossRef]

13. Bruno, D.; Grimaldi, A. Delamination failure of layered composite plates loaded in compression. *Int. J. Solids Struct.* **1990**, *26*, 313–330. [CrossRef]

14. Allix, O.; Ladevèze, P. Interlaminar interface modeling for the prediction of delamination. *Compos. Struct.* **1992**, *22*, 235–242. [CrossRef]

15. Corigliano, A. Formulation, identification and use of interface models in the numerical analysis of composite delamination. *Int. J. Solids Struct.* **1993**, *30*, 2779–2811. [CrossRef]

16. Allix, O.; Ladevèze, P.; Corigliano, A. Damage analysis of interlaminar fracture specimens. *Compos. Struct.* **1995**, *31*, 61–74. [CrossRef]

17. Point, N.; Sacco, E. A delamination model for laminated composites. *Int. J. Solids Struct.* **1996**, *33*, 483–509. [CrossRef]

18. Bruno, D.; Greco, F. Mixed mode delamination in plates: a refined approach. *Int. J. Solids Struct.* **2001**, *38*, 9149–9177. [CrossRef]

19. Qiao, P.; Wang, J. Mechanics and fracture of crack tip deformable bi-material interface. *Int. J. Solids Struct.* **2004**, *41*, 7423–7444. [CrossRef]

20. Szekrényes, A. Improved analysis of unidirectional composite delamination specimens. *Mech. Mater.* **2007**, *39*, 953–974. [CrossRef]

21. Bennati, S.; Valvo, P.S. Delamination growth in composite plates under compressive fatigue loads. *Compos. Sci. Technol.* **2006**, *66*, 248–254. [CrossRef]

22. Bennati, S.; Colleluori, M.; Corigliano, D.; Valvo, P. S. An enhanced beam-theory model of the asymmetric double cantilever beam (ADCB) test for composite laminates. *Compos. Sci. Technol.* **2009**, *69*, 1735–1745. [CrossRef]

23. Bennati, S.; Fisicaro, P.; Valvo, P.S. An enhanced beam-theory model of the mixed-mode bending (MMB) test—Part I: Literature review and mechanical model. *Meccanica* **2013**, *48*, 443–462. [CrossRef]

24. Bennati, S.; Fisicaro, P.; Valvo, P.S. An enhanced beam-theory model of the mixed-mode bending (MMB) test—Part II: Applications and results. *Meccanica* **2013**, *48*, 465–484. [CrossRef]

25. Xie, J.; Waas, A.; Rassaian, M. Analytical predictions of delamination threshold load of laminated composite plates subject to flexural loading. *Compos. Struct.* **2017**, *179*, 181–194. [CrossRef]

26. Dimitri, R.; Tornabene, F.; Zavarise, G. Analytical and numerical modeling of the mixed-mode delamination process for composite moment-loaded double cantilever beams. *Compos. Struct.* **2018**, *187*, 535–553. [CrossRef]

27. Park, K.; Paulino, G.H. Cohesive Zone Models: A Critical Review of Traction-Separation Relationships Across Fracture Surfaces. *Appl. Mech. Rev.* **2013**, *64*, 060802. [CrossRef]

28. Dimitri, R.; Trullo, M.; De Lorenzis, L.; Zavarise, G. Coupled cohesive zone models for mixed-mode fracture: A comparative study. *Eng. Fract. Mech.* **2015**, *148*, 145–179. [CrossRef]

29. Borg, R.; Nilsson, L.; Simonsson, K. Simulation of delamination in fiber composites with a discrete cohesive failure model. *Compos. Sci. Technol.* **2001**, *61*, 667–677. [CrossRef]

30. Camanho, P.P.; Dávila, C.G.; de Moura, M.F. Numerical simulation of mixed-mode progressive delamination in composite materials. *J. Compos. Mater.* **2003**, *37*, 1415–1438. [CrossRef]

31. Yang, Q.; Cox, B. Cohesive models for damage evolution in laminated composites. *Int. J. Fract.* **2005**, *133*, 107–137. [CrossRef]

32. Parmigiani, J.P.; Thouless, M.D. The effects of cohesive strength and toughness on mixed-mode delamination of beam-like geometries. *Eng. Fract. Mech.* **2007**, *74*, 2675–2699. [CrossRef]

33. Turon, A.; Dávila, C.G.; Camanho, P.P.; Costa, J. An engineering solution for mesh size effects in the simulation of delamination using cohesive zone models. *Eng. Fract. Mech.* **2007**, *74*, 1665–1682. [CrossRef]

34. Harper, P.W.; Hallett, S.R. Cohesive zone length in numerical simulations of composite delamination. *Eng. Fract. Mech.* **2008**, *75*, 4774–4792. [CrossRef]

35. Sørensen, B.F.; Jacobsen T.K. Characterizing delamination of fiber composites by mixed mode cohesive laws. *Compos. Sci. Technol.* **2009**, *69*, 445–456. [CrossRef]

36. Xie, J.; Waas, A.; Rassaian, M. Closed-form solutions for cohesive zone modeling of delamination toughness tests. *Int. J. Solids Struct.* **2016**, *88–89*, 379–400. [CrossRef]

37. Wu, C.; Gowrishankar, S.; Huang, R.; Liechti, K.M. On determining mixed-mode traction–separation relations for interfaces. *Int. J. Fract.* **2016**, *202*, 1–19. [CrossRef]

38. Wu, C.; Huang, R.; Liechti, K.M. Simultaneous extraction of tensile and shear interactions at interfaces. *J. Mech. Phys. Solids* **2019**, *125*, 225–254. [CrossRef]

39. da Silva, L.F.M.; das Neves, P.J.C.; Adams, R.D.; Spelt, J.K. Analytical models of adhesively bonded joints—Part I: Literature survey. *Int. J. Adhes. Adhes.* **2009**, *29*, 319–330. [CrossRef]

40. Bigwood, D.A.; Crocombe, A.D. Elastic analysis and engineering design formulae for bonded joints. *Int. J. Adhes. Adhes.* **1989**, *9*, 229–242. [CrossRef]

41. Krenk, S. Energy release rate of symmetric adhesive joints. *Eng. Fract. Mech.* **1992**, *43*, 549–559. [CrossRef]

42. Penado, F.E. A closed form solution for the energy release rate of the double cantilever beam specimen with an adhesive layer. *J. Compos. Mater.* **1993**, *27*, 383–407. [CrossRef]

43. Bennati, S.; Taglialegne, L.; Valvo, P.S. Modelling of interfacial fracture of layered structures. In Proceedings of the ECF 18—18th European Conference on Fracture, Dresden, Germany, 30 August–3 September 2010; ISBN 978-3-00-031802-3.

44. Liu, Z.; Huang, Y.; Yin, Z.; Bennati, S.; Valvo, P.S. A general solution for the two-dimensional stress analysis of balanced and unbalanced adhesively bonded joints. *Int. J. Adhes. Adhes.* **2014**, *54*, 112–123. [CrossRef]

45. Jiang, W.; Qiao, P. An improved four-parameter model with consideration of Poisson's effect on stress analysis of adhesive joints. *Eng. Struct.* **2015**, *88*, 203–215. [CrossRef]

46. Zhang, Z.W.; Li, Y.S.; Liu, R. An analytical model of stresses in adhesive bonded interface between steel and bamboo plywood. *Int. J. Solids Struct.* **2015**, *52*, 103–113. [CrossRef]

47. Blackman, B.R.K.; Hadavinia, H.; Kinloch, A.J.; Williams, J.G. The use of a cohesive zone model to study the fracture of fiber composites and adhesively-bonded joints. *Int. J. Fract.* **2003**, *119*, 25–46. [CrossRef]

48. Li, S.; Thouless, M.D.; Waas, A.M.; Schroeder, J.A.; Zavattieri, P.D. Mixed-mode cohesive-zone models for fracture of an adhesively bonded polymer–Matrix composite. *Eng. Fract. Mech.* **2006**, *73*, 64–78. [CrossRef]

49. Jones, R.M. *Mechanics of Composite Materials*, 2nd ed.; Taylor & Francis Inc.: Philadelphia, PA, USA, 1999; ISBN 978-1560327127.

50. Schapery, R.A.; Davidson, B.D. Prediction of energy release rate for mixed-mode delamination using classical plate theory. *Appl. Mech. Rev.* **1990**, *43*, S281–S287. [CrossRef]

51. Valvo, P.S. On the calculation of energy release rate and mode mixity in delaminated laminated beams. *Eng. Fract. Mech.* **2016**, *165*, 114–139. [CrossRef]

52. Tsokanas, P.; Loutas, T. Hygrothermal effect on the strain energy release rates and mode mixity of asymmetric delaminations in generally layered beams. *Eng. Fract. Mech.* **2019**, *214*, 390–409. [CrossRef]

53. Taglialegne, L. Modellazione Meccanica Della Frattura Interlaminare di Provini in Composito Non Simmetrici. Master's Thesis, University of Pisa, Pisa, Italy, 2014. Available online: http://etd.adm.unipi.it/theses/available/etd-09172014-092832 (accessed on 23 July 2019).

54. Davidson, B.D.; Hu, H.; Schapery, R.A. An Analytical Crack-Tip Element for Layered Elastic Structures. *J. Appl. Mech.* **1995**, *62*, 294–305. [CrossRef]

55. Wang, J.; Qiao, P. Interface crack between two shear deformable elastic layers. *J. Mech. Phys. Solids* **2004**, *52*, 891–905. [CrossRef]

56. Timoshenko, S.P. *Strength of Materials, Vol. 1: Elementary Theory And Problems*, 3rd ed.; D. Van Nostrand: New York, NY, USA, 1955; ISBN 978-0442085414.

57. Rice, J.R. A path independent integral and the approximate analysis of strain concentrations by notches and cracks. *J. Appl. Mech.* **1968**, *35*, 379–386. [CrossRef]

58. Arouche, M.M.; Wang, W.; Teixeira de Freitas, S.; de Barros, S. Strain-based methodology for mixed-Mode I+II fracture: A new partitioning method for bi-material adhesively bonded joints. *J. Adhes.* **2019**, *95*, 385–404. [CrossRef]

59. Vannucci, P.; Verchery, G. A special class of uncoupled and quasi-homogeneous laminates. *Compos. Sci. Technol.* **2001**, *61*, 1465–1473. [CrossRef]

60. Garulli, T.; Catapano, A.; Fanteria, D.; Jumel, J.; Martin, E. Design and finite element assessment of fully uncoupled multi-directional layups for delamination tests. *J. Compos. Mater.* **2019**. [CrossRef]

61. Samborski, S. Numerical analysis of the DCB test configuration applicability to mechanically coupled Fiber Reinforced Laminated Composite beams. *Compos. Struct.* **2016**, *152*, 477–487. [CrossRef]

62. Bennati, S.; Valvo, P.S. An experimental compliance calibration strategy for mixed-mode bending tests. *Proc. Mater. Sci.* **2014**, *3*, 1988–1993. [CrossRef]

63. Sørensen, B.F.; Jørgensen, K.; Jacobsen, T.K.; Østergaard, R.C. DCB-specimen loaded with uneven bending moments. *Int. J. Fract.* **2006**, *141*, 163–176. [CrossRef]
64. Davidson, B.D.; Gharibian, S.J.; Yu, L. Evaluation of energy release rate-based approaches for predicting delamination growth in laminated composites. *Int. J. Fract.* **2000**, *105*, 343–365. [CrossRef]
65. Davidson, B.D.; Bialaszewski, R.D.; Sainath, S.S. A non-classical, energy release rate based approach for predicting delamination growth in graphite reinforced laminated polymeric composites. *Compos. Sci. Technol.* **2006**, *66*, 1479–1496. [CrossRef]
66. Harvey, C.M.; Wang, S. Experimental assessment of mixed-mode partition theories. *Compos. Struct.* **2012**, *94*, 2057–2067. [CrossRef]
67. Harvey, C.M.; Eplett, M.R.; Wang, S. Experimental assessment of mixed-mode partition theories for generally laminated composite beams. *Compos. Struct.* **2015**, *124*, 10–18. [CrossRef]
68. Bednarcyk, B.A.; Aboudi, J.; Yarrington, P.W.; Collier, C.S. Simplified Shear Solution for Determination of the Shear Stress Distribution in a Composite Panel From the Applied Shear Resultant. In Proceedings of the 49th AIAA/ASME/ASCE/AHS/ASC Structures, Structural Dynamics, and Materials Conference, Schaumburg, IL, USA, 7–10 April 2008. [CrossRef]
69. Kumar, R.; Nath, J.K.; Pradhan, M.K.; Kumar, R. Analysis of Laminated Composite Using Matlab. In Proceedings of the ICMEET 2014—1st International Conference on Mechanical Engineering: Emerging Trends for Sustainability, MANIT, Bhopal, India, 29–31 January 2014.
70. Wallner-Novak, M.; Augustin, M.; Koppelhuber, J.; Pock, K. *Cross-Laminated Timber Structural Design*; proHolz: Vienna, Austria, 2018; Volume II, ISBN 978-3-902320-96-4.

applied sciences

Article

Wedge-Splitting Test on Carbon-Containing Refractories at High Temperatures

Martin Stückelschweiger [1,*], Dietmar Gruber [2], Shengli Jin [2] and Harald Harmuth [2]

[1] K1-MET GmbH, Linz 4020, Austria
[2] Chair of Ceramics, Montanuniversitaet Leoben 8700, Austria
* Correspondence: martin-josef.stueckelschweiger@alumni.unileoben.ac.at; Tel.: +43-3842-402-3207

Received: 4 July 2019; Accepted: 5 August 2019; Published: 8 August 2019

Abstract: The mode I fracture behavior of ordinary refractory materials is usually tested with the wedge-splitting test. At elevated temperatures, the optical displacement measurement is difficult because of the convection in the furnace and possible reactions of refractory components with the ambient atmosphere. The present paper introduces a newly developed testing device, which is able to perform such experiments up to 1500 °C. For the testing of carbon-containing refractories a gas purging, for example, with argon, is possible. Laser speckle extensometers are applied for the displacement measurement. A carbon-containing magnesia refractory (MgO–C) was selected for a case study. Based on the results obtained from tests, fracture mechanical parameters such as the specific fracture energy and the nominal notch tensile strength were calculated. An inverse simulation procedure applying the finite element method yields tensile strength, the total specific fracture energy, and the strain-softening behavior. Additionally, the creep behavior was also considered for the evaluation.

Keywords: high-temperature wedge splitting test; fracture parameters; reducing condition; carbon-containing refractories; strain-softening; fracture energy

---

1. Introduction

The wedge-splitting test is performed on notched prismatic specimens that enable stable crack formation for relatively large specimens [1,2]. The reasons are the action of the wedge and the relatively high fracture surface to sample volume ratio. The specimen length is 100 mm, the height is also 100 mm, and the thickness 75 mm. A three-dimensional schematic representation of the sample including the load transmission parts is shown in Figure 1. The wedge-splitting test (WST) at room temperature is performed with a loading rate of 0.5 mm/min. During the test, a video extensometer detects the horizontal displacement at the measuring points. The displacement is measured at the height where the rollers contact the load transmission elements.

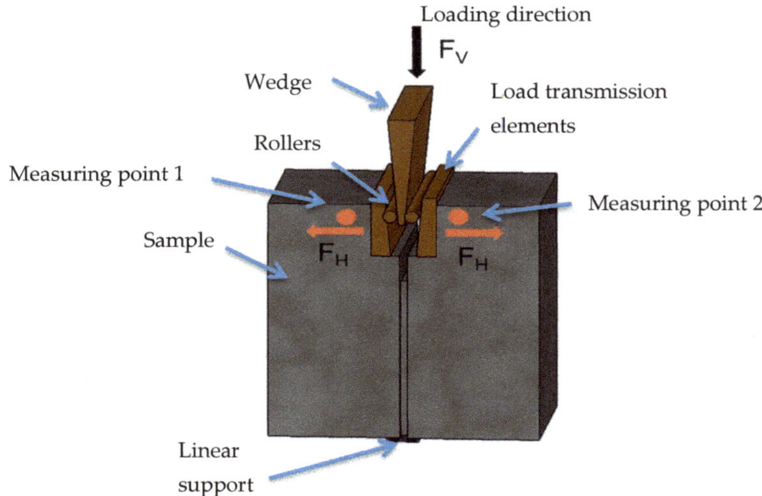

Figure 1. Schematic representation of the wedge-splitting test.

The sample rests on a linear support ($3 \times 3 \times 90$ mm^3) to allow for the free vertical displacement of the sample except for the support. The linear support, as well as the wedge and the rollers, are made of corundum to withstand the testing temperatures.

From the results, the specific fracture energy G_f' and the nominal notch tensile strength σ_{NT}, can be calculated directly from the load-displacement curve (Equations (2) and (3)). The horizontal force F_H, which is needed for Equations (2) and (3), is calculated from the geometry of the wedge and the measured vertical force F_V (Equation (1)):

$$F_H = \frac{F_V}{2 tan\frac{\alpha}{2}} \tag{1}$$

In this equation, α is the wedge angle (10°).

$$G_{f'} = \frac{1}{A} \int_0^\delta F_H d\delta_H \tag{2}$$

The specific fracture energy G_f' is the area below the load-displacement curve divided by the ligament area A. The measured horizontal displacement is defined with δ_H, and the horizontal displacement at 15% of the maximum load is δ.

Here G_f' (Equation (2)) is the specific fracture energy directly evaluated from the load-displacement curve. This parameter includes only the major part of the specific fracture energy because a premature termination of the test is necessary to avoid the contact of the wedge with the ligament. This means that the WST is stopped before the force reaches zero.

The nominal notch tensile strength σ_{NT} is calculated according to Equation (3):

$$\sigma_{NT} = \frac{F_{H,max}}{bh}\left(1 + \frac{6y}{h}\right) \tag{3}$$

Here, $F_{H,max}$ is the maximum horizontal load, b and h are the width and the height of the ligament, and y is the vertical distance from the loading position to the center of the ligament.

The simplest possibility for the displacement measurement is to make use of the crosshead travel. Unfortunately, this is not accurate because deformations of the testing machine influence the results as well as the adjustment of the load transmission parts during load application. Out of these reasons, it is advantageous to measure the displacement directly on the sample.

At elevated temperatures, one possibility is to measure the displacement with an optical dilatometer. Here a halogen lamp illuminates the sample on one side and on the other side the shadow cast by the sample against the light source is recorded by a high-resolution complementary metal-oxide-semiconductor (CMOS) camera [3]. With this method, the displacement is measured on one side of the sample only. Although the performance of the wedge splitting test at room temperature with optical displacement measurement is state of the art [4], at elevated temperatures, the optical measurement is far more difficult due to the radiation and convection in the furnace chamber.

High-temperature testing above 600 °C of carbon-containing refractories shall be carried out in non-oxidizing atmospheres. Otherwise, the oxidation of carbon $2C_{(g)} + O2_{(g)} = 2CO_{(g)}$ takes place [5], and the microstructure and the properties will be changed significantly [6].

To counteract these problems, a new device was developed to perform the WST on carbon-containing materials up to 1500 °C in an argon atmosphere and to measure the displacement on the front and rear side of the sample. The horizontal displacements are measured with two laser speckle extensometers (LSE). In the present paper, the features of the new device and the fracture behavior of a resin-bonded magnesia carbon material at elevated temperatures are discussed.

2. Material and Sample Preparation

The tested resin bonded magnesia carbon brick is composed of fused magnesia, large crystal magnesia clinker, and graphite. A typical application is the refractory lining of a steel ladle in the secondary steel metallurgy. The residual carbon content after cooking under reducing conditions was 10%. The composition showed a CaO/SiO_2 (C/S) ratio of 3.8. Bricks were heat-treated at 1000 °C for 5 h in a coke bed prior to testing; this enhanced the carbon bond and reduced the emission of volatiles during testing. These emissions caused problems for the displacement measurement by laser speckle extensometers.

3. Device

The wedge splitting test device (universal testing device with displacement measurement) was developed in close cooperation with the company ZwickRoell Testing Systems GmbH (Furstenfeld, Austria). An existing universal testing device with a rigid frame and two rotating spindles were equipped with several components in order to perform the wedge-splitting test at high temperatures. A schematic drawing of the device is illustrated in Figure 2. The load cell is integrated in the lower piston and the maximum load capacity is 300 kN at room temperature. The vertical load rate is set to 0.5 mm/min for the wedge-splitting test.

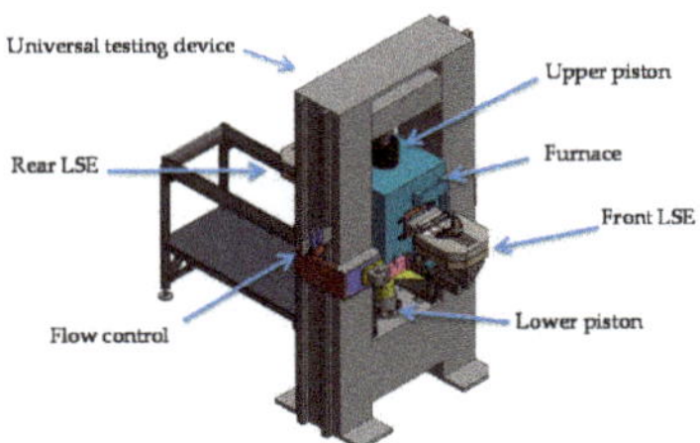

Figure 2. High-temperature testing device (courtesy of Messphysik).

In this device, a gas-tight furnace was installed (Figure 2, Figure 3). The furnace has two windows (front and rear) in order to measure the displacement during the test. The maximum testing temperature is 1500 °C. In order to protect the carbon in the samples from oxidation, the furnace chamber can be purged with an inert gas. The purging rate depends on the material and on the testing temperature and can be adjusted between 8 and 70 L/h. The adjustment of the purging rate was done with a valve at the digital flowmeter.

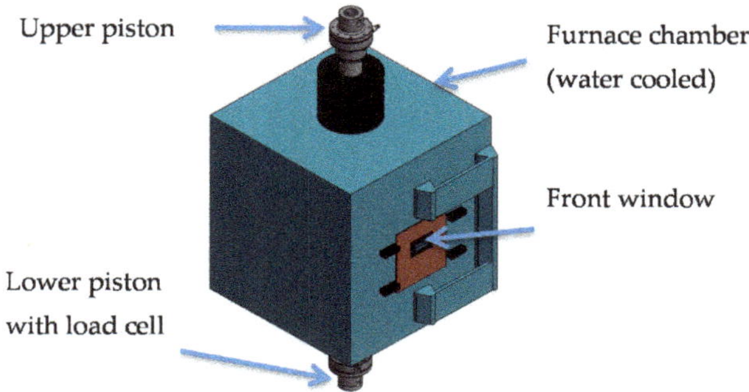

Figure 3. Gas-tight furnace (courtesy of Messphysik).

For the displacement measurement during the wedge-splitting test, the device was equipped with two laser-speckle extensometers. One is situated in the front and the other one in the rear of the furnace. The sample was illuminated with green laser light (wavelength: 532 nm), creating a speckled pattern on the sample surface. The movement of the speckled pattern was recorded with two full-screen digital cameras. In each evaluation area, a displacement measurement of the speckled pattern was performed. The increase in the distance between the measurement points is calculated. It is called horizontal displacement in this paper [7].

The setup of the furnace including the laser-speckle extensometers is shown in Figure 4 (all units in mm).

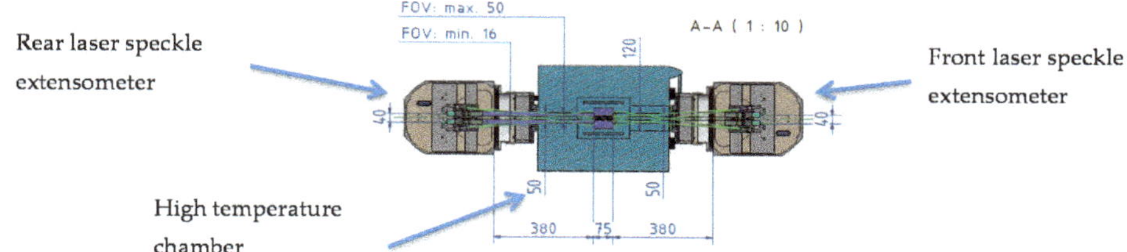

Figure 4. Furnace with two laser-speckle extensometers (courtesy of Messphysik).

Each extensometer has two lasers and two cameras. The green lines in Figure 4 show the course of the laser beam. The distance of the measuring points was set to 40 mm. This initial measurement length can be adjusted according to the sample geometry. The distance between the position of the camera and the sample surface is 380 mm.

The advantage of a laser-speckle-extensometer compared to a mechanical extensometer is that the sample and the extensometer fingers are not in contact. Contacting extensometer fingers could be damaged during the cracking of the sample, furthermore, a possible deformation of the extensometers could influence the results. An advantage compared to a standard laser extensometer is that an application of measuring marks to increase the contrast is not necessary because of the characteristic speckled pattern. The resolution of the strain measurement is 0.11 µm according to the specifications from ZwickRoell. The scattering of the measurement signal during a high-temperature test was below 3 µm. This value was determined together with the company ZwickRoell in the course of preliminary investigations. The horizontal displacement was measured in the front and rear side, the average value was used for the evaluation. Figure 5 shows the configuration after a test.

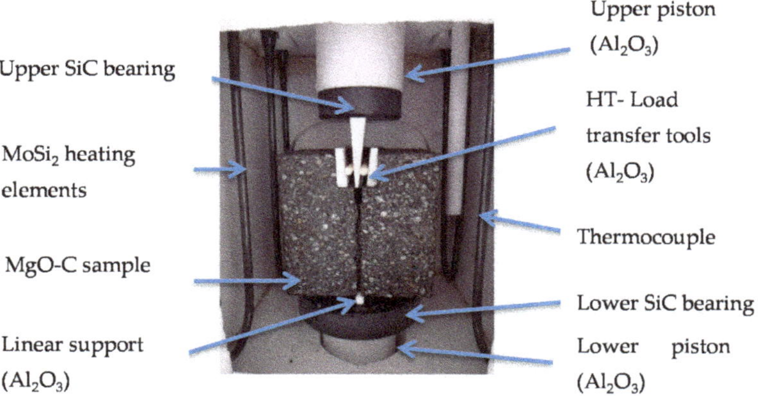

Figure 5. Wedge splitting test setup and sample after testing.

The alignment of the rolls and the wedge is essential to achieve symmetrical displacement on the front and rear side of the sample.

4. Results and Discussion

This section is divided into three subsections. It provides a concise and precise description of the experimental results, their interpretations, and the conclusions that can be drawn.

4.1. Experiments and Evaluation

Tests with the newly developed device were carried out successfully between room temperature and 1470 °C. During the high-temperature experiments, the furnace chamber was purged with argon and a purging rate up to 70 L/h. Figure 6 shows the load-displacement diagrams from room temperature to 1470 °C.

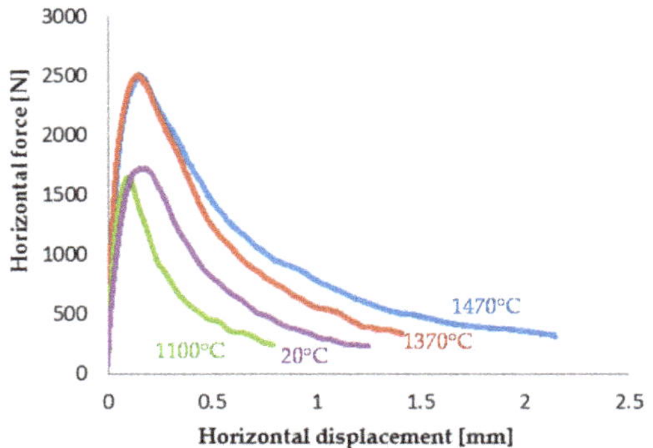

Figure 6. Load/displacement diagram of the magnesia carbon material measured at different temperatures.

As the oxidation of carbon in the sample increases with increasing temperature, a purging rate was defined for each individual temperature. The purging rate at 1100 °C was set to 10 L/h, at 1370 °C 40 L/h and to 70 L/h for the test at 1470 °C.

The Table 1 below shows the maximum horizontal force $F_{H,max}$, the specific fracture energy Gf' and the nominal notch tensile strength σ_{NT} according to the Equations (1)–(3).

Table 1. Fracture mechanical parameters determined from tests performed at room temperature (RT), 1100 °C, 1370 °C, and 1470 °C.

	20 °C (RT)	1100 °C	1370 °C	1470 °C
F_{Vmax} [N]	301	288	439	440
F_{Hmax} [N]	1720	1646	2508	2510
Specific fracture energy G_f' [N/m]	240	125	362	513
Nominal notch tensile strength σ_{NT} [N/mm^2]	2.71	3.13	4.02	4.35
G_f'/σ_{NT} [mm]	0.09	0.04	0.09	0.12

For the high-temperature tests the specific fracture energy G_f' and the nominal notch tensile strength σ_{NT} show an increase with increasing temperature. The test at room temperature shows the smallest nominal notch tensile strength but the specific fracture energy G_f' is higher than that at 1100 °C and lower than that at 1370 °C. The ratio between the specific fracture energy G_f' and the notch tensile strength σ_{NT} are applied as an indicator for brittleness [8].

The following figure illustrates a sample after the experiment at 1470 °C. The oxidation of the carbon in the sample is relatively small, only slight traces of oxidation are visible on the surface. The ligament area of the sample (the area inside the red frame) is without any visible oxidation (Figure 7).

Figure 7. Sample after a wedge-splitting test at 1470 °C.

The weight loss after the test at 1470 °C is below 1% (18 g out of 2250 g of a sample in virgin state) for a purging rate of 70 L/h. The reaction of C(s) with O_2(g) decreases with an increase in the argon purging rate [9]. The weight loss after the test at 1370 °C is 0.8%. For the test at 1100 °C, the weight loss decreases to 0.5%.

4.2. Inverse Evaluation

In order to determine other parameters of interest, such as the tensile strength f_t and the total specific fracture energy G_f and to consider the influence of the creep behavior, an inverse evaluation method was applied on a two-dimensional finite element model of the wedge splitting test developed earlier [10]. The commercial software package ABAQUS (6.13, Dassault Sys Simulia Corp, Providence, RI, USA) was used for modeling. The mode I fracture was considered with an exponential strain-softening behavior applied in the ligament area (Figure 8) (Equation (4)).

$$\sigma = f_t \left\{ 1 - \left(\frac{1 - e^{-\varphi \frac{x}{x_{ult}}}}{1 - e^{-\varphi}} \right) \right\} \tag{4}$$

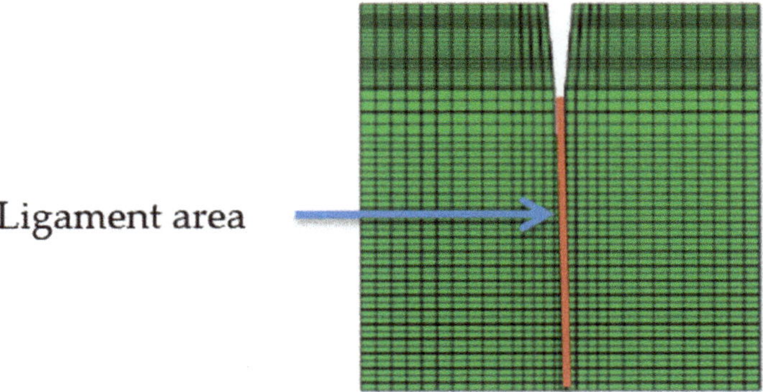

Figure 8. Finite element model geometry and mesh.

In this equation f_t is the tensile strength of the material, the parameter x_{ult} is the ultimate crack opening and φ is a non-dimensional parameter that defines the rate of damage evolution [11]. Cohesive elements were applied to the interfaces of the two half specimens to simulate the mode I fracture [11]. Through an adaptive nonlinear least-squares algorithm, the fracture parameters are inversely calculated by fitting the results of finite element simulations to the results of laboratory wedge splitting tests.

The following Figures 9–11 show the load-displacement diagrams and the determined strain-softening behavior for experiments between 1100 °C and 1470 °C. The green curves in the left diagrams (a) are the measured ones from the wedge splitting tests. The blue curves (a) are the curves simulated by finite element simulations. The red curves in the diagrams show the results with consideration of the creep behavior and are explained in the following chapter. The (b) figures show the strain-softening behavior determined via inverse evaluation from parameters f_t, x_{ult}, and φ. Results from (b) correspond to (a).

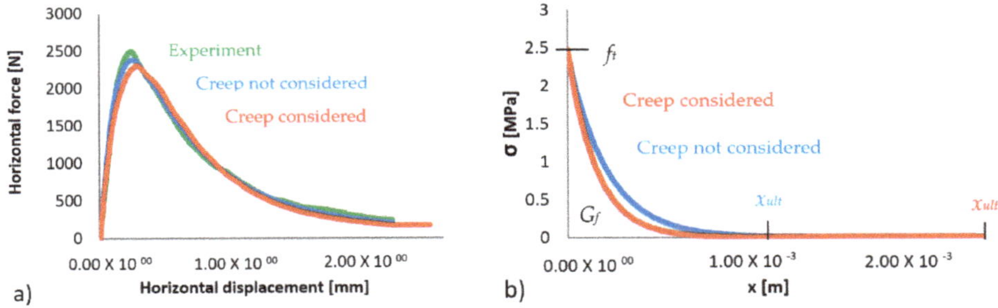

Figure 9. Load-displacement diagram (**a**) and strain-softening behavior (**b**) at 1470 °C.

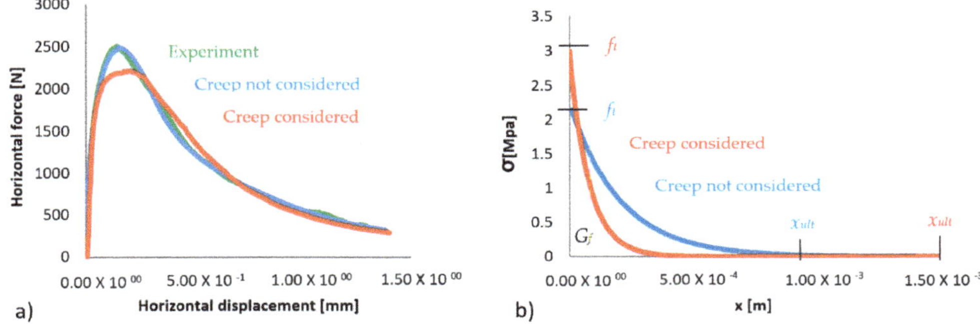

Figure 10. Load-displacement diagram (**a**) and strain-softening behavior (**b**) at 1370 °C.

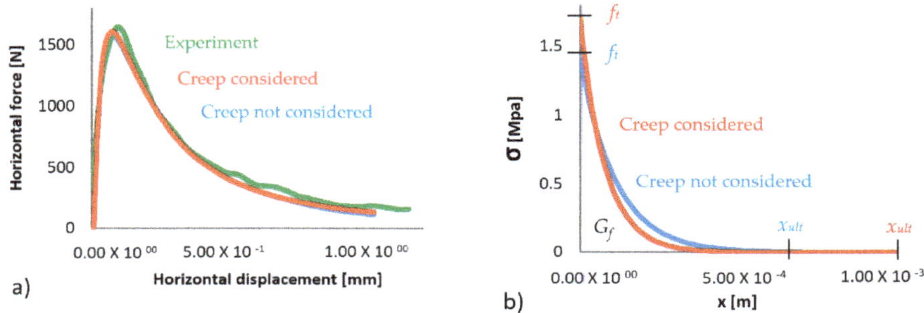

Figure 11. Load-displacement diagram (**a**) and strain-softening behavior (**b**) at 1100 °C.

The inversely calculated curve (a) shows only a slight deviation from the tested one. The minimization procedure is automatically terminated when a minimum of residuals is reached [10]. The tensile strength and x_{ult} show a decrease with decreasing temperature when the creep is not considered.

Table 2 shows the inversely evaluated parameters and brittleness indicators.

Table 2. Fracture mechanical parameters and brittleness indicators at different temperatures calculated from ABAQUS simulations.

Parameter	1470 °C	1370 °C	1100 °C
Total specific fracture energy G_f [N/m]	563	453	162
Tensile strength f_t [N/mm^2]	2.40	2.26	1.42
σ_{NT}/f_t	1.83	1.77	2.2
G_f/f_t [mm]	0.23	0.20	0.11

With increasing temperature, the specific fracture energy G_f increases. The tensile strength follows the same trend. At 1470 °C and 1370 °C, the maximum horizontal forces are similar but the specific fracture energy is lower at 1370 °C because of lower horizontal displacement.

Additionally, the ratio between σ_{NT} and f_t was calculated. The difference between σ_{NT} and f_t is expected. Reasons are the action of the notch on the one hand and on the other hand the dependence on the brittleness number [12]. As well as G_f'/σ_{NT}, the ratio between the real specific fracture energy G_f and the tensile strength f_t is an indicator of the brittleness. This ratio decreases with decreasing temperature.

4.3. Inverse Evaluation with the Consideration of Creep

In this chapter, the tensile creep behavior was considered in the inverse evaluation. The creep behavior of this material was already investigated by the authors [13]. It is clear that creep has an influence on the service behavior and furthermore on the results of the high-temperature wedge splitting test. The aim of these simulations is to calculate the specific fracture energy, which is caused by crack propagation only. The contribution of creep depends on the loading rate. For low loading rates, the influence of creep is comparatively high and vice versa for high loading rates. The Norton–Bailey creep equation was applied (Equation (5)).

$$\dot{\varepsilon}_{cr} = K(T)\sigma^n \, \varepsilon_{cr}^a \tag{5}$$

Here, σ is the applied load, $\dot{\varepsilon}_{cr}$ is the creep strain rate, ε_{cr} is the creep strain, $K(T)$ is a temperature-dependent function and the parameters n and a are the stress and strain exponents, respectively.

These parameters were implemented in the simulation of the wedge-splitting test and the creep behavior was considered in the whole specimen volume. The diagrams (Figures 9–11) in Section 4.2 shows the results in detail. The left diagrams (a) at each temperature show the experimental curve and the fits, both, with and without consideration of creep. The strain-softening behavior, which is plotted on the right side in Figures 9b, 10b and 11b, shows the change of the energy and the tensile strength of the material at each temperature.

Table 3 shows the change of the parameter with respect to temperature when creep is considered. The inverse calculated parameter G_f shows an increase of the specific fracture energy with increasing temperature. The ratio between the specific fracture energy and the tensile strength that is related to the brittleness shows, for both cases (with and without consideration of creep), an increase with increasing temperature. The difference between G_f with consideration of creep and G_f without consideration of creep (Table 2), ΔG_f, shows a maximum temperature of 1370 °C and is between 15% and 47% for the high-temperature tests.

Table 3. Fracture mechanical parameters at different temperatures calculated considering creep and influence on G_f.

Parameter	1470 °C	1370 °C	1100 °C
Specific fracture energy G_f, creep considered [N/m]	417	238	137
ΔG_f [N/m]	146 (26%)	215 (47%)	25 (15%)
Tensile strength f_t, creep considered [N/mm^2]	2.47	2.99	1.71
G_f/f_t, creep considered [mm]	0.17	0.08	0.08

5. Conclusions

A new testing device to perform the WST was introduced in this paper. This new device allows fracture mechanical characterization of carbon-containing refractories up to 1500 °C under reducing conditions. Tests on resin bonded magnesia carbon material were successfully carried out. The oxidation of the carbon in the sample during the high-temperature test can be avoided by inert gas purging during the whole test. The displacement measurement directly on the sample shows a resolution below 3 µm and works stably over the whole temperature range. Based on the data obtained from laboratory experiments, fracture parameters can be determined and implemented in finite element simulations with an inverse estimation procedure. Besides, simulations including the tensile creep behavior were performed and the fracture parameters under pure mode I failure were determined. The influence of creep on the results obtained from the inverse evaluation for the specific fracture energy is between 15% (1100 °C) and 47% (1370 °C).

Author Contributions: Conceptualization, M.S. and D.G.; Methodology, M.S.; Software, S.J.; Validation, D.G., S.J. and H.H.; Formal Analysis, M.S. and. S.J.; Investigation, M.St.; Resources, M.S. and D.G.; Data Curation, D.G.; Writing—Original Draft Preparation, M.S.; Writing—Review & Editing, D.G.; Visualization, M.S.; Supervision, H.H.; Project Administration, D.G.; Funding Acquisition, H.H.

Acknowledgments: This research was funded by K1-MET GmbH, metallurgical competence center. The research program of the competence center K1-MET is supported by COMET (Competence Centre for Excellent Technologies), the Austrian program for competence centers. COMET is funded by the Federal Ministry for Transport, Innovation and Technology, the Federal Ministry for Science, Research and Economy, the province of Upper Austria, Tyrol, and Styria, the Styrian Business Promotion Agency. The authors gratefully thank the company ZwickRoell Testing Systems GmbH (former Messphysik Materials Testing GmbH) for the good cooperation and manufacturing of the newly developed device.

Conflicts of Interest: The authors declare no conflict of interest

References

1. Harmuth, H.; Rieder, K.; Krobath, M.; Tschegg, E. Investigation of the nonlinear fracture behaviour of ordinary ceramic refractory materials. *Mater. Sci. Eng. A Struct.* **1996**, *214*, 53–61. [CrossRef]

2. Tschegg, E. Equipment and Appropriate Specimen Shapes for Tests To Measure Fracture Values. Austria Patent 390328, 31 January 1986.

3. Brochen, E.; Dannert, C. Thermo-mechanical characterisation of magnesia-carbon refractories by means of wedge splitting test under controlled atmosphere at high-temperature. In Proceedings of the 13th Biennial Worldwide Congress UNITECR 2013, Victoria, BC, Canada, 10–13 September 2013.

4. Dai, J.; Gruber, D.; Harmuth, H. Determination of the fracture behaviour of MgO-refractories using multi-cycle wedge splitting test and digital image correlation. *J. Eur. Ceram. Soc.* **2017**, *37*, 5035–5043. [CrossRef]

5. Zhu, T.; Li, Y.; Luo, M.; Sang, S.; Wang, Q.; Zhao, L.; Li, Y.; Li, S. Microstructure and mechanical properties of MgO-C refractories containing graphite oxide nanosheets (GONs). *Ceram. Int.* **2013**, *39*, 3017–3025. [CrossRef]

6. Mahato, S.; Shantanu, K.; Behera, K. Oxidation resistance and microstructural evolution in MgO–C refractories with expanded graphite. *Ceram. Int.* **2016**, *42*, 7611–7619. [CrossRef]

7. Song, J.; Yang, J.; Liu, F.; Lu, K. High temperature strain measurement method by combining digital image correlation of laser speckle and improved RANSAC smoothing algorithm. *Opt. Lasers Eng.* **2018**, *111*, 8–18. [CrossRef]

8. Harmuth, H.; Bradt, R.C. Investigation of refractory brittleness by fracture mechanical and fractographic methods. *InterCeram Int. Ceram. Rev.* **2010**, *62*, 6–10.

9. Faghihi-Sani, M.; Yamaguchi, A. Oxidation kinetics of MgO–C refractory bricks. *Ceram. Int.* **2002**, *28*, 835–839. [CrossRef]

10. Jin, S.; Gruber, D.; Harmuth, H. Determination of Young's modulus, fracture energy and tensile strength of refractories by inverse estimation of a wedge splitting procedure. *Eng. Fract. Mech.* **2014**, *116*, 228–236. [CrossRef]

11. Abaqus 6.13 Documentation. Available online: http://dsk.ippt.pan.pl/docs/abaqus/v6.13/index.html (accessed on 4 July 2019).

12. Auer, T.; Harmuth, H. Numerical simulation of a fracture test for brittle disordered materials. In Proceedings of the 16th European Conference of Fracture, Alexandroupolis, Greece, 3–7 July 2006.

13. Stueckelschweiger, M.; Gruber, D.; Jin, S.; Harmuth, H. Creep testing of carbon containing refractories under reducing conditions. *Ceram. Int.* **2019**, *45*, 9776–9781. [CrossRef]

Article

How Soft Polymers Cope with Cracks and Notches

Andrea Spagnoli *, Michele Terzano, Roberto Brighenti, Federico Artoni and Andrea Carpinteri

Department of Engineering and Architecture, University of Parma, Parco Area delle Scienze 181/A, 43124 Parma, Italy; michele.terzano@studenti.unipr.it (M.T.); roberto.brighenti@unpr.it (R.B.); federico.artoni@studenti.unipr.it (F.A.); andrea.carpinteri@unipr.it (A.C.)
* Correspondence: spagnoli@unipr.it; Tel.: +39-0521-905-927

Received: 5 February 2019; Accepted: 9 March 2019; Published: 14 March 2019

Abstract: Soft matter denotes a large category of materials showing unique properties, resulting from a low elastic modulus, a very high deformation capability, time-dependent mechanical behavior, and a peculiar mechanics of damage and fracture. The flaw tolerance, commonly understood as the ability of a given material to withstand external loading in the presence of a defect, is certainly one of the most noticeable attributes. This feature results from a complex and highly entangled microstructure, where the mechanical response to external loading is mainly governed by entropic-related effects. In the present paper, the flaw tolerance of soft elastomeric polymers, subjected to large deformation, is investigated experimentally. In particular, we consider the tensile response of thin plates made of different silicone rubbers, containing defects of various severity at different scales. Full-field strain maps are acquired by means of the Digital Image Correlation (DIC) technique. The experimental results are interpreted by accounting for the blunting of the defects due to large deformation in the material. The effect of blunting is interpreted in terms of reduction of the stress concentration factor generated by the defect, and failure is compared to that of traditional crystalline brittle materials.

Keywords: soft materials; polymers; strain rate; defect tolerance; digital image correlation; stress concentrators; notch blunting

---

1. Introduction

A class of materials, which are relevant from the point of view of advanced applications, is represented by the so-called soft or highly deformable materials, such as elastomeric polymers, colloids, liquid-crystals polymers, gels, foams, as well as biological materials—such as soft tissues—that are roughly governed by the same mechanical principles [1]. Typically, these materials have mechanical properties falling within the following range; elastic modulus of 0.1–1.5 MPa, tensile strengths of 1–10 MPa, ultimate tensile strains up to 2000%, and fracture energy of 100–1000 J/m^2. Soft materials are endowed with unique features and mechanical properties, explaining the great attention that they have been receiving from the scientific community in the last decades. In natural systems, soft tissues are a fertile source of inspiration for advanced applications, with mechanics and biology going hand-in-hand to formulate the underlying mechanical principles and develop new optimized structural materials [2–6].

The peculiar properties exhibited by soft materials are a direct consequence of their complex and entangled molecular structure, which involves millions or billions of atoms forming linear chains. We owe the fundamentals of the chemical- and physics-based mechanical behavior of this class of materials to the research work of Paul J. Flory and P.J. De Gennes [7–9]. In fully amorphous materials, the structure at the nanoscale level consists of a three-dimensional network of polymer chains, linked together at several discrete points identified as cross-links. The mechanical response of these materials at the meso- or macroscale is heavily affected by the amount of entanglement and the number of existing cross-links per unit volume, rather than by the bonding strength existing between the atoms, as happens in fully crystalline materials (e.g., metals, ceramics, etc.). Upon the application of load, the deformation induces an alignment and unentanglement of the polymeric chains, with the initial amorphous conformation turned into a semicrystalline-like one. Such a phenomenon usually occurs at quite large deformation, thanks to the highly-oriented arrangement of the chains along the tensile direction [10–12]. This so-called strain-induced crystallization is responsible for the typical stiffening behavior which is noticed in the stress–strain curve of polymers at high strain levels [13,14].

Soft materials normally can deform several times their original length, without being damaged or ruptured. If macroscopic flaws are introduced in the form of cuts, cracks or notches, the stretchability can be affected in different measure, depending on the flaw sensitivity of the material. The concept of a material length scale, separating flaw-sensitive from flaw-insensitive rupture, has been initially proposed for hard materials [15] and later extended to soft polymers [16]. Differently from traditional crystalline solids, rupture of soft polymers occurs at large deformation, when the rearrangement of the polymeric chains leads to flaw reshaping and strengthening around the highest strained region. In particular, an existing initial sharp crack blunts significantly before propagation, with the effect of relieving the strain concentration around the tip [17–19]. The resulting feature is an enhanced defect tolerance, with some tough soft materials showing an insensitivity to flaws up to a few millimeters long, in contrast to the typical nanometer scale for hard brittle solids [20]. For instance, biological tissues such as skin are known for their extreme resistance to flaws, which makes virtually impossible to propagate tearing in a stretched sample. Experimental observations on rabbit skin showed that a small notch does not propagate but progressively blunts, due to straightening and stretching of the collagen fibers [21]. However, other soft materials, such as the so-called hydrogels, are often prone to premature fracture and have low fatigue resistance. In such materials, the rupture process is strongly influenced by the fluid interaction [22], and it has been discovered that increased toughness and fatigue resistance are obtained through the development of polymer networks containing chains of different lengths (such as the double-network hydrogels, containing both a short- and a long-chain highly stretchable network). In this fashion, the shortest chains act as sacrificial elements, while the longer ones provide the material with a further elastic behavior [23,24]. From this perspective, it appears that the polymer network characteristics play a crucial role in defining the macroscopic behavior of the material. Furthermore, soft materials are sensible to time-dependent effects, with the fracture energy depending on the rate of application of the external loads, because of the viscous energy dissipation occurring in the crack tip region [25–27]. Sometimes, under a constant applied load, a so-called delayed fracture has been observed, depending on the network structure of the soft matter [28].

In this work we present a comprehensive investigation into the defect tolerance of flawed specimens of rubber-like polymers, with a detailed summary of experimental findings recently published by the authors [29–31]. Various configurations with cracks and notches are examined in order to evaluate the macroscopic mechanical response in relation to the flaw severity. Moreover, the effect of the applied strain rate is taken into account phenomenologically. All the experimental tests are conducted under strain control, and the kinematically-related quantities are measured through a contactless Digital Image Correlation (DIC) technique. Due to the local high deformations in proximity of the defects, a severe defect remodeling with evident blunting is noticed. The main purpose is to show how crack blunting affects the rupture behavior of soft materials. Through a simplified analytical model, the increase in the curvature radius at the notch root with the remote applied load is described. Such a model is applied to the experimental results, putting into evidence that it is the blunting effect which controls the rupture process; in particular, the grade of blunting appears to determine the transition from the typical small-scale yielding failure to rupture at a constant theoretical strength.

The paper is organized as follows. Section 2 presents a collection of the results obtained from the experiments, with details on the materials and methods employed. Accurate images taken from the DIC elaboration are included. In Section 3, a detailed discussion is developed in order to give a comprehensive interpretation of the experiments, with specific reference to the notch blunting effect, which is computed through a simplified analytical model. Finally, Section 4 sets out the conclusions.

2. Experimental Tests on Macroscopically Flawed Thin Plates

In this section, we present the results of tensile tests on thin plates made of various elastomeric polymers, carried out under displacement control up to complete failure. The plates contain flaws with different geometrical features, ranging from cracks to notches, denoted here as geometric discontinuities with a finite radius of curvature. Rubber-like polymers have been chosen because their behavior is highly representative of a vast range of soft materials. For instance, silicone rubber is often employed as a substitute of human skin, since it does not show the strain-induced stiffening of natural rubbers, maintaining comparable values of tensile and tearing strength [32]. The stress–strain curves obtained from tensile testing of silicone rubbers are typical of a hyperelastic, almost incompressible behavior. Several models have been proposed [33], which usually show good agreement at low to moderate stretches (generally below 1.5). During our experiments we have found that the one-parameter neo-Hookean model offers a satisfactory approximation in the stretch range of interest (see, for instance, Figure 1).

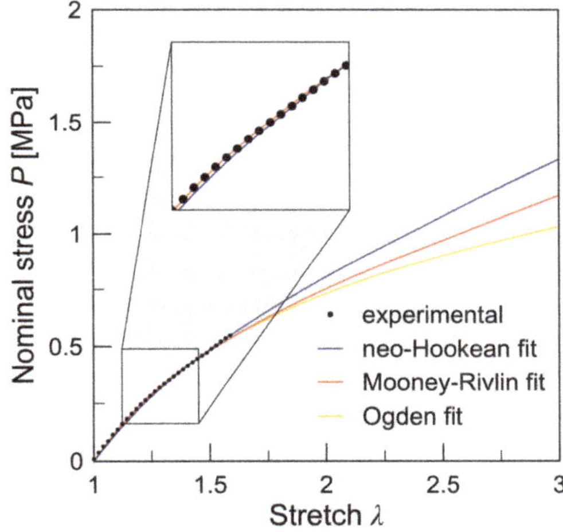

Figure 1. Typical curve obtained from tensile testing of a silicone rubber (experimental and fitted data). The enlarged inlet shows the region $\lambda = 1.2$–1.4, with excellent agreement between the different models.

The purpose of the experiments herein described is to evaluate the effect of initial flaws on the tensile strength of the material, considering both the effect of the flaw size and of the applied strain rate. The response of the specimens during the experimental tests is monitored by measuring the applied force and corresponding displacements, and by using the Digital Image Correlation (DIC) technique. The DIC is a contactless technique widely used to get full-field displacement and strain maps in experiments, through numerical reconstruction of the kinematic field shown by the surface points of the samples. For an optimal use of such a technique, the surface of the specimens needs to be covered with speckle patterns before testing. Several parameters affect measurement accuracy and spatial resolution, included optical measures connected to the camera and lens resolution, image magnification, mean size of the speckle pattern, and factors depending on the correlation algorithm, such as the image subset size and the gray-level interpolation. In particular, the resolution of the displacement measurement is governed by the subset size, whose lower bound is limited by the speckle size and, consequently, by the available pixels [34,35]. In the present work, the images are acquired with a high-resolution digital camera (maximum resolution of 5184 × 3456 pixels) mounted on a tripod, and lights are used to ensure a uniform illumination of the samples. The sequence of images is treated by means of the freeware software Ncorr [36], developed in MATLAB environment, for monitoring the displacement and the strain fields within the specimen.

2.1. Plates Containing Elliptical Flaws

The first set of experiments is carried out on elastomeric sheets under tension, containing elliptical flaws of length $2a$, characterized by a finite notch root radius (Figure 2a). The plates are made of Sylgard®, a common silicone polymer having the following elastic parameters: initial Young modulus $E = 0.84$ MPa and Poisson ratio $\nu = 0.37$. Three types of specimens have been prepared, containing a centered elliptical flaw with different values of the root radius ρ and same length $2a$. The plate aspect ratio is kept constant at $L/W \approx 2$. The geometric characteristics of the specimens are reported in Table 1. The flawed plates are subjected to tensile loading along the y-axis, applied at a constant strain rate of $\dot{\varepsilon}_0 = \dot{\delta}/(2L) = 4.8 \cdot 10^{-3}\ s^{-1}$. The tests have been interrupted before failure, since an evident notch remodeling was noticed. The response during the experimental tests is monitored by means of the DIC technique.

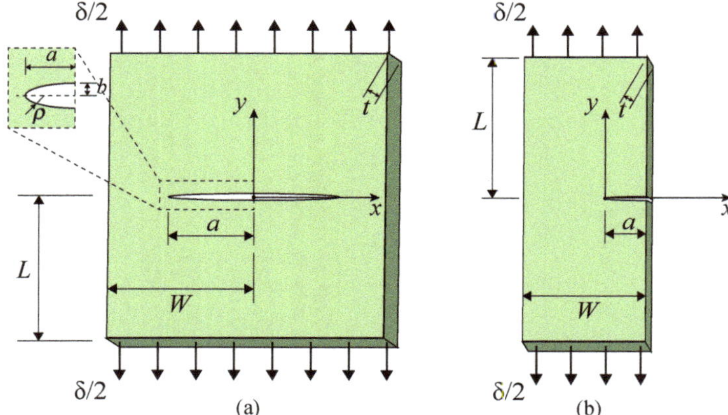

Figure 2. (**a**) Sketch of a thin plate containing a centered internal flaw. The enlarged view shows the case of an elliptical notch, with semi-axes a and b. ρ is the tip radius of the notch ($\rho = 0$ in the case of a crack-like flaw). (**b**) Sketch of an edge-cracked plate. The plates are subjected to a remote displacement δ applied along the y-axis.

Table 1. Geometric characteristics of the thin plates with elliptical notches. The rightmost column contains the values of the initial notch root radius $\rho = b^2/a$.

Specimen ID	W (mm)	a (mm)	b (mm)	t (mm)	a/W (-)	ρ (mm)
El1	58.5	20	5	2.0	0.342	1.250
El2	58.5	20	2.5	2.0	0.342	0.3125
El3	58.5	20	1	2.0	0.342	0.0050

Figure 3 shows the initial (undeformed) and the generic stretched shapes of the specimen El1 at two increasing levels of the remote applied stretch. The corresponding strain patterns obtained from the DIC analysis are illustrated in Figure 4, specifically, the Green–Lagrange strain E_{yy} parallel to the loading direction and the strain E_{xx} transversal to such a direction. Three different levels of the applied remote stretch are considered, defined as

$$\lambda_0 = 1 + \frac{\delta}{2L} \tag{1}$$

where δ is the applied displacement.

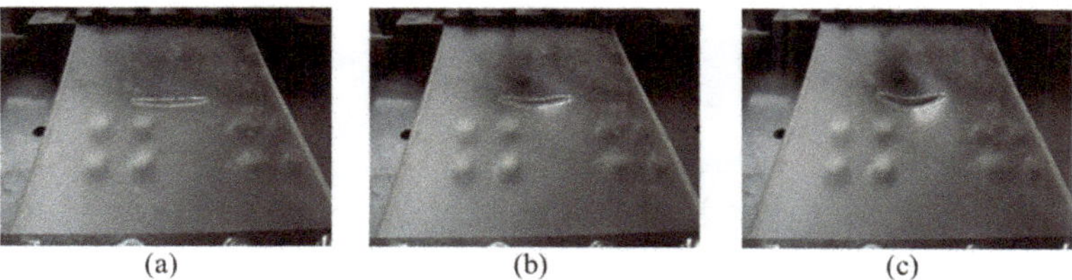

Figure 3. Qualitative view of the (**a**) undeformed and (**b,c**) increasingly deformed configurations, in the flawed specimen El1.

The DIC plots show that the material is able to comply with very high deformations, leading to a severe defect remodeling characterized by an evident blunting of the notch. A compressed region just in front and behind the elliptical hole is observed, due to the contraction effect arising in the direction normal to the applied load (see also Figure 3c). This phenomenon has a beneficial effect in terms of the strain concentration, with a sort of augmented notch blunting due to the local flexural instability of the thin plates in the compressed zones [29].

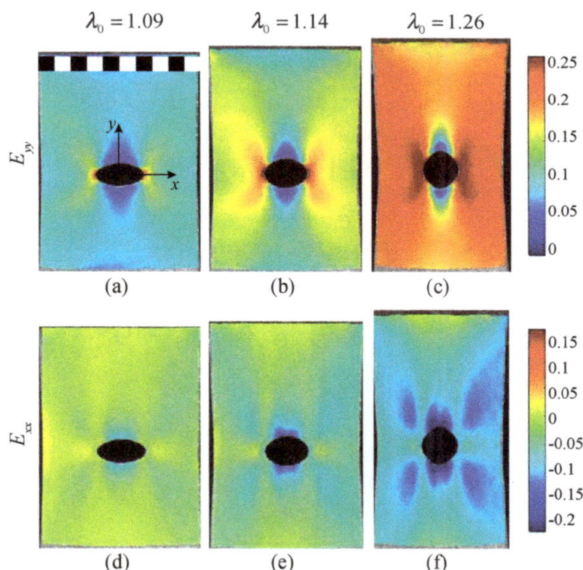

Figure 4. Strain maps obtained from the Digital Image Correlation (DIC) analyses on the elliptically notched specimen El1, for three increasing values of the applied remote stretch λ_0. (**a–c**) Green–Lagrange strain E_{yy}; (**d–f**) corresponding maps for the strains E_{xx}. In (**a**), the length scale is expressed in cm.

2.2. Plates Containing Internal Crack-Like Flaws

A second set of experiments deals with plates containing a centered crack of length $2a$ (Figure 2a with $b = \rho = 0$). The samples are made of a commercial silicone rubber (TSE3478T by Momentive). The desired crack is obtained by manually cutting the samples with a sharp blade. From tensile tests on sound specimens, the initial Young modulus of the material is found approximately equal to $E = 1.12$ MPa and the Poisson ratio to $\nu = 0.42$. Four series of specimens containing a centered crack are here considered, characterized by different values of the relative crack length a/W, whereas the plate aspect ratio is kept constant at $L/W \approx 2$. The geometric characteristics of the specimens are reported in Table 2. The plates are subjected to tensile loading along the y-axis up to failure, applied at a constant strain rate of $\dot{\varepsilon}_0 = 5.8 \cdot 10^{-3}\ \mathrm{s}^{-1}$. The response during the experimental tests is monitored by means of the DIC technique.

Table 2. Geometric characteristics of the specimens containing a centered crack. The rightmost column contains the ultimate remote stretch before failure.

Specimen ID	W (mm)	a (mm)	t (mm)	a/W	λ_U
CC1	56	10	2.75	0.179	1.75
CC2	56	15	3.00	0.268	1.61
CC3	56	20	2.75	0.357	1.55
CC4	56	25	2.85	0.446	1.32

Figure 5 shows the deformed patterns of the specimens, for three different stages of the applied loading: at the beginning of the test (undeformed, left column), at an intermediate stage (remote stretch $\lambda_0 = 1.29$, central column), and at the final stage before failure (ultimate remote stretch, right column). The strain patterns obtained from the DIC analyses are also shown, specifically, the Green–Lagrange strain E_{yy} (parallel to the loading direction) at the intermediate and the ultimate stages. It can be noticed that the initial crack-like shape tends to blunt under loading and the applied ultimate stretch generally decreases for an increasing relative crack size a/W. At the intermediate stage, the DIC maps clearly show a strain concentration typical of an elliptical notch, where the maximum strain values occur in the locations corresponding to the original crack tips. At incipient failure, the strain maps exhibit a complex distribution due to the failure mechanisms developing in the vicinity of the notch roots. It is worth noticing that out-of-plane displacements occur in two limited regions close to the crack edges (the blue regions in Figure 5), because of the contraction effect arising in the direction normal to the applied load, i.e., in the x-direction.

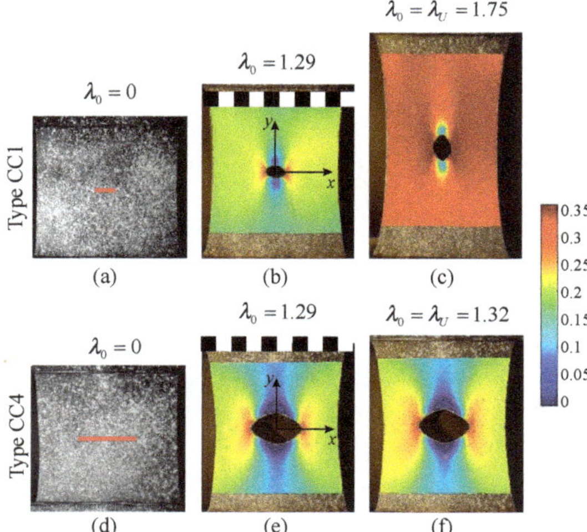

Figure 5. Images of the precracked specimens at the initial stage (left column), and with maps of the Green–Lagrange strain E_{yy} at an intermediate stage (central column) and final stage at incipient failure (right column). The remote stretch is shown on the top of each plot. (**a–c**) Specimens type CC1 ($a/W = 0.179$); (**d–f**) specimens type CC4 ($a/W = 0.446$). In (b,e), the length scale is expressed in cm.

The distribution of the Green–Lagrange strain E_{yy} along the x-axis is shown in Figure 6.

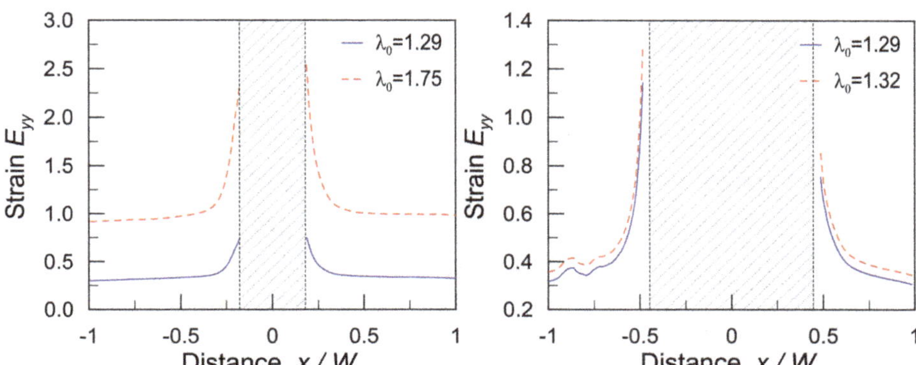

Figure 6. Green–Lagrange strain E_{yy} distribution along the cracked section of the plate, at an intermediate (continuous lines) and ultimate values (dashed lines) of the remote applied stretch. (**a**) Specimens type CC1 (a/W = 0.179) and (**b**) specimens type CC4 (a/W = 0.446).

2.3. Plates Containing Edge Crack-Like Flaws

The last set of experiments that we review is related to plates containing edge crack-like defects (Figure 2b), where we consider the effects of the relative crack size a/W and of the applied strain rate. A first group of edge-cracked plates (specimens EC1–EC4 in Table 3) has been prepared with the commercial silicone RTV 615 (Elantas Europe Srl). From tensile tests on sound specimens, the initial Young modulus of the material is equal to E = 1.50 MPa and the Poisson ratio is equal to ν = 0.42. The plates are subjected to tensile loading along the y-axis up to failure, applied with three different strain rates: $\dot{\varepsilon}_0^{(1)} = 1.9 \cdot 10^{-3}$ s^{-1}, $\dot{\varepsilon}_0^{(2)} = 0.48 \cdot 10^{-3}$ s^{-1}, and $\dot{\varepsilon}_0^{(3)} = 0.16 \cdot 10^{-3}$ s^{-1}. The response during the experimental tests is monitored by means of the DIC technique.

Table 3. Geometric characteristics of the edge-cracked plates. The rightmost column contains the ultimate remote stretch before failure. In specimens EC1–EC4, the fastest rate $\dot{\varepsilon}_0^{(1)} = 1.9 \cdot 10^{-3}$ s^{-1} is considered.

Specimen ID	W (mm)	a (mm)	t (mm)	a/W	λ_u
EC1	26.3	1	2.3–3.5	0.038	1.26
EC2	26	2	2.3–3.2	0.077	1.18
EC3	26	5	2.3–2.9	0.192	1.11
EC4	26	8	2.5–2.9	0.308	1.07
EC5	25	1.8	4.2	0.072	1.42
EC6	25.5	3	4.2	0.118	1.36
EC7	26	4	4.3	0.154	1.28
EC8	25	5	4.6	0.202	1.26

A second group of edge-cracked plates (EC5-EC8 in Table 3) has been prepared using a different silicone rubber (Elite Double 32 by Zhermack Dental), with an estimated Young modulus equal to E = 1.36 MPa and the Poisson ratio is equal to ν = 0.42. The plates are subjected to tensile loading along the y-axis up to failure; such a loading is applied with a strain rate $\dot{\varepsilon}_0 = 3.8 - 4.2 \cdot 10^{-3}$ s^{-1}. Four series of specimens have been prepared for each group, with different values of the relative crack length a/W, while the plate aspect ratio is kept constant at $L/W \approx 1.5$.

The results obtained from the DIC analyses in the specimens EC2–EC4 are illustrated in Figure 7, specifically, the Green–Lagrange strain E_{yy} (parallel to the loading direction) at incipient failure for the highest strain rate among the three considered. From a qualitative observation, it can be noticed that crack tip blunting is limited (compared, for instance, to the case of an internal crack in Figure 5).

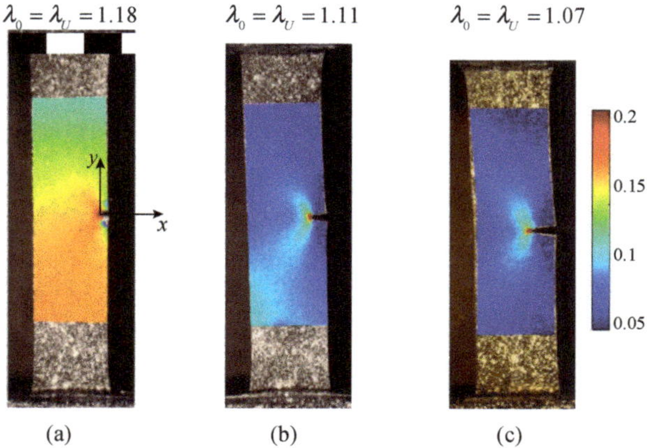

Figure 7. Map of the Green–Lagrange strain E_{yy} at incipient failure, at the applied rate $\dot{\varepsilon}_0{}^{(1)}$. The remote stretch is shown on the top of each plot. (**a**) Specimens type EC2 ($a/W = 0.077$), (**b**) type EC3 ($a/W = 0.192$) and (**c**) type EC4 ($a/W = 0.308$). In (**a**), the length scale is expressed in cm.

The results show that specimens with only one ligament zone are more sensible to defects with respect to the ones containing a central crack (CC1–CC4). The eccentric load effect, which increases with the deformation, plays a crucial role in intensifying the stress close to the crack tip, leading to a premature failure of the polymer network chains within the crack process region.

2.4. Effects of Intrinsic Material Defects

The group of experiments on edge-cracked plates has also involved samples with intrinsic defects, in the form of microvoids (D1–D4 in Table 4). For this purpose, we have prepared samples with the same material and sizes of the group EC1–EC4, but following a different treatment during the preparation of the silicone mixture. To prepare the samples, 50 g of component A (matrix) are thoroughly mixed with 5 g of component B (curing agent). At this point, the mixture for the material without defects is carefully degassed in vacuo and subsequently mechanically spread into the custom-made aluminum mold. This stage is followed by a second degassing, then the mixture is cured in oven at 60 °C overnight and finally mechanically removed from the mold. In order to obtain microbubbles embedded in the material, the silicone mixture is directly cured in oven without the degassing stages.

Table 4. Geometric characteristics of the edge-cracked plates with intrinsic defects. The rightmost column contains the ultimate remote stretch before failure.

Specimen ID	W (mm)	a (mm)	t (mm)	a/W	λ_U
D1	26.3	1	2.7–3.3	0.038	1.23
D2	26	2	2.7–3.1	0.077	1.15
D3	26	5	2.8–3.0	0.192	1.12
D4	26	8	2.8–3.5	0.308	1.09

A comparison of the materials, with and without the microvoids, is shown in Figure 8a,b. The average void radius is equal to $r_v = 0.35$ mm, and its distribution is described by adopting a normal probability function (variance equal to 0.021 mm^2). The plot of the probability function (Figure 8c) shows a good agreement with the measured distribution, and ~80% of the voids have a radius within half of the smallest edge crack length considered.

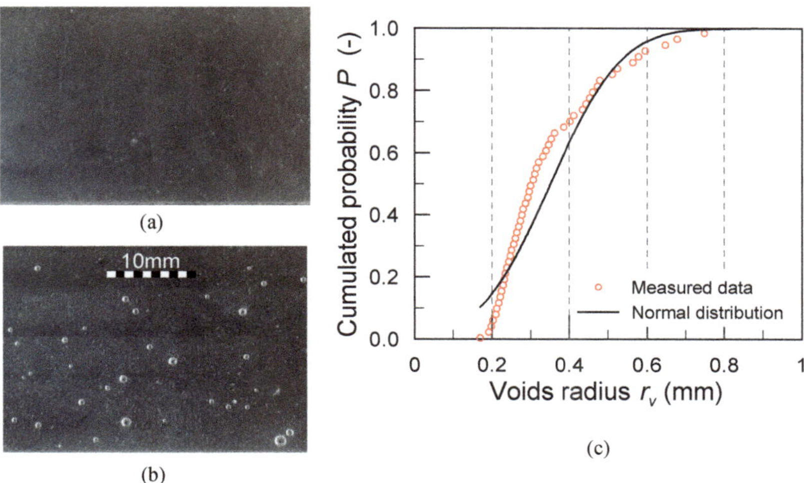

Figure 8. Images of the elastomeric material employed in specimens EC1–EC4 and D1–D4. Detail of the material (**a**) without and (**b**) with intrinsic defects. (**c**) Distribution of the void size cumulated probability.

Figure 9 illustrates the failure behavior of edge-cracked specimens EC1–EC4 and D1–D4, in terms of the ultimate stretch vs the relative crack size. As was expected, smaller ultimate stretches are attained for increasing values of the relative crack size a/W. The decreasing trend, with a roughly quadratic pattern, confirms that the material is sensitive to the presence of the initial flaws.

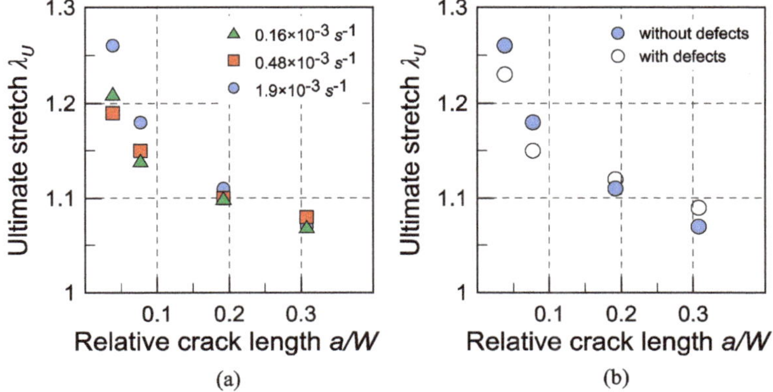

Figure 9. Ultimate stretch vs the initial relative crack size for the edge-cracked samples. (**a**) Effect of the applied strain rate in the specimens EC1–EC4 and (**b**) influence of the material defects, for a strain rate $\dot{\varepsilon}_0 = 1.9 \cdot 10^{-3}$ s^{-1} (results from samples EC1–EC4 are shown as solid circles, and D1–D4 as hollow circles).

The effect of the strain rate is investigated in Figure 9a. The highest strain rate provides the highest ultimate stretch, although the effect tends to reduce at lower ultimate stretches, suggesting that the influence of flaw size on the ultimate stretch decreases if the deformation is applied sufficiently slow. Figure 9b compares the failure in specimens with and without intrinsic material defects. It can be inferred that microdefects tend to anticipate failure, although the results are not so clear, probably due to the uneven distribution of the microbubbles in the samples.

3. Discussion

3.1. Model for Notch Blunting

Cracks and notches in soft materials become distorted under loading due to large strain effects, with an evident tip blunting, as is clearly shown in the figures obtained from the DIC on the experimental tests. In order to quantify the flaw severity and account for the geometrical effect of blunting, we resort to the concept of the stress concentration factor. Considering a notched sample under remote uniform stress acting parallel to the y-axis, the stress concentration factor can be defined as

$$K_t = \sigma_{max}/\sigma_0 \tag{2}$$

where $\sigma_{max} = \sigma_{yy,max}$ is the maximum notch root stress and σ_0 is the corresponding remote stress, measured with respect to a uniformly stressed (gross) section of the sample. In the following, we propose a simplified analytical model, capable of relating the stress concentration factor with the tip radius ρ of the blunted notch and explore the variation of $K_t(\rho)$ under increasing applied loading. Such an approach, conceived for the analysis of blunting of elliptical notches [29], is extended to the tip blunting of cracks (Section 3.2).

The starting point is the solution for an elliptical notch in an infinite elastic plate, having semi-axes a and b. Its equation can be written as follows

$$y(x) = \frac{b}{a}\sqrt{a^2 - x^2} \tag{3}$$

and the radius of curvature at the tips ($x = \pm a$) is equal to

$$\rho = b^2/a \tag{4}$$

The stress concentration factor K_t for an elliptical notch is obtained from the renowned Inglis' solution [37]:

$$K_t(\rho) = \left(1 + 2\sqrt{\frac{a}{\rho}}\right) Y\left(\frac{a}{W}, \frac{a}{b}\right) \tag{5}$$

where Y is a corrective factor introduced to account for the finite width of the specimens [38] (considering the initial geometry of the tested notched specimens, the values of Y range from 1.09 to 1.10). For the sake of simplicity, this factor is taken equal to unity in the following equations.

In order to describe the change of the stress concentration factor with the increasing deformation of the blunted tip under loading, we need to consider the variation of the radius of curvature. Such a variation depends on the deformation state in proximity of the notch root, which has to be evaluated in the deformed configuration (see Figure 10a). Let us consider a small square element of edge length equal to h, located in the proximity of the notch root, with edges inclined by an angle α with respect to the horizontal axis (Figure 10b). The small strain tensor ε' in the local reference system 1–2 (with components $\varepsilon_{11}, \varepsilon_{22}, \varepsilon_{12}$) is related to the corresponding tensor ε (with components $\varepsilon_{tt}, \varepsilon_{nn}, \varepsilon_{tn}$) in the reference system t-n, with its origin located at the notch root, through the well-known relationships [33]:

$$\varepsilon_{11} = c^2\varepsilon_{tt} + s^2\varepsilon_{nn}, \varepsilon_{22} = s^2\varepsilon_{tt} + c^2\varepsilon_{nn}, \varepsilon_{12} = cs\varepsilon_{tt} - cs\varepsilon_{nn} \tag{6}$$

with $c = \cos\alpha, s = \sin\alpha$. Assuming plane stress conditions and the material governed by the generalized Hooke model, the strain tensor components are

$$\varepsilon_{tt} = -\nu\sigma_{nn}/E, \varepsilon_{nn} = \sigma_{nn}/E, \varepsilon_{tn} = 0 \tag{7}$$

If the angle α is sufficiently small, the only non-zero component of the stress tensor is

$$\sigma_{nn} = K_t(\rho)\sigma_0 \tag{8}$$

The radius of curvature at the notch root, in the undeformed state (point A in Figure 10a,b), is approximated by the radius of the local osculating circle:

$$\rho \simeq \frac{h}{2\cos\beta} \tag{9}$$

with $\beta = \pi/2 - \alpha$, and the increased radius of curvature in the deformed state (point A' in Figure 10a,b) is approximated as

$$\rho' = \frac{h(1 + \varepsilon_{22})}{2\cos(\beta + \gamma)} \tag{10}$$

where $\gamma = 2\varepsilon_{12}$.

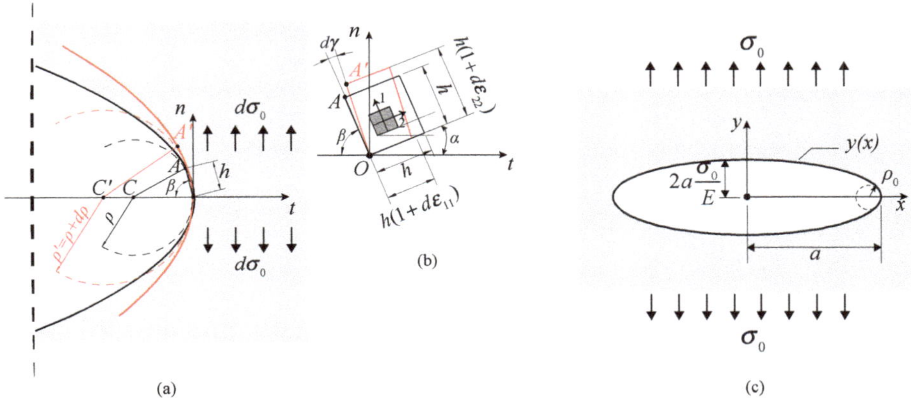

Figure 10. (**a**) Model of notch blunting. Schematic of the notch profile modification due to the large strains occurring in the material during an increment of the remote stress $d\sigma_0$. The radius of curvature at point A (undeformed) and A' (deformed) is shown. (**b**) Deformation of a small material element in the vicinity of the notch root. (**c**) Elliptical notch, with the equivalent semi-axes obtained from the crack flank displacement under tensile stress σ_0.

The model here described is nonlinear, because the local strains depend on K_t through Equations (6)–(8), where, in turn, ρ is also a function of the local strains through Equation (10). Following an incremental procedure, firstly the increments of strain are evaluated for an applied remote stress variation $d\sigma_0$ through Equations (6)–(8); then, the updated radius of curvature is obtained from Equation (10) and the stress concentration factor is computed from Equation (5). In other words, following a stepwise updated Lagrangian approach, at the first increment the increased radius of curvature ρ' of Equation (10) is calculated with respect to the reference configuration on the undeformed notch, whilst at successive increments the reference configuration, i.e., ρ from Equation (9), corresponds to the radius of the notch root under the current stress level. The results applied to one of the elliptically-notched plates tested experimentally are illustrated in Figure 11.

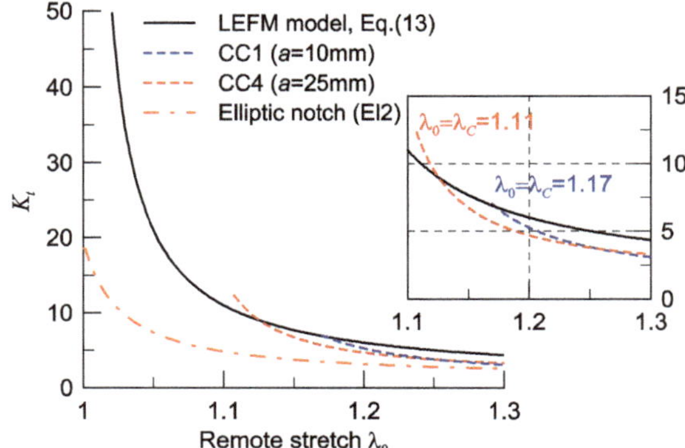

Figure 11. Stress concentration factor as a function of the remote applied stretch. Prediction of the 'pure' LEFM model (continuous line) and results obtained from the 'modified' model, in the center-cracked specimens CC1 and CC4. The dash-dot line refers to the elliptically-notched plate El2. The enlarged view shows the region $1.1 < \lambda < 1.3$.

3.2. Crack Tip Blunting

Linear Elastic Fracture Mechanics (LEFM) is grounded on the hypothesis of infinitesimal displacements, with the near-tip region being described by the K-dominated fields, which imply, for instance, a square root singularity in the strain and stress components at the crack tip. In soft materials, the large strain region near the tip can be relevant, and the local fields need to be defined within the framework of finite strain elastostatic, showing for instance stress singularities different from the well-known square-root [39]. However, tip blunting can still be described according to the simplified model introduced in Section 3.1, provided that the local strain is not too large.

According to LEFM, the deformed shape of a generic crack of semilength a, subjected to a uniform stress σ_0, is described by the following expression of the crack flank displacement [40].

$$y(x) = 2\frac{\sigma_0}{E}\sqrt{a^2 - x^2} \tag{11}$$

which is the equation of an ellipse, having the major semi-axis equal to a and the minor semi-axis equal to $b_0 = 2a\frac{\sigma_0}{E}$ (Figure 10c). Retrieving the expression of the tip radius of an ellipse from Equation (4), the equivalent radius of curvature is

$$\rho_0 = 4a\left(\frac{\sigma_0}{E}\right)^2 \tag{12}$$

A direct relationship between the applied remote stretch and the stress concentration factor predicted by the Inglis solution, Equation (5), is derived as

$$K_t(\lambda_0) = 1 + 2\sqrt{\frac{a}{\rho_0}} = \frac{\lambda_0}{\lambda_0 - 1} \tag{13}$$

where we have used the standard relation for linear elastic materials: $\lambda_0 = 1 + \frac{\sigma_0}{E}$. The noticeable feature of such an expression is that it does not depend on the initial length of the crack. According to LEFM, the ultimate condition at failure occurs when the remote stress equals

$$\sigma_C = \frac{1}{Z\left(\frac{a}{W}\right)}\sqrt{\frac{G_C E}{\pi a}} \tag{14}$$

where $Z\left(\frac{a}{W}\right)$ is a corrective factor for the finite width of the cracked specimens [41] (considering the initial geometry of the tested notched specimens, the values of Z range from 1.02 to 1.69) and G_C is the fracture resistance of the material. The critical stretch, omitting the corrective factor Z, is then obtained as

$$\lambda_C = 1 + \sqrt{\frac{G_C}{\pi E}}\, a^{-\frac{1}{2}} \tag{15}$$

Up to this point, we have not considered the deformation of the blunted crack under loading. The predicted stress concentration factor, as is described from this 'pure' LEFM model in Equation (13), is illustrated by the continuous line in Figure 11.

The effects of the blunted tip deformation under increasing loading and the resulting stress concentration factor are described adopting the 'modified' model introduced in Section 3.1, where the current tip radius is computed on the deformed configuration from Equation (10). We assume that the initial configuration of the crack is the critical condition of LEFM, Equation (14), when $\sigma_0 = \sigma_C$ and the corresponding value of the radius of curvature, from Equation (12), is

$$\rho_C = \frac{4}{\pi}\frac{G_C}{E} \tag{16}$$

The results of the 'modified' model are also shown in Figure 11. We have considered the center-cracked specimens CC1 and CC4, which present two different lengths of the crack: we notice that, contrary to the LEFM model results, there is a dependence of K_t on a.

3.3. Application to Experimental Data

The model is applied to account for the effect of blunting in the failure of the specimens tested during the experiments. The data collected in Figure 12 illustrate the ultimate condition, when the remote stretch is equal to $\lambda_0 = \lambda_U$, with respect to the initial length a of the defect. In this plot we have also added the predicted trend of LEFM, as obtained from Equation (15), where an average value of the fracture resistance for silicone rubbers is taken equal to $G_c = 1$ kJ/m^2.

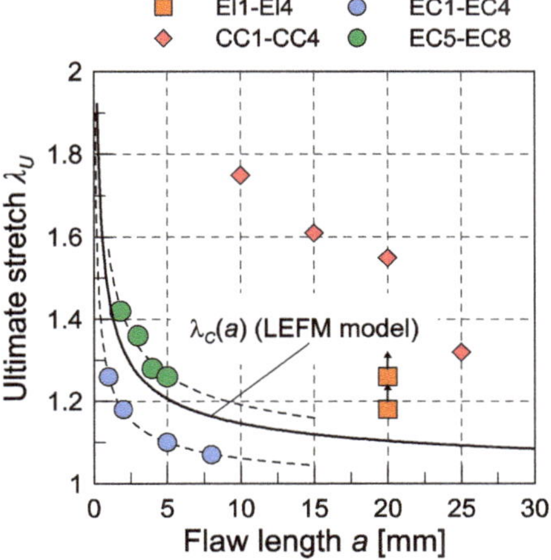

Figure 12. Ultimate remote stretches at failure λ_U as a function of the initial crack length a. In the case of the elliptical notches, the upward arrows indicate that failure has not occurred within the considered range of applied stretch. The continuous line corresponds to the prediction of the 'pure' LEFM solution, Equation (15). The dashed lines correspond to best-fit curves with a λ_U vs. a power law dependence with exponent -0.5. Results for EC1–EC4 for a strain rate $\dot{\varepsilon}_0 = 1.9 \cdot 10^{-3}$ s^{-1}.

It should be noticed that the plot in Figure 12 is by no means a representation of the flaw sensitivity in the canonical form, primarily because the smallest size of the flaw in the tested specimens is larger than the critical flaw size (see [20]). The discriminant here seems to be the grade of crack blunting:

the edge-cracked specimens, which fail at smaller stretches, are well approximated by the 'pure' LEFM model, where the ultimate stretch decreases with the square root of a. On the contrary, the plates containing central cracks (specimens CC1–CC4) resist to larger stretches and do not agree well with the power law trends, suggesting that crack blunting plays a fundamental role.

To the latter group of specimens, we have applied the 'modified' model previously described. The results are summarized in Figure 13, where the quantity $K_t(\lambda_U - 1)$ is plotted as a function of the crack length a. Note that the quantity $\frac{1}{E}(K_t(\lambda_U - 1)E)$ represents the normalized true stress at the notch root at incipient failure, and can be therefore considered a material property, independent of the presence of the flaw. From the observation of the results reported in Figure 13, it can be noticed that such a local stress seems to be an appropriate parameter for quantifying the material failure in the case of the centered-cracked plates CC1–CC4, but not in the others. In other words, where crack blunting allows rupture at larger stretches, failure can be predicted by a local stress quantity rather than by a fracture mechanics-related one.

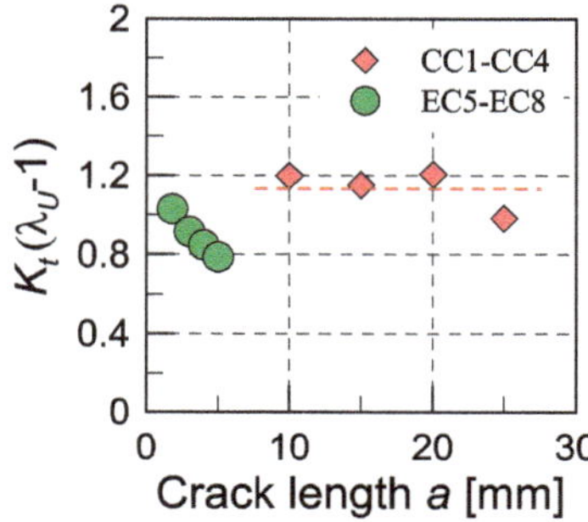

Figure 13. Normalized notch root true stress at incipient failure. The effect of crack blunting is enhanced for the specimens CC1–CC4, with the failure stress being approximately constant (dashed line).

4. Conclusions

The present paper investigates the defect tolerance capability of flawed silicone specimens subjected to tensile loading, adopting both a theoretical and an experimental approach. Silicone rubber has been chosen as illustrative of the typical response of soft matter in general, including other polymers, gels, and biological tissues. Different flaw geometries, sizes, and rates of the applied strain are examined in the experimental tests, with the aim of understanding their influence on the macroscopic mechanical response of the samples. The presence of internal defects, in the form of microbubbles, is also considered. The experimental response is monitored by measuring the applied force and using the Digital Image Correlation (DIC) technique to obtain precise two-dimensional displacement and full-field strain maps.

Flawed samples of soft materials subjected to tensile loading undergo a remarkable notch blunting prior to failure, which tends to reduce the stress concentration due to the presence of the flaw. Moreover, when notches are contained in nonconfined thin elements, a further notch blunting occurs as consequence of the local buckling of the material in the compressed zones, which arises normally to the loading direction. Experimental tests on thin silicone plates with elliptical notches showed that, irrespective of the initial size of the notch, very high remote stresses are supported thanks to the favorable notch profile evolution under load. With respect to the strain rate effects, we can observe that the flaw sensitivity is augmented at higher rates while, if a sufficiently slow strain rate is applied, the ultimate strain before failure is less affected by the size of the initial flaw. Indeed, slow deformation rates allow the internal microstructure of the material in high strained regions to be rearranged and to fail locally, with the consequence to smooth out the peak strain arising close to the geometric discontinuities.

A simple analytical model is proposed to account for the effect of crack blunting, in terms of the increase of the tip radius with the remote applied stretch. Although based on the assumption of linear elastic behavior, the application of the model to the experimental results has provided an interesting insight into the defect tolerance of cracked thin plates under tensile loading. Specifically, we have observed a transition from a typical flaw-size-dependent failure, as predicted by linear elastic fracture mechanics, to rupture occurring at a constant theoretical strength of the material. Our results show that, even at flaw lengths larger than the critical size found in other studies [20], silicone rubber specimens can withstand large stretches thanks to the flaw reshaping allowed by their peculiar microstructure. We are confident that the obtained results might be applied profitably to the evaluation of the safety levels of notched soft structural components, commonly found in numerous advanced applications.

Author Contributions: Conceptualization, A.S.; Data curation, R.B. and F.A.; Methodology, R.B.; Supervision, A.S.; Visualization, M.T.; Writing–original draft, M.T.; Writing–review & editing, A.C.

Funding: This research received no external funding.

Conflicts of Interest: The authors declare no conflict of interest.

References

1. Doi, M. *Soft Matter Physics*; OUP Oxford: Oxford, UK, 2013.
2. Bruck, H.; Evans, J.J.; Peterson, M. The Role of Mechanics in Biological and Biologically Inspired Materials. *Exp. Mech.* **2002**, *42*, 361–371. [CrossRef]
3. Bao, G.; Suresh, S. Cell and Molecular Mechanics of Biological Materials. *Nat. Mater.* **2003**, *2*, 715–725. [CrossRef] [PubMed]
4. Buehler, M.J. Multiscale Mechanics of Biological and Biologically Inspired Materials and Structures. *Acta Mech. Solida Sin.* **2010**, *23*, 471–483. [CrossRef]
5. Rogers, J.A.; Someya, T.; Huang, Y. Materials and Mechanics for Stretchable Electronics. *Science* **2010**, *327*, 1603–1607. [CrossRef]
6. Mirzaeifar, R.; Dimas, L.S.; Qin, Z.; Buehler, M.J. Defect-Tolerant Bioinspired Hierarchical Composites: Simulation and Experiment. *ACS Biomater. Sci. Eng.* **2015**, *1*, 295–304. [CrossRef]
7. Flory, P.J. *Principles of Polymer Chemistry*; Baker Lectures 1948; Cornell University Press: Menasha, WI, USA, 1953.
8. de Gennes, P.G. Reptation of a Polymer Chain in the Presence of Fixed Obstacles. *J. Chem. Phys.* **1971**, *55*, 572–579. [CrossRef]
9. Flory, P.J. *Statistical Mechanics of Chain Molecules*; Hanser: Cincinnati, OH, USA, 1989.
10. Rao, I.J.; Rajagopal, K.R. A Study of Strain-Induced Crystallization of Polymers. *Int. J. Solids Struct.* **2001**, *38*, 1149–1167. [CrossRef]
11. Tosaka, M.; Kohjiya, S.; Ikeda, Y.; Toki, S.; Hsiao, B.S. Molecular Orientation and Stress Relaxation during Strain-Induced Crystallization of Vulcanized Natural Rubber. *Polym. J.* **2010**, *42*, 474–481. [CrossRef]
12. Candau, N.; Laghmach, R.; Chazeau, L.; Chenal, J.-M.; Gauthier, C.; Biben, T.; Munch, E. Strain-Induced Crystallization of Natural Rubber and Cross-Link Densities Heterogeneities. *Macromolecules* **2014**, *47*, 5815–5824. [CrossRef]
13. Kuhn, W. Dependence of the Average Transversal on the Longitudinal Dimensions of Statistical Coils Formed by Chain Molecules. *J. Polym. Sci.* **1946**, *1*, 380–388. [CrossRef]
14. Treloar, L.R.G. The Elasticity and Related Properties of Rubbers. *Rep. Prog. Phys.* **1973**, *36*, 755–826. [CrossRef]
15. Gao, H.; Ji, B.; Jager, I.L.; Arzt, E.; Fratzl, P. Materials Become Insensitive to Flaws at Nanoscale: Lessons from Nature. *Proc. Natl. Acad. Sci. USA* **2003**, *100*, 5597–5600. [CrossRef]
16. Talamini, B.; Mao, Y.; Anand, L. Progressive Damage and Rupture in Polymers. *J. Mech. Phys. Solids* **2018**, *111*, 434–457. [CrossRef]
17. Zhifei, S.; Gao, Y.-C. Stress-Strain Field near the Notch Tip of a Rubber Sheet. *Acta Mech. Sin.* **1995**, *11*, 169–177. [CrossRef]
18. Hui, C.-Y.; Jagota, A.; Bennison, S.J.; Londono, J.D. Crack Blunting and the Strength of Soft Elastic Solids. *Proc. R. Soc. A Math. Phys. Eng. Sci.* **2003**, *459*, 1489–1516. [CrossRef]
19. Goldman Boué, T.; Harpaz, R.; Fineberg, J.; Bouchbinder, E. Failing Softly: A Fracture Theory of Highly-Deformable Materials. *Soft Matter* **2015**, *11*, 3812–3821. [CrossRef]

20. Chen, C.; Wang, Z.; Suo, Z. Flaw Sensitivity of Highly Stretchable Materials. *Extrem. Mech. Lett.* **2017**, *10*, 50–57. [CrossRef]
21. Yang, W.; Sherman, V.R.; Gludovatz, B.; Schaible, E.; Stewart, P.; Ritchie, R.O.; Meyers, M.A. On the Tear Resistance of Skin. *Nat. Commun.* **2015**, *6*, 6649. [CrossRef]
22. Mao, Y.; Anand, L. A Theory for Fracture of Polymeric Gels. *J. Mech. Phys. Solids* **2018**, *115*, 30–53. [CrossRef]
23. Zhao, X. Designing Toughness and Strength for Soft Materials. *Proc. Natl. Acad. Sci. USA* **2017**, *114*, 8138–8140. [CrossRef]
24. Zhang, W.; Liu, X.; Wang, J.; Tang, J.; Hu, J.; Lu, T.; Suo, Z. Fatigue of Double-Network Hydrogels. *Eng. Fract. Mech.* **2018**, *187*, 74–93. [CrossRef]
25. Bergstrom, J. Constitutive Modeling of the Large Strain Time-Dependent Behavior of Elastomers. *J. Mech. Phys. Solids* **1998**, *46*, 931–954. [CrossRef]
26. Roland, C.M.M. Mechanical Behavior of Rubber at High Strain Rates. *Rubber Chem. Technol.* **2006**, *79*, 429–459. [CrossRef]
27. Lavoie, S.R.; Long, R.; Tang, T. A Rate-Dependent Damage Model for Elastomers at Large Strain. *Extrem. Mech. Lett.* **2016**, *8*, 114–124. [CrossRef]
28. Bonn, D. Delayed Fracture of an Inhomogeneous Soft Solid. *Science* **1998**, *280*, 265–267. [CrossRef]
29. Brighenti, R.; Spagnoli, A.; Carpinteri, A.; Artoni, F. Notch Effect in Highly Deformable Material Sheets. *Thin-Walled Struct.* **2016**, *105*, 90–100. [CrossRef]
30. Brighenti, R.; Spagnoli, A.; Carpinteri, A.; Artoni, F. Defect Tolerance at Various Strain Rates in Elastomeric Materials: An Experimental Investigation. *Eng. Fract. Mech.* **2017**, *183*, 79–93. [CrossRef]
31. Brighenti, R.; Carpinteri, A.; Artoni, F.; Domenichelli, I. Defect Sensitivity of Highly Deformable Polymeric Materials with Different Intrinsic Qualities at Various Strain Rates. *Fatigue Fract. Eng. Mater. Struct.* **2018**, *41*, 806–820. [CrossRef]
32. Shergold, O.A.; Fleck, N.A. Experimental Investigation Into the Deep Penetration of Soft Solids by Sharp and Blunt Punches, With Application to the Piercing of Skin. *J. Biomech. Eng.* **2005**, *127*, 838–848. [CrossRef]
33. Holzapfel, G.A. *Nonlinear Solid Mechanics: A Continuum Approach for Engineering*; Wiley: Hoboken, NJ, USA, 2000.
34. Bornert, M.; Brémand, F.; Doumalin, P.; Dupré, J.-C.; Fazzini, M.; Grédiac, M.; Hild, F.; Mistou, S.; Molimard, J.; Orteu, J.-J.; et al. Assessment of Digital Image Correlation Measurement Errors: Methodology and Results. *Exp. Mech.* **2009**, *49*, 353–370. [CrossRef]
35. Reu, P.L.; Sweatt, W.; Miller, T.; Fleming, D. Camera System Resolution and Its Influence on Digital Image Correlation. *Exp. Mech.* **2015**, *55*, 9–25. [CrossRef]
36. Blaber, J.; Adair, B.; Antoniou, A. Ncorr: Open-Source 2D Digital Image Correlation Matlab Software. *Exp. Mech.* **2015**, *55*, 1105–1122. [CrossRef]
37. Inglis, C.E. Stresses in a Plate Due to the Presence of Cracks and Sharp Corners. *Trans. Inst. Nav. Archit.* **1913**, *55*, 219–230.
38. Budynas, R.G.; Young, W.C.; Sadegh, A.M. *Roark's Formulas for Stress and Strain*, 8th ed.; McGraw-Hill Education: New York, NY, USA, 2011.
39. Long, R.; Hui, C.-Y. Crack Tip Fields in Soft Elastic Solids Subjected to Large Quasi-Static Deformation—A Review. *Extrem. Mech. Lett.* **2015**, *4*, 131–155. [CrossRef]
40. Janssen, M.; Zuidema, J.; Wanhill, R.J.H. *Fracture Mechanics*; Delft Academic Press: Delft, The Netherlands, 2006.
41. Hiroshi, T. A Note on the Finite Width Corrections to the Stress Intensity Factor. *Eng. Fract. Mech.* **1971**, *3*, 345–347. [CrossRef]

Article

Analytical Study of Reinforced Concrete Beams Tested under Quasi-Static and Impact Loadings

Sayed Mohamad Soleimani [1],* and Sajjad Sayyar Roudsari [2]

[1] Department of Civil Engineering, Australian College of Kuwait, P.O. Box 1411, Safat 13015, Kuwait
[2] Department of Computational Science and Engineering, North Carolina A&T State University, 1601 E. Market Street, Greensboro, NC 27411, USA
* Correspondence: s.soleimani@ack.edu.kw; Tel.: +965-2537-6111

Received: 28 May 2019; Accepted: 13 July 2019; Published: 16 July 2019

Abstract: During dynamic events (such as impact forces), structures fail to absorb the incoming energy and catastrophic collapse may occur. Impact and quasi-static tests were carried out on reinforced concrete beams with and without externally bounded sprayed and fabric glass fiber-reinforced polymers. For impact loading, a fully instrumented drop-weight impact machine with a capacity of 14.5 kJ was used. The drop height and loading rate were varied. The load-carrying capacity of reinforced concrete beams under impact loading was obtained using instrumented anvil supports (by summing the support reactions). In quasi-static loading conditions, the beams were tested in three-point loading using a Baldwin Universal Testing Machine. ABAQUS FEA software was used to model some of the tested reinforced concrete beams. It was shown that the stiffness of reinforced concrete beams decreases with increasing drop height. It was also shown that applying sprayed glass fiber-reinforced polymers (with and without mechanical stiffeners) and fabric glass fiber-reinforced polymers on the surface of reinforced concrete beams increased the stiffness. Results obtained from the software analyses were in good agreement with the laboratory test results.

Keywords: reinforced concrete beam; impact and quasi-static loading; retrofitting

---

1. Introduction

Concrete structures are subjected to a variety of different loading conditions in day-to-day situations, with the two most ubiquitous of these conditions being quasi-static and impact loadings. With regard to the former, there exists an abundance of research studying the behavior of reinforced concrete (RC) elements under quasi-static load. Conversely, structures such as high-rise buildings are often exposed to damaging impact forces, including the incidental impact of objects, explosions, sudden collapse of cranes, and unregulated motion of heavy machinery. As such, it is also imperative to study the behavior of RC beams under these dynamic loading conditions. To this end, there have been numerous studies conducted on impact loading and its effects on RC structures, with the first known dynamic test being conducted on concrete in 1917 [1]. After years of inactivity, there has been a resurgence in various impact loading experiments performed on concrete over the past 50 years. Researchers such as Atchley and Furr [2], Scott et al. [3], Dilger et al. [4], Mlakar et al. [5], and Soroushian et al. [6] studied the behavior of RC elements under dynamic conditions and discovered that with incremental increases in loading rate, the ultimate stress and strain of concrete increases by about 25%. Moreover, with respect to the compressive strength of RC elements, experiments carried out by Wastein [7], and Malvar and Ross [8] suggest that the compressive strength of concrete under dynamic loads increased by 85–100%. However, the existing literature on the bending capacity of RC beams in response to impact loading remains comparatively scarce. Bertero et al. conducted experiments on simply supported RC beams [9] and noted that the flexural capacity and rigidity of RC beams show an incremental increase at high rates of strain. Similarly, Wakabayashi et al. [10]

ran dynamic experiments on beams under strain rates of 0.001 s^{-1} and discovered that the loading capacity of beams increased by 30% in high strain rates. Comparable experiments have also been carried out by Fujikake et al. [11] to experimentally and analytically evaluate the effect of hammer height and longitudinal reinforcement on the behavior of RC beams, using a mass-spring damper system to simulate the impact response. Furthermore, Pham and Hao [12] performed an accurate numerical simulation (less than 12% average absolute error) using an artificial neural network to predict the response of RC beams to impact loading, with concrete compressive strength, hammer velocity, and mass of hammer being considered as variables. Another important variable—namely, contact stiffness—was studied experimentally and numerically by Pham et al. [13], with results indicating that this variable can significantly influence the peak of load–displacement diagram. Moreover, contact stiffness significantly influenced the bending moment and shear force of the RC beams.

Given the advancing age of RC structures, especially those in developed countries, the importance of using advanced materials—such as fiber-reinforced polymers (FRP)—for rehabilitation purposes encompassing both quasi-static and impact loadings should be emphasized. The use of externally bonded FRP as a strengthening technique for RC structures has gained popularity over the past three decades, with a notable surge in its use being observed in recent years. As such, it is necessary to study the behavior of FRP-strengthened RC elements under the aforementioned loading conditions. With respect to quasi-static loading, there has been an increasing trend in the number of studies conducted in recent years [14,15], establishing the role of externally bonded FRP in augmenting the load-carrying capacity of RC elements. On the other hand, researchers have found that the behavior of RC elements strengthened by externally bonded FRP under impact loading conditions is slightly different compared to that under quasi-static loading, with the significance of the correlation between effective FRP-concrete bonding and performance under load being highlighted [16–21].

Considering the existing body of literature, there exists the possibility of further development pertaining to the analytical study of RC beams—specifically, those tested under impact loading, as well as those strengthened by means of sprayed glass fiber-reinforced polymer (GFRP) under quasi-static loading. This paper will attempt to develop upon these topics in order to illuminate the behavior of RC beams treated with sprayed GFRP under quasi-static loading conditions. Greater insight into the use of sprayed GFRP in strengthening RC elements affords several noteworthy advantages over externally bonded fabric FRP, including ease of implementation, cost reduction, and time efficiency. As such, the novelty of this paper lies in two primary domains—the finite element modeling of RC beams strengthened for shear by sprayed GFRP, which remains an understudied area of research, in addition to the presence of mechanical fasteners (i.e., through bolts) in associated modeling schemes.

2. Materials and Methods

A total of 17 RC beams were tested in this study and the experimental results were evaluated using ABAQUS software. Nine RC beams were tested under quasi-static loading including two control and seven retrofitted beams. The other eight beams were used to study the behavior of RC beams under impact loading. Different retrofitting schemes using sprayed GFRP (with and without mechanical fasteners) were employed for shear strengthening of the RC beams tested under quasi-static loading. Fabric GFRP was employed for flexural strengthening. Samples previously experimented in Soleimani's doctoral dissertation [22] were employed for the modeling and analysis of the behavior of RC beams under quasi-static and impact loads. Tables 1–3 show the characteristics of RC beams as well as the properties of sprayed GFRP containing randomly distributed chopped fibers and unidirectional fabric GFRP.

2.1. Quasi-Static Loading

Nine beams (with and without stirrups) were tested using three-point loading under quasi-static condition using a 400 kip Baldwin Universal Testing Machine. The bending capacity of these beams is higher than their shear capacity, meaning that the failure will always be in shear. Two beams (C-NS and C-S-2) were tested as control samples without shear strengthening. The rest were strengthened for shear using sprayed GFRP to evaluate the effectiveness in increasing the shear capacity of the beams. Dimensions and reinforcement details of the beams tested under quasi-static loading are shown in Figure 1. Three Linear Variable Differential Transformers (LVDT) were used to measure the displacement of the beams during testing, as shown in Figure 1. Various parameters, such as the presence of stirrups and through-bolts as mechanical fasteners, were considered as shown in Figure 2. Details of these beams are tabulated in Table 4.

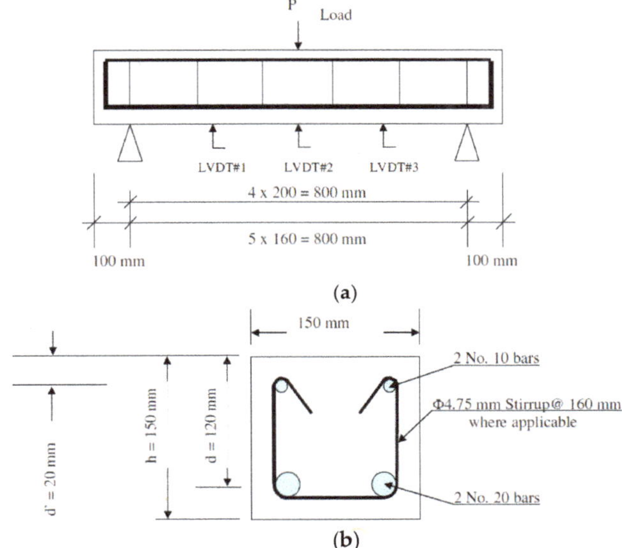

Figure 1. Reinforced concrete (RC) beams tested under quasi-static loading: (**a**) elevation; (**b**) cross-section [22].

Table 1. Characteristics of reinforced concrete beams.

Parameter	Definition	Value	Unit
f'_c	Specified compressive strength of concrete	44	MPa
f_y	Specified yield strength of tension reinforcement	M-10: 474; M-20: 440	MPa
f_{ys}	Specified yield strength of shear reinforcement	600	MPa
f_u	Specified ultimate strength of tension reinforcement	M-10: 720; M-20: 695	MPa
f_{us}	Specified ultimate strength of shear reinforcement	622	MPa
A_s	Area of reinforcement (M-10 and M-20 for tension and φ4.75 for shear)	M-10: 100; M-20: 300; φ4.75: 18.1	mm^2

Table 2. Sprayed glass fiber-reinforced polymer (GFRP) properties.

Properties	Value	Unit
Ultimate tensile strength	69	MPa
Tensile modulus	14	GPa
Ultimate rupture strain	0.63	%

Table 3. Fabric GFRP properties.

Properties	Value	Unit
Ultimate tensile strength	1517	MPa
Tensile modulus	72.4	GPa
Ultimate tensile strength per unit width	0.536	KN/mm/ply
Tensile modulus per unit width	25.6	KN/mm/ply
Ultimate rupture strain	2.1	%

2.2. Impact Loading

A drop weight impact machine with a capacity of 14.5 kJ was used, as shown in Figure 3. A mass of 591 kg (including the striking tup) can be dropped from as high as 2.5 m. During a test, the hammer is raised to a certain height above the specimen using a hoist and chain system. At this position, air brakes are applied on the steel guide rails to release the chain from the hammer. By releasing the breaks, the hammer falls and strikes the specimen. Instrumented anvil supports were used to record the bending loads during the impact [22–24].

Table 4. Designations of RC beams tested under quasi-static loading as shown in Figures 1 and 2.

Beam's Designation	Stirrups?	Sprayed GFRP on Two Sides or Three?	Width of Sprayed GFRP on the Sides (mm)	Thickness of Sprayed GFRP (mm)	No. of Through Bolts as Mechanical Fastener
C-NS	No	N/A	N/A	N/A	N/A
B2-NS	No	2	100	4	N/A
B2-4B-NS-3	No	2	100	4	4
B2-6B-NS-1	No	2	100	3.5	6
C-S-2	Yes	N/A	N/A	N/A	N/A
B2-S-1	Yes	2	150	3.5	N/A
B2-4B-S-1	Yes	2	150	3.5	4
B2-6B-S-1	Yes	2	150	4	6
B3-S-2	Yes	3	150	4	N/A

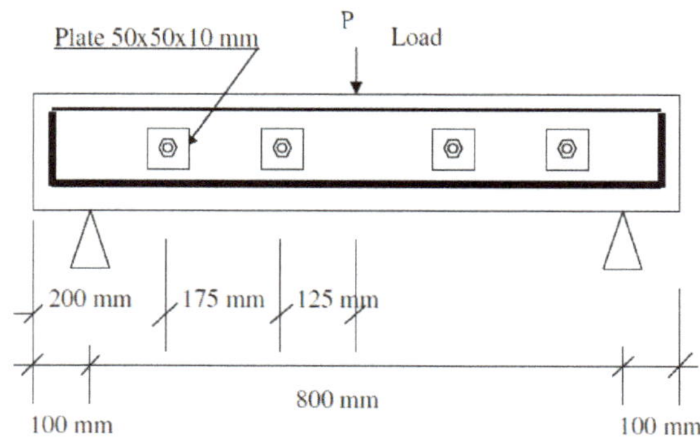

(a) Beams: B2-4B-NS-3 and B2-4B-S-1.

Figure 2. *Cont.*

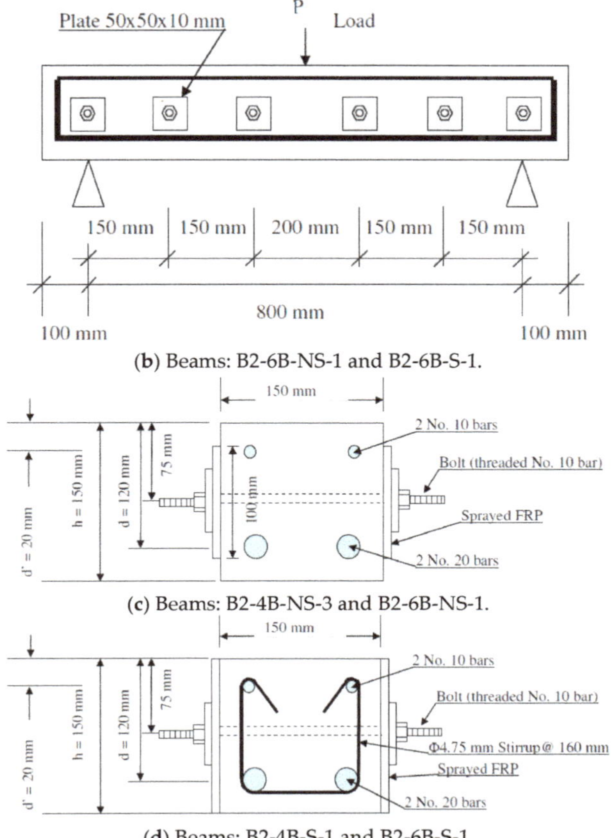

(**b**) Beams: B2-6B-NS-1 and B2-6B-S-1.

(**c**) Beams: B2-4B-NS-3 and B2-6B-NS-1.

(**d**) Beams: B2-4B-S-1 and B2-6B-S-1.

Figure 2. Bolt configuration of RC beams with sprayed GFRP: (**a**) four bolts; (**b**) six bolts; (**c**) cross-sections with no stirrups; (**d**) cross-sections with stirrups [22].

Figure 3. Impact machine.

Dimensions and reinforcement details of the beams tested under impact loading are shown in Figure 4. Enough stirrups were provided to make sure that the beam would fail under bending. Two beams (BS and BS-GFRP) were tested under quasi-static loading and the rest under impact loading with various impact velocities ranging from 2.80 to 6.26 m/s (see Table 5). It is worth mentioning that BS-GFRP and BI-600-GFRP are identical beams strengthened by fabric GFRP. One layer of unidirectional GFRP fabric with a total thickness of about 1.2 mm, length of 750 mm, and width of 150 mm was applied longitudinally on the tension (bottom) side of the beam for flexural strengthening. An extra layer with fibers perpendicular to the fiber direction of the first layer was applied on three sides (two sides and the tension side) for shear strengthening. It is important to note that while the control RC beam (beam "BS") failed in flexure, the strengthened RC beam failed in shear, indicating that shear strengthening was not as effective as flexural strengthening and perhaps more layers of GFRP were needed to overcome the deficiency of shear strength in these beams. Beam displacement was measured via an accelerometer affixed to the bottom face of the beam in impact experiments (refer to Figure 4).

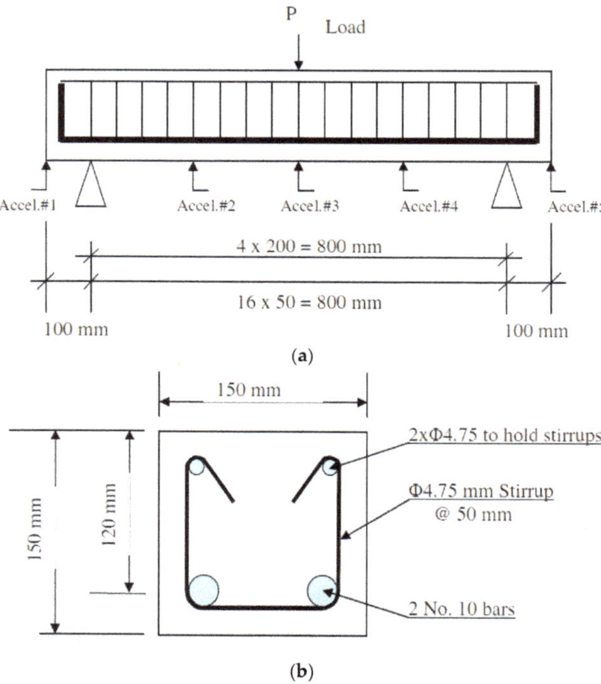

Figure 4. Elevation and cross-section of RC beams tested under impact loading; (**a**) elevation; (**b**) cross-section [22].

Table 5. Designations and details of RC beams tested under impact loading as shown in Figure 4.

Beam's Designation	Quasi-Static or Impact?	Drop Height (mm)	Impact Velocity (m/s)
BS	Quasi-static (control)	NA	NA
BI-400	Impact	400	2.8
BI-500	Impact	500	3.13
BI-600	Impact	600	3.43
BI-1000	Impact	1000	4.43
BI-2000	Impact	2000	6.26
BS-GFRP	Quasi-static (control)	NA	NA
BI-600-GFRP	Impact	600	3.43

3. Finite Element Model (FEM)

ABAQUS FEA software, a commercially available finite-element analysis program, was used to model the RC beams with and without GFRP strengthening schemes. For concrete modeling, 3D eight-node linear isoparametric elements with reduced integration (C3D8R) were utilized. For longitudinal and transverse bars, truss elements (T3D2) were applied. FRP were modeled using shell elements (S4R). Embedded region coupling was utilized to simulate the bond between longitudinal and transverse reinforcements with concrete. This coupling enables the user to identify one region as the host, and another one as embedded. In this paper, reinforcements represented the embedded region, and concrete was the host region [25]. In order to prevent the scattering of results, coupling restraint was applied. This coupling restraint is formed through a reference point (RP) situated at the center of the support on the bottom of the beam.

For impact, the loading is exerted via a hammer at the mid-span of the beam, and a face-to-face constraint was utilized. Therefore, one could obtain force displacement results at one point with minimal errors. Since the beams were simply supported, U1, U2, and U3 degrees of freedom were set to zero. The arrangement of reinforcing bars, RC beams, supports, and hammer in the software are presented in Figure 5. It is noteworthy that the analyses conducted for these models were quasi-static and dynamic. The BS and BS-GFRP samples shown in Table 5 and all samples mentioned in Table 4 were modeled using quasi-static loading, and BI samples in Table 5 were modeled using dynamic/impact analyses. Dynamic explicit analysis was employed for the impact tests, while dynamic implicit analysis was used for quasi-static analyses. To simulate the behavior of steel reinforcing bars in ABAQUS under a high strain rate, a linear kinematic hardening model was used [25]. In order to model the behavior of confined concrete, a concrete damage plasticity (CDP) model was utilized based on the work of Sayyar Roudsari et al. [26].

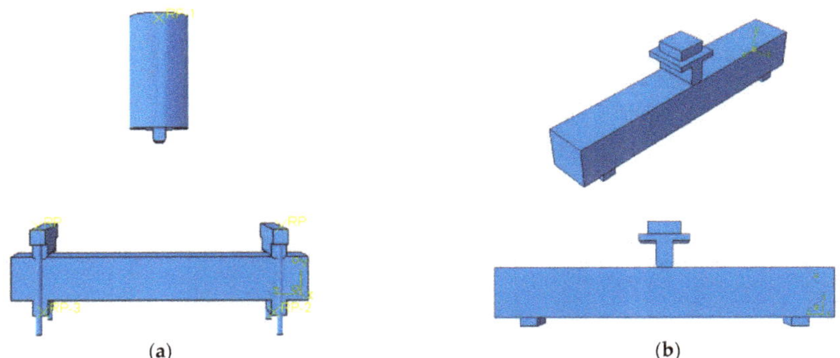

(a)   (b)

Figure 5. RC beam assembly simulations; (**a**) impact, (**b**) quasi-static.

4. Results and Discussion

Load vs. mid-span displacement curves from laboratory experiments (EXP) and finite element models (FEM) in ABAQUS are illustrated in Figures 6–22.

Beams in Figures 6–13 have the same reinforcement details as those shown in Figure 4. Figure 6 shows the load vs. mid-span displacement curve of the RC beam under quasi-static loading, and Figures 7–11 demonstrate similar curves for RC beams tested under impact loading. Results obtained for RC beams strengthened with fabric GFRP under quasi-static and impacts loading are illustrated in Figures 12 and 13, respectively. Results from the finite element models prove tangible correspondence with those of the laboratory experiments, so much so that the maximum difference between the laboratory and software results was found to be no more than 20% (ranging between 0.1% for BS in Figure 6, and 20% for BI-400 in Figure 7. It is worth mentioning that the difference between the laboratory and software results are negligible except for BI-400. It can be inferred that a minimum drop height of around 500 mm is required to make the RC beam fail, which was not met for BI-400 as its drop height was only 400 mm. This is the most probable cause of the 20% difference. Comparing the results, the beam BS-GFRP that is strengthened with GFRP and tested under quasi-static loading showed a 29.3% increase in load-carrying capacity in the laboratory experiment and a 30.0% increase in the software model.

Figures 14–22 present the load vs. mid-span displacement curves under quasi-static loading based on Table 4. Results from the FEM analysis are in very good agreement with the experimental results, with a maximum difference of 5%. In effect, the FEM analysis is seen to accurately predict the rate of load-carrying capacity vs. displacement before and after the maximum load.

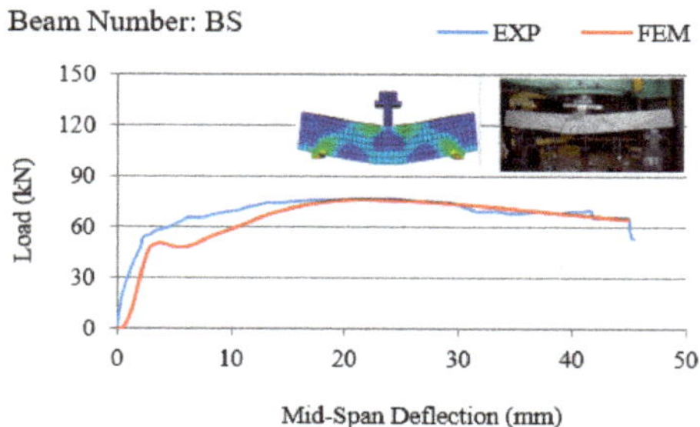

Figure 6. Load vs. mid-span deflection for BS; EXP: experimental; FEM: finite element modeling.

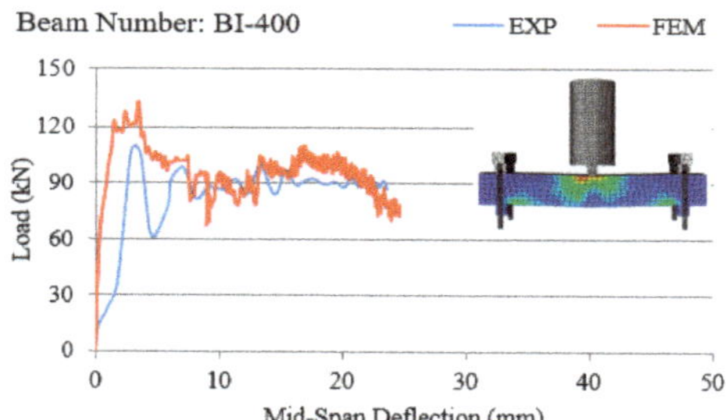

Figure 7. Load vs. mid-span deflection for BI-400.

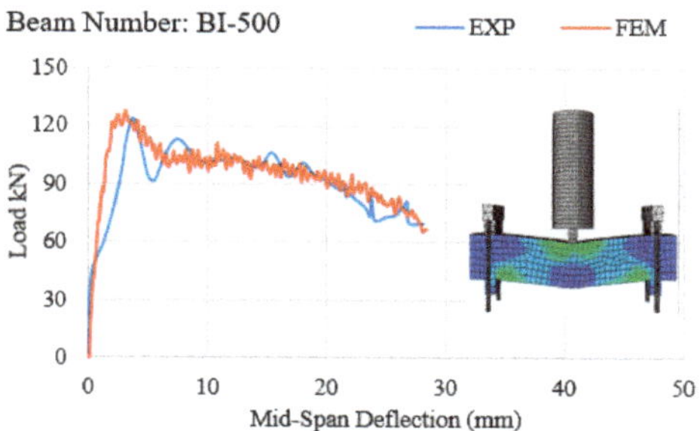

Figure 8. Load vs. mid-span deflection for BI-500.

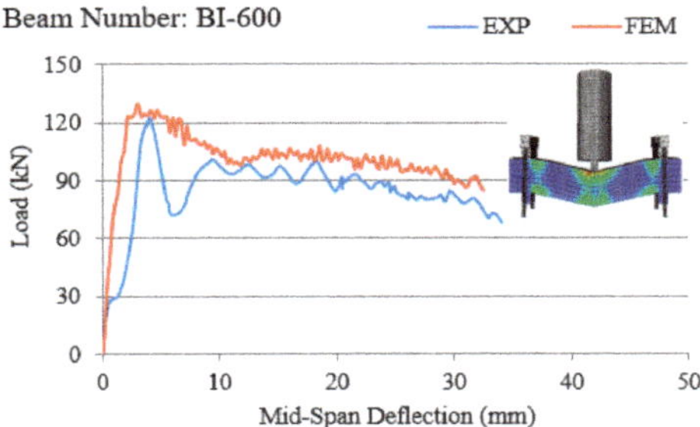

Figure 9. Load vs. mid-span deflection for BI-600.

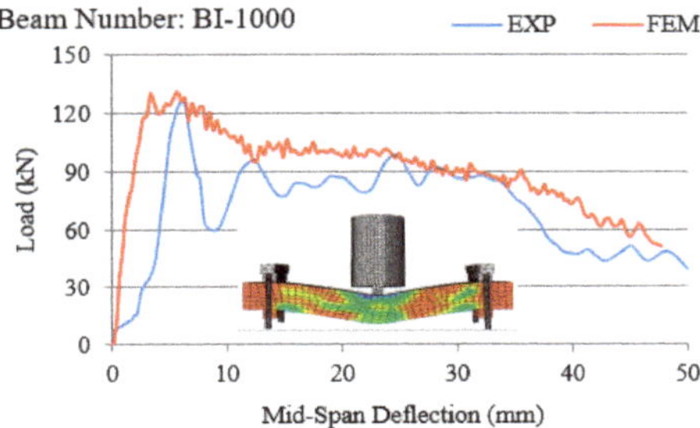

Figure 10. Load vs. mid-span deflection for BI-1000.

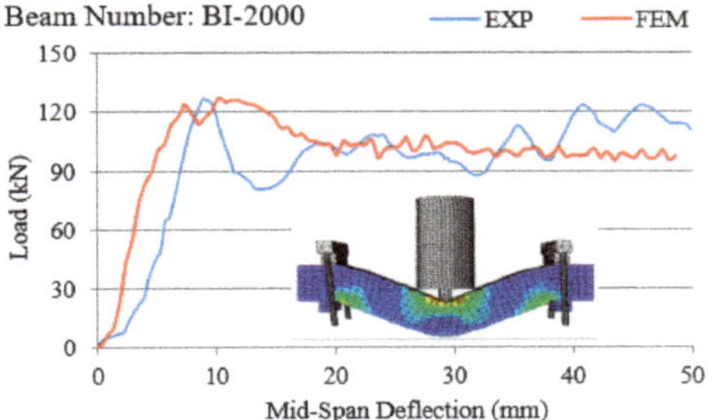

Figure 11. Load vs. mid-span deflection for BI-2000.

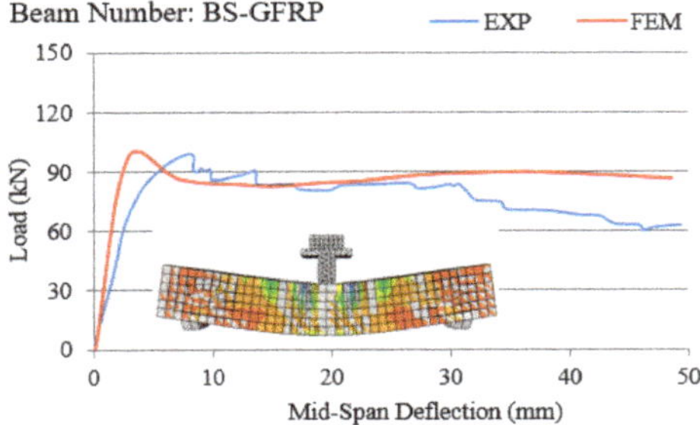

Figure 12. Load vs. mid-span deflection for BS-GFRP.

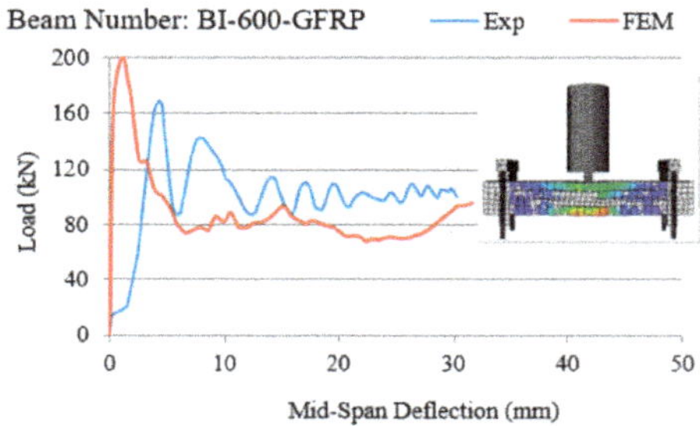

Figure 13. Load vs. mid-span deflection for BI-600-GFRP.

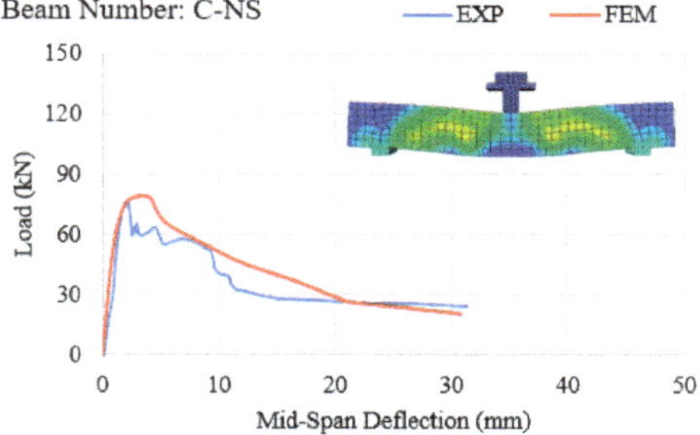

Figure 14. Load vs. mid-span deflection for C-NS.

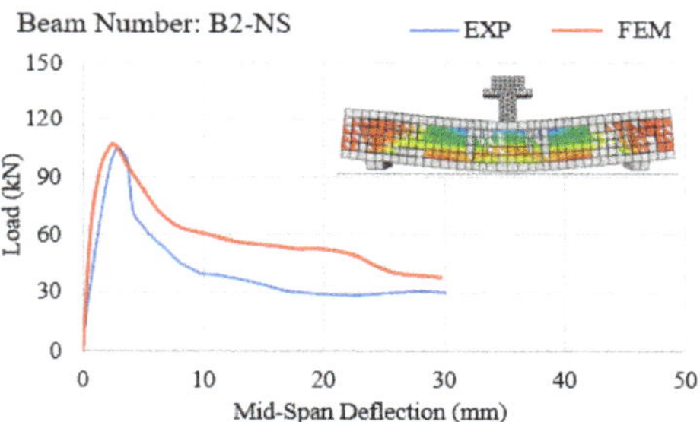

Figure 15. Load vs. mid-span deflection for B2-NS.

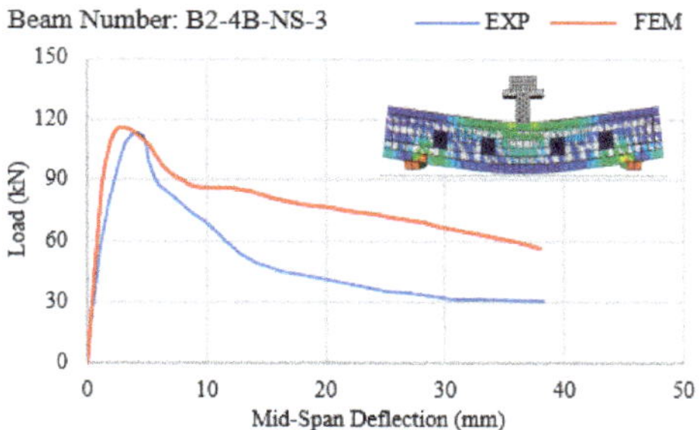

Figure 16. Load vs. mid-span deflection for B2-4B-NS-3.

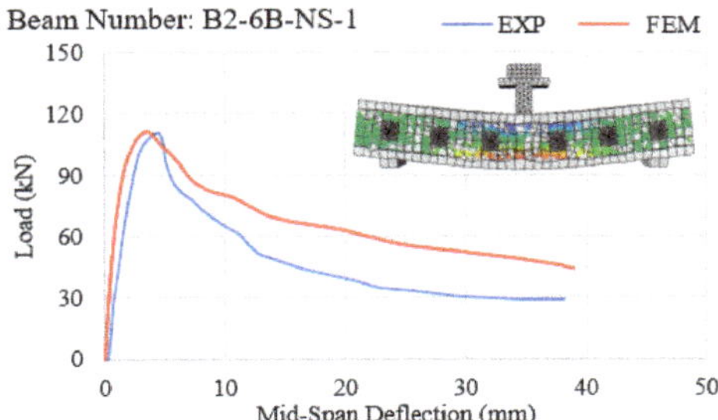

Figure 17. Load vs. mid-span deflection for B2-6B-NS-1.

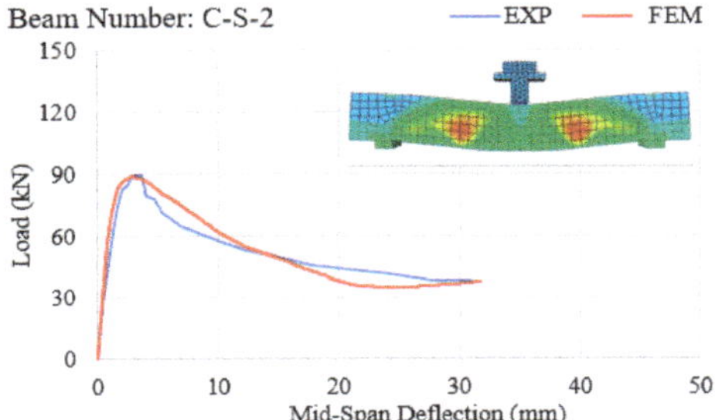

Figure 18. Load vs. mid-span deflection for C-S-2.

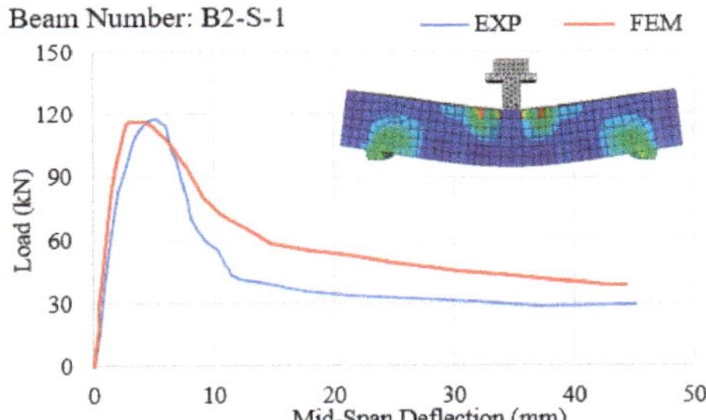

Figure 19. Load vs. mid-span deflection for B2-S-1.

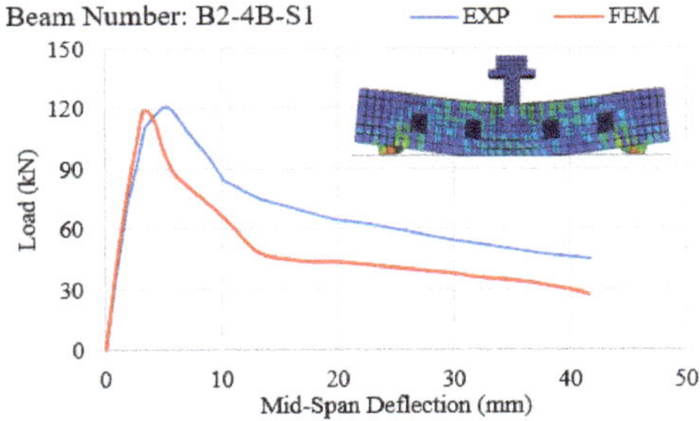

Figure 20. Load vs. mid-span deflection for B2-4B-S-1.

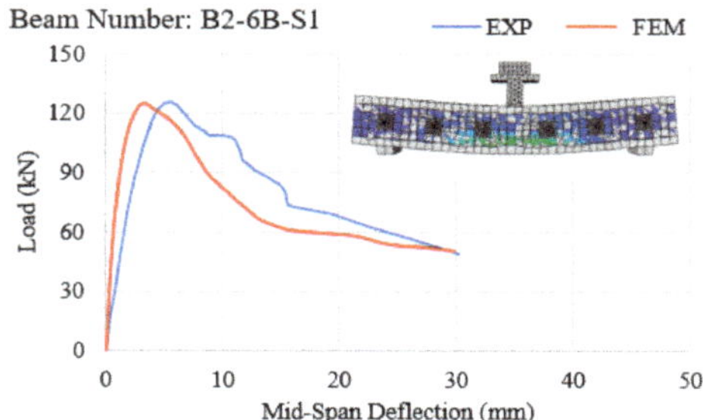

Figure 21. Load vs. mid-span deflection for B2-6B-S-1.

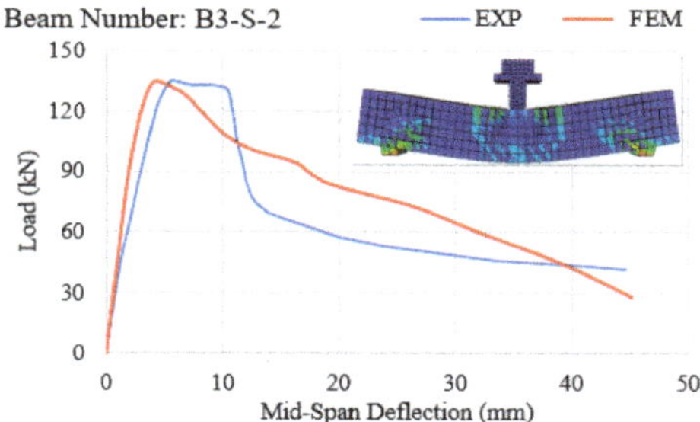

Figure 22. Load vs. mid-span deflection for B3-S-2.

The load–displacement curve can be idealized by a bilinear graph similar to Figure 23. The stiffness of the beam can then be calculated using Equation (1): [27]

$$\text{Stiffness} = V_y/\Delta_y. \tag{1}$$

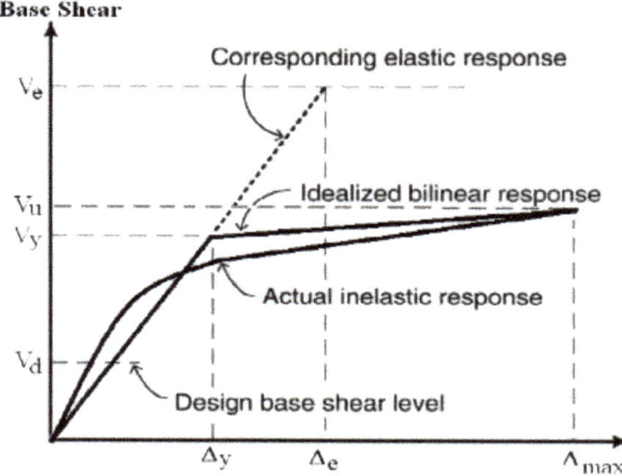

Figure 23. Load vs. displacement bilinear graph [27].

The load–displacement curve of C-NS beam is idealized in Figure 24, and the stiffness was calculated. Following the same procedure, the stiffness of all beams was calculated, and the values are illustrated in Figures 25 and 26.

As shown in Figure 25, it is apparent that the stiffness of the beam under impact loading increased with respect to its stiffness under quasi-static loading. Also, increasing the impact velocity reduced the stiffness (the stiffness increased from BS to BI-400 and then decreased from BI-400 to BI-2000). The same trend was observed in FEM as well as experimental results. Fabric GFRP increased the stiffness of the beam under both quasi-static and impact loadings. The increase in stiffness under impact loading was more apparent in BS and BS-GFRP compared to BI-600 and BI-600-GFRP. One can conclude that fabric GFRP is quite effective in enhancing the stiffness of an RC beam under impact loading.

As shown in Figure 26, sprayed GFRP is more effective than steel stirrups in enhancing the stiffness of the beam (C-NS vs. C-S-2 as well as C-NS vs. B2-NS, B2-4B-NS-3, and B2-6B-NS-1). Shear strengthening of RC beams using sprayed FRP effectively increased the stiffness of the beam, either with or without stirrups. This was apparent from the results obtained from both FEM and experiments. Through-bolts are more effective in increasing the stiffness when they are employed for strengthening of the beams with steel stirrups. The thickness of the sprayed GFRP plays a role in increasing the stiffness (4 mm thickness is more effective than 3.5 mm, even if the number of through-bolts is greater in the thinner layer of sprayed FRP).

Comparing Figures 25 and 26, one can conclude that the FEM are more successful in predicting the stiffness of RC beams under quasi-static loading. Tanarslan et. al. [28] tested RC beams strengthened by prefabricated ultra-high-performance fiber RC laminates and also reported the increase of stiffness.

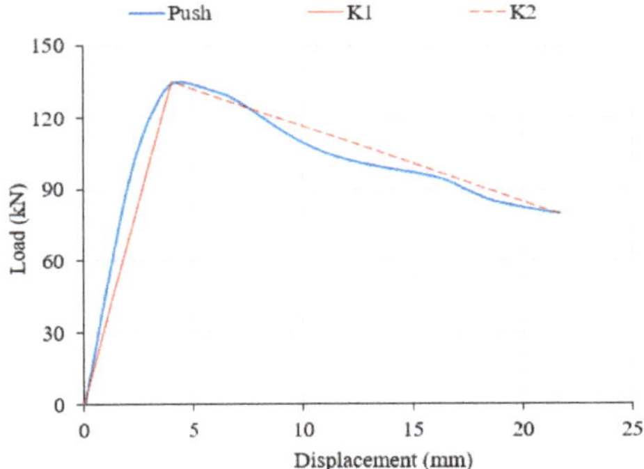

Figure 24. Bilinear graph for C-NS beam.

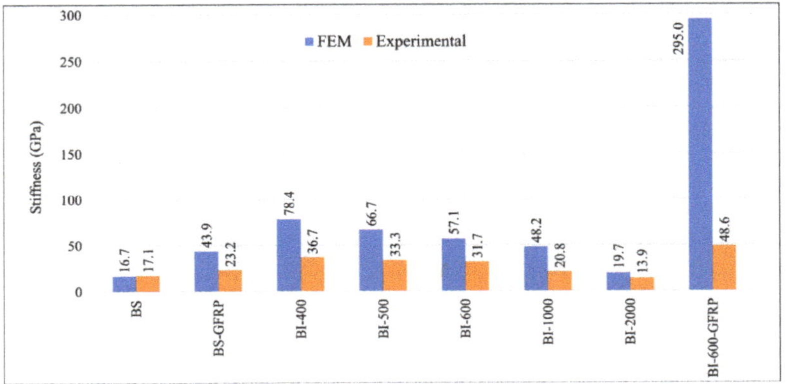

Figure 25. Comparison of beam stiffness under impact loading: finite element model (FEM) vs. experimental results.

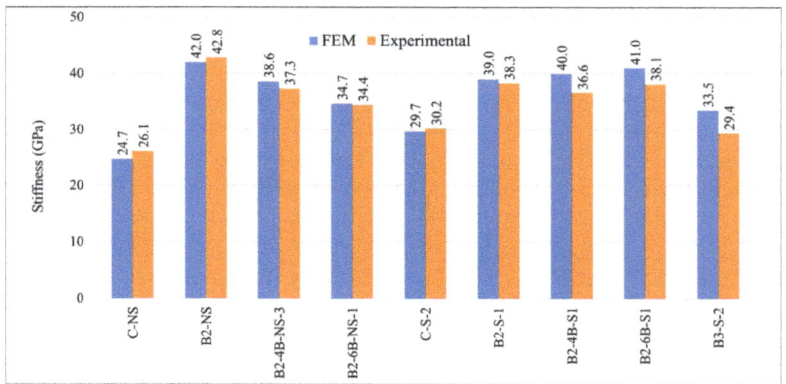

Figure 26. Comparison of beam stiffness under quasi-static loading: FEM vs. experimental results.

5. Conclusions

Based on the results obtained from the experiment and verified by the finite element analyses, the following conclusions can be drawn from this study:

- Load vs. mid-span displacement curves obtained from finite element models generated by ABAQUS software, for both quasi-static and impact loadings, suggest close correspondence with laboratory results.

- In quasi-static loading on an RC beam (Figure 6) with an adequate number of stirrups, both ABAQUS software and laboratory results verified that the stress in longitudinal tension reinforcements exceeded the yield stress of steel, and the load-carrying capacity of the beam increased subsequent to the yield of tensile steel. This implies that the failure of the RC beam presented in Figure 6 is flexural.

- In all experiments conducted on RC beams under impact loading (Figures 7–11 and Figure 13), the beams failed in shear as opposed to flexure. Thus, there is a correlation between the application of fabric GFRP and the increased shear strength capacity of RC beams. The RC beams strengthened by fabric GFRP depicted a higher load-carrying capacity in both quasi-static and impact loadings compared to that of non-strengthened RC beams (Figure 7, Figure 9, Figure 12, and Figure 13). Fabric GFRP effectively increased the beam's stiffness under both quasi-static and impact loadings (Figure 25).

- The load-carrying capacity of the beam was substantially increased under impact loading, and the impact velocity for the range tested in this study does not affect this increment (Figures 6–13).

- In quasi-static loading on an RC beam (Figure 14 with no stirrups and Figure 18 with an inadequate number of stirrups), both ABAQUS software and laboratory results verified that the RC beam presented in these figures would fail in shear.

- Although the beam's stiffness under impact loading is increased as opposed to that under quasi-static loading, stiffness under impact loading will be reduced by increasing impact velocity (Figure 25).

- Sprayed GFRP (with and without through-bolts) increases the beam's stiffness (Figure 26). The through-bolts are more effective in increasing the beam's stiffness when some steel stirrups are present.

- Using FEM to predict the stiffness of RC beams was more effective for beams tested under quasi-static conditions as compared to those tested under impact loading.

Considering these results and their implications, this paper has highlighted the possibility of further research with respect to the analytical study of RC beams strengthened by sprayed GFRP under impact loading. Research into this field is not as developed as compared to externally bonded fabric FRP, and it is imperative that this gap be narrowed by future works.

Author Contributions: S.S.R. conducted the finite element analysis and prepared the original draft; S.M.S. supervised the project, checked the results, reviewed and edited the paper.

Funding: This research received no external funding.

Conflicts of Interest: The authors declare no conflict of interest.

References

1. Abrams, D.A. Effects of rate of application of load on the compressive strength of concrete. In Proceedings of the American Society for Testing and Materials, Atlantic City, NJ, USA, 26–29 June 1917; Volume 17, Part II, pp. 364–377.
2. Atchley, B.L.; Furr, H.L. Strength and energy absorption capabilities of plain concrete under dynamic and static loadings. *Am. Concr. Inst. J.* **1967**, *64*, 745–756.
3. Scott, B.D.; Park, R.; Priestley, M.J.N. Stress-strain behaviour of concrete confined by overlapping hoops at low and high strain rates. *J. Am. Concr. Inst.* **1982**, *79*, 496–498.
4. Dilger, W.; Kand, R.; Kowalczyk, R. Ductility of plain and confined concrete under different strain rates. *Am. Concr. Inst. J.* **1984**, *81*, 73–81.
5. Mlakar, P.F.; Vitaya-Udom, K.P.; Cole, R.A. Dynamic tensile-compressive behaviour of concrete. *Am. Concr. Inst. J.* **1986**, *86*, 484–491.
6. Soroushian, P.; Choi, K.-B.; Alhamad, A. Dynamic constitutive behaviour of concrete. *Am. Concr. Inst. J.* **1986**, *83*, 251–259.
7. Watstein, D. Effect of straining rate on the compressive strength and elastic properties of concrete. *Am. Concr. Inst. J.* **1953**, *24*, 729–744.
8. Malvar, L.J.; Ross, C.A. Review of strain rate effects for concrete in tension. *Mater. J.* **1998**, *95*, 735–739.
9. Bertero, V.V.; Rea, D.; Mahin, S.; Atalay, M.B. Rate of loading effects on uncracked and repaired reinforced concrete members. In Proceedings of the 5th World Conference on Earthquake Engineering, Rome, Italy, 25–29 June 1973; Volume II, pp. 1461–1470.
10. Wakabayashi, M.; Nakamura, T.; Yoshida, N.; Iwai, S.; Watanabe, Y. Dynamic loading effects on the structural performance of concrete and steel materials and beams. In Proceedings of the 7th World Conference on Earthquake Engineering, Istanbul, Turkey, 8–13 September 1980.
11. Fujikake, K.; Li, B.; Soeun, S. Impact response of reinforced concrete beam and its analytical evaluation. *J. Struct. Eng.* **2009**, *135*, 938–950. [CrossRef]
12. Pham, T.M.; Hao, H. Prediction of the impact force on RC beams from a drop weight. *Adv. Struct. Eng.* **2016**. [CrossRef]
13. Pham, T.M.; Hao, Y.; Hao, H. Sensitivity of impact behavior of RC beams to contact stiffness. *Int. J. Impact Eng.* **2017**. [CrossRef]
14. Jabr, A.; El-Ragaby, A.; Ghrib, F. Effect of the fiber type and axial stiffness of FRCM on the flexural strengthening of RC beams. *Fibers* **2017**, *5*, 2. [CrossRef]
15. Peng, H.; Shang, S.; Jin, Y.-J.; Wang, M. Experimental study of reinforced concrete beam with prestressed CFRP plate. *Eng. Mech.* **2008**, *25*, 142–151.
16. White, T.; Soudki, K.; Erki, M. Response of RC beams strengthened with CFRP laminates and subjected to a high rate of loading. *J. Compos. Constr.* **2001**, *5*, 153–162. [CrossRef]
17. Tang, T.; Saadatmanesh, H. Behavior of concrete beams strengthened with fiber-reinforced polymer laminates under impact loading. *J. Compos. Constr.* **2003**, *7*, 209–218. [CrossRef]
18. Tang, T.; Saadatmanesh, H. Analytical and experimental studies of fiber-reinforced polymer strengthened concrete beams under impact loading. *ACI Struct. J.* **2005**, *102*, 139–149.
19. Hamed, E.; Rabinovitch, O. Dynamic behavior of reinforced concrete beams strengthened with composite materials. *J. Compos. Constr.* **2005**, *9*, 429–440. [CrossRef]
20. Pham, T.M.; Hao, H. Impact behavior of FRP-strengthened RC beams without stirrups. *J. Compos. Constr.* **2016**. [CrossRef]

21. Meola, C.; Boccardi, S.; Carlomagno, G.M. Infrared thermography for inline monitoring of glass/epoxy under impact and quasi-static bending. *Appl. Sci.* **2018**, *8*, 301. [CrossRef]
22. Soleimani, S.M. Sprayed Glass Fiber Reinforced Polymers in Shear Strengthening and Enhancement of Impact Resistance of Reinforced Concrete Beams. Ph.D. Thesis, The University of British Columbia, Vancouver, BC, Canada, 2007.
23. Soleimani, S.M.; Banthia, N. A novel drop weight impact setup for testing reinforced concrete beams. *J. Exp. Tech.* **2014**, *38*, 72–79. [CrossRef]
24. Soleimani, S.M.; Banthia, N. Reinforced concrete beams under impact loading and influence of a GFRP coating. *Int. Rev. Civ. Eng.* **2010**, *1*, 68–77.
25. ABAQUS Analysis User's Guide. Available online: http://dsk.ippt.pan.pl/docs/abaqus/v6.13/books/usb/default.htm (accessed on 16 July 2019).
26. Sayyar Roudsari, S.; Hamoush, S.A.; Soleimani, S.M.; Madandoust, R. Evaluation of large-size reinforced concrete columns strengthened for axial load using fiber reinforced polymers. *Eng. Struct.* **2019**, *178*, 680–693. [CrossRef]
27. Mahmoudi, M.; Zaree, M. Determination the response modification factors of buckling restrained braced frames. *Procedia Eng.* **2013**, *54*, 222–231. [CrossRef]
28. Tanarslan, M.; Alver, N.; Jahangiri, R.; Yalçınkaya, Ç.; Yazıcı, H. Flexural strengthening of RC beams using UHPFRC laminates: Bonding techniques and rebar addition. *Constr. Build. Mater.* **2017**, *155*, 45–55. [CrossRef]

Article

Curable Area Substantiation of Self-Healing in Concrete Using Neutral Axis

Choonghyun Kang [1] and Taewan Kim [2,*]

[1]   Department of Civil and Environmental Engineering, Chonnam National University, Yeosu, Chonnam-do 59626, Korea; kangcivil@gmail.com

[2]   Department of Civil Engineering, Pusan National University, Busan 46241, Korea

*   Correspondence: ring2014@naver.com; Tel.: +82-10-6357-8142

Received: 4 March 2019; Accepted: 11 April 2019; Published: 13 April 2019

Featured Application: **This study substantiates a curable area with self-healing in concrete material using a flexural test. The neutral axis was estimated as tension strains from attached successive strain gauge, and this before and after healing specified the curable area.**

Abstract: The self-healing nature of concrete has been proved in many studies using various methods. However, the underlying mechanisms and the distinct area of self-healing have not been identified in detail. This study focuses on the limits of the area of self-healing. A bending specimen with a notch is used herein, and its flexural strength and stiffness before and after healing are compared and used for self-healing assessment. In addition, the neutral axis of the specimen was measured using successive strain gauges attached to the crack propagation part. Although the strength and stiffness of the concrete recovered after self-healing, the change in the location of the neutral axis before and after healing was insignificant, which indicates that physical recovery did not occur for once-opened crack areas.

Keywords: neutral axis; self-healing; successive strain gauge; flexural test

---

1. Introduction

Self-healing of concrete is attracting attention as a solution for reducing maintenance costs and environmental issues. Conventionally, studies in the literature on concrete development have focused mainly on improving the properties of concrete such as strength; durability; resistance against acid, chlorides, corrosion, etc. [1]. Ultra-high-strength concrete of over 100 MPa has been developed [2], and the strength of the concrete used in actual applications has been increased [3,4]. Moreover, various admixtures are being applied to improve the durability of concrete [5–9]. These efforts ultimately aim to reduce construction costs by reducing the section areas of a framework. By contrast, there is a debate about reducing maintenance costs of an increasing cumulative number of old concrete structures [10–12]. In addition, research on mitigating environmental problems is required in the concrete industry based on the notion that environmental issues should be assessed at any cost [13]. Self-healing of concrete presents advantages in terms of reducing costs and solving environmental problems. The main aim is to reduce maintenance costs through automatic crack detection and recovery in general, and multiple self-healing methods are available to this end. Moreover, the increased lifetime of concrete structures due to self-healing reduces the overall usage of cement in the long term, thereby gradually mitigating the environmental harm arising from the CO_2 generated during cement production.

Self-healing of concrete has been studied from various viewpoints over the past decade. The most famous study involved the active method of using capsules containing an adhesive [14–20]; this method is classified as active because it uses additives that are not used in normal concrete. Another famous active method was proposed based on the use of microorganisms [21–24], which help deposit crack-restoring substances. Moreover, many passive methods exist, for example, improvement of self-healing environment using fibers that prevent the occurrence of a single large crack and induce multiple cracks of small width [25–28]. The recovery due to self-healing is not only general strength recovery but also durability recovery such as decreased water permeability [29–33]. Furthermore, self-healing techniques have been studied in various ways depending on the types and shapes of cracks that can occur in a structure, and various methods for verifying the healing effect have been proposed. However, existing studies have generally measured the degree of healing by an indirect method because direct observation of cracks and post-healing conditions is not easy and has various limitations.

The most important part of self-healing assessment is to measure efficiently and quantify accurately the healing effects. Indirect measurements, such as strength measurement, permeability estimation, and ultrasonic pulse velocity measurement, can easily quantify the healing effect by comparing the measured values before and after healing. However, these measurements are limited to relative comparison because the generated cracks are not directly measured and compared after healing. In addition, it is difficult to exclude the influence of the portion with no cracking because the measurement range covers the entire specimen as opposed to only the area around the generated crack. Especially, the age effect, which is necessarily accompanied by healing time, makes it difficult to evaluate only the healing effect accurately. Therefore, to distinguish the self-healing effect more clearly, a more direct measure is required.

2. Experimental Setup

2.1. Three-Point Bending Test with A Notch

In this study, the three-point bending test with a notch was performed for assessment of the self-healing property of concrete. The self-healing measurement process generally proceeds in the order of crack inducement, performance measurement before healing, healing, and performance measurement after healing. Crack inducement involves determining the depth or area of a crack; in other words, the recoverable area is determined in this process. It is important to maintain constant width and depth of the crack for an accurate comparison of the healing effect, and the three-point bending test with a notch is advantageous in this respect. The notch given just below the load point limits the crack evolution to only a single crack, and the crack mouth opening displacement (CMOD) can be measured easily by installing the instrument at the bottom of the notch. Crack propagation is correlated directly to the CMOD in the three-point bending test; therefore, it can be controlled easily during the loading process. In addition, performance before healing can be measured by investigating the behavior of the specimen during the reloading process (2nd loading) just after crack inducement. After healing, the same scheme of loading (3rd loading) is applied to measure performance after healing. The 2nd loading and the 3rd loading follow the same loading schemes at different times, so the comparison before and after healing is easy and clear. Furthermore, the specimens used in this test can be fabricated easily without the use of any specialized equipment. Owing to the abovementioned reasons, we used the three-point bending test with a notch for assessments in this study. We performed the experiment by referring to the Réunion Internationale des Laboratoires et Experts des Matériaux (RILEM) and Japan Concrete Institute (JCI) standards [34,35]. Figure 1 shows a schematic of the experimental setup.

The degree of crack inducement was set to CMOD 0.05 mm. It is necessary to standardize the depth of the crack in all specimens to quantify the healing effect. The crack depth was controlled constantly by advancing the CMOD. Granger et al. used a certain percentage (60%) of the peak load as a standard for the crack depth [36]. However, it was confirmed that the behaviors of the specimens after the peak load differed, even in the case of specimens of the same age. Therefore, a constant CMOD of 0.05 mm was applied to the 1st loading of all cases by referring to the literature [23,37], and unloading was performed until the load decreased to zero. The 2nd loading was also conducted until a

CMOD of 0.05 mm was achieved. The crack depth based on the CMOD was confirmed to be relatively even in all specimens.

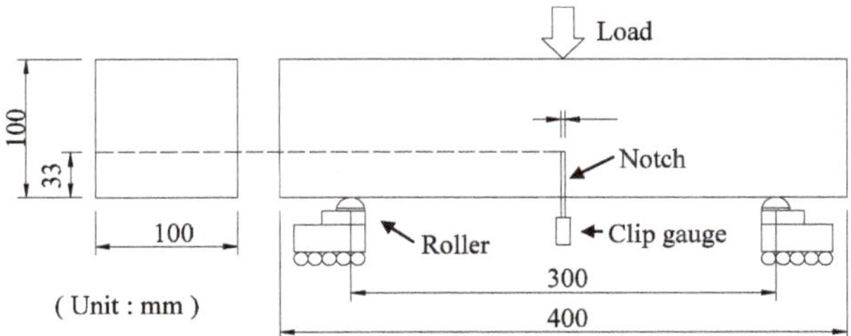

Figure 1. Schematic of the experimental setup.

2.2. Specimen

A prismatic specimen measuring L400 × W100 × D100 mm was used in this study, as specified in the JCI standard (JCI-S-001-2003). RILEM specifies a longer specimen for the three- or four-point bending test, but small specimens are convenient from the viewpoint of the most essential curing process in self-healing [38]. Ordinary Portland cement was used as the base material, and additional materials that can aid the healing process were not included because we sought to focus on the crack itself. The water–cement ratio (W/C) was set to 45%. Detailed specifications of the concrete mix are listed in Table 1.

Table 1. Mixing proportions of the specimen.

W/C (%)	Unit Content (kg/m^3)			
	Water	Cement	Sand	Gravel
45	175	389	780	879

All specimens were fabricated at the same time and with the same materials, and the notch was also created at the same time and cured for seven days. All specimens were cast in steel molds, and they were demolded and cured in fresh water at 20 °C for the following 24 h. Five different cases were considered in this study, and the faster loading process (1st loading) of these cases was planned on the seventh day after fabrication. Therefore, the notches in all specimens were sawed on the seventh day of the curing process. A wet concrete cutter with a diamond blade was used for sawing the notch.

The notch depth depends on the maximum load capacity of the specimens. If the notch is too deep, there would not be sufficient space to measure the neutral axis. Considering these aspects, the notch depth was set to 33 mm, which is one-third of the specimen depth, and the notch width was set to 3 ± 0.5 mm. Cracks occur near the peak load and begin to propagate, and the crack propagation speed is generally higher with a higher peak load. As a preliminary experiment, 25% of the load specified in RILEM or 30% of the load specified in the JCI standard was found to be too high as peak loads from the viewpoint of controlling crack propagation easily.

These were three loading processes per case. The 1st loading induced the crack in each specimen. The 2nd loading just followed the 1st loading. The 2nd loading was limited to the extent that the crack induced in the 1st loading was no longer advanced, and it measured performance before healing. After the 2nd loading, specimens entered the healing process. The healing process was done in different healing conditions (freshwater and air) and in different periods of time (21 days, 42 days and 63 days). The 3rd loading was performed after the healing process. We were able to determine the degree of healing effect by comparing the 2nd loading and the 3rd loading performed before and after the healing process.

Five different cases were planed; 7-28 case, 7-49 case, 7-70 case, 28-49 case and 49-70 case. The first number of case names indicates when the 1st loading and the 2nd loading took place. Both of them was done at the same days because the 2nd loading was just followed the 1st loading. The second number of case names indicates the date when the 3rd loading was made, and the healing period could be calculated from the difference between two numbers in the case name. For example, 28-70 case means that the 1st loading and the 2nd loading was done on the 28th days and the 3rd loading was done on the 70th day with a 42-day healing period. Finally, the letters W and A at the end of the case name represent the healing condition; healing in freshwater at 20 °C and healing in air in a curing room maintained at 20 °C.

The effect of the healing period would be investigated comparing 7-28 case, 7-49 case and 7-70 case. These cases took place on the 7th day of the same day as the 1st loading and the 2nd loading, only different healing period. And, the effect of the crack inducement time would be investigated comparing 7-28 case, 28-49 case and 49-70 case. The 1st loading and the 2nd loading were performed on different dates in these cases, but the healing period is the same as 28-day. More than three specimens were prepared for every case. Table 2 describes the test program

Table 2. Test program.

Case	Condition	______ Days after Casting ______			
		7	28	49	70
7-28	Water / Air	1st loading 2nd loading healing ▶▶▶▶	3rd loading		
7-49	Water / Air	1st loading 2nd loading	healing ▶▶▶▶	3rd loading	
7-70	Water / Air	1st loading 2nd loading		healing ▶▶▶▶	3rd loading
28-49	Water / Air		1st loading 2nd loading healing ▶▶▶▶	3rd loading	
49-70	Water / Air			1st loading 2nd loading healing ▶▶▶▶	3rd loading

3. Self-Healing Assessment

3.1. Strength

The flexural strengths of the specimens at four points were compared in this study. Strength is the most widely used and most comparable criterion in the case of a specimen under a given condition. When using flexural strength, it is important to fix the points that are the subjects of comparison, because of which the degree of evaluation stabilizes accordingly. First, the load at the unloading point of the 1st loading was compared to the maximum load of the 3rd loading. If the latter was found to be larger than the former, the healing effect was considered to occur, although the age effect of the undamaged part may have played a role as well. That is, the increase in flexural strength after healing can be ascribed not only to the healing effect but also the age effect. Second, the load at CMOD 0.05 mm of the 3rd loading was compared to the load at the same point of the 2nd loading. This comparison made it relatively straightforward to compare the flexural strength recovery before and after healing because these values were measured with the same degree of damage. This comparison may be less affected by the age effect because it was made at the point at which the crack does not advance.

3.2. Flexural Stiffness

The initial flexural stiffness of the specimen was used for the self-healing assessment instead of the flexural strength. The healing area was limited to the crack tip because van Breugel reported that the recoverable limit of a crack was under 0.2 mm [21,22], and this crack tip may be involved in the initial behavior of the crack re-opening. By contrast, Jenq et al. reported that the modulus of elasticity of the specimen material is directly related to its flexural stiffness, which is described by the compliance between the CMOD and the load [39]. Moreover, Shah et al. reported that the flexural stiffness of the material is related to the critical stress intensity factor [40]. In addition, the initial part of flexural stiffness may explain the crack re-opening behavior. This is the reason why the initial flexural stiffness was compared in the present study.

The initial flexural stiffness of the 2nd loading before healing and the 3rd loading after healing was compared in this study. The behaviors of the 2nd loading and the 3rd loading were different, as shown in Figure 2. Especially, the initial parts of these behaviors differed significantly according to the healing time and the healing condition. In the load-CMOD relationship, the slope from 0.004 mm to 0.005 mm of CMOD from the starting point of reloading (2nd and 3rd loading) which is a relatively straight line, was calculated as the initial flexural stiffness, as shown in Figure 3. The slopes of the 2nd and the 3rd loading were obtained and the ratio of these two values was used for the quantification of the healing effect.

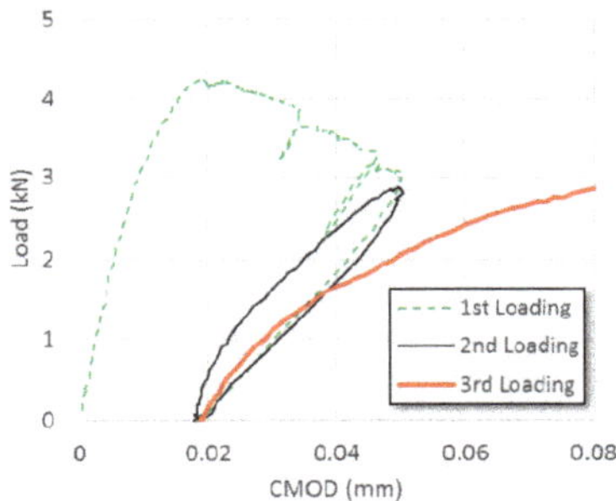

Figure 2. Typical Load-CMOD (crack mouth opening displacement) curves (Case: 7-28-A).

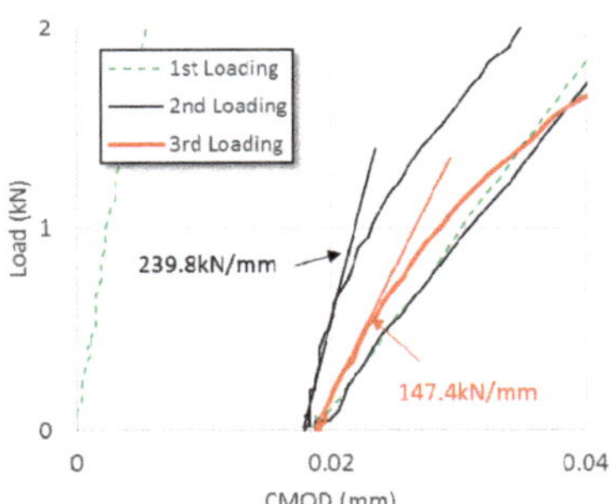

Figure 3. Initial flexural stiffness of the 2nd and 3rd loading.

3.3. Neutral Axis Estimation

The strain distribution of the area in which crack propagation was expected to occur was investigated by the attached successive strain gauges. In the preliminary experiment, seven strain gauges were attached at intervals of 10 mm from 5 mm above the notch end, as shown in Figure 4, and the vertical strain distribution in the lower part of the load point was obtained, as shown in Figure 5. The crack induced at the end of the notch can obviously be expected to advance past this part in the three-point bending test with a notch. The strain showed a relatively linear distribution under the initial low load but it gradually changed to a nonlinear distribution with increasing load. The nonlinearity of the strain distribution may be ascribed to stress concentration due to the notch and tensile softening of the tensile part. The exact stress distribution could not be estimated directly from this strain distribution, even if the elastic modulus were to be known, because the stress–strain relationship of the tensile failure part remains unknown. However, it is known that in parts wherein the strain is zero, the stress is zero as well. In other words, the position of the neutral axis can be estimated.

Figure 4. Seven strain gauges for investigation of the vertical strain distribution.

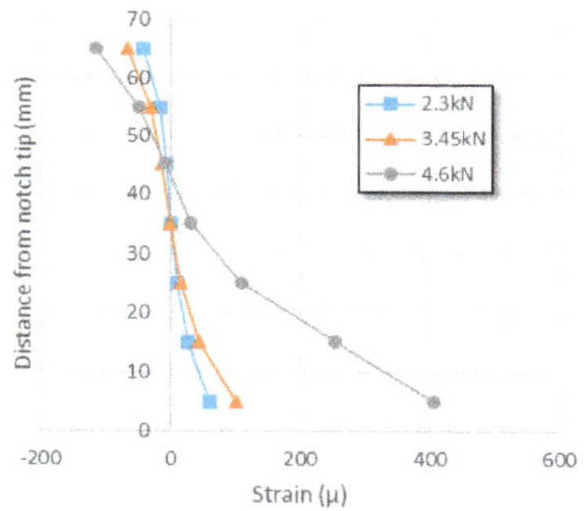

Figure 5. Strain distribution with different loads.

The neutral axis clarifies the quantification of the healing effect. The neutral axis is probably located at the center of the cross-section at the beginning of the loading. It moves along with the crack's propagation. In the bending test, the crack propagation is limited to the tensile side below the neutral axis. This indicates that the neutral axis explicitly indicates the area wherein self-healing is possible. Meanwhile, the neutral axis does not move without further crack propagation, so it is expected to not move in the 2nd loading conducted just after the crack inducement (1st loading). Thereafter, if this stress neutral axis changes after the healing period, such change can be ascribed to the self-healing effect as a physical recovery. The movement of the neutral axis depends on the physical recovery, and it is relatively unaffected by changes in material properties owing to the age effect.

The position of the neutral axis was estimated by seven successive strains gauges installed at intervals of 10 mm on the specimen in this study. Physical measurements at intervals less than 10 mm were not possible, and the interval selected herein seemed to be sufficient to reflect any changes in strain in the entire area. A few scattered data of the strain gauge attached underneath the loading point were detected, but the neutral area showed continuous data. The neutral axes were estimated by quadratic polynomial regression of the strain results obtained from the seven strain gauges at each point in time, as shown in Figure 6 and Table 3. As can be seen, the neutral axis moves upward the cross-section as the load increases and the crack propagates.

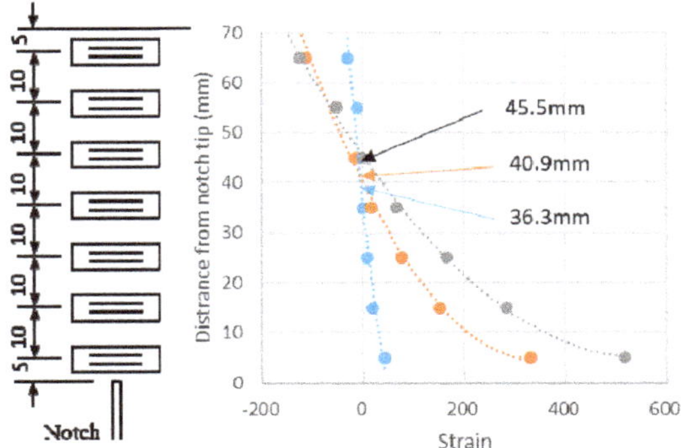

Figure 6. Estimation of a neutral axis.

Table 3. Estimated neutral axis with different load level.

CMOD (mm)		0.005	0.026	0.034
Load (kN)		1.70	4.53	4.55
Measured strain (μ)	5 mm	43.6	330.8	518.5
	15 mm	19.0	153.4	284.1
	25 mm	8.5	76.8	166.8
	35 mm	0.9	18.0	66.9
	45 mm	−5.7	−15.2	1.0
	55 mm	−11.4	−50.2	−51.2
	65 mm	−29.4	−111.8	−125.1
Neutral axis		36.3	40.9	45.5

4. Results and Discussion

4.1. Behavior Change after Healing

The behaviors in the 1st loading and the 2nd loading were almost equivalent, regardless of the case. Figure 7 shows the typical curves of the specimens with different healing periods, and Figure 8 shows the typical curves of the specimens with different crack inducement times. The dotted line indicates the crack inducement process, and the solid line and the thick solid line indicate the 2nd loading and the 3rd loading, respectively. The peak load increased with age, and the average peak loads were 4.5 kN on the 7th day, 4.9 kN on the 28th day, and 5.7 kN on the 49th day. The load at the unloading point increased with age as well. Therefore, the average ratio of these two values was almost similar: 0.86 on the 7th day, 0.88 on the 28th day, and 0.88 on the 49th day. These results indicate that the damage or the crack inducement due to the 1st loading was similar, regardless of age. The curves were drawn downward to the left during unloading, and the residual CMODs averaged around 0.017, regardless of the case. The 2nd loading was performed as soon as the unloading was completed, and the initial flexural stiffness of the 2nd loading was gentler (smaller) than that of the 1st loading. These two values were very different for each specimen; hence, the ratio of these two values was not constant. The 2nd loading was conducted as the CMOD reached 0.05 mm again. The maximum load of the 2nd loading was always less than the load at the unloading point, and the average ratio of these two values was 0.96, regardless of the case. The unloading curve of the 2nd loading followed the traces of the unloading curve of the 1st loading, and the residual CMOD remained almost unchanged. Table 4 shows the average values of the cases.

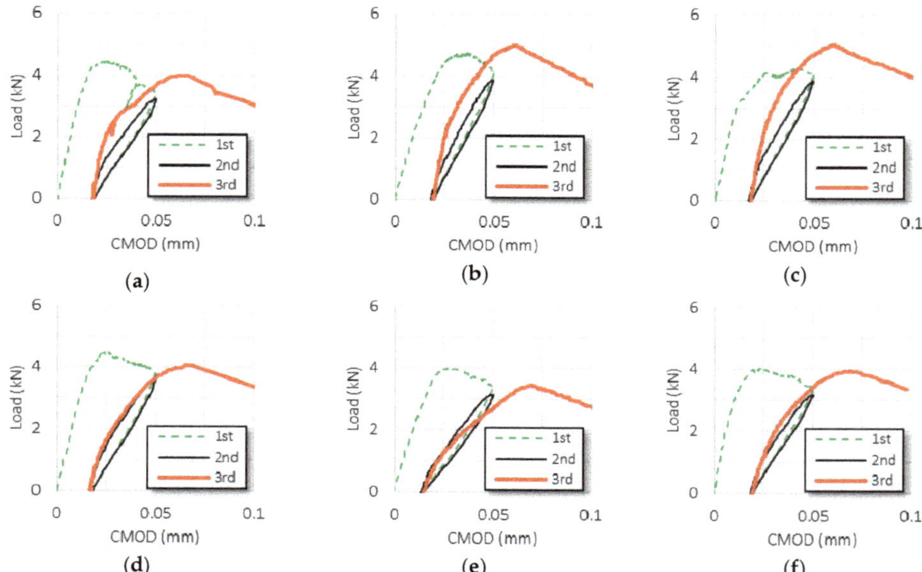

Figure 7. Typical load-CMOD curves for different healing periods: (**a**) 7-28W; (**b**) 7-49W; (**c**) 7-70W; (**d**) 7-28A; (**e**) 7-49A; (**f**) 7-70A.

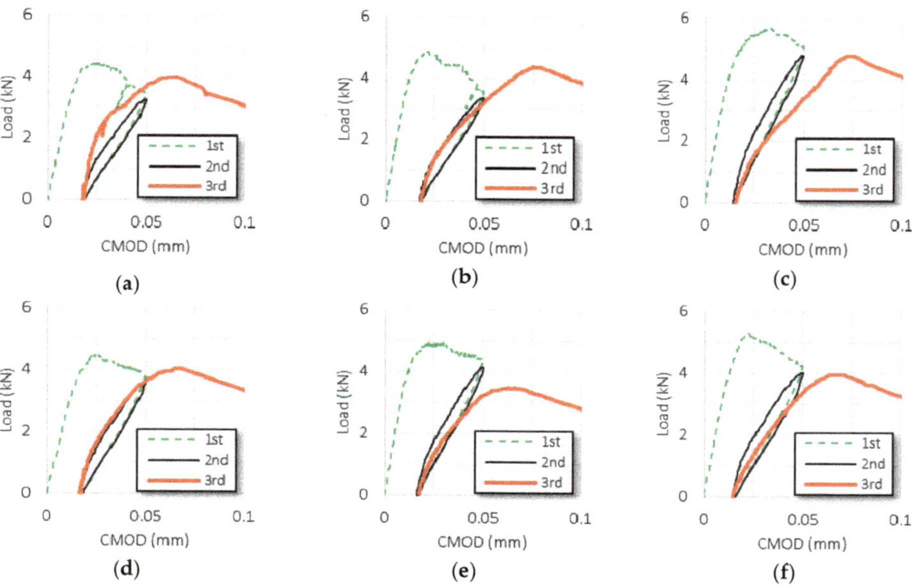

Figure 8. Typical load-CMOD curves for different crack inducement times: (**a**) 7-28W; (**b**) 28-49W; (**c**) 49-70W; (**d**) 7-28A; (**e**) 28-49A; (**f**) 49-70A.

The behaviors associated with the 3rd loading after healing were different for each case. They could be classified into two main groups. One group of specimens seemed to be healed; the initial flexural stiffness associated with the 3rd loading was steeper than that associated with the 2nd loading, and the maximum load in case of the 3rd loading was higher than the load at the unloading point. These phenomena could be ascribed to the effect of healing. However, another group of specimens seemed to be not healed; in the cases of specimens from this group, the initial flexural stiffness was gentle, and the maximum load was low. As such, the initial flexural stiffness and the maximum load seemed to be related closely, as discussed separately in the subsequent sections.

Table 4. Average load and ratio.

Case	7-28		7-49		7-70		28-49		49-70	
Healing condition	Water	Air	Water	Air	Water	Air	Water	Air	Water	Air
ⓐ Peak load of 1st loading	4.48	4.43	4.61	4.46	4.85	4.26	4.88	4.92	5.83	5.63
ⓑ Unloading point	3.83	3.70	4.19	3.72	4.36	3.63	4.25	4.38	5.21	4.91
ⓒ Maximum load of 3rd loading	4.46	3.88	5.01	3.88	5.08	4.20	4.32	4.26	4.83	4.53
ⓑ/ⓐ	0.85	0.83	0.91	0.84	0.90	0.85	0.87	0.89	0.89	0.87
ⓒ/ⓓ	1.16	1.04	1.19	1.04	1.16	1.17	1.03	0.98	0.93	0.92

4.2. Self-Healing Assessment Based on the Flexural Strength

The maximum load in case of the 3rd loading and the load at the unloading point of the same specimen were compared. Figure 9 shows the average ratios of cases for different healing periods: the blue square box denotes the results of the specimens cured in water, and the red triangle denotes the results of the specimens cured in air. As the figure indicates, all ratios are more than 1.0, which means the maximum load in case of the 3rd loading was higher than the load at the unloading point. This phenomenon in which the peak load of the 3rd loading after the healing process is higher than the previous peak load has been reported in previous studies [23,36]. The average ratio appears to increase as the age increases, but it does not increase considerably.

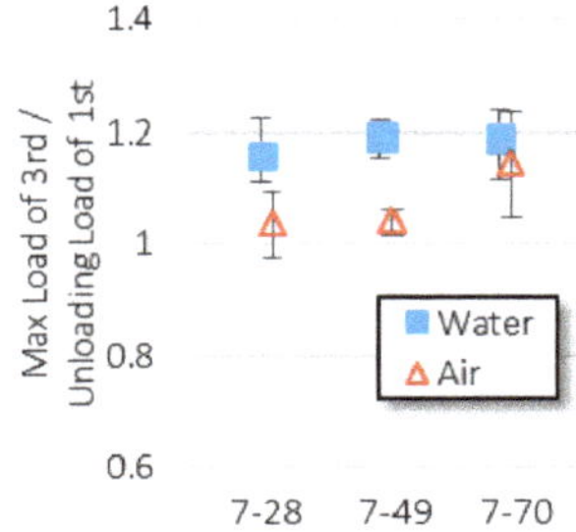

Figure 9. Ratios between the maximum load in the case of 3rd loading and unloading load in case of 1st loading (different healing periods).

The loads of the 3rd loading and the 2nd loading at CMOD = 0.05 mm of the same specimen were compared. The maximum load of the 3rd loading must be accompanied by the age effect because it involves the uncracked part as well. To exclude the age effect to the extent as much as possible, the loads under the same damage conditions must be compared. Figure 10 shows the average ratios of the 2nd loading and the 3rd loading, and these load values were obtained when CMOD was 0.05 mm. In this comparison, the increasing trend was more evident when the aging period was longer, especially in the case of a specimen cured in air. However, the ratios of the cases 7-28-A and 2-49-A show values than lower 1.0; the maximum load after healing in air was lower than the load at the unloading point.

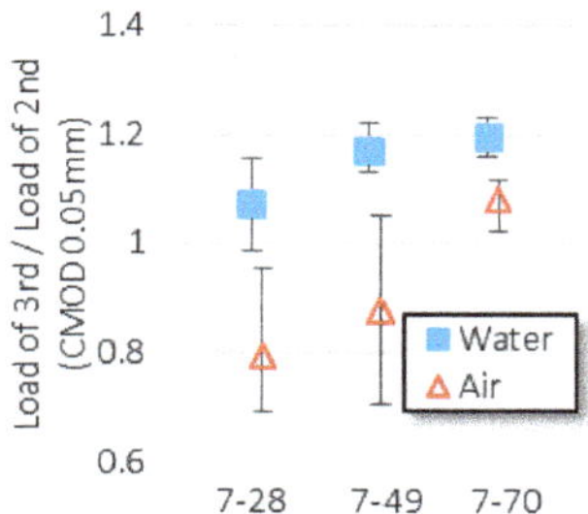

Figure 10. Ratios between the load of the 3rd loading and the load of the 2nd loading with CMOD = 0.05 mm (different healing periods).

Figure 11 shows the average ratios of the maximum load of the 3rd loading and the load at the unloading point with different crack inducement times. As can be seen, the ratios decrease as the crack inducement time increases, regardless of the curing condition. However, the average ratios of the 3rd loading and the 2nd loading at CMOD = 0.05 mm do not show any significant decrease trend, as in Figure 12.

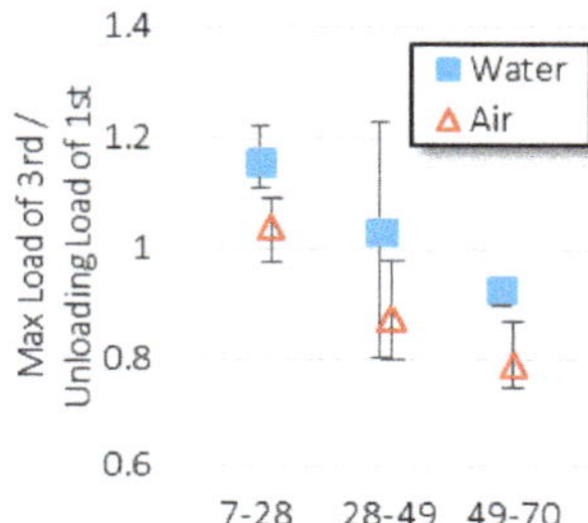

Figure 11. Ratios between the maximum load of the 3rd loading and the unloading load of the 1st loading (different crack inducement times).

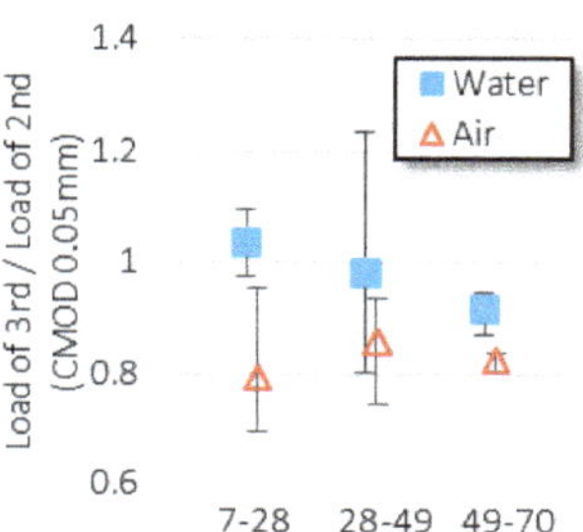

Figure 12. Ratios between load of 3rd loading and load of 2nd loading with CMOD = 0.05 mm (different crack inducement time).

Compared to the results in Figures 9 and 11, the results in Figures 10 and 12 show more clearly the changes in the healing conditions and periods. A clearer difference is thought to be due to comparing flexural strength under certain conditions as the same CMOD, which is different from similar previous studies [23,36]. Three-point bending test with a notch is considered to be more advantageous in controlling comparative conditions than experiments of other criteria such as permeability or chloride transport [32,41–43].

4.3. Self-Healing Assessment Based on Initial Flexural Stiffness

A comparison of the initial flexural stiffness between the 2nd loading and the 3rd loading directly and quantitatively showed the changes before and after healing. The reloading after the crack inducement process degraded the performance of the specimen, even if the crack did not propagate further, because the microcracking zone accumulated damage and gradually turned into a bridging zone. Both the maximum strength and the initial flexural stiffness in case of the 2nd loading decreased slightly as a result of this damage accumulation. These two parameters varied depending on the healing period and the crack inducement time, and the self-healing assessment was quantified by comparing the degree of variation. As described above, the maximum strength clearly reflects not only the healing effect but also the age effect, so only the initial flexural stiffness was compared. The ratio of the slope of the 3rd loading to the slope of the 2nd loading could be expected to be greater than 1.0 if there was a positive effect during the healing period. A value less than or equal to 1.0 means that there was no positive effect, not even the age effect. This value was considered to represent the healing effect in this study and used as a quantitative marker in self-healing assessments.

The initial flexural stiffness in case of the 3rd loading increased as the length of the healing period increased. For curing in freshwater, in Cases 7-28-W, 7-49-W, and 7-70-W, it can be seen that all the initial flexural stiffness values in case of the 3rd loading are steeper than those in case of the 2nd loading, as shown in Figure 7a–c; these results represent a positive effect. However, in Cases 7-28-A and 7-49-A, with healing in air, the slope in case of the 3rd loading decreases. Only in Case 7-70-A, in which the specimen was healed for 63 days, a positive effect was observed. This finding is similar to the results of Granger et al. [36], but a direct comparison of the same specimen before and after the healing makes our results more concrete. This is evident from the quantitative comparison of the slopes, as shown in Figure 13. Although there are deviations, the average healing effect increased with increasing healing time, especially the healing effect was greater than 1.0 in the cases of all specimens cured in water. The healing effect in case of the specimens cured in air was lower than 1.0 for 21 days of healing period (7-28), but it increased remarkably and reached a level similar to that of the specimens cured in water for 63 days (7–70). These results indicate that the healing effect depends on the healing period, regardless of the healing circumstance.

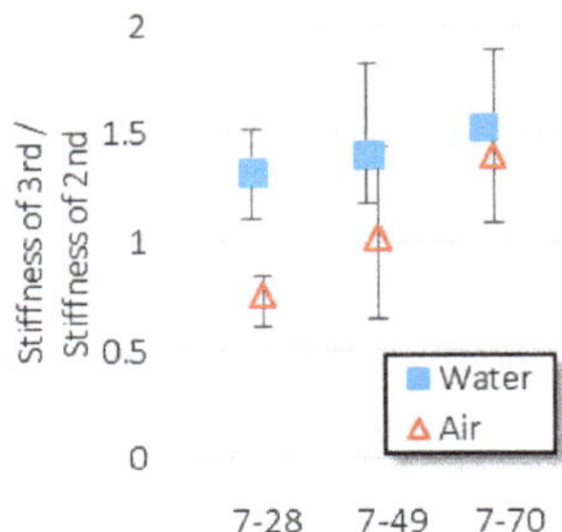

Figure 13. Ratios between flexural stiffness of 3rd loading and that of 2nd loading (different healing periods).

The initial flexural stiffness in case of the 3rd loading decreased with delayed crack inducement time. The curve of Case 7-28-W shows a positive effect, but the curves of Cases 28-49-W and 49-70-W are below the 2nd loading curve despite the fact that the specimens were cured in water. Furthermore, all specimens cured in air showed no positive effect, i.e., the 3rd loading curves are below the 2nd loading curves. Figure 14 shows the quantitative changes according to the crack inducement time. Despite curing in water, the healing effect decreased significantly as the crack inducement time was delayed. A more unexpected phenomenon can be seen in the results of specimens cured in air; all average healing effects with different crack inducement times were almost the same and less than 1.0.

These results indicate that the healing effect is strongly influenced by the crack inducement time and the curing circumstance if the healing period is constant.

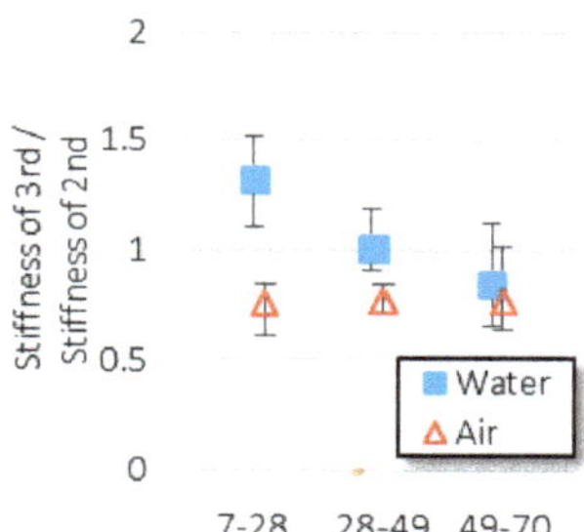

Figure 14. Ratios between flexural stiffness of 3rd loading and that of 2nd loading (different crack inducement times).

The results of the flexural strength comparison and the initial flexural stiffness comparison are similar if we look at the results themselves. However, considering that healing occurs primarily around the end of a crack in self-healing of concrete, it is deemed more reasonable to compare the stiffness of the moment when the healed crack is reopened than to compare the strength, which is the total ability of the specimen. Both the flexural strength and the initial flexural stiffness healed to a greater extent in case of the specimens cured in water. Regardless of the healing period and the crack inducement time, the healing effects were stronger in cases of the specimens cured in water compared to those in cases of the specimens cured in air. Although in a few cases it was observed that the specimens cured in air could be healed if the healing period was sufficiently long, the water condition was better for the self-healing of concrete. The healing tendency according to the healing condition is consistent with the results of many other studies [14–33,36,37,41,43]. It is a reasonable fact that the underwater environment is advantageous for self-healing of concrete. Importantly, these results were similar when obtained in the environment of the hydration reaction. The hydration reaction is time-dependent, and the longer the reaction time, the more effective is the reaction. However, the strength of the reaction decreases noticeably over time after casting. The results of the healing effects obtained in this study and those of the hydration reaction share a few commonalities. Consequently, the healing effect observed in this study can be said to be highly related to the re-hydration reaction, which is one of the two major mechanisms of self-healing of concrete.

4.4. Crack Depths and Neutral Axises

The real crack depth with CMOD = 0.05 mm was observed to be about 20 mm. It was observed by means of an ink injection after the crack inducement, and several surplus specimens of seven days of age were used for this purpose because the largest number of specimens was prepared for the seven cases. A cellophane adhesive tape was attached to both sides of the specimens to prevent ink leakage, and red ink was injected from the notch. Next, after sufficient time provision for ink injection and drying (24 h), reloading was applied to split the specimens. Figure 15 shows the cracked area stained by red ink injection. The crack depth was calculated by dividing the area stained with red ink by the specimen width, and the average crack depth of the specimens of seven days of age was about 20 mm. Five additional specimens of 28 days of age showed a crack depth of about 20 mm as well, although the deviation in their crack depths was larger than that in case of the seven-day-old specimens.

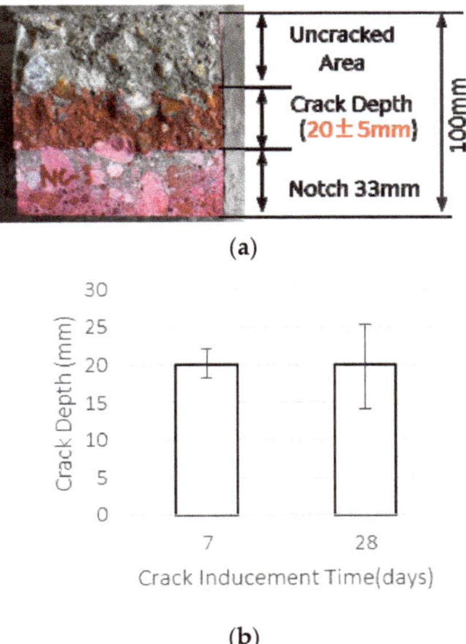

(a)

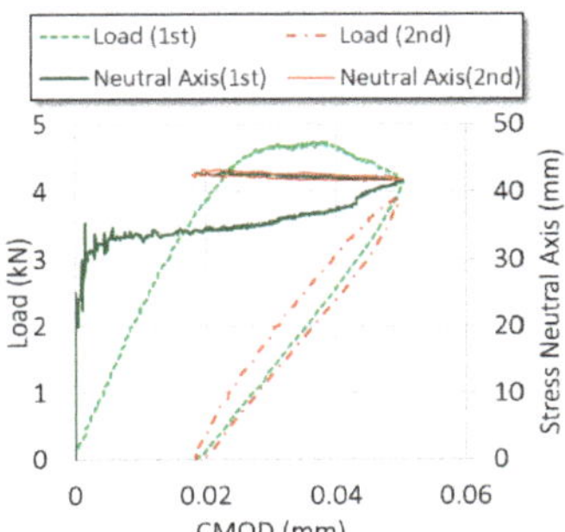

(b)

Figure 15. Investigation of crack depth by ink injection. (a) Stained crack part; (b) measured crack depth.

Meanwhile, the neutral axis ascended to about 44 mm. Figure 16 shows the change in the position of the neutral axis with changes in the CMOD. The neutral axis was estimated from seven strains measured by successive strain gauges, so slight scattering can be seen at the beginning of load introduction. Initially, the neutral axis was located at about 33 mm from the notch tip, which corresponds to the center of the un-notched section. It moved upward gradually with increasing CMOD and was located at 40–50 mm at CMOD = 0.05 mm. It changed insignificantly during the unloading process, but in the case of a few specimens, the neutral axis moved slightly upward. There was a minor difference among each of the specimens, and the differences in the position of the neutral axis when fully unloaded averaged 44 mm. The movement of the neutral axis in the 2nd loading almost followed the trail of the unloading, and it could not be distinguished. This result confirms that the 2nd loading did not induce additional crack propagation.

Figure 16. The behavior of the neutral axis with CMOD.

Accordingly, there was a difference of about 24 mm between the neutral axis and the actual crack depth when CMOD = 0.05 mm. Nanakorn and Horii mentioned that the fracture process zone of concrete comprises the microcracking zone and the bridging zone [44]. In the microcracking zone, the concentrated stress near the crack tip causes the existing defects and pores to expand. Moreover, the interface between aggregates and the cement matrix is weakened, and the two components began to separate. These deteriorations develop into the bridging zone, and the stress in this area starts to dissipate. This stress dissipation in the bridging zone is expressed as the tension-softening relationship of concrete. A fully separated section is considered a stress-free crack. The stress-free crack area was obviously stained with the injected red ink, and a portion of the bridging zone was also likely stained. Consequently, it can be expected that the microcracking zone and the bridging zone are located at about 24 mm between the neutral axis and the actual crack depth, as shown in Figure 17.

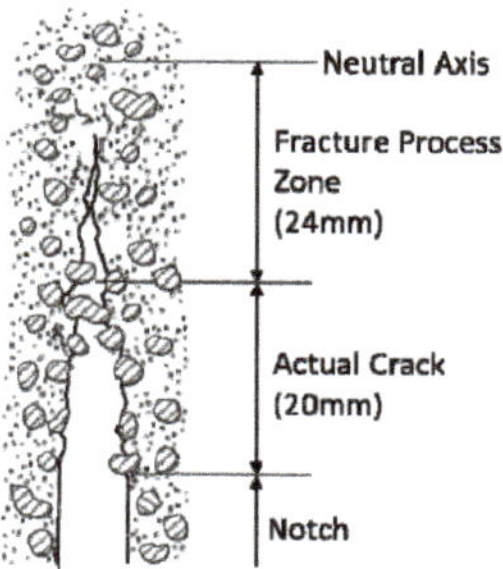

Figure 17. The concept of actual crack depth and neutral axis.

The neutral axis of the 3rd loading followed the neutral axis trace of the 2nd loading almost equally, regardless of the case. Figure 18 shows the typical stress-neutral axis curves of the specimens with different healing periods, and Figure 19 shows the typical stress-neutral axis curves of the specimens with different crack inducement times. Compare to Case 7-70-W (Figure 18c) and Case 49-70-A (Figure 19f), which are the strongest and the weakest healing effects in terms of the flexural strength and the initial flexural stiffness. As can be seen, the changes in the neutral axis in both cases are almost the same as that in the case of the 2nd loading until CMOD = 0.05 mm. Similar results were obtained in all other cases as well. This means there was no physical recovery to the extent that the neutral axis changed. In order words, in all cases, the cracks once opened were never reattached, even over a long healing period.

Consequently, the healing effect seems to have occurred between the neutral axis and the opened crack area, as opposed to over the opened crack area. In the introduction, we expected that recovery would occur through self-healing, as shown in Figure 20b. However, in practice, a certain area from the crack tip is closed, and this part is attached again, which restores the stiffness of the specimen. As a result, the neutral axis must move downward. However, the neutral axis shows no sign of movement, which means stiffness recovery did not occur in the once-opened cracked part. The initial flexural stiffness increased or decreased depending on the healing condition, and it can be stated confidently that this phenomenon depended on the area between the neutral axis and the crack tip, as opposed to the opened crack part, as shown in Figure 20c. In other words, the healing effect was limited to the fracture process zone, and the once-opened crack area did not recover physically.

The changes in the neutral axis can explain the dependence on the re-hydration reaction. Contrary to expectations outlined in the introduction, the neutral axis of the 3rd loading did not show any signs of healing in the once-opened crack area, and this indicated that physical recovery did not occur. The precipitation of calcium hydroxide, which is another of the two major mechanisms of self-healing and can be expected to occur in the opened crack area, was not helpful for restoring the physical properties of the opened crack area.

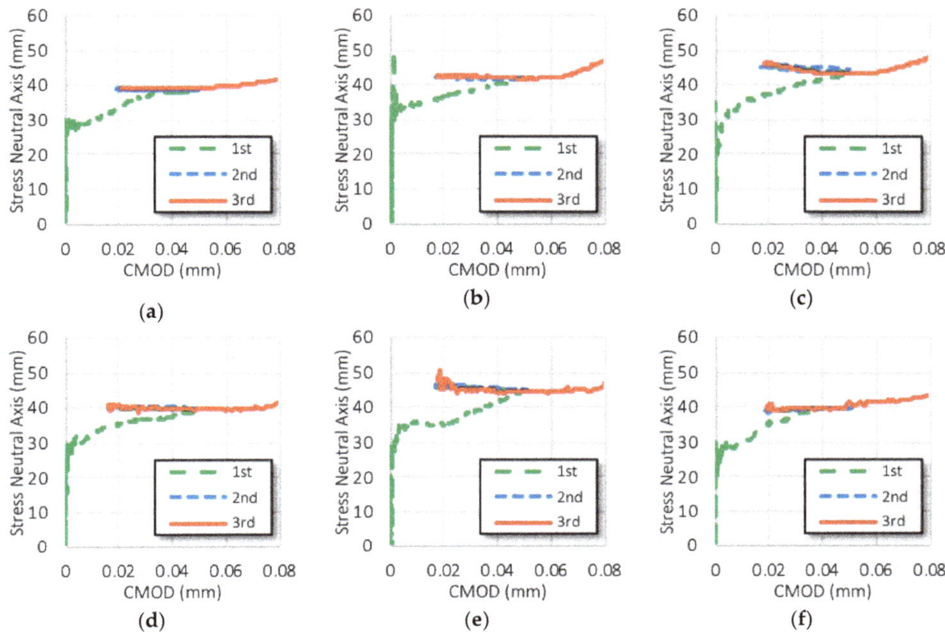

Figure 18. Typical stress neutral axis-CMOD curves for different healing periods. (**a**) 7-28W; (**b**) 7-49W; (**c**) 7-70W; (**d**) 7-28A; (**e**) 7-49A; (**f**) 7-70A.

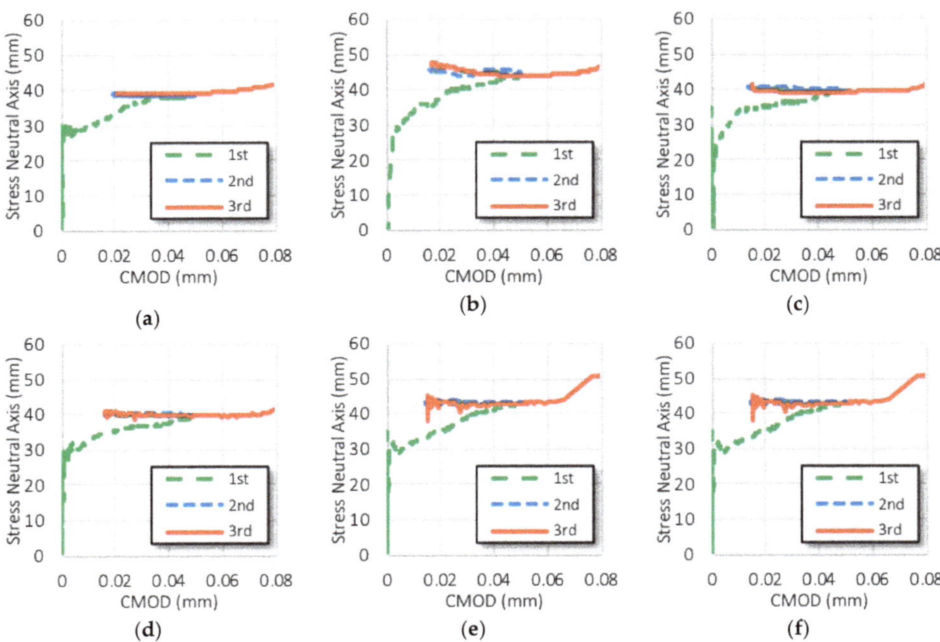

Figure 19. Typical stress neutral axis-CMOD curves for different crack inducement times. (**a**) 7-28W; (**b**) 28-49W; (**c**) 49-70W; (**d**) 7-28A; (**e**) 28-49A; (**f**) 49-70A.

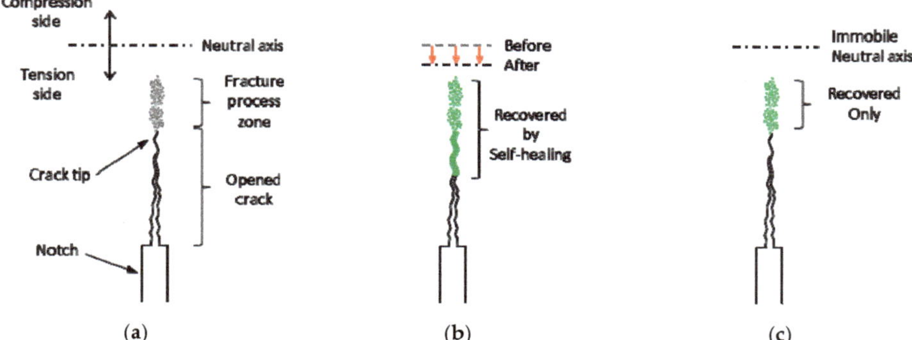

Figure 20. Schematic of neutral axis movement before and after healing. (**a**) Unloaded state; (**b**) assumed neutral axis movement; (**c**) measured result.

5. Conclusions

In this study, the three-point bending test with a notch was performed for assessing the self-healing of concrete, and the neutral axis was estimated to investigate the curable area by self-healing. Our results are as follows:

(1)   Within the scope of this study, the longer the healing period, the higher the healing effect. In the same healing period (21-day), the later the crack was induced, the less effective the healing effect was. These results are generally consistent with previous studies.

(2)   The method for estimating the neutral axis using successive strain gauges attached to the crack propagation part was proposed in this study. The estimated neutral axis showed proper behavior in comparison with Load-CMOD curve; the neutral axis moved slightly upward with crack propagation, but there was an insignificant move when the crack was closed or reopened.

(3)   Recovery of the flexural strength and the initial flexural stiffness before and after healing was confirmed, but no major fluctuations in the neutral axis were confirmed. Consequently, the curable area appears to be limited to the area between the neutral axis and the crack tip, as opposed to over the opened crack area.

The conclusions were judged to be valid within the scope of this study. We noted that all results in this study show a certain trend while data is statistically insufficient. It is considered that further experiments will produce more significant results.

Author Contributions: Conceptualization, C.K. and T.K.; methodology, T.K.; software, C.K.; validation, C.K. and T.K.; formal analysis, C.K.; investigation, T.K.; writing—original draft preparation, C.K.; writing—review and editing, T.K.; visualization, T.K.; supervision, T.K.; project administration, C.K.; funding acquisition, C.K.

Funding: This research was funded by the Basic Science Research Program through the National Research Foundation of Korea(NRF) funded by the Ministry of Education, grant number NRF-2017R1D1A1B03034470.

Conflicts of Interest: The funders had no role in the design of the study; in the collection, analyses, or interpretation of data; in the writing of the manuscript, and in the decision to publish the results.

References

1.   Aguado, A.; Gettu, R.; Shah, S. Concrete technology: New trends, industrial applications. In *Proceedings of the International RILEM Workshop*; CRC Press: Boca Raton, FL, USA, 1994.

2.   Zhang, L.X.B.; Hsu, T.T. Behavior and analysis of 100 MPa concrete membrane elements. *J. Struct. Eng.* **1998**, *124*, 24–34. [CrossRef]

3.   Aïtcin, P.C. *High Performance Concrete*; CRC Press: Boca Raton, FL, USA, 2011.

4. Kim, C.S.; Park, H.G.; Chung, K.S.; Choi, I.R. Eccentric axial load testing for concrete-encased steel columns using 800 MPa steel and 100 MPa concrete. *J. Struct. Eng.* **2011**, *138*, 1019–1031. [CrossRef]

5. Caldarone, M.A.; Gruber, K.A.; Burg, R.G. High reactivity metakaolin (HRM): A new generation mineral admixture for high performance concrete. *Concr. Int.* **1994**, *16*, 37–41.

6. Hassan, K.E.; Cabrera, J.G.; Maliehe, R.S. The effect of mineral admixtures on the properties of high-performance concrete. *Cem. Concr. Compos.* **2000**, *22*, 267–271. [CrossRef]

7. Poon, C.S.; Kou, S.C.; Lam, L. Compressive strength, chloride diffusivity and pore structure of high performance metakaolin and silica fume concrete. *Constr. Build. Mater.* **2006**, *20*, 858–865. [CrossRef]

8. Shi, H.S.; Xu, B.W.; Zhou, X.C. Influence of mineral admixtures on compressive strength, gas permeability and carbonation of high performance concrete. *Constr. Build. Mater.* **2009**, *23*, 1980–1985. [CrossRef]

9. Barbudo, A.; de Brito, J.; Evangelista, L.; Bravo, M.; Agrela, F. Influence of water-reducing admixtures on the mechanical performance of recycled concrete. *J. Clean. Prod.* **2013**, *59*, 93–98. [CrossRef]

10. Frangopol, D.M.; Lin, K.Y.; Estes, A.C. Life-cycle cost design of deteriorating structures. *J. Struct. Eng.* **1997**, *123*, 1390–1401. [CrossRef]

11. Frangopol, D.M.; Liu, M. Maintenance and management of civil infrastructure based on condition, safety, optimization, and life-cycle cost. *Struct. Infrastruct. Eng.* **2007**, *3*, 29–41. [CrossRef]

12. Kim, K.W.; Yun, S.H. A case study of life cycle cost analysis on apartment houses and Han-Ok. *J. Korea Inst. Build. Constr.* **2010**, *10*, 1–6. [CrossRef]

13. Guggemos, A.A.; Horvath, A. Comparison of environmental effects of steel-and concrete-framed buildings. *J. Infrastruct. Syst.* **2005**, *11*, 93–101. [CrossRef]

14. White, S.R.; Sottos, N.R.; Geubelle, P.H.; Moore, J.S.; Kessler, M.R.; Sriram, S.R. Autonomic healing of polymer composites. *Nature* **2001**, *409*, 794. [CrossRef]

15. Brown, E.N.; Sottos, N.R.; White, S.R. Fracture testing of a self-healing polymer composite. *Exp. Mech.* **2002**, *42*, 372–379. [CrossRef]

16. Mookhoek, S.D.; Blaiszik, B.J.; Fischer, H.R.; Sottos, N.R.; White, S.R.; Van Der Zwaag, S. Peripherally decorated binary microcapsules containing two liquids. *J. Mater. Chem.* **2008**, *18*, 5390–5394. [CrossRef]

17. Dry, C. Three design for the internal release of sealants, adhesives and waterproofing chemical into concrete to release. *Cem. Concr. Res.* **2000**, *30*, 1969–1977. [CrossRef]

18. Joseph, C.; Jefferson, A.D.; and Cantoni, M.B. Issues relating to the autonomic healing of cementitious materials. In Proceedings of the 1st International Conference on Self-Healing Materials, Noordwijk aan Zee, The Netherlands, 18–20 April 2007; pp. 1–8.

19. Joseph, C.; Jefferson, A.D.; Isaacs, B.; Lark, R.; Gardner, D. Experimental investigation of adhesive-based self-healing of cementitious materials. *Mag. Concr. Res.* **2010**, *62*, 831–843. [CrossRef]

20. Sun, L.; Yu, W.Y.; Ge, Q. Experimental research on the self-healing performance of micro-cracks in concrete bridge. *Adv. Mater. Res.* **2011**, *250*, 28–32. [CrossRef]

21. van Breugel, K. Is there a market for self-healing cement based materials? In Proceedings of the 1st International Conference on Self Healing Materials, Noordwijk aan Zee, The Netherlands, 18–20 April 2007; pp. 1–9.

22. van Breugel, K. Self-healing en vloeistofdichtheid. *Cement* **2003**, *7*, 85–89.

23. Schlangen, E.; Heide, N.; van Breugel, K. Crack healing of early age cracks in concrete. In *Measuring, Monitoring and Modeling Concrete Properties*; Konsta-Gdoutos, M.S., Ed.; Springer: Dordrecht, The Netherlands, 2006; pp. 273–284.

24. Kim, W.J.; Kim, S.T.; Park, S.J.; Ghim, S.Y.; Chun, W.Y. A study on the development of self-healing smart concrete using microbial biomineralization. *J. Korea Concr. Inst.* **2009**, *21*, 501–511. [CrossRef]

25. Li, V.C.; Stang, H.; Krenchel, H. Micromechanics of crack bridging in fibre-reinforced concrete. *Mater. Struct.* **1993**, *26*, 486–494. [CrossRef]

26. Stang, H.; Li, V.C.; Krenchel, H. Design and structural applications of stress-crack width relations in fibre reinforced concrete. *Mater. Struct.* **1995**, *28*, 210–219. [CrossRef]

27. Kunieda, M.; Choonghyun, K.; Ueda, N.; Nakamura, H. Recovery of protective performance of cracked ultra high performance-strain hardening cementitious composites (UHP-SHCC) due to autogenous healing. *J. Adv. Concr. Technol.* **2012**, *10*, 313–322. [CrossRef]

28. Williams, G.; Trask, R.; Bond, I. A self-healing carbon fibre reinforced polymer for aerospace applications. *Compos. Part A Appl. Sci. Manuf.* **2007**, *38*, 1525–1532. [CrossRef]

29. Homma, D.; Mihashi, H.; Nishiwaki, T. Self-healing capability of fiber reinforced cementitious composites. *J. Adv. Concr. Technol.* **2009**, *7*, 217–228. [CrossRef]

30. Hosoda, A.; Kishi, T.; Arita, H.; Takakuwa, Y. Self healing of crack and water permeability of expansive concrete. In Proceedings of the 1st International Conference on Self Healing Materials, Noordwijk aan Zee, The Netherlands, 18–20 April 2007.

31. Jacobsen, S.; Sellevold, E.J. Self healing of high strength concrete after deterioration by freeze/thaw. *Cem. Concr. Res.* **1996**, *26*, 56–62. [CrossRef]

32. Jacobsen, S.; Marchand, J.; Boisvert, L. Effect of cracking and heating on chloride transport in OPC concrete. *Cem. Concr. Res.* **1996**, *26*, 869–881. [CrossRef]

33. Wang, G.; Guo, Z.Q.; Yu, J.; Shui, Z. Effect of self-healing materials on steel reinforcement corrosion of concrete. In Proceedings of the First International Conference on Self Healing Materials, Noordwijk aan Zee, The Netherlands, 18–20 April 2007.

34. Barr, B.I.; Lee, M.K.; de Place Hansen, E.J.; Dupont, D.; Erdem, E.; Schaerlaekens, S.; Schnütgen, B.; Stang, H.; Vandewalle, L. Round-robin analysis of the RILEM TC 162-TDF beam bending test: Part 3—Fibre distribution. *Mater. Struct. Mat. Constr.* **2003**, *36*, 631–635. [CrossRef]

35. *JCI-S-001 Method of Test for Fracture Energy of Concrete by Use of Notched Beam*; JCI Standards for Test of FRC; Japan Concrete Institute: Tokyo, Japan, 2003; pp. 1–13.

36. Granger, S.; Loukili, A.; Pijaudier-Cabot, G.; Chanvillard, G. Experimental characterization of the self-healing of cracks in an ultra high performance cementitious material, Mechanical test and acoustic emission analysis. *Cem. Concr. Res.* **2007**, *37*, 519–527. [CrossRef]

37. Matsushita, H.; Sue, Y.; Kiyosaki, R. A study of the strength recovery of mortar with an initiated crack. *Proc/JCI* **2003**, *25*, 605–610.

38. TC162-TDF, RILEM. Test and design methods for steel fibre reinforced concrete: Bending test. *Mater. Struct.* **2000**, *33*, 3–5. [CrossRef]

39. Jenq, Y.; Shah, S.P. Two parameter fracture model for concrete. *J. Eng. Mech.* **1985**, *111*, 1227–1241. [CrossRef]

40. Shah, S.P. Determination of fracture parameters (KIcs and CTODc) of plain concrete using three-point bend tests. *Mater. Struct.* **1990**, *23*, 457–460. [CrossRef]

41. Reinhardt, H.W.; Joosss, M. Permeability and self-healing of cracked concrete as a function of temperature and crack width. *Cem. Concr. Res.* **2003**, *33*, 981–985. [CrossRef]

42. Edvardsen, C. Water permeability and autogenous healing of cracks in concrete. *Mater. J.* **1999**, *96*, 448–454.

43. Yamada, K. Crack self healing properties of expansive concretes with various cements and admixture. In Proceedings of the 1st International Conference on Self Healing Materials, Noordwijk, The Netherlands, 18–20 April 2007.

44. Nanakorn, P.; Horii, H. A finite element with embedded displacement discontinuity. *Build. 21st Century* **1995**, *1*, 33–38.

 *applied sciences*

Article

Research on the Blow-Off Impulse Effect of a Composite Reinforced Panel Subjected to Lightning Strike

Senqing Jia, Fusheng Wang *, Weichao Huang and Bin Xu

School of Mechanics, Civil Engineering and Architecture, Northwestern Polytechnical University, Xi'an 710129, China; sqjia@mail.nwpu.edu.cn (S.J.); HWC2018201315@163.com (W.H.); xubin@nwpu.edu.cn (B.X.)
* Correspondence: fswang@nwpu.edu.cn; Tel.: +86-29-8843-1000

Received: 24 January 2019; Accepted: 13 March 2019; Published: 19 March 2019

Abstract: The blow-off impulse effect of a composite reinforced panel subjected to lightning strike is studied combing electric-thermal coupling with explicit dynamic methods. A finite element model of a composite reinforced panel is established under the action of 2.6/10.5 µs impulse current waveform with current peak 60 kA. Blow-off impulse elements are selected according to numerical results of electric-thermal coupling analysis. Elements failure, pressure, and von Mises stress distribution are discussed when blow-off impulse analysis is completed. The results show that the blow-off impulse effect can alter the damage forms of a composite reinforced panel and causes the damage distribution to deviate from the initial fiber direction in each layer. Elements failure modes around the blow-off impulse area are similar to that around the attachment area of the lightning strike. The blow-off impulse effect can well model the internal damage, concave pit, and bulge phenomenon around the attachment area. Additionally, pressure contours are not presented as an anisotropic characteristic but an isotropic characteristic under the blow-off impulse effect, which indicates that the mechanical behavior of composite materials presents as an anisotropic characteristic in low pressure while as an isotropic characteristic in high pressure. This method is suitable to evaluate shock damage of a composite reinforced panel induced by lightning strike.

Keywords: lightning strike; composite reinforced panel; blow-off impulse; electric-thermal coupling

1. Introduction

With the rapid development of aircraft industry, carbon fiber/epoxy reinforced composite materials have been widely used in aircraft design in recent decades for its advantages such as lightweight, high specific modulus, high specific strength and designability, etc. However, composite materials have poor electric and thermal conductivity compared with traditional metal materials such as aluminum alloy and titanium alloy, which make aircraft structures more vulnerable to catastrophic damage in a lightning environment because of weak anti-lightning ability.

Both military aircraft and civil aircraft will inevitably fly in thunderstorm weather, and probably encounter lightning strike in the process of normal service. Relevant reports show that an aircraft may encounter one lightning strike per 1000–1500 h of flight and this is roughly equivalent to once a year for regular airliner aircraft. Thermal damage induced by lightning strike attributes to ablation, phase-transition, thermal shock, and the blow-off impulse effect, etc. While traditional thermal loading such as fire does not include the dynamical effects of thermal shock and blow-off impulse, etc. When high-energy lightning current attaches to the surface of a composite structure, tremendous Joule heat will be transmitted to the composite structure immediately in the form of conduction and radiation, which will generate great energy deposition and resulting in the temperature to rise rapidly around the attachment area. Furthermore, when the temperature exceeds the critical value of molten, vaporization and decomposition of material, a series of physic-chemical changes will be generated around the attachment area and three-phase transition of solid-liquid-vapor will also be occurred instantly [1]. Temperature is unevenly distributed due to the anisotropic characteristic of composite materials, and then leads to the uneven expansion of materials. Therefore, the gas generated by matrix vaporization is easily surrounded by the non-vaporized matrix and fibers, which will cause a rapid rise in internal pressure. The vapor spatter phenomenon which leads to a thermal explosion will occur when the internal pressure exceeds the constraint strength of surrounding materials, thus resulting in a reverse impact effect on the composite structure. This impact effect can be called the blow-off impulse effect [2]. The huge impact generated by the thermal explosion will cause more serious damage to the composite structure. Therefore, the blow-off impulse effect should be considered when the direct effects of lightning strike are analyzed.

At present, many scholars have investigated the thermo-dynamic response of composite materials and there have been abundant achievements on lightning damage of a composite structure. The representative studies on this experiment are as follows: Hirano et al. [3] carried out the lightning strike experiment of IM600/133 composite laminates, finding that damage modes of composite materials mainly include fiber fracture, matrix crack, and intra-laminar delamination etc. Deierling et al. [4] conducted an experiment to study the electric-thermal behavior of carbon fiber/epoxy composite materials subject to high-lightning current. The results reveal that lightning currents lead to a significant temperature rise around the attachment area, which is a result of the intense Joule heat effect generated in electric conductive fibers. Feraboli, Minller, and Kawakami et al. [5,6]. conducted research on composite specimens using simulated lightning strike, with the fundamental damage responses of specimens studied and the damage mechanism of composite materials subject to three different current peaks compared. Dong, Li, and Yin et al. [7–10] all reported a series of lightning strike experiments, which indicate that electric conductivity exerts a heavier effect on damage degree than thermal conductivity does. Furthermore, boundary conditions also have an obvious effect on the damage degree of composite materials during experiments.

There are also representative studies on numerical simulation of the thermo-dynamic response of composite materials. Ogasawara et al. [11] analyzed the temperature distribution in composite laminates from the perspective of electric-thermal coupling. The results indicate that Joule heat influences lightning strike damage significantly. Specifically, intra-laminar delamination is caused by the decomposition of resin and a concave pit is formed due to the sublimation of fibers. Abdelal et al. [12] predicted the thermal damage of composite panels subjected to lightning strike through electric-thermal coupling element. Meanwhile, the temperature-dependence material properties were considered as well. The results show that this simulation method is capable of capturing the damage size and the temperature profile in composite panels exactly. Naghipour et al. [13] studied the intra-laminar delamination of CFRP laminates induced by lightning strike using temperature-dependence interface elements. Wang et al. [14–16] has further conducted a series of studies on the thermo-dynamic response and the residual strength of composite materials after lightning strike, with fruitful results being achieved. Numerous studies on lightning strike protection have been conducted by many scholars [17–21]. Protective performances of different designs were compared and the best design scheme was proposed.

In general, the above studies of composite materials induced by lightning strike have important reference value and guiding significance. But previous studies mainly focused on the ablation analysis of composite materials, the blow-off impulse effect caused by the thermal explosion was rarely studied. Nevertheless, the structural response of composite materials subjected to lightning strike involves complex damage types, such as thermal shock wave, phase transition and thermal explosion, etc. Therefore, lightning strike response cannot be analyzed only by ablation damage and the blow-off impulse effect should be considered. The vaporized gas enclosed in materials will lead to a thermal explosion when thermal pressure continues to increase, then the inner explosion phenomenon will be formed and result in blow-off impulse damage. Therefore, it is necessary to study the blow-off impulse effect of composite materials under high temperature, high pressure, and high energy. However, there are few studies on the blow-off impulse effect of composite materials subjected to lightning strike so far, and the related reports are rare too. Only a small amount of studies about the thermal shock wave effect of composite materials under radiation conditions such as laser and X-Ray are reported. At present, studies concerning the blow-off impulse effect are mainly presented as follows: For example, Tang et al. [22–24] have conducted research about multi-physics effects on the surface of composite materials radiated by pulse, the material spatter caused by the pulse is called blow-off impulse. Huang et al. [25,26] studied the propagation rules of thermal shock wave in anisotropic material induced by X-Ray, damage characteristics of anisotropic material under strong radiation were also discussed. The results indicate that thermal shock waves exhibit different shapes under the radiation of soft and hard X-Ray, great differences exist in the form mechanisms of thermal shock wave, wave peak, penetration depth, gasification phenomenon, tensile intensity and so on.

In this paper, a method which integrates electric-thermal coupling with an explicit dynamic is put forward to study the blow-off impulse effect of a composite reinforced panel induced by lightning strike. The dynamic failure model of a composite reinforced panel is established and the temperature distribution in the benchmark skin is analyzed. Blow-off elements are obtained according to the temperature distribution in benchmark skin. The blow-off impulse effect of a composite reinforced panel subjected to lightning strike is then investigated. Finally, element failure, pressure, and von Mises stress distribution around the blow-off impulse area are discussed. The research achievements can be applied to the analysis of the damage mechanism of composite materials under the action of lightning strike, which has great engineering significance.

2. Material Properties and Calculation Model

2.1. Main Material Parameters

The temperature changes rapidly around the attachment area of a composite reinforced panel subjected to lightning strike. Related studies show that the local temperature may reach 10,000 °C and the high temperature will cause the variation of material properties, so the material parameters show obvious temperature-dependence in this case [12]. The influence of material properties that vary with temperature is considered in order to improve the accuracy and reliability of calculation results. The material type adopted in this research is an IM600/133 composite material. Mechanical, electric, and thermal properties of the IM600/133 composite material at different temperatures are given in Tables 1–4.

Table 1. Mechanical properties of IM600/133 composite material.

Temperature/°C	E_x/GPa	E_y/GPa	E_z/GPa	μ_{xy}	μ_{yz}	μ_{xz}	G_{xy}/GPa	G_{yz}/GPa	G_{xz}/GPa
25	137	8.2	8.2				4.36	3	4.36
200	137	6.56	6.56				3.488	2.4	3.488
260	137	0.082	0.082	0.02	0.34	0.02	0.03488	0.024	0.03488
600	137	0.0041	0.0041				0.001744	0.0012	0.001744
3316	137	0.0041	0.0041				0.001744	0.0012	0.001744

Table 2. Electric resistances of IM600/133 composite material [12,27,28].

Temperature/°C	Electric Resistances/$\Omega \cdot$m		
	Longitudinal	Transverse	Thickness
27	6.224×10^{-5}	0.3558	
127	5.948×10^{-5}	0.3362	
227	5.676×10^{-5}	0.3195	
327	5.429×10^{-5}	0.3043	
427	5.2×10^{-5}	0.2906	
457	5.139×10^{-5}		
527	4.994×10^{-5}		
627	4.801×10^{-5}	0.0547	
727	4.627×10^{-5}		
827	4.459×10^{-5}		
3316	13.442×10^{-5}		

Table 3. Thermal expansion coefficients of IM600/133 composite material.

Temperature/°C	Thermal Expansion Coefficients/°C^{-1}		
	Longitudinal	Transverse	Thickness
25	1.8×10^{-8}	2.16×10^{-5}	
200	5.4×10^{-8}	3.78×10^{-5}	
260	5.4×10^{-8}	3.78×10^{-5}	
600	5.4×10^{-8}	3.78×10^{-5}	
3316	5.4×10^{-8}	3.78×10^{-5}	
3317	5.4×10^{-8}	3.78×10^{-5}	

Table 4. Thermal conductivity, specific heat, and density of IM600/133 composite material [29,30].

Temperature/°C	Thermal Conductivity/W·m^{-1}·°C^{-1}			Specific Heat/J·kg^{-1}·°C^{-1}	Density/kg/m^3
	Longitudinal	Transverse	Thickness		
25	11.8	0.609		1065	
330	6.02	0.31		2050	1520
360	5.46	0.28		4250	
500	2.8	0.14		4200	
525	2.33	0.12		1800	1170
815	1.4	0.072		1850	
3316	1.4	0.072		2300	

2.2. Structure and Finite Element Models of a Composite Reinforced Panel

The size of the composite reinforced panel is 500 × 250 mm, with the height and width of the T stripper 38 and 50 mm, respectively. The reinforced core is filled with a mixture of fiber and resin, an adhesive of J-116B-δ0.15 is used to glue the benchmark skin and T stripper. The cross-section is shown in Figure 1, 24 layers of which are in the benchmark skin and the thickness of each layer is 0.15 mm, with a total thickness 3.6 mm and stacking sequence [45°/0°/−45°/90°/−45°/0°/45°/0°/45°/90°/−45°/0°]$_S$. Stacking sequence of the left side in the T stripper is [45°/0°/−45°/0°/90°/0°/−45°/0°/90°/0°/45°/0°] and the right side is [−45°/0°/45°/0°/90°/0°/45°/0°/90°/0°/−45°/0°], with a thickness of 1.8 mm, respectively. Stacking sequence of the bottom layer in the T stripper is [45°/0°], with a thickness of 0.3 mm. Stacking sequences in each component of the composite reinforced panel are given in Table 5.

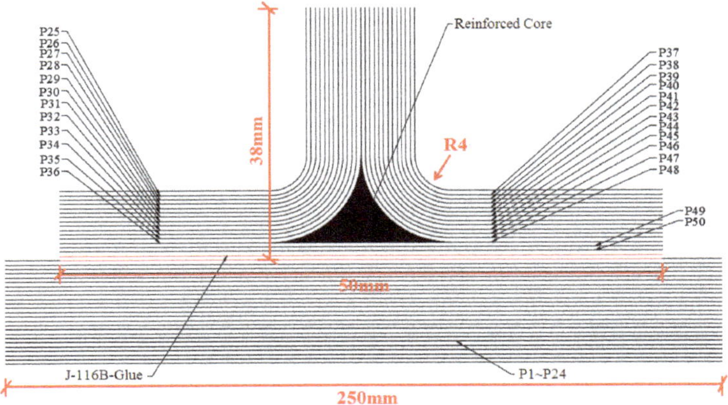

Figure 1. Cross-section of the composite reinforced panel.

Table 5. Stacking sequences in each component of the composite reinforced panel.

Benchmark Skin		Angle/°	Left Side of T Stripper		Right Side of T Stripper		Bottom of T Stripper	Angle/°
			Layer	Angle/°	Layer	Angle/°		
P1	P24	45	P25	45	P37	−45	P49	45
P2	P23	0	P26	0	P38	0	P50	0
P3	P22	−45	P27	−45	P39	45		
P4	P21	90	P28	0	P40	0		
P5	P20	−45	P29	90	P41	90		
P6	P19	0	P30	0	P42	0		
P7	P18	45	P31	−45	P43	45		
P8	P17	0	P32	0	P44	0		
P9	P16	45	P33	90	P45	90		
P10	P15	90	P34	0	P46	0		
P11	P14	−45	P35	45	P47	−45		
P12	P13	0	P36	0	P48	0		

This research is studied in ANSYS software and is divided into two modules. Namely, an electric-thermal coupling module and blow-off impulse module. Firstly, electric-thermal coupling analysis is performed and the temperature distribution in the composite reinforced panel is obtained. Blow-off impulse elements are then selected according to the temperature distribution. Finally, blow-off impulse analysis is performed according to the distribution of blow-off impulse elements. Electric-thermal element SOLID69 is adopted in the electric-thermal coupling module, this element type has 8 nodes and 2 degrees of freedom per node. Therefore, the model can be divided into hexahedral elements. Boundary conditions are set as follows: thermal radiation coefficient ε is equal to 0.9 in the surface of benchmark skin and surrounding sides of the composite reinforced panel. The T stripper and the surrounding sides are grounded, so electric potential U is assumed to be 0. The bottom surface of the benchmark skin and T stripper are adiabatic. Element dimension is approximately $8.33 \times 7.0 \times 0.15$ mm. There are 2640 elements in each layer of benchmark skin, 720 elements in each layer of the T stripper and 1440 elements in the bottom of the T stripper. The finite element model and boundary conditions of the composite reinforced panel are shown in Figure 2.

Lightning current is applied to the center-node in the benchmark skin of the composite reinforced panel. Current waveform applied in this research is a double exponential waveform, which can be expressed in the form:

$$I(t) = I_0\left(e^{-\alpha t} - e^{-\beta t}\right) \tag{1}$$

where, I_0 represents current constant; $I(t)$ is transient current; α is reciprocal of wave-tail time; β is reciprocal of wave-front time; t is time.

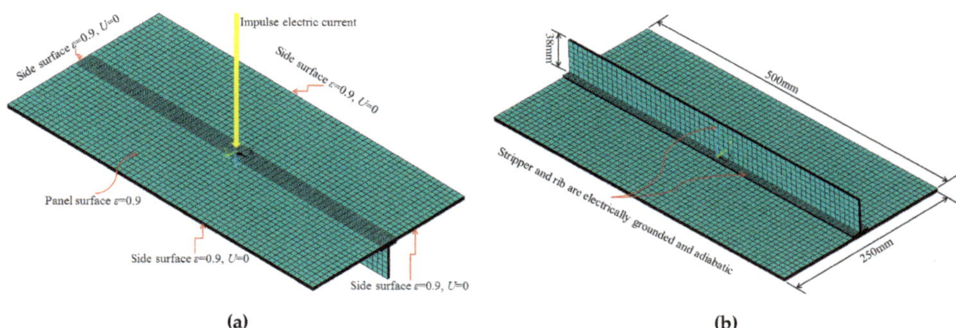

(a) (b)

Figure 2. Finite element model and boundary conditions of the composite reinforced panel. (**a**) Boundary conditions in benchmark skin; (**b**) boundary conditions in the T stripper.

Pulse current waveform is defined through a pair of parameters t_1/t_2 and current peak I_p [3,9,31]. Where, t_1 is the time from 10% to 90% of the maximum current and t_2 is the time from 10% to 50% through 90% of the maximum current. The relationship of t_1/t_2 is shown in Figure 3a. However, the most common waveform parameters used are I_p, t_1, and t_2 in an actual lightning strike experiment. The main parameters of lightning current adopted in this research and the waveform are shown in Figure 3b. Current duration is 80 μs and it is divided into 12 steps to load during the calculation, sub-time step is equal to 10 in each load step. The time and corresponding current values in each load step are given in Table 6.

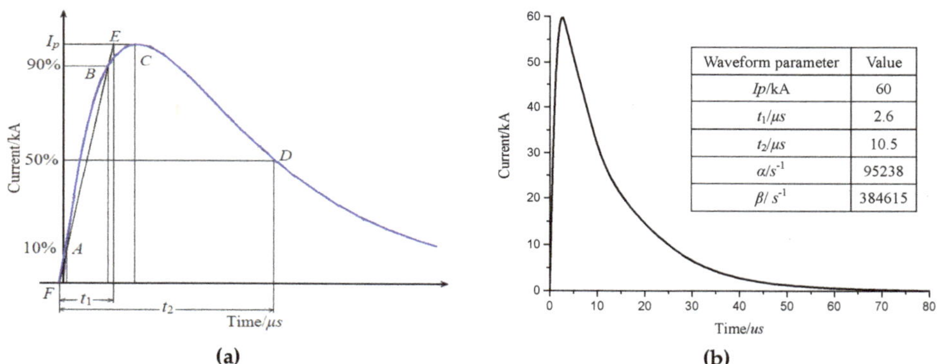

Waveform parameter	Value
I_p/kA	60
$t_1/\mu s$	2.6
$t_2/\mu s$	10.5
α/s^{-1}	95238
β/s^{-1}	384615

(a) (b)

Figure 3. Double exponential impulse current waveform. (**a**) Relationship of t_1/t_2; (**b**) waveform of 2.6/10.5 μs-0 kA.

Table 6. The time and current value in each load step of 2.6/10.5 μs-60 kA.

Load step	1	2	3	4	5	6	7	8	9	10	11	12
Time/μs	1	2	2.6	3	7	10	10.5	11	15	30	50	80
Load/kA	46.3	58.9	60	59.5	42.6	31.5	30.3	28.5	19.1	4.2	0.57	0.028

2.3. Blow-Off Impulse Model of a Composite Reinforced Panel

According to the temperature distribution of the composite reinforced panel, blow-off impulse elements are obtained after the electric-thermal coupling analysis is done. SOLID69 is then replaced by the explicit dynamic element SOLID164 for blow-off impulse analysis in ANSYS software. This element type has 8 nodes and 3 degrees of freedom per node. The meshing method and number of meshes are the same as that in Section 2.2. At the same time, the initial composite materials model in the blow-off impulse area is replaced by a high-explosive material model which is described by Jones–Wilkins–Lee equation of state (JWL EOS). The pressure of JWL EOS is defined as follows:

$$p = A\left(1 - \frac{\omega}{R_1 V}\right)e^{-R_1 V} + B\left(1 - \frac{\omega}{R_2 V}\right)e^{-R_2 V} + \frac{\omega E_0}{V} \tag{2}$$

where, p is the pressure of the high-explosive element. V is the initial relative volume. E_0 is the initial explosion energy per unit volume. A, B, R_1, R_2, and ω are material constants.

The material type adopted for the area which expects blow-off elements is the 59# constitutive model. This represents that the material type of Mat_composite_failure_solid_model ranks No.59 in keyword user's manual of ANSYS/LS-DYNA. The surrounding sides of the composite reinforced panel are fixed and the unified system of unit kg-m-s is adopted. The constitutive model 59# can be defined as follows:

$$f = \frac{4\left[\sigma_1 - \frac{X_t - X_c}{2}\right]^2}{(X_t + X_c)^2} + \frac{4\left[\sigma_2 - \frac{Y_t - Y_c}{2}\right]^2}{(Y_t + Y_c)^2} + \frac{4\left[\sigma_3 - \frac{Z_t - Z_c}{2}\right]^2}{(Z_t + Z_c)^2} + \frac{\sigma_{12}^2}{S_{12}^2} + \frac{\sigma_{13}^2}{S_{13}^2} + \frac{\sigma_{23}^2}{S_{23}^2} - 1 \tag{3}$$

where, X_t and X_c are longitudinal tensile and compressive strengths. Y_t and Y_c are transverse tensile and compressive strengths. Z_t and Z_c are normal tensile and compressive strengths. S_{12} is shear strength in-plane. S_{13} and S_{23} are transverse shear strengths. f is an ellipsoidal function. The whole calculation process is shown in Figure 4. The main parameters of the high-explosive model, JWL EOS and 59# constitutive model are given in Table 7.

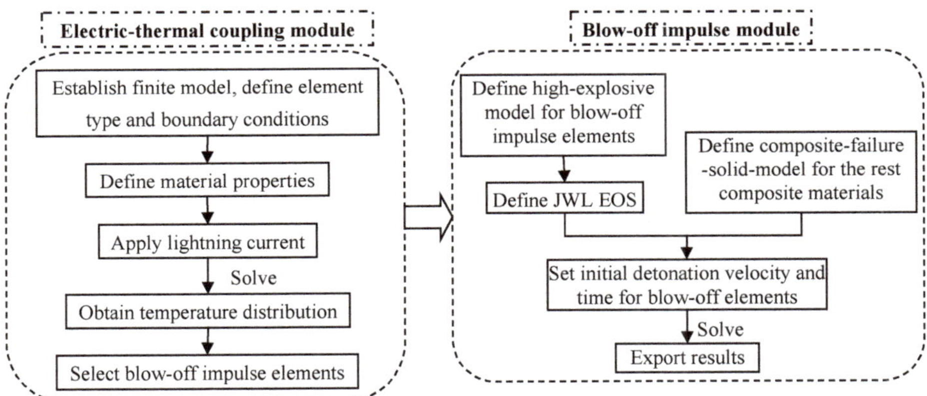

Figure 4. Calculation process of blow-off impulse analysis.

Table 7. Main parameters of the high-explosive model, JWL EOS, and 59# constitutive model.

Parameters	ρ/kg/m^3	D/m/s	P_{CJ}/GPa	A/GPa	B/GPa	R_1	R_2	ω	E_0/kJ
Value	1520	6718	18.5	540.9	9.4	4.5	1.1	0.35	8×10^6
Parameters	X_c/MPa	X_t/MPa	Y_c/MPa	Y_t/MPa	Z_c/MPa	Z_t/MPa	S_{12}/MPa	S_{13}/MPa	S_{23}/MPa
Value	1281	1708	192	34	280	52	128	128	96

3. Results and Discussion

3.1. Analysis of Electric-Thermal Coupling

Due to the fact that electric conductivity and thermal conductivity of fiber are much higher than that of the matrix and the anisotropic characteristic of carbon fiber/epoxy composite materials, composite materials present low electric conductivity and thermal conductivity as a whole. When lightning current attaches to the surface of composite materials, it will conduct internally along the attachment points and huge Joule heat will be generated during the conduction process within composite materials. Temperature distribution around the attachment area of the composite reinforced panel is shown in Figure 5 when the calculation is completed.

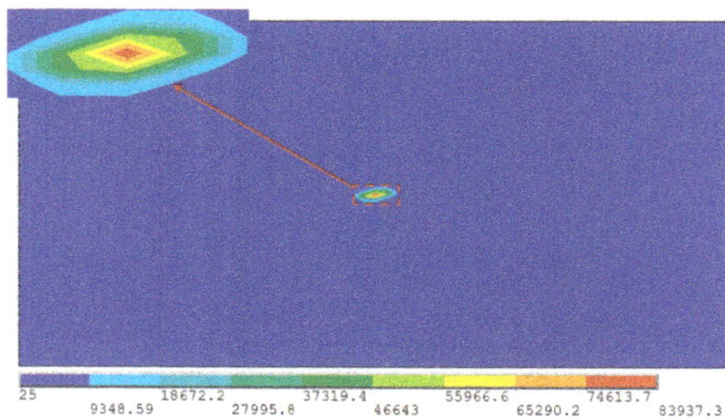

Figure 5. Temperature distribution around the attachment area of the composite reinforced panel. (Unit: °C).

It can be seen from Figure 5 that the temperature profile around the attachment area resembles an ellipse, and the long axis of the ellipse is along fiber direction in P1 layer. The highest temperature is 83937.3 °C, which is much higher than the ablation and sublimation temperatures of carbon fiber. However, the temperature drops sharply along the thickness direction due to the low electric conductivity and thermal conductivity in the thickness direction. Figure 6 presents the temperature distribution in the cross-section and Figure 7 presents the temperature change in the top 17 layers of the benchmark skin. It can be seen from Figures 6 and 7 that the temperatures around the attachment area mainly focus on the top three layers of benchmark skin, with very a small temperature rise in the P4–P16 layers. Temperatures in P16–P24 layers and the T stripper have almost no change. Therefore, it can be concluded that the Joule heat generated by the lightning current mainly causes significant thermal damage in the top three layers around the attachment area, with very little damage in other areas.

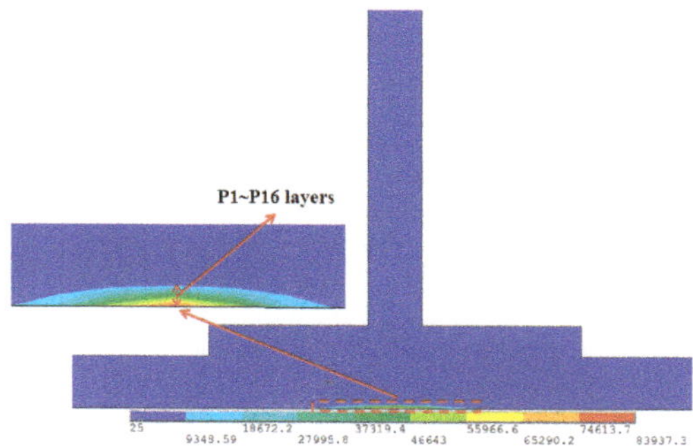

Figure 6. Temperature distribution in the cross-section of the composite reinforced panel. (Unit: °C).

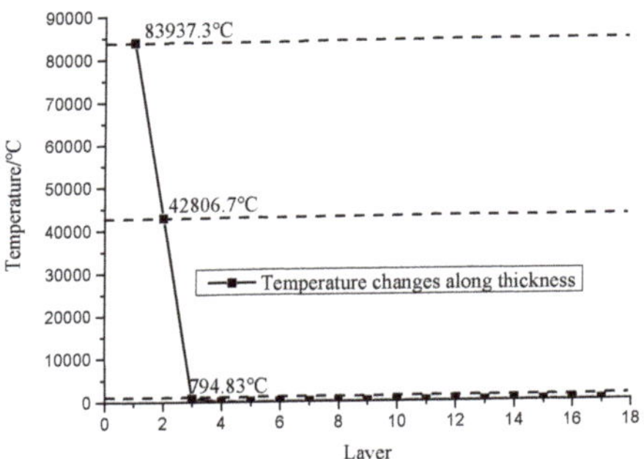

Figure 7. Temperature change along the thickness direction.

Figure 8 presents the temperature distribution in the top 17 layers of benchmark skin. It can be seen that the closer to the top layer, the higher the temperature will be. The temperature profile in the P1 layer is similar to an ellipse shape and the long axis of the ellipse is roughly along the fiber direction (45°). The influence between each layer is very significant, which can be verified from the temperature contours in P2 layer, P3 layer, P5 layer, P6 layer, and P8 layer. Stacking sequences in these layers are 0°, −45°, −45°, 0°, and 0°, respectively, but the temperature distribution is not along the fiber direction in each layer. These serious effects are mainly caused by the low thermal conductivity in the thickness and transverse direction. The Joule heat effect mainly concentrates on the top three layers of benchmark skin, which is due to the fact that the electric conductivity and thermal conductivity in the thickness direction are much lower than that in the fiber direction. Lightning current mainly conducts along fiber direction in P1 layer, P2 layer, and P3 layer, only a small amount of lightning currents conduct along the thickness direction. The high electric resistance blocks the conduction of lightning current along the thickness direction, which can be confirmed through the temperature distribution in the cross-section, as shown in Figure 6. For example, the highest temperatures in P1 layer and P2 layer are 83,937.3 and 42,806.7 °C, respectively. While the temperature in P3 layer has dropped to 794.83 °C, indicating that only the top two layers will be ablated around the attachment area under this current waveform. The temperature in P4 layer is just 83.0886 °C, which is lower than the molten temperature of resin. Temperatures around the attachment area are between 25 and 37 °C from P5 layer to P17 layer, which is in the range of the environment temperature. Since then, temperatures in inner layers of the benchmark skin as well as the T stripper are not on the rise, agreeing well with Figures 6 and 7. At the same time, although the thermal conductivity in the fiber direction is much greater than that in the other two directions, the difference is far smaller than that of electric conductivity in the fiber direction and the other two directions. Therefore, the influence of thermal conductivity in the thickness direction on the temperature distribution of each layer can almost be ignored, except for the top three layers with high temperature.

3.2. Analysis of the Blow-Off Impulse

The elements with temperatures exceeding 3316 °C are not defined as failures in the electric-thermal coupling module, but selected as the blow-off impulse elements in the blow-off impulse module. Therefore, element failure is not considered in the electric-thermal coupling module. Figure 9 presents the finite element model of blow-off impulse, it can be seen from Figure 9 that there are 38 blow-off impulse elements in the center area of the composite reinforced panel, while the rest areas are non-blow-off impulse elements. These 38 blow-off impulse elements all concentrate on P1 layer and P2 layer of the benchmark skin. Solution time is 2 µs, the step length factor is set as 0.6, and the output step number is 22. Keyword file is output and submitted to LS-DYNA solver after it is modified in LS-PrePost. Failure elements are defined according to the maximum failure strain. As expressed in Equation (3), yield function can be built through strength parameters and stress tensor of composite materials. Elements will enter the plastic phase when *f* is greater than zero. Elements will then deform continuously subject to external load and the stiffness of composite materials will be reduced too. The maximum failure strain is defined in the keyword file of ANSYS/LS-DYNA. If the equivalent strains of elements are greater than the maximum failure strain, the elements are defined as failures and will be deleted.

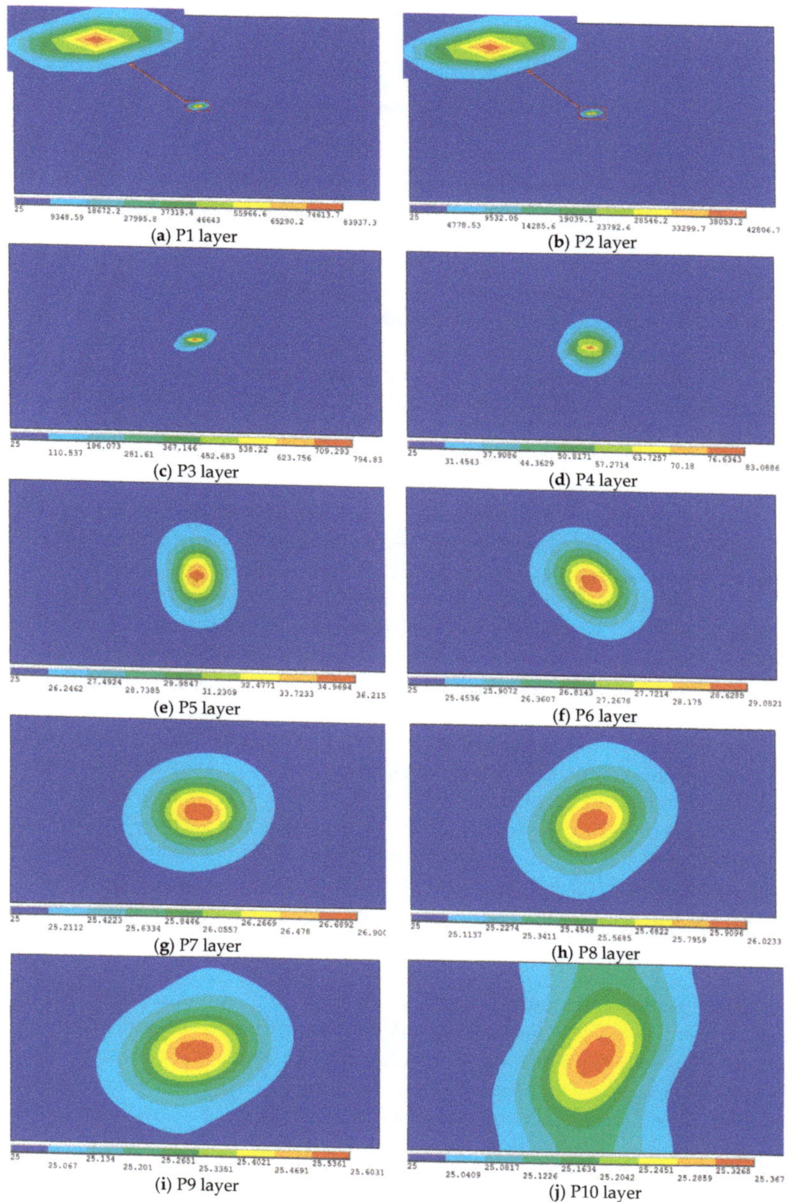

Figure 8. *Cont.*

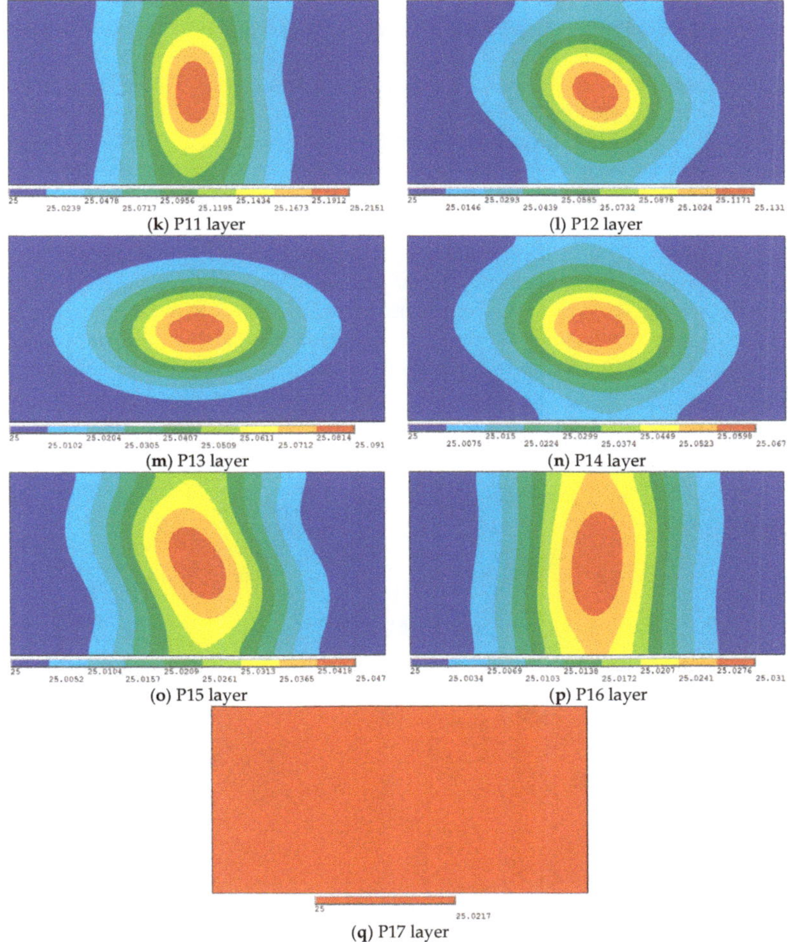

Figure 8. Temperature distribution of the top 17 layers in the benchmark skin of the composite reinforced panel. (Unit: °C).

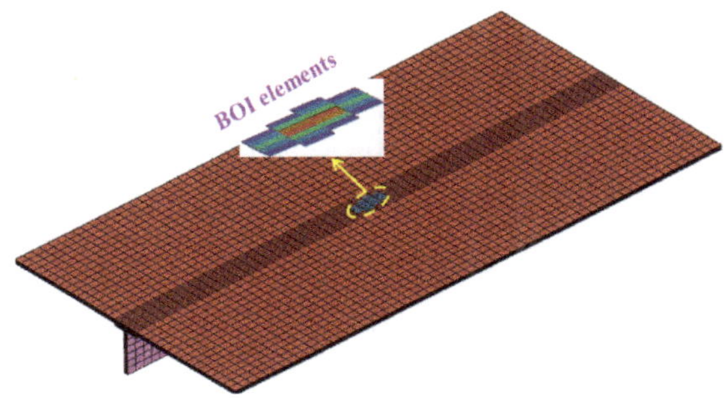

Figure 9. Finite element model of the blow-off impulse of the composite reinforced panel.

Figure 10 presents the contours of von Mises stress in the top seven layers when the blow-off impulse analysis is completed. It can be seen that the damage caused by blow-off impulse mainly concentrates on the center areas of each layer, and the element deletion phenomenon only appears in the top six layers of the benchmark skin. P1 layer is the most seriously damaged, with 38 elements deleted and size of the failure area is about 50×21 mm. However, the contours of von Mises stress is not along $45°$ in P1 layer but deviates from the fiber direction. In P2 layer and P3 layer, 27 and 40 elements are removed, respectively. The sizes of the failure areas in these two layers are almost the same. There are four elements deleted in P4 layer, P5 layer, and P6 layer, respectively, and no elements deleted after P7 layer. However, there is a large stress concentration phenomenon in the center area of P7 layer. Additionally, the influence on stress distribution between each layer is very significant. For example, stacking sequences of P2 layer and P3 layer are $0°$ and $-45°$, respectively, but the contours of von Mises stress are not along the fiber direction in each layer. It can be concluded that the damage forms caused by the blow-off impulse effect alter the initial damage distribution caused by the electric-thermal coupling effect, which makes the damage distribution t deviate from the initial fiber direction.

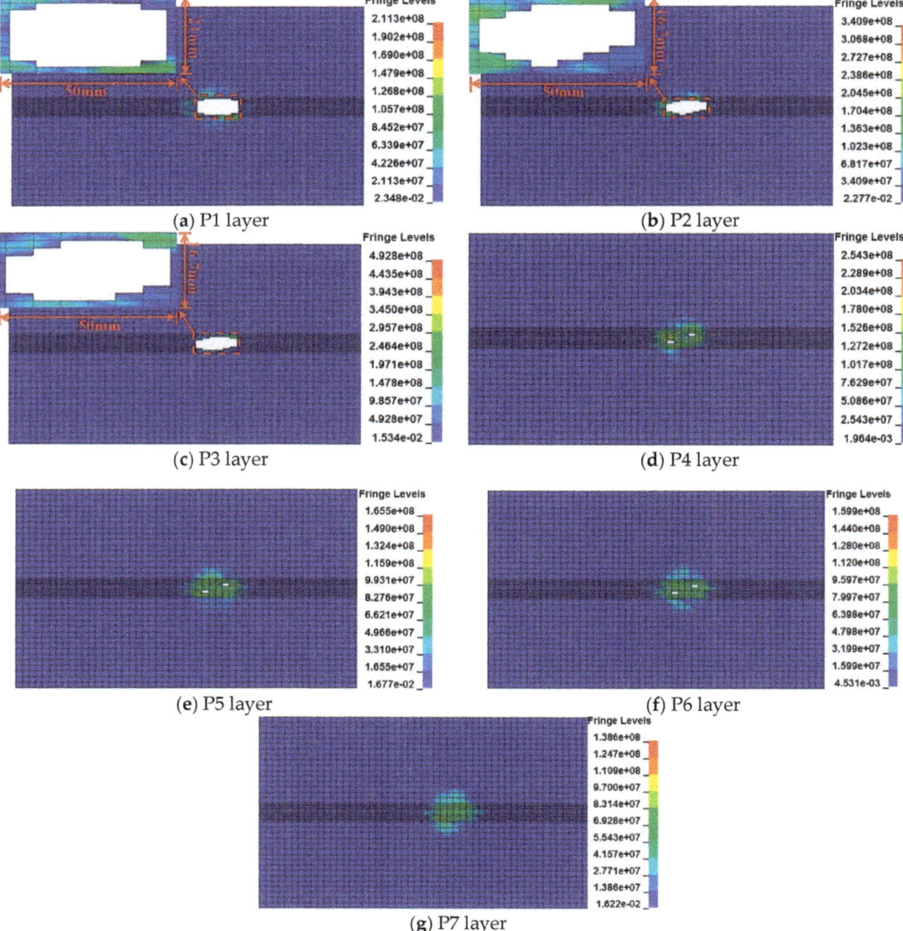

Figure 10. Contours of von Mises stress on the top seven layers of the benchmark skin.

Figure 11 presents the pressure contours of the composite reinforced panel in several typical moments. It can be seen that the implosion effect starts to appear when time is equal to 0.099 µs and great pressure is formed immediately. The maximum pressure is 4.443×10^8 N, but no elements are deleted at this moment. The element deletion phenomenon starts to appear on the top three layers when time is equal to 0.4 µs. Six layers exhibit element deletion phenomenon when time is equal to 1.5 µs, and an obvious concave pit is formed around the blow-off impulse area. But the pressure is not as large as when the implosion effect starts, the maximum pressure is 2.846×10^8 N at this moment. The blow-off impulse effect is completed when time is equal to 2 µs, and the maximum damage is reached at this point. A large concave pit is formed around the blow-off impulse area and the depth of the concave pit is about 1.088 mm, as shown in Figure 12. Additionally, it also can be seen from Figure 12 that the element deletion phenomenon also occurs in the T stripper under the action of the blow-off impulse, indicating that the shock wave may cause some inner damage that cannot be seen on the surface. The element bulge phenomenon appears around the edges of element failure and the bottom of the concave pit is uneven. It can be concluded that the blow-off impulse effect not only causes external damage in the benchmark skin but also causes internal damage in the T stripper.

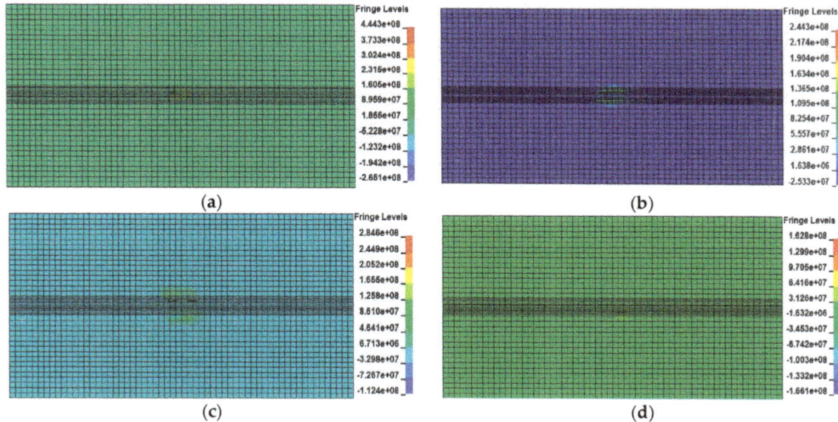

Figure 11. Pressure contours in several typical moments. (**a**) $t = 0.099$ µs; (**b**) $t = 0.4$ µs; (**c**) $t = 1.5$ µs; (**d**) $t = 2$ µs.

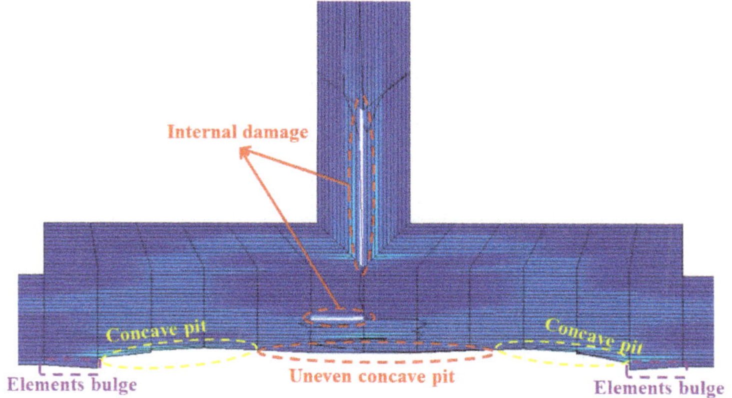

Figure 12. Damage forms in the cross-section of the composite reinforced panel.

Figure 13 presents the pressure contours in blow-off impulse elements at the beginning, middle, and end of the calculation, respectively. Figure 14 presents the pressure change of blow-off impulse elements at different moments. It can be seen from Figures 13 and 14 that the pressure increases sharply when the explosion begins and the maximum pressure is 4.181×10^9 N. The pressure then decreases sharply and the decrease-rate becomes slow and eventually tends to balance. However, the volume of blow-off impulse elements expands rapidly in the process of the explosion, then the volume reaches the maximum when the blow-off impulse analysis is completed, which can reflect the bulge phenomenon around the attachment area of lightning strike.

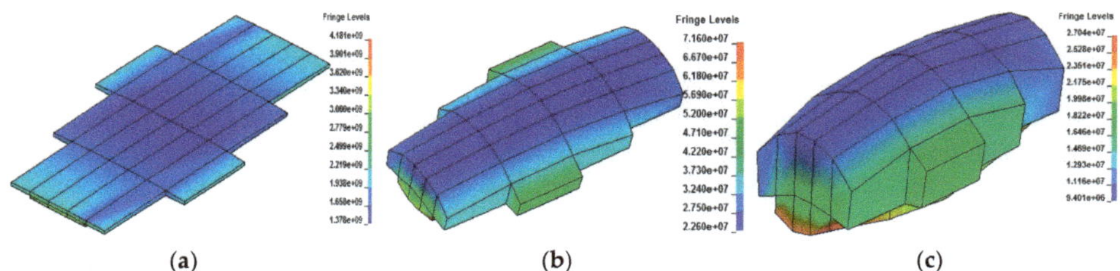

(a)   (b)   (c)

Figure 13. Pressure contours in blow-off impulse elements. (**a**) $t = 0.0096$ μs; (**b**) $t = 1$ μs; (**c**) $t = 2$ μs.

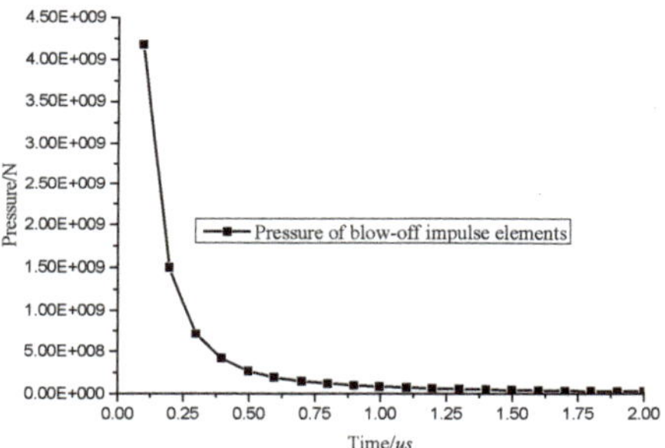

Figure 14. Pressure change of blow-off impulse elements.

Figure 15 presents the overall pressure contours of the composite reinforced panel after the blow-off impulse analysis is completed. It can be seen that an obvious bulge phenomenon appears around the blow-off impulse area, and the element failure mode around the blow-off impulse area is similar to that around the attachment area of lightning strike. Therefore, the blow-off impulse effect can reflect the damage forms of the composite reinforced panel induced by lightning strike and should be considered. Additionally, the pressure around the blow-off impulse area presents as isotropic rather than anisotropic, which agrees well with the fact that the mechanical properties of composite materials present as anisotropic in low pressure and isotropic in high pressure [25,26]. Figure 16 presents the von Mises stress and pressure in the T stripper. It can be seen that there is no element deletion on the surface of the T stripper, but some internal elements are deleted as shown in Figure 12. This indicates that the blow-off impulse effect has a serious influence on the center areas in the top six layers of the benchmark skin, while the influences on the T stripper and other areas are relatively less serious.

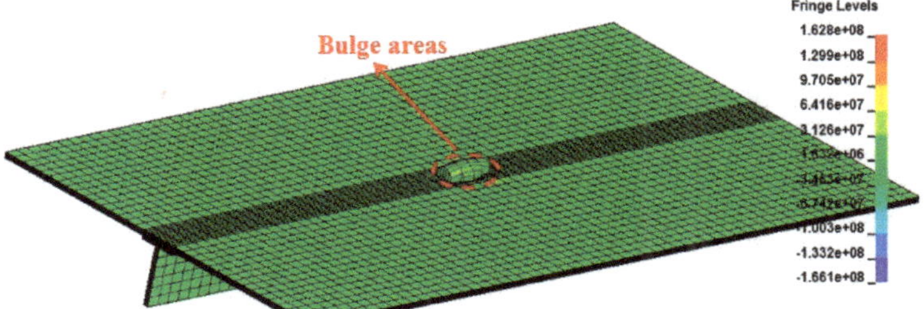

Figure 15. Overall pressure contours of the composite reinforced panel.

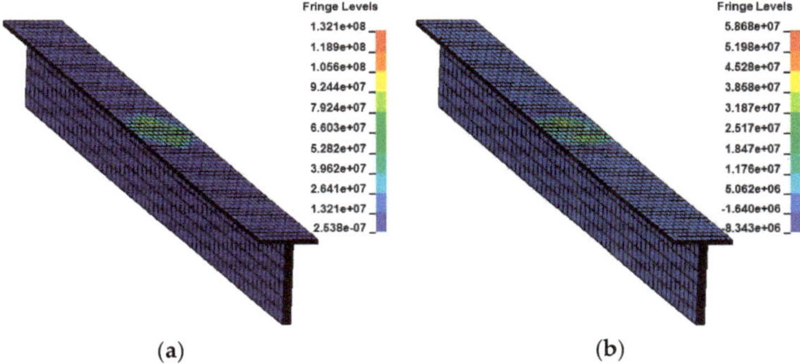

(a)   (b)

Figure 16. Contours of von Mises and pressure in the T stripper. (**a**) Contours of von Mises stress; (**b**) pressure contours.

Mesh quality and element dimensions have a great influence on the calculation results. In order to study the influence of mesh quality on calculation results, the damage degree of a composite reinforced panel with five kinds of element dimensions are compared. Calculation results for different element dimensions are given in Table 8. It can be seen that the error will increase with the increase of element dimensions. Although failure area errors change greatly with the increase of the element dimension, damage depth errors change little. However, there will be no blow-off impulse elements around the attachment area if the element dimension is larger than 10.0mm in length direction. Therefore, the element dimension should be as small as possible to enable better calculation results.

Table 8. Comparison of calculation results for different element dimensions.

Serial Number	Element Dimension/mm	Damage Depth/mm	Error	Failure Area/mm^2	Error
Mesh-1	8.33 × 7.0 × 0.15	1.088	-	1050	-
Mesh-2	8.47 × 7.0 × 0.15	1.080	7.30%	936	10.8%
Mesh-3	8.62 × 7.0 × 0.15	0.962	11.58%	864	17.7%
Mesh-4	8.92 × 7.0 × 0.15	0.615	43.47%	670	36.2%
Mesh-5	10.0 × 7.0 × 0.15	0.550	49.48%	500	52.3%

4. Conclusions

Based on the anti-lightning strike background of composite materials widely used in aircraft structure design, the thermal explosion phenomenon of composite materials is rarely studied. The blow-off impulse effect of a composite reinforced panel induced by lightning strike is studied through numerical simulation in this paper. A method integrating electric-thermal coupling with an explicit dynamic is proposed in order to study the blow-off impulse effect of a composite reinforced panel. The damage mechanism of a composite reinforced panel caused by the blow-off impulse effect is discussed. The conclusions can be summarized as follows:

1. The blow-off impulse effect alters the damage distribution caused by the electric-thermal coupling effect, which makes the damage distribution deviate from the initial fiber direction in each layer.
2. The blow-off impulse effect could well present the internal damage, concave pit, and bulge phenomenon around the attachment area of lightning strike, and the failure modes in the blow-off impulse area are similar to the damage forms caused by lightning strike.
3. Pressure increases sharply when explosion begins, and then decreases gradually with the increase of time and tends to balance in the end. The pressure of composite materials is not presented as anisotropic but isotropic, agreeing well with the observed characteristic that mechanical behavior of composite materials exhibits anisotropic in low pressure while isotropic in high pressure.

The results obtained in this study can reflect the dynamical damage behavior of composite materials induced by lightning strike to some extent. However, phase transition and delamination are other important damage modes of composite materials, which have not been involved in this study. Therefore, this field needs to be further considered in future investigations.

Author Contributions: Data curation, S.J. and W.H.; formal analysis, S.J., F.W., W.H. and B.X.; methodology, S.J., F.W. and B.X.; writing—original draft, S.J. and F.W.

Funding: This research was funded by the National Natural Science Foundation of China, grant number: 51875463 & 51475369, the Natural Science Basic Research Plan in Shaanxi Province of China, grant number: 2018JM1001.

Acknowledgments: This study is supported by the National Natural Science Foundation of China (No.: 51875463 & No.: 51475369), the Natural Science Basic Research Plan in Shaanxi Province of China (No.: 2018JM1001).

References

1. Katunin, A.; Krukiewicz, K.; Turczyn, R.; Sul, P.; Lasica, A. Synthesis and characterization of the electrically conductive polymeric composite for lightning strike protection of aircraft structures. *Compos. Struct.* **2017**, *159*, 773–783.
2. Liu, Z.Q.; Yue, Z.F.; Wang, F.S.; Ji, Y.Y. Combining analysis of coupled electrical-thermal and Blow-off impulse effects on composite laminate induced by lightning strike. *Appl. Compos. Mater.* **2015**, *22*, 189–207. [CrossRef]
3. Hirano, Y.; Katsumata, S.; Iwahori, Y.; Todoroki, A. Artificial lightning testing on graphite/epoxy composite laminate. *Compos. Part A Appl. Sci. Manuf.* **2010**, *41*, 1461–1470. [CrossRef]
4. Deierling, P.E.; Zhupanska, O.I. Experimental study of high electric current effects in carbon/epoxy composites. *Compos. Sci. Technol.* **2011**, *71*, 1659–1664. [CrossRef]
5. Feraboli, P.; Minller, M. Damage resistance and tolerance of carbon/epoxy composite coupons subjected to simulated lightning strike. *Compos. Part A Appl. Sci. Manuf.* **2009**, *40*, 954–967.
6. Kawakami, H.; Feraboli, P. Lightning strike damage resistance and tolerance of scarf-repaired mesh-protected carbon fiber composites. *Compos. Part A Appl. Sci. Manuf.* **2011**, *42*, 1247–1262.
7. Dong, Q.; Guo, Y.L.; Sun, X.C.; Jia, Y.X. Coupled electrical-thermal-pyrolytic analysis of carbon fiber/epoxy composites subjected to lightning strike. *Polymer* **2015**, *56*, 385–394. [CrossRef]
8. Dong, Q.; Guo, Y.L.; Chen, J.L.; Yao, X.L.; Yi, X.S.; Ping, L.; Jia, Y.X. Influencing factor analysis based on electrical-thermal-pyrolytic simulation of carbon fiber composites lightning damage. *Compos. Struct.* **2016**, *140*, 1–10. [CrossRef]
9. Li, S.L.; Yin, J.J.; Yao, X.L.; Chang, F. Damage analysis for carbon fiber/epoxy composite exposed to simulated lightning current. *J. Reinf. Plast. Compos.* **2016**, *35*, 1201–1213.
10. Yin, J.J.; Li, S.L.; Yao, X.L.; Chang, F. Lightning strike ablation damage characteristic analysis for carbon fiber/epoxy composite laminate with fastener. *Appl. Compos. Mater.* **2016**, *23*, 821–837.
11. Ogasawara, T.; Hirano, Y.; Yoshimura, A. Coupled thermal-electrical analysis for carbon fiber/epoxy composites exposed to simulated lighting current. *Compos. Part A Appl. Sci. Manuf.* **2010**, *41*, 973–983. [CrossRef]

12. Abdelal, G.; Murphy, A. Nonlinear numerical modeling of lightning strike effect on composite panels with temperature dependent material properties. *Compos. Struct.* **2014**, *109*, 268–278. [CrossRef]

13. Naghipour, P.; Pineda, E.J.; Arnold, S.M. Simulation of lightning-induced delamination in un-protected CFRP laminates. *Appl. Compos. Mater.* **2016**, *23*, 523–535. [CrossRef]

14. Wang, F.S.; Ji, Y.Y.; Yu, X.S.; Chen, H.; Yue, Z.F. Ablation damage assessment of aircraft carbon fiber/epoxy composite and its protection structures suffered from lightning strike. *Compos. Struct.* **2016**, *145*, 226–241. [CrossRef]

15. Wang, F.S.; Ding, N.; Liu, Z.Q.; Ji, Y.Y.; Yue, Z.F. Ablation damage characteristic and residual strength prediction of carbon fiber/epoxy composite suffered from lightning strike. *Compos. Struct.* **2014**, *117*, 222–233. [CrossRef]

16. Wang, F.S.; Yu, X.S.; Jia, S.Q.; Li, P. Experimental and numerical study on the residual strength of carbon/epoxy composite after lightning strike. *Aerosp. Sci. Technol.* **2018**, *75*, 304–314. [CrossRef]

17. Gagné, M.; Therriault, D. Lightning strike protection of composites. *Prog. Aerosp. Sci.* **2014**, *64*, 1–16. [CrossRef]

18. Wang, B.; Duan, Y.G.; Xin, Z.B.; Yao, X.L.; Abliz, D.; Ziegmann, G. Fabrication of an enriched graphene surface protection of carbon fiber/epoxy composites for lightning strike via a percolating-assisted resin film infusion method. *Compos. Sci. Technol.* **2018**, *158*, 51–60. [CrossRef]

19. Dhanya, T.M.; Chandra, S.Y. Lightning strike effect on carbon fiber reinforced composites-effect of copper mesh protection. *Mater. Today. Commun.* **2018**, *16*, 124–134. [CrossRef]

20. Soulas, F.; Espinosa, C.; Lachaud, F.; Guinard, S. A method to replace lightning strike tests by ball impacts in the design process of lightweight composite aircraft panels. *Int. J. Impact. Eng.* **2018**, *11*, 165–176. [CrossRef]

21. Vipin, K.; Tomohiro, Y.; Takao, O.; Hirano, Y.; Goto, T.; Takahashi, T.; Hassen, A.A.; Ogasawara, T. Polyaniline-based all-polymeric adhesive layer: An effective lightning strike protection technology for high residual mechanical strength of CFRPs. *Compos. Sci. Technol.* **2019**, *172*, 49–57.

22. Zhang, K.; Tang, W.H.; Fu, K.K. Modeling of dynamic behavior of carbon fiber-reinforced polymer (CFRP) composite under X-ray radiation. *Materials* **2018**, *11*, 143. [CrossRef]

23. Tang, W.H.; Wang, D.R.; Huang, X. The sublimation energy versus temperature and pressure and its influence on Blow-off Impulse. *Int. Conf. Appl. Math. Mech. Phys.* **2011**, *5*, 1492–1495.

24. Chen, H.; Tang, W.H.; She, J.H.; Ran, X.W.; Xu, Z.H. The PUFF equation of state parameters for synthetic rubber. *Mech. Mater.* **2011**, *43*, 69–74. [CrossRef]

25. Huang, X.; Tang, W.H.; Jiang, B.H.; Guo, X. The numerical simulation of 2-D thermal shock wave induced by X-Ray in anisotropic material. *Chin. J. High. Pres. Phys.* **2011**, *1*, 41–47. (In Chinese)

26. Huang, X.; Tang, W.H.; Jiang, B.H.; Ran, X.W. Anisotropic constitutive model and its application in simulation of thermal shock wave propagation for cylinder shell composite. *Int. Conf. Appl. Math. Mech. Phys.* **2011**, *9*, 1823–1828.

27. Abry, J.C.; Bochard, S.; Chateauminois, A.; Salvia, M.; Giraud, G. In situ detection of damage in CFRP laminates by electrical resistance measurements. *Compos. Sci. Technol.* **1999**, *59*, 925–935. [CrossRef]

28. Lago, F.; Gonzalez, J.J.; Freton, P.; Uhlig, F.; Piau, G. A numerical modeling of an electric arc and its interaction with the anode: part III. Application to the interaction of a lightning strike and an aircraft in flight. *J. Phys. D Appl. Phys.* **2006**, *39*, 2294–2310. [CrossRef]

29. Chen, J.K.; Sun, C.T.; Chang, C.I. Failure analysis of a graphite/epoxy laminate subjected to combined thermal and mechanical loading. *J. Compos. Mater.* **1985**, *19*, 408–423. [CrossRef]

30. Zaman, W.; Li, K.Z.; Ikram, S.; Li, W.; Zhang, D.S.; Guo, L.J. Morphology, thermal response and anti-ablation performance of 3D-four directional pitch-based carbon/carbon composites. *Corros. Sci.* **2012**, *61*, 134–142. [CrossRef]

31. *Aircraft Lightning Environment and Related Test Waveforms*; Aerospace Recommended Practice ARP 5412; SAE: Warrendale, PA, USA; Troy, MI, USA, 2005.

Article

Numerical–Experimental Correlation of Impact-Induced Damages in CFRP Laminates

Andrea Sellitto [1,*], Salvatore Saputo [1], Francesco Di Caprio [2], Aniello Riccio [1], Angela Russo [1] and Valerio Acanfora [1]

[1] Department of Engineering, University of Campania Luigi Vanvitelli, via Roma 29, 81031 Aversa, Italy; salvatore.saputo@unicampania.it (S.S.); aniello.riccio@unicampania.it (A.R.); angela.russo@unicampania.it (A.R.); valerio.acanfora@unicampania.it (V.A.)

[2] C.I.R.A.—Italian Aerospace Research Centre, via Maiorise snc, 81043 Capua, Italy; f.dicaprio@cira.it

[*] Correspondence: andrea.sellitto@unicampania.it; Tel.: +39-081-5010-407

Received: 8 April 2019; Accepted: 5 June 2019; Published: 11 June 2019

Featured Application: Numerical investigation on the damage behavior of CFRP laminates subjected to low velocity impacts.

Abstract: Composite laminates are characterized by high mechanical in-plane properties and poor out-of-plane characteristics. This issue becomes even more relevant when dealing with impact phenomena occurring in the transverse direction. In aeronautics, Low Velocity Impacts (LVIs) may occur during the service life of the aircraft. LVI may produce damage inside the laminate, which are not easily detectable and can seriously degrade the mechanical properties of the structure. In this paper, a numerical-experimental investigation is carried out, in order to study the mechanical behavior of rectangular laminated specimens subjected to low velocity impacts. The numerical model that best represents the impact phenomenon has been chosen by numerical–analytical investigations. A user defined material model (VUMAT) has been developed in Abaqus/Explicit environment to simulate the composite intra-laminar damage behavior in solid elements. The analyses results were compared to experimental test data on a laminated specimen, performed according to ASTM D7136 standard, in order to verify the robustness of the adopted numerical model and the influence of modeling parameters on the accuracy of numerical results.

Keywords: CFRP; Low Velocity Impacts; Cohesive Zone Model (CZM); Finite Element Analysis (FEA); VUMAT; inter-laminar damage; intra-laminar damage

---

1. Introduction

Composite materials are characterized by high mechanical properties values if compared to classic metallic materials. However, their damage behavior is difficult to predict and control. As a matter of fact, impact events on a composite structure are still a "hot topic" for research, nowadays. The different impact phenomena can be divided into three main groups, based on the velocity of the impactor: low velocity impacts, ballistic impacts, and hyper-velocity impacts. In particular, Low Velocity Impacts (LVIs) often result in Barely Visible Impact Damages (BVIDs), which can be hardly detected by the naked eye, but may relevantly decrease the material strength [1–3]. Damages occurring as a consequence of low velocity impacts include both intra-laminar (fiber and/or matrix failure) and inter-laminar damages (delaminations). Among the others, delaminations, which can growth under service loads, can seriously reduce the load carrying capability of the overall structure [4,5]. For this reason, a large number of studies, focused on the prediction of the damages (especially delaminations) induced by LVI, can be found in the literature. In these studies, different approaches, both analytical–numerical and experimental, were considered [6–9]. Different techniques are available and presented in the literature, able to predict and evaluate the damage and the residual strength during LVI phenomena. In [10], an acousto–ultrasonic approach was used to characterize the material impact strength. Low velocity analytical impact prediction models available in the literature are often characterized by a large degree of simplification. However, despite the simplifications in the models, good results can be achieved. In particular, the study in Reference [11] describes a numerical model for predicting the onset and the propagation of damages arising from low velocity impacts induced by large and small impact masses. A further evolution of the analytical model for predicting the behavior of a thin-walled composite structure during a low velocity-high mass impact phenomenon was presented in [12]. An additional analytical methodology based on the energy balance for the development of damage during a LVI phenomenon with the help of a localized strain field was introduced in [13]. However, analytical approaches, which can only be adopted for few particular conditions [1,6,7], cannot be generally considered for a complete and detailed study of the impact event, due to inherent complexity of the impact phenomenon itself [9,14–16]. A more comprehensive study of the impact phenomenon can be surely obtained by means of numerical methods. Indeed, FEM (Finite Element Method) codes can be adopted to model a low velocity impact event on a composite specimen, taking into account both intra-laminar and inter-laminar damages [9,17–19]. In order to evaluate the intra-laminar and inter-laminar damage onset and propagation during an impact phenomenon, progressive damage methods have been presented by several authors, showing good results compared to the experimental data. In [20], LVI tests were performed and the Puck criteria were adopted to analyze the composite structure mechanical response. In the same work, an inter-laminar matrix crack density parameter was introduced, in order to mitigate the computational cost, while decreasing, on the other side, the analysis accuracy. In [21–23], a model capable of predicting the inter-laminar damage in a unidirectional composite laminate was introduced. In particular, the proposed model took into account a matrix non-linear shear behavior, focusing on the element characteristic length in order to alleviate the mesh dependency. Actually, fracture mechanic is widely used to model inter-laminar damage evolution through FE-based procedures such as Virtual Crack Closure Technique (VCCT) and Cohesive Zone Model theory (CZM) [17,24–29]. In [29–31], the influence of intra-laminar damages on the inter-laminar ones is assessed adopting CZM, while, in [26,31–33], the mesh-dependency issue of both inter-laminar and intra-laminar prediction models, which may lead to an unrealistic inter-laminar damage prediction, is emphasized. Low velocity impact events on composite specimens, with different stacking sequences, were studied in several works, which implement the intra-laminar and inter-laminar damages, respectively, by means of Continuum Damage Mechanics (CDM) and CZM approaches [9,18,19]. In these works, good results, in terms of correlation between numerical outputs and experimental data, were obtained for various impact energy levels.

In this work, we propose an analytical model to preliminary evaluate the response of a unidirectional composite plate subjected to a low velocity impact. The analytical model has been developed in accordance with [1,2,33]. Then, the results of the analytical model were compared with those from a numerical model, which uses the analytical boundary conditions. Indeed, in the frame of this first step, the introduced analytical model, based on Classical Laminated Plate Theory (CLPT), was adopted as a benchmark for appropriately selecting the main numerical modeling parameters to be used for the numerical procedure (element type, mesh density, element size, etc.). Subsequently, the numerical model was improved, introducing the boundary conditions of the experimental test. Finally, a further numerical model, which takes into account inter-laminar and intra-laminar damages, was implemented and the results were compared with the experimental test ones. This final numerical test was performed by adopting the numerical model, from the previous step, which best represents the impact event. The low velocity impact phenomenon was numerically simulated by considering the boundary conditions and the specimen size according to the ASTM (American Society for Testing and Materials) D7136 standard [34]. The composite specimens were subjected to a 10 J low velocity impact. A reduction of the computational time has been achieved by introducing global/local techniques [35] between the impacted area and the rest of the specimen. The results, in terms of intra-laminar and inter-laminar damages, have been compared with the experimental ones, obtained by means of Ultrasonic C-Scan tests. Further numerical–experimental correlations on the impactor displacement and velocity, the internal energy, and the force exerted during the impact have been performed as well.

Numerical results, obtained considering different Abaqus element types, were compared to each other. The Finite Element Model has been discretized by means of SC8R Continuum shell elements and C3D8R Solid elements [36]. Inter-laminar damages were modeled by a Cohesive Zone Model based approach, while intra-laminar damage was modeled by means of Hashin's failure criteria and gradual material properties degradation rules. However, the intra-laminar damage model was only provided for Continuum shell elements in Abaqus. Since no intra-laminar damage model was provided for Solid elements, a user-defined material model, able to take into account the intra-laminar damage, was implemented in a user subroutine VUMAT.

In Section 2, the theoretical background on the analytical model and on the intra-laminar damage model implemented in the VUMAT are presented. In Section 3, the test cases are introduced, while in Section 4 the results are presented and discussed.

2. Theoretical background

2.1. Classical Laminated Plate Theory

A multi-degree of freedom analytical model [1,2,37] was introduced in order to study the impact event. A simply supported plate was analytically solved by using the Classical Laminated Plate Theory. The solutions for the analytical approach were obtained in terms of contact force, maximum displacement, impactor velocity, and kinetic energy. Then, a system of ordinary differential equations (ODE), representing the simply supported plate, coupled with a differential equation representing the impactor motion, was obtained.

A laminated composite plate, with in-plane dimensions a and b, respectively, along the x- and y-directions, was considered. The plate was considered simply supported along the four edges, and a concentrated force was applied at the center ($x_0 = a/2$, $y_0 = b/2$). Under these assumptions, the equation of motion along the transverse direction can be expressed as the following system of ordinary differential equations [1]:

$$\ddot{\alpha}_{mn} + \omega_{mn}^2 \alpha_{mn} = \frac{4F}{abI_1} \sin\left(\frac{m\pi x_0}{a}\right)\sin\left(\frac{n\pi y_0}{b}\right), \tag{1}$$

In Equation (1), F is a concentrated applied force, while ω_{mn}, which is the natural frequency of the system for the strain modes m,n, is given by:

$$\omega_{mn} = \sqrt{\frac{\pi^4}{a^4 I_1}(D_{11}m^4 + 2(D_{12} + 2D_{66})m^2n^2r^2 - 4D_{16}m^3nr - 4D_{26}mn^3r^3 + D_{22}n^4r^4)}, \tag{2}$$

where D_{ij} are the terms of the bending stiffness matrix, I_1 is the mass per unit length, and $r = a/b$.

As reported in [1], when both values of m and n are negative, the effects of deflection in the plate become negligible. The choice of high positive values for both values is related to the phenomenon being analyzed: as the values of m and n increase, the rotary inertia and shear deformation effects becomes significant.

Moreover, the following Equation (3), governing the impactor dynamic, has been considered:

$$M\ddot{z}(t) + F_c(t) = 0, \tag{3}$$

where M and $\ddot{z}(t)$ are, respectively, the impactor mass and acceleration, $F_c(t)$ is the contact force, and t is the time variable.

A proper contact law has been chosen to couple Equations (3) and (1). In particular, the contact force $F_c(t)$ has been related to the indentation $\alpha = z(t) - w_0(x_0,y_0,t)$, expressed as the difference between the impactor displacement $z(t)$ and the laminate central point displacement $w_0(x_0,y_0,t)$. In this work, a linearized form of the Hertz Contact Law was adopted [37], therefore the contact force history, exerted during the impact phenomenon, can be expressed as:

$$F_c(t) = k_y \alpha, \tag{4}$$

where k_y represents the linearized contact stiffness.

The boundary conditions of the system of ODEs are:

$$\alpha_{mn}(0) = \dot{\alpha}_{mn}(0) = 0, \tag{5}$$

$$\begin{aligned} z(0) &= 0 \\ \dot{z}(0) &= V \end{aligned} \tag{6}$$

where V is the initial velocity of the impactor.

2.2. Composites Damage Criteria

In order to account for the damage behavior of the laminate, a User Material was developed in the ABAQUS FEM environment capable to consider the fiber and matrix damage mechanisms in tension and compression [4,19,29]. Each failure mode consisted of two different phases. Initially, a linear mechanical behavior is considered up to the damage initiation threshold. Then, the damage evolution up to the complete damage is evaluated.

The criteria used for evaluate the damage initiation are based on Hashin's (fiber and matrix tension and fiber compression) and Puck–Shurmann [38] (matrix compression) failure criteria, as reported in Equations (7)–(10).

$$F_{ft} = \left(\frac{\sigma_1}{X_T}\right) \geq 1 \quad if \quad \sigma_1 \geq 0, \tag{7}$$

$$F_{fc} = \left(\frac{\sigma_1}{X_C}\right) \geq 1 \quad if \quad \sigma_1 < 0, \tag{8}$$

$$F_{mt} = \left(\frac{\sigma_2}{Y_T}\right)^2 + \left(\frac{\tau_{12}}{S_{12}}\right)^2 + \left(\frac{\tau_{23}}{S_{23}}\right)^2 \geq 1 \quad if \quad \sigma_2 \geq 0, \tag{9}$$

$$F_{mc} = \left(\frac{\tau_{nt}}{S_{23}^A - \mu_{nt}\sigma_{nn}}\right)^2 + \left(\frac{\tau_{nl}}{S_{12} - \mu_{nl}\sigma_n}\right)^2 \geq 1 \quad if \quad \sigma_2 < 0, \tag{10}$$

where X_T and X_C are the fiber tensile and compressive strengths, Y_T is the matrix tensile strength, and S_{12} and S_{23} are respectively the longitudinal and transverse shear strengths. Moreover, the parameters introduced in Equation (10) have been defined with respect to the potential fracture plane; in particular, S^A_{23} is the transverse shear strength in the potential fracture plane, μ_{nt} and μ_{nl} are respectively the friction coefficients in the transverse and longitudinal directions, σ_{nn} is the normal stress, and τ_{nt} and τ_{nl} are respectively the shear stresses in the transverse and longitudinal directions. Indeed, experiments have demonstrated that unidirectional laminates experience shear fail [39] when subjected to transverse compressive loads, with a fracture plane oriented with an angle $\theta_f = 53° \pm 2°$ [21]. Hence, to correctly evaluate the damage status of the laminate, the stress/strain components in the θ_f oriented fracture plane have been considered, instead of the nominal 0° oriented nominal plane.

Therefore, the stress components in the general fracture plane L-N-T oriented with a fracture angle θ_f, as in Figure 1, can be obtained as a function of the stress components defined in the lamina coordinate system 1-2-3:

$$\begin{cases} \sigma_n = \sigma_2 \cos^2 \theta_f + \sigma_3 \sin^2 \theta_f + 2\tau_{23} \cos \theta_f \sin \theta_f \\ \tau_{nl} = \tau_{23} \cos \theta_f + \tau_{31} \cos \theta_f \\ \tau_{nt} = (\sigma_3 - \sigma_2) \cos \theta_f \sin \theta_f + \tau_{23}\left(\cos^2 \theta_f - \sin^2 \theta_f\right) \end{cases} \tag{11}$$

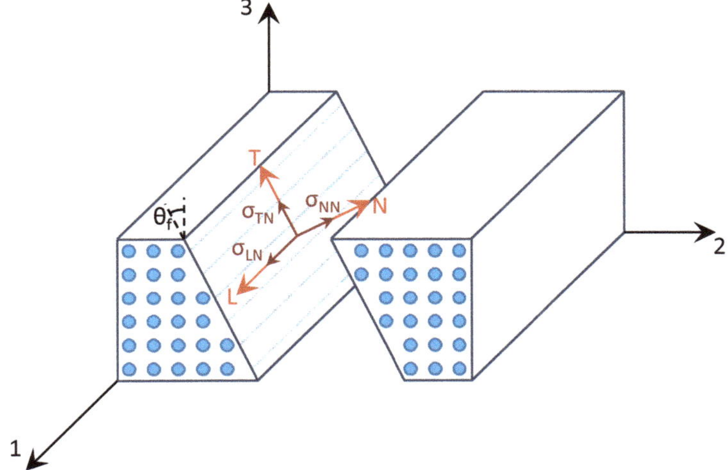

Figure 1. Compressive matrix failure and fracture plane.

Hence, the parameters related to the fracture angle introduced in Equation (10) can be defined [40]:

$$\phi = 2 \cdot \theta_f - 90°, \tag{12}$$

$$\mu_{nt} = \tan \phi, \tag{13}$$

$$S_{23}^A = \frac{Y_C(1 - \sin \phi)}{2 \cos \phi}, \tag{14}$$

$$\mu_{nl} = \mu_{nt} \cdot \frac{S_{12}}{S_{23}^A}, \tag{15}$$

where Y_T is the matrix compressive strength.

Each mode is characterized by the failure behavior reported in Figure 2. In particular, a linear elastic behavior with an initial stiffness K can be observed up to location A, which identifies the damage onset. Then, each location B on the segment AC identifies a partial damage condition characterized by a degraded stiffness K_d, up to location C, where the complete failure can be observed.

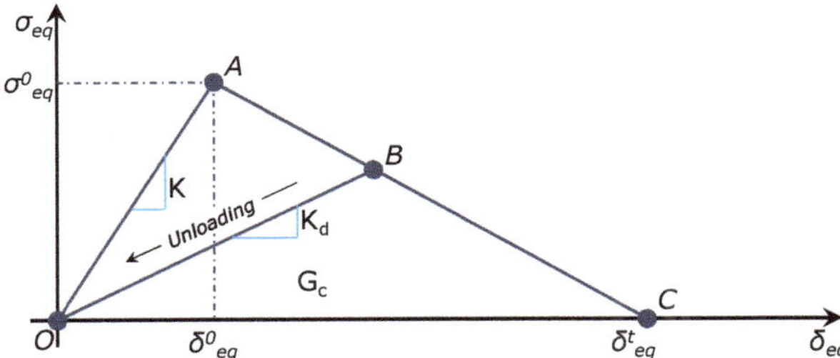

Figure 2. Constitutive relation adopted for fiber and matrix failure modes in tension and compression.

To take into account the material stiffness degradations in locations B, the damage coefficients d_i [36] are introduced for each failure mode:

$$d_i = \frac{\delta_{i,eq}^t\left(\delta_{i,eq} - \delta_{i,eq}^0\right)}{\delta_{i,eq}\left(\delta_{i,eq}^t - \delta_{i,eq}^0\right)}; \quad \delta_{i,eq}^0 \le \delta_{i,eq} \le \delta_{i,eq}^t; \quad i \in (fc, ft, mc, mt). \tag{16}$$

Indeed, the damaged stiffness matrix $\mathbf{C_D}$ can be expressed as a function of the damage coefficients as:

$$\mathbf{C_D} = \frac{1}{D}\begin{bmatrix} \left(1-d_f\right)E_1 & \left(1-d_f\right)(1-d_m)v_{21}E_1 & 0 \\ \left(1-d_f\right)(1-d_m)v_{12}E_2 & (1-d_m)E_2 & 0 \\ 0 & 0 & D(1-d_s)G_{12} \end{bmatrix}, \tag{17}$$

where $D = 1 - (1 - d_f)(1 - d_m)v_{12}v_{21}$, and d_f, d_m, and d_s are respectively the fiber, matrix, and shear damage coefficients:

$$d_f = \begin{cases} d_{ft} & \text{if } \hat{\sigma}_{11} \ge 0 \\ d_{fc} & \text{if } \hat{\sigma}_{11} < 0 \end{cases}$$
$$d_m = \begin{cases} d_{mt} & \text{if } \hat{\sigma}_{22} \ge 0 \\ d_{mc} & \text{if } \hat{\sigma}_{22} < 0 \end{cases} \tag{18}$$
$$d_s = 1 - \left(1 - d_{ft}\right)\left(1 - d_{fc}\right)(1 - d_{mt})(1 - d_{mc})$$

However, this methodology has been found strongly dependent on the domain discretization. Hence, to reduce mesh dependencies, the element characteristic length L_C has been introduced. In this work, the solution proposed by Bazant and Oh [39] and reported in Equation (19) has been adopted:

$$L_C = \frac{\sqrt{A_{ip}}}{\cos\theta}, \tag{19}$$

where A_{ip} is the Area of the element corresponding to the ip-th integration point and θ is the fracture angle. Then, the equivalent displacements was computed as a function of the element characteristic length L_C:

$$\delta_{mt,eq}^0 = \frac{\sqrt{(\varepsilon_2 \cdot L_C)^2 + (\gamma_{12} \cdot L_C)^2 + (\gamma_{23} \cdot L_C)^2}}{\sqrt{F_{mt}}}, \tag{20}$$

$$\delta_{mc,eq}^0 = \frac{\sqrt{(\gamma_{ln} \cdot L_C)^2 + (\gamma_{tn} \cdot L_C)^2}}{\sqrt{F_{mc}}}, \tag{21}$$

$$\delta_{i,eq}^0 = \frac{\sqrt{(\varepsilon_1 \cdot L_C)^2}}{\sqrt{F_i}}; \quad i \in (fc, ft), \tag{22}$$

$$\delta_{ft,eq}^t = \frac{2G_{IC}^T}{\sigma_{ft,eq}^0}, \tag{23}$$

$$\delta_{fc,eq}^t = \frac{2G_{IC}^C}{\sigma_{fc,eq}^0}, \tag{24}$$

$$\delta_{mt,eq}^t = \frac{2G_{IIC}^T}{\sigma_{mt,eq}^0}, \tag{25}$$

$$\delta^t_{mc,eq} = \frac{2G^C_{IIC}}{\sigma^0_{mc,eq}}, \tag{26}$$

where G^T_{IC} and G^C_{IC} are the tensile and compressive Mode I fracture toughness, G^T_{IIC} and G^C_{IIC} are the tensile and compressive Mode II fracture toughness, and:

$$\gamma_{nt} = \gamma_{12} \cos \theta_f + \gamma_{13} \sin \theta_f, \tag{27}$$

$$\gamma_{nl} = -\varepsilon_2 \cos \theta_f \sin \theta_f + \varepsilon_3 \cos \theta_f \sin \theta_f + \gamma_{23}\left(2 \cos^2 \theta_f - 1\right). \tag{28}$$

3. Test Cases Descriptions

3.1. Analytical Set-Up

The analyzed test case is a simply supported laminated plate impacted by a 3.58 kg mass steel sphere. A 10 J impact energy is considered, leading to an impactor initial velocity equal to 2.364 m/s. The plate is made by a unidirectional carbon-fiber laminate, with in-plane dimensions in the x- and y-directions, respectively, a = 150 mm and b = 100 mm. The stacking sequence of the laminate is $[45/-45/0/0/90/0]_{2S}$, while the mechanical properties of the adopted material system and the ply thickness are reported in Table 1.

Table 1. Lamina mechanical properties.

Property [unit]	Value
ρ [kg/m^3]	1600
E_{11} [MPa]	149,500
$E_{22} = E_{33}$ [MPa]	8430
$G_{12} = G_{13}$ [MPa]	4200
G_{23} [MPa]	2520
$\upsilon_{12} = \upsilon_{13}$ [-]	0.3
υ_{23} [-]	0.45
υ_{21} [-]	0.0186
X_T [MPa]	2143
X_C [MPa]	1034
Y_T [MPa]	75
Y_C [MPa]	250
$S_{12} = S_{13}$ [MPa]	108
S_{23} [MPa]	95
G^T_{1c} [kJ/m^2]	30.72
G^C_{1c} [kJ/m^2]	7.15
G^T_{2c} [kJ/m^2]	0.667
G^C_{2c} [kJ/m^2]	7.41
Ply thickness [mm]	0.186

A spherical impactor with an 8 mm in radius was considered. The linear contact stiffness was set equal to $k_y = 7.8755 \times 10^3$ N/mm, according to the equations reported in [34]. In these analyses, aimed to preliminary assess the elastic behavior of the structure under impact loading conditions, only the elastic behavior has been taken into account. Therefore, no damage has been considered.

3.2. Experimental Set-Up

The experimental impact test was performed according to the ASTM D7136 standard [34], as shown in Figure 3.

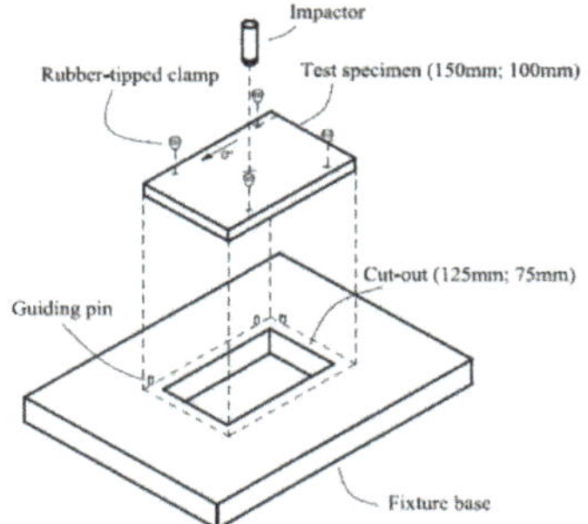

Figure 3. Experimental set-up.

The three tested specimens were cut from a single CFRP plate by using the waterjet technique. The CFRP plate is composed of prepreg high modulus carbon fiber and thermoset epoxy resin, cured in autoclave. The test was performed with the CEAST Fractovis plus drop tower test machine. The plates were normally impacted at the center. The impactor was a 16 mm diameter hemisphere made of hardened steel, with a mass of 3.58 kg. A 10 J impact energy was considered. Ultrasonic inspections were performed on the specimen prior to the impact test to assure the lack of manufacturing defects, and after the experimental test to evaluate the damaged area.

4. Results and Discussion

4.1. Preliminary Analyses

The analytical approach results expressed in terms of impactor displacement, contact force, impactor velocity, and kinetic energy versus time are shown respectively in Figures 4–7. The results were obtained by considering an increasing number of modes.

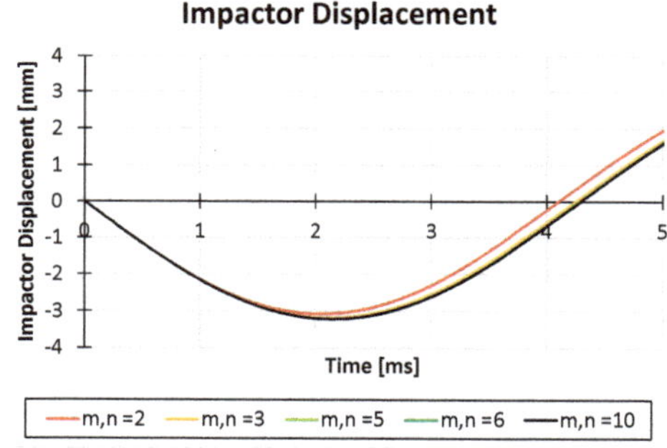

Figure 4. Analytical solution—impactor displacement.

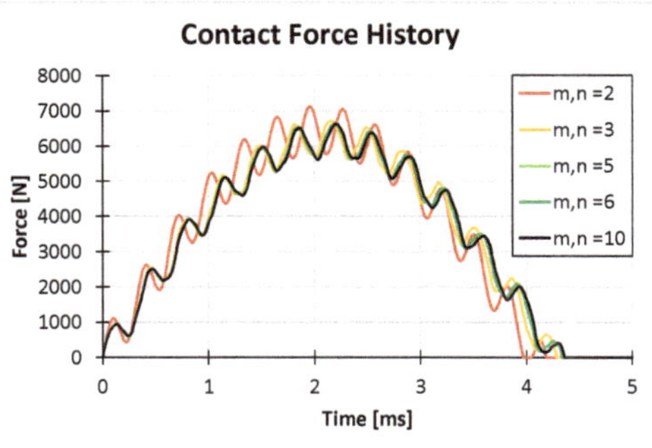

Figure 5. Analytical solution—contact force history.

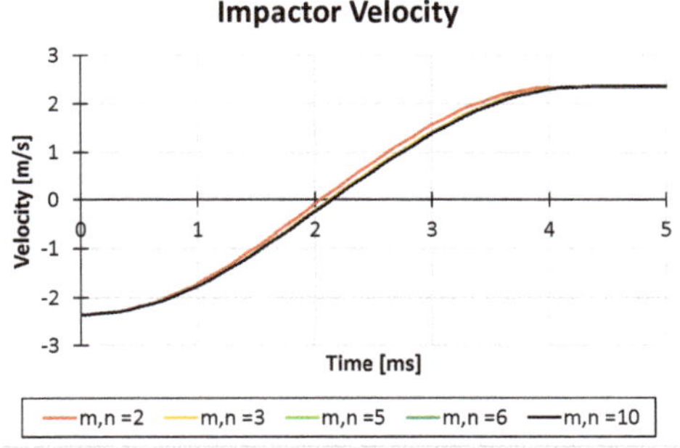

Figure 6. Analytical solution—impactor velocity.

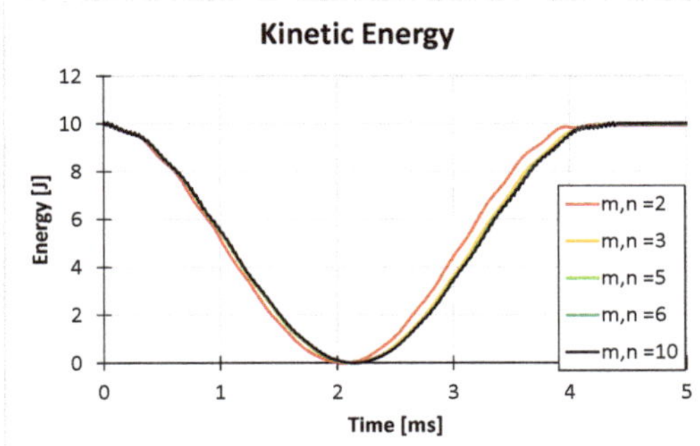

Figure 7. Analytical solution—kinetic energy.

Figures 4 and 5 highlight that lower values of m and n results in higher values of stiffness. Indeed, increasing the m and n values led to a decrease of the peak force and to an increase in the maximum deflection of the plate central point. Additionally, according to Figures 4 and 5, the panel response in terms of impact force and maximum deflection tended to converge as the values of m and n increased. Figure 6 shows the trends of the impact velocity considering different m and n values. Although the velocity had the same initial and final values, different m and n led to different velocity trends. Moreover, as a direct consequence of the variation of the impact velocity, the kinetic energy, shown in Figure 7, showed different behavior when low values of m and n were considered. However, as m and n increase, the graphs shown in Figures 4–7 appear to be perfectly overlapped.

Then, the numerical results were compared with the analytical ones. The $m,n = 10$ analytical results were chosen as the reference analytical solution.

In order to define the numerical model which best-fit the analytical results, several parameters were investigated, such as the element type (reduced integration continuum shell elements SC8R and reduced integration solid elements C3D8R), the in-plane element size (1 mm, 2 mm, and 4 mm), and the number of elements along the thickness direction. In particular, Continuum Shell elements were discretized with 1, 12, and 24 elements in the thickness direction, while Solid elements were discretized with 24 elements in the thickness direction (one element per ply). Hence, more than one orientation is associated to the continuum shell elements belonging to the models characterized by less than 24 elements in the thickness direction.

In the numerical analyses, the plate was considered simply supported on the edges, as reported in Figure 8. In the framework of preliminary analyses, aimed to compare the analytical model with the numerical ones to determine the numerical parameters which best represent the mechanical behavior of the plate subjected to low velocity impact, both the inter-laminar and intra-laminar damages were neglected.

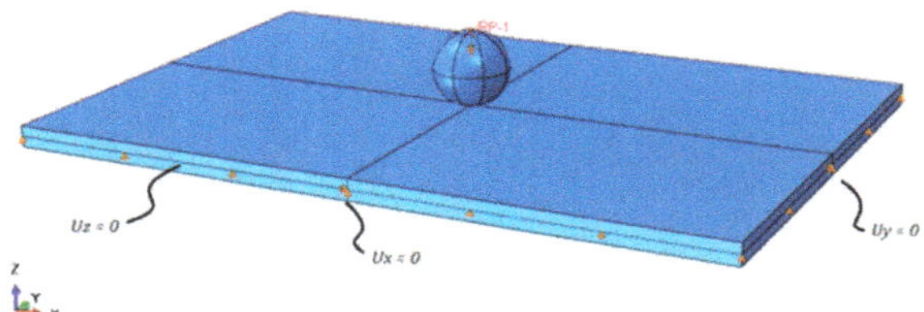

Figure 8. Shell model boundary conditions.

In Figures 9–13, the numerical and analytical results, in terms of displacement of the impactor and contact force, are compared. In particular, the results shown in Figure 9 were obtained by using a shell element formulation, while the results in Figures 10–12 were obtained by using a continuum shell formulation considering, respectively, 1, 12, and 24 elements in the thickness direction. Finally, in Figure 13, the results obtained by using a solid element formulation, with 24 elements in the thickness direction, are introduced.

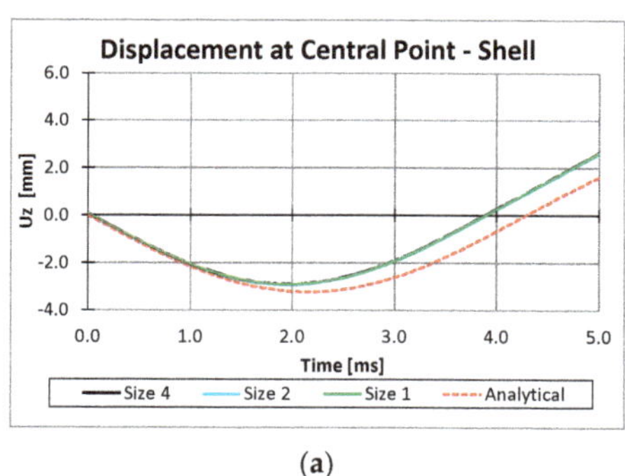

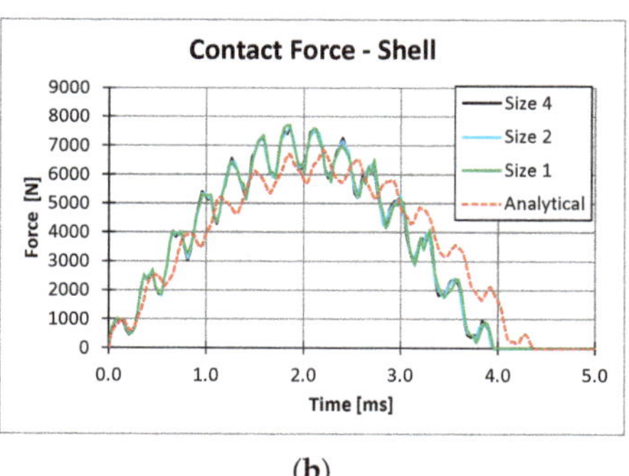

(a)   (b)

Figure 9. Shell elements: analytical–numerical comparison. (a) Central point displacement; (b) contact force.

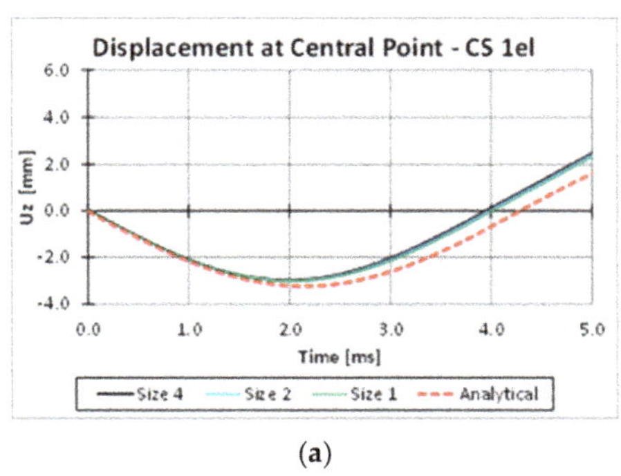

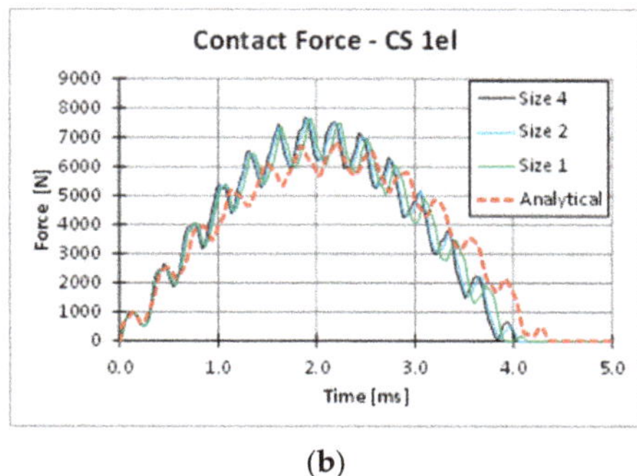

(**a**)   (**b**)

Figure 10. Continuum shell elements with one element in the thickness direction: analytical–numerical comparison. (**a**) Central point displacement; (**b**) contact force.

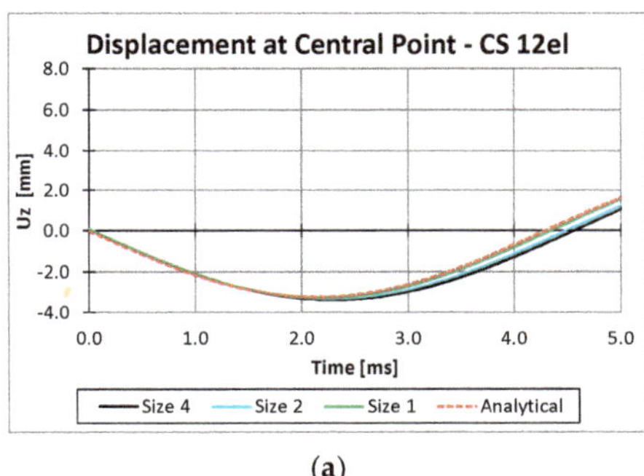

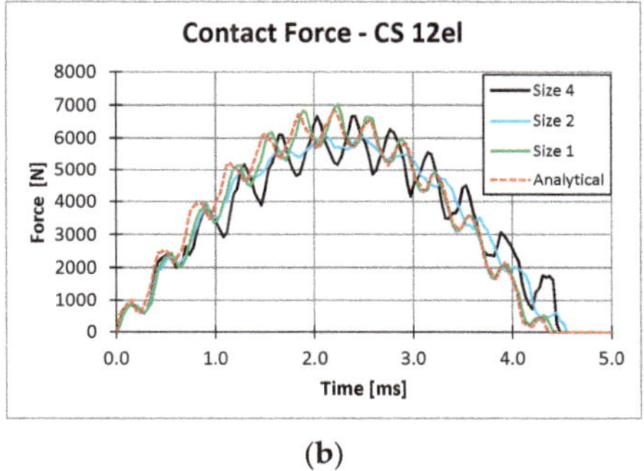

(**a**)   (**b**)

Figure 11. Continuum shell elements with 12 elements in the thickness direction: analytical–numerical comparison. (**a**) Central point displacement; (**b**) contact force.

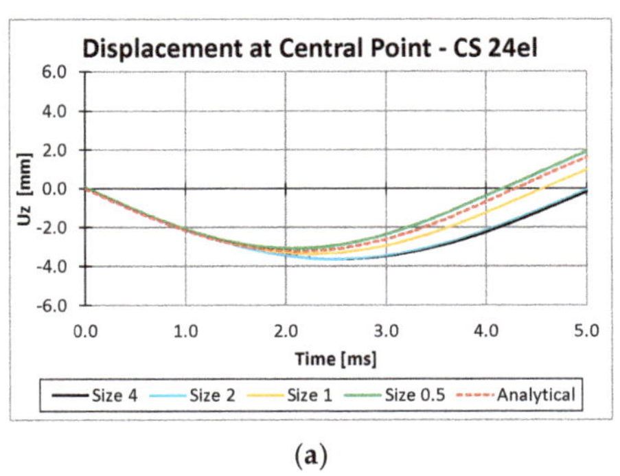

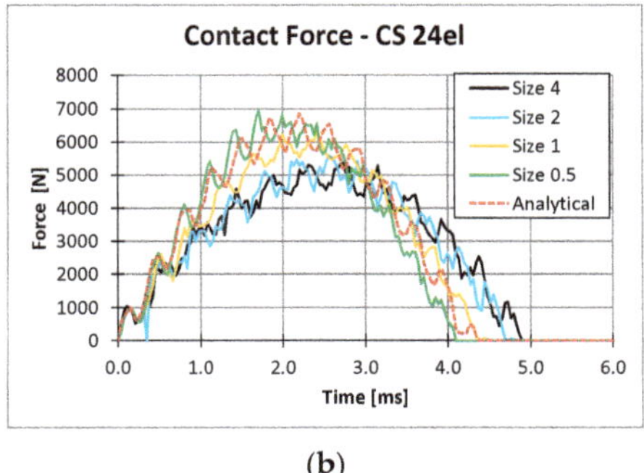

(**a**)   (**b**)

Figure 12. Continuum shell elements with 24 elements in the thickness direction: analytical–numerical comparison. (**a**) Central point displacement; (**b**) contact force.

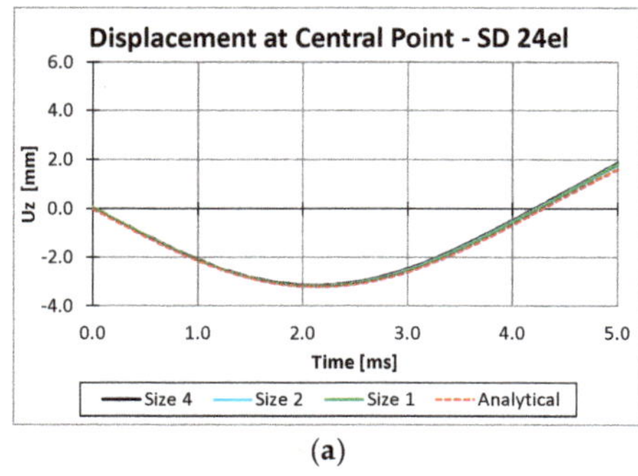
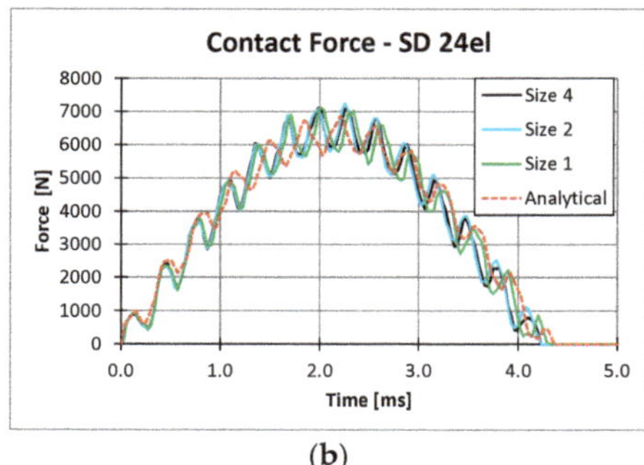

Figure 13. Solid elements with 24 elements in the thickness direction: analytical–numerical comparison. (**a**) Central point displacement; (**b**) contact force.

According to the Figures 9–13, Shell and Solid elements' models were not particularly influenced by the in-plane element size. On the other hand, Continuum shell elements' models were more sensitive to the in-plane element size as the number of elements in the thickness direction increased. In particular, the numerical results of the Continuum shell model with 12 and 24 elements in the thickness direction associated to coarser mesh (size 2 mm or 4 mm) differed noticeably from the analytical results. This is mainly due to hourglass problems, which can be relevant for the reduced integration scheme elements. Unfortunately, the reduced integration scheme is mandatory in Abaqus/Explicit when Continuum shell elements were adopted. To alleviate hourglass deformation problems that could arise in the numerical models, the default hourglass controls available in Abaqus/Explicit [36] was adopted.

In general, the results highlight that the best results were obtained by the Solid element model discretized with one element per ply and an in-plane element size equal to 1 mm; however, a good numerical–analytical correlation was obtained also by the Continuum shell mode, discretized by 12 elements in the thickness direction and an in-plane element size equal to 1 mm. Indeed, this last configuration was chosen for the analysis presented hereafter due to its excellent compromise between low computational cost and agreement with the analytical solution.

Despite their simplifications, the analytical models available in the literature well describe the behavior of a CFRP plate subjected to impact loading conditions. Hence, the numerical model was validated by comparisons with the analytical one, and the used modeling approach, neglecting in this preliminary phase the damages, has been assessed. Once the accuracy of the impact numerical model is verified, some considerations should be made about the relevant difference between the boundary conditions implemented in the analytical model and the ones implemented in the experimental test. Actually, simply supported conditions have been considered in the analytical and preliminary numerical models; however, the boundary conditions of the experimental test, suggested by the ASTM standards and reported in Figure 14, are substantially different. These boundary conditions are a consequence of a support fixture whose degrees of freedom are restrained and four rigid clamps, also modeled as rigid bodies, which hold the specimen still during impact and avoid any rebound.

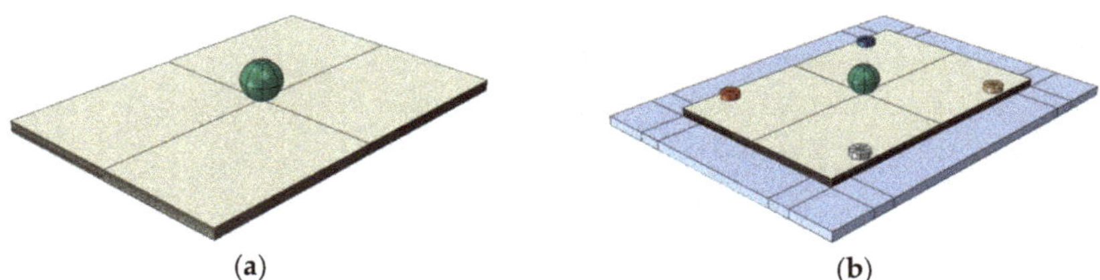

Figure 14. Boundary conditions. (**a**) Analytical; (**b**) experimental.

A further numerical model obtained with the implementation of the Experimental boundary condition (BC-E) was introduced to understand the effects of boundary conditions on the numerical results. In Figure 15, the analytical, experimental and numerical results of the shell model obtained by considering both analytical (BC-A) and experimental (BC-E) boundary conditions were compared.

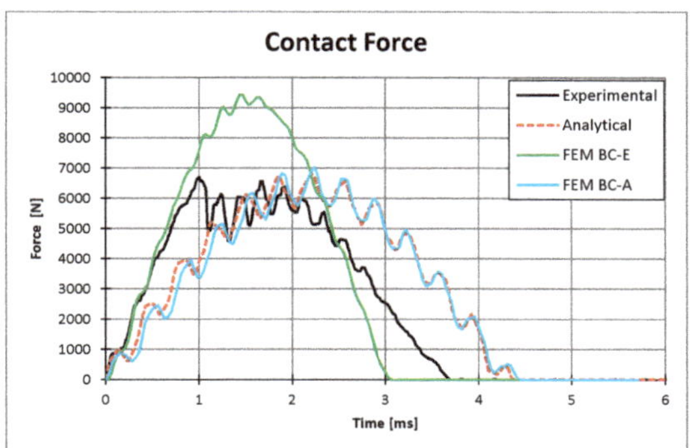

Figure 15. Contact force comparison between analytical and numerical boundary conditions.

In Figure 15, a comparison between experimental, analytical, and numerical results, in terms of force as a function of the time, is reported. The comparison shows that in the first part of the curve both numerical and analytical models were in agreement with the experimental results. However, after a first phase, the analytical model and the FEM BC-A numerical model greatly differed from the experimental result. On the other hand, the FEM BC-E numerical model was characterized by the same mechanical response in terms of stiffness of the laminate, up to the maximum force peak. Indeed, neglecting the damages in the numerical model resulted in grater force peak, if compared with the experiment. Once the accuracy of the numerical model, in terms of elastic response, was established, the numerical models which provided the best results compared to the analytical solution was considered for the final impact test: final Continuum shell model and Final 3D solid model. Since Abaqus does not provide any intra-laminar damage models for Solid elements, a VUMAT was introduced. On the other side, Abaqus built-in Hashin's criteria were used for model discretized by means of Continuum shell elements. For both the final continuum shell model and the final solid FEM model, gradual material properties degradation rules have been adopted for intra-laminar damage progression and inter-laminar damage were implemented through cohesive elements layers placed at plies interfaces.

4.2. Final Solid Model

This Final Solid model was created by introducing two different meshed areas connect by a global–local conditions. Indeed, in the specimen impacted area, where most of the damage is supposed to develop (Figure 16) a ply-by-ply refined discretization was considered with each ply represented by a single layer of 3D solid elements C3D8R with a reduced integration scheme. On the contrary, the external global domain has been modeled with a coarser continuum shell mesh. With this discretization scheme, an accurate study can be performed by reducing, at the same time, the computational effort [40].

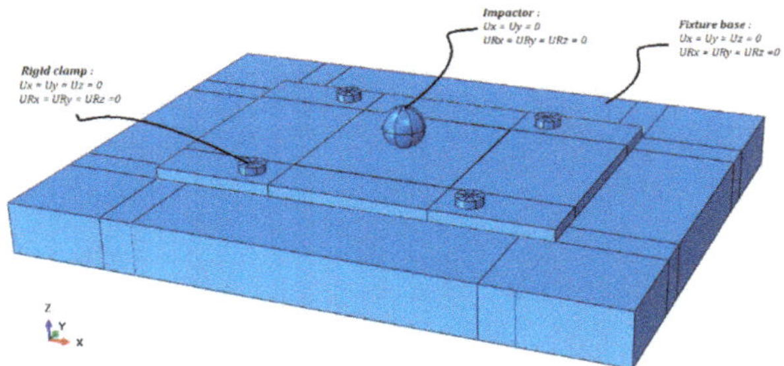

Figure 16. Domain and boundary conditions.

For 3D solid elements, the intra-laminar damage model introduced in Section 2.2 was applied by means of a VUMAT. Moreover, since the inter-laminar damages are likely to occur only between plies with different fiber orientation, zero-thickness cohesive layers were placed only in these locations. As highlighted in Figure 17, cohesive layers were connected to the adjacent layer by means of tie-constraint. Table 2 reports the mechanical properties of the cohesive material system, which were experimentally evaluated by means of three-point bending (E_n, E_t, E_s), double cantilever beam (N_{max}, G_{Ic}), and end notched flexure tests (T_{max} and S_{max}, G_{IIc} and G_{IIIc}).

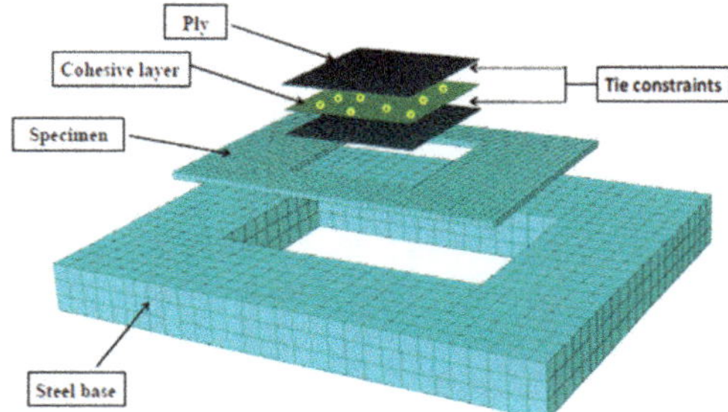

Figure 17. Exploded model (detail).

Table 2. Cohesive mechanical properties.

Property [unit]	Value
E_n [*MPa/mm*]	1.155×10^6
E_t [*MPa/mm*]	6×10^5
E_s [*MPa/mm*]	6×10^5
N_{max} [*MPa*]	62.3
T_{max} [*MPa*]	92.3
S_{max} [*MPa*]	92.3
G_{Ic} [*kJ/m²*]	0.18
G_{IIc} [*kJ/m²*]	0.5
G_{IIIc} [*kJ/m²*]	0.5

Both solid and cohesive elements have been discretized by 1×1 mm² elements. The external global domain, which has been modeled by one layer of Continuum shell *SC8R* elements along thickness direction, has been discretized with an element in-plane size of 4 mm. The specimen is constrained by a fixture and four clamps, modeled with R3D4 rigid elements. The impactor was modeled as a sphere of 16 mm diameter modeled with 3D Solid C3D8R elements. A surface-to-surface contact was used between the impactor and the specimen. During the analysis, the completely damaged elements were removed from the model to increase the computational efficiency. In particular, when cohesive elements were removed, plies can touch each other at the interfaces. Hence, a general contact on the whole model was defined (except for the already defined impact surfaces) to take into account this behavior. A friction coefficient with a value of 0.5 was introduced for both impactor-to-ply contact and ply-to-ply contact. Enhanced hourglass control was set to decrease hourglass problems related to the reduced integrated elements.

4.3. Final Continuum Shell Model

In the Final Continuum shell model, 12 continuum shell elements along thickness direction were used to model the specimen (Figure 18). For each element, a composite layup composed of two plies was defined.

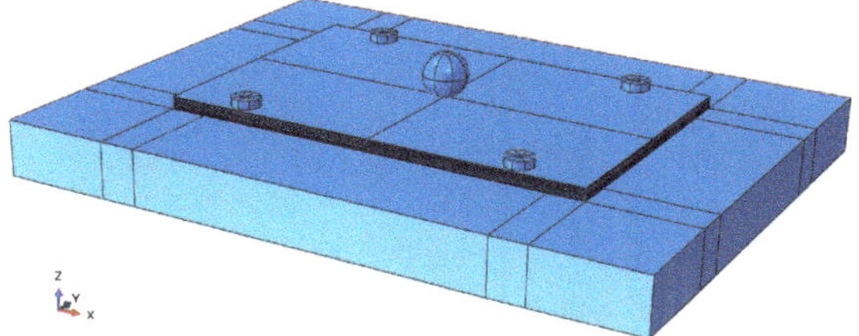

Figure 18. Continuum shell elements domain.

The intra-laminar damages have been accounted by means of Hashin's failure criteria, while cohesive layers placed at plies interfaces and introduced in Table 2 were used to model inter-laminar damage. The contact laws introduced in Section 4.2 were adopted.

4.4. Numerical–Experimental Correlation

In this section, a comparison between the experimental results and the numerical ones for the both investigated configurations (continuum shell element and solid elements) is performed considering both intra-laminar and inter-laminar damages. Three 10 J impact experimental tests were performed. The damaged area was acquired by means of the ultrasonic C-Scan. Figure 19 shows the C-scan of one representative specimen.

Figure 19. C-scan of the damaged area.

A linear pattern obtained by means of Abaqus/Viewer was used to evaluate the experimental damaged area. The tested specimens were comparable in terms of force, displacement, and kinetic energy as a function of the time, and inter-laminar damages. In particular, the experimental damaged area was about 250 mm^2, while the numerical damaged area, evaluated by using different numerical models, was in the range 166–609 mm^2.

The continuum shell model uses a lower number of interfaces between the plies, if compared with the solid element discretization model, and two plies with different orientations are discretized with only one element through the thickness. This assumption could influence the plate mechanical response, leading to different results in terms of delaminated area.

In Figures 20 and 21, the numerical results of the final Continuum shell model in terms of inter-laminar and intra-laminar damages are respectively reported.

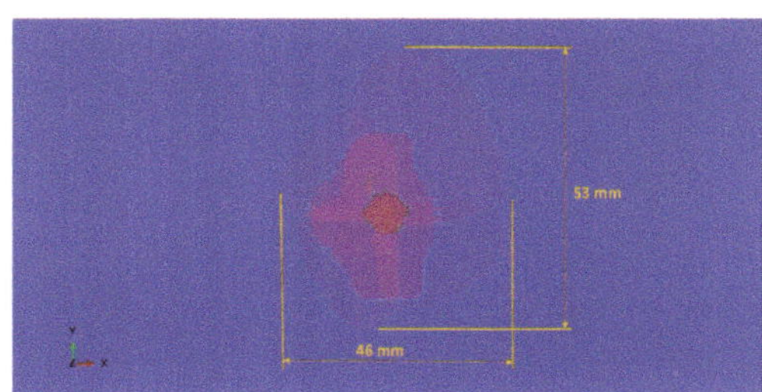

Figure 20. Shell element model: delamination.

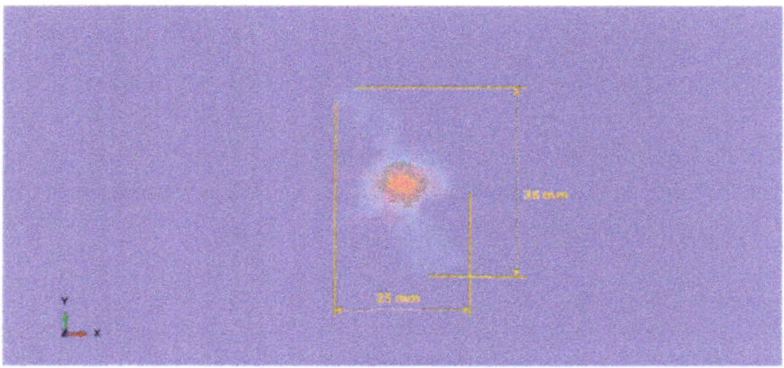

Figure 21. Shell element model: intra-laminar damages (envelope).

According to Figure 20, the simulation over predicts the size of the delamination. This over-prediction can be related to the shell formulation, which is not able to correctly predict the forces acting in the out-of-plane direction, and to the discretization in the thickness direction. Indeed, assuming 12 elements in the thickness direction reduces the number of interfaces where the delamination can occur, leading to larger inter-laminar damages in the remaining interfaces.

Figures 22 and 23 show the predicted and measured contact force, impactor displacement, internal energy, and impactor velocity as a function of the time.

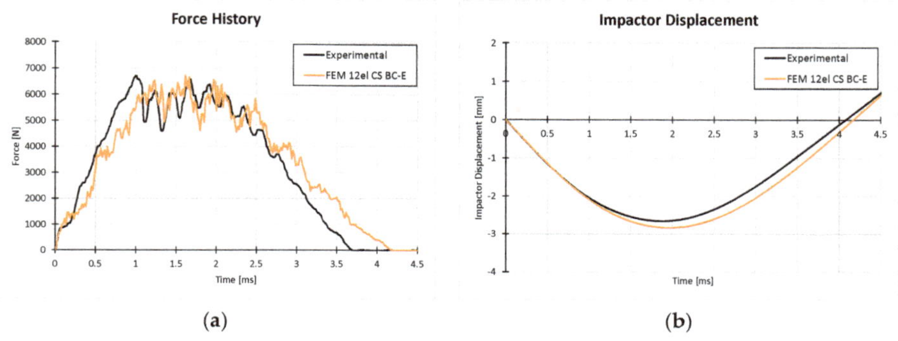

Figure 22. Shell element model. (**a**) Contact force history; (**b**) impactor displacement.

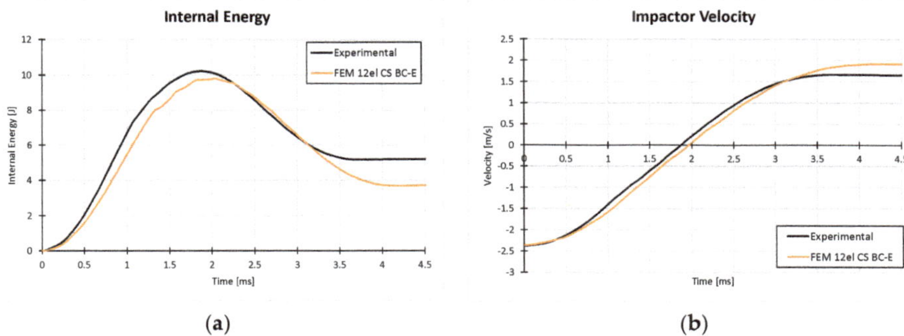

Figure 23. Shell element model. (**a**) Internal energy; (**b**) impactor velocity.

According to Figure 22a, the numerical curve deviates from the experimental one at about 0.3 ms due to anticipated numerical prediction of the onset of delaminations, resulting in a longer contact duration. This effect can be due to the lower number of interfaces. Moreover, the numerically overestimated impactor displacement (6.77% as reported in Figure 22b) and the underestimated residual internal energy of the numerical model (−28.52% as reported in Figure 23a), confirm the over-prediction of damages observed in the simulation. Hence, in the adopted numerical model, significant differences can be observed between the numerical and experimental inter-laminar and intra-laminar damages, as shown in Figures 20 and 21. In particular, the stiffness of the continuum shell element is underestimated if compared with experimental results. Indeed, the maximum displacement of the numerical result is lower than the experimental result in conjunction with a lower peak force.

The numerical results of the solid element model in terms of inter-laminar and intra-laminar damages are reported in Figures 24–27.

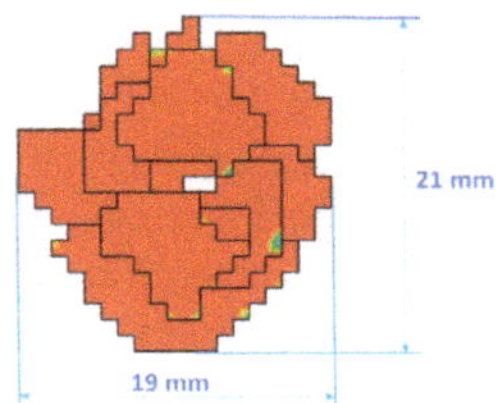

Figure 24. Solid element model: delamination.

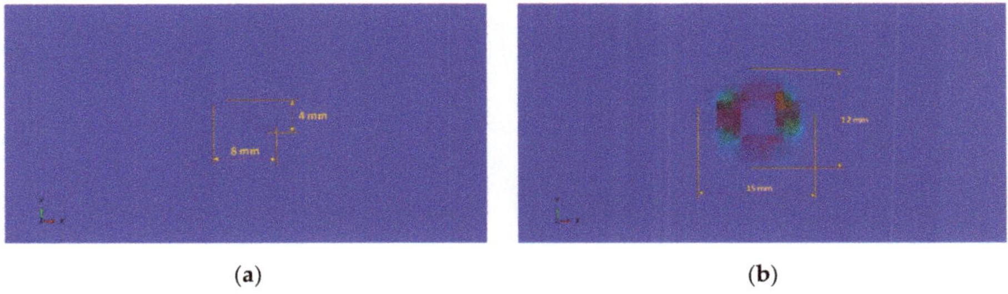

Figure 25. (**a**) Fiber compressive damage; (**b**) matrix tensile damage.

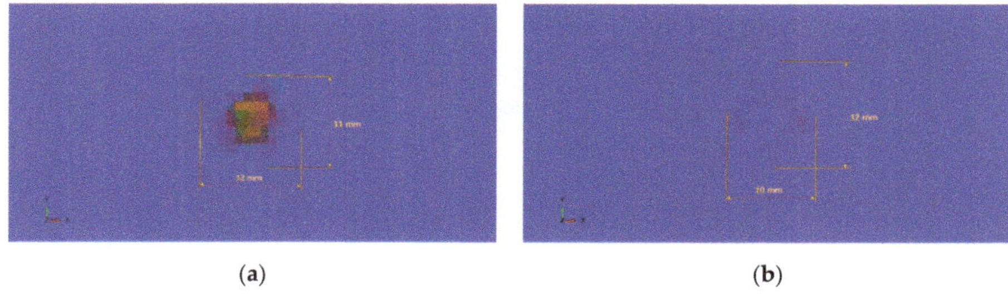

(a) **(b)**

Figure 26. (**a**) Matrix compressive damage; (**b**) longitudinal shear damage.

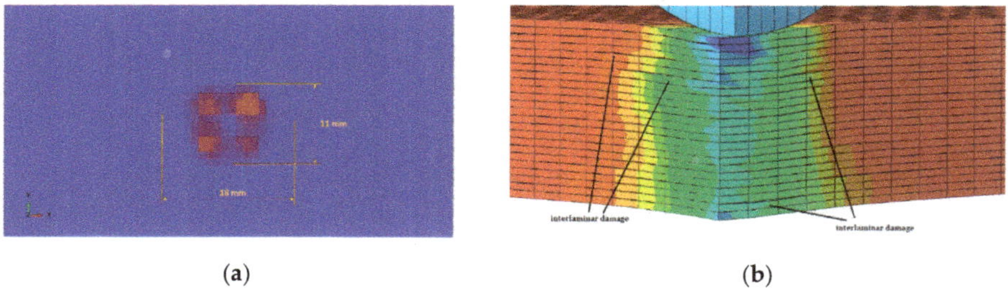

(a) **(b)**

Figure 27. (**a**) Transversal Shear damage; (**b**) specimen section detail.

Figure 24 shows a good agreement between the predicted numerical delaminated area (267 mm^2) and the experimental one. Moreover, the delaminations through the specimen sections are reported in Figure 27b. A good agreement between numerical and experimental results has been reached in terms of displacement-time, energy-time, and impact velocity-time as highlighted in Figure 28a,b and Figure 29b. The numerical model with solid element approach is able to predict with good approximation the maximum peak force, with a deviation of −8.90% with respect to the experiment, although it is not able to accurately predict the entire load history. Moreover, the proposed numerical model better predicts the maximum displacement (deviation of 0.37%) and reproduces the overall stiffness of the whole plate. According with Figure 29a, a good correlation in terms of internal energy vs time has been reported. In particular, the solid model compared to the shell one reproduces, in a good way, the amount of energy dissipated in the form of friction, viscosity, and especially due to breakages, is better computed by the solid model (−15.41%) respect to the shell one (−28.52%).

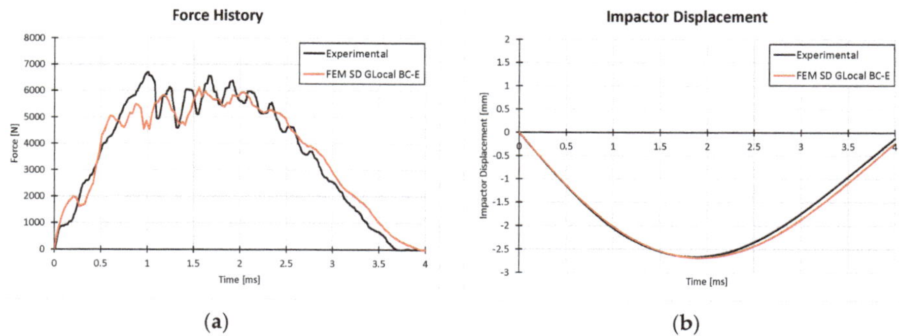

(a) **(b)**

Figure 28. (**a**) Contact force history; (**b**) impactor displacement.

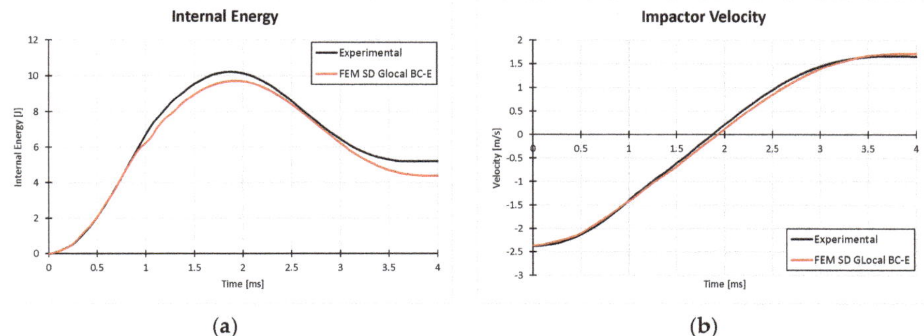

Figure 29. (**a**) Internal energy; (**b**) impactor velocity.

5. Conclusions

In this work, the influence of modeling parameters on the accuracy of results in simulating the impact response of a composites plate subjected to low velocity impact was investigated. Indeed, a numerical procedure has been introduced, able to assess the relevance of modeling parameters for the simulation of intra-laminar and inter-laminar damage onset and progression. In the first stage, an analytical procedure for the prediction of the impact event has been introduced. The results have been used to calibrate the numerical models in terms of element types, element size, and element discretization along the thickness. Continuum shell and solid elements formulations have been analyzed. After this first screening based on analytical–numerical comparisons, continuum shell element model (12 elements through the thickness) and a solid element model (24 elements through the thickness, therefore one element for ply) were selected as the most suitable combination to represent the impact behavior of the specimen. The built-in Abaqus intra-laminar damage model has been used for Continuum Shell elements, while a user defined material model has been developed to simulate the intra-laminar damages in Solid elements, by means of Hashin's failure criteria and gradual material properties degradation rules. Cohesive zone model elements were adopted to simulate the inter-laminar damage for both the FEM models. The results obtained highlight some important aspects. The firstly introduced analytical method provides a very fast approach in terms of computational costs to preliminary assess the impact behavior of the composite plate. In the secondly introduced numerical model, which neglects the damages and uses a plain strain formulation, the best ratio between accuracy of the results and computational costs is represented by a reduced number of elements in the thickness direction. Then, Abaqus built-in failure criteria for intra-laminar and inter-laminar damage onset and propagation have been used to preliminary assess the damage behavior of the impacted panel. Finally, the solid element formulation, considering both intra-laminar and inter-laminar damages, has been introduced. In particular, user-defined intra-laminar failure criteria (Hashin's for fiber and tensile matrix, Puck–Shurmann for compressive matrix, and a non-linear behavior of the matrix shear) have been considered, allowing a more accurate prediction of the impact response of the plate. All the presented models provide an increasing degree of accuracy, starting from the analytical to the numerical solid model, being characterized, however, by increasing computational cost.

The results demonstrate the superior capabilities of the solid element model in conjunction with the user material model to simulate the low velocity impact phenomenon on the composite plate, with respect to the shell element model which substantially anticipate the damage onset and overestimates the damages. In conclusion, the solid elements model gives a more realistic representation of the impact behavior of the panel. Indeed, the shell elements formulation does not allow to accurately evaluate the stresses in the normal direction of the plane (which is the predominant direction of deformation during an impact phenomenon) and neglects the effects of the shear.

Author Contributions: All authors contributed equally to this work.

Funding: This research received no external funding.

Conflicts of Interest: The authors declare no conflict of interest.

References

1. Abrate, S. *Impact on Composite Structures*; Cambridge University Press: Cambridge, UK, 1998.
2. Abrate, S. *Impact Engineering of Composites Structures*; Springer: Wien, Austria; Udine, Italy, 2011.
3. Zhang, X. *Low Velocity Impact and Damage Tolerance of Composite Aircraft Structures*; Cranfield University: Cranfield, UK, 2003.
4. Soutis, C.; Curtis, P. Prediction of the post-impact compressive strength of CFRP laminated composites. *Compos. Sci. Technol.* **1996**, *56*, 677–684. [CrossRef]
5. Wisnom, M. The role of delamination in failure of fibre-reinforced composites. *Philos. Trans. R. Soc.* **2012**, *370*, 1850–1870. [CrossRef] [PubMed]
6. Abrate, S. Modeling of impacts on composite structures. *Compos. Struct.* **2001**, *51*, 129–138. [CrossRef]
7. Christoforou, A.; Swanson, S. Analysis of impact response in composite plates. *Int. J. Solid Struct.* **1991**, *27*, 161–170. [CrossRef]
8. Carvalho, A.; Guedes Soares, C. Dynamic response of rectangular plates of composite materials subjected to impact loads. *Compos. Struct.* **1996**, *34*, 55–63. [CrossRef]
9. Gonzàlez, E.; Maimì, P.; Camanho, P.; Turon, A.; Mayugo, J. Simulation of drop-weight impact and compression after impact tests on composite laminates. *Compos. Struct.* **2012**, *94*, 3364–3378. [CrossRef]
10. Barile, C.; Casavola, C.; Pappalettera, G.; Vimalathithan, P.K. Acousto-ultrasonic evaluation of interlaminar strength on CFRP laminates. *Compos. Struct.* **2019**, *208*, 796–805. [CrossRef]
11. Olsson, R. Analytical prediction of large mass impact damage in composite laminates. *Compos.-Part A Appl. Sci. Manuf.* **2001**, *32*, 1207–1215. [CrossRef]
12. Olsson, R. Closed form prediction of peak load and delamination onset under small mass impact. *Compos. Struct.* **2003**, *59*, 341–349. [CrossRef]
13. Olsson, R. Analytical prediction of damage due to large mass impact on thin ply composites. *Compos.-Part A Appl. Sci. Manuf.* **2015**, *72*, 184–191. [CrossRef]
14. Huang, K.Y.; De Boer, A.; Akkerman, R. Analytical modeling of impact resistance and damage tolerance of laminated composite plates. *AIAA J.* **2008**, *46*, 2760–2772. [CrossRef]
15. Tita, V.; deCarvalho, J.; Vandepitte, D. Failure analysis of low velocity impact on thin composite laminates: Experimental and numerical approaches. *Compos. Struct.* **2008**, *83*, 413–428. [CrossRef]
16. Li, C.; Hu, N.; Yin, Y.; Sekine, H.; Fukunaga, H. Low-velocity impact-induced damage of continuous fiber-reinforced composite laminates. Part I: An FEM numerical model. *Composites* **2002**, *33*, 1055–1062. [CrossRef]
17. Elder, D.; Thomson, R.; Nguyen, M.; Scott, M. Review of delamination predictive methods for low speed impact of composite laminates. *Compos. Struct.* **2004**, *66*, 677–683. [CrossRef]
18. Riccio, A.; Caputo, F.; Di Felice, G.; Saputo, S.; Toscano, C.; Lopresto, V. A Joint Numerical-Experimental Study on Impact Induced Intra-laminar and Inter-laminar Damage in Laminated Composites. *Appl. Compos. Mater.* **2016**, *23*, 219–237. [CrossRef]
19. Riccio, A.; Raimondo, A.; Saputo, S.; Sellitto, A.; Battaglia, M.; Petrone, G. A numerical study on the impact behavior of natural fibres made honeycomb cores. *Compos. Struct.* **2018**, *202*, 909–916. [CrossRef]
20. Li, N.; Chen, P.H. Micro–macro FE modeling of damage evolution in laminated composite plates subjected to low velocity impact. *Compos. Struct.* **2016**, *147*, 111–121. [CrossRef]
21. Faggiani, A.; Falzon, B.G. Predicting low-velocity impact damage on a stiffened composite panel. *Compos.-Part A Appl. Sci. Manuf.* **2010**, *41*, 737–749. [CrossRef]
22. Falzon, B.G.; Apruzzese, P. Numerical analysis of intralaminar failure mechanisms in composite structures. Part I: FE implementation. *Compos. Struct.* **2011**, *93*, 1039–1046. [CrossRef]
23. Falzon, B.G.; Apruzzese, P. Numerical analysis of intralaminar failure mechanisms in composite structures. Part II: Applications. *Compos. Struct.* **2011**, *93*, 1047–1053. [CrossRef]
24. Tay, T. Characterization and analysis of delamination fracture in composites: An overview of developments from 1990 to 2001. *Appl. Mech. Rev.* **2003**, *56*, 1–32. [CrossRef]

25. Wisnom, M. Modeling discrete failures in composites with interface elements. *Compos. Part A* **2010**, *41*, 795–805. [CrossRef]

26. Camanho, P.; Davila, C. *Mixed-Mode Decohesion Finite Elements for the Simulation of Delamination in Composite Materials*; NASA/TM-2002-211737; National Aeronautics and Space Administration, Langley Research Center: Hampton, VA, USA, 2002; pp. 1–37.

27. Riccio, A.; Russo, A.; Sellitto, A.; Raimondo, A. Development and application of a numerical procedure for the simulation of the "Fibre Bridging" phenomenon in composite structures. *Compos. Struct.* **2017**, *168*, 104–119. [CrossRef]

28. Riccio, A.; Ricchiuto, R.; Di Caprio, F.; Sellitto, A.; Raimondo, A. Numerical investigation of constitutive material models on bonded joints in scarf repaired composite laminates. *Eng. Fract. Mech.* **2017**, *173*, 91–106. [CrossRef]

29. Aymerich, F.; Dore, F.; Priolo, P. Simulation of multiple delaminations in impacted cross-ply laminates using a finite element model based on cohesive interface elements. *Compos. Sci. Technol.* **2009**, *69*, 1699–1709. [CrossRef]

30. DeMoura, M.; Gonçalves, J. Modeling the interaction between matrix cracking and delamination in carbon–epoxy laminates under low velocity impact. *Compos. Sci. Technol.* **2004**, *64*, 1021–1027. [CrossRef]

31. Zhang, Y.; Zhu, P.; Lai, X. Finite element analysis of low-velocity impact damage in composite laminated plates. *Mater. Des.* **2006**, *27*, 513–519. [CrossRef]

32. deBorst, R.; Remmers, J.J.; Needleman, A. Mesh-independent discrete numerical representations of cohesive-zone models. *Eng. Fract. Mech.* **2006**, *73*, 160–177.

33. Maimì, P.; Camanho, P.; Mayugo, J.; Davila, C. A continuum damage model for composite laminates: Part I—Constitutive model. *Mech. Mater.* **2007**, *39*, 897–908. [CrossRef]

34. *ASTM D7136/D-Standard Test Method for Measuring the Damage Resistance of a Fiber-Reinforced Polymer Matrix Composite to a Drop-Weight Impact Event*; ASTM International: West Conshohocken, PA, USA, 2005.

35. Sellitto, A.; Borrelli, R.; Caputo, F.; Riccio, A.; Scaramuzzino, F. Application to plate components of a kinematic global-local approach for non-matching finite element meshes. *Int. J. Struct. Integr.* **2012**, *3*, 260–273. [CrossRef]

36. *Dassault System Abaqus 2016 User's Manual*; Dassault Systèmes Simulia Corp.: Providence, RI, USA, 2016.

37. Christoforou, A.; Yigit, A. Effect of flexibility on low velocity impact response. *J. Sound Vib.* **1998**, *217*, 563–578. [CrossRef]

38. Puck, A.; Schurmann, H. Failure analysis of FRP laminates by means of physically based phenomenological models. *Compos. Sci. Technol.* **1998**, *58*, 1045–1067. [CrossRef]

39. Bazant, Z.P.; Oh, B.H. Crack band theory for fracture of concrete. *Mater. Struct.* **1983**, *16*, 155–177.

40. Davies, G.A.O.; Olsson, R. Impact on composite structures. *Aeronaut. J.* **2004**, *108*, 541–563. [CrossRef]

Article

Failure Probability Prediction of Thermally Stable Diamond Composite Tipped Picks in the Cutting Cycle of Underground Roadway Development

Yong Sun *, Xingsheng Li and Hua Guo

CSIRO Mineral Resources, PO Box 883, Kenmore, QLD 4069, Australia
* Correspondence: yong.sun@csiro.au

Received: 17 July 2019; Accepted: 6 August 2019; Published: 11 August 2019

Featured Application: Pick failure risk assessment in mining and civil industries.

Abstract: The Thermally Stable Diamond Composite (TSDC) tipped pick has been developed to replace Tungsten Carbide (WC) tipped picks for hard rock cutting. Due to the material properties of TSDC, a major failure mode of TSDC tipped picks during rock cutting is random failures caused by excessive bending force acting on the cutting tips. A probabilistic approach has been proposed to estimate the failure probability of picks with this failure mode. However, there are two limitations in existing research: only one drum revolution is considered, and the variation of rock thickness is ignored. This study aims to extend the current approach via removing these limitations, based on the failure probability analysis of picks over a full cutting cycle in the underground coal mining roadway development process. The research results show that both drum advance direction and the variation of rock thickness have significant impacts on pick failure probability. The extended approach can be used to estimate pick failure probability for more realistic scenarios in real applications with improved accuracy. Although the study focused on TSDC tipped picks, the developed approach can also be applied to other types of picks.

Keywords: failure probability; diamond composite; material failure characteristics; reliability; rock cutting picks; mining; civil engineering

---

1. Introduction

The Thermally Stable Diamond Composite (TSDC) material is made using ceramic-based silicon to bind synthetic diamond grains together [1,2]. Research on TSDC's properties and its applications can be found in [3–8]. In brief, as a type of diamond composite, the wear resistance of TSDC is several hundred times better than Tungsten Carbide (WC) [7,8]. Due to the use of silicon as the binder, the mechanical properties of TSDC remain stable at temperatures up to 1200 °C [2]. The thermal stability of TSDC is much higher than that of ordinary polycrystalline diamond (PCD), which employs metallic cobalt as the binding material [2]. PCD is generally suitable for operational temperature below 750 °C [9]. Higher thermal stability is critical for hard rock cutting tools because research has revealed that the temperature at the contact area between cutting tip and rock surface can reach as high as 1100 °C [10]. High wear resistance and high thermal stability make TSDC a potentially superior material for hard and abrasive rock cutting. One type of TSDC tipped picks, namely SMART*CUT picks, have been specifically developed to tackle the challenge of high abrasiveness and high temperature observed in hard rock cutting [3–5].

Picks are a type of rock cutting tool which are broadly equipped on excavation machines such as continuous miners and roadheaders to break rock in the mining and construction industries. Understanding the performance and reliability of picks is important for industries to increase production efficiency, improve production safety and reduce production cost [11,12]. Many efforts including laboratory experiments and field trials have been made to study picks from various aspects [3–24]. An important aspect of these studies is pick force analysis [4,5,12–19]. During the course of rock cutting, a pick is generally subjected to three orthogonal forces—cutting force, normal force and lateral force [13]. These forces can be affected by various factors including depth of cut (DOC) [5,12–15], rock strength [13–15] and attack angle [5]. Another important aspect of pick performance study is failure analysis [11,12,17,20–22,24,25], including identification of typical pick failure modes [17], investigation of the causes of high consumption of picks [12], protection of pick body by use of a cap tip [12,20] and understanding of the body bending failures of picks [21].

Although existing studies largely focused on WC tipped picks, studies with a focus on the performance and failure characteristics of TSDC tipped picks have also been carried out (see [3–6,11,25]). These studies showed that while TSDC tips were able to cut hard rock with a uniaxial compressive strength of 260 MPa [24], they were less capable to bear bending and impact forces than WC tips due to their lower toughness [5]. Failures caused by excessive impact or bending forces were a major concern for TSDC tipped picks [11]. Therefore, it is important to predict the failure risk of TSDC tipped picks which are subjected to excessive bending or impact forces in real industrial applications. As indicated above, the forces acting on a pick are affected by multiple factors during the rock cutting process. Some of these factors often have considerable uncertainties (i.e., their values vary randomly within a considerable range) in the real world. For example, there are often uncertainties in rock properties and machine operations during the mining production process. Due to these uncertainties, the forces acting on a pick consequently change randomly during the production process. Additionally, the material properties of TSDC cutting tips often have significant random variations [25]. In addition, the geometry of TSDC cutting tips with the same design may also vary considerably. As a result, the sudden failures of TSDC tips caused by excessive bending force or impact force during normal production often occur in a random manner.

Currently, a probabilistic approach has been developed to assess the failure risk of picks with random sudden failures based on underground coal mine roadway development with a case study on the TSDC tipped picks in [11]. Note that although the approach proposed in [11] was developed based on the scenario of underground roadway development in the coal mining industry, it is a general approach and can be applied to other rock cutting scenarios. This approach consists of three steps: firstly, the continuous rock cutting process of a pick over one drum revolution is discretized into a series of small segments; secondly, the failure probability of a pick for cutting each of these segments is estimated; and finally the failure probability of the pick over the whole cutting process is estimated. In [11], the influences of multiple major factors including pick tip geometry, Brazilian Tensile Strength (BTS) of pick tip, the attack angle of the pick, DOC and rock BTS on the failure probability of picks were modelled. Later on, a reliability-based-approach (RBP) [6] was developed to link the failure probability of pick tips subject to sudden bending failures for cutting a segment of rock to the length of the rock segment, so that the pick failure probability estimation accuracy can be improved.

This study aims to improve the usefulness of existing approach via removing two critical limitations in existing models. The first limitation is that only one drum revolution was considered in the modelling process. The second limitation is that the variation of rock thickness is ignored. However, in the actual production process, the drum advance distance is generally much larger than the drum's advance distance in one revolution. For example, in underground coal mine roadway development, the sump-in depth is typically about the half of the drum's cutting diameter which is the advance distance of the drum over multiple revolutions. Over the sump-in stage, the drum often involves cutting the roadway roof with different rock thickness. Therefore, to make the failure analysis approach more applicable to the real-world scenario, these two limitations should be removed.

To address the above issue, the existing approach will be extended to take into account a more realistic rock cutting production process. Given that there are various rock cutting production processes utilized in industry, the extended approach will be developed based on the scenario of an underground coal mine roadway development using continuous miners with TSDC tipped picks, in order to remain consistent with [11]. However, the extended approach will also be applicable to other rock cutting scenarios and other types of picks.

Development of an underground roadway normally consists of many cutting cycles. A cutting cycle is the process in which a continuous miner implements a full face cut (an example is given in the next section). Extension of the analysis approach to a full cutting cycle in mining production could face a number of challenges, e.g., determination of the drum revolutions and the thickness of hard rock cut in each revolution. However, the extension of the approach from one cutting cycle to multiple cutting cycles will be relatively straightforward. Therefore, this paper focuses on the extension of the approach to a full cutting cycle. More realistic scenarios, the corresponding challenges and solutions will be discussed. A case study on the failure probability assessment of TSDC tipped picks in a full cutting cycle will be presented to demonstrate the application of the extended approach.

2. Extended Pick Failure Probability Assessment Approach

2.1. A Typical Cutting Cycle

Figure 1 shows a schematic view of a typical cutting cycle in underground roadway development with a continuous miner. As an example, Figure 2 shows the locus of a pick tip on the cutting drum of a continuous miner over such a cutting cycle.

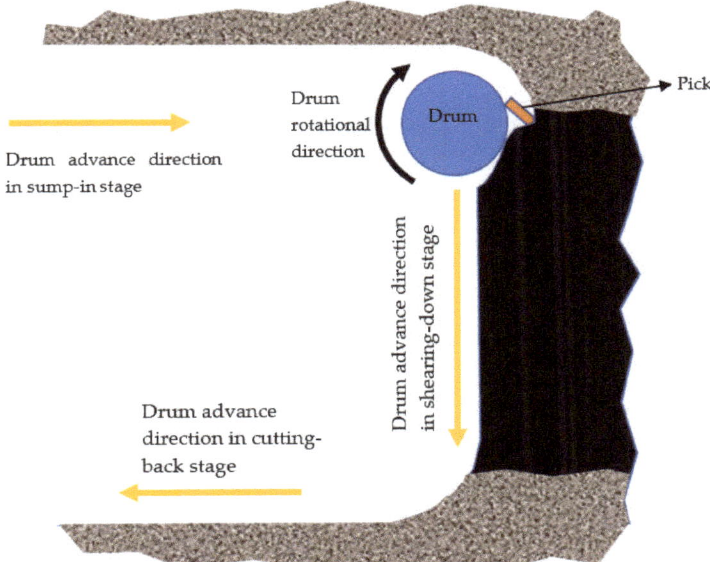

Figure 1. A schematic view of a typical cutting cycle in underground roadway development.

A drum is normally equipped with a large number of picks which are attached to the drum on the external surface. The reason that only a pick is shown in Figure 1 is to make the figure clear.

The roadway face can be divided into three sections: the roof rock, the floor rock and the coal seam. The cases where coal exists inside a rock section or rock bands are included inside the coal seam are not considered in this study. A 'full cutting cycle' is the complete process for cutting these three sections sequentially. In this cycle, the drum first cuts into the face along the roof for a half of the drum to form the roof, then shears down to the floor, and finally cuts back to clear the floor. Therefore, a full cutting cycle can be divided into three stages: sump-in (cutting-into), shearing-down and cutting-back.

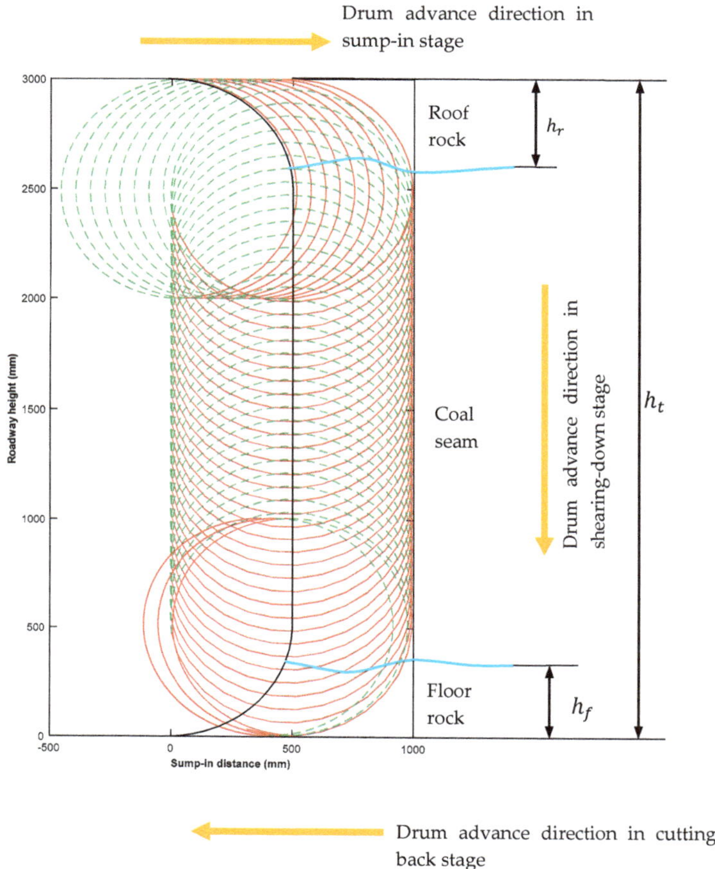

Figure 2. Locus of a typical pick tip in a typical cutting cycle in underground roadway development.

In Figure 2, h_t, h_r and h_f are respectively the height of the roadway, the thickness of the rock removed to form the roadway roof (roof rock) and the thickness of the rock removed to form the roadway floor (floor rock). The thickness of the roof rock and that of the floor rock are measured from the roadway roof and the roadway floor respectively. The parameters used to develop Figure 2 are shown in Table 1.

Table 1. Parameters used to draw Figure 2.

Parameter	Unit	Value
Roadway height	mm	3000
Drum's tip-to-tip diameter	mm	1000
Drum rotational speed	rpm	50
Drum advance speed	m/min	3
Drum sump-in depth	mm	500
Drum cutting-back distance	mm	150

Other information used to draw Figure 2 is as follows:

A full cutting cycle involves multiple drum revolutions which are depicted by the locus of a pick tip in Figure 2. In order to show the drum advance directions more clearly, the locus of the pick tip in one drum revolution is divided into two equal parts: the half involving rock/coal cutting is marked in red and the other half is marked in green. The black line represents the face profile before this cutting cycle. The two blue curves respectively indicate the interface between the roof rock and the coal seam, and that between the coal seam and the floor rock. At the beginning of the face cutting cycle, i.e., the initial position, the circle formed by the cutting tips of the drum just touches the face profile before this cutting cycle and the pick tip in consideration is located at the top most position, which is the start point to measure the angular position of the pick in the sump-in stage.

The aim of this study is to estimate the failure probability of a pick over a full cutting cycle. The pick failures in the following analysis have the same meaning as those defined in [11], that is, the sudden failure of the tip and/or body of a pick due to excessive bending or impact force acting on the pick tip. Other failure modes such as tip fatigue failure will be studied in the future. In this study, it is assumed that rock breakout is effective. When a pick is installed on a drum, it is randomly selected from a given batch of picks. Once being installed on the drum, the pick will be in service continuously until it has failed.

From [11], it is known that the failure of a pick is a combination of its tip failure and body failure. As the failure probability analysis approach of body failures is the same as that of tip failures and it is assumed that the tip failures are independent of body failures, the following analysis mainly focuses on the pick tip failure probability estimation. In addition, as the failure probability of a pick tip over a cutting cycle consisting of three stages is a combination of its failure probability over these three stages, the pick tip failure probability over each individual stage will be estimated initially.

2.2. Pick Tip Failure Probability during the Sump-in Stage

In the sump-in stage, the drum normally cuts into the face along the roof for a sump-in depth of about the cutting radius (i.e., the half of the tip-to-tip diameter) of the drum. To implement the required sump-in depth, the drum needs to rotate multiple revolutions. The pick tip failure probability in this stage is a combination of the tip failure probability over individual drum revolutions. If the roof rock thickness is less than the tip-to-tip diameter of the drum (which is quite common during real production), the drum will partially cut roof hard rock and partially cut the coal seam over a single revolution. As the coal is generally much softer and weaker than the rock, the pick failure probability due to cutting coal can be ignored [11].

To estimate the pick tip failure probability over each drum revolution, the thickness of the roof rock in each individual drum revolution needs to be determined. In reality, the roof rock thickness generally fluctuates randomly, and it is normally difficult to obtain the distribution of roof rock thickness. For simplicity, instead of using a continuous distribution, the roof rock thickness can be approximated by several mean values with their corresponding occurrence likelihoods in industrial applications. In this case, based on the total probability theorem, the failure probability of a pick tip with a given allowable bending force x_a^t in a drum revolution can be estimated by

$$F_{ri} = \sum_{j=1}^{n_{hr}} F_{ri}\left(h_{rj}, \, x_a^t\right) p\left(h_{rj}\right) \tag{1}$$

where F_{ri} is the pick tip failure probability at the i-th drum revolution in the sump-in stage; n_{hr} is the number of mean roof rock thickness values in consideration; $F_{ri}\left(h_{rj}, \, x_a^t\right)$ is the pick tip failure probability at the i-th drum revolution in the sump-in stage when the mean roof rock thickness is h_{rj} (mm) and tip allowable bending force is x_a^t; and $p\left(h_{rj}\right)$ is the likelihood (%) that the mean roof rock thickness is h_{rj}. Obviously,

$$\sum_{j=1}^{n_{hr}} p\left(h_{rj}\right) = 100\% \tag{2}$$

The pick failure probability $F_{ri}\left(h_{rj},\ x_a^t\right)$ can be estimated using the models given in [11].

Nevertheless, when using the models given in [11], one needs to consider the following points:

The first issue is the determination of the cutting sector of the drum, α_{rc} (degree) (see Figure 3). In [11], since the impact of cutting soft coal is ignored, the cutting sector indicates the hard rock cutting sector and is given by

$$\alpha_{rrj} = \arccos\left(1 - \frac{2h_{rj}}{D_d}\right) * \frac{180}{\pi} \tag{3}$$

where, α_{rrj} is the hard rock cutting sector of the drum (in degrees) when the mean roof rock thickness is h_{rj} (mm). D_d is the tip-to-tip diameter of the drum (mm).

However, determination of the cutting sector in this way would be inconvenient when multiple drum revolutions are involved, because of the uncertainty of the rock thickness in each individual drum revolution. To resolve this problem, the cutting sector of a drum in one revolution is defined as the central angle of the drum engaging coal and/or rock cutting. With this definition, the cutting sector is between 90 degrees and 180 degrees in the sump-in stage (see Figure 3). The cutting sector of the drum at the i-th revolution can be obtained by solving the following equation:

$$\frac{1000v_a}{60}\left[\frac{60(i-1)}{n} + \frac{\alpha_{rci}}{6n}\right] + \frac{D_d}{2}\sin\frac{\pi\alpha_{rci}}{180} - \frac{D_d}{2} = 0 \tag{4}$$

where, α_{rci} is the cutting sector of the drum at the i-th revolution (in degrees), n is the drum rotational speed (rpm), and v_a is the drum advance speed (m/min).

As the hard rock cutting sector can be calculated using Equation (3) and the failure probability for cutting coal is ignored, the cutting sector in the sump-in stage can be set to 180 degrees in the calculation if the roof rock thickness is not greater than $\frac{D_d}{2}$.

In shearing-down stage, the cutting sector is 90 degrees when the distance between the center of the drum and the floor is not less than D_d. The estimation of the cutting sector when the distance between the center of the drum and the floor is shorter than D_d will be discussed in the next section. Obviously, the hard rock cutting sector for a given revolution cannot be greater than the full cutting sector of this revolution.

The second issue for consideration is the calculation of the DOC of the pick at the first revolution. According to [11,18], the maximum DOC (nominal DOC) can be calculated by

$$D_{max} = \frac{v_n}{m} \tag{5}$$

where m is the number of starts of the drum and v_n is the drum advance distance per revolution (mm/r):

$$v_n = \frac{1000v_a}{n} \tag{6}$$

As the cutting starts from the initial position on the first revolution, the maximum DOC of the pick in consideration in this revolution is only a quarter of that in other revolutions because the pick rotates only a quarter of a revolution to reach the maximum advance distance in this revolution $v_n/4$ (see Figures 2 and 3). For example, with the parameters in Figure 2 and the number of the starts being 2, the maximum DOC of this pick in the first revolution is only 7.5 mm, while it is 30 mm in the other revolutions.

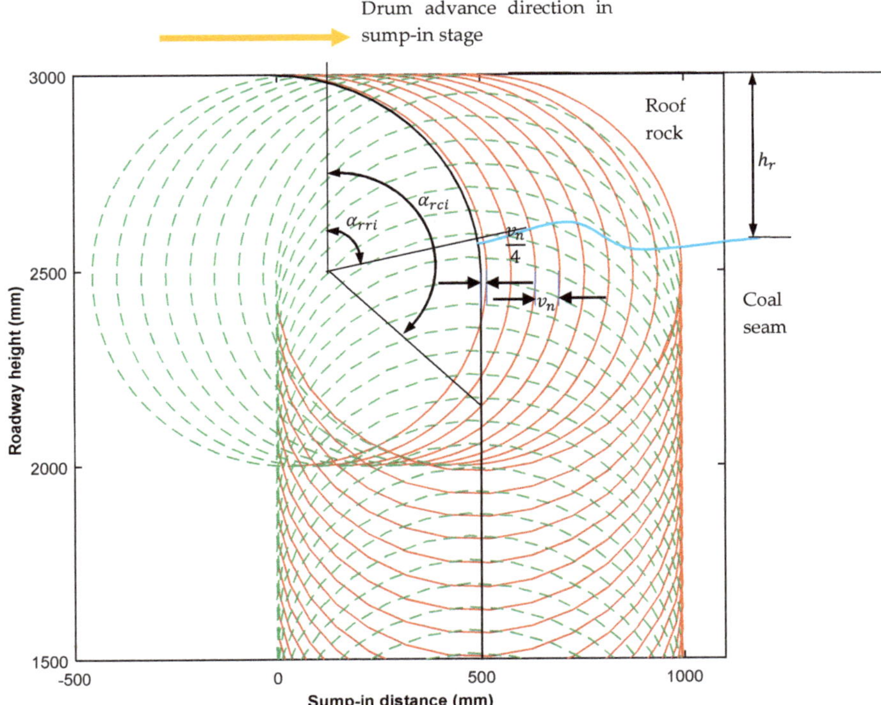

Figure 3. Cutting sector in sump-in stage.

The final point is the estimation of the pick tip failure probability of the last revolution. On this revolution, the pick may not complete a full cut of the roof rock before the drum moves to shearing-down stage. However, as the drum will continue cutting when its cutting mode changes from sump-in to shearing-down, and the variations of the DOC during the transition period can be ignored, the calculation of the pick failure probability of the last revolution can be treated as the same as that in the other revolutions (see Figure 2).

Once the failure probability of the pick tip in each drum revolution in the sump-in stage has been estimated, its failure probability in the sump-in stage, $F_1^t\left(x_a^t\right)$, is

$$F_1^t\left(x_a^t\right) = 1 - \prod_{i=1}^{n_{dr}}\left[1 - \sum_{j=1}^{n_{hr}} F_{ri}\left(h_{rj},\ x_a^t\right) p\left(h_{rj}\right)\right] \tag{7}$$

where n_{dr} is the number of drum revolutions for implementing the sump-in cutting process, which is given by

$$n_{dr} = \text{ceil}\left(\frac{D_s}{v_n}\right) \tag{8}$$

In Equation (8), D_s is the sump-in distance (mm).

In reality, the sump-in depth cannot be accurately controlled due to limitations of the equipment control systems, resulting in the number of the drum revolutions for sump-in cutting fluctuating randomly. This issue will be studied in due course. In this paper, only fixed sump-in depth is considered.

2.3. Pick Failure Probability during the Shearing-Down Stage

When the operation of the drum shifts from the sump-in stage to the shearing-down stage, its advance direction changes from horizontal (along the roof) to vertical (towards the floor). In the shearing-down stage, the drum primarily cuts the coal seam, as well as some floor rock (Figure 2). This stage may also involve cutting roof rock if the thickness of the roof rock is greater than the cutting radius of the drum (Figure 2). However, the roof rock thickness is generally less than two thirds of the tip-to-tip diameter of the drum, and the coal is much softer than the rock. Therefore, in the following analysis, the impact of cutting roof rock and the coal seam on the pick tip failure probability in the shearing-down stage is ignored, and only the impact of cutting the floor rock on the pick tip failure probability is considered. The scenario where the roof rock thickness can take any value will be studied in due course.

Three issues need to be addressed in the analysis of the pick tip failure probability during the shearing-down stage. The first issue is that although the location of the pick at the drum transition point from the sump-in stage to the shearing-down stage can be calculated, it is not convenient to use this location to count the drum revolutions. To address this, the following simplification is possible if the roof rock thickness is less than the 2/3 of the tip-to-tip diameter of the drum. As the rock cutting in the last cycle at the sump-in stage has been fully considered, it is acceptable to assume that the first revolution in the shearing-down stage always starts from the point with the maximum sump-in distance (rightmost point) at the last revolution in the sump-in stage.

The second issue is that similarly to the roof rock, the thickness of the floor rock also generally varies from place to place. As a result, both the number of the drum revolutions involving floor hard rock cutting and the rock length cut in each drum revolution can be uncertain (i.e., their values can change randomly). However, this process is different from the sump-in stage. To solve this problem, the calculation of the pick tip failure probability always starts from the first drum revolution in the shearing-down stage, because

$$F_{fi} = 0, \text{ when } n_{di}v_n \le h_t - D_d - h_{fmax} \tag{9}$$

In Equation (9), h_{fmax} is the largest mean thickness of the floor rock (mm), F_{fi} is the pick tip failure probability over the i-th drum revolution in the shearing-down stage, h_t is the height of the roadway (mm), and n_{di} is the number of revolutions from the 1st drum revolution to the n_{di}-th drum revolution in the shearing-down stage.

Equation (9) holds because the drum will not start to cut the floor rock until the n_{di}-th drum revolution satisfies the following condition:

$$n_{di} > \frac{\left(h_t - D_d - h_{fmax}\right)}{v_n} \tag{10}$$

The third issue is that the thickness of the floor rock can vary over a drum revolution in the shearing-down stage, and thus it is required to identify whether a segment being cut is rock or not. Figure 4 shows an arbitrary segment (the s-th segment) in the shearing-down stage. It can be seen from this figure that a segment being cut is rock only when its distance from the floor is shorter than h_{fj} and its angular position (central angle) is less than the cutting sector. Note that, unlike in the sump-in stage where the angular position of a segment is measured from vertical, the angular position of a segment in the shearing-down stage is the angle between the horizontal direction and the line to the middle point of the segment. The cutting sector in the shearing-stage is also measured from the horizontal direction.

As previously mentioned, the cutting sectors of the revolutions in the shearing-down stage are not all the same. If the distance between the centre of the drum and the floor is not less than the drum's tip-to-tip diameter, the cutting sector of the drum in a revolution is 90 degrees; otherwise, the cutting sector will be greater than 90 degrees. From Figure 4, it can be found that the drum in the last revolution has the largest cutting sector. As the thickness and the length of any segment are very small compared with the drum's cutting radius, the influence of the segment on the cutting sector can be ignored. As such, triangle $\Delta O_{l1}O_{l2}A_l$ is an equilateral triangle. This means that the cutting sector in the last revolution is 120 degrees. In other words, the end angle of the cutting sector in the last revolution in the shearing-down stage, α_{fcl}, is

$$\alpha_{fcl} = 120 \text{ (degree)} \tag{11}$$

The end angles of cutting sectors in other revolutions in the shearing-down stage will be between 90 degrees and 120 degrees. However, accurate calculation of the cutting sector in an arbitrary drum revolution is not further discussed as it is not the focus of this study. For this paper, the cutting sector in the i-th revolution, α_{fci}, is approximated to be

$$\alpha_{fci} = \begin{cases} 90 \, , \, 0 \leq n_{di} \leq n_{db} \\ \quad \vdots \\ 90 + \dfrac{60v_n(n_{di}-n_{db})}{D_d} , \, n_{db} < n_{di} \leq \dfrac{h_t-D_d}{v_n} \end{cases} \tag{12}$$

where

$$n_{db} = floor\left(\frac{2h_t - 3D_d}{2v_n}\right) \tag{13}$$

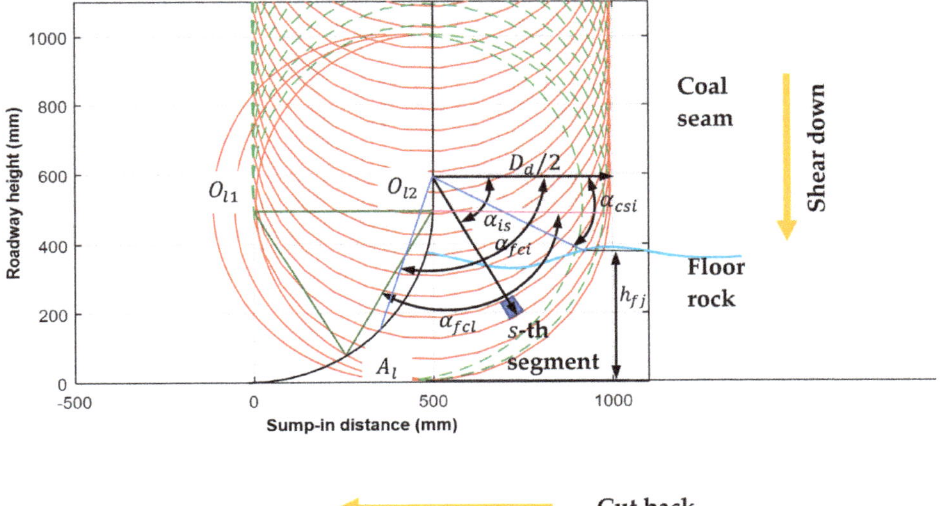

Figure 4. A segment in the shearing-down stage.

Given these constraints, the pick tip failure probability over the shearing-down stage can be estimated using a similar approach for calculating the pick tip failure probability over the sump-in stage, with a consideration of the differences between these two stages. The first difference in the tip failure probability calculation is the segment rock condition. In the sump-in stage, if the n_i-th segment in the i-th revolution is not rock, then any k_i-th ($k_i > n_i$) segment in the same revolution will not be rock; while in the shearing-down stage, any segments within the drum's cutting sector in a drum revolution may or may not be rock. However, once a segment in a drum revolution is rock, all segments at the same angular position in the following drum revolutions will also be rock. The second difference is the rock cutting start angle. While the start angle of rock cutting sector in the sump-in stage is generally equal to zero, the start angle of rock cutting sector in the shearing-down stage is often non-zero (see Figure 4). As the DOC is normally much smaller than the drum's tip-to-tip diameter, the start angle of rock cutting sector in the shearing-down stage, α_{csi}, can be estimated by

$$\alpha_{csi} = \mathrm{H}(L_i)\arcsin\left(\frac{2L_i}{D_d}\right), \text{ when } L_i \leq \frac{D_d}{2} \tag{14}$$

where $\mathrm{H}(\cdot)$ is the Heaviside step function, and

$$L_i = h_t - \frac{D_d}{2} - v_n n_{di} - h_{fj} \tag{15}$$

As h_{fj} is a random variable, both L_i and α_{csi} are also random variables. If $L_i > \frac{D_d}{2}$, the drum will not cut the floor rock.

Due to the above-mentioned differences between the sump-in stage and the shearing-down stage, there are some variations in the approach to estimating the failure probability of pick tips in the shearing-down stage. In the sump-in stage, the pick tip failure probability for each individual drum revolution is estimated first, and then the pick tip failure probability over the whole stage is estimated. However, this approach is not suitable for handling the segment rock conditions in the shearing-down stage. To address this issue, the pick tip failure probability for cutting the s-th segments in all the revolutions in the shearing-down stage will be estimated first, and then the pick tip failure probability over the whole shearing-down stage is estimated.

With a consideration of the variation of rock thickness, the pick tip failure probability for cutting the s-th segments in all the revolutions over the shearing-down stage is given by

$$F_{fs} = \sum_{j=1}^{n_{hf}}\left[1 - \prod_{i=1}^{n_{df}}(1 - F_{fis}(h_{fj}, x_a^t))\right]p(h_{fj}) \tag{16}$$

where F_{fs} is the failure probability of the pick tip for cutting the s-th segments in all the revolutions in the shearing-down stage, n_{hf} is the number of mean floor rock thickness values in consideration, $F_{fis}(h_{fj}, x_a^t)$ is the pick tip failure probability for cutting the s-th segment in the i-th drum revolution in the shearing-down stage when the mean floor rock thickness is h_{fj} (mm), $p(h_{fj})$ is the likelihood (%) that the mean floor rock thickness is h_{fj} (mm), and n_{df} is the number of drum revolutions in the shearing-down stage:

$$n_{df} = \mathrm{ceil}\left(\frac{h_t - D_d}{v_n}\right) \tag{17}$$

Then, the failure probability of the pick tip over the shearing-down stage, $F_2^t(x_a^t)$, is

$$F_2^t(x_a^t) = 1 - \prod_{s=1}^{n_{sf}}(1 - F_{fs}) \tag{18}$$

where n_{sf} is the number of segments in a revolution in the shearing-down stage:

$$n_{sf} = \mathrm{ceil}\left(\frac{360}{\alpha_{sf}}\right) \tag{19}$$

In Equation (19), α_{sf} is the central angle of a segment in the shearing-down stage (in this study, all drum revolutions are equally segmented).

From Equations (16) and (18), it can be seen that to estimate the failure probability of the pick tip in the shearing-down stage, $F_{fis}(h_{fj}, x_a^t)$ must be calculated first. To calculate $F_{fis}(h_{fj}, x_a^t)$, it is necessary to determine whether an individual segment in the shearing-down stage is hard rock. To address this issue, the mid-point method [26] applied in [11] is applied again. If the midpoint of a segment is within the floor hard rock, the whole segment is regarded as the same hard rock. Based on this concept, the pick tip failure probability for cutting the s-th segment in the i-th drum revolution with the rock thickness of h_{fj} is given by

$$F_{fis}\left(h_{fj}, x_a^t\right) = \begin{cases} P_{is}\left(x_a^t\right), & \text{when } h_{fj} > D_{is} \text{ and } \alpha_{fci} \geq \alpha_{is} \\ 0, & \text{others} \end{cases} \tag{20}$$

where, $P_{is}\left(x_a^t\right)$ is the probability that actual bending force is greater than the allowable bending force x_a^t of the pick tip, and α_{is} is the angular position of the s-th segment in the i-th drum revolution (degree).

In Equation (20), D_{is} is the distance of the s-th segment in the i-th drum revolution from the roadway floor:

$$D_{is} = h_t - v_n\left(n_{di} - 1 + \frac{\alpha_{is}}{360}\right) - \frac{D_d}{2}\left[1 + \sin\left(\frac{\pi\alpha_{is}}{180}\right)\right] \tag{21}$$

The computation based on Equations (20) and (21) can often be simplified, because in the real world, the chord of a drum engaging in rock cutting cannot exceed its tip-to-tip diameter which is usually about 1000 mm. The rock thickness normally does not change dramatically over this short distance. In this case, it is reasonable to assume that once drum starts to cut rock in a revolution, it will continue to cut the rock over the rest of its cutting sector in the same revolution. This means that although rock thickness changes randomly, its mean value over a drum revolution remains constant. Then the failure probability of the pick tip for cutting the s-th segment in the i-th drum revolution with the rock thickness of h_{fj} can be estimated using the following equation:

$$F_{fis}\left(h_{fj}, x_a^t\right) = \begin{cases} P_{is}\left(x_a^t\right), & \text{when } \alpha_{fci} \geq \alpha_{is} \geq \alpha_{csi} \\ 0, & \text{others} \end{cases} \tag{22}$$

Correspondingly, the pick tip failure probability over the shearing-down stage can be estimated using the following equation to replace Equations (16) and (18):

$$F_2^t\left(x_a^t\right) = \sum_{j=1}^{n_{hf}}\left[1 - \prod_{s=1}^{n_{sf}}\prod_{i=1}^{n_{df}}\left(1 - F_{fis}\left(h_{fj}, x_a^t\right)\right)\right]p\left(h_{fj}\right) \tag{23}$$

2.4. Pick Tip Failure Probability during the Cutting-Back Stage and over the Full Cutting Cycle

The methods for analyzing the failure probability of the pick tip over the last stage in a cutting cycle and the accumulation of the failure probability of the pick tip over the full cutting cycle are presented in this section. The pick failure probability over the full cutting cycle is also analyzed in this section.

2.4.1. Pick Tip Failure Probability over the Cutting-Back Stage

The analysis method used for the pick tip failure probability over the sump-in stage can be adopted for analyzing the pick tip failure probability over the cutting-back stage. However, from Figure 2, it can be seen that the amount of hard rock cutting involved in this stage is tiny, the bending force from the cutting process is generally smaller than the allowance bending force of the pick tips. As a result, it can be assumed that

$$F_3^t\left(x_a^t\right) = 0 \tag{24}$$

where, $F_3^t\left(x_a^t\right)$ is the failure probability of the pick tip over the cutting-back stage.

2.4.2. Pick Tip Failure Probability over the Full Cutting Cycle

Once the pick tip failure probabilities over all three stages in a cutting cycle have been obtained, the tip's failure probability of the given pick over the full cutting cycle, $F(x_a^t)$, can be estimated by

$$F^t(x_a^t) = 1 - \prod_{i=1}^{3}\left[1 - F_i^t(x_a^t)\right] \tag{25}$$

For an arbitrarily selected pick, its cutting tip's failure probability over the full cutting cycle is given by

$$F^t = \int_{-\infty}^{\infty}\left(1 - \prod_{i=1}^{3}\left[1 - F_i(x_a^t)\right]\right)p(x_a^t)dx_a^t \tag{26}$$

where $p(x_a^t)$ is the probability density function of the allowable forces of the pick tips.

2.4.3. Pick Failure Probability over the Full Cutting Cycle

As indicated previously, the failure of a pick is the combination of its tip failure and its body failure. The pick body failure probability over a cutting cycle can also be estimated using the same approach for estimating the pick tip failure probability. If the body failure probability of a pick over the abovementioned full cutting cycle is F^b, and the body failure of the pick is independent of its tip failure, then failure probability of the pick over the full cutting cycle, F is

$$F = 1 - (1 - F^t)(1 - F^b). \tag{27}$$

In the application of Equation (25) or Equation (26), it should be noted that the three stages occur sequentially. To take the sequence of the stages into account, it is better to introduce a variable of time. This is discussed in the next section.

2.5. Change of the Pick Failure Probability over Time

The failure probability prediction models given in the above sections show the relationship between the pick failure probability and drum revolutions. It may be desired to understand the change of the pick failure probability over time during mining production. This can be done by linking the revolution to time in the above models. As all revolutions are equally segmented, the relationship between the pick failure probability and time can be expressed using the following piece-wise function:

$$F(t) = \begin{cases} 0, & t = 0 \\ F_{s1}, & 0 < t \le T_s \\ \quad \vdots \\ F_{sk}, & (k-1)T_s < t \le kT_s \end{cases} \tag{28}$$

In Equation (28), $F(t)$ is the pick failure probability over time t, F_{s1} is the pick failure probability for cutting the first segment, and F_{sk} is the pick failure probability for sequentially cutting k segments starting from the first. This calculation needs to consider the cutting stages and the variation of rock thickness and can be done by using the models developed in the previous sections. Parameter T_s is the time required for a pick to cut one segment (sec) which is given by

$$T_s = \frac{a_s}{6n}, \tag{29}$$

where α_s is the central angle of a segment (in degrees).

3. A Case Study

A case study is presented in this section to demonstrate the application of the extended approach. The relevant parameters and values in the case study shown in [11] were adopted in this case study. Detailed conditions of this case study are as follows:

It is assumed that a continuous miner was used to develop a roadway in an underground coal mine. The roadway consists of three sections as shown in Figure 2. The total height of the roadway was 3000 mm. The roof rock and the floor rock were both sandstone with the same material properties as shown in Table 1 in [11].

The continuous miner was equipped with a 2-start drum which has a tip-to-tip diameter of 1000 mm. The rotational speed and the advance speed of the drum were respectively 50 rpm and 2.35 m/min. In a full face cutting cycle, the drum first cut into the roof of the roadway with a sump-in depth of 500 mm, then sheared down 2000 mm, and finally cut back for 150 mm. The rock breakout during all the cutting processes was effective. According to Equations (8) and (17), it can be calculated that the number of drum revolutions for the sump-in of 500 mm is 11 and for the shearing-down of 2000 mm it is 44.

A full set of TSDC tipped picks were installed on the drum. These picks were identical to those TSDC tipped picks studied in [11]. The bottom diameter of the tips was 12 mm, and the arm length for the bending force acting on the tips was 4.6 mm. The attack and tilt angles of the picks in consideration were 55 degrees and 0 (zero) degree respectively. Each pick cuts one line.

In addition, the following assumptions which are made in the case study in [11] are also adopted in this case study:

- Only sudden failures caused by excessive bending force are considered;
- Tip failure of a pick is independent of its body failure;
- Each small segment covers a central angle of 10 degrees.

Given that the influences of drum operating parameters, pick geometry, and the material properties of pick and rock have been analyzed in [11], and rock variations and drum advance directions are two new factors introduced in this extended approach, this study focused on the investigation of the influences of these two new factors on the pick failure probability. To this end, only average TSDC tips with the mean allowable force of 9.745 kN were analysed. Furthermore, a series of what-if analyses were carried out based on the following three scenarios:

Scenario 1: The thickness of the roof rock can be represented by two mean values, (410 mm, 460 mm), with their corresponding likelihoods of (35%, 65%). The thickness of the floor rock can be represented by three mean values, (450 mm, 500 mm, 610 mm), with their corresponding likelihoods of (25%, 50%, 25%).

Scenario 2: The thickness of the roof rock and the floor rock can be represented by the same set of two mean values, (410 mm, 460 mm), with their corresponding likelihoods of (35%, 65%).

Scenario 3: The thickness of the roof rock and the floor rock can be represented by a single value of 442.5 mm (=410 × 35% + 460 × 65%), which is the expected value of the random variable "rock thickness" in Scenario 2.

3.1. Results

Figure 5 shows the pick tip failure probability changing with cutting time in a cutting cycle for Scenario 1.

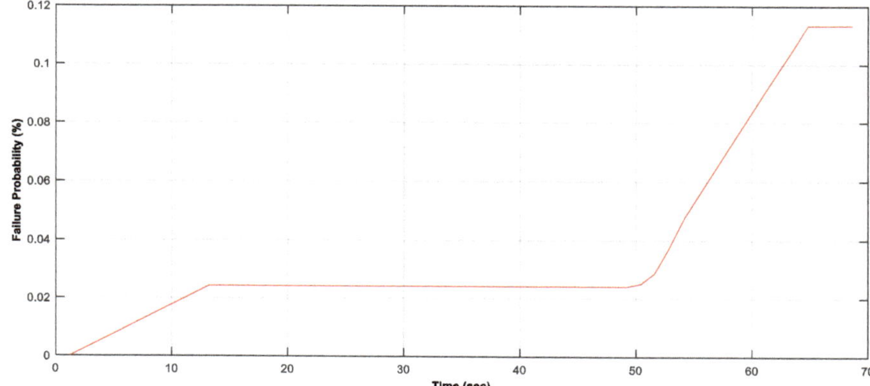

Figure 5. Thermally Stable Diamond Composite (TSDC) tip failure probability over a cutting cycle–Scenario 1.

Figures 6 and 7 illustrate the pick tip failure probabilities over the sump-in stage and the shearing-down stage for Scenario 2, respectively. As the tip failure probability in the cutting-back stage is regarded to be zero, the cutting back stage is no longer considered in the analysis for Scenarios 2 and 3.

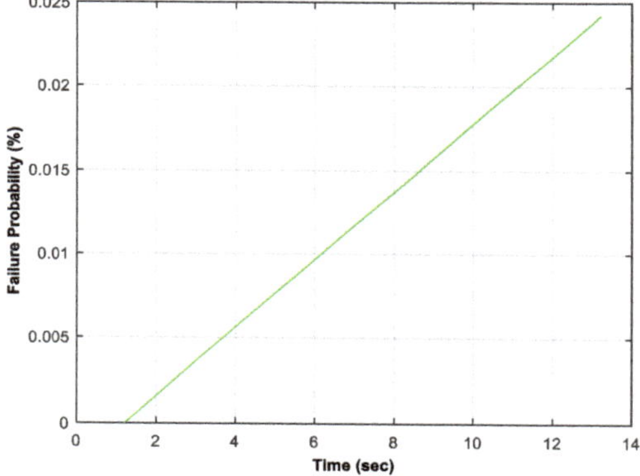

Figure 6. TSDC tip failure probability in the sump-in stage–Scenario 2.

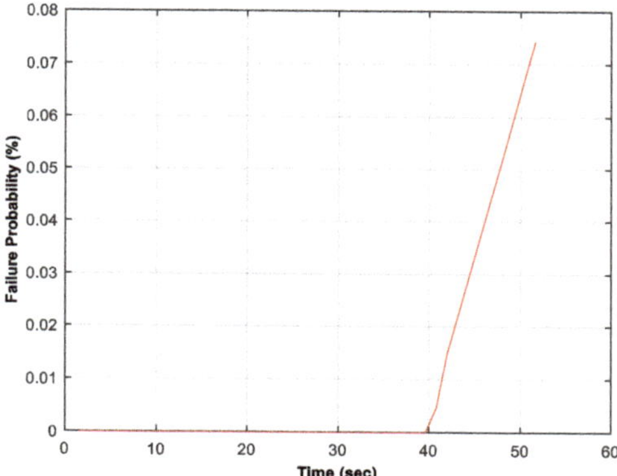

Figure 7. TSDC tip failure probability in the shearing-down stage–Scenario 2.

Figures 8 and 9 respectively depict the pick tip failure probabilities over the sump-in stage and the shearing-down stage for Scenario 3.

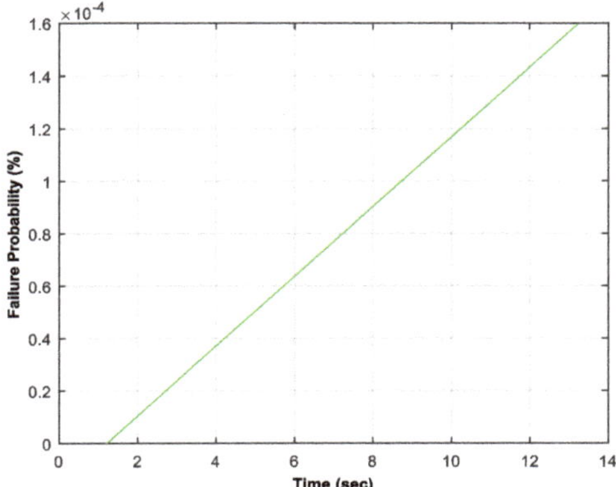

Figure 8. TSDC tip failure probability in the sump-in stage–Scenario 3.

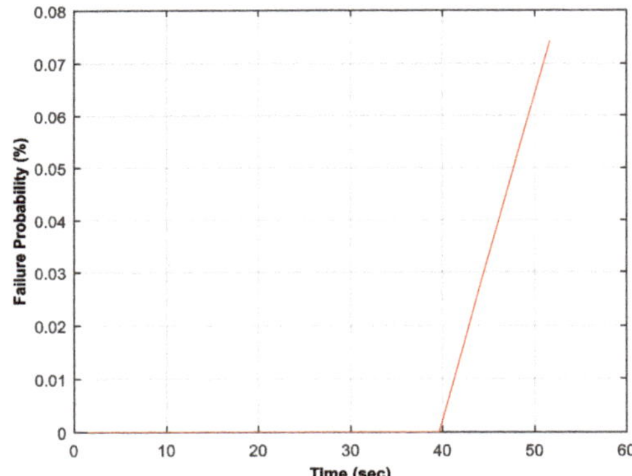

Figure 9. TSDC tip failure probability in the shearing-down stage–Scenario 3.

3.2. Discussion

Figure 5 shows that in a cutting cycle, the pick tip failure probability changes over time in a style of constant–increasing–constant–increasing–constant. The first constant period (zero period) occurs in the sump-in stage and lasts about 1.2 s, mainly due to the small DOC in the first revolution as mentioned in Section 2.2. In the first revolution, the maximum DOC is only 5.875 mm. The bending force generated from cutting the given sandstone with this small DOC will not damage the tips in the consideration. The first increase period happens in the sump-in stage due to hard rock cutting. The second constant period and the second increase period both occur in the shearing-down stage. In this stage, the tip cuts the coal seam first and then cuts hard floor rock. When cutting the coal seam only, the tip has a failure probability of 0 (zero) because as assumed, tips will not be subjected to bending-caused failure when they are cutting coal. As a result, the tip failure probability over the coal cutting period remains constant. Once the pick involves floor rock cutting, its tip failure probability increases over time again. The third constant period corresponds to the cutting-back stage. As discussed in Section 2.4.1, the failure probability of the tip at this stage can be ignored (i.e., it is regarded to be zero).

A comparison between Figures 6 and 8 indicates that the variation in the rock thickness can significantly affect the pick tip failure probability in the sump-in stage. Although the rock thickness in Scenario 3 is the expected value of the rock thickness in Scenario 2, the tip failure probability over the sump-in stage in Scenario 2 is much higher than that in Scenario 3. The reason is that there is 65% chance for the roof rock mean thickness to be 460 mm, which is much higher than the expected value of 442.5 mm. As all the rock thickness values are less than the drum's cutting radius (500 mm), a larger value of rock thickness means that the pick needs to cut rock segments with larger DOC. According to the formula given by Goktan [14] and Equation (20) in [11], the bending force acting on the tip is proportional to the square of DOC, and therefore a larger DOC will result in a higher failure probability of tips.

However, a comparison between Figures 7 and 9 shows that the variation in the rock thickness has little influence on the pick tip failure probability in the shearing-down stage. The reason for this phenomenon is that the drum advance direction in the sharing-down stage is perpendicular to that in the sump-in stage. When the drum advances towards the floor rock, an increase in the floor rock thickness will change the number of drum revolutions involving hard rock cutting, rather than the DOC of rock segments being cut. The changed number of drum revolutions involving hard rock cutting will also change the tip failure probability, but the magnitude of the change in the tip failure probability heavily depends on the ratio of the rock thickness variation to the average rock thickness. In the above examples, the variation of the maximum rock thickness in Scenario 2 (460 mm) to the expected mean rock thickness in Scenario 3 (442.5 mm), and that of the minimum rock thickness in Scenario 2 (410 mm) to the expected mean rock thickness in Scenario 3 are respectively 17.5 mm and -32.5 mm. As the absolute values of the variations are both less than the drum's advance distance per revolution (47 mm), the difference between drum revolutions involving hard rock cutting in Scenario 2 and Scenario 3 is less than 1 (one) revolution. This small change in the number of the drum revolutions involving hard rock cutting has a trivial influence on the tip failure probability, because the ratios of the variations of the maximum and the minimum rock thickness to the expected value are small (3.95% and -7.34% respectively). As a result, the pick tip failure probabilities over the shearing-down stage in both scenarios are nearly the same.

The influences of the drum advance direction on the pick tip failure probability can be further explored by comparing Figure 7 with Figure 6 and comparing Figure 9 with Figure 8. From Figures 6 and 7, it can be found that the tip failure probability in the shearing-down stage is much higher than that in the sump-in stage although the rock thickness distribution in both stages are the same. This finding is also evidenced by the comparison between Figures 8 and 9. In Scenario 3, the rock removed in the sump-in stage is slightly more than that in the shearing stage, but the tip failure probability over the sump-in stage is $1.6 \times 10^{-4}\%$, much lower than the tip failure probability over the shearing-down stage which is 0.07433%. A major cause for this failure probability difference is the different maximum DOC due to the different drum advance directions in these two stages. In the shearing-down stage, the drum advance direction is perpendicular to the floor. If the drum involves rock cutting in a revolution, it will often cut the rock with the maximum DOC. By contrast, in the sump-in stage, the drum advances along the roof. When the rock thickness is less than the drum's cutting radius, the DOC of the rock cut by the drum will be less than the maximum DOC. As abovementioned, a small change in DOC can result in a large variation in pick tip failure probability.

From the above analysis, it is noted that the prediction accuracy of the bending force on pick tip directly affects the accuracy of pick tip failure probability estimation. Currently, the bending force was analyzed based on the formula given by Goktan [14] and laboratory rock cutting tests [11]. However, laboratory rock cutting tests are often costly and time-consuming. Additionally, measuring forces on individual picks in situ is difficult. Therefore, new methods such as inverse analysis [27,28] may be needed to help force prediction.

4. Conclusions

In this paper, an extended probabilistic approach is proposed to remove the following two critical limitations in existing research on the estimation of the probability of pick random failures due to excessive bending force on pick cutting tips:

- Only one drum revolution is considered;
- The variation of rock thickness is ignored.

As such, this extended approach can be used to estimate the pick failure probability for a cutting process which consists of multiple drum revolutions and involves different rock thickness.

The research results show that drum advance direction can dramatically influence pick tip failure probability. Pick tip failure probability when the drum advances along the rock panel (like cutting a roadway roof in the sump-in stage) is lower than that when the drum advances towards the rock panel (like cutting the floor rock in the shearing-down stage), if rock thickness is less than the drum's cutting radius. The variation of rock thickness can also impact the pick tip failure probability, but this influence depends upon drum advance direction, the rock mean thickness, and the relative variation of the rock thickness to its mean value. In general, the influence of the variation in rock thickness on the pick tip failure probability when the drum advances along the rock panel is much greater than that when the drum advances towards the rock panel.

This extended approach is developed based on the failure probability analysis of picks over a full cutting cycle in the underground roadway development of coal mine. Nevertheless, it can be extended to other rock cutting applications straightforwardly. The research results can be used to optimize the design and operation of cutting drums with both TSDC picks and other type of picks to increase productivity and reduce production costs.

However, in the current approach, it is assumed that the roof rock thickness is less than two thirds of the drum's tip-to-tip diameter, the sump-in depth is constant, and failures are independent of each other. Further studies are needed to address these issues.

Author Contributions: Y.S. conceptualized the study and developed the original methodology. X.L. and H.G. contributed to the improvement of the analysis. Y.S. wrote the original draft of the manuscript. X.L., H.G. and Y.S. reviewed and improved the manuscript.

Funding: This research received no external funding.

Acknowledgments: The research was sponsored by the Commonwealth Scientific and Industrial Research Organization (CSIRO).

Conflicts of Interest: The authors declare no conflict of interest.

References

1. Mlungwane, K.; Herrmann, M.; Sigalas, I. The low-pressure infiltration of diamond by silicon to form diamond–silicon carbide composites. *J. Eur. Ceram. Soc.* **2008**, *28*, 321–326. [CrossRef]
2. Boland, J.N.; Li, X.S. Microstructural characterisation and wear behaviour of diamond composite materials. *Materials* **2010**, *3*, 1390–1419. [CrossRef]
3. Li, X.S.; Sun, Y. Development of hard rock cutting tool with advanced diamond composites. *Adv. Mater. Res.* **2013**, *690–693*, 1831–1835. [CrossRef]
4. Shao, W.; Li, X.S.; Sun, Y.; Huang, H. Laboratory comparison of SMART*CUT picks with WC picks. *Adv. Mater. Res.* **2014**, *1017*, 323–328. [CrossRef]
5. Shao, W.; Li, X.S.; Sun, Y.; Huang, H. Parametric study of rock cutting with SMART*CUT picks. *Tunn. Undergr. Space Technol.* **2017**, *61*, 134–144. [CrossRef]
6. Sun, Y.; Li, X.S. Failure probability of thermally stable diamond composite tips for cutting rock segments. *Mater. Sci. Forum* **2018**, *936*, 192–197. [CrossRef]
7. Li, X.S.; Boland, J.N. The wear characteristics of superhard composite materials in abrasive cutting operations. *Wear* **2005**, *259*, 1128–1136. [CrossRef]

8. Li, X.S.; Boland, J.N.; Guo, H. A comparison of wear and cutting performance between diamond composite and tungsten carbide tools. *Ind. Diam. Rev.* **2008**, *68*, 51–54.

9. Vogt, D. A review of rock cutting for underground mining: Past, present and future. *J. S. Afr. Inst. Min. Metall.* **2016**, *116*, 1011–1026. [CrossRef]

10. Shao, W.; Li, X.S.; Sun, Y.; Huang, H.; Tang, J.Y. An experimental study of temperature at the tip of point-attack pick during rock cutting process. *Int. J. Rock Mech. Min. Sci.* **2018**, *107*, 39–47. [CrossRef]

11. Sun, Y.; Li, X.S. A probabilistic approach for assessing failure risk of cutting tools in underground excavation. *Tunn. Undergr. Space Technol.* **2017**, *70*, 299–308. [CrossRef]

12. Prokopenko, S.; Li, A.; Kurzina, I.; Sushko, A. Improved operating performance of mining machine picks. *IOP Conf. Ser. Mater. Sci. Eng.* **2016**, *142*, 1–8. [CrossRef]

13. Evans, I. A theory of the cutting force for point-attack picks. *Int. J. Min. Eng.* **1984**, *2*, 63–71. [CrossRef]

14. Goktan, R. A suggested improvement on Evan's cutting theory for conical bits. In *Proceedings of the Fourth International Symposium on Mine Mechanization and Automation, Brisbane, Australia, 6–9 July 1997*; Gurgenci, H., Hood, M., Eds.; Cooperative Research Centre for Mining Technology and Equipment: Kenmore, Australia, 1997; Volume I, pp. 57–61.

15. Nishimatsu, Y. The mechanics of rock cutting. *Int. J. Rock Mech. Min. Sci.* **1972**, *9*, 261–270. [CrossRef]

16. Goktan, R.M.; Gunes, N. A semi-empirical approach to cutting force prediction for point-attack picks. *J. S. Afr. Inst. Min. Metall.* **2005**, *105*, 257–263.

17. Sun, Y.; Li, X.S. Ineffective rock breaking and its impacts on pick failures. In *Proceedings of the 31st International Symposium on Automation and Robotics in Construction and Mining, Sydney, Australia, 9–11 July 2014*; Ha, Q., Shen, X.S., Akbarnezhad, A., Eds.; University of Technology, Sydney: Sydney, Australia, 2014; pp. 754–760.

18. Hekimoglu, O.Z.; Fowell, R.J. Theoretical and practical aspects of circumferential pick spacing on boom tunnelling machine cutting heads. *Min. Sci. Technol.* **1991**, *13*, 257–270. [CrossRef]

19. Sun, Y.; Li, X.S. Slant angle and its influence on rock cutting performance. *Adv. Civ. Eng.* **2018**, *2018*, 1–11. [CrossRef]

20. McNider, T.; Grygiel, E.; Haynes, J. Reducing frictional ignitions and improving bit life through novel pick and drum design. In *Proceedings of the 3rd Mine Ventilation Symposium, University Park, State College, PA, USA, 12–14 October 1987*; Mutmansky, J.M., Ed.; Society of Mining Engineers: Littleton, CO, USA, 1987; pp. 119–125.

21. Sun, Y.; Li, X.S. Experimental investigation of pick body bending failure. *Int. J. Mech. Eng. Robot. Res.* **2018**, *7*, 184–188. [CrossRef]

22. Hurt, K.G.; MacAndrew, K.M. Cutting efficiency and life of rock-cutting picks. *Min. Sci. Technol.* **1985**, *2*, 139–151. [CrossRef]

23. Sun, Y.; Li, X.S. Comparison of the methods for calculating instant depth of cut. *Adv. Mater. Res.* **2013**, *634–638*, 3313–13320. [CrossRef]

24. Li, X.S.; Tiryaki, B.; Cleary, P.W. Hard rock cutting with SMART/CUT technology. In *Proceedings of the 22nd World Mining Congress and Expo, Istanbul, Turkey, 11–16 September 2011*; Eskikaya, S., Ed.; Aydoğdu Ofset: Ankara, Turkey, 2011; pp. 725–732.

25. Sun, Y.; Li, X.S. Modelling the property variation of diamond composite and its impact on the reliability of cutting tools. *Adv. Mater. Res.* **2012**, *565*, 448–453. [CrossRef]

26. Vu, K.A.T.; Stewart, M.S. Predicting the likelihood and extent of reinforced concrete corrosion-induced cracking. *J. Struct. Eng.* **2005**, *131*, 1681–1689. [CrossRef]

27. Ning, J.; Liang, S.Y. Model-driven determination of Johnson-Cook material constants using temperature and force measurements. *Int. J. Adv. Manu. Tech.* **2018**, *97*, 1053–1060. [CrossRef]

28. Ning, J.; Liang, S.Y. Inverse identification of Johnson-Cook material constants based on modified chip formation model and iterative gradient search using temperature and force measurements. *Int. J. Adv. Manu. Tech.* **2019**, *102*, 2865–2876. [CrossRef]

Article

Triaxial Loading and Unloading Tests on Dry and Saturated Sandstone Specimens

Diyuan Li [1], Zhi Sun [1], Quanqi Zhu [1] and Kang Peng [2,3,*]

[1]   School of Resources and Safety Engineering, Central South University, Changsha 410083, China;
       diyuan.li@csu.edu.cn (D.L.); zhi.sun@csu.edu.cn (Z.S.); quanqi_zhu@csu.edu.cn (Q.Z.)
[2]   State Key Laboratory of Coal Mine Disaster Dynamics and Control, Chongqing University,
       Chongqing 400044, China
[3]   College of Resources and Environmental Science, Chongqing University, Chongqing 400044, China
*   Correspondence: pengkang@cqu.edu.cn; Tel.: +86-159-7426-9965

Received: 16 February 2019; Accepted: 10 April 2019; Published: 24 April 2019

Featured Application: The specific application or a potential application of the work is related to deep mining and tunneling engineering.

Abstract: The brittle failure of hard rock due to the excavation unloading in deep rock engineering often causes serious problems in mining and tunneling engineering, and the failure process is always affected by groundwater. In order to investigate the effects of stress paths and water conditions on the mechanical properties and failure behavior of rocks, a series of triaxial compression tests were conducted on dry and saturated sandstones under various loading and unloading paths. It was found that when the sandstone rock samples are saturated by water, the cohesion, the internal friction angle and the Young's modulus will decrease but the Poisson's ratio will increase. The fracturing characteristics of the sandstone specimens are related to the initial confining pressure, the stress paths and the water conditions from both macroscopic and microscopic viewpoints. The failure of sandstone in unloading test is more severe than that under loading test, particularly for dry sandstone samples. In unloading test, the energy is mainly consumed for the circumferential deformation and converted into kinetic energy for the rock bursts. The sandstone is more prone to produce internal cracks under the effect of water, and the absorbed energy mainly contributes to the damage of rock. It indicates that the possibility of rockburst in saturated rock is lower than the samples in dry condition. It is important to mention that water injection in rock is an effective way to prevent rockburst in deep rock engineering.

Keywords: triaxial compression test; sandstone; rock mechanics; rock fracture; energy evolution

1. Introduction

The excavation of underground rock engineering is often affected by groundwater [1,2]. Under high geo-stress, high groundwater pressure is prone to cause engineering geological disasters, such as water inrush. It seriously affects the construction progress and personnel safety for underground rock engineering [3]. The excavation of underground engineering, in fact, is the triaxial loading and unloading processes of the surrounding rock masses. It is therefore important to carry out the triaxial loading and unloading tests with different water contents of the rocks in laboratory.

Considerable efforts have been devoted to the effects of water on rock failure. For the influence of water on the mechanical properties of rocks, it has been found that a small increase in the water content may significantly lower the strength and stiffness of the rocks [4–9]. The statistical and numerical analysis on 14 kinds of the rocks with different water contents were also carried out, and it was found a negative exponential relationship between rock strength and moisture content [10].

Meanwhile, Li et al., [11] conducted the triaxial compression tests and found that the internal friction angle decreases while the cohesion increases when the meta-sedimentary rocks was saturated by water. Most of previous related studies were focused on the influence of water on the uniaxial or conventional triaxial strength of rocks [5–7,10]. However, few experimental and numerical studies are reported to discuss the effect of water content on the strength and deformability of the rocks under triaxial stress conditions [12–14]. In the practical engineering, the existence of groundwater not only affects the mechanical properties of rock, but also influences the damage degree and even the failure mode of the rock [15]. Zhou et al. [16] investigated the effects of water and joints on the properties of rock masses using the data from the Shirengou iron mine. Their numerical simulation results indicated that water is the critical factor for rock damage pattern. Several scholars also conducted experiments on deep rock and successfully showed a layered failure for saturated rock in tensile tests [17]. Spalling behaviors were studied in ture-triaxial unloading conditions and scanning electron microscope (SEM) observations revealed the distribution of microcracks in the fragments [18]. Besides, there are many studies related to rock uniaxial compression and tensile failure affected by water [19–22]. The deformation and failure of rock are energy-driven processes which include energy absorption, evolution, dissipation, and release [23–26]. Based on uniaxial and triaxial compression tests, Hua and You [27] studied the characteristics of rock energy evolution during unloading failure and concluded that the strain energy stored in rock material is sufficiently large to cause failure when it is released. Furthermore, water injection to reduce rockburst occurrences has also been discussed mainly on numerical simulation [28–30], and partially for experiments [2,31]. The numerical modeling showed that the peak stress can be reduced as water injection, which causes a significant reduction of internal energy stored within the rock, and the possibility of a rockburst occurrence may be reduced [28]. Moreover, the effects of water on rock energy evolution have also been studied [32–37], though it mainly concerns on uniaxial compression [34,35] and axial loading-unloading experiments [32,33,36,37]. Undoubtedly, the above study enriches our knowledge of the effect of water on rock failure and energy evolution. However, there are a few studies concern on the effect of water on rocks under triaxial tests, especially for the influence of unloading high confining pressure conditions, which represents a more realistic site environment.

Therefore, based on deep underground engineering such as mining and tunnel engineering under high geo-stress and water-rich conditions, this study is tried to investigate the combining effects of water content and high stress on the mechanical properties, failure and energy evolution of rocks in different stress paths. It is benefited the exploitation of deep resources and construction of deep rock engineering. Moreover, this study will be helpful to understand the effect of water on rock failure in high stress.

2. Materials and Methods

2.1. Specimen Preparation

Red sandstone samples which were obtained from a quarry in Yunnan Province, were used to test in the present study. A thin section analysis was carried out to examine the microstructure of the sandstone. It was identified as fine-grained sandstone by petrographic microscopy (Leica DM2700P, Leica Microsystems Inc., Wetzlar, German). The microstructure of the sandstone under plane polarization light (PPL) and cross polarization light (CPL) are illustrated in Figure 1. The mineral composition and grain size distribution of the sandstone specimen are listed in Table 1.

(a)   (b)

Figure 1. Polarized light micrographs of sandstone specimen: (a) plane polarization light (PPL); (b) cross polarization light (CPL). (Note: The letters Qz, Kfs and Cc represent quartz, potassium feldspar and calcite, respectively).

Table 1. Mineral composition and grain size distribution of sandstone specimen.

Rock Composition	Grain Size (mm)	Mineral Content (%)
Quartz	0.03 ~ 0.2	77
Potassium feldspar	0.05 ~ 0.2	10
Calcite	0.01 ~ 0.08	7
Sericite	0.01 ~ 0.02	4
Others	0.01 ~ 0.1	2

Both oven-dried and water-saturated specimens were prepared to study the influence of water on triaxial failure of the sandstone samples. In drying process, the sandstone specimens were placed in an oven at 105 °C for 24 h, and then they were removed to a desiccator and weighed after cooling to room temperature. In saturation process, the specimens were placed in water and allowed to absorb water for 48 h under atmospheric pressure, and then specimens were taken out and weighed after removing surface moisture of the rock. These specimens were considered as saturated samples in this study. The natural water content of specimen was about 2.5% and the saturated water content was about 3.3%. The diameter (D) of the cylindrical specimens was 50 mm and the specimen height (H) was 100 mm. The density and the P-wave velocity of each testing specimen are provided in Table 2.

Table 2. Test scheme and physical parameters of sandstone specimens.

Group	Specimen No.	Density (kg/m^3)	P-Wave Velocity (m/s)	Initial Confining Pressure (MPa)	Unloading Point (MPa)	Loading Rate	Unloading Rate (MPa/min)
TC-D	TC-D-0	2387.3	2951	0		0.1 mm/min	
	TC-D-10	2380.4	2979	10		0.1 mm/min	
	TC-D-20	2387.5	2916	20		0.1 mm/min	
	TC-D-30	2398.3	3063	30		0.1 mm/min	
	TC-D-40	2382.1	3077	40		0.1 mm/min	
TC-S	TC-S-0	2456.8	3381	0		0.1 mm/min	
	TC-S-10	2459.8	3363	10		0.1 mm/min	
	TC-S-20	2464.1	3362	20		0.1 mm/min	
	TC-S-30	2459.7	3393	30		0.1 mm/min	
	TC-S-40	2450.0	3392	40		0.1 mm/min	

Table 2. *Cont.*

TU-D	TU-D-10	2380.7	3299	10	$0.8\sigma_{(10)}$	0.1 mm/min-1.5 MPa/min	3
	TU-D-20	2391.3	3292	20	$0.8\sigma_{(20)}$	0.1 mm/min-1.5 MPa/min	3
	TU-D-30	2387.8	3315	30	$0.8\sigma_{(30)}$	0.1 mm/min-1.5 MPa/min	3
	TU-D-40	2391.3	3313	40	$0.8\sigma_{(40)}$	0.1 mm/min-1.5 MPa/min	3
TU-S	TU-S-10	2454.7	3396	10	$0.8\sigma_{(10)}$	0.1mm/min-1.5 MPa/min	3
	TU-S-20	2457.7	3347	20	$0.8\sigma_{(20)}$	0.1 mm/min-1.5 MPa/min	3
	TU-S-30	2457.8	3349	30	$0.8\sigma_{(30)}$	0.1 mm/min-1.5 MPa/min	3
	TU-S-40	2457.7	3399	40	$0.8\sigma_{(40)}$	0.1 mm/min-1.5 MPa/min	3

2.2. Test Scheme

The triaxial tests including conventional triaxial compression tests (abbreviate to loading test) and unloading confining pressure tests (abbreviate to unloading test), were conducted by a servo-controlled material testing machine (MTS 815) (MTS Systems Corporation, Minnesota, USA) at laboratory of Central South University. Both of axial and circumferential strains can be recorded by extensometers.

Each test includes two experimental groups of dry and saturated sandstones respectively. Thus, the specimens are divided into four groups of dry and saturated sandstones in different loading and unloading tests (see in Table 2). The initial confining pressures (σ_3^0) were set as 10, 20, 30, and 40 MPa, respectively. Meanwhile, the unloading point for the confining pressure in unloading tests was set as the axial stress reaching 80% of the corresponding triaxial compression strength (such as $0.8\sigma_{(10)}$ and so on). Then, the axial loading method was changed from displacement control (0.1 mm/min) to loading control (1.5 MPa/min) at unloading point. In general, four sets of specimens were prepared, i.e., TC-D, TC-S, TU-D, TU-S. Specimens of set TC-D and TC-S, with the dry and saturated state respectively, were used in the conventional triaxial compression test. Specimens of set TU-D and TU-S, with the dry and saturated state respectively, were used in the unloading confining pressure test. And the number of specimen set represents for the corresponding confining pressure. For example, as shown in Table 2, TU-D-10 represents the unloading confining pressure test of dry specimen under the confining pressure of 10 MPa.

The detailed test procedures of the triaxial unloading test are listed as follows:

Step 1: Apply hydrostatic pressure on the specimen to an initial confining pressure (σ_3^0).

Step 2: Keep σ_3^0, and increase axial stress to 80% of the corresponding triaxial compression strength by a displacement control method at 0.1 mm/min.

Step 3: Reach the unloading point, change the axial stress loading method from displacement control to load control at the unloading point.

Step 4: Increase the axial stress at 1.5 MPa/min and also reduce the confining pressure simultaneously by load control method at a specified unloading rate (3.0 MPa/min) until the failure of the rock specimen.

3. Results

3.1. Mechanical Properties

Figure 2 shows the variation of the peak strength with the initial confining pressure of dry and saturated sandstone specimens. It can be seen that the peak strength of sandstone specimen increases as the initial confining pressure increases, but the increase rate slows down with the increase of the initial confining pressure. The difference of the peak strength between dry and saturated sandstones grows with the increasing of initial confining pressure. It means that the effect of water on rock strength is more remarkable under high confining pressure.

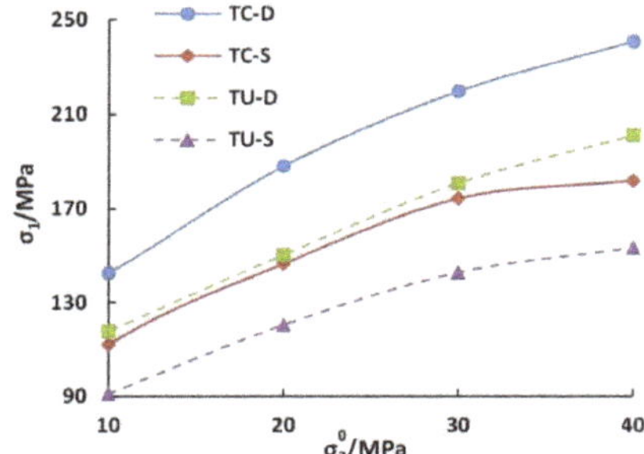

Figure 2. Peak strength of sandstone specimens.

Figure 3 illustrates the strength parameters of different groups of specimens under triaxial loading and unloading tests. The strength parameters of sandstone specimens are differed in different stress paths. The strength parameters of C (cohesion) and φ (internal friction angle) under loading conditions are higher than those of unloading tests with the same water condition. Meanwhile, both C and φ values of saturated sandstones are lower than those of dry sandstones, which indicate that water has a deep impact in decreasing strength parameters (C and φ) of sandstone samples. When the sandstone samples are saturated by water, their shear strength parameters are reduced, where changes in results of φ are more remarkable. The weakening of strength parameters further affects the ultimate bearing capacity of sandstone.

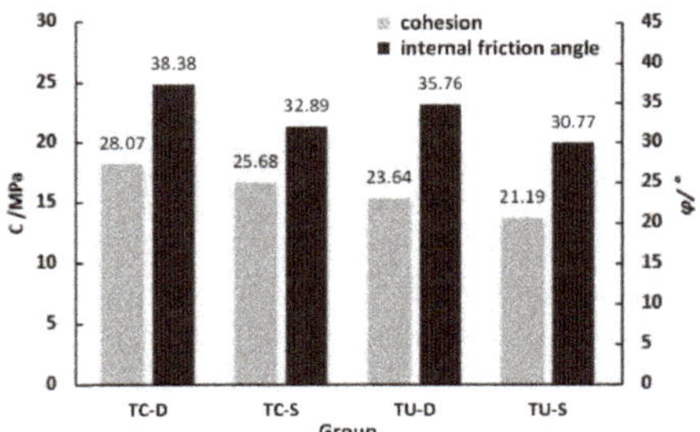

Figure 3. Strength parameters of sandstone specimens.

The values of the axial strain (ε_1, ε_1'), circumferential strain (ε_3, ε_3'), and volumetric strain (ε_v, ε_v') at the unloading point (σ_3^0) and the critical failure point (σ_3^f) are listed in Table 3. It can be seen that the absolute value of volumetric strain for rock in unloading test is greater than that of rock samples in conventional triaxial compression test, particularly for saturated samples. It means that the characteristics of volume expansion of rock under unloading conditions are more pronounced than the rock samples in loading tests. In addition, the incremental rates of circumferential strain are several times larger than those of axial strain from unloading point to critical failure point. The rock failure occurs mainly for the circumferential expansion under unloading conditions. Furthermore, except for the initial confining pressure of 20 MPa, the confining pressures of saturated sandstones at failure point (σ_3^f) are higher than that for dry sandstones which indicate that saturated sandstones fail earlier than dry sandstones. This is due to the fact that the structures of rock samples are weaker when they are saturated by water as compared to structures of dry rock samples.

Table 3. Critical strain values of sandstone specimens at the unloading point and the critical failure point of sandstone specimens.

Group	Specimen Number	Unloading Point				Critical Failure Point			
		σ_3^0	ε_1	ε_3	ε_v	σ_3^f	ε_1'	ε_3'	ε_v'
TC-D	TC-D-10	10	-	-	-	10	0.0110	−0.0084	−0.0058
	TC-D-20	20	-	-	-	20	0.0136	−0.0085	−0.0034
	TC-D-30	30	-	-	-	30	0.0156	−0.0118	−0.008
	TC-D-40	40	-	-	-	40	0.0218	−0.0166	−0.0114
TC-S	TC-S-10	10	-	-	-	10	0.0118	−0.0124	−0.013
	TC-S-20	20	-	-	-	20	0.0137	−0.0122	−0.0107
	TC-S-30	30	-	-	-	30	0.0180	−0.0104	−0.0028
	TC-S-40	40	-	-	-	40	0.0185	−0.0100	−0.0015
TU-D	TU-D-10	10	0.0063	−0.0012	0.0039	1.52	0.0077	−0.0080	−0.0083
	TU-D-20	20	0.0082	−0.0025	0.0032	14.82	0.0137	−0.0116	−0.0095
	TU-D-30	30	0.0088	−0.0019	0.005	15.51	0.0123	−0.0104	−0.0085
	TU-D-40	40	0.0096	−0.0022	0.0052	20.01	0.0145	−0.0135	−0.0125
TU-S	TU-S-10	10	0.0063	−0.0017	0.0029	4.56	0.0085	−0.0095	−0.0105
	TU-S-20	20	0.0086	−0.0020	0.0046	10.56	0.0103	−0.0134	−0.0165
	TU-S-30	30	0.0082	−0.0025	0.0032	19.73	0.0118	−0.0100	−0.0082
	TU-S-40	40	0.0082	−0.0018	0.0046	21.61	0.0137	−0.0136	−0.0135

The relationship between the deformation parameters (i.e., the Young's modulus, E, and the Poisson's ratio, μ) and the initial confining pressure (σ_3^0) is shown in Figure 4. As shown in Figure 4a, E values of sandstone firstly increase and then decrease as the initial confining pressure increases. Meanwhile, E values of dry sandstones are significantly higher than that of saturated sandstones in both loading and unloading tests. In Figure 4b, μ values of sandstone increase with the increase of initial confining pressure. Because the sandstone specimens are more prone to produce cracks paralleling to axial loading direction with the release of circumferential constraint in unloading test, the Poisson's ratio of the sandstone in unloading test is higher than those values in loading test. Meanwhile, the Poisson's ratio of the saturated specimen is higher than that of dry specimen under the same stress path. The bearing capacity decreases, and the Young's modulus of the sandstone decreases while the Poisson's ratio increases.

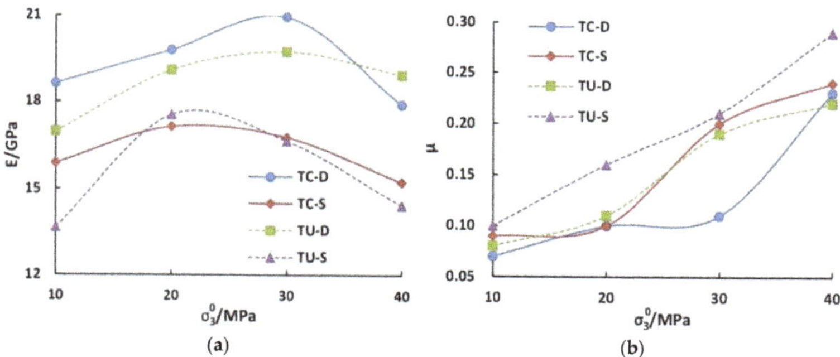

Figure 4. Variation of the deformation parameters of sandstone with the initial confining pressure (**a**) Young's modulus; and (**b**) Poisson's ratio.

3.2. Fracturing Characteristics

Figure 5 shows the failure modes of dry and saturated sandstones in conventional triaxial compression tests. It is found from Figure 5a,b that the spalling and tensile cracks are visible under uniaxial compression. When the initial confining pressure increases to 10 MPa, axial or radial tensile cracks are often produced along the shear plane, and then tensile cracks develop and form a macroscopic failure surface, which even breaks earlier than the main shear fracture, as shown in Figure 5d. It indicates that the failure mode of rock specimens shows a combined tension and shear failure under low initial confining pressure condition ($\sigma_3^0 \leq 10$ MPa). However, under high initial confining pressures ($\sigma_3^0 > 10$ MPa), a main shear fracture passes through the entire specimen usually extend to the end of specimen. The specimens are cut into two triangular vertebral bodies by the main fracture, and the main failure surface of sandstone develops from the end of specimen to the side surface as the increase of initial confining pressure. Shear failure occurs under high initial confining pressures ($\sigma_3^0 > 10$ MPa), and a single shear fracture dominates the failure processes. Besides, all dry sandstone specimens show structural failure while some saturated sandstone specimens fail partly, as shown in Figure 5, implying that the failure characteristics of dry sandstones are more severe than those of saturated sandstones.

Figure 5. Failure modes of dry and saturated sandstones in conventional triaxial compression tests: (a) TC-D-0; (b) TC-S-0; (c) TC-D-10; (d) TC-S-10; (e) TC-D-20; (f) TC-S-20; (g) TC-D-30; (h) TC-S-30; (i) TC-D-40; (j) TC-S-40.

Figure 6 shows the failure modes of dry and saturated sandstones in unloading confining pressure tests. When the initial confining pressure is 10 MPa, the rock tends to produce longitudinal tensile cracks approximately parallel to the loading direction, as the reduction of circumferential constraints in unloading test. Tensile cracks propagate and coalesce with the shear crack, leading to the final failure of rock specimens. When the initial confining pressure is 20 MPa, there are numerous rock fragments caused by shear friction on the rock failure plane, implying that the rock specimen undergoes shear failure and forms a macroscopic shear surface. Meanwhile, longitudinal (Figure 6c) and radial (Figure 6d) tensile cracks are often produced at the lower part of the triangular vertebral body. Combined shear and tension failure occurs for the rock specimen. Under high initial confining pressures ($\sigma_3^0 > 20$ MPa), shear failure dominates the failure process in unloading tests, which is the same as the conventional triaxial compression test. The shear friction of the rock is severe, and a lot of rock fragments are observed on the failure surface. Besides, by comparing failed rock fragments, it can be found that the failure characteristics of dry sandstones are more violent than those of saturated sandstones under unloading test.

Figure 6. Failure modes of dry and saturated sandstones in unloading confining pressure tests: (**a**) TU-D-10; (**b**) TU-S-10; (**c**) TU-D-20; (**d**) TU-S-20; (**e**) TU-D-30; (**f**) TU-S-30; (**g**) TU-D-40; (**h**) TU-S-40.

The phenomenon of volume expansion occurs significantly in the triaxial tests, particularly for the saturated rock in unloading confining pressure test. The failure characteristics of sandstone in unloading test are more violent than those under triaxial loading test.

In order to compare and analyze the microscopic and macroscopic failure characteristics of the rocks under different confining pressures, stress paths and water contents, the microscopic analysis of dry and saturated sandstones with an initial confining pressure of 10 and 40 MPa were carried out using the scanning electron microscopy (SEM) method. Figure 7 shows the SEM photos of sandstone fractures with the initial confining pressure 10 and 40 MPa in conventional triaxial compression test. In Figure 7a,c, the spalling fractures occurs in microscopic view under the initial confining pressure of 10 MPa, which is similar to the macroscopic spalling failure of rock specimen. The microscopic fractures retain sharp and smooth crystal surfaces. Meanwhile, the tearing traces (left arrows) are formed in Figure 7b, which are the typical microscopic characteristics of tensile failure [25,38]. Besides, there are tiny rock fragments (as shown at two sides of Figure 7a,d) and slip scratches (as shown in Figure 7d) left behind by shear friction [39]. It means that the micro-cracks of sandstone are shown as a combined tension and shear failure under low initial confining pressures ($\sigma_3^0 = 10$ MPa represents low initial confining pressure). It is consistent with the macroscopic failure characteristics of sandstone. However, under high initial confining pressures ($\sigma_3^0 = 40$ MPa), the microscopic fractures of the rock are stepped and serrated as shear effect [40]. In addition, there are lots of rock fragments left in the low-lying area, as illustrated in Figure 7e–h. Furthermore, the slip scratches along the crystal are also existed in Figure 7e. The rock crystals were cut off and angular edges were flattened in Figure 7h. The fractures almost have no traces of tension failure on the microscopic section of sandstone. Under high initial confining pressure conditions, the microscopic failure of sandstone is dominated by shear, which is consistent with the macroscopic failure characteristics of sandstone in the conventional triaxial compression test.

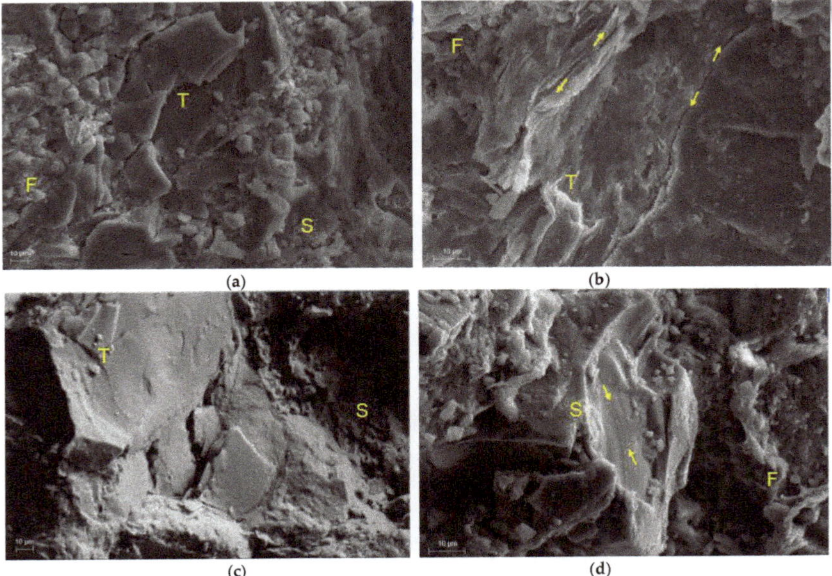

Figure 7. *Cont.*

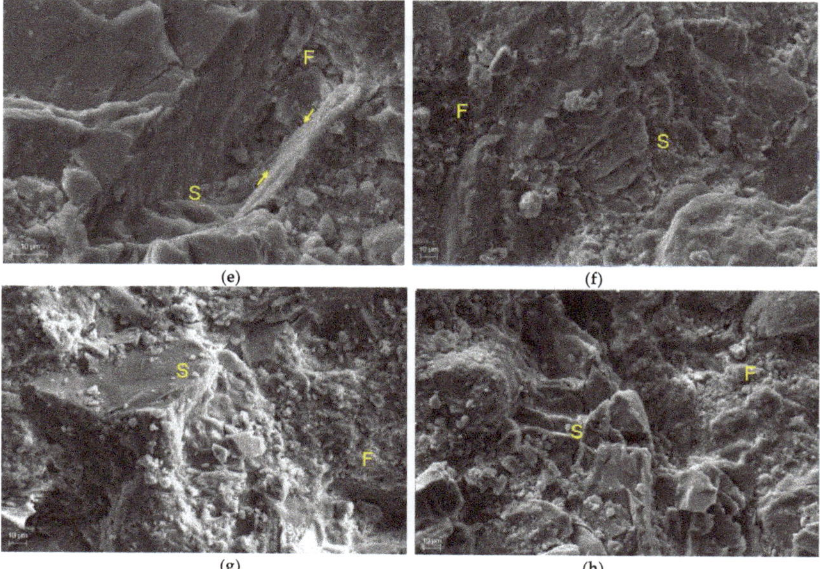

Figure 7. Typical scanning electron microscopy (SEM) images of sandstone fractures with the initial confining pressure 10 and 40 MPa in conventional triaxial compression tests: (**a,b**) TC-D-10; (**c,d**) TC-S-10; (**e,f**) TC-D-40; (**g,h**) TC-S-40 (Note: Opposite arrows and inverse arrows represent slip trace and tensile trace respectively; and the letters F, S and T represent fragments, shear crack and tensile crack, respectively).

Figure 8 shows the SEM photos of sandstone fractures with the initial confining pressure 10 and 40 MPa in unloading confining pressure tests. It can be seen clearly from Figure 8 that there are visible tensile cracks in dry and saturated sandstones under different confining pressures. In Figure 8a,c,e,g,h, the microscopic section shows sharp angular edges and smooth crystal surfaces. Moreover, there are evident traces of tension and spalling fractures in microscopic view. The microscopic section exhibit a typical tensile failure. In addition, as shown in Figure 8b,d,f, there are a large number of rock fragments left behind on the roughly microscopic section. The scratches caused by shear friction appear on the right part of Figure 8h. The microscopic failure section also shows shear failure characteristics. Thus, the microscopic failure of sandstone mainly shows a combined tension and shear failure in unloading test. It is consistent with macroscopic failure of sandstone under low initial confining pressure, but it is different from macroscopic failure of sandstone under high initial confining pressure, where the specimen only shows shear failure in macroscopic view.

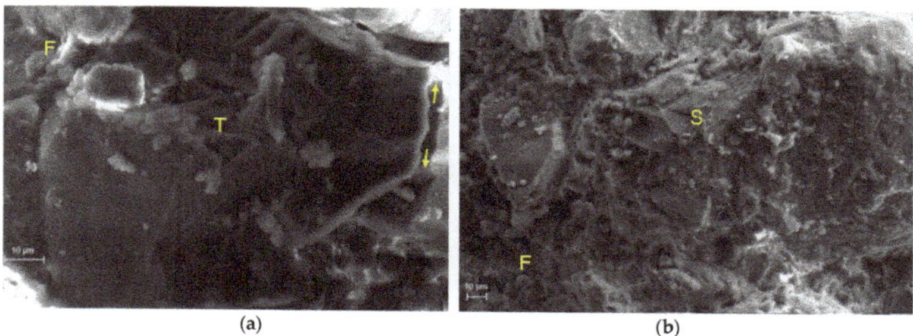

Figure 8. *Cont.*

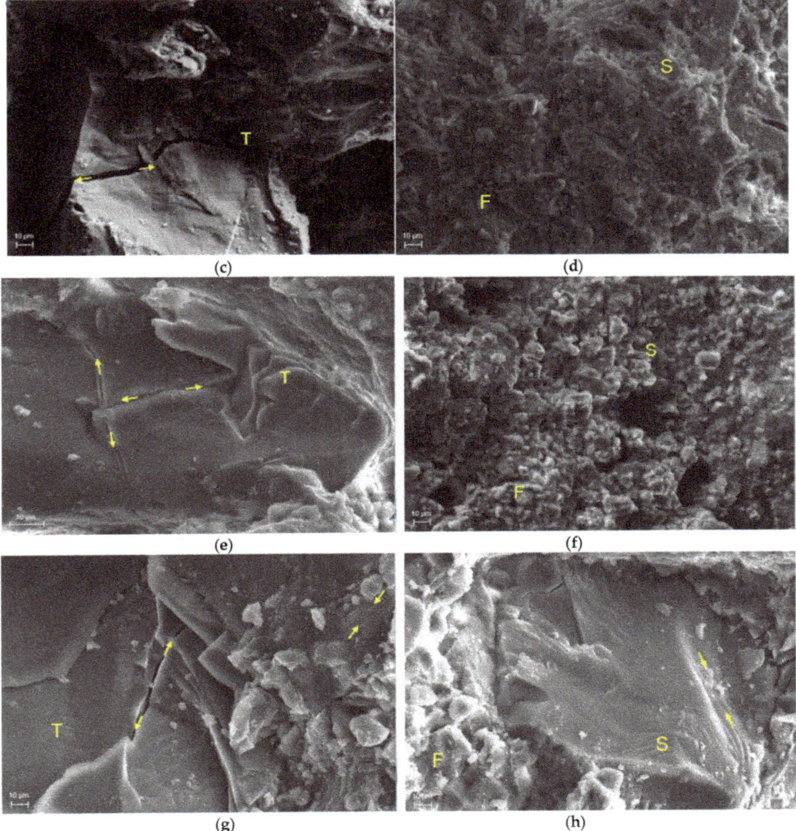

Figure 8. Typical SEM images of sandstone fractures with the initial confining pressure 10 and 40 MPa in unloading confining pressure tests: (**a**,**b**) TC-D-10; (**c**,**d**) TC-S-10; (**e**,**f**) TC-D-40; (**g**,**h**) TC-S-40 (Opposite arrows and inverse arrows represent slip trace and tensile trace respectively; and the letters F, S and T represent fragments, shear crack and tensile crack, respectively).

3.3. Energy Evolution of Rock Failure

Based on the relevant strain energy calculation formula and the experimental data, the strain energy evolution curves can be obtained with loading time [41,42]. The typical time history curves of strain energy for dry and saturated sandstone specimens under different initial confining pressures are shown in Figure 9. The initial confining pressures of 10 MPa and 40 MPa represent low and high confining pressures respectively. The strain energy curves for other specimens under low or high confining pressures are similar to these curves in each testing group.

For conventional triaxial compression tests, as shown in Figure 9a–d, several typical stages can be divided corresponding to the points in the stress-time curves. When the rock specimen is under low initial confining pressure ($\sigma_3^0 = 10$ MPa), as shown in Figure 9a,c, the curves of total strain energy (U) and elastic strain energy (U_e) almost overlap, and the dissipative strain energy (U_d) is relatively low in the micro-cracks compaction stage (OA) and the elastic deformation stage (AB). It indicates that the energy absorbed from the test machine is basically converted into U_e. However, while the rock specimen is under high initial confining pressure ($\sigma_3^0 = 40$ MPa), particularly for the saturated sandstone as shown in Figure 9d, the curves of U and U_e begin to separate gradually and there are few increases of U_d in the elastic deformation stage (AB). In the crack initiation and expansion stage (BC), when the rock is under low initial confining pressure, the curves of U and U_e separate gradually at point B. Meanwhile, U_d starts to increase from this point, owing to the development of plastic deformation and propagation of micro-cracks of the specimens. When the rock is under high initial confining pressure, U, U_e, and U_d increase steadily where the increase rate of U_d is the minimum among them. During the unstable crack development up to the failure stage (CD), the increase rate of U_e slows down, and U_e reaches the elastic energy storage limit at the peak strength point D. U_d increases rapidly in this stage, illustrating accelerated growth of micro-cracks, and the failure approaches. In the post-failure stage (DE), U_e releases quickly to a small value, and U_d increases rapidly to a large value. The absorbed strain energy is basically transformed into U_d, which contributes to the development of internal cracks and a large shear deformation along the fracture surface. However, U still increases rapidly as the large axial deformation at this stage. Finally, the specimen shows a brittle failure.

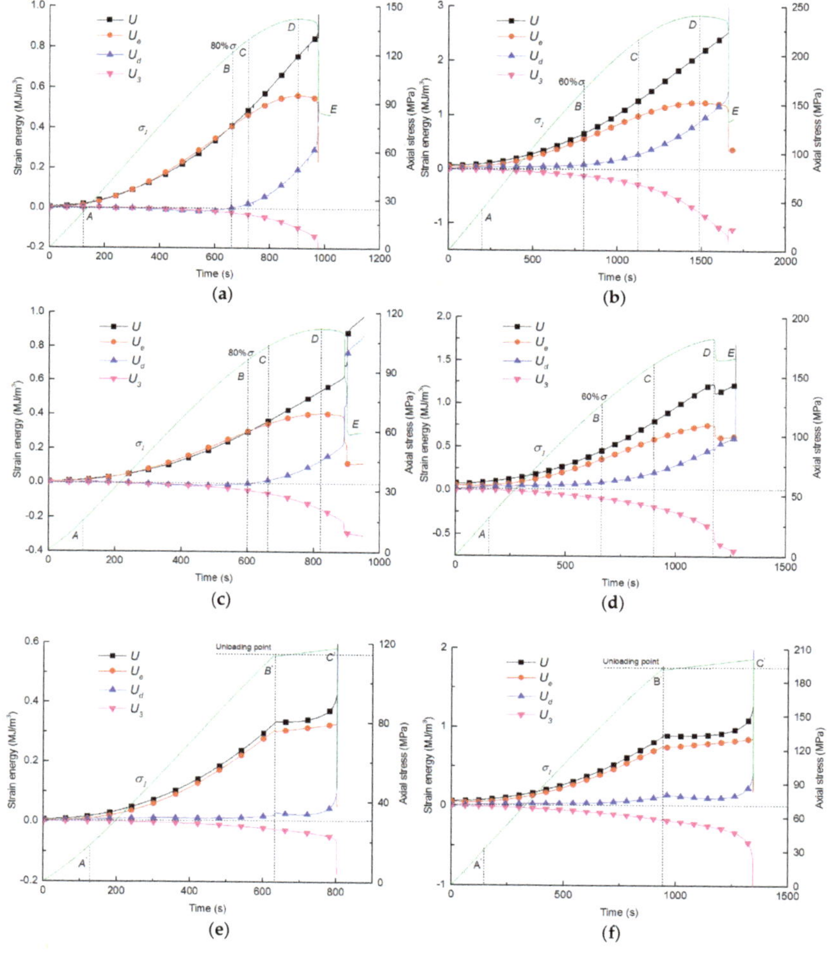

Figure 9. *Cont.*

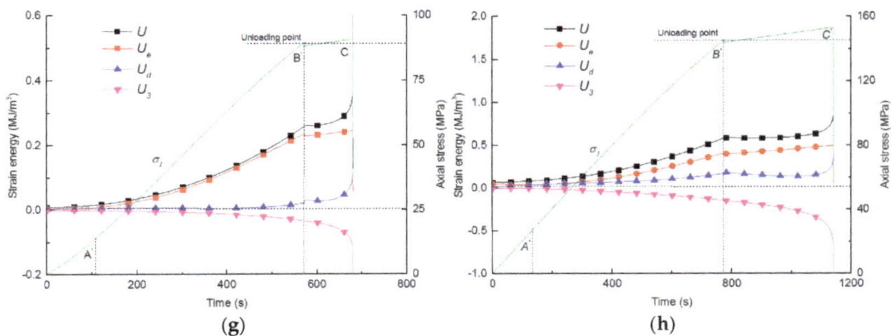

Figure 9. Typical strain energy and axial stress time curves: (**a**) TC-D-10; (**b**) TC-D-40; (**c**) TC-S-10; (**d**) TC-S-40; (**e**) TU-D-10; (**f**) TU-D-40; (**g**) TU-S-10; (**h**) TU-S-40.

For the unloading confining pressure tests, the strain energy time curves of sandstone specimens are shown in Figure 9e–h. The characteristics of strain energy curves in unloading tests are quite different from conventional triaxial compression tests, especially from the unloading point. Before the unloading point, in the micro-cracks compaction stage and the elastic stage (OB′), the curves of U and U_e almost overlap with a low initial confining pressure (10MPa), while the curves of U and U_e begin to separate under a high initial confining pressure (40MPa). The variation trend is similar with the loading test. However, after unloading point (B′C′), U and U_d increase steadily while U_e shows a slow rate of increase. Meanwhile, since the rock is still under the effect of axial loading in this stage, the energy is still absorbed from the test machine. The absorbed energy is mainly converted into U_d, which is used for the development and expansion of micro-cracks within the rock. When reaching the failure point (C′), U_d and circumferential strain energy (U_3) grow rapidly, and the stress-time curve falls down immediately. At this moment, U_e releases rapidly because of the propagation of rock cracks. Following the point C′, almost all of the releasable strain energy is released. Meanwhile, U_d and U_3 increase to the maximum value. Finally, the rock specimen fails violently due to the unloading of confining pressure.

Undoubtedly, there are some differences in the energy evolution laws of the rocks under different confining pressure conditions, particularly in the elastic deformation stage in triaxial loading and unloading process. During the elastic deformation stage, the curves of U and U_e almost overlap when the rock specimen is under low initial confining pressure, and almost all of the absorbed energy is converted into U_e. However, when the rock is under high initial confining pressure, the curves of U and U_e begin to separate, and the absorbed energy is basically stored as U_e, and partially converted as U_d. It is manifested that the internal damage has occurred earlier under the high initial confining pressure conditions, and the internal micro-cracks of rock begin to develop and propagate during the elastic deformation stage. Especially for the saturated sandstone under high initial confining pressures, the curves of U and U_e separate significantly in Figure 9h, which signifies that the water effect on rock energy evolution are nonnegligible, and it has a significant influence on the deformation and failure process of rock under high initial confining pressures. In addition, the curves separate at the point where the axial stress is about 80% of the peak strength for low initial confining pressure condition, while the curves separate at the point where the axial stress is about 60% of the peak strength under high initial confining pressure. It also indicates that the internal damage and micro-cracks develop earlier under high initial confining pressure than that under low confining pressure conditions.

The dissipation of energy leads to the development and propagation of internal cracks, which is the main reason of rock failure. It has been proved that the use of U_d/U is more favorable to analyze rock deformation and failure process instead of dissipative strain energy (U_d) [25,43]. The variation curves of U_d/U with the loading time under different confining pressures are shown in Figure 10.

The U_d/U exhibits some typical evolution stages as shown in Figure 10. In the conventional triaxial compression test, it increases initially, then decreases, and then increases slowly; finally, it increases rapidly at the failure stage of the rock. However, the curve increases initially, and then it decreases to a nearly smooth platform; finally, it increases sharply in unloading tests. During the failure stage, the elastic strain energy accumulated within the rock releases instantaneously, prompting rock fractures and damages. Meanwhile, it has been testified that the faster dissipation of rock energy causes the crack to propagate and penetrate faster, and then the failure of the rock occurs more suddenly [42,44]. In the conventional triaxial compression test, the energy dissipation curve increases along a skew line, showing a gradual increase trend. In contrast, the energy dissipation curve shows a nearly vertical increase for the triaxial unloading test, indicating a sudden and violent brittle failure of rock specimens.

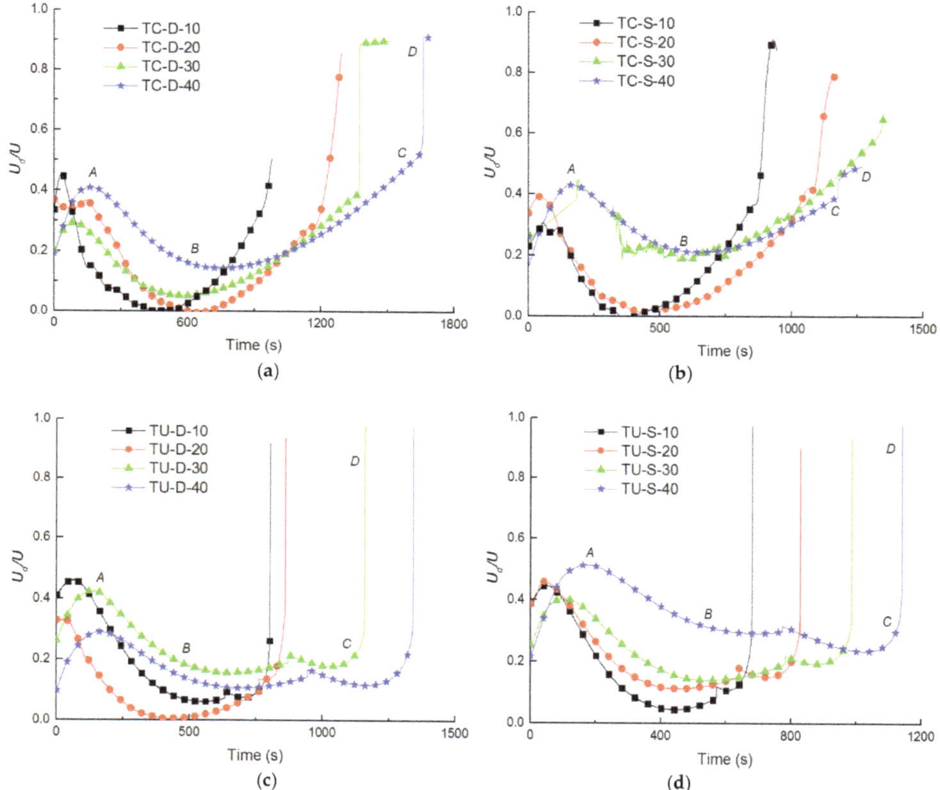

Figure 10. Evolution of U_d/U with loading time under triaxial loading and unloading tests for dry and saturated sandstone specimens (**a**) TC-D; (**b**) TC-S; (**c**) TU-D; (**d**) TU-S.

The relationship between the U_d/U and initial confining pressure at the corresponding critical failure point is plotted in Figure 11. The U_d/U values increase as σ_3^0 increases. When the rock specimen is under high initial confining pressures, it has more time for specimen deformation and damage. Thus the dissipative strain energy takes a large proportion of the total absorption energy in the rock. The deformation and failure process of rock in loading tests includes the micro-cracks compaction stage (OA), the elastic deformation stage (AB), the crack initiation and propagation stage (BC), the unstable crack development up to the failure stage (CD), the post-failure stage (DE) and so on. However, the rock in unloading tests enters the failure stage almost directly from the elastic deformation stage.

Thus, relatively fewer cracks develop inside the rock, resulting in less dissipation of rock energy and the energy is mainly consumed for the circumferential deformation and converted into kinetic energy for the failure of rock, such as rockbursts. The U_d/U of rock in unloading test is significantly lower than that of conventional triaxial compression test, as shown in Figure 11. Furthermore, the U_d/U of saturated rock is greater than dry rocks in the deformation and failure process of triaxial tests. Saturated by water, the internal friction strength of the rock is weakened as the water immersion, which results the rocks more prone to be deformed and damaged. With the accumulated damage within the rock, the energy account for the dissipation increases and the energy release decreases, leading to a relatively moderate failure of saturated rock, and there are less possibilities for rockbursts. It is also one of the basic theoretical foundations for the rockburst prevention through water injection in engineering. In addition, comparing the influence of stress paths and water conditions of the rock in Figure 11, it can be found that the effect of stress paths on energy evolution is more pronounced than the water conditions of the rock.

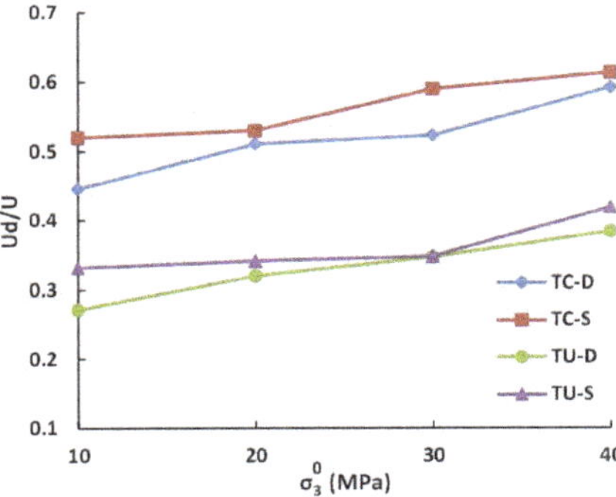

Figure 11. Variation of the U_d/U values with the initial confining pressure for dry and saturated sandstone specimens at the critical failure point.

4. Discussion

When a rock specimen approaches the failure point, U_e can be quickly released and it may be transformed into U_d. In the conventional triaxial compression test, the internal cracks develop completely within the rock specimen. However, the unloading failure usually occurs abruptly, and the rock specimens fail violently with loud fracturing noise, which has the characteristics of rockburst as observed in deep rock excavation engineering.

Table 4 lists the average strain energy values of sandstone specimens at the unloading point and the critical failure point in present triaxial tests. In the conventional triaxial compression test, U_d is larger than U_3 at critical failure point of the rock, implying that the energy consumption is mainly used for the micro-crack initiation and internal damage of the rock. In contrast, under unloading tests, the energy consumption during the failure process of rock specimens mainly contributes to the circumferential strain (ε_3) as the reduction of confining pressure, and thus U_3 is larger than U_d as shown in Table 4. Moreover, the following energy incremental value can be calculated based on the data in Table 4.

$$\Delta U_e = U'_e - U_e, \Delta U_d = U'_d - U_d, \Delta U_1 = U'_1 - U_1, \Delta U_3 = U'_3 - U_3 \tag{1}$$

It can be inferred that $\Delta U_e < 0$, $\Delta U_d > 0$ in unloading tests, and U_e converts into U_d for the development and propagation of internal cracks from the unloading point to the failure point. Meanwhile, the present study discovered that $\Delta U_d > \Delta U_e$. The dissipative energy includes not only the energy converted from U_e, but also the newly absorbed energy from the test machine. In unloading test, there is the following relationship between the dry and saturated rock:

$$\Delta U_e^{TU-D} \geq \Delta U_e^{TU-S}, \Delta U_d^{TU-D} \geq \Delta U_d^{TU-S} \tag{2}$$

It implies that the failure of dry specimens is more severe than the saturated specimen. Furthermore, $\Delta U_1 + |\Delta U_e| - |\Delta U_3| - \Delta U_d > 0$ at the critical failure point also exists in unloading test. The energy absorbed by the rock from test machine is mainly consumed for the development of internal fractures, axial deformation, and circumferential deformation, while partly remained. The residual energy $(\Delta U_1 + |\Delta U_e| - |\Delta U_3| - \Delta U_d)$ may account for the kinetic energy due to rockbursts, which may occur in the unloading process such as rock excavation with high in-situ stresses in hard rock. Besides, it can be obtained that:

$$0 < \Delta U_1^{TU-S} + \left|\Delta U_e^{TU-S}\right| - \left|\Delta U_3^{TU-S}\right| - \Delta U_d^{TU-S} < \Delta U_1^{TU-D} + \left|\Delta U_e^{TU-D}\right| - \left|\Delta U_3^{TU-D}\right| - \Delta U_d^{TU-D} \tag{3}$$

There is more residual energy for dry rock than saturated rock. It implies that the rock burst is more likely to occur in dry rock, and water infusion is effective to reduce the possibility of rockbursts in the excavation of deep underground rock engineering.

Table 4. Strain energy of sandstone specimens at the unloading point and the critical failure point in triaxial test.

Group	σ_3^0 (MPa)	Unloading Point					Critical Failure Point				
		U(MJ/m³)	U_1(MJ/m³)	U_e(MJ/m³)	U_d(MJ/m³)	U_3(MJ/m³)	U'(MJ/m³)	$U_1{}'$(MJ/m³)	$U_e{}'$(MJ/m³)	$U_d{}'$(MJ/m³)	$U_3{}'$(MJ/m³)
TC-D	10	-	-	-	-	-	0.85	1.01	0.41	0.43	−0.17
	20	-	-	-	-	-	1.31	1.62	0.73	0.59	−0.33
	40	-	-	-	-	-	1.80	2.54	0.84	0.95	−0.79
	60	-	-	-	-	-	2.52	3.75	1.02	1.50	−1.30
TC-S	10	-	-	-	-	-	0.66	0.90	0.32	0.34	−0.25
	20	-	-	-	-	-	0.95	1.40	0.44	0.50	−0.48
	40	-	-	-	-	-	1.60	1.98	0.56	0.94	−0.82
	60	-	-	-	-	-	1.62	2.40	0.62	0.99	−0.85
TU-D	10	0.33	0.35	0.31	0.02	−0.02	0.45	0.50	0.33	0.12	−0.06
	20	0.59	0.66	0.53	0.06	−0.10	0.83	1.13	0.55	0.27	−0.36
	40	0.75	0.82	0.61	0.13	−0.11	1.01	1.34	0.66	0.35	−0.37
	60	0.89	1.01	0.74	0.15	−0.17	1.35	1.98	0.83	0.52	−0.69
TU-S	10	0.26	0.28	0.24	0.02	−0.03	0.36	0.48	0.24	0.12	−0.12
	20	0.38	0.43	0.32	0.06	−0.08	0.52	0.85	0.34	0.18	−0.35
	40	0.54	0.64	0.41	0.10	−0.15	0.71	1.15	0.46	0.25	−0.48
	60	0.59	0.66	0.42	0.17	−0.15	0.85	1.48	0.50	0.36	−0.70

5. Conclusions

(1) The peak strength difference between dry and saturated sandstones grows with the increasing of initial confining pressure, and the effect of water on rock strength is more remarkable under high initial confining pressures. As a result, when the sandstone specimens are saturated by water, the values of cohesion, internal friction angle and Young's modulus are decreased whereas the Poisson's ratio is increased.

(2) In the conventional triaxial compression test, both macroscopic and microscopic failure of the rocks show a combined tension and shear fracture under low initial confining pressure ($\sigma_3^0 \leq 10$ MPa), but it shows shear fracture under high initial confining pressure ($\sigma_3^0 > 10$ MPa). The microscopic failure of the rock in unloading test mainly shows a combined tension and shear fracture, which is consistent with the macroscopic failure of sandstone under low initial confining pressure ($\sigma_3^0 \leq 20$ MPa). However, under high initial confining pressure conditions ($\sigma_3^0 > 20$ MPa), it mainly shows shear fracture in macroscopic view. The failure of sandstone in unloading test is more violent than that the failure under triaxial loading test, particularly for dry sandstones.

(3) The energy evolution processes confirm that there are some differences for rock specimens under different confining pressures, particularly for the elastic deformation stage. During this stage,

the total energy (U) and elastic strain energy (U_e) curves are almost overlapped when the rock specimen is under low initial confining pressures. However, when the rock is under high initial confining pressures, U and U_e begin to separate, and the absorbed energy is basically stored as U_e, partially as the dissipative strain energy (U_d). The internal damage has occurred and the internal cracks of the rock develop earlier under the high initial confining pressure conditions.

(3)   The energy consumption during the failure process mainly contributes to crack initiation and internal damage in loading tests. In unloading test, the energy is mainly consumed for the circumferential deformation and converted into kinetic energy for rock failure. The failure of the saturated rock is relatively moderate because the absorbed energy is mainly used for internal damage and only a part of energy is used for release. Rockburst phenomenon is tended to occur for dry rocks, especially under triaxial unloading conditions. It also indicates that the water injection of the rock may be effective to prevent rockbursts under high in-situ stresses conditions.

Author Contributions: Conceptualization, D.L. and Z.S.; methodology, Z.S.; software, Z.S. and Q.Z.; validation, D.L., Z.S. and K.P.; formal analysis, Z.S.; investigation, Z.S. and Q.Z.; resources, D.L.; data curation, Z.S.; writing—original draft preparation, Z.S.; writing—review and editing, D.L.; visualization, Z.S.; supervision, D.L.; project administration, D.L. and K.P.; funding acquisition, D.L. and K.P.

Funding: This research was funded by the National Natural Science Foundation of China (51474250, 51774058) and Chongqing Basic Research and Frontier Exploration Project (cstc2018jcyjA3320, cstc2016jcyjA1861).

Conflicts of Interest: The authors declare no conflict of interest.

References

1.   Al-Shalabi, E.W.; Sepehrnoori, K. A comprehensive review of low salinity/engineered water injections and their applications in sandstone and carbonate rocks. *J. Pet. Sci. Eng.* **2016**, *139*, 137–161. [CrossRef]

2.   Zhou, Z.; Cai, X.; Cao, W.; Li, X.; Xiong, C. Influence of Water Content on Mechanical Properties of Rock in Both Saturation and Drying Processes. *Rock Mech. Rock Eng.* **2016**, *49*, 3009–3025. [CrossRef]

3.   Wang, Y.; Meng, F.; Geng, F.; Jing, H.; Zhao, N. Investigating Water Permeation through the Soil-Rock Mixture in Underground Engineering. *Pol. J. Environ. Stud.* **2017**, *26*, 1777–1788. [CrossRef]

4.   Vasarhelyi, B. Statistical analysis of the influence of water content on the strength of the miocene limestone. *Rock Mech. Rock Eng.* **2005**, *38*, 69–76. [CrossRef]

5.   Talesnick, M.; Shehadeh, S. The effect of water content on the mechanical response of a high-porosity chalk. *Int. J. Rock Mech. Min. Sci.* **2007**, *44*, 584–600. [CrossRef]

6.   Erguler, Z.A.; Ulusay, R. Water-induced variations in mechanical properties of clay-bearing rocks. *Int. J. Rock Mech. Min. Sci.* **2009**, *46*, 355–370. [CrossRef]

7.   Török, Á.; Vásárhelyi, B. The influence of fabric and water content on selected rock mechanical parameters of travertine, examples from Hungary. *Eng. Geol.* **2010**, *115*, 237–245. [CrossRef]

8.   Yilmaz, I. Influence of water content on the strength and deformability of gypsum. *Int. J. Rock Mech. Min. Sci.* **2010**, *47*, 342–347. [CrossRef]

9.   Çelik, M.Y.; Ergül, A. The influence of the water saturation on the strength of volcanic tuffs used as building stones. *Environ. Earth Sci.* **2015**, *74*, 3223–3239. [CrossRef]

10.   Shi, X.; Cai, W.; Meng, Y.; Li, G.; Wen, K.; Zhang, Y. Weakening laws of rock uniaxial compressive strength with consideration of water content and rock porosity. *Arab. J. Geosci.* **2016**, *9*, 369. [CrossRef]

11.   Li, D.; Wong, L.N.Y.; Liu, G.; Zhang, X. Influence of water content and anisotropy on the strength and deformability of low porosity meta-sedimentary rocks under triaxial compression. *Eng. Geol.* **2012**, *126*, 46–66. [CrossRef]

12.   Rathnaweera, T.; Ranjith, P.; Perera, M.; De Silva, V. Development of a laboratory-scale numerical model to simulate the mechanical behaviour of deep saline reservoir rocks under varying salinity conditions in uniaxial and triaxial test environments. *Measurement* **2017**, *101*, 126–137. [CrossRef]

13.   Bejarbaneh, B.Y.; Armaghani, D.J.; Amin, M.F.M. Strength characterisation of shale using Mohr–Coulomb and Hoek–Brown criteria. *Measurement* **2015**, *63*, 269–281. [CrossRef]

14.   Mckean, S.H.; Priest, J.A.; Priest, J. Multiple failure state triaxial testing of the Montney Formation. *J. Pet. Sci. Eng.* **2019**, *173*, 122–135. [CrossRef]

15. Zhu, W.; Wong, T.-F.; Baud, P.; Wong, T. Failure mode and weakening effect of water on sandstone. *J. Geophys. Res. Biogeosci.* **2000**, *105*, 16371–16389.

16. Zhou, J.; Wei, J.; Yang, T.; Zhu, W.; Li, L.; Zhang, P. Damage analysis of rock mass coupling joints, water and microseismicity. *Tunn. Undergr. Space Technol.* **2018**, *71*, 366–381. [CrossRef]

17. Duan, H.F.; Jiang, Z.Q.; Zhu, S.Y.; Xiao, W.G.; Li, D.L. Micro-mechanism of water stability and characteristics of strength softening of rock in deep mines. *Chin. J. Geotech. Eng.* **2012**, *34*, 1636–1645.

18. Zhao, X.G.; Wang, J.; Cai, M.; Cheng, C.; Ma, L.K.; Su, R.; Zhao, F.; Li, D.J. Influence of Unloading Rate on the Strainburst Characteristics of Beishan Granite Under True-Triaxial Unloading Conditions. *Rock Mech. Rock Eng.* **2014**, *47*, 467–483. [CrossRef]

19. Kundu, J.; Mahanta, B.; Sarkar, K.; Singh, T.N. The Effect of Lineation on Anisotropy in Dry and Saturated Himalayan Schistose Rock Under Brazilian Test Conditions. *Rock Mech. Rock Eng.* **2018**, *51*, 5–21. [CrossRef]

20. Liu, D.; Wang, Z.; Zhang, X.; Wang, Y.; Zhang, X.; Li, D. Experimental investigation on the mechanical and acoustic emission characteristics of shale softened by water absorption. *J. Nat. Gas Sci. Eng.* **2018**, *50*, 301–308. [CrossRef]

21. Lyu, Q.; Long, X.; Ranjith, P.; Tan, J.; Kang, Y. Experimental investigation on the mechanical behaviours of a low-clay shale under water-based fluids. *Eng. Geol.* **2018**, *233*, 124–138. [CrossRef]

22. Xiao, T.L.; Li, X.P.; Lu, Y.I. Structural plane of rock mass strength characteristics and its impact on Groundwater, in Trends in Civil Engineering. *Adv. Mater. Res.* **2012**, *446–449*, 476–479. [CrossRef]

23. Wasantha, P.; Ranjith, P.; Shao, S. Energy monitoring and analysis during deformation of bedded-sandstone: Use of acoustic emission. *Ultrasonics* **2014**, *54*, 217–226. [CrossRef]

24. Peng, K.; Liu, Z.P.; Zou, Q.L.; Zhang, Z.Y.; Zhou, J.Q. Static and dynamic mechanical properties of granite from various burial depths. *Rock Mech. Rock Eng.* **2019**. [CrossRef]

25. Li, D.; Sun, Z.; Xie, T.; Li, X.; Ranjith, P.G. Energy evolution characteristics of hard rock during triaxial failure with different loading and unloading paths. *Eng. Geol.* **2017**, *228*, 270–281. [CrossRef]

26. Zhang, D.; Li, S.; Bai, X.; Yang, Y.; Chu, Y. Experimental Study on Mechanical Properties, Energy Dissipation Characteristics and Acoustic Emission Parameters of Compression Failure of Sandstone Specimens Containing En Echelon Flaws. *Appl. Sci.* **2019**, *9*, 596. [CrossRef]

27. Hua, A.Z.; You, M.Q. Rock failure due to energy release during unloading and application to underground rock burst control. *Tunn. Undergr. Space Technol.* **2001**, *16*, 241–246. [CrossRef]

28. Lu, T.K.; Zhao, Z.J.; Hu, H.F. Improving the gate road development rate and reducing outburst occurrences using the waterjet technique in high gas content outburst-prone soft coal seam. *Int. J. Rock Mech. Min. Sci.* **2011**, *48*, 1271–1282. [CrossRef]

29. Song, D.; Wang, E.; Liu, Z.; Liu, X.; Shen, R. Numerical simulation of rock-burst relief and prevention by water-jet cutting. *Int. J. Rock Mech. Min. Sci.* **2014**, *70*, 318–331. [CrossRef]

30. Bautista, J.; Taleghani, A.D. Prediction of formation damage at water injection wells due to channelization in unconsolidated formations. *J. Pet. Sci. Eng.* **2018**, *164*, 1–10. [CrossRef]

31. Guo, W.Y.; Tan, Y.L.; Yang, Z.L.; Zhao, T.B.; Hu, S.C. Effect of Saturation Time on the Coal Burst Liability Indexes and Its Application for Rock Burst Mitigation. *Geotech. Geol. Eng.* **2018**, *36*, 589–597. [CrossRef]

32. Zhang, Z.Z.; Feng, G. Experimental investigation on the energy evolution of dry and water-saturated red sandstones. *Int. J. Min. Sci. Technol.* **2015**, *25*, 383–388. [CrossRef]

33. Hong-qing, Z.; Min-bo, Z.; Shuai-hu, Z.; Bei-fang, G.; Xiang, S. Effect of triaxial compression on damage deformation of coal rock under pulsed pore water pressure. *Rock Soil Mech.* **2015**, *36*, 2137–2143.

34. Cai, C.; Gao, F.; Li, G.; Huang, Z.; Hou, P. Evaluation of coal damage and cracking characteristics due to liquid nitrogen cooling on the basis of the energy evolution laws. *J. Nat. Gas Sci. Eng.* **2016**, *29*, 30–36. [CrossRef]

35. Zhu, Q.; Li, D.; Han, Z.; Li, X.; Zhou, Z. Mechanical properties and fracture evolution of sandstone specimens containing different inclusions under uniaxial compression. *Int. J. Rock Mech. Min. Sci.* **2019**, *115*, 33–47. [CrossRef]

36. Zhuang, D.; Tang, C.; Liang, Z.; Ma, K.; Wang, S.; Liang, J. Effects of excavation unloading on the energy-release patterns and stability of underground water-sealed oil storage caverns. *Tunn. Undergr. Space Technol.* **2017**, *61*, 122–133. [CrossRef]

37. Wang, S.; Hagan, P.; Zhao, Y.; Chang, X.; Song, K.I.; Zou, Z. The Effect of Confining Pressure and Water Content on Energy Evolution Characteristics of Sandstone under Stepwise Loading and Unloading. *Adv. Civ. Eng.* **2018**, *2018*. [CrossRef]

38. He, Z.; Li, G.; Tian, S.; Wang, H.; Shen, Z.; Li, J. SEM analysis on rock failure mechanism by supercritical CO_2 jet impingement. *J. Pet. Sci. Eng.* **2016**, *146*, 111–120. [CrossRef]

39. Mancktelow, N.S.; Pennacchioni, G. The influence of grain boundary fluids on the microstructure of quartz-feldspar mylonites. *J. Struct. Geol.* **2004**, *26*, 47–69. [CrossRef]

40. Storti, F.; Billi, A.; Salvini, F. Particle size distributions in natural carbonate fault rocks: Insights for non-self-similar cataclasis. *Earth Planet. Sci. Lett.* **2003**, *206*, 173–186. [CrossRef]

41. Xie, H.; Li, L.; Peng, R.; Ju, Y. Energy analysis and criteria for structural failure of rocks. *J. Rock Mech. Geotech. Eng.* **2009**, *1*, 11–20. [CrossRef]

42. Li, Y.; Huang, D.; Li, X. Strain Rate Dependency of Coarse Crystal Marble under Uniaxial Compression: Strength, Deformation and Strain Energy. *Rock Mech. Rock Eng.* **2014**, *47*, 1153–1164. [CrossRef]

43. Ning, J.; Wang, J.; Jiang, J.; Hu, S.; Jiang, L.; Liu, X. Estimation of Crack Initiation and Propagation Thresholds of Confined Brittle Coal Specimens Based on Energy Dissipation Theory. *Rock Mech. Rock Eng.* **2018**, *51*, 119–134. [CrossRef]

44. Peng, R.; Ju, Y.; Wang, J.G.; Xie, H.; Gao, F.; Mao, L. Energy Dissipation and Release during Coal Failure under Conventional Triaxial Compression. *Rock Mech. Rock Eng.* **2015**, *48*, 509–526. [CrossRef]

Article

Optimization of Shape Design of Grommet through Analysis of Physical Properties of EPDM Materials

Young Shin Kim [1], Eui Seob Hwang [2] and Euy Sik Jeon [3,*]

[1] Industrial Technology Research Institute, Kongju National University, Kongju 31080, Korea; people9318@gmail.com
[2] Company-affiliated R&D center/New Business Manager, DAESUNG HI-TECH Co., Ltd., Cheonan 31025, Korea; hus5174@dscmc.com
[3] Department of Mechanical Engineering, Industrial Technology Research Institute, Kongju National University, Kongju 31080, Korea
* Correspondence: osjun@kongju.ac.kr; Tel.: +82-41-521-9284

Received: 10 December 2018; Accepted: 27 December 2018; Published: 2 January 2019

Abstract: Ethylene propylene diene monomer (EPDM) has superior mechanical properties, water resistance, heat resistance, and ozone resistance. It can be applied to various products owing to its low hardness and high slip resistance properties. A grommet is one of the various products made using EPDM rubber. It is a main component of automobiles, in which it protects wires throughout the inside and outside of a vehicle body. The grommet, made of EPDM, has different mounting performance depending on the process parameters and the shape of the grommet. This study conducted optimization to improve the mounting performance of a grommet using EPDM materials. The physical properties of the main molding materials were investigated according to process parameters. A grommet was fabricated according to the process parameters of fabrication. Insertion force and separation force were examined through experiments. Nonlinear material constants were determined through uniaxial and biaxial tensile tests. The nonlinear analysis of the grommet was conducted, and a compound design that incorporated the shape parameters for the minimum load of each part was derived. Then, additional nonlinear analysis was performed. This was followed by a comparative analysis of the actual model through experimental evaluation.

Keywords: Ethylene-propylene diene monomer rubber EPDM; grommet; physical properties; optimization of shape design

---

1. Introduction

Ethylene propylene diene monomer (EPDM) rubber is a terpolymer in which ethylene, propylene, and diene are irregularly bonded. Compared to general rubbers, it has superior mechanical properties, water resistance, heat resistance, and ozone resistance. In addition, it has high inheritance and corona discharge resistance because of limited in its chemical structure [1]. Moreover, it can be applied to various products owing to its low hardness and high slip resistance properties [2,3]. Numerous studies have been conducted based on the diverse applications of EPDM, with a focus on the reliability of EPDM-based products [4–6]. Studies have been conducted to analyze the physical characteristics of composite materials [7,8].

A grommet is one of the various products made using EPDM rubber [9]. It is a main component of automobiles, in which it protects wires throughout the inside and outside of a vehicle body. Unlike conventional plastic, EPDM rubber is characterized by high flexibility, high elasticity, and high tensile strength. It is fabricated through an injection molding process. Grommets are also produced through the injection molding process of EPDM [7]. The parameters of this process, such as time and temperature, not only change the physical properties of raw materials but also affect the insertion force and separation force generated during mounting a grommet on a body when molding a grommet product. Accordingly, in this study, to analyze the physical properties of raw materials according to process parameters, we set the main factors of molding process parameters using a design of an experimental method (DOE) [10–12]. Specimens were prepared according to process conditions. Tensile strength and elongation were measured, and the correlation was analyzed.

The process parameters of EPDM raw materials were set, and the experimental design was established by applying factorial designs from among experimental design methods. The physical properties of the raw materials were tested using the standard test method for soft vulcanized rubber (KS M 6518) [13] for confirming the changes in the physical properties according to process parameters. The physical properties of EPDM were checked and reflected during grommet vulcanization [14]. The experimental design was established using factorial designs among the bellows type, cable type, and cable-less type. Then, we compared and analyzed the maximum insertion force and maximum separation force generated during mounting. It was confirmed that an insertion force and separation force tended to not occur depending on process parameters. We conducted the nonlinear analysis of EPDM to improve grommet design for mounting performance. For this purpose, stress–strain rate information was obtained through uniaxial and biaxial tensile tests, and the nonlinear material constants required for the analysis were determined. The shape parameters of the grommet were set, and a mounting performance simulation was conducted for various shapes. Based on the analytical results, we derived the dimensions for optimizing grommet mounting performance. Additional analysis was conducted with the derived dimensions. An actual grommet was manufactured and analyzed to verify its feasibility.

2. Experimental Analysis Based on Molding Process Conditions

2.1. Analysis of Physical Properties Based on Molding Process Conditions

The physical properties of EPDM rubber, which is a raw material for making grommet, were analyzed according to the conditions of the injection process. The controllable factors that were expected to affect the physical properties were set, and specimens were fabricated according to the KS M 6518 [13] standard for each experimental condition. The physical properties were set as tensile strength and elongation, which were measured using a universal testing machine. Table 1 shows the main process parameters for injection molding. The experiment was conducted 16 times. The table shows the experimental conditions for each process value.

Table 1. Experimental conditions for different process values (DOE).

No.	Temp. (°C)	Time (s)	Degassing	Strength (Mpa)	Elongation (%)	No	Temp (°C)	Time (s)	Degassing	Strength (Mpa)	Elongation (%)
1	160	200	O	11.2	969.0	9	160	200	X	10.5	968.3
2	160	200	O	11.1	960.4	10	160	200	X	10.8	944.8
3	160	600	O	13.4	853.9	11	160	600	X	13.5	851.5
4	160	600	O	12.9	815.2	12	160	600	X	13.0	821.2
5	180	200	O	13.2	840.5	13	180	200	X	12.8	838.7
6	180	200	O	13.1	836.0	14	180	200	X	13.2	844.8
7	180	600	O	13.4	775.0	15	180	600	X	13.0	749.7
8	180	600	O	13.7	792.3	16	180	600	X	13.3	767.1

Figure 1 shows the main effects and interactions of the factors that affect tensile strength. Temperature and time affect tensile strength, while degassing does not. Additionally, interactions occur according to time and temperature and no interactions occur, owing to degassing.

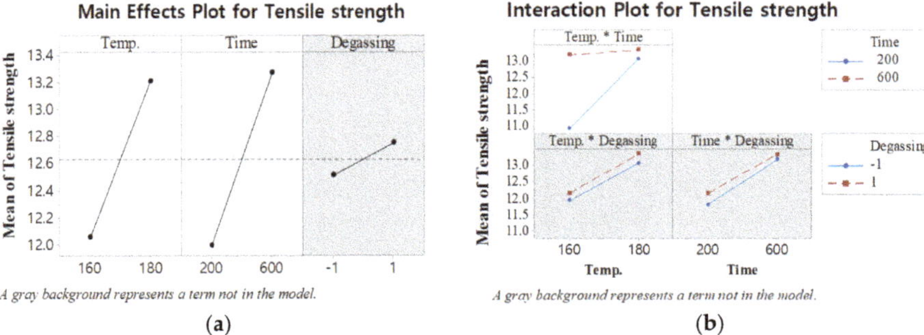

Figure 1. Main effects plot and interaction plot for tensile strength. (**a**) Main effects plot for tensile strength; and (**b**) Interaction plot for tensile strength.

Table 2 shows the analysis of variance (ANOVA) result for the factors that affect tensile strength. ANOVA is a collection of statistical models and their associated estimation procedures (such as the "variation" among and between groups) used to analyze the differences among group means in a sample. 'Adj. SS' represents the sum of squares, 'Adj. MS' is the mean squares, and '*F*-Value' is the value of the adj. The SS of each factor divided by the mean squares error. '*p*-value' was derived based on *F* value. *p* values larger than 0.05 are pooled as error terms, and only significant factors are shown. The regression Equation (1) was derived through ANOVA analysis. The R^2 value of the regression equation is 94.59% and the adj. R^2 value is 93.24%.

$$tensile\ strenght\ =\ -15.59 + 0.1585 Temp + 0.04592 Time - 0.00025 Temp \times Time \tag{1}$$

Table 2. ANOVA result for tensile strength.

Source	DF	Adj SS	Adj MS	*F*-Value	*p*-Value
Model	3	15.95	5.31661	69.95	0.000
Linear	2	11.91	5.95374	78.33	0.000
Temp (°C)	1	5.380	5.38007	70.78	0.000
Time (s)	1	6.5274	6.52741	85.88	0.000
2-way interactions	1	4.0423	4.04234	53.18	0.000
T (°C) × Time (s)	1	4.0423	4.04234	53.18	0.000
Error	12	0.9121	0.0760		
Total	15	16.8619			

2.2. Elongation according to Time and Gas Removal Conditions

Figure 2 shows the plot of the major factors that determine elongation. Time and temperature affect elongation, and gas removal has a minor effect on elongation. Figure 2b shows the diagram of the interactions. It can be confirmed that temperature and time affect each other.

Table 3 shows the ANOVA results for the analysis of the factors that affect elongation. Based on the analysis of tensile strength according to process parameters, temperature and time affect tensile strength and elongation, while degassing does not. However, a few specimens without degassing did exhibit pores. We set the degassing parameter in additional experiments. The regression Equation (2) was derived through ANOVA analysis. The R^2 value of the regression equation is 96.69% and the adj. R^2 value is 95.86%.

$$elongation\ =\ 2213 - 7.436 Temp - 1.437 Time - 0.00702 Temp \times Time \tag{2}$$

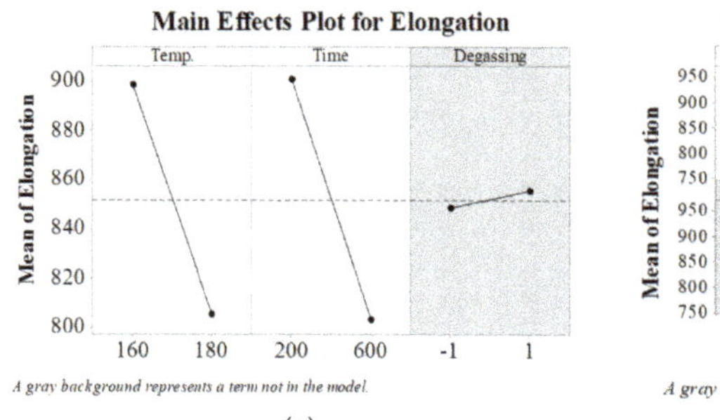
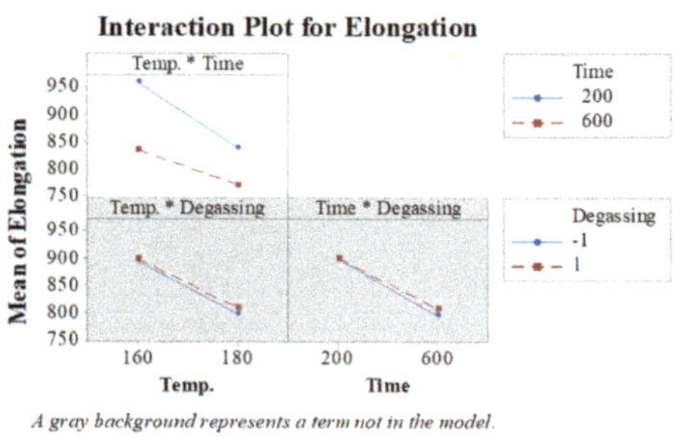

(a) (b)

Figure 2. Main effects plot and interaction plot for elongation. (**a**) Main effects plot for elongation; and (**b**) Interaction plot for elongation.

Table 3. ANOVA result for elongation.

Source	DF	Adj SS	Adj MS	*F*-Value	*p*-Value
Model	3	75,096.2	25,032.1	116.87	0.000
Linear	2	71,937.7	35,968.9	167.93	0.000
Temp. (°C)	1	34,243.5	34,243.5	159.88	0.000
Time (s)	1	37,694.2	37,694.2	175.99	0.000
2-way interactions	1	3158.4	3158.4	14.75	0.034
T (°C) × Time (s)	1	3158.4	3158.4	14.75	0.002
Error	12	2570.2	214.2		
Total	15	77,666.4			

2.3. Measurement of Grommet Mounting Performance according to Molding Process Parameters

We derived the process parameters that increase mounting performance by measuring mounting performance according to grommet shape. Based on the results described in the previous section, temperature and time were set as the process parameters because they affect tensile strength and elongation. We set the maximum and minimum values for each factor according to grommet shape and fabricated the grommet. Here, degassing was applied in the fabrication of all products. The mounting performance of the products was analyzed by measuring the insertion force required for fastening the grommet and the required separation force. Insertion force and separation force were measured using a universal tensile tester when the grommet was inserted into or removed from a panel fixing jig. The experimental speed was set as 50 mm/min, and the experiment was repeated twice. We employed three widely used types of shapes for the grommet. The temperature and time parameters for molding the grommet of each shape in the initial test are shown in Table 4. Figure 3 shows the types of grommet shape.

Table 4. Experimental conditions.

Type	Temp. (°C)		Time (s)	
	Min(−1)	Max(1)	Min(−1)	Max(1)
Bellows	160	180	200	600
Cable	170	190	400	800
Blank	180	200	300	900

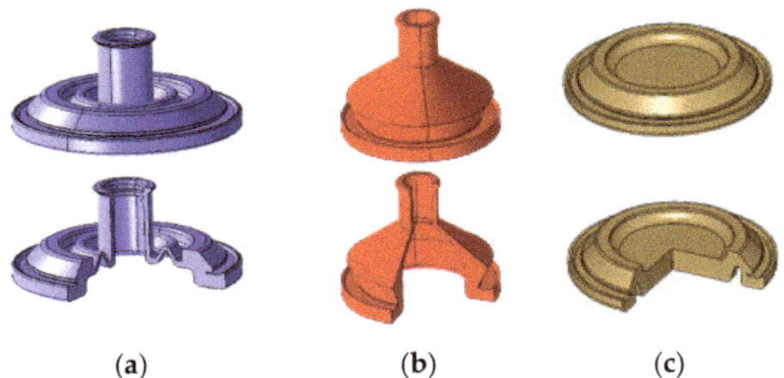

(a)   (b)   (c)

Figure 3. Grommet shapes. (**a**) Bellows type; (**b**) Cable type; and (**c**) Cable-less type.

The insertion force and separation force for different grommet shape types are given in Table 5. ANOVA was used to analyze the insertion force and separation force for each type of shape, and it was confirmed that there was no difference between insertion force and separation force according to grommet process parameters. As shown in Figure 4, even though there is no difference between insertion force and separation force according to process parameters, the times at which the maximum insertion force and maximum separation force occur vary depending on process parameters. This appears to be because the elongation rate changes according to process parameters. Moreover, it was confirmed that the change in insertion force and separation force was more influenced by the changes in the shape of the grommet.

Table 5. Experimental results.

Process Value		Bellows Type		Cable Type		Cable-Less Type	
Temp.	Time	Insertion Force	Separation Force	Insertion Force	Separation Force	Insertion Force	Separation Force
−1	−1	94.1	85.3	226.5	120.6	99.0	73.5
1	1	94.1	88.3	268.7	119.6	97.1	65.7

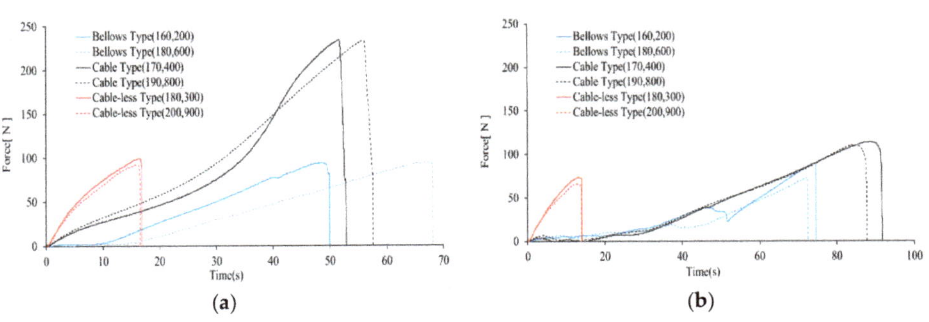

Figure 4. Insertion force and separation force experimental results according to process parameters by grommet type. (**a**) Insertion force results according to grommet shape type; and (**b**) separation force results according to grommet shape type.

3. Nonlinear Analysis Using FEM

3.1. Parameter Settings According to Shape

This study considered that the factors that influenced insertion force and separation force were more affected by the shape of the product than by the process parameters of the product. Therefore, the shape of the product was parameterized to analyze insertion force and separation force according to the changes in shape. Figure 5 shows the shape parameters of the main part of the grommet, and Table 6 shows the values of each shape parameter.

Table 6. Parameters according to shape.

Level	a (mm)	b (mm)	c (mm)	d (mm)	e (°C)
1	4	10	1	5	135
2	4.5	11	1.5	6	145
3	5	12	2	7	155

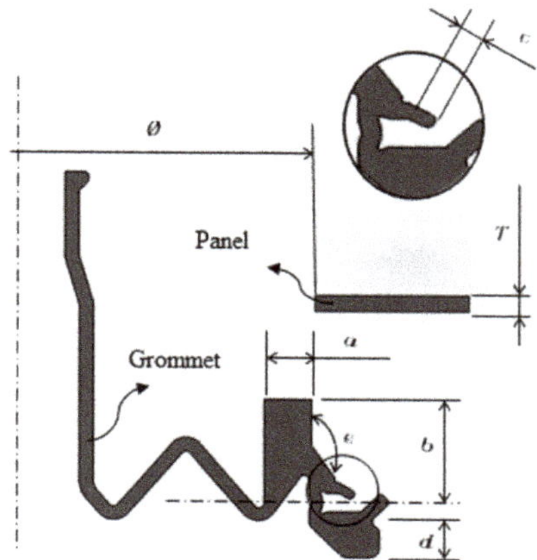

Figure 5. Grommet shape parameter settings.

3.2. Nonlinear Analysis by Setting Material Constants

Unlike metals, rubber retains its elasticity even under large strain. As rubber has hyper-elastic properties that exhibit nonlinearity between load and strain, it is important to understand its nonlinear properties [15]. We conducted uniaxial and biaxial tensile tests of EPDM rubber to obtain stress–strain rate information. Then, we determined the nonlinear material constants required for finite element analysis. The uniaxial and biaxial tensile tests were performed using an EPDM 50 material to obtain stress–strain data, as shown in Figure 6. We determined the material constants required for nonlinear analysis. The tensile test was performed using the KS standard dumbbell-type three test [13]. The Mullins effect is observed in EPDM materials, such as rubber, in which the initial molecular structure is rearranged upon repeated loads [16–19]. As shown in Figure 6a,b, as strain range gradually increases, if a strain larger than the previously applied strain is received, a certain permanent strain occurs and strain does not become zero, even if stress is zero. In addition, while the gauge distance of a specimen increases in the repeated loading process, the cross-sectional area decreases.

The nonlinear material constants for finite element analysis were obtained through the curve fitting of the stress-strain data obtained in the uniaxial and biaxial tensile tests. The relationship between stress and strain was determined to obtain the final nonlinear material constants considering the change in the cross-sectional area under the repeated loading of rubber, as shown in Figure 7.

The Ogden model was used for nonlinear analysis. An Ogden model is a hyper elastic material model that can be used for predicting the nonlinear stress-strain behavior of materials such as rubber or polymer. Ogden model was introduced by Ogden in 1972, and the strain energy density function for an Ogden material is as follows (3)

$$w = \sum_{k=1}^{n} \mu_k \left(\frac{\lambda_1^{a_k} + \lambda_2^{a_k} + \lambda_3^{a_k} - 3}{a_k} \right) \tag{3}$$

where w: Strain energy density; μ_k, a_k: Ogden constants; and λ: Stretch ratio; $n = 3$.

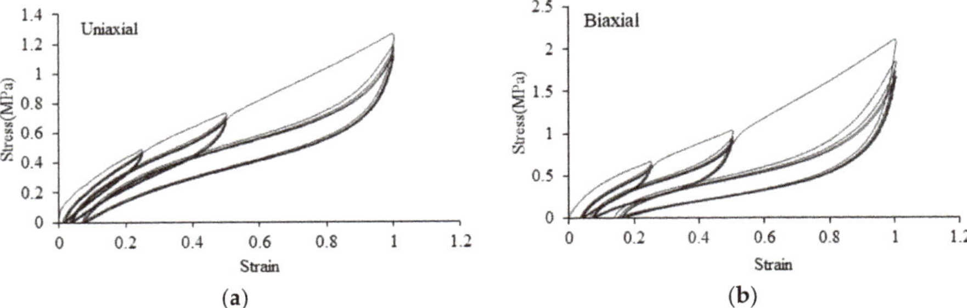

Figure 6. Experimental stress–strain curves for compound EPDM. (**a**) Uniaxial tension; (**b**) Biaxial tension.

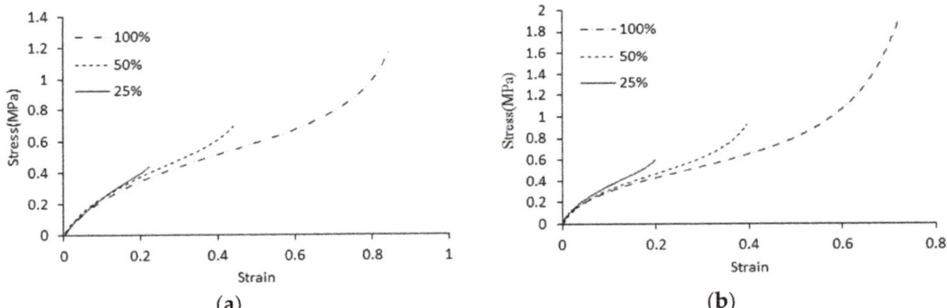

Figure 7. Stable stress–strain curves for compound EPDM. (**a**) Uniaxial tension; and (**b**) biaxial tension.

Table 7 shows the coefficients of the Ogden model with a strain range of 100%, which is a material model used to describe nonlinear material constants. The ABAQUS commercial software was used [20–22].

Table 7. Ogden model 3rd constant values.

Material	Ogden Model 3rd Constant Values					
	μ1	μ2	μ3	α1	α2	α3
EPDM50 (100%)	3.557	8.004	4.550×10^{-1}	2.000×10^{-2}	4.000×10^{-3}	2.381

Figure 8 shows an image of the analysis results. Figure 8a shows the initial state before analysis and (b) shows graphically one of the various analysis results. The maximum value of insertion force and separation force was confirmed through the analysis results. The results for insertion force and separation force were obtained for 32 conditions through the nonlinear analysis, as shown in Table 8. ANOVA was performed for insertion force and separation force according to shape design parameters. Results showed that all variables except d were significant among the variables that affected insertion force, while shape parameters b and e affected separation force.

Table 8. Experimental conditions for different dimension values (DOE).

No.	a [mm]	b [mm]	c [mm]	d [mm]	e [o]	Insertion Force (N)	Separation Force (N)	No.	a [mm]	b [mm]	c [mm]	d [mm]	e [o]	Insertion Force (N)	Separation Force (N)
1	4.5	11	1.5	6	145	62.37	69.72	17	4	10	2	7	155	64.13	70.60
2	4	10	1	7	135	62.66	64.52	18	4.5	11	2	6	145	62.95	70.21
3	4	10	2	5	135	61.78	64.33	19	5	12	2	5	135	62.37	70.21
4	5	10	1	7	155	63.25	70.41	20	4.5	11	1.5	6	145	62.37	69.72
5	4.5	11	1.5	6	145	62.86	70.01	21	4.5	11	1	6	145	62.27	69.23
6	5	11	1.5	6	145	62.56	70.80	22	5	12	2	7	155	63.84	78.25
7	5	12	1	7	135	62.07	69.82	23	4	12	1	7	155	63.44	74.33
8	4.5	11	1.5	6	155	63.15	73.05	24	4.5	11	1.5	5	145	62.07	69.33
9	5	10	1	5	135	61.58	66.19	25	4	12	1	5	135	61.58	67.27
10	4	10	1	5	155	62.27	69.03	26	4.5	11	1.5	6	135	62.95	67.27
11	4.5	11	1.5	6	145	62.37	69.72	27	4.5	11	1.5	6	145	62.37	69.72
12	4.5	11	1.5	6	145	62.37	69.72	28	4.5	11	1.5	7	145	63.35	70.31
13	4	12	2	5	155	62.76	74.72	29	4	11	1.5	6	145	62.37	68.54
14	5	10	2	5	155	70.31	62.76	30	4	12	2	7	135	62.95	68.15
15	5	10	2	7	135	62.86	66.88	31	4.5	10	1.5	6	145	62.27	67.37
16	4.5	12	1.5	6	145	62.37	71.78	32	5	12	1	5	155	63.15	73.15

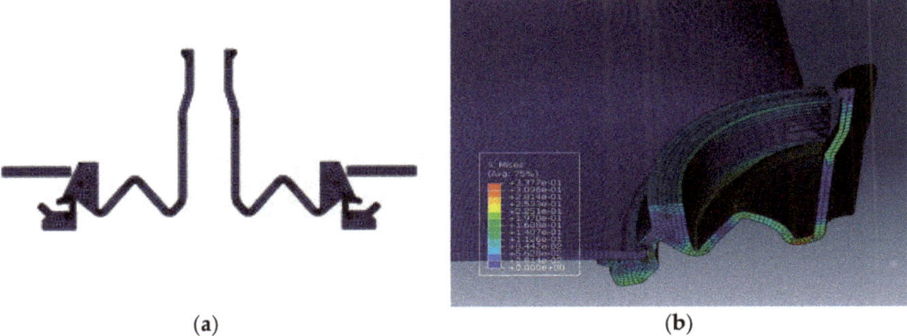

(a)   (b)

Figure 8. Analysis results. (**a**) Sectional view for grommet analysis; and (**b**) analysis result confirmation of stress and maximum stress distribution by the position of grommet.

4. Optimization of Shape Parameters

4.1. Derivation of Optimal Shape Parameters

We derived the shape parameters for minimizing insertion force and maximizing separation force based on the results obtained from the nonlinear analysis. Large values were obtained for shape parameters a, b, d, and e and a value of 1.313 was obtained for c. Insertion force and separation force were predicted as 62.46 N and 76.98, respectively. Figure 9 is an optimization graph that shows insertion force and separation force according to the values of each shape parameter obtained using the response surface optimization [23].

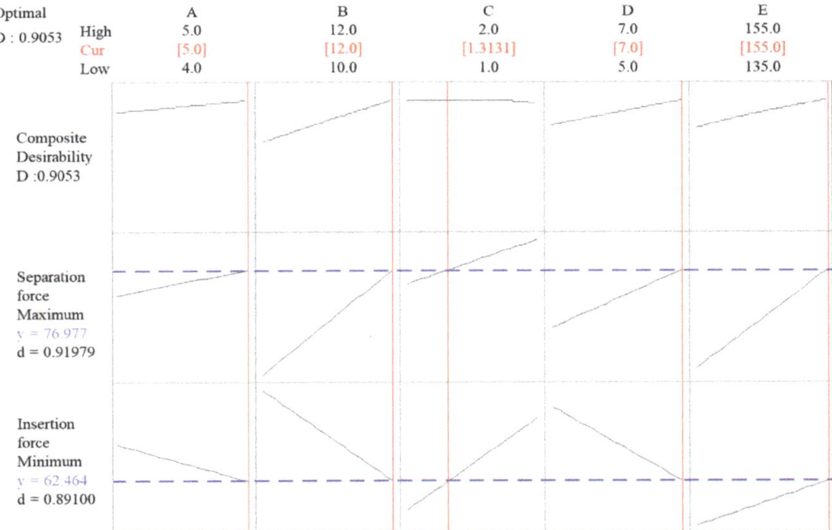

Figure 9. Shape parameter optimization results using response optimization tool.

4.2. Design Verification

As shown in Table 9, nonlinear analysis was conducted to verify the feasibility of the shape parameters and predicted values were derived using the response optimization tool. To verify the feasibility of the predicted values, nonlinear analysis was conducted by modeling the values of the derived shape parameters. An insertion force of 62.76 N was derived from the results of the additional nonlinear analysis; the difference from the predicted value (62.46 N) was 0.29 N.

The predicted value of separation force was confirmed to be 77.08 N using the response optimization tool. The value obtained through the additional nonlinear analysis was 76.98 N, and the difference between the predicted and analysis values was 0.10 N. Table 9 shows the predicted values obtained from the response optimization tool and the values obtained via the additional analysis.

Table 9. Comparison of predicted and analysis values.

Classification	Predicted Value (N)	Analysis Value (N)	Difference (N)
Insertion force	62.46	62.76	0.29
Separation force	76.98	77.08	0.10

4.3. Verification of Effectiveness

A grommet was fabricated to experimentally test insertion force and separation force using the predicted shape parameters. Based on the process variables set in the previous test, the temperature was set to 170 °C and the time was set to 300 s to produce a grommet. The experiments were conducted as shown in Figure 10. Insertion force was 50.0 N, and separation force was 85.3 N. The predicted and experimental values are different because, when the grommet is fabricated, a protrusion is formed to reduce the friction between the grommet and the mounting part in the insertion part. This causes insertion force to be smaller than the predicted value. Table 10 show the Predicted and measured insertion force and separation force.

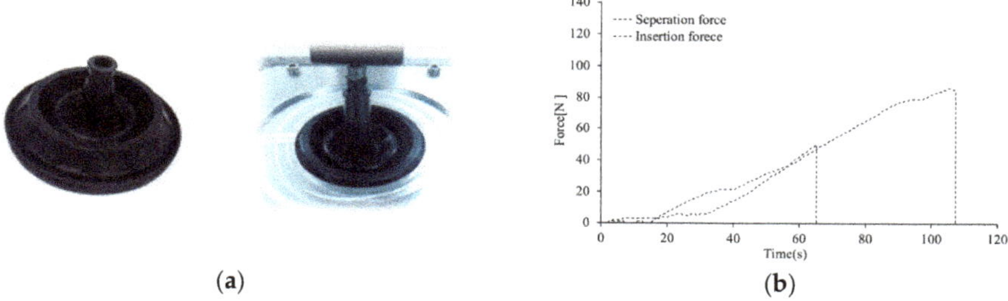

(a)   (b)

Figure 10. Comparison of simulated and experimental values. (**a**) Insertion force measurement; and (**b**) comparison of insertion and separation forces.

Table 10. Predicted and measured insertion force and separation force.

Classification	Predicted Value (N)	Simulated Value (N)	Experimental Value (N)
Insertion force	62.5	62.8	50.0
Separation force	77.0	77.1	85.3

5. Conclusions

This study conducted optimization to improve the mounting performance of a grommet using EPDM materials. The physical properties of the main molding materials were investigated according to process parameters. A grommet was fabricated according to the process parameters of fabrication. Insertion force and separation force were examined through experiments. Nonlinear material constants were determined through uniaxial and biaxial tensile tests. The nonlinear analysis of the grommet was conducted, and a compound design that incorporated the shape parameters for the minimum load of each part was derived. Then, additional nonlinear analysis was performed. This was followed by a comparative analysis of the actual model through experimental evaluation.

1. The physical properties of EPDM materials were analyzed according to molding parameters. Tensile strength and elongation were measured. Tensile strength increased with temperature and time.

2. A grommet was fabricated by applying the process parameters that affected the properties of specimens. Experiments were conducted to measure the insertion force and separation force of the fabricated grommet. We confirmed that the maximum load did not change with tensile strength and elongation. Moreover, differences in insertion time occurred owing to differences in elongation.

3. Uniaxial and biaxial elongation tests of the EPDM materials were conducted to perform the nonlinear analysis of the grommet, and physical property data were derived through the Ogden model. The grommet model was set for each shape parameter and analyzed for various cases. The influence of insertion force and separation force was confirmed through the set shape parameters, and the dimensions for minimizing insertion force and maximizing separation force were derived.

4. Additional analysis was performed for comparing the results of the optimization and experiments to verify the feasibility of the derived dimensions.

Author Contributions: Y.S.K., E.S.H. and E.S.J. conceived and designed the experiments; Y.S.K., E.S.H. and E.S.J. performed the experiments; Y.S.K., E.S.H. and E.S.J. analyzed the data; Y.S.K., E.S.H. and E.S.J. contributed reagents/materials/analysis tools; Y.S.K., E.S.H. and E.S.J. wrote the paper.

Funding: This research was financially supported by the Ministry of SMEs and Startups (MSS), Korea, under the "Convergence and Integration R&D(P0005112)" supervised by the Korea Institute for Advancement of Technology (KIAT).

Conflicts of Interest: The authors declare no conflict of interest.

References

1. Abou-Helal, M.O.; El-Sabbagh, S.H. A study on the compatibility of NR–EPDM blends using electrical and mechanical techniques. *J. Elastomers Plast.* **2005**, *37*, 319–346. [CrossRef]

2. Kondyurin, A. EPDM rubber modified by nitrogen plasma immersion ion implantation. *Materials* **2018**, *11*, 657. [CrossRef] [PubMed]

3. Ginic-Markovic, M.; Choudhury, N.R.; Dimopoulos, M.; Matisons, J. Adhesion between polyurethane coating and EPDM rubber compound. *J. Adhes. Sci. Technol.* **2004**, *18*, 575–596. [CrossRef]

4. Kwak, S.B.; Choi, N.S.; Choi, Y.J.; Shin, S.M. Nondestructive characterization of degradation of EPDM rubber for automotive radiator hoses. *Key Eng. Mater.* **2006**, *326–328*, 565–568. [CrossRef]

5. Zhao, Q.; Li, X.; Gao, J. Aging of ethylene–propylene–diene monomer (EPDM) in artificial weathering environment. *Polym. Degrad. Stab.* **2007**, *92*, 1841–1846. [CrossRef]

6. Ismail, H.; Pasbakhsh, P.; Ahmad Fauzi, M.N.; Bakar, A.A. Morphological, thermal and tensile properties of halloysite nanotubes filled ethylene propylene diene monomer (EPDM) nanocomposites. *Polym. Test.* **2008**, *27*, 841–850. [CrossRef]

7. Gatos, K.G.; Thomann, R.; Karger-Kocsis, J. Characteristics of ethylene propylene diene monomer rubber/organoclay nanocomposites resulting from different processing conditions and formulations. *Polym. Int.* **2004**, *53*, 1191–1197. [CrossRef]

8. Ismail, H.; Shaari, S.M. Curing characteristics, tensile properties and morphology of palm ash/halloysite nanotubes/ethylene-propylene-diene monomer (EPDM) hybrid composites. *Polym. Test.* **2010**, *29*, 872–878. [CrossRef]

9. Kim, Y.S.; Jeon, E.S.; Hwang, E.S. Effects of injection molding process conditions on physical properties of EPDM using design of experiment method. *MATEC Web Conf.* **2018**, *167*, 1–6. [CrossRef]

10. Maamoun, A.; Xue, Y.; Elbestawi, M.; Veldhuis, S. Effect of selective laser melting process parameters on the quality of Al alloy parts: Powder characterization, density, surface roughness, and dimensional accuracy. *Materials* **2018**, *11*, 2343. [CrossRef] [PubMed]

11. Ezedine, F.; Linares, J.M.; Chaves-Jacob, J.; Sprauel, J.M. Measurement parameters optimized for sequential multilateration in calibrating a machine Tool with a DOE Method. *Appl. Sci.* **2016**, *6*, 313. [CrossRef]

12. López, A.; Aisa, J.; Martinez, A.; Mercado, D. Injection moulding parameters influence on weight quality of complex parts by means of DOE application: Case study. *Measurement* **2016**, *90*, 349–356. [CrossRef]

13. KS M 6518:2016. *Physical Test Methods for Vulcanized Rubber*; Korean Industrial Standards: Seoul, Korea, 2016.

14. Kumar, C.S.S.R.; Nijasure, A.M. Vulcanization of rubber. *Resonance* **1997**, *2*, 55–59. [CrossRef]

15. Woo, C.S.; Kim, W.D.; Choi, S.S. Material Characteristics Evaluation and Useful Life Prediction by Heating Aging of Rubber Materials for Electronic Component. In Proceedings of the KSME 2005 Spring Annual Meeting, Busan, Korea, 25–27 May 2005; pp. 130–135.

16. Jang, W.J.; Han, C.Y.; Lee, S.B. Material Tests and Prediction for Finite Element Analysis of EPDM. In Proceedings of the KSME 2008 Autumn Annual Meeting, Pyeongchang, Korea, 5–7 November 2008; pp. 65–70.

17. Kowalik, M.; Pyrzanowska, J.; Piechal, A.; Blecharz-Klin, K.; Widy-Tyszkiewicz, E.; Suprynowicz, K.; Pyrzanowski, P. Determination of mechanical properties of rat's artery using optimization based method and Ogden's model. *Mater. Today Proc.* **2017**, *4*, 5849–5854. [CrossRef]

18. Wu, Y.; Wang, H.; Li, A. Parameter identification methods for hyperelastic and hyper-viscoelastic models. *Appl. Sci.* **2016**, *6*, 386. [CrossRef]

19. Kim, W.D.; Kim, W.S.; Kim, D.J.; Woo, C.S.; Lee, H.J. Mechanical testing and nonlinear material properties for finite element analysis of rubber components. *Trans. KSME* **2004**, *28*, 848–859.

20. Bhowmick, S.; Liu, G.R. Three dimensional cs-fem phase-field modeling technique for brittle fracture in elastic solids. *Appl. Sci.* **2018**, *8*, 2488. [CrossRef]

21. Ullah, S.N.; Hou, L.F.; Satchithananthan, U.; Chen, Z.; Gu, H. A 3D RITSS approach for total stress and coupled-flow large deformation problems using ABAQUS. *Comput. Geotech.* **2018**, *99*, 203–215. [CrossRef]

22. Kim, B.; Lee, S.B.; Lee, J.; Cho, S.; Park, H.; Yeom, S.; Park, S.H. A comparison among Neo-Hookean Model, MooneyRivlin Model, and Ogden Model for chloroprene rubber. *Int. J. Precis. Eng. Manuf.* **2012**, *13*, 759–764. [CrossRef]

23. Chu, Z.; Zheng, F.; Liang, L.; Yan, H.; Kang, R. Parameter determination of a minimal model for brake squeal. *Appl. Sci.* **2018**, *8*, 37. [CrossRef]

Article

Fatigue Life of RC Bridge Decks Affected by Non-Uniformly Dispersed Stagnant Water

Eissa Fathalla [1], Yasushi Tanaka [2] and Koichi Maekawa [3,*]

[1] Department of Civil Engineering, The University of Tokyo, 7-3-1 Hongo, Bunkyo-ku, Tokyo 113-8656, Japan; eissa.tokyo.concrete@gmail.com

[2] Department of Civil and Environmental Engineering, Kanazawa Institute of Technology, 7-1 Ohgigaoka, Nonoichi, Ishikawa 921-8501, Japan; ytanaka@neptune.kanazawa-it.ac.jp

[3] Department of Civil Engineering, Graduate School of Urban Innovation, Yokohama National University, 79-1 Tokiwadai, Hodogaya, Yokohama 240-8501, Japan

* Correspondence: maekawa-koichi-tn@ynu.ac.jp; Tel.: +81-45-339-4155

Received: 19 January 2019; Accepted: 9 February 2019; Published: 12 February 2019

Featured Application: This study aims to achieve the deterioration rank of the site-inspected wetting locations of road bridge decks, where the wetting locations can be detected by non-destructive testing technology. Thus, rational life assessment of bridge decks is secured.

Abstract: Stagnant water on reinforced concrete (RC) decks reduces their life significantly compared to the case of dry states. Fully submerged states have been investigated as the most severe case, which is however rarely experienced in reality. Currently, it is possible to simulate concrete–water interactions for lifetime prediction of RC decks. In this study, fatigue lifetime is systematically computed for various locations of stagnant water at the upper layer of RC decks. It is found that the patterns of wet and dry areas have a great influence on the remaining fatigue life even though the same magnitude of cracking develops. Then, a hazard map for the wetting locations with regard to the remaining fatigue life is presented based on the systematically arranged simulation. Finally, a nonlinear correlation is introduced for fatigue life prediction based upon site inspected wetting locations, which can be detected by non-destructive testing technology.

Keywords: multi-scale simulation; fatigue loading; road bridge decks; stagnant water

---

1. Introduction

Reinforced concrete (RC) bridge decks suffer from high deteriorations due to environmental attacks besides traffic loading, where corrosion, freeze and thawing, alkali silica reaction, and shrinkage and thermal cracking were reported to be significant on the reduction of life of RC decks [1–9].

On the other hand, in high seismic risk countries like Japan, thicknesses of RC decks were aimed to be thinner in order to reduce the inertia forces at earthquakes to satisfy earthquake resistant design requirements since bridge slabs are main source of seismic loads to bridges. These limited-thicknesses of decks (less than 200 mm) were constructed in 1960–70s, where enormous numbers of highway bridges were built. After around half a century, degradation of bridge decks has been observed from accumulated loads of daily traffics, where these deteriorations may reduce the safety of users.

Previous research reported that RC slabs exposed to moving loads are extremely deteriorated compared to those exposed to fixed-point pulsating ones. The reversal cyclic-shear along crack planes induced by moving wheel type loading rapidly deteriorates the shear transfer of aggregates interlock along concrete cracks [10]. Finally, RC slabs speedily lose their stiffness until total failure. Thus, we have common issues in view of bridge deterioration where mechanical fatigue loads and environmental actions develop together with more or less interaction.

It should be noted that the performance of RC decks can be upgraded by utilizing pre-stressing techniques [11] for newly constructed bridges. In fact, traffic-induced cracking is suppressed and water may not come inside crack gaps of concrete. However, we have to face serious problems such as that damaged RC decks of many bridges cannot be easily replaced since it will directly disturb the traffic flow leading to social problems. Therefore, old RC decks shall be retrofitted and/or limitedly replaced for extending the lifetime of RC decks on the basis of reliable maintenance plans.

Moreover, stagnant water from rainfall may remain on RC bridge decks due to imperfect waterproofing works and/or insufficient maintenance. Stagnant water is well recognized to seriously influence fatigue life [12–17], where performance of concrete is weakened and disintegration between aggregates and cement binder has been reported from experiments and site investigations, as shown in Figure 1. Accelerated wearing of the surface of cracks is demonstrated experimentally by cyclic shear tests in pure water, and the loss of the strength of concrete is qualitatively explained by changing the surface energy of Calcium-Silicate-Hydrate (CSH) binders. Under a high deformational rate, condensed water cannot easily disperse leading to a sharp rise and/or fall of the pore water pressure owing to its viscosity [18]. Fatigue loading experiments of water supply on RC decks demonstrate the negative impacts on their lives, and it was reported that the reduction in fatigue life is around 1/200 of the dry states [12,13].

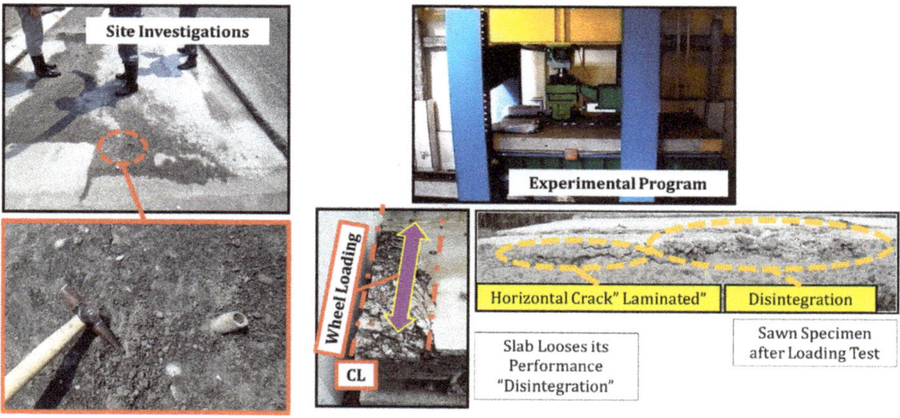

Figure 1. Site and experimental investigations for the disintegration phenomenon.

In recent research [17], predictive models were proposed for fatigue life of RC decks based on site inspected cracks under stagnant water, where artificial neural network model and mechanics-based correlation are introduced on the basis of an enormous number of investigated crack patterns. However, in the mentioned research, the RC decks are assumed to be in fully submerged states for safer and conservative assessment of RC decks at the current moment. It was found that cracks with the existence of stagnant water have more negative impacts on fatigue life of RC decks than the cracks in dry states, but it should be noted that the negative impacts of stagnant water on fatigue life are much higher than the one of the cracks as an overall.

Most of these research usually has targeted RC decks in fully submerged conditions, but little attention has been directed to more realistic moisture states where the wet area is scattered in space. Nowadays, some multi-scale simulation can deal with the crack to water interaction on the basis of upgraded constitutive laws for fatigue and Biot's theory [19–21], where spatially non-uniform wetting and dry regions can be numerically reproduced. On the other hand, in the past few decades, there were no quick-feasible technologies for detecting the wetting regions in real RC decks on site. However, currently, Mizutani et al. [22] are developing a fast detection technique for wetting locations in the upper layer by signal processing of UHF-band ground penetrating radar. This technique has been validated at the site and partially utilized in bridge maintenance in Japan.

In this study, the authors investigate the remaining fatigue life of various wetting regions in the upper layer of RC decks by utilizing the multi-scale simulation. It is empirically known that locations of stagnant water substantially affect the remaining fatigue life and that the wheel-loading path is the highest risk area that may reduce the fatigue life. Then, a detailed hazard map on wetting locations in between pavement and the concrete decks is aimed to meet the challenge of rational life assessment. Finally, a predictive correlation is proposed for fatigue life of RC decks based on site inspected wetting locations with high accuracy.

2. Methodology for Predicting the Remaining Fatigue Life

Figure 2 shows the methodology for predicting the remaining fatigue life of RC decks based upon site inspected wetting locations on the surface of decks as follows:

1. Ground penetrating radar is installed on a vehicle that runs about 80 km/h over the bridge, where signal responses of the hidden information of RC decks below the pavement layers are detected [22].
2. These signals are processed to achieve sound locations of the water especially at the upper surface layer of RC decks.
3. These wetting locations are induced into the finite element model by utilizing the multi-scale simulation program.
4. Travelling wheel load is applied until the failure of RC decks based on a failure criterion that will be stated in a later section.
5. Finally, remaining fatigue life of RC decks is computed.

The life-simulation is based upon the multi-scale thermo-hygral analysis [21], where the constitutive laws are upgraded for high cycle fatigue loading and the concrete–water interaction [20,21] as shown in Figure 3. The micro-fatigue model with cyclic pore pressure was integrated into the multi-scale simulation, where the average stiffness degrades with the disintegration of the aggregates-cement composites during the fatigue simulation of concrete with condensed water. The disintegration phenomenon between aggregates and cement composites is reproduced by integrating the damage evolved with increasing local pressure between aggregates and cement paste. This local pressure, which is also computed by the constitutive model for multi-directional cracked concrete, degrades the bond between cement matrix and aggregates leading to erosion of the composite system [20].

On the basis of previous research of the freeze-thawing mechanism [23] and rate-type of fatigue modelling [24], Equation (1) [21] expresses the overall stresses of disintegrated reinforced concrete, where it is computed by integrating three constitutive models: (1) non-eroded concrete with and without cracking, (2) steel reinforcement and (3) assembly of aggregates. When there is no erosion caused by internal water impact, stresses of the element are rooted in (1) and (2). If full disintegration develops finally, the total stresses come solely from (3). Here, the erosion factor denoted by K is defined to interpolate these two extreme states. Then, we have

$$\sigma_{ij} = K \cdot \sigma_{c,ij} + \sqrt{K} \cdot \sigma_{s,ij} + (1 - K) \cdot \sigma_{agg,ij}$$
$$K = e^{-Z} \tag{1}$$
$$dZ = -10^{n} \cdot (1 + f_n) \cdot p_{ampl}{}^{f_n} \cdot dp$$

where σ_{ij} is total compressive stress tensor, $\sigma_{c,ij}$ and $\sigma_{s,ij}$ are stress tensors carried by cracked concrete and steel reinforcement, $\sigma_{agg,ij}$ is the stress tensor representing the fictitious aggregate particles, Z is set forth as an accumulated damage of concrete in micro-pore structure, (n, f_n) are coefficients related to the intersection and the slope of S–N diagrams and equal to (2.0, 0.4), respectively, p_{ampl} is amplitude of pore water pressure. The concrete and steel stresses ($\sigma_{c,ij}$ and $\sigma_{s,ij}$) are computed from the given space-averaged strain of finite element by the non-orthogonal multi-directional cracked concrete constitutive model and the aggregate stresses are computed by the multi-spring soil model [21].

During evolution of the disintegration, the cement binder is eroded and the effective stress of the concrete and steel parts is degraded until reaching zero at the same particular time of full erosion (K = 0). Here, the assembly of the remaining aggregates still can sustain compressive stresses from volumetric contraction similar to the behavior of soil particles in the geotechnical field, where the stiffness of the aggregates assembly without cement binder is estimated to be $1/100$ of the normal concrete [25]. Finally, by that model, the deterioration from the elevated pressure of crack–water interaction during fatigue loading can be simulated.

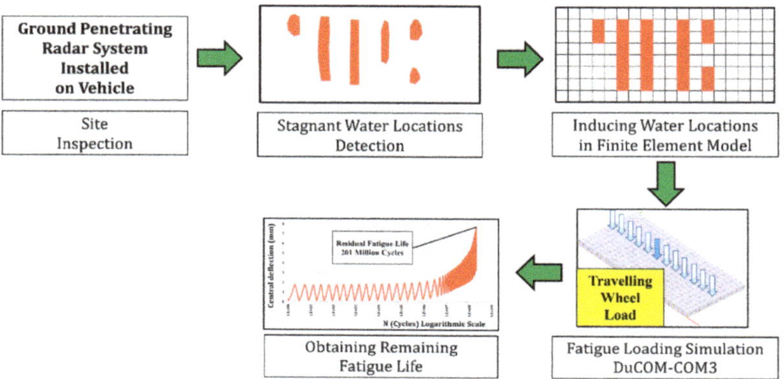

Figure 2. Overview of remaining fatigue life prediction methodology.

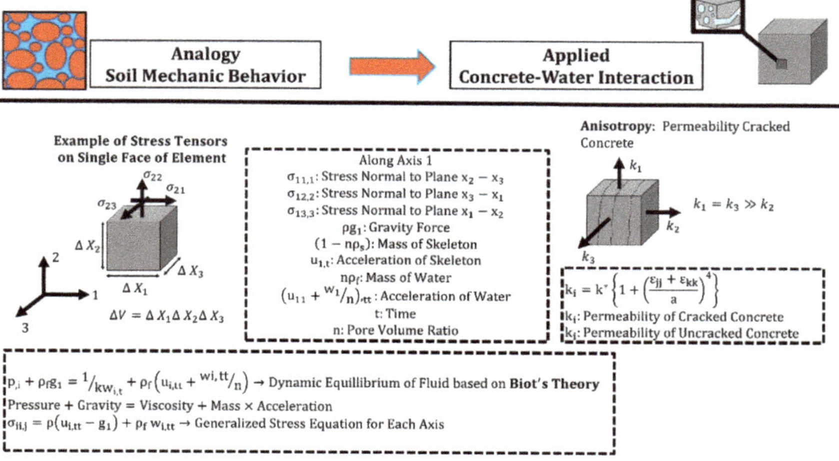

Figure 3. Constitutive laws of concrete–water interaction.

3. Specifications for the Parametric Study

3.1. Referential Reinforced Concrete Deck

In this study, we select a referential RC deck in reference to typical old bridge decks in Japan. Figure 4 shows the dimensions and the reinforcement arrangement of the targeted deck. A large amount of RC decks, which are currently problematic in maintenance due to short fatigue life, were designed as one-way slab supported by side girders, while their length generally depends on several conditions such as the span of adjacent girders. In a previous research study [26], 6.0 meters' length was selected as a reference to represent a large amount of bridge infra-stocks in Japan. Thus, the authors select the same target as well.

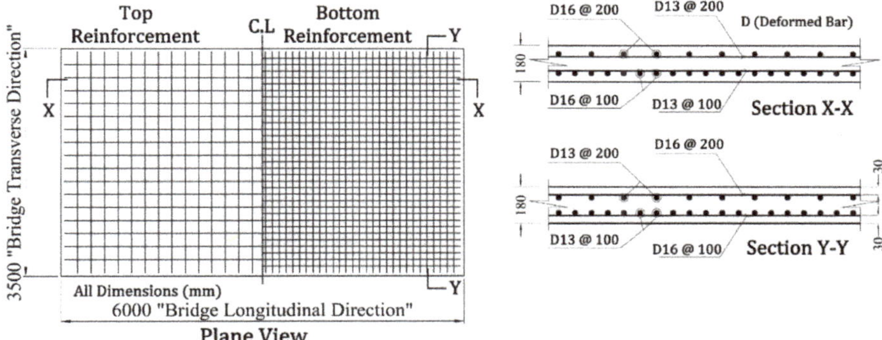

Figure 4. Dimensions and reinforcement of the reinforced concrete (RC) deck discussed.

3.2. Material Properties

Material properties of concrete and steel of the studied deck are shown in Table 1 on the basis of general design values used in the past construction practice of highway bridge decks.

Table 1. Material properties of the slab for analysis.

Material Type		Concrete	Steel Reinforcement
Young's Modulus	N/mm^2	24,750	205,000
Compressive Strength	N/mm^2	30	295
Tensile Strength	N/mm^2	2.2	295
Specific Weight	kN/m^3	24	78

3.3. Wheel-Type Moving Load

Referring to the specification for highway bridges-Part III [27], the deck is subjected to travelling wheel-type design load of 98 kN as shown in Figure 5. Running speed of the wheel is chosen as 60 km/h, which is used to be the legal speed limit for national routes. The dimensions of the wheel are 500 and 250 mm in reference to the contact area of wheel tires.

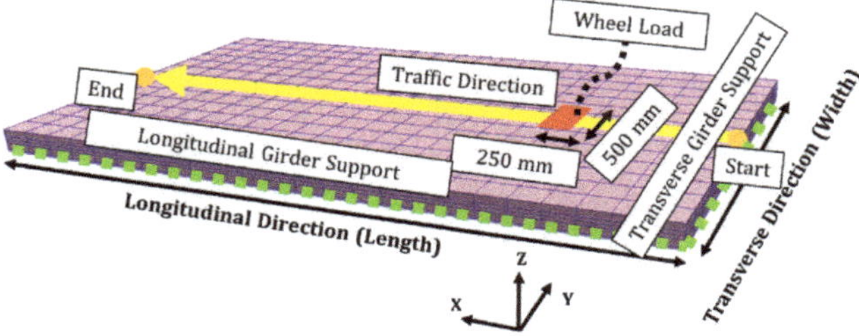

Figure 5. Standardized state of the RC deck discussed.

3.4. Failure Criterion

The fatigue limit state was specified according to the central live load deflection on the basis of past experiments and experiences [28,29]. When the live load deflection defined by Equation (2) reaches the limit state, which is equivalent to the loss of bond in flexure between concrete and steel (see Figure 6), it is judged as the fatigue failure criterion. We also accept this failure criterion in order to refer to our past research works. As this limit state live load deflection, from the experimental results, is equal to ($\cong$3) times its initial value, the authors directly apply the criterion denoted by Equation (3).

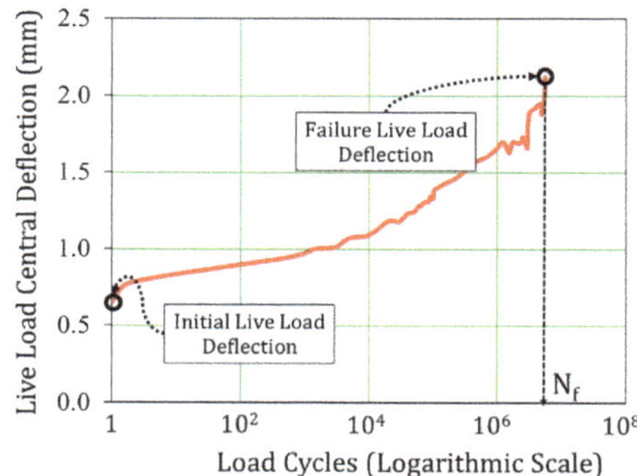

Figure 6. Experimentally obtained live load deflection with equivalent number of cycles.

Generally, fatigue life of RC decks is dependent on dimensions, material properties, and boundary conditions of the specimens. However, this fatigue failure criterion was empirically found to be rational since the properties of RC decks are rooted in the mere values of live load deflection. Moreover, it should be noted that the fatigue rupture of steel rebar, under moving loading, does not occur before failure of concrete [30]. Therefore, this criterion is thought to be feasible for the assessment of the failure of RC decks.

$$\delta_{L,N} = \delta_{1,N} - \delta_{2,N}, \tag{2}$$

$$\delta_{L,N}/\delta_{L,0} \geq 3.0, \tag{3}$$

where $\delta_{L,N}$ is central live load deflection at the Nth of cycles, $\delta_{1,N}$ is central total deflection at the Nth of cycles at the loading step, $\delta_{2,N}$ is central total deflection at N^{th} of cycles at unloading step, $\delta_{L,0}$ is initial live load deflection, and N_f is the failure number of cycles corresponding to ($\delta_{L,N}$ from Equation (3)).

3.5. Numerical Model

The simulation model is discretized with finite elements by using the open code program "FABriS" [26], which is specialized for the wheel-type moving load on RC bridge decks. Mesh size is chosen to be 250 × 250 mm in the *x*–*y* plane, and the number of layers in the *z*-direction is four as shown previously in Figure 5. The slab is supported by hinges at the boundaries of the RC deck. The studied cases of the effect of wetting locations on the fatigue life of the RC deck are based on varying the wetting locations among the top layer of the RC deck, as shown in Figure 7, since the radar system technique can efficiently obtain this kind of information from site [22]. The simulation explained as above has been validated by experiments and site inspection data [17].

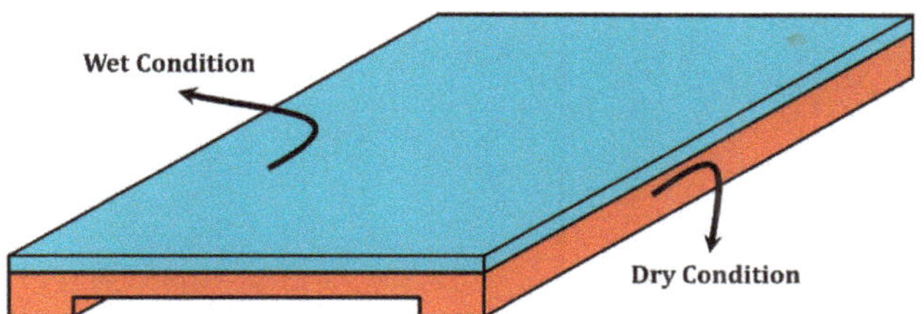

Figure 7. Standardized states for stagnant water studied cases.

4. Dry and Wet Ambient Conditions

The fatigue life of the referential RC deck in dry conditions and fully submerged was investigated in previous research [17,26,31]. However, this study targets the effect of wetting locations only at the upper entire surface layer of the RC deck. Figure 8 shows the relation of load cycles and the central live load deflection of three cases: dry, upper mesh layer "wet", and full submerged conditions, where their remaining fatigue lives are 221.49, 8.16, and 2.93 million cycles, respectively, as shown in Table 2.

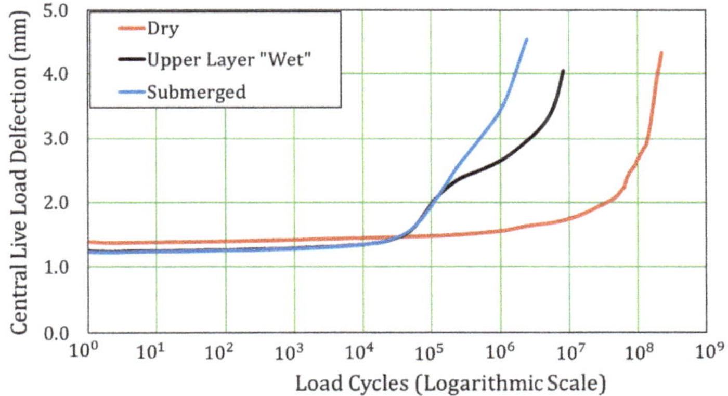

Figure 8. Load cycles and central live load deflection of dry and wet ambient conditions.

Table 2. Remaining fatigue life of dry and wet ambient conditions.

Case	Remaining Fatigue Life (Million Cycles)	Reduction in Life (Compared to Dry Case)
Dry Condition	221.49	1.0
Upper Layer (Wet)	8.16	1/27
Submerged	2.93	1/75

The pore-mechanics based simulation can indicate the profile of erosion by K factor (Equation (1)) at the central zone of the referential RC deck (upper layer "wet" in Table 2). Concrete erosion is severe around the path of the moving wheel as shown in Figure 9, and the concrete is fully eroded at 70,000 cycles as shown in Figure 10. The computed erosion profile matches the reality of the RC decks in the laboratory. Figure 11 shows the rise and decay of the pore water pressure. When the concrete composite is more or less sound as solid, pore pressure may rise according to the external load. However, when the erosion evolves much, the pore water pressure cannot rise because the water may easily pass through an assembly of cracks. In accordance with the erosion, we have varying compressive stresses of concrete at the top layer central elements and tensile stresses of reinforcement at the bottom ones as shown in Figures 12 and 13, respectively.

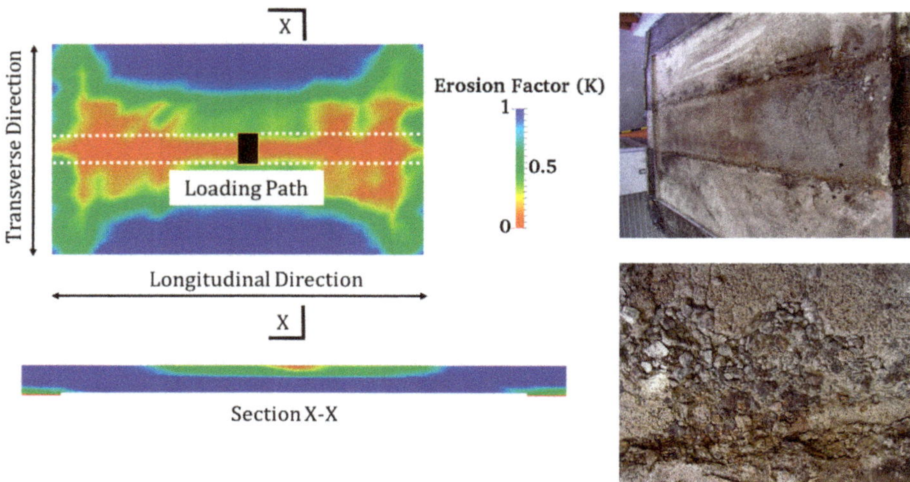

Figure 9. Multi-scale simulation of RC deck under fatigue loading in upper layer "wet" case.

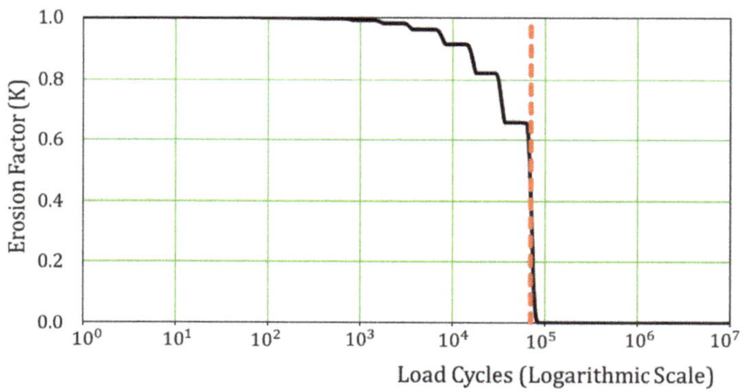

Figure 10. Load cycles and the erosion factor of the central zone of the RC deck.

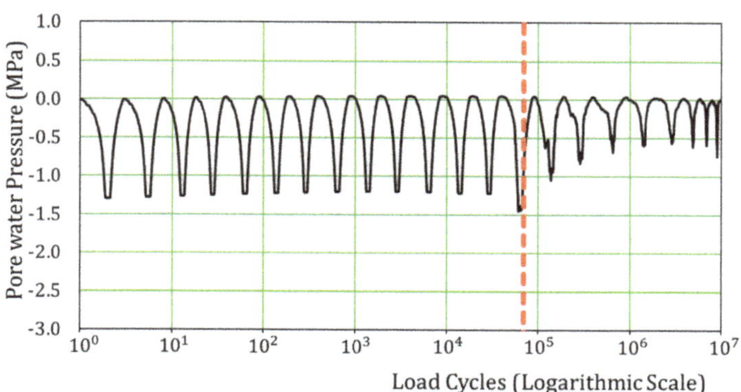

Figure 11. Load cycles and the pore water pressure of the central zone of the RC deck.

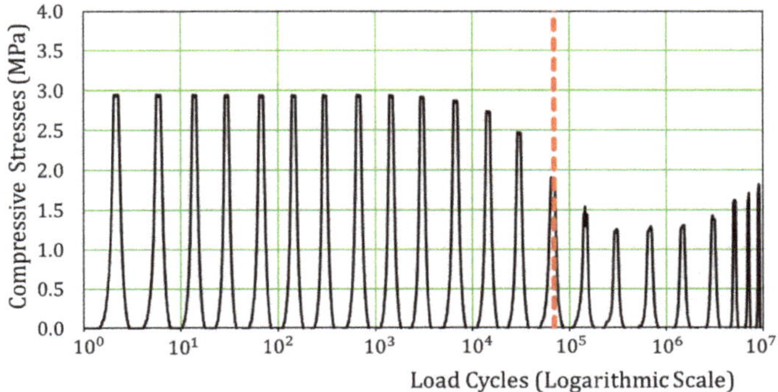

Figure 12. Load cycles and the top compressive stresses of the transverse direction of the central zone of the RC deck.

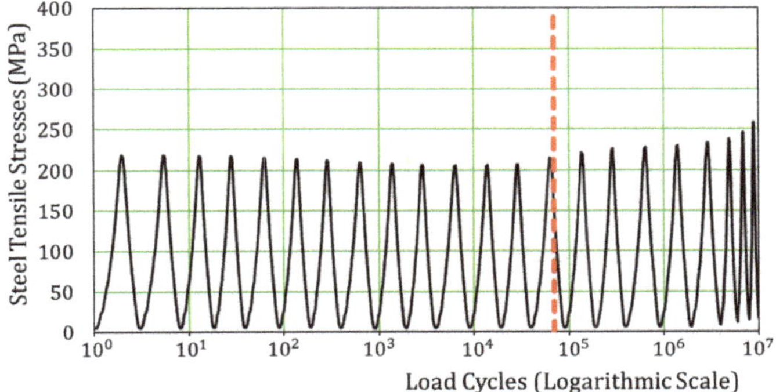

Figure 13. Load cycles and the bottom-steel tensile stresses of the central zone of the RC deck.

5. Case Study

Figure 14 shows the non-uniform patterns of the locations of stagnant water (30 cases) at the top surface of studied RC deck. The relation of the wetting rate at the top surface (WR) expressed in Equation (4), and the computed fatigue lives is shown in Figure 15, where their detailed remaining fatigue lives, normalized fatigue life by the fully dry condition, and WR indices are listed in Table 3. The simulation results prove the wide range of fatigue life despite of the same WR aiming at the significance of wetting patterns on the fatigue life. It is clear that the fatigue life is significantly reduced when the wetting locations are at the central zone of the wheel loading path as the following cases: (3), (13), (16), (20), (27), and (29).

On the other hand, the impact of stagnant water is significantly reduced, when the wetting locations are near the sides of the deck, away from the wheel loading path, as the following cases: (4–9), (17–19), and (26). It is demonstrated that the negative impact of the stagnant water on the fatigue life is reduced as the wetting locations start to be farther from the central zone of the wheel-loading path, as shown in Figure 16, where X is the longitudinal distance of the central wetting location to the center of the RC deck.

$$\text{WR\%} = \frac{\sum_{k=1}^{k=n}(E_k)}{n} \times 100, \tag{4}$$

where WR% is the wetting rate among the top mesh layer of the RC deck, k is the kth element at the top surface, E_k is environmental condition of the kth element (0 for dry, 1 for wet), and n is the total number of elements at the top surface of the deck (336 is this study).

For further investigation for the wetting pattern, a comparison has been made between case (5) that has low water impact and case (16) that has high impact, where the difference in their fatigue lives is around 1.6 times despite of the quite close value of WR. Figure 17 shows the development of the central live load deflection with loading cycles for cases (5) and (16), where their remaining fatigue lives are 160.0 and 99.8 million cycles, respectively. By investigating the rate of erosion of the concrete at the wetting locations for both cases (see Figure 18), it is obvious that the erosion starts earlier in case (16) than case (5). This is also clear in the rise of the pore water pressure for each case at the same particular time of reaching full erosion (K = 0), as shown in Figure 19. This can be explained by checking the average principal strains at the bottom surface of the RC deck just below their wetting locations, as shown in Figure 20.

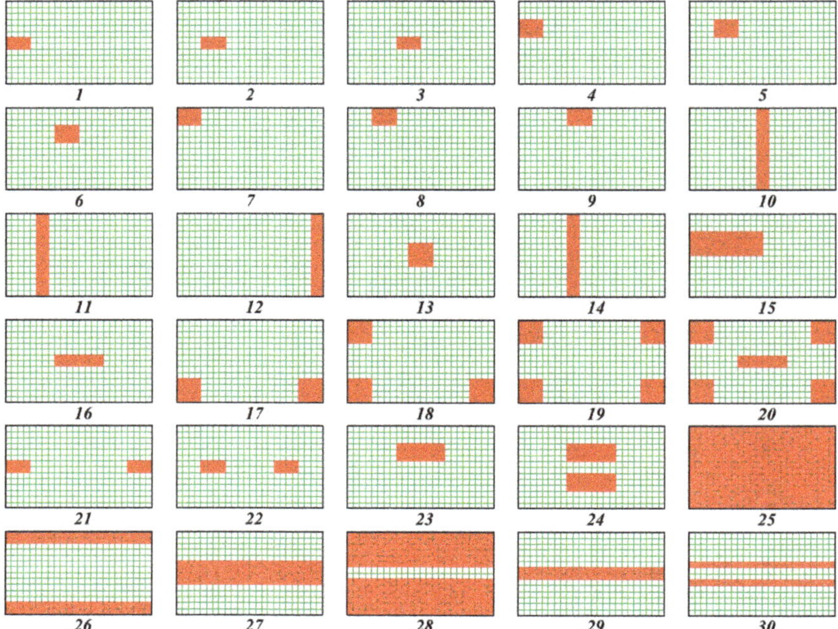

Figure 14. Location patterns of stagnant water.

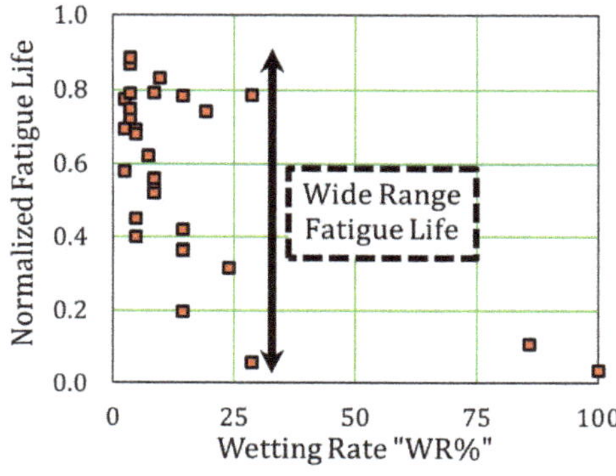

Figure 15. Relation of wetting rate at the top layer of RC deck and the fatigue life.

Table 3. Studied remaining fatigue lives and wetting rates.

Case	Life "Million Cycles"	Normalized "Dry"	WR%	Case	Life "Million Cycles"	Normalized "Dry"	WR%
1	171.7	0.78	2.4	16	99.8	0.45	4.8
2	153.8	0.69	2.4	17	184.0	0.83	9.5
3	128.3	0.58	2.4	18	173.5	0.78	14.3
4	174.9	0.79	3.6	19	164.5	0.74	19.0
5	160.0	0.72	3.6	20	69.7	0.31	23.8
6	165.7	0.75	3.6	21	153.1	0.69	4.8
7	193.6	0.87	3.6	22	150.6	0.68	4.8
8	192.6	0.87	3.6	23	137.6	0.62	7.1
9	196.5	0.89	3.6	24	81.0	0.37	14.3
10	115.5	0.52	8.3	25	8.2	0.04	100.0
11	121.7	0.55	8.3	26	174.0	0.79	28.6
12	175.7	0.79	8.3	27	13.1	0.06	28.6
13	88.8	0.40	4.8	28	24.2	0.11	85.7
14	124.2	0.56	8.3	29	43.6	0.20	14.3
15	93.1	0.42	14.3	30	80.6	0.36	14.3

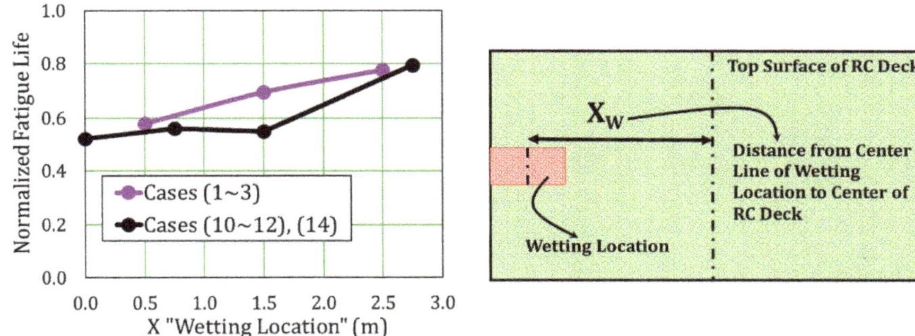

Figure 16. Effect of wetting locations on the fatigue life of RC decks.

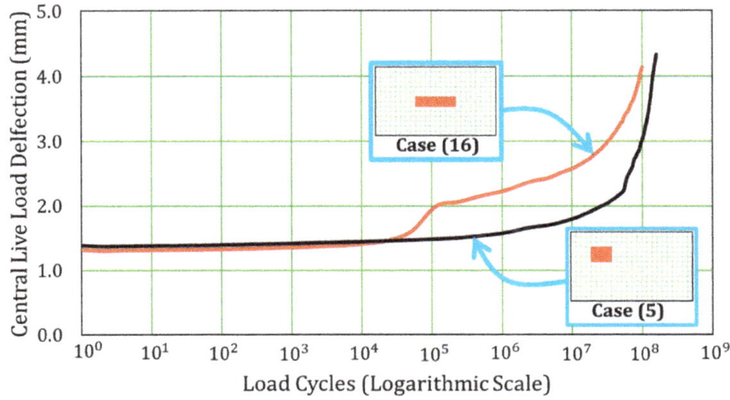

Figure 17. Load cycles and the central live load deflection of cases (5) and (16).

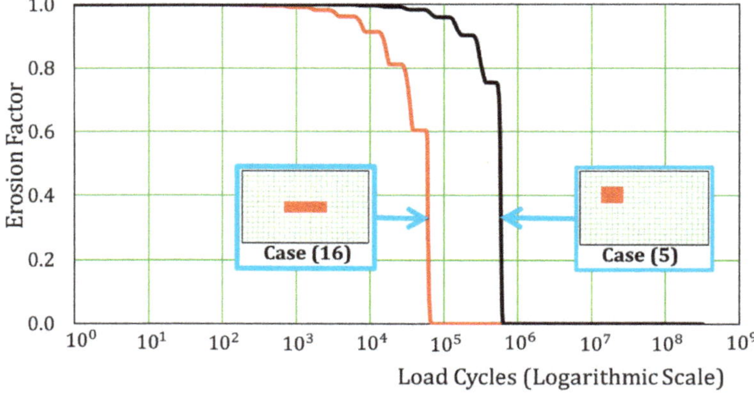

Figure 18. Load cycles and the erosion factor of the upper zone of cases (5) and (16) at their wetting locations.

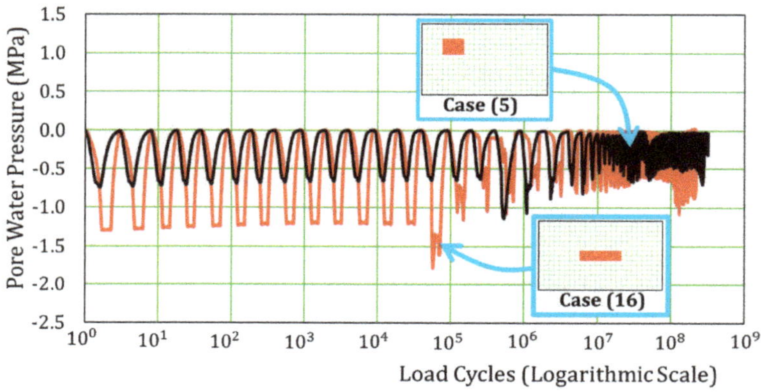

Figure 19. Load cycles and the pore water pressure of the upper zone of cases (5) and (16) at their wetting locations.

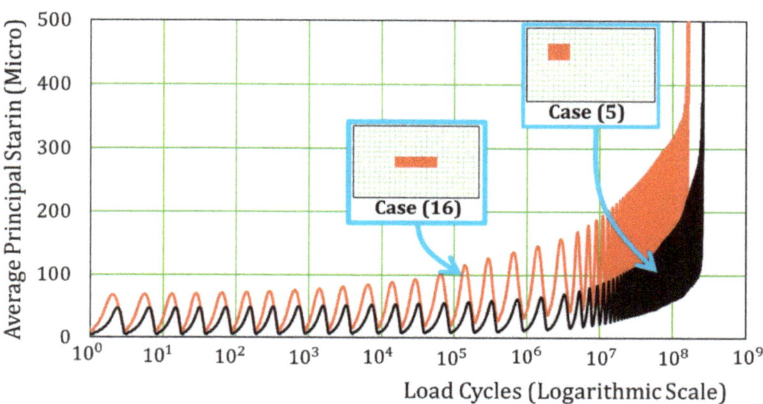

Figure 20. Load cycles and the average principal strains of the bottom zone of cases (5) and (16) at their wetting locations.

It is found that the amplitude of the principal strains for case (16) is higher than that of case (5), since it is the location of maximum bending moment, where the flexure cracks occur with great strain. On the other hand, flexural cracks are dispersing as we go farther from the central zone of the RC deck, where the reduction of principal strains for case (5) is clearly seen. The segregation of aggregates and cement paste is dependent on the opening and closure of the cracks, where the pore water pressure increases at the closure of the cracks leading to the damage of the local interface between the aggregates and the cement binder, as shown in Figure 21. Thus, more flexure cracks and deformation amplitudes of case (16) lead to the negative impacts on the remaining fatigue life compared to case (5).

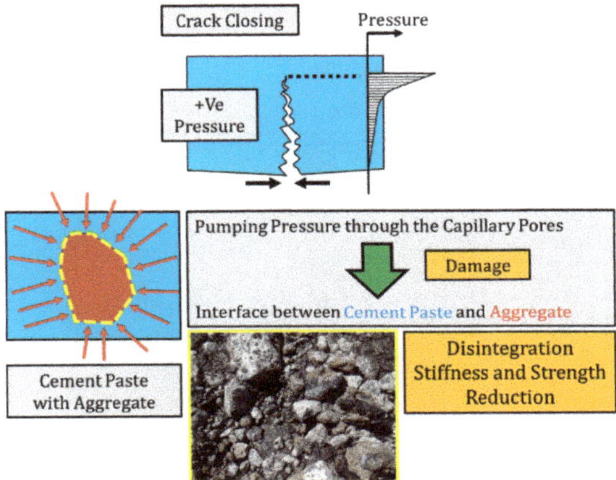

Figure 21. Mechanisms of disintegration of concrete under high cycle fatigue.

6. Hazard Map for Engineering Practice

In this section, the authors try to build a hazard map of water for the wetting locations based on the simulation results. Cases (1~9), besides the laws of symmetry, are utilized to build the hazard map since these cases cover one quarter of the top surface of the RC deck without overlapping, as shown in Figure 22. The normalized fatigue life compared to the full dry state of the reference case is normalized between 0 and 1 to achieve 2D contour lines of the hazard map with the upper bound of (1) and the lower one of (0), which denote the lowest and the highest hazard regions, respectively, as shown in Table 4. Finally, a hazard map for the wetting locations is introduced, as shown in Figure 23. It is demonstrated that the central zone of the wheel-loading path is highly hazardous. The introduced hazard map allows the bridges' inspectors to evaluate the risk of the RC decks by utilizing the visual and/or non-destructive inspection [22,32].

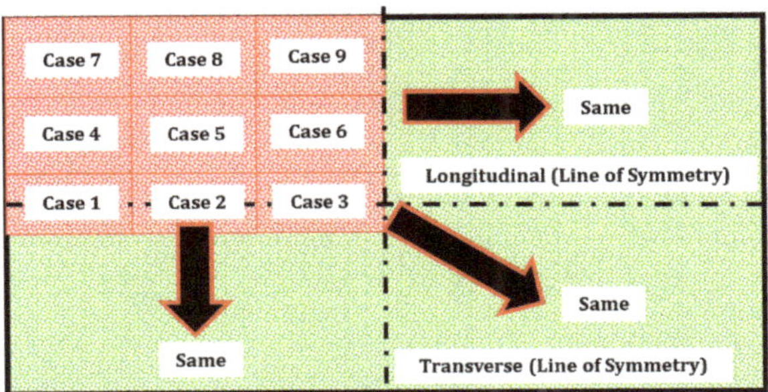

Figure 22. Methodology for achieving the hazard map.

Table 4. Utilized cases for building the hazard map.

Case	Life "Million Cycles"	Normalized "Dry"	Normalization (0 and 1)
1	171.7	0.78	0.65
2	153.8	0.69	0.35
3	128.3	0.58	0.00
4	174.9	0.79	0.68
5	160.0	0.72	0.45
6	165.7	0.75	0.55
7	193.6	0.87	0.94
8	192.6	0.87	0.94
9	196.5	0.89	1.00

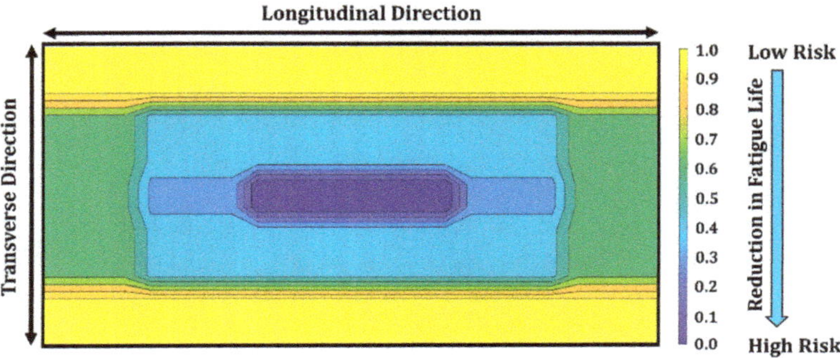

Figure 23. Hazard map for the wetting locations of higher risk.

7. Predictive Correlation

On the basis of the proposed hazard map, a predictive correlation for fatigue life of RC decks, based upon site inspected wetting locations, is introduced. The top surface of the RC deck is divided into four regions (see Figure 24), where WR index (Equation (4)) is calculated for each zone. Then, a predictive damage index (D.I.) is introduced by integrating the WR index for the four regions, where each region is assigned to different weight based on its risk impact, as shown in Equation (5). Finally, nonlinear correlation (Equation (6)) is introduced for the remaining fatigue life as a function of the proposed damage index. Figure 25 shows the relation of the damage index and the fatigue life of the studied cases, where the regression coefficient, coefficient of variation (C.O.V), and prediction interval of variance of 95% (P.I.) of the proposed correlation are 0.91, 21%, and 17%, respectively. The proposed correlation offers to analyze the remaining fatigue life of RC deck based on site inspected wetting locations with high accuracy:

$$\text{D.I.} = 0.55 \cdot \text{WR}_1 + 0.23 \cdot \text{WR}_2 + 0.17 \cdot \text{WR}_3 + 0.05 \cdot \text{WR}_4, \tag{5}$$

$$\text{Life} = 1.49 + 1.43 \cdot \tanh(-0.36 - 0.24 \cdot \text{D.I.}), \tag{6}$$

where D.I. is the damage index of water impact, and (WR$_1$, WR$_2$, WR$_3$, WR$_4$) are wetting rates for the regions shown in Figure 24 and calculated from Equation (4).

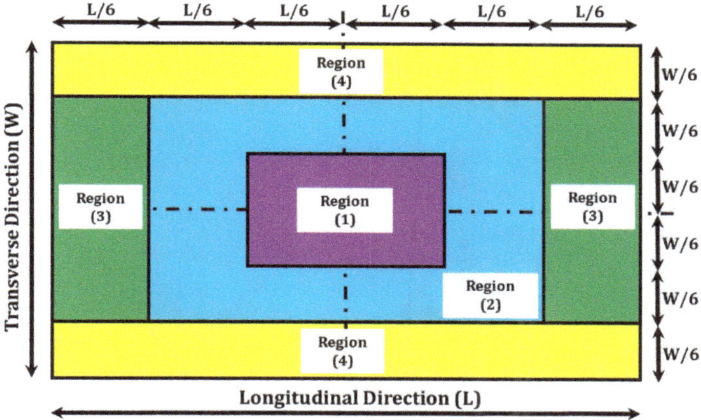

Figure 24. Damage index concept.

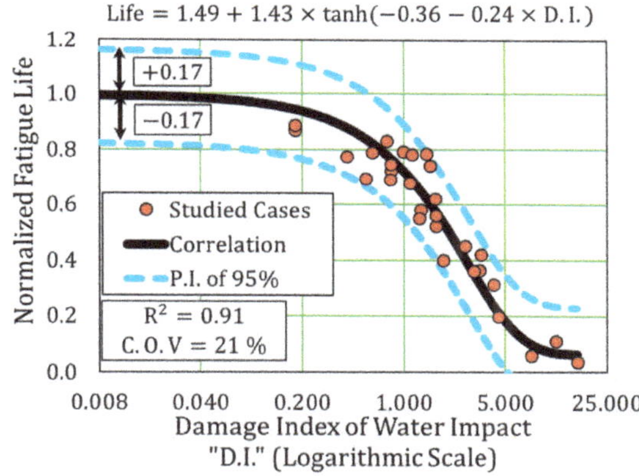

Figure 25. Relation of the damage index of water impact and the normalized fatigue life.

8. Conclusions

The impact of wetting by the stagnant water on the remaining fatigue life of the RC decks is studied by utilizing the multi-scale simulation, which was previously validated by experiments and site experiences, and the following conclusions are earned.

1. The patterns of wet and dry areas have a great influence on the remaining fatigue life of the RC decks.
2. It is found that the central zone of the wheel-loading path is the wetting location of higher risk on the fatigue life.
3. The negative impacts of stagnant water reduce gradually as the wetting locations go farther from the central zone of the wheel-loading path, where these impacts tend to significantly reduce at the sides of the RC deck away from the wheel-loading path.
4. A hazard map for the wetting locations of higher risk is proposed based on the simulation results, which is beneficial for bridges' inspectors.
5. A predictive correlation is proposed for fatigue life prediction of RC deck based on site inspected wetting locations with high accuracy, which fulfill the engineers' needs to conduct the risk assessment of RC decks during maintenance.

It should be noted that the scope of the proposed predictive correlation is to achieve the risk magnitude of inspected wetting locations with regard to the reference RC deck. For other RC decks, whose dimensions and material properties differ from the referential RC deck, the sensitivity of the proposed correlation will be checked in the near future regarding its generalization to deal with any RC deck. Then, as a next future step, the authors aim to couple the introduced predictive model in this study with our previous proposed artificial intelligence (AI) models that deal with fatigue life prediction based on site-inspected cracks [17,26,31].

Author Contributions: E.F. conducted and analyzed the numerical studied cases; Y.T and K.M. supervised over the analytical process and developed the simulation program; E.F. and K.M. wrote the paper.

Funding: This study was financially supported by the Council for Science, Technology, and Innovation, "Cross-ministerial Strategic Innovation Promotion Program (SIP), Infrastructure Maintenance, Renovation, and Management" granted by the Japan Science and Technology Agency (JST).

Acknowledgments: The authors extend their appreciation to Yozo Fujino of Yokohama National University and the University of Tokyo, for his valuable advice and encouragement to bridge the multi-disciplines in engineering. Many thanks to Takahiro Yamaguchi of the University of Tokyo for his great discussions and valuable comments about ground penetrating radar system technology.

Conflicts of Interest: The authors declare no conflict of interest.

References

1. Val, D.V.; Stewart, M.G.; Melchers, R.E. Effect of reinforcement corrosion on reliability of highway bridges. *Eng. Struct.* **1998**, *20*, 1010–1019. [CrossRef]

2. Vassie, P. Reinforcement corrosion and the durability of concrete bridges. In *Proceedings of Institution of Civil Engineers*; Institution of Civil Engineers: London, UK, 1984; Part 1; Volume 76.

3. Cady, P.D.; Weyers, R.E. Chloride penetration and the deterioration of concrete bridge decks. *Cem. Concr. Aggreg.* **1983**, *5*, 81–87.

4. Freyermuth, C.L.; Klieger, P.; Stark, D.C.; Wenke, H.N. Durability of concrete bridge decks—A review of cooperative studies. *Highw. Res. Rec.* **1970**, *328*, 50–60.

5. Lindquist, W.D.; Darwin, D.; Browning, J.; Miller, G.G. *Effect of Cracking on Chloride Content in Concrete Bridge Decks*; American Concrete Institute: Farmington Hills, MI, USA, 2006.

6. Shah, S.P.; Weiss, W.J.; Yang, W. Shrinkage Cracking—Can It Be Prevented? *Concr. Int.* **1998**, *20*, 51–55.

7. Sakulich, A.R.; Bentz, D.P. Increasing the service life of bridge decks by incorporating phase-change materials to reduce freeze-thaw cycles. *J. Mater. Civ. Eng.* **2011**, *24*, 1034–1042. [CrossRef]

8. Detwiler, R.J.; Kojundic, T.; Fidjestol, P. Evaluation of bridge deck overlays. *Concr. Int. Detroit* **1997**, *19*, 43–46.

9. Ulm, F.J.; Coussy, O.; Kefei, L.; Larive, C. Thermo-chemo-mechanics of ASR expansion in concrete structures. *J. Eng. Mech.* **2000**, *126*, 233–242. [CrossRef]

10. Perdikaris, P.C.; Beim, S. RC Bridge Decks Under Pulsating and Moving Load. *ASCE J. Struct. Eng.* **1988**, *114*, 591–607. [CrossRef]

11. Poston, R.W.; Carrasquillo, R.L.; Breen, J.E. Durability of post-tensioned bridge decks. *Mater. J.* **1987**, *84*, 315–326.

12. Matsui, S. Fatigue Strength of RC-Slabs of Highway Bridge by Wheel Running Machine and Influence of Water on Fatigue. In *Proceedings of Japan Concrete Institute*; Japan Concrete Institute: Tokyo, Japan, 1987; Volume 9, pp. 627–632.

13. Matsui, S. Lifetime prediction of bridge. *J. Jpn. Soc. Civ. Eng.* **1996**, *30*, 432–440.

14. Waagaard, K. Fatigue strength evaluation of offshore concrete structures. *ACI-SP* **1982**, *75*, 373–397.

15. Fujino, Y. Infrastructure Maintenance Renovation and Management, Cross-Ministerial Strategic Innovation Promotion Program. 2017. Available online: http://www.jst.go.jp/sip/dl/k07/pamphlet_2017_en.pdf (accessed on 17 May 2018).

16. Otsu, K. The current situation and maintenance of expressways in Japan. In Proceedings of the 3rd International Conference on Sustainable Construction Materials and Technology, Kyoto, Japan, 18–21 August 2013.

17. Fathalla, E.; Tanaka, Y.; Maekawa, K.; Sakurai, A. Quantitative deterioration assessment of road bridge decks based on site inspected cracks under stagnant water. *J. Adv. Concr. Technol.* **2019**, *17*, 16–33. [CrossRef]

18. Solwik, V.; Saouma, E.V. Water pressure in propagating concrete cracks. *J. Struct. Eng. Am. Soc. Civ. Eng.* **2000**, *126*, 235–242. [CrossRef]
19. Biot, M.A. General theory of three-dimensional consolidation. *J. Appl. Phys.* **1941**, *12*, 155–164. [CrossRef]
20. Maekawa, K.; Fujiyama, C. Crack water interaction and fatigue life assessment of RC bridge decks. In Proceedings of the Poromechanics V: American Society of Civil Engineers, Vienna, Austria, 10–12 July 2013; pp. 2280–2289.
21. Maekawa, K.; Ishida, T.; Chijiwa, N.; Fujiyama, C. Multiscale coupled-hygromechanistic approach to the life-cycle performance assessment of structural concrete. *J. Mater. Civ. Eng. Am. Soc. Civ. Eng.* **2015**, *27*, A4014003. [CrossRef]
22. Mizutani, T.; Nakamura, N.; Yamaguchi, T.; Tarumi, M.; Hara, I. Bridge slab damage detection by signal processing of uhf-band ground penetrating radar data. *J. Disaster Res.* **2017**, *12*, 415–421. [CrossRef]
23. Powers, T.C. A working hypothesis for further studies of frost resistance of concrete. *J. Proc.* **1945**, *41*, 245–272.
24. El-Kashif, K.F.; Maekawa, K. Cyclic cumulative damaging of reinforced concrete in post-peak regions. *J. Adv. Concr. Technol.* **2004**, *2*, 257–271.
25. Amini, F.; Sama, K.M. Behavior of stratified sand–silt–gravel composites under seismic liquefaction conditions. *Soil Dyn. Earthq. Eng.* **1999**, *18*, 445–455. [CrossRef]
26. Fathalla, E.; Tanaka, Y.; Maekawa, K. Remaining fatigue life assessment of in-service road bridge decks based upon artificial neural networks. *Eng. Struct.* **2018**, *171*, 602–616. [CrossRef]
27. Japan Road Association. *Specification for Highway Bridges-Part III Concrete Bridges*; Japan Road Association: Tokyo, Japan, 2012.
28. Maeshima, T.; Koda, Y.; Tsuchiya, S.; Iwaki, I. Influence of corrosion of rebars caused by chloride induced deterioration on fatigue resistance in RC road deck. *J. Jpn. Soc. Civ. Eng.* **2014**, *70*, 208–225.
29. Tanaka, Y.; Maekawa, K.; Maeshima, T.; Iwaki, I.; Nishida, T.; Shiotani, T. Data assimilation for fatigue life assessment of RC bridge decks coupled with hygro-mechanistic model and nondestructive inspection. *J. Disaster Res.* **2017**, *12*, 422–431. [CrossRef]
30. Matsui, S.; Maeda, Y. A rational valuation method for deterioration of highway bridge decks. *J. Jpn. Soc. Civ. Eng.* **1986**, *374*, 419–426.
31. Fathalla, E.; Tanaka, Y.; Maekawa, K.; Sakurai, A. Quantitative deterioration assessment of road bridge decks based on site inspected cracks. *Appl. Sci.* **2018**, *8*, 1197. [CrossRef]
32. Kakuma, K.; Sato, K.; Nishi, H. Field survey on the detection of stagnant water on highway bridge RC slabs using electromagnetic wave radar method. In Proceedings of the Hokkaido Chapter of JSCE, Sapporo, Japan, 4–5 February 2017; Volume 73, p. A-36.

Article

Effect of Crack Orientation on Fatigue Life of Reinforced Concrete Bridge Decks

Eissa Fathalla [1], Yasushi Tanaka [2] and Koichi Maekawa [3,*]

[1] Department of Civil Engineering, The University of Tokyo, 7-3-1 Hongo, Bunkyo-ku, Tokyo 113-8656, Japan; eissa.tokyo.concrete@gmail.com

[2] Department of Civil and Environmental Engineering, Kanazawa Institute of Technology, 7-1 Ohgigaoka, Nonoichi, Ishikawa 921-8501, Japan; ytanaka@neptune.kanazawa-it.ac.jp

[3] Department of Civil Engineering, Graduate School of Urban Innovation, Yokohama National University, 79-1 Tokiwadai, Hodogaya, Yokohama 240-8501, Japan

* Correspondence: maekawa-koichi-tn@ynu.ac.jp; Tel.: +81-45-339-4155

Received: 26 February 2019; Accepted: 15 April 2019; Published: 20 April 2019

Abstract: In visual inspection of bridges at sites, much attention is given to the density and width of cracks of concrete, but little attention is paid to crack orientation for the diagnosis of bridge performance. In this research, the effect of crack orientation on the remaining fatigue life of reinforced concrete (RC) bridge decks is investigated for crack patterns with a wide range of possible crack orientations. The data assimilation technology of multi-scale simulation and the pseudo-cracking method, which are widely validated for fatigue-lifetime simulation, are utilized in this study. The impact of the crack direction on fatigue life is found to be associated with the coupled flexure-shear mode of failure, and the mechanism to arrest shear cracking by preceding cracks is quantitatively estimated. This mechanism is similar to the stop-hole to prevent fatigue cracks in steel structures, and it enables us to enhance the fatigue life of RC decks. It is demonstrated that the crack orientations that approximate the longitudinal and transverse directions of RC decks are the ones that most extend remaining fatigue life. Finally, the higher risk cracking locations on the bottom surface of RC decks are discussed, presenting information of use to site inspectors.

Keywords: fatigue loading; bridge decks; pseudo-cracking method; data assimilation

---

1. Introduction

Bridges are essential infrastructure for transportation and trade. However, throughout the world, fatigue-inducing repetitive traffic loads and environment-related defects, such as corrosion, thermal and shrinkage cracking, and freezing and thawing, lead to degradation of reinforced concrete (RC) bridge decks [1–9]. Particularly in areas of higher seismic activity, where the reduced weight of bridge viaducts is beneficial, the fatigue damage of thinner RC decks whose thickness is less than 200 mm is a common issue. Moreover, the capacity of concrete declines significantly compared to its ultimate short-term strength [10–16]. In Japan, the preservation plan for Japanese highways [17] calls for more than 50% of the total maintenance budget to be spent on the renewal and repair of old RC decks, which were designed and constructed before the 1970s. Thus, the remaining fatigue life of RC decks needs to be rationally estimated for reliable social stock management.

When assessing the health of bridges, inspectors direct their attention to the density and width of the cracks appearing on the concrete surfaces of RC decks, as shown in Figure 1. RC is designed to allow cracking under service loads, and the orientation of cracks in RC is related to its performance [18–22]. The fact that crack-to-crack interactions may cause behavioral changes in structural members has been proven, and this has been investigated for beams and panels, where crack propagation can be observed by the naked eye. In fact, crack interactions have been investigated in view of the serviceability and ultimate limit states [20–22].

In the case of bridge decks, however, it is hard to detect crack interactions since cracks develop out of view inside the structure. Thus, it is of practical importance to apply knowledge of structural mechanics to site inspections in terms of crack orientation and crack-to-crack interactions. This study focuses on the effect of crack orientation on the remaining fatigue life of RC decks with consideration of a wide range of crack orientations. The core technologies employed in this study are multi-scale cracked concrete simulation [23,24] and the pseudo-cracking method [25,26], which have been validated to effectively assess the remaining fatigue life of RC decks [20–22,27].

First, the authors built a dataset of various crack patterns to study bridge performance by using a randomized artificial crack pattern program (RACP) to overcome the biased real cracks at sites and cover the entire range of crack orientations. The RACP was made to produce fictitious but probable cracks and was applied to build the training datasets of neural network models [21,22]. Crack density and width were kept fixed in the scheme of the RACP, and only crack direction was changed in this study. The effect of crack orientation on the remaining fatigue life of RC decks was discussed, and crack orientation was found to be associated with the coupled flexure-shear action, where pre-cracks tend to arrest the propagation of post-shear cracks. Finally, after capturing the general trend of the crack orientation effect on fatigue life, the impact of crack location was studied again in accordance with the most sensitive crack orientations.

The pre-crack arrest mechanism in RC members closely approximates that of the stop-holes that are drilled in steel members to stop the growth of fatigue cracks [28–31]. The pre-cracks in RC members act as stop-holes, stopping the propagation of post-shear cracks and resulting in extended fatigue life. Thus, this arrest mechanism enables us to upgrade the fatigue life of RC decks and to conduct a fair assessment of RC decks that may present a rough appearance, but have mechanically enhanced fatigue life.

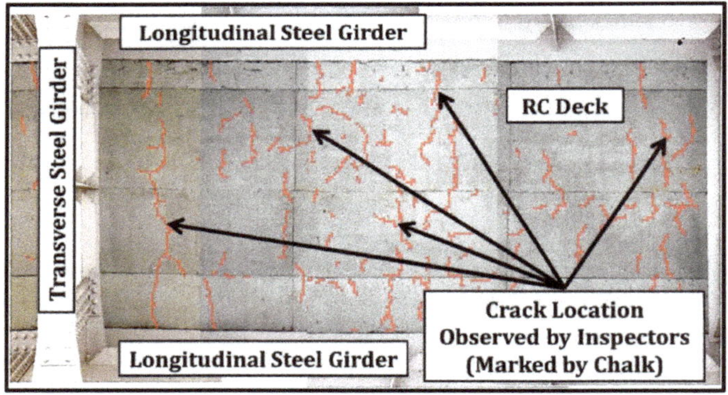

Figure 1. Site inspection for bottom surface cracks of reinforced concrete (RC) decks.

2. Methodology and Studied Cases

2.1. Data Assimilation Technology

For fatigue-lifetime simulation of RC decks based on site inspected cracks, the data assimilation technology of the multi-scale simulation program and the pseudo cracking method, which are broadly validated [23–26], were utilized. The constitutive laws of the multi-scale simulation were upgraded in the last few decades to deal with high cycle fatigue, so they can account for fatigue damage of concrete by a decrease in both strength and stiffness, and increase in time dependent deformations on the basis of the direct path integral method [24]. As concrete is a cementitious composite, a single crack opening (Mode-I) though cement paste solid exhibits tension softening and the fracture energy is much less than that of crack shear transfer of Mode-II (see Figure 2) owing to the interlocking of rough crack surfaces. Then, blunt multiple cracks are consequently dispersed, and the direction of subsequently propagating cracks is assumed to coincide with that of the updated principal stress, which satisfies the equilibrium with non-orthogonal preceding cracks. It means the minimum fracture energy of the whole analysis domain associated with the shear transfer along crack planes and tension softening normal to cracks [24]. The high shear transfer along preceding crack planes of concrete is a fatigue resistant mechanism which differs from the case of smooth preceding crack planes without shear transfer [32].

Generally, the existing crack will propagate if its propagation is associated with a decrease in the total energy of the system [33,34]. In the case of RC members, the post-shear cracks propagate through the preceding cracks when the shear can transfer from one crack's plane to another. Therefore, the system may easily reach the point of minimum potential energy. On the other hand, when the preceding crack's arrest mechanism occurs, the system is in a stable state, however, more energy is required to create independent shear cracks until forming a failure path. These are so-called cracks interactions, which are explained in detail in later sections.

On the other hand, the pseudo cracking method [26] is a numerical technique for fatigue life of RC decks based upon site-inspected cracks on the surfaces of RC decks. Internal unknown cracks are generated in the finite element model in the early cycles of fatigue loading by the corrector-predictor approach based on energy principles in order to satisfy dynamic equilibrium and deformational compatibility [26].

The multi-scale simulation program can successfully capture the response of a multi-cracked element under multi-stress conditions [18,23], where the global response is an assembly of local crack responses of all the cracks inside the element. The loading conditions of each crack must satisfy the deformational compatibility and equilibrium in its local direction. Through this process, some cracks are activated, while others are idle, resulting in a mechanism called anisotropic crack interaction. The activated cracks dominate the overall behavior, while other cracks are dormant. The activation of pre-cracks reduces the stress concentration in the diagonal direction, where it may affect the response of new cracks. Thus, crack interactions (activation or dormancy) may govern the global structural behavior of RC members. It should be noted that the kinematics of cracks of concrete including opening/closing, slip, and anisotropic interactions are already integrated into the multi-scale simulation program [18,23].

Figure 3 shows examples of the validation of data assimilation technology. Figure 3a shows an actual RC deck from a site with current damage [26] that was tested experimentally to check its remaining fatigue life. Next, it was analyzed by the data assimilation technology for validation. It is clear that the simulation results are in good agreement with the experimental ones. Figure 3b shows another example of validation for an experimental specimen [25], where the data assimilation technology succeeded to predict its remaining fatigue life at different time intervals of damage.

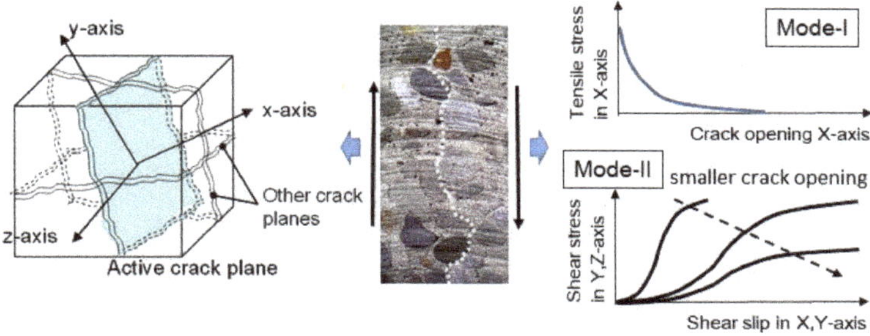

Figure 2. Tension and shear stress transfer across crack planes of concrete.

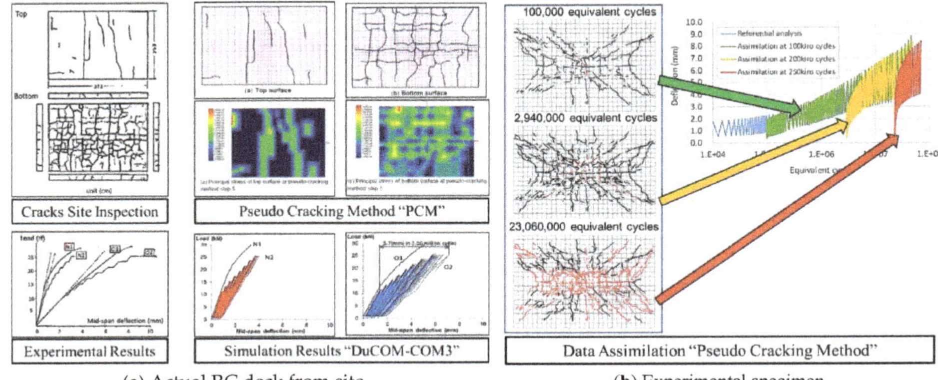

(**a**) Actual RC deck from site.   (**b**) Experimental specimen.

Figure 3. Validations of data assimilation technology.

2.2. Studied RC Deck

The RC decks of road steel-girder bridges are targeted in this study. Such decks are vertically supported by main steel girders in the longitudinal direction (traffic direction), as shown in Figure 4. These main girders are connected laterally by transverse girders to secure their lateral stability. Generally, RC decks are subjected to a heavy regimen of repeated traffic loads as the result of daily use, leading to deterioration, as stated in the introduction section.

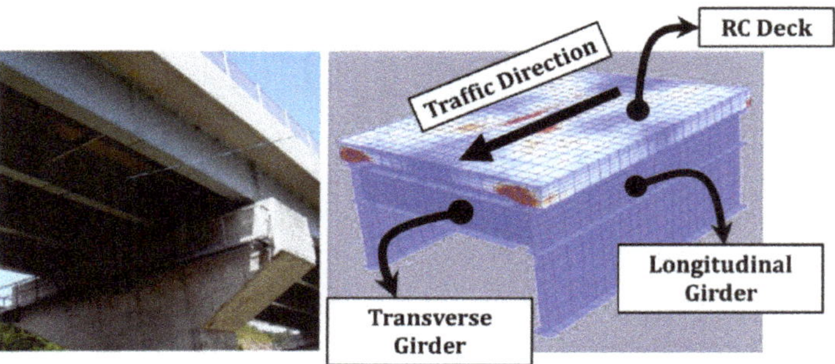

Figure 4. RC deck of road steel-girder bridges.

2.2.1. Dimensions and Reinforcement

Here, let us focus on fatigue loading in the dry state [20,21]. The targeted deck slab is a thin RC plate of a type that was typically built in the past and is still in service. Figure 5 shows the dimensions and the arrangement of reinforcing bars. Generally, bridge decks follow the conceptual design of one-way slabs supported by longitudinal girders, while the length of RC decks depends on several aspects such as the type of bridge. In a previous study [20], the authors chose the length of 6.0 m as the optimum length for fulfilling engineering practices at sites and effective analysis of RC decks, and thus, this length was also selected in this research. This referential target is basically the same as the ones selected in the referenced studies [20–22,27] for the purpose of knowledge integration.

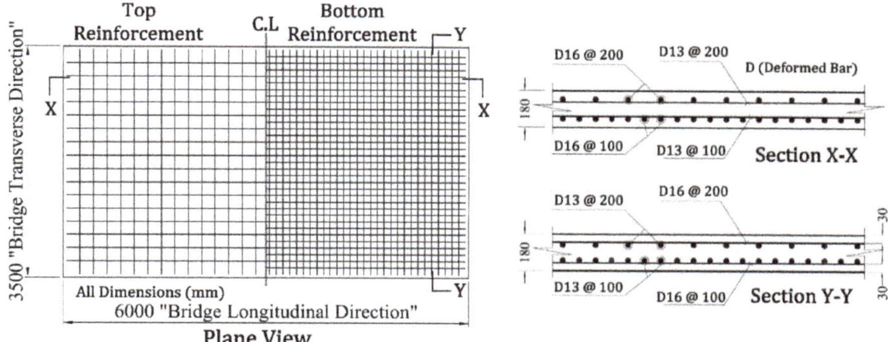

Plane View

Figure 5. Dimensions and reinforcement arrangement of the studied slab.

2.2.2. Material Properties

The material properties of the concrete and steel of the referential RC slab are listed in Table 1 [20,21]. These values are followed by general designs used in past construction of highway bridge decks.

Table 1. Material properties of the referential slab.

Material Type		Concrete	Steel Reinforcement
Young's Modulus	N/mm^2	24,750	205,000
Compressive Strength	N/mm^2	30	295
Tensile Strength	N/mm^2	2.2	295
Specific Weight	kN/m^3	24	78

2.2.3. Referential Loads & Boundary Conditions

On the basis of Japan's Specifications for Highway Bridges-Part III [35], we set up the deck to be subjected to a traveling wheel-type load of 98 kN, as shown in Figure 6. The speed of the running wheel for simulation was specified to be 60 km/h, which is the legal speed limit for national routes. The wheel load length and width used in the simulation were 500 mm and 250 mm, respectively, in reference to the width of vehicle tires, as shown in Figure 6. Hinged supports that allow free rotation were chosen, but the translation movement was restrained with reference to real conditions. Compared to real bridge conditions, this boundary may lead to a somewhat shorter fatigue life to be on the safe side. These are the same dimensions as those adopted by the past laboratory experiments against which the present fatigue simulation is examined and validated.

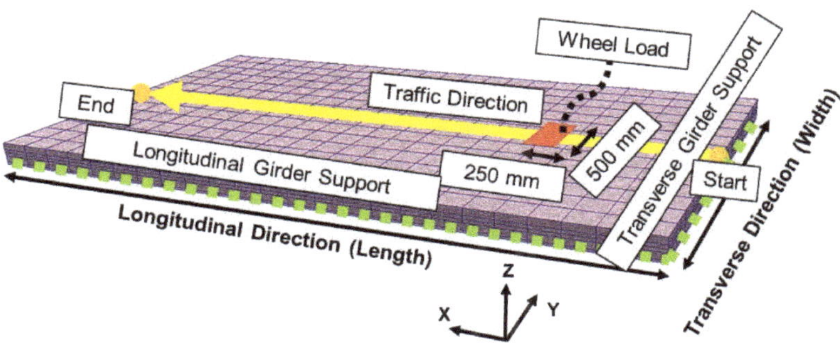

Figure 6. Wheel traveling load on the studied deck.

2.2.4. Failure Criteria

Based on the past experiments [25], the fatigue limit state was specified by the central live load deflection [36]. When the live load deflection defined by Equation (1) reaches the limit state deflection, which corresponds to no bond between the concrete and main reinforcement, it is judged that the slab comes up to the fatigue failure. It has been confirmed that the specific deflection associated with no bond is on the order of 2.5 times to 3.5 times its initial value [37], and the authors chose to apply the same criterion, denoted by Equation (2), as in past studies [20,21].

In fact, the fatigue life of RC decks depends on several aspects such as dimensions, material properties, reinforcement ratio, and load values. The limit state criterion was found to be rational as the properties of decks are included explicitly in the mean values of live load deflection. In almost all cases, out-of-plane shear failure of concrete occurs before the fatigue rupture of reinforcing bars [38]. In fact, no failure of reinforcement has been reported in past laboratory tests nor has been observed in real bridges. This can be attributed to the weakness of thinner concrete slabs rather than the fatigue strength of reinforcement. Thus, the limit state criterion is expected to be practical for the development and implementation of effective maintenance plans.

$$\delta_{L,N} = \delta_{1,N} - \delta_{2,N}, \tag{1}$$

$$\delta_{L,N}/\delta_{L,0} \geq 3.0, \tag{2}$$

where $\delta_{L,N}$ is the central live load deflection at N cycles, $\delta_{1,N}$ is the central total deflection at N cycles at the loading stage, $\delta_{2,N}$ is the central total deflection at N cycles at the unloading stage, $\delta_{L,0}$ is the initial live load deflection, and N_f is the number of cycles corresponding to $\delta_{L,N}$ (Equation (2)).

2.2.5. Fatigue Life of Non-Damaged Deck

Figure 7 shows the relation of the wheel load cycles and the total central deflection for the referential deck (initially un-cracked), which is obtained by the multi-scale simulation program [20,21]. According to the failure criteria mentioned in Section 2.2.4, the fatigue life of this model is estimated at 221 million cycles. The live load deflection at the first cycle (A) is 1.4 mm, while that at failure (B) is 4.3 mm. Finally, the total deflection of 7.8 mm was specified as the value for failure for all the studied crack patterns.

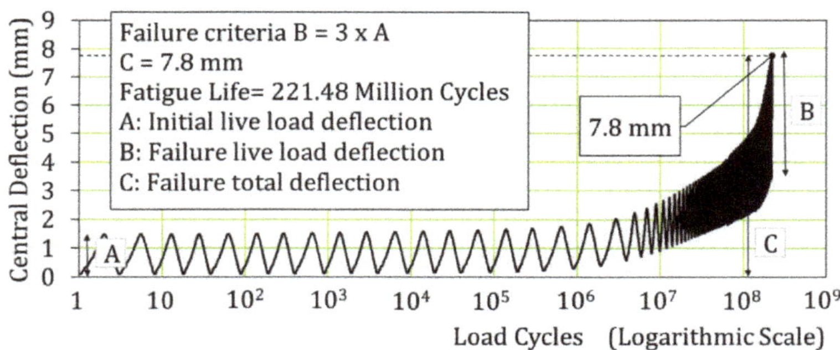

Figure 7. Load cycles versus central total deflection for the referential RC deck.

2.3. Randomized Artificial Crack Patterns (RACP)

To complement missing crack patterns over real decks, randomized artificial crack patterns are utilized. Figure 8 shows the basic RACP procedure concept. Here, the random variable is the number of elements (1–336). To avoid the overlap of case studies and reduce the effect of crack locations as much as possible, the randomization range of elements was limited to the first quarter of the deck.

The other quarters were provided with the same cracks obtained from the first quarter by using the laws of symmetry in both the longitudinal and transverse directions, as shown in Figure 9.

The crack angles were arranged in sequential order (each crack pattern has a single crack orientation) to cover all possible ranges from 0 degrees to 180 degrees with the step-interval of 15 degrees. The cracks with angles close to 0 degrees, 90 degrees, 180 degrees most extend the remaining fatigue life. Therefore, the smaller step interval of two degrees was further applied for the cases near 0 degrees, 90 degrees, and 180 degrees. The RACP produces continuous cracks, and three shapes of continuity are produced: x-direction, y-direction, or diagonal, on the basis of the sequential angle.

Both the crack width (0.3 mm) and crack density (average strain = 0.05%) were kept fixed in the RACP scheme to allow independent study of the effect of crack orientation, with the average strain parameter calculated by Equation (3). Although the RACP is based on randomization of elements, it can avoid repetition of cracking patterns. If a crack pattern is repeated, the RACP will continue randomization until a unique one is obtained.

It should be noted that the induced pre-crack is set to reach the upper mesh layer whose size is almost the same as that of the neutral axis of the studied RC deck [26].

$$\varepsilon_{avg.}(A.S) = \frac{\sum_{k=1}^{k=n}\left(\varepsilon_{xx} + \varepsilon_{yy}\right)}{n}, \tag{3}$$

where $\varepsilon_{avg.}$ is the average strain on the bottom surface of the RC deck, ε_{xx} is the concrete's normal strain in x-direction for the kth element, ε_{yy} is the one in y-direction, and n is the total number of elements (336 in this study).

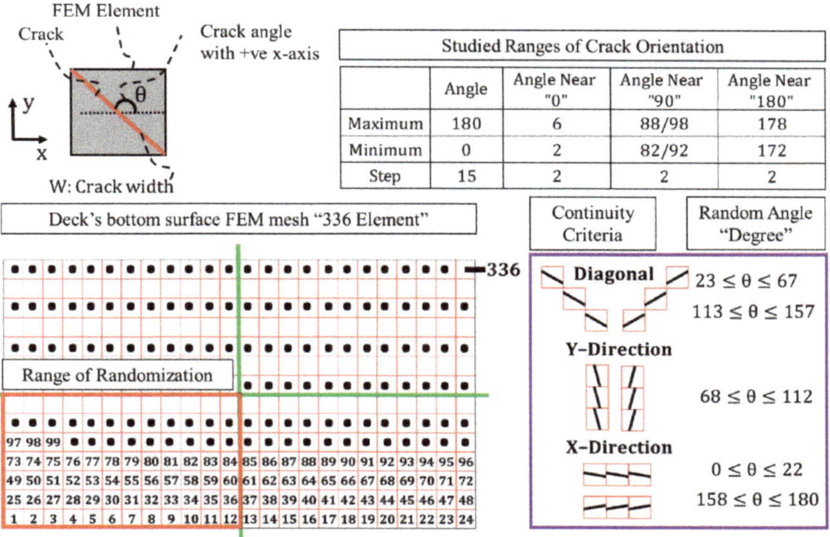

Figure 8. Randomized artificial crack pattern (RACP) scheme.

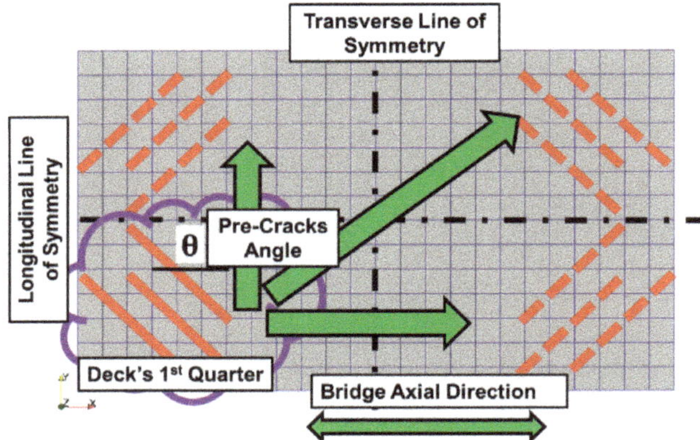

Figure 9. Utilization of laws of symmetry in RACP scheme.

3. Crack Patterns & Simulation Results

By utilizing the RACP, 148 crack patterns with a wide range of crack orientations (θ) were artificially produced. Figure 10 shows an example of the studied dataset. These crack patterns were analyzed for remaining fatigue life by the integrated multi-scale simulation program [23,24] and the pseudo-cracking method [25,26]. In spite of the fixed crack density, a wide range of fatigue life from 0.45 to 2.0 normalized by the non-cracked reference was obtained, as shown in Figure 11. The crack orientation was confirmed to be substantial.

Figure 12 shows the relation of the crack orientations and the remaining fatigue life of the crack patterns. It is clear that the cracks that are nearly parallel to the longitudinal or transverse direction (close to 0 degrees, 90 degrees, 180 degrees) most extend the remaining fatigue life, even beyond the undamaged sound condition. Here, the pre-cracks arrest the propagation of the post-shear cracks [20,21]. For other crack orientations, however, the remaining fatigue life remains almost unchanged and the effect of the crack orientations on the remaining fatigue life is moderate.

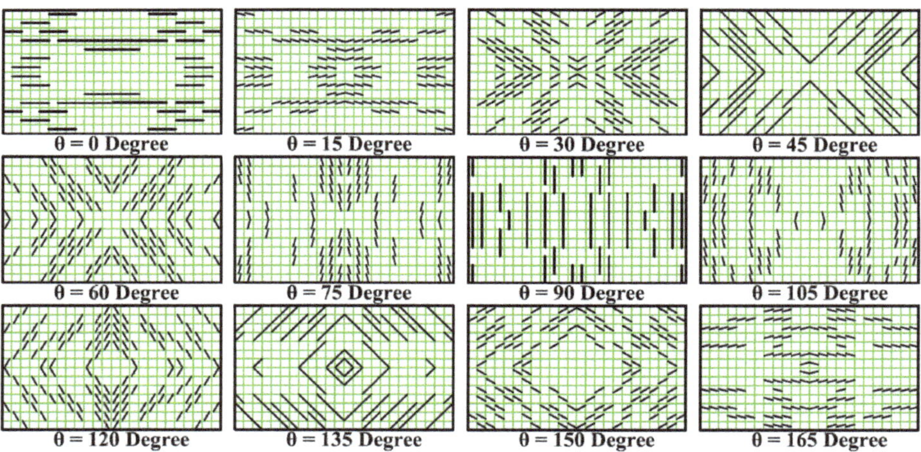

Figure 10. Example of the crack patterns for the studied dataset.

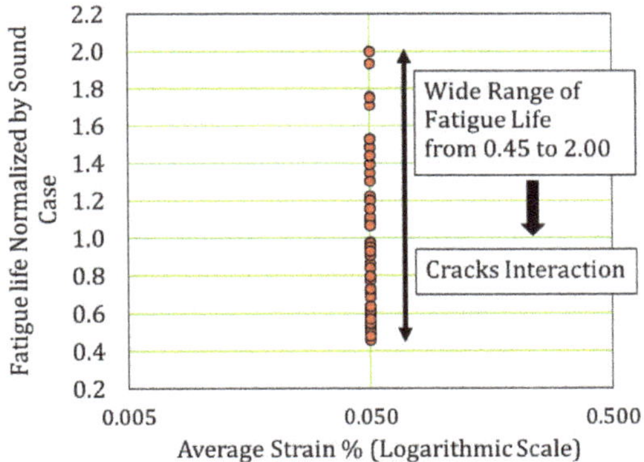

Figure 11. Relation of the average strain and the normalized fatigue life of the studied crack patterns.

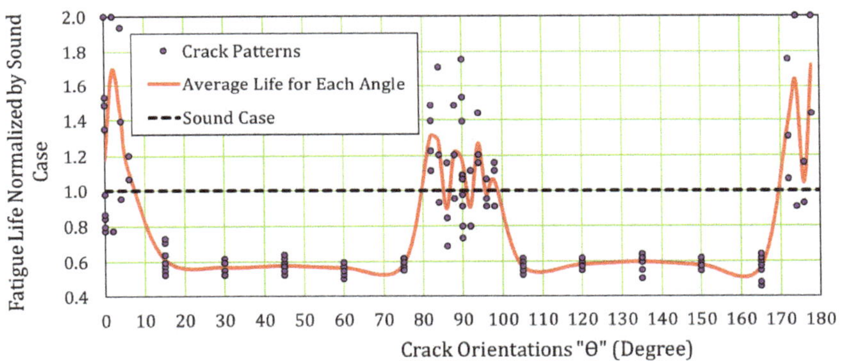

Figure 12. Relation of the crack orientations and the remaining fatigue life.

4. Discussions

4.1. Coupled Flexure-Shear Mode–Crack Arrest Mechanism

The crack arrest mechanism under shear was previously investigated for the case of RC beams [18] in which pre-cracks were introduced by flexural actions. Afterwards, the shear forces were applied to examine the propagation of diagonal shear cracks. It was reported that pre-cracks stop the propagation of post-shear diagonal cracks, leading to upgraded static capacity even greater than the sound condition of no pre-cracks. Once the stop mechanism arises, shear cracks are arrested by the preceding cracks and more energy is required to merge the post-shear cracks together with the preceding ones.

This phenomenon is highly dependent on the crack width, as shown in Figure 13. If pre-cracks hardly open, the diagonal shear crack can easily propagate across the pre-cracks. In this case, the shear stress is fairly transferred along crack planes by the aggregate interlock [18,20]. On the other hand, if the crack width of pre-cracks is larger, the local shear stress cannot be successfully transferred along pre-cracks' planes. This reduced shear stiffness along cracks causes stress release, which may not produce other diagonal cracks. Then, the crack arrest mechanism causes an elevated static capacity of RC members in shear. On that basis, RC decks also exhibit the crack arrest mechanism, and enhanced fatigue life can be realized as shown in Figure 12 [20,21].

The crack arrest mechanism in RC members is similar to the drilled stop-holes in steel members. It was reported that stop-holes stimulate the retardation of crack growth, where the crack tip is transferred to notch and the stresses at the crack tip reduce significantly, followed by extended fatigue life [28–31]. Based on this analogy (see Figure 14), it can be said that the pre-cracks in concrete are like the stop-holes drilled through steel members.

It is clearly demonstrated in the previous section that the cracks that approximate the longitudinal and transverse directions of the deck most extend fatigue life. In consideration of the diagonal shear crack inside RC decks, the crack arrest mechanism is highly dependent on the crossing angle (ϕ) of the pre-cracks and the post-shear cracks (see Figure 15). The crack arrest mechanism may be active when less shear stress is transferred across crack planes. In fact, it was demonstrated that the crack arrest mechanism comes up when pre-cracks and post-cracks intersect non-orthogonally [18]. Figure 16 shows the horizontal projections of pre-cracks and post-shear cracks in the x-y plane of the RC deck, where the post-shear cracks should mechanically be nearly diagonal in the horizontal projection.

If the pre-cracks are nearly diagonal, the intersection of the post-shear cracks is either orthogonal or nearly parallel. Therefore, the crack arrest mechanism hardly acts. In the case of the parallel intersection, the pre- and post-cracks may divert from each other without interaction. However, if the pre-cracks approximate the transverse or longitudinal direction of the RC deck, the non-orthogonality of the cracks' intersection is satisfied, and the crack arrest mechanism works, leading to extended fatigue life [18].

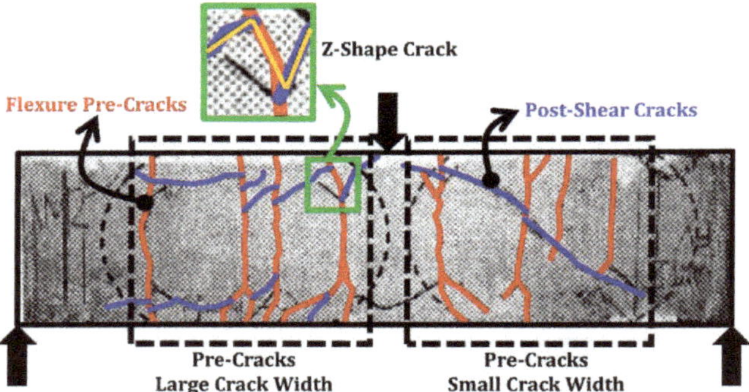

Figure 13. Pre-cracks stopping mechanism for RC beams.

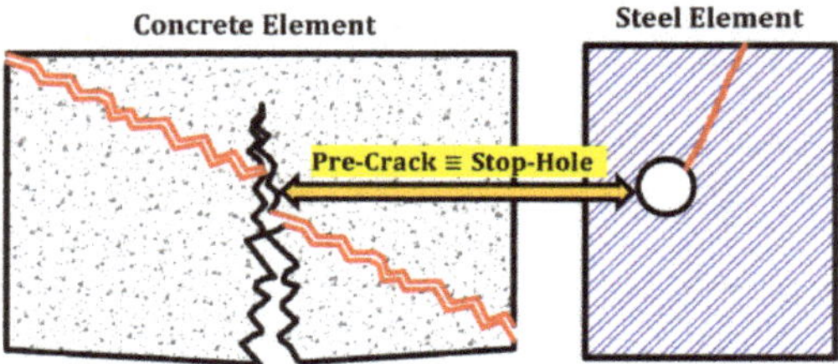

Figure 14. Stopping mechanism of crack growth in concrete and steel elements.

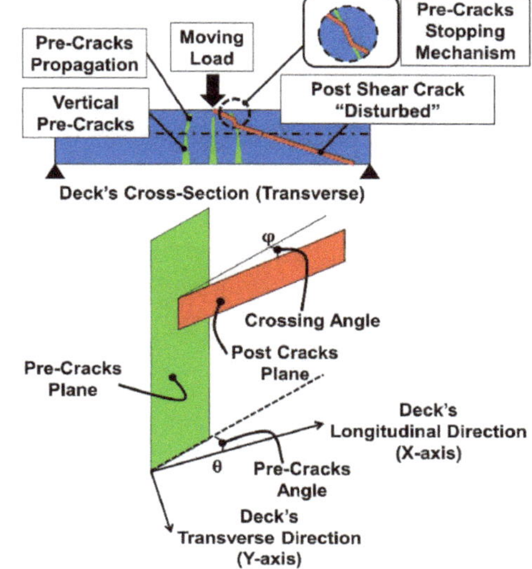

Figure 15. Interaction of pre-cracks and post-shear cracks.

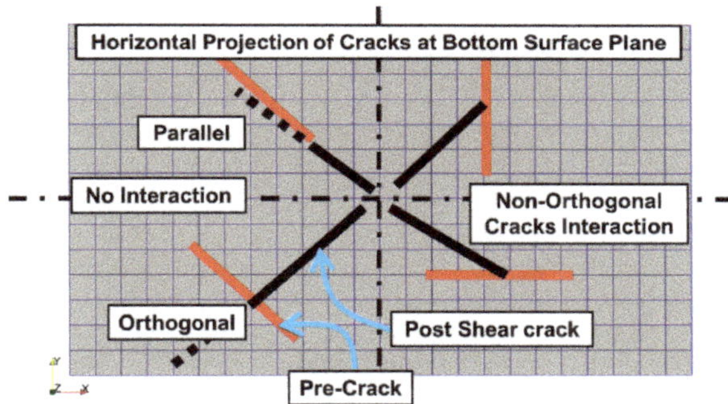

Figure 16. Horizontal projection of pre- and post-shear cracks at bottom surface of RC deck.

4.2. Effect of Cracks Location on Coupled Flexure-Shear Mode

Studied in this section is the effect of cracks location on the remaining fatigue life associated with cracks parallel to the transverse direction of the deck ($\theta = 90°$, see Figure 17). Here, we have a wide range of fatigue lives from 0.73 to 1.75 despite the same global crack density. It is found that there is a clear correlation between the average strain of the cracks in the central zone of the loading path (A.S*, see Figure 18) and the remaining fatigue life, as shown in Figure 19. When cracks are not present in the central zone (A.S* = 0), the crack arrest mechanism does not arise to enhance fatigue life.

On the other hand, by the increase in A.S*, the remaining fatigue life is extended, and the crack arrest mechanism becomes more effective. If the cracks pre-exist in the central zone, arrested cracks become dormant since the diagonal shear cracks start from the loading point, as shown in Figure 20.

When cracks exist outside the central zone of the wheel-loading path, the crack arrest mechanism is put off and the RC deck starts losing global stiffness prior to the limit for fatigue failure. For further investigation of this phenomenon, two crack patterns were focused on to obtain their static capacity: crack pattern (6) with (A.S*) equal to 0%, and crack pattern (11) with maximum (A.S*) equal to 0.084%. Figure 21 shows the relation of the deck's central deflection and the static load, where crack pattern

(11) has higher static capacity than that of the crack pattern (6) with about 14% despite their similar initial stiffness. The simulation results demonstrate that the crack arrest mechanism is effective when pre-cracks exist in the central zone.

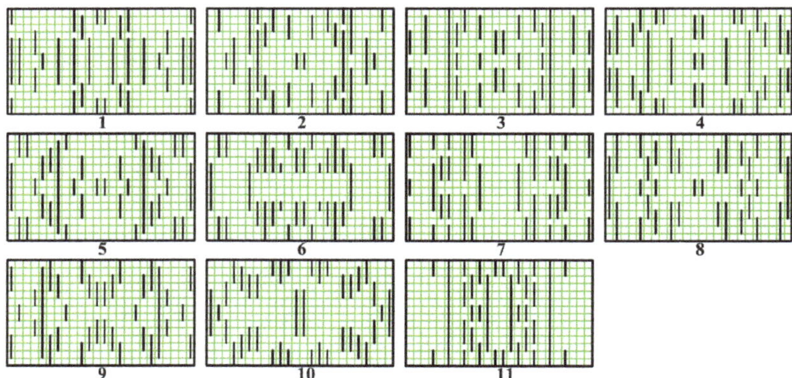

Figure 17. Artificial crack patterns with θ (90°).

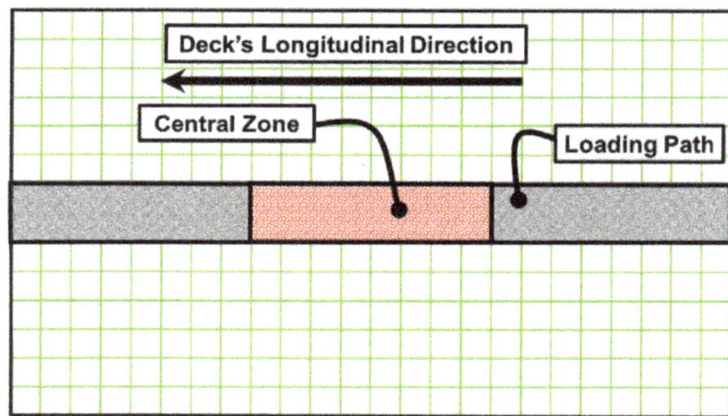

Figure 18. Finite element mesh of the bottom surface of the deck.

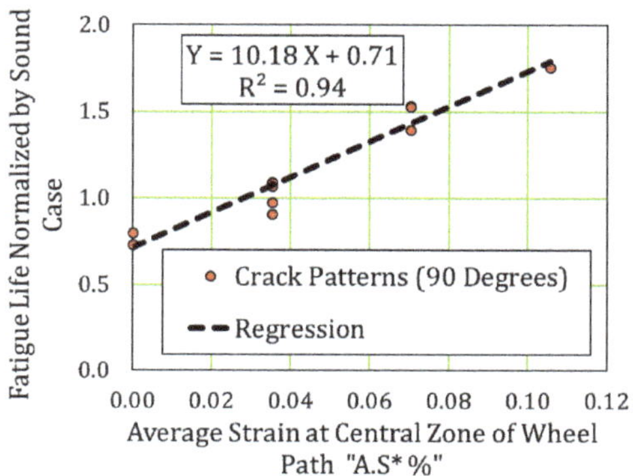

Figure 19. Relation of the average strain of the central zone of the loading path and the remaining fatigue life.

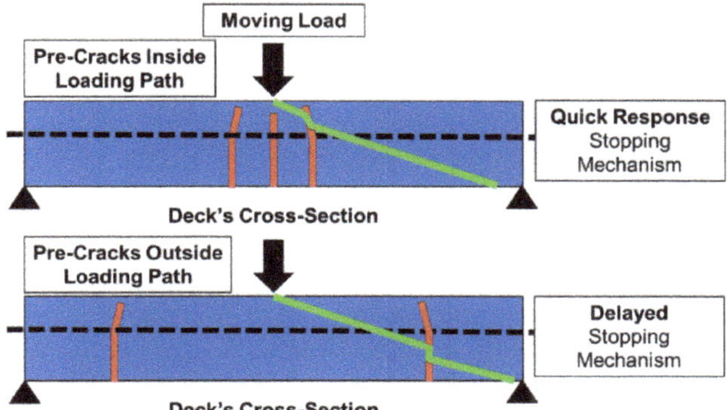

Figure 20. Pre-cracks stopping mechanism inside and outside loading path.

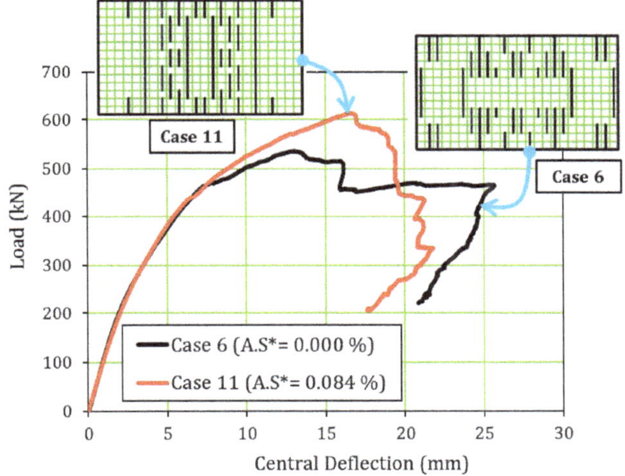

Figure 21. Central deflection of the RC deck and the static load for crack patterns (6 and 11).

4.3. High Risk Location of Cracking

In the previous research [21], a hazard map was achieved for the high-risk crack locations, i.e., the central zone and the corners of the RC deck, as shown in Figure 22, and highly coupled flexure and shear action was not explicitly included for safer and conservative assessment of remaining fatigue life. Thus, taking into consideration the uncertainty of cracks' depth on site, the depth of the cracks was previously intended to be shallow (non-penetrated cracks).

In this study, we targeted RC decks with penetrated cracks that are close to their top fibers in order to check different scenarios that may occur on site. In Section 4.2, we demonstrated that the central zone of the loading path is the location where the crack arrest mechanism is activated, and the fatigue life is elevated. This is beneficial in view of sustainable maintenance. Although the central zone of RC decks shall be designated as a higher-risk cracking region [21], the pre-crack arrest mechanism can be expected to lead to enhanced fatigue life.

On the other hand, the corners are also considered as higher-risk cracking locations, and the crack arrest mechanism hardly works there, as shown in Figure 22. Thus, cracks in the corners of RC decks are riskier than cracks in the central zone, where their depth is close to the top fibers of the RC deck. By combining the findings of the previous research [21] and those of this study, the upshot is that more attention should be given to cracks at the corners than in the central zone of the RC deck during a bridge inspection.

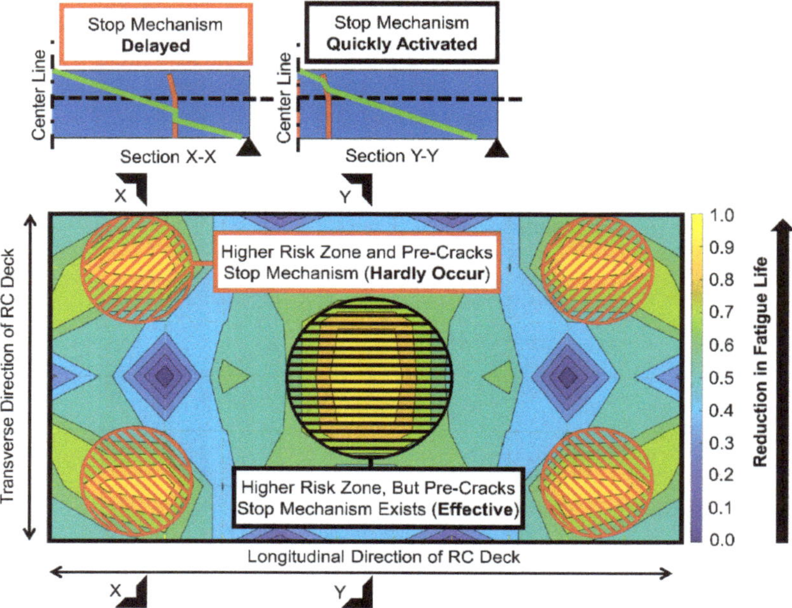

Figure 22. Hazard map for the locations of higher risk cracking.

5. Conclusions

The effect of crack orientation on the fatigue life of RC decks is investigated by using the multi-scale simulation and the pseudo-cracking method with respect to the most sensitive crack direction. On the basis of large numbers of reviewed crack patterns, the following conclusions are drawn.

i.   The crack orientation is highly associated with the coupled flexure-shear mode of failure, where pre-cracks tend to arrest the ensuing propagation of post-shear cracks. As a result, enhanced fatigue life is obtained.

ii.   The pre-crack arrest mechanism enables bridge maintenance managers to conduct a fair assessment of RC decks that present a rough appearance but have mechanically enhanced fatigue life.

iii.   The cracks that approximate the deck's longitudinal or transverse direction are understood to most extend the remaining fatigue life of RC decks.

iv.   The central zone of the loading path is found to be the location where the crack arrest mechanism is effective, and the remaining fatigue life is elevated.

v.   During inspections, careful attention shall be paid to cracks at the corners of RC decks since the stopping mechanism that arrests diagonal cracking does not effectively function there.

Author Contributions: E.F. conducted and analyzed the numerical studied cases; Y.T. and K.M. supervised the analytical process and developed the simulation program; E.F. and K.M. wrote the paper.

Funding: This study was financially supported by the Council for Science, Technology, and Innovation, "Cross-ministerial Strategic Innovation Promotion Program (SIP), Infrastructure Maintenance, Renovation, and Management" granted by the Japan Science and Technology Agency (JST).

Acknowledgments: The authors extend their appreciation to Yozo Fujino of Yokohama National University and The University of Tokyo, for his valuable advice and encouragement to bridge multiple disciplines in engineering.

Conflicts of Interest: The authors declare no conflict of interest.

References

1. Val, D.V.; Stewart, M.G.; Melchers, R.E. Effect of reinforcement corrosion on reliability of highway bridges. *Eng. Struct.* **1998**, *20*, 1010–1019. [CrossRef]
2. Vassie, P. Reinforcement corrosion and the durability of concrete bridges. In *Proceedings of Institution of Civil Engineers*; Institution of Civil Engineers: London, UK, 1984; Volume 76, pp. 713–723.
3. Cady, P.D.; Weyers, R.E. Chloride penetration and the deterioration of concrete bridge decks. *Cem. Concr. Aggreg.* **1983**, *5*, 81–87.
4. Freyermuth, C.L.; Klieger, P.; Stark, D.C.; Wenke, H.N. Durability of concrete bridge decks—A review of cooperative studies. *Highw. Res. Rec.* **1970**, *328*, 50–60.
5. Lindquist, W.D.; Darwin, D.; Browning, J.; Miller, G.G. Effect of Cracking on Chloride Content in Concrete Bridge Decks. *Aci Mater. J.* **2006**, *103*, 467–473.
6. Shah, S.P.; Weiss, W.J.; Yang, W. Shrinkage Cracking—Can It Be Prevented? *Concr. Int.* **1998**, *20*, 51–55.
7. Sakulich, A.R.; Bentz, D.P. Increasing the service life of bridge decks by incorporating phase-change materials to reduce freeze-thaw cycles. *J. Mater. Civ. Eng.* **2011**, *24*, 1034–1042. [CrossRef]
8. Detwiler, R.J.; Kojundic, T.; Fidjestol, P. Evaluation of bridge deck overlays. *Concr. Int. Detroit* **1997**, *19*, 43–46.
9. Ulm, F.J.; Coussy, O.; Kefei, L.; Larive, C. Thermo-chemo-mechanics of ASR expansion in concrete structures. *J. Eng. Mech.* **2000**, *126*, 233–242. [CrossRef]
10. Schläfli, M.; Brühwiler, E. Fatigue of existing reinforced concrete bridge deck slabs. *Eng. Struct.* **1998**, *20*, 991–998. [CrossRef]
11. Lee, M.K.; Barr, B.I.G. An overview of the fatigue behaviour of plain and fibre reinforced concrete. *Cem. Concr. Compos.* **2004**, *26*, 299–305. [CrossRef]
12. Sparks, P.R.; Menzies, J.B. The effect of rate of loading upon the static and fatigue strengths of plain concrete in compression. *Mag. Concr. Res.* **1973**, *25*, 73–80. [CrossRef]
13. Cornelissen, H.A.W.; Reinhardt, H.W. Uniaxial tensile fatigue failure of concrete under constant-amplitude and program loading. *Mag. Concr. Res.* **1984**, *36*, 216–226. [CrossRef]
14. Tepfers, R.; Kutti, T. Fatigue Strength of Plain, Ordinary, and Lightweight Concrete. *ACI J.* **1979**, *76*, 635–652.
15. Kolluru, S.V.; O'Neil, E.F.; Popovics, J.S.; Shah, S.P. Crack propagation in flexural fatigue of concrete. *J. Eng. Mech.* **2000**, *126*, 891–898. [CrossRef]
16. Slowik, V.; Plizzari, G.A.; Saouma, V.E. Fracture of concrete under variable amplitude fatigue loading. *ACI Mater. J.* **1996**, *93*, 272–283.
17. NEXCO-Japan. Regarding Large-Scale Renewal and Large-Scale Repair on the Expressway Managed by Eastern. Central and West Japan Expressway Co., Ltd. 2014. Available online: https://www.c-nexco.co.jp/koushin/pdf/about.pdf (accessed on 17 May 2018).
18. Pimanmas, A.; Maekawa, K. Shear failure of RC members subjected to pre-cracks and combined axial tension and shear. *J. Mater. Concr. Struct. Pavements JSCE* **2001**, *53*, 159–174. [CrossRef]
19. Tanaka, Y.; Kishi, T.; Maekawa, K. Experimental research on the structural mechanism of RC members containing artificial crack in shear. *JSCE* **2005**, *802*, 109–121.
20. Fathalla, E.; Tanaka, Y.; Maekawa, K. Remaining fatigue life assessment of in-service road bridge decks based upon artificial neural networks. *Eng. Struct.* **2018**, *171*, 602–616. [CrossRef]
21. Fathalla, E.; Tanaka, Y.; Maekawa, K.; Sakurai, A. Quantitative deterioration assessment of road bridge decks based on site inspected cracks. *Appl. Sci.* **2018**, *8*, 1197. [CrossRef]
22. Fathalla, E.; Tanaka, Y.; Maekawa, K.; Sakurai, A. Quantitative deterioration assessment of road bridge decks based on site inspected cracks under stagnant water. *J. Adv. Concr. Technol.* **2019**, *17*, 16–33. [CrossRef]
23. Fathalla, E.; Tanaka, Y.; Maekawa, K. Fatigue life of RC bridge decks affected by non-uniformly dispersed stagnant water. *Appl. Sci.* **2019**, *9*, 607. [CrossRef]
24. Maekawa, K.; Ishida, T.; Chijiwa, N.; Fujiyama, C. Multiscale coupled-hygromechanistic approach to the life-cycle performance assessment of structural concrete. *J. Mater. Civ. Eng. Am. Soc. Civ. Eng.* **2015**, *27*, A4014003. [CrossRef]

25. Maekawa, K.; Toongoenthong, K.; Gebreyouhannes, E.; Kishi, T. Direct path-integral scheme for fatigue simulation of reinforced concrete in shear. *J. Adv. Concr. Technol.* **2006**, *4*, 159–177. [CrossRef]
26. Tanaka, Y.; Maekawa, K.; Maeshima, T.; Iwaki, I.; Nishida, T.; Shiotani, T. Data assimilation for fatigue life assessment of RC bridge decks coupled with hygro-mechanistic model and nondestructive inspection. *J. Disaster Res.* **2017**, *12*, 422–431. [CrossRef]
27. Fujiyama, C.; Tang, X.J.; Maekawa, K.; An, X. Pseudo-cracking approach to fatigue life assessment of RC bridge decks in service. *J. Adv. Concr. Technol.* **2013**, *11*, 7–21. [CrossRef]
28. Ayatollahi, M.R.; Razavi, S.M.J.; Chamani, H.R. Fatigue Life Extension by Crack Repair Using Stop-hole Technique under Pure Mode-I and Pure mode-II Loading Conditions. *Procedia Eng.* **2014**, *74*, 18–21. [CrossRef]
29. Song, P.S.; Shieh, Y.L. Stop drilling procedure for fatigue life improvement. *Int. J. Fatigue* **2004**, *26*, 1333–1339. [CrossRef]
30. Mikkelsen, O.; Rege, K.; Hemmingsen, T.; Pavlou, D.G. Numerical estimation of the stop holes-induced fatigue crack growth retardation in offshore structures taking into account the corrosion effect. In Proceedings of the 27th International Ocean and Polar Engineering Conference, International Society of Offshore and Polar Engineers, San Francisco, CA, USA, 25–30 June 2017.
31. Razavi, S.M.J.; Ayatollahi, M.R.; Sommitsch, C.; Moser, C. Retardation of fatigue crack growth in high strength steel S690 using a modified stop-hole technique. *Eng. Fract. Mech.* **2017**, *169*, 226–237. [CrossRef]
32. Pavlou, D.G.; Labeas, G.N.; Vlachakis, N.V.; Pavlou, F.G. Fatigue crack propagation trajectories under mixed-mode cyclic loading. *Eng. Struct.* **2003**, *25*, 869–875. [CrossRef]
33. Bažant, Z.P.; Estenssoro, L.F. Surface singularity and crack propagation. *Int. J. Solids Struct.* **1979**, *15*, 405–426. [CrossRef]
34. Irwin, G.R. *Fracture Mechanics, in the Book of Structural Mechanics*; Pergamon Press: New York, NY, USA, 1958; pp. 557–594.
35. Japan Road Association. *Specifications for Highway Bridges—Part III Concrete Bridges*; Japan Road Association: Tokyo, Japan, 2012.
36. Matsui, S. Lifetime prediction of bridge. *J. Jpn. Soc. Civ. Eng.* **1996**, *30*, 432–440.
37. Maeshima, T.; Koda, Y.; Tsuchiya, S.; Iwaki, I. Influence of corrosion of rebars caused by chloride induced deterioration on fatigue resistance in RC road deck. *J. Jpn. Soc. Civ. Eng.* **2014**, *70*, 208–225.
38. Matsui, S.; Maeda, Y. A rational valuation method for deterioration of highway bridge decks. *J. Jpn. Soc. Civ. Eng.* **1986**, *374*, 419–426.

applied sciences

Article

Fatigue Assessment of Prestressed Concrete Slab-Between-Girder Bridges

Eva O.L. Lantsoght [1,2,*], Rutger Koekkoek [3], Cor van der Veen [2] and Henk Sliedrecht [4]

[1]   Politécnico, Universidad San Francisco de Quito, Quito 17015EC, Ecuador
[2]   Concrete Structures, Delft University of Technology, 2628 CN Delft, The Netherlands; C.vanderveen@tudelft.nl
[3]   BAM Infraconsult, 2801 SC Gouda, The Netherlands; rutger.koekkoek@bam.com
[4]   Rijkswaterstaat, Ministry of Infrastructure and the Environment, 3526 LA Utrecht, The Netherlands; henk.sliedrecht@rws.nl
*   Correspondence: elantsoght@usfq.edu.ec or e.o.l.lantsoght@tudelft.nl; Tel.: +593-2-297-1700 (ext. 1186)

Received: 25 April 2019; Accepted: 1 June 2019; Published: 5 June 2019

Featured Application: The results of this work can be used in the evaluation of existing prestressed concrete slab-between-girder bridges for fatigue.

Abstract: In the Netherlands, the assessment of existing prestressed concrete slab-between-girder bridges has revealed that the thin, transversely prestressed slabs may be critical for static and fatigue punching when evaluated using the recently introduced Eurocodes. On the other hand, compressive membrane action increases the capacity of these slabs, and it changes the failure mode from bending to punching shear. To improve the assessment of the existing prestressed slab-between-girder bridges in the Netherlands, two 1:2 scale models of an existing bridge, i.e., the Van Brienenoord Bridge, were built in the laboratory and tested monotonically, as well as under cycles of loading. The result of these experiments revealed: (1) the static strength of the decks, which showed that compressive membrane action significantly enhanced the punching capacity, and (2) the Wöhler curve of the decks, showed that the compressive membrane action remains under fatigue loading. The experimental results could then be used in the assessment of the most critical existing slab-between-girder bridges. The outcome was that the bridge had sufficient punching capacity for static and fatigue loads and, therefore, the existing slab-between-girder bridges in the Netherlands fulfilled the code requirements for static and fatigue punching.

Keywords: assessment; bridge evaluation; compressive membrane action; concrete bridges; fatigue; fatigue assessment; live loads; prestressed concrete; punching shear; scale model

1. Introduction

The majority of the bridges in the Dutch highway bridge stock were built in the decades following World War II, and this post-war period was an era of rapid and extensive expansion of the Dutch road network. These bridges were designed for the live loads of that era, which resulted in lower demands on the bridges compared to the recently introduced Eurocode live loads from the NEN-EN 1991-2:2003 standard [1]. In terms of capacity, the design capacities for shear and punching in the previously used Dutch codes (e.g., VBC 1995-NEN 6723 [2]) were larger than the capacities determined using the recently introduced Eurocode for concrete structures NEN-EN 1992-1-1:2005 [3]. With these higher demands and lower capacities according to the Eurocodes, the outcome of a bridge assessment is often that existing bridges will not meet the code requirements for brittle failure modes, such as shear [4] and punching [5]. This problem is not limited to the Netherlands, as similar discussions are taking place in Germany [6], Sweden [7], Switzerland [8], and other European countries, as well as in the United States [9], where bridge construction peaked during the 1930s (the New Deal) and between 1956 and 1992 (i.e., the construction of the Interstate Highway System). As one can see, the methods used for an accurate assessment of existing bridges are becoming increasingly important, as the safety of the traveling public should be protected, and at the same time, unnecessary bridge replacements or strengthening actions should be avoided [10].

The preliminary assessment of existing bridges in the Netherlands according to the new Eurocodes was based on hand calculations (Quick Scans [11,12]), where the categories of bridge types that required further study were identified. One such category is the prestressed slab-between-girder bridges; this subset contains about 70 bridges [5]. The structural system of these bridges is a combination of prestressed girders with the deck slab cast in between the girders and is transversely prestressed. As a result, the top of the flange of the girders is flush with the top of the deck. Additionally, prestressed diaphragm beams provide stiffness to the overall system. Upon assessment, the thin deck slabs did not fulfil the code requirements for punching shear. One mechanism that is not considered in the codes, but that enhances the capacity of these thin decks, is the compressive membrane action [13–20]. Additionally, the fatigue capacity of the thin decks is still subject to discussion, as it remains unknown whether progressive cracking and damage accumulation affect the capacity-enhancing effect of the compressive membrane action [21].

This work summarized the experimental results from testing 1:2 scale models of prestressed slab-between-girder bridges, and then it applied these results to the punching and fatigue assessment of an existing bridge. We demonstrated how compressive membrane action improves the assessment for punching shear, and how the Wöhler curve from the fatigue tests could be used for the assessment of the bridge deck under fatigue. The summarized experiments are unique in nature, as the tested specimens give us insights into the behavior of slab-between-girder bridges as a structural system. Most fatigue testing in the past has focused on testing small specimens [22,23] or structural elements [24–31], instead of the structural systems. The insights from these experiments are reported in this study for the first time within the context of bridge assessment. This analysis showed that, based on the experimental evidence, we found that the existing slab-between-girder bridges in the Netherlands satisfied the safety requirements of the code, and in particular, the requirements for punching shear under static and fatigue live loading.

2. Materials and Methods

2.1. Description of Case Study Bridge

Of the 70 slab-between-girder bridges in the Netherlands, the bridge that has the most critical slab geometry (largest span to depth ratio of 3.6 m/0.2 m = 18) is the approach bridge of the Van Brienenoord Bridge in Rotterdam, see Figure 1a. The approach spans are 50 m in length and consist of thin, transversely post-tensioned decks that are cast between simply supported post-tensioned girders, see Figure 1b [13]. The clear span of the slab is 2100 mm. The transverse prestressing level is 2.5 MPa. The duct spacing on the deck is 650 mm at the center, and at some positions it is increased to 800 mm at the center. Table 1 gives the main properties of the geometry and reinforcement of the decks. Post-tensioned crossbeams are built at the end of the spans and post-tensioned diaphragm beams are provided at 1/3 and 2/3 of the span length.

At the time of construction, the design concrete compressive strength of the deck was B35 ($f_{ck,cube}$ = 35 MPa) and B45 ($f_{ck,cube}$ = 45 MPa) for the girders. Testing of the cores taken from the deck slab resulted in an average $f_{cm,cube}$ = 98.8 MPa ($f_{ck,cube}$ = 84.6 MPa), as a result of the continued cement hydration. For the assessment calculations, we conservatively assumed that the mean compressive cylinder strength f_{cm} = 65 MPa on the deck. The associated characteristic concrete compressive strength was f_{ck} = 53 MPa.

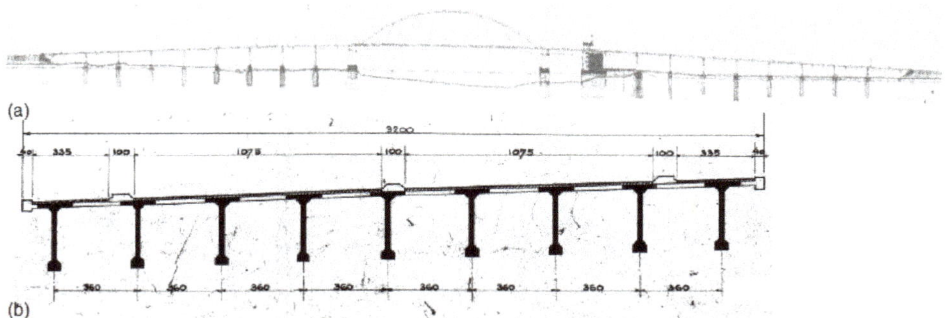

Figure 1. Van Brienenoord Bridge: (**a**) sketch of the elevation of the entire bridge structure, showing the approach slabs, as well as the steel arch; (**b**) cross-section of the slab-between-girder approach bridge. Dimensions in cm.

Table 1. Main properties of geometry and reinforcement of the decks of the Van Brienenoord Bridge.

Dimension	Value
Thickness h	200 mm
Concrete cover c	30 mm
Longitudinal reinforcement	ϕ8 mm–250 mm
Effective depth longitudinal d_l	166 mm
Area of longitudinal reinforcement $A_{s,l}$	201.1 mm^2/m
Longitudinal reinforcement ratio ρ_l	0.12%
Transverse reinforcement	ϕ8 mm–200 mm
Effective depth transverse d_t	158 mm
Area of transverse reinforcement $A_{s,t}$	251.3 mm^2/m
Transverse reinforcement ratio ρ_t	0.16%
Average effective depth d	162 mm
Average reinforcement ratio ρ_{avg}	0.14%
Prestressing reinforcement	462 mm^2–800 mm
Area of prestressing steel A_{sp}	0.5775 mm^2/mm

2.2. Live Load Models

We used two live load models for the assessment of the Van Brienenoord Bridge: Load model 1, for the assessment of the punching capacity, and fatigue load model 1, for the fatigue assessment, both adapted from the NEN-EN 1991-2:2003 [1].

Live load model 1 combines a distributed lane load with a design tandem. The design tandem has the following characteristics: (1) wheel print of 400 mm × 400 mm, (2) axle distance of 1.2 m, and (3) transverse spacing between wheels of 2 m. The magnitude of the axle load is α_{Q1} × 300 kN in the first lane, α_{Q2} × 200 kN in the second lane, and α_{Q3} × 300 kN in the third lane [12]. For the Netherlands, the values of all the α_{Qi} = 1, with i = 1 ... 3. The uniformly distributed load acts over the full width of the notional lane of 3 m width, and it equals α_{q1} × 9 kN/m^2 for the first lane, and α_{qi} × 2.5 kN/m^2 for all the other lanes. In the Netherlands, for bridges with three or more notional lanes, the value of α_{q1} = 1.15 and α_{qi} = 1.4, with i > 1. Figure 2 shows a sketch of the live load model 1.

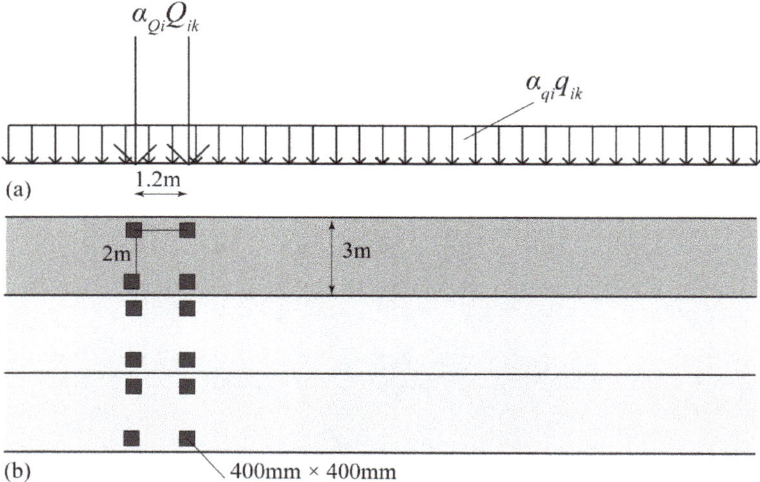

Figure 2. Live load model 1 from NEN-EN 1991-2:2003 [1]: (**a**) elevation; (**b**) top view. Edited from Reference [12], and reprinted with permission.

Fatigue load model 1 has the same configuration as load model 1, with $0.7Q_{ik}$ for the axle loads, and $0.3q_{ik}$ for the distributed lane loads. In other words, the axle load becomes 0.7×300 kN = 210 kN, and the load per wheel print becomes 105 kN. The distributed lane load is $0.3 \times 1.15 \times 9$ kN/m^2 = 3.105 kN/m^2. The fatigue load model has as a reference load of 2 million trucks per year. In the Netherlands, the guidelines for the assessment of bridges (RBK [32]) uses a higher number of passages: 2.5 million trucks per year. Over a lifespan of 100 years, the result is 250 million truck passages.

In the Netherlands, assessment is carried out using both a wheel print of 400 mm × 400 mm (as prescribed by the Eurocode 1 NEN-EN 1991-2:2003 [1]) and a wheel print of 230 mm × 300 mm (used for the fatigue evaluation of joints, but also often used as an additional check in assessments).

2.3. Description of Experiments

We built two 1:2 scale models of an existing bridge in the laboratory, which we tested monotonically, as well as under cycles of loading. Full descriptions of the first series of static tests [5,13], first series of fatigue tests [33–35], and second series of fatigue tests [36,37] can be found elsewhere. The description in this paper was limited to the information necessary to interpret the test results for application in assessment of the case study bridge.

The first 1:2 scale model (6.4 m × 12 m, see Figure 3) used four prestressed concrete T-girders with a center-to-center spacing of 1.8 m, length $l = 10.95$ m, and height $h = 1.3$ m; two post-tensioned crossbeams ($b = 350$ mm, $h = 810$ mm), and three transversely post-tensioned decks with $h = 100$ mm and $b = 1050$ mm between the girders. The choice of the size of the scale model and number of girders was determined as a function of the available test floor space in the laboratory. The post-tensioning of the deck was applied through prestressing bars placed in 30 ducts with a 40 mm diameter, and spaced 400 mm apart. To increase the number of experiments that could be carried out on this scale model, the middle deck was removed after testing and a new deck was cast. Therein, one segment of the new deck contained ducts of diameter 30 mm that were spaced 300 mm apart to study the influence of the duct spacing.

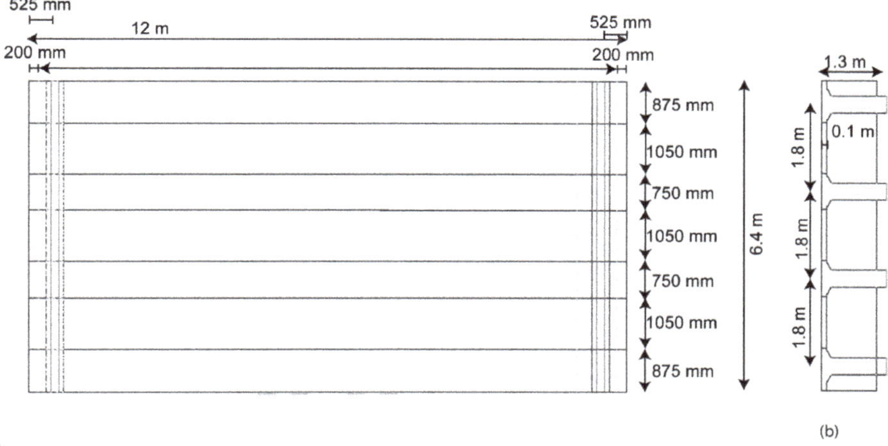

Figure 3. Dimensions of first 1:2 scale model: (**a**) top view; (**b**) cross-section view. Figure adapted from Reference [34]. Reprinted with permission. This figure was originally published in Vol. 116 of the ACI Structural Journal.

The second 1:2 scale model (4.6 m × 12 m, see Figure 4) used three prestressed concrete bulb T-girders and two post-tensioned decks. The dimensions of the girders, crossbeams, and decks were similar to the dimensions for the first 1:2 scale model, with the exception of the shape of the girders (T-girders in the first scale model and bulb T-girders in the second scale model). For the second scale model, the top flange of the girders was cast in the laboratory, monolithically with the deck. The advantage of this approach was that the weight of the girders was reduced, which facilitated transportation and handling.

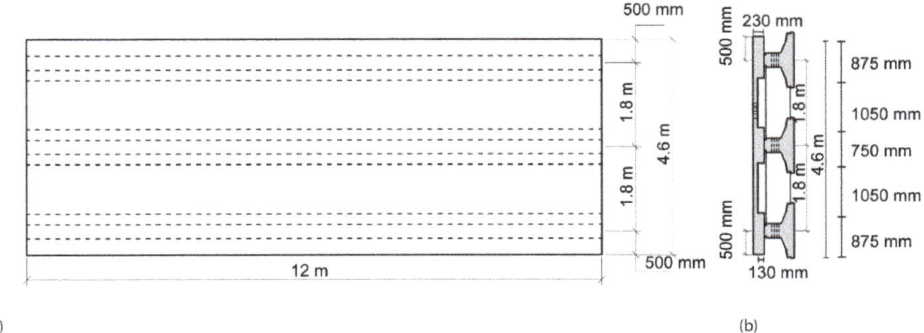

Figure 4. Overview of the second 1:2 scale setup: (**a**) top view; (**b**) cross-section view. Figure adapted from Reference [37]. Reprinted with permission. This figure was originally published in Vol. 116 of the ACI Structural Journal.

Standard cube specimens were used to determine the concrete compressive strength for the concrete of the different casts. The results for the 28 days strength were as follows: $f_{cm,cube}$ = 75 MPa for the original slab in setup 1, $f_{cm,cube}$ = 68 MPa for the newly cast slab in setup 1, $f_{cm,cube}$ = 81 MPa for the first cast of setup 2, and $f_{cm,cube}$ = 79 MPa for the second cast of setup 2.

Mild steel reinforcement is used for the longitudinal and transverse reinforcement in the deck slabs. In setup 1, the longitudinal reinforcement was φ = 6 mm at 200 mm o.c. top and bottom, and the transverse reinforcement was φ = 6 mm at 250 mm o.c. top and bottom. In setup 2, the longitudinal reinforcement was φ = 8 mm at 200 mm o.c. top and bottom, and the transverse reinforcement was φ = 8 mm at 240 mm o.c. top and bottom. The clear cover to the reinforcement was 7 mm. The mild steel reinforcement in the setups was B500B steel, except for the bars of a 6 mm diameter, for which B500A steel was used. Stress–strain curves of the mild steel for all the bar diameter were measured in the laboratory, see References [33,36].

The prestressing steel in the girders were Y1860S tendons, and the prestressing steel in the crossbeams and slabs were Y1100H prestressing bars with a diameter of 15 mm. The transverse prestressing in the deck resulted in an axial compressive stress of 2.5 MPa.

The size of the concentrated load in the experiments was 200 mm × 200 mm for the experiments on the original first setup, which was a 1:2 scale of the wheel print of 400 mm × 400 mm from the design tandem of load model 1 in NEN-EN 1991-2:2003 [1]. For all the other experiments, the size of the loading plate was 115 mm × 150 mm, or the 1:2 scale wheel print of 230 mm × 300 mm that was used in the Netherlands for the assessment of bridge joints for fatigue.

The load was applied using a hydraulic jack mounted on a steel frame test setup. Figure 5 shows an overview photograph of the test setup. For the static tests, the load was applied in a stepwise monotonic loading protocol. In two experiments, a loading protocol with three cycles per load levels was used. For the static tests and the tests with three cycles per load level, the load was applied in a displacement-controlled way. For the fatigue tests, the load was cycled between a lower limit and an upper limit, with the lower limit F_{min} being 10% of the upper limit. A sine function was used with a frequency of 1 Hz. In the fatigue tests, the load was applied in a force-controlled way. If fatigue failure did not occur after a large number of cycles, the upper load level was increased (and the associated lower limit of 10% of the upper limit was adjusted as well).

Figure 5. Overview of the test setup, showing setup 1 with the new middle deck.

3. Results

3.1. Results of the Experiments

The complete results of all the experiments can be consulted in Reference [5] for the static tests on the first setup, in Reference [34] for the fatigue tests on the first setup, and in Reference [37] for the tests on the second setup. In this study, only the results that were relevant for the assessment of the case study bridge were summarized.

Table 2 gives an overview of the relevant static tests from the first setup (BB tests) and the second setup (FAT tests). For the BB series, all the experiments were consecutively numbered. For the FAT series, the test number provides information about the experiment: FAT (fatigue testing series of experiments on setup 2), followed by the test number, and then the S (static test) or D (dynamic test), and 1 (load applied through one loading plate representing a single wheel load) or 2 (load applied through two loading plates representing a double wheel load). The table gives the size of the loading plate used for testing, the load at failure P_{max}, the age of the concrete of the slab at the moment of testing, and the concrete cube compressive strength $f_{cm,cube}$ determined at the day of testing the slab.

Table 2. Overview of the static tests used for assessment of the case study bridge.

Test Number	Size Load (mm × mm)	P_{max} (kN)	Age (days)	$f_{cm,cube}$ (MPa)
BB1	200 × 200	348.7	96	80.0
BB2	200 × 200	321.4	99	79.7
BB7	200 × 200	345.9	127	80.8
BB19	200 × 200	317.8	223	79.9
FAT1S1	150 × 115	347.8	94	82.2
FAT7S1	150 × 115	393.7	240	88.8
FAT8S2	2 of 150 × 115	646.1	245	88.6

Table 3 gives an overview of the fatigue tests. Here, all the tests were considered relevant for the fatigue assessment, since all the fatigue tests were used to derive the Wöhler curves. The test number is given, with BB being the experiments on the first setup and FAT being the experiments on the second setup. Then, the number of the setup was listed, with "1, new" for the experiments that were carried out on the newly cast deck in the first setup. Next, the size of the loading plate used to apply the load on the slab was reported, followed by the "wheel", which can be S (single wheel print) or D (double wheel print). Then, the upper load level used in the test, F/P_{max} (with P_{max} from the static test) was given, as well as N, the number of cycles. For the variable amplitude fatigue tests, N was the number of cycles for the associated load level F/P_{max}. After N cycles at load level F/P_{max}, given in one row of Table 3, the test was continued with N cycles at another load level F/P_{max}, given in the next row. The column "age" gives the age of the slab at the age of testing, and $f_{cm,cube}$ gives the associated cube concrete compressive strength. For fatigue tests that lasted several days, a range of ages was given in the column "age", indicating the age of the concrete in the slab at the beginning of testing and at the end of testing. Similarly, a range of compressive strengths was given for $f_{cm,cube}$, representing the strength determined at the beginning and the end of testing.

Table 3. Overview of the punching fatigue experiments.

Test Number	Setup	Size Load (mm × mm)	Wheel	F/P_{max}	N	Age (days)	$f_{cm,cube}$ (MPa)
BB17	1	200 × 200	S	0.80	13	147	82.6
BB18	1	200 × 200	S	0.85	16	56	82.6
BB23	1	200 × 200	S	0.60	24,800	301	79.9
BB24	1	200 × 200	S	0.45	1,500,000	307–326	79.9
BB26	1, new	150 × 115	S	0.48	1,405,337	35–59	70.5–76.7
BB28	1, new	150 × 115	S	0.48	1,500,000	68–97	76.8–77.1
				0.58	1,000,000	97–113	77.1–77.3
				0.70	7144	113	77.3
BB29	1, new	150 × 115	S	0.58	1,500,000	117–136	77.3–77.5
				0.64	264,840	136–139	77.5–77.6

Table 3. *Cont.*

Test Number	Setup	Size Load (mm × mm)	Wheel	$F/P_{\max}$	N	Age (days)	$f_{cm,cube}$ (MPa)
BB30	1, new	150 × 115	D	0.58	100,000	143–144	77.6
				0.50	1,400,000	144–162	77.6–77.8
				0.58	750,000	162–171	77.8–77.9
				0.67	500,000	171–177	77.9–78.0
				0.75	32,643	177	78.0
BB32	1, new	150 × 115	S	0.70	10,000	184	78.1
				0.58	272,548	185–187	78.1
FAT2D1	2	150 × 115	S	0.69	100,000	102–144	82.6–84.6
				0.58	2,915,123		
				0.69	100,000		
				0.75	150,000		
				0.81	20,094		
FAT3D1	2	150 × 115	S	0.69	200,000	149–168	84.9–85.8
				0.58	1,000,000		
				0.69	100,000		
				0.75	300,000		
				0.81	6114		
FAT4D1	2	150 × 115	S	0.58	1,000,000	169–190	85.8–86.8
				0.69	200,000		
				0.75	100,000		
				0.81	63,473		
FAT5D1	2	150 × 115	S	0.71	10,000	192–217	91.6–89.6
				0.51	1,000,000		
				0.61	100,000		
				0.66	1,000,000		
				0.71	1424		
FAT6D1	2	150 × 115	S	0.71	10,000	219–239	89.6–88.8
				0.51	1,000,000		
				0.61	100,000		
				0.71	160,000		
				0.51	410,000		
				0.71	26,865		
FAT9D2	2	150 × 115	D	0.59	500,000	246–255	88.5–88.2
				0.65	209,800		
FAT10D2	2	150 × 115	D	0.63	100,000	260–284	90.2–91.3
				0.56	1,000,000		
				0.63	950,928		
FAT11D2	2	150 × 115	D	0.67	100,000	288–315	91.5–92.8
				0.60	1,000,000		
				0.67	1,100,000		
				0.75	1720		
FAT12D1	2	150 × 115	S	0.89	30	318	85.9
FAT13D1	2	150 × 115	S	0.86	38	319	85.8

3.2. Resulting Wöhler Curve

To find the Wöhler curve of the fatigue experiments, the relation between the logarithm of the number of cycles N and the applied load ratio $F/P_{\max}$ was plotted, see Figure 6. For this curve, we interpreted the variable amplitude loading tests as follows: if N_1 cycles at load level F_1 are applied, followed by N_2 cycles at load level F_2, and then N_3 cycles to failure at F_3, with increasing load levels $F_1 < F_2 < F_3$, it is conservative to assume that the slab can withstand $N_1 + N_2 + N_3$ cycles at the load level F_1, $N_2 + N_3$ cycles at load level F_2, and N_3 cycles at load level F_3. This approach led to three datapoints for one variable amplitude fatigue test. As a result of this approach, we obtained 16 datapoints on the first setup and 28 datapoints on the second setup, resulting in 44 datapoints in Figure 6. The average value of the Wöhler curve is shown as "mean" in Figure 6, and it is described with the following expression, using S for the load ratio and N for the number of cycles to failure:

$$S = -0.062 \log N + 0.969 \tag{1}$$

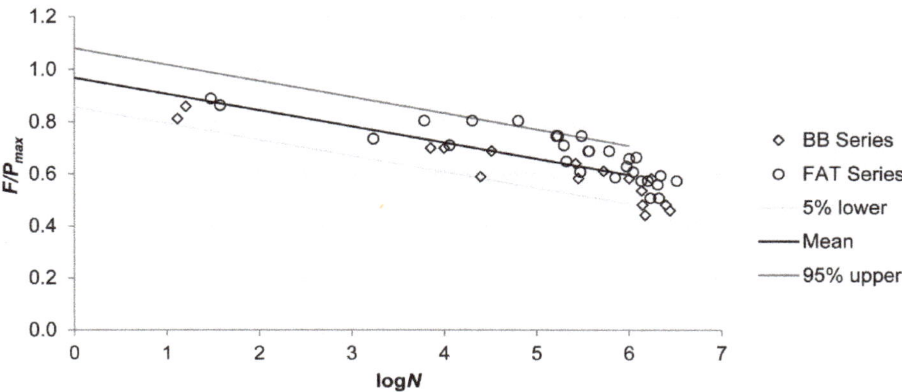

Figure 6. Relation between the number of cycles N and the applied load ratio F/P_{max} in all the fatigue experiments, adapted from Reference [37]. Reprinted with permission. This figure was originally published in Vol. 116 of the ACI Structural Journal.

The current approach was based on a linear fit for the Wöhler curve. For an improved approach, two- and three-parameter Weibull distribution models could be used as in Reference [38]. In [39–41], the methodology for selecting the Weibull distribution models and the compatibility requirements over the whole S-N field are given. This approach may be suitable in ascertaining the predictive fatigue life assessment.

For the current approach, the goodness-of-fit was calculated using the chi-squared test. For all the fatigue tests (datapoints in Figure 6), the value equals 1. As such, the approach was considered satisfactory for our purposes.

Since the assessment was carried out separately for one and two wheel prints, it was interesting to examine the difference in the Wöhler curve for the experiments with one and two wheel prints. Figure 7 gives these results, with Figure 7a showing the datapoints from the FAT series for the single wheel print, and Figure 7b showing the datapoints for the double wheel print. The markers in Figure 7 are different for the datapoints obtained at a number of cycles that resulted in failure and a number of cycles that were calculated using the previously mentioned conservative assumption. The Wöhler curve for the datapoints with a single wheel load is:

$$S = -0.066 \log N + 1.026 \tag{2}$$

The 5% lower bound (characteristic value) of this expression, which can be used for the assessment, is:

$$S_{char} = -0.066 \log N + 0.922 \tag{3}$$

The Wöhler curve for the datapoints with a double wheel load is:

$$S = -0.045 \log N + 0.885 \tag{4}$$

The 5% lower bound of this expression is:

$$S_{char} = -0.045 \log N + 0.825 \tag{5}$$

The slope of the Wöhler curve for the case with two wheel loads is lower compared to the case with a single wheel load. However, for the case with a double wheel load, no low-cycle fatigue experimental results are available. For one load cycle Equation (2) gives a load ratio of 1.026, and Equation (4) gives a load ratio of 0.885. The difference between the two Wöhler curves for one cycle is significant. However, for 1 million load cycles, Equation (2) gives a load ratio of 0.63 and Equation (4) gives a load ratio of 0.62. Therefore, for a large number of load cycles, the difference between the two Wöhler curves becomes smaller. For the assessment of existing bridges, a large number of cycles need to be considered.

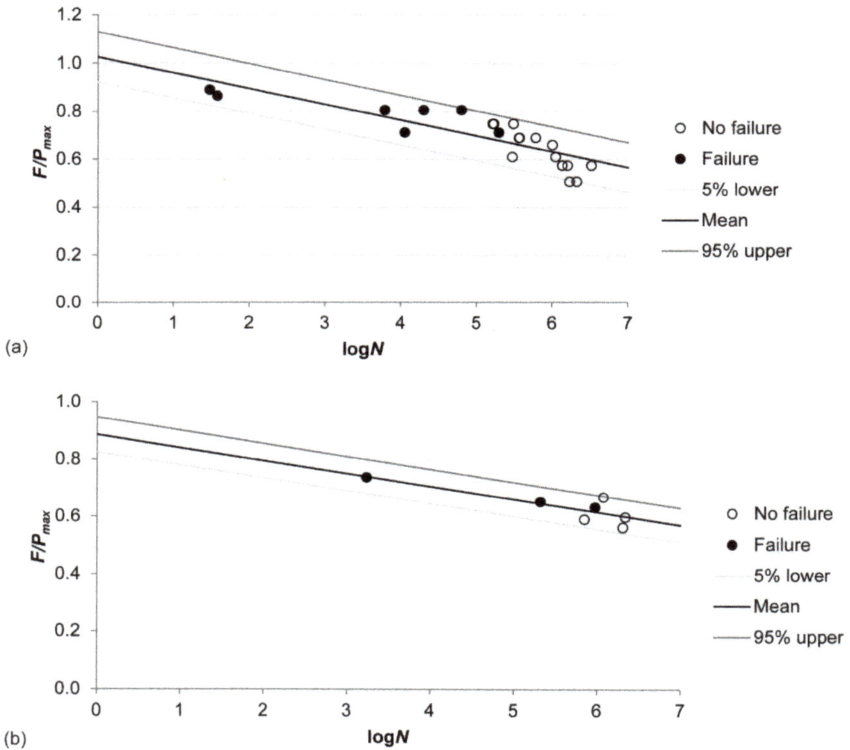

Figure 7. Relation between the number of cycles N and applied load level F/P_{max} for (**a**) a single wheel load; and (**b**) a double wheel load, from Reference [37]. Reprinted with permission. This figure was originally published in Vol. 116 of the ACI Structural Journal.

3.3. Assessment of the Case Study Bridge for Punching

First, the capacity of the thin slab for punching was evaluated based on the experimental results. The shear capacity according to NEN-EN 1992-1-1:2005 [3] was calculated:

$$v_{Rd,c} = C_{Rd,c}k\left(100\rho_{avg}f_{ck}\right)^{1/3} + k_1\sigma_{cp} \geq v_{min} + k_1\sigma_{cp} \tag{6}$$

With

$$k = 1 + \sqrt{\frac{200\text{ mm}}{d}} \leq 2 \tag{7}$$

And

$$\rho_{avg} = \sqrt{\rho_l \times \rho_t} \tag{8}$$

$$\sigma_{cp} = \frac{\sigma_{cx} + \sigma_{cy}}{2} \tag{9}$$

The recommended value for $k_1 = 0.1$, for $C_{Rd,c} = 0.18/\gamma_c$ with $\gamma_c = 1.5$, and for $v_{\min}$:

$$v_{\min} = 0.035k^{3/2}\sqrt{f_{ck}} \tag{10}$$

Using the properties in Table 1, we found that $k = 2$ and the punching shear stress capacity of the case study bridge equals:

$$v_{Rd,c} = \frac{0.18}{1.5} \times 2 \times (100 \times 0.001388 \times 53.3\text{ MPa})^{1/3} + 0.1 \times 1.25\text{ MPa} = 0.572\text{ MPa} \tag{11}$$

To find the maximum punching force, we calculated the punching perimeter around the 400 mm wheel print as sketched in Figure 8:

$$u = 4 \times 400\text{ mm} + 2\pi \times 2 \times 162\text{ mm} = 3636\text{ mm} \tag{12}$$

For the 230 mm × 300 mm wheel print, the punching perimeter length became:

$$u = 2 \times (230\text{ mm} + 300\text{ mm}) + 2\pi \times 2 \times 162\text{ mm} = 3096\text{ mm} \tag{13}$$

The maximum punching force for these two wheel prints then became:

$$V_{Rd,c} = 0.572\text{ MPa} \times 3636\text{ mm} \times 162\text{ mm} = 336.8\text{ kN} \tag{14}$$

$$V_{Rd,c} = 0.572\text{ MPa} \times 3096\text{ mm} \times 162\text{ mm} = 286.8\text{ kN} \tag{15}$$

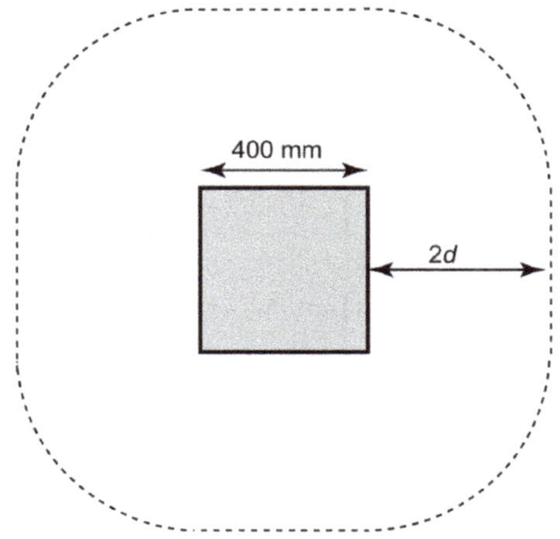

Figure 8. Punching perimeter around the wheel print.

The load that the deck has to resist is a combination of the concentrated live load and the distributed live load. The axle load of 300 kN results in a wheel load of 150 kN. The distributed lane load was 1.15×9 kN/m² = 10.35 kN/m². The contributions of the self-weight and asphalt were 25 kN/m³ × 200 mm = 5 kN/m² and 23 kN/m³ × 120 mm = 2.8 kN/m², respectively. The area over which these loads were considered was the area within the punching perimeter, $A_u = (400 \text{ mm})^2 + 4 \times 162 \text{ mm} \times 400 \text{ mm} + \pi(162 \text{ mm}/2)^2 = 439{,}812 \text{ mm}^2 = 0.4398 \text{ m}^2$. The corresponding loads for the distributed lane load, self-weight, and asphalt then became 4.55 kN, 2.2 kN, and 1.23 kN, respectively, when the Eurocode wheel print was considered. For the smaller wheel print, the area within the punching perimeter became $A_u = 0.2613 \text{ m}^2$, resulting in loads of 2.7 kN, 1.3 kN, and 0.7 kN, respectively, for the distributed lane load, the self-weight, and the asphalt.

The load combination for the assessment of existing bridges in the Netherlands depends on the required safety level, as prescribed by NEN 8700:2011 [42] and the RBK (Richtlijnen Beoordeling Kunstwerken = Guidelines for the Assessment of Existing Bridges) [32]. The highest level was the "design" level (associated reliability index β = 4.3), which gave the following load combination: $U = 1.25DL + 1.25DW + 1.50LL$, with DL being the dead load, DW the superimposed dead load, and LL the live load. The resulting factored concentrated load for evaluation then became 236 kN, for the 400 mm × 400 mm wheel print, and 232 kN, for the 230 mm × 300 mm wheel print.

The assessment was carried out based on the Unity Check, where the Unity Check is the ratio of design demand to design capacity. In this case, for punching, the Unity Check was the ratio of the factored concentrated load acting on the wheel print to the design punching shear force capacity. To fulfil the code requirements, the Unity Check has to be smaller than 1. Table 4 gives an overview of the resulting Unity Checks for the different wheel prints that were studied. We observed that assessment of the deck using the Eurocode shows that it already fulfils the code's requirements. In the introduction, we stated that there was discussion about the punching capacity of the decks in the existing slab-between-girder bridges. The reason why this assessment showed that the deck met the code requirements was the higher punching capacity established based on the results of the drilled cores.

Table 4. Overview of the resulting Unity Checks according to the Eurocode.

Wheel Print	V_{Ed} (kN)	$V_{Rd,c}$ (kN)	Unity Check
400 mm × 400 mm	236	337	0.70
230 mm × 300 mm	232	287	0.81

In a next step of the assessment, the maximum loads obtained in the static tests were applied to the assessment of the Van Brienenoord Bridge. When assessing the bridge based on the results of the experiments, we could replace the design capacity according to the Eurocode $V_{Rd,c}$, using the capacity obtained in the tests. To translate the capacity obtained in the test to a representative design capacity of the case study bridge, we had to consider the following (see Annex D of NEN-EN 1990:2002 [43]):

- The laboratory setup was a 1:2 scale of the case study bridge, resulting in a factor 2^2;
- Considering scaling laws, a scale factor of 1.2 [13] had to be included in the capacity;
- The partial factor derived from the experiments γ_T had to be included.

First, we derived the partial factor from the experiments γ_T. To calculate this factor, we compared the punching capacity obtained in the static experiments with the average punching stress capacity $v_{R,c}$ according to NEN-EN 1992-1-1:2005 [3]. The expression for $v_{R,c}$ was given in the background report of Eurocode 2 [44] as follows:

$$v_{R,c} = 0.18 \times k \times \left(100 \times \rho_{avg} \times f_{cm}\right)^{1/3} + 0.08\sigma_{cp} \tag{16}$$

To find the punching shear capacity $V_{R,c}$, the stress $v_{R,c}$ was then multiplied with $u \times d$, where u was determined as shown in Figure 8, for the considered wheel print. Table 5 combines the experimental results V_{exp} and the predicted capacities $V_{R,c}$, as well as the ratio of the tested to predicted capacity

$V_{exp}/V_{R,c}$. The average value of $V_{exp}/V_{R,c}$ was 2.61, with a standard deviation of 0.296 and a coefficient of variation of 11%. This information led to the derivation of γ_T as determined in Annex C of NEN-EN 1990:2002 [43]:

$$\gamma_T = \frac{\mu}{B_{Rd}} \tag{17}$$

With

$$B_{Rd} = \mu(1 - \alpha \times \beta \times COV) = 2.61(1 - 0.8 \times 4.3 \times 0.11) = 1.622 \tag{18}$$

where $\alpha = 0.8$ is the factor for considering the experimental results, and β is the target reliability index. The value for γ_T then becomes:

$$\gamma_T = \frac{\mu}{B_{Rd}} = \frac{2.61}{1.622} = 1.61 \tag{19}$$

As for the influence of the difference in scale between the test setup in the laboratory and the case study bridge, the experimental result V_{exp} could be scaled to the capacity of the bridge V_{BB} as follows:

$$V_{BB} = V_{exp} \times \frac{2^2}{1.2} \tag{20}$$

where the factor 2^2 corrects for the 1:2 scale, and 1.2 is the scaling factor. The design capacity based on the test results is then:

$$V_{BB,d} = \frac{V_{BB}}{\gamma_T} \tag{21}$$

Table 6 shows the results for the V_{BB} according to Equation (20) and $V_{BB,d}$ according to Equation (21), as well as the demand V_{Ed} that corresponds to the wheel print in the experiment under consideration (see Table 4). The average value of $V_{BB,d}/V_{Ed} = 3.06$, which meant that the margin of safety was 3.23, or that the Unity Check was the inverse, $UC = 0.33$. When comparing this value based on the experiments to the values in Table 4, we observed the beneficial effect of compressive membrane action on the capacity of the thin, transversely prestressed concrete slabs.

Table 5. Comparison between the mean predicted punching capacity and punching capacity in experiment.

Test Number	Wheel Print (mm $\times$ mm)	V_{exp} (kN)	$V_{R,c}$ (kN)	$V_{exp}/V_{R,c}$
BB1	200 $\times$ 200	348.7	141.9	2.458
BB2	200 $\times$ 200	321.4	141.9	2.266
BB7	200 $\times$ 200	345.9	141.9	2.438
BB19	115 $\times$ 150	317.8	121.6	2.613
FAT1S1	115 $\times$ 150	347.8	124.4	2.795
FAT7S1	115 $\times$ 150	393.7	127.4	3.091

Table 6. Determination of the safety factor for the deck of Van Brienenoord Bridge.

Test Number	V_{exp} (kN)	V_{BB} (kN)	$V_{BB,d}$ (kN)	V_{Ed} (kN)	$V_{BB,d}/V_{Ed}$
BB1	348.7	1162.3	721.9	236.0	3.06
BB2	321.4	1071.3	665.4	236.0	2.82
BB7	345.9	1153.0	716.1	236.0	3.03
BB19	317.8	1059.3	658.0	232.0	2.84
FAT1S1	347.8	1159.3	720.1	232.0	3.10
FAT7S1	393.7	1312.3	815.1	232.0	3.51

3.4. Assessment of Case Study Bridge for Fatigue

The results of the experiments and the developed Wöhler curve could be interpreted for the assessment of fatigue. Given the geometry of the deck (see Figure 1), only two wheels (one of each axle) out of the four wheels of the tandem could act together on the deck. The clear span was 2.1 m whilst the width of the design tandem was 2.4 m in total, and 2.0 m center-to-center. For the interpretation of the test results, this meant that the outcome of the tests with a double wheel print (Wöhler curve in Figure 7b) should be evaluated for the case study bridge over 250 million cycles, and that the outcome of the tests with a single wheel print (Wöhler curve in Figure 7b) should be evaluated for the case study bridge over 2 × 250 million cycles = 500 million cycles.

To use the Wöhler curves derived in the experiments for the assessment of the Van Brienenoord Bridge for fatigue, we scaled the fatigue load model to the 1:2 size of the test setup. Note that this approach differed from the assessment for punching, where we scaled up the capacity from the laboratory setup to the capacity of the case study bridge. Here, we used the opposite approach to avoid having to change the Wöhler curve. The concentrated load of the fatigue load model was 105 kN. Scaling this load down to the 1:2 scale model, we used a factor $2^2 = 4$, so that the concentrated load became 26.25 kN. The distributed lane load of the fatigue load model was 3.105 kN/m^2. For the 1:2 scale model, the distributed lane load became 0.776 kN/m^2.

In the 1:2 scale model, only the concentrated loads were used, so the load that represented the concentrated load, as well as the distributed lane load, should be determined. To determine the region over which the distributed lane load should be considered, the cracking patterns in the experiments were studied. The cracking pattern extended over 1.2 m for the experiments with a single wheel load, and over 2 m for the experiments with a double wheel load. To find the equivalent point load, we first determined the bending moment caused by the distributed load, considering that the slab spanned over 1.8 m:

$$M_{dist,1wheel} = \frac{1}{8}\left(0.776\ \frac{\text{kN}}{\text{m}^2} \times 1.2\ \text{m}\right)(1.8\ \text{m})^2 = 0.38\ \text{kNm} \tag{22}$$

$$M_{dist,2wheel} = \frac{1}{8}\left(0.776\ \frac{\text{kN}}{\text{m}^2} \times 2\ \text{m}\right)(1.8\ \text{m})^2 = 0.63\ \text{kNm} \tag{23}$$

The equivalent concentrated load was then:

$$F_{eq} = \frac{4M_{dist}}{l_{span}} \tag{24}$$

which resulted in $F_{eq} = 0.83$ kN for a single wheel load, and $F_{eq} = 1.40$ kN for a double wheel load. Then, the total load was $F = 27.08$ kN for a single wheel load and $F = 27.65$ kN for a double wheel load.

The punching shear capacity of setup 2 is given in Table 5 for FAT1S1 or cast 1 of the concrete as 124.4 kN, and for FAT7S1 or cast 2 as 127.4 kN based on the Eurocode punching provisions. Recall that the design value of the enhancement factor was $B_{Rd} = 1.622$. As such, the design capacity of the punching resistance with the punching perimeter around one wheel load, including the enhancing effect of compressive membrane action became 1.622 × 124.4 kN = 201.8 kN, for the most critical case (lowest capacity $V_{Rd,c}$ as a result of the lowest concrete compressive strength). To determine the capacity for punching with the case of a double wheel print, one could expect a double capacity. However, the results in Table 2 show that the capacity in the FAT8S2 was 1.64 times the capacity in FAT7S1. This ratio was used to determine the punching shear capacity. The capacity was now 1.64 × 201.8 kN = 331.0 kN.

The load ratio could now be determined. For a single wheel load, the load ratio was 27.08 kN/201.8 kN = 0.134, and for a double wheel load, the load ratio was 2 × 27.65 kN/331.0 kN = 0.167.

For the evaluation for one wheel load, Equation (3) was used with $N = 500$ million cycles. The resulting ratio was $S_{char} = 0.348$. For two wheel loads, using Equation (5) with $N = 250$ million cycles gave $S_{char} = 0.447$. The outcome of the assessment was that the margin of safety for one wheel print was 0.348/0.134 = 2.60, or that inversely, the UC = 0.39. For the case with two wheel prints, the margin of safety was 0.447/0.167 = 2.68 or inversely UC = 0.37. Thus, the results for one and two wheel prints were very similar. The conclusion of the assessment was that based on the experimental results, we found that the case study bridge met the code requirements for fatigue.

4. Discussion

In the previous two paragraphs, we calculated the Unity Checks for static punching ($UC = 0.31$), for fatigue punching of one wheel load after 500 million cycles of the single load ($UC = 0.391$), and for fatigue punching of two wheel loads after 250 million cycles of the axle ($UC = 0.37$). Comparing these Unity Checks leads to the conclusion that the most critical case is punching fatigue for a single wheel load. However, the difference between the punching fatigue Unity Check for one and two wheel loads was negligible. In addition, the Unity Checks were small, and significantly smaller than the limiting value of 1.0. This analysis shows the beneficial effect of considering compressive membrane action.

All resulting Unity Checks were smaller than the limiting value of 1.0. This result means that the code requirements for static and fatigue punching were met for the case study bridge. This outcome directly shows the benefit of testing a scaled version of the Van Brienenoord Bridge in the laboratory.

In addition to the conclusion that the Van Brienenoord Bridge met the code requirements for static and fatigue punching, we need to recall that this case study bridge was selected since it had the most critical geometry (largest span to depth ratio for the slab) of the existing slab-between-girder bridges in the Netherlands. As such, the conclusion becomes that all slab-between-girder bridges in the Netherlands, which form a well-defined subset of bridges in the Dutch bridge stock, fulfil the Eurocode requirements. Drawing this conclusion is valid, since these bridges were all built in the same time period, with the same materials, and using the same execution techniques—and are thus all very similar, with only small variations in the geometry and material properties.

However, a side note that we should place with the conclusion that all slab-between-girder bridges in the Netherlands meet the requirements for static and fatigue punching, is that this conclusion is only valid for bridges without material degradation or other forms of damage. To ensure this premise, routine inspections remain necessary. Inspections are an important tool within the bridge management toolbox. If during an inspection indications of material degradation or damage are found, the bridge requires further analysis, and it should be evaluated to check that the conclusion that was based on an undamaged structure is still valid.

For this research, the outcome was twofold: (1) the small resulting Unity Checks based on the experimental results, and (2) the fact that with this approach, the existing slab-between-girder bridges have been shown to fulfil the code requirements. This result also shows that constructing the 1:2 scale setups in the laboratory has been beneficial in the assessment of existing slab-between-girder bridges. Whilst building a 1:2 scale bridge in the laboratory may be considered expensive and time-consuming, testing such a setup gives unique insights on the overall structural behavior of a structural system. Testing at the component level cannot provide such insights. Therefore, the cost-benefit analysis of these experiments is in favor of testing a structural system. Taking this approach is not common, but it may be become an interesting approach for ministries or departments of transportation when they are confronted with a problem for an entire category of bridges.

5. Conclusions

A number of existing slab-between-girder bridges in the Netherlands do not fulfil the requirements of the newly introduced Eurocodes when these bridges are independently evaluated for punching (both static and for cycles of loading). The Eurocode model for determining the punching shear capacity is an empirical model, derived from the results of (mostly concentric) slab-column connection tests [44]. The structural behavior of the thin slabs in slab-between-girder bridges is different from that of slab–column connections. In particular, the development of compressive membrane action increases the capacity significantly.

To study the structural behavior of slab-between-girder bridges, we selected as a case study the Van Brienenoord Bridge because it has the most critical slab geometry (largest span-to-depth ratio for the slabs) of this subset of bridges in the Dutch bridge stock. Based on the geometry of the case study bridge, we built two setups in the laboratory at a 1:2 scale and carried out static and dynamic tests.

The outcome of the static tests could be used for assessing the static punching strength of the Van Brienenoord Bridge. Using the method given in the Eurocode for design by testing, a factor for converting mean values in the design values of 1.53 was derived. Using this approach, the resulting Unity Check for punching shear of the Van Brienenoord Bridge became 0.31.

The outcome of the fatigue tests could be used to derive the Wöhler curve for thin slabs in slab-between-girder bridges. Analyzing the fatigue live load model, we selected two critical loading cases for the fatigue assessment: the case with a single wheel load, and the case with two wheel loads (one of each axle). For both cases, we obtained the results of fatigue tests, and thus a Wöhler curve. The assessment was then carried out based on a service life of 100 years, which led to 500 million cycles for the single wheel load, and 250 million cycles for the double wheel load. Considering the factor to convert the mean values to design values of 1.622 as derived from the static tests, we compared the applied load ratio to the load ratio resulting from the characteristic (5% lower bound) Wöhler curve. Comparing these values gives a Unity Check of 0.39, for the case with a single wheel print, and of 0.37, for the case with a double wheel print.

Evaluating the results of the Unity Checks, we could identify the most critical case, which was (by a small margin) the case of fatigue punching under a single wheel load. However, the resulting Unity Checks were much smaller than the limiting value of 1.0. As such, the conclusion is that the Van Brienenoord Bridge meets the Eurocode requirements for static punching and fatigue. Since the case study bridge was selected based on the most critical geometry, we could say that by extension, all other slab-between-girder bridges in the Netherlands meet the Eurocode requirements for static and fatigue punching. However, this final conclusion is only valid for bridges without deterioration and material degradation. Routine inspections remain an important bridge management tool to identify bridges that require further study.

Author Contributions: Conceptualization, C.v.d.V. and H.S.; methodology, C.v.d.V., R.K., and E.O.L.L.; validation, E.O.L.L.; formal analysis, R.K. and E.O.L.L.; investigation, R.K. and C.v.d.V.; resources, H.S.; data curation, E.O.L.L. and R.K.; writing—original draft preparation, E.O.L.L.; writing—review and editing, R.K., C.v.d.V., and H.S.; visualization, R.K. and E.O.L.L.; supervision, C.v.d.V. and H.S.; project administration, C.v.d.V. and H.S.; funding acquisition, C.v.d.V.

Funding: This research was funded by Rijkswaterstaat, Ministry of Infrastructure and the Environment. The APC was funded by the Delft University of Technology.

Acknowledgments: The authors wish to express their gratitude and sincere appreciation to the Dutch Ministry of Infrastructure and the Environment (Rijkswaterstaat) for financing this research work. We are deeply indebted to our colleague Albert Bosman for his work in the laboratory and the meticulous reporting of the first series of experiments. We would also like to thank our former colleagues Sana Amir and Patrick van Hemert for their contributions to the beginning of the experimental work.

Conflicts of Interest: The authors declare no conflict of interest. The funders had no role in the design of the study; in the collection and analyses of data. The funders were involved with the practical interpretation of the data, writing of the manuscript, and the decision to publish the results.

List of Notations

b	width
c	concrete cover
d	average effective depth
d_l	effective depth to the longitudinal reinforcement
d_t	effective depth to the transverse reinforcement
$f_{ck,cube}$	characteristic cube concrete compressive strength

$f_{cm,cube}$	average cube concrete compressive strength
f_{ck}	characteristic cylinder concrete compressive strength
f_{cm}	average cylinder concrete compressive strength
h	height
k	size effect factor
k_1	factor on effect of axial stresses
l	length
l_{span}	span length
q_{ik}	distributed lane load
u	punching perimeter length
v_{min}	lower bound of shear capacity
$v_{R,c}$	mean capacity for punching shear
$v_{Rd,c}$	design capacity for punching shear
$A_{s,l}$	longitudinal reinforcement area
A_{sp}	area of prestressing steel
$A_{s,t}$	transverse reinforcement area
A_u	area within punching perimeter
B_{Rd}	design capacity derived from statistical results of experiments
COV	coefficient of variation
$C_{Rd,c}$	constant in punching capacity equation
DL	dead load
DW	superimposed dead load
F	applied load
F_{eq}	equivalent load
F_{min}	lower limit of the load as used in the fatigue tests
LL	live load
$M_{dist,1wheel}$	bending moment caused by distributed lane load for influence area of one wheel load
$M_{dist,2wheel}$	bending moment caused by distributed lane load for influence area of two wheel loads
N	number of cycles
P_{max}	load at failure
Q_{ik}	axle load of design tandem
S	load ratio
S_{char}	characteristic value of load ratio (5% lower bound Wöhler curve)
U	load combination
UC	Unity Check
V_{BB}	average capacity of deck of Van Brienenoord Bridge based on experiments
$V_{BB,d}$	design capacity of deck of Van Brienenoord Bridge based on experiments
$V_{R,c}$	mean value of the punching shear capacity
$V_{Rd,c}$	design value of the punching shear capacity
V_{Ed}	design value of punching shear demand
V_{exp}	experimental punching capacity
α	factor that considered effect of experiments
α_{qi}	factor on distributed lane loads
α_{Qi}	factor on design tandem
β	reliability index
γ_T	partial factor derived from experiments
μ	mean value of experimental results
ρ_{avg}	average reinforcement ratio
ρ_l	longitudinal reinforcement ratio
ρ_t	transverse reinforcement ratio
σ_{cp}	average axial stress
σ_{cx}	longitudinal axial stress
σ_{cy}	transverse axial stress

References

1. CEN. *Eurocode 1: Actions on Structures—Part 2: Traffic Loads on Bridges*; Nen-en 1991-2:2003; Comité Européen de Normalisation: Brussels, Belgium, 2003; p. 168.
2. Code Committee 351001. *NEN 6720 Technical Foundations for Building Codes, Concrete Provisions Tgb 1990—Structural Requirements and Calculation Methods (VBC 1995)*; Civil Engineering Center for Research and Regulation, Dutch Normalization Institute: Delft, The Netherlands, 1995; p. 245. (In Dutch)
3. CEN. *Eurocode 2: Design of Concrete Structures—Part 1-1 General Rules and Rules for Buildings*; NEN-EN 1992-1-1:2005; Comité Européen de Normalisation: Brussels, Belgium, 2005; p. 229.
4. Lantsoght, E.O.L.; van der Veen, C.; de Boer, A.; Walraven, J.C. Recommendations for the shear assessment of reinforced concrete slab bridges from experiments. *Struct. Eng. Int.* **2013**, *23*, 418–426. [CrossRef]
5. Amir, S.; Van der Veen, C.; Walraven, J.C.; de Boer, A. Experiments on punching shear behavior of prestressed concrete bridge decks. *ACI Struct. J.* **2016**, *113*, 627–636. [CrossRef]
6. Teworte, F.; Herbrand, M.; Hegger, J. Structural assessment of concrete bridges in germany—Shear resistance under static and fatigue loading. *Struct. Eng. Int.* **2015**, *25*, 266–274. [CrossRef]
7. Bagge, N.; Nilimaa, J.; Puurula, A.; Täljsten, B.; Blanksvärd, T.; Sas, G.; Elfgren, L.; Carolin, A. Full-scale tests to failure compared to assessments—Three concrete bridges. In *High Tech Concrete: Where Technology and Engineering Meet*; Springer: Cham, Switzerland, 2018.
8. Brühwiler, E.; Vogel, T.; Lang, T.; Luechinger, P. Swiss standards for existing structures. *Struct. Eng. Int.* **2012**, *22*, 275–280. [CrossRef]
9. Kong, J.S.; Frangopol, D.M. Probabilistic optimization of aging structures considering maintenance and failure costs. *J. Struct. Eng. ASCE* **2005**, *131*, 600–616. [CrossRef]
10. Frangopol, D.M.; Sabatino, S.; Dong, Y. Bridge life-cycle performance and cost: Analysis, prediction, optimization and decision making. In *Maintenance, Monitoring, Safety, Risk and Resilience of Bridges and Bridge Networks*; Bittencourt, T.N., Frangopol, D.M., Beck, A., Eds.; CRC Press: Foz do Iguacu, Brazil, 2016; pp. 2–20.
11. Vergoossen, R.; Naaktgeboren, M.; Hart, M.; De Boer, A.; Van Vugt, E. Quick scan on shear in existing slab type viaducts. In Proceedings of the International IABSE Conference, Assessment, Upgrading and Refurbishment of Infrastructures, Rotterdam, The Netherlands, 6–8 May 2013; p. 8.
12. Lantsoght, E.O.L.; van der Veen, C.; de Boer, A.; Walraven, J. Using eurocodes and aashto for assessing shear in slab bridges. *Proc. Inst. Civ. Eng. Bridge Eng.* **2016**, *169*, 285–297. [CrossRef]
13. Amir, S. Compressive Membrane Action in Prestressed Concrete Deck Slabs. Ph.D. Thesis, Delft University of Technology, Delft, The Netherlands, 2014.
14. Collings, D.; Sagaseta, J. A review of arching and compressive membrane action in concrete bridges. *Inst. Civ. Eng. Bridge Eng.* **2015**, *169*, 271–284. [CrossRef]
15. Eyre, J.R. Direct assessment of safe strengths of rc slabs under membrane action. *J. Struct. Eng. ASCE* **1997**, *123*, 1331–1338. [CrossRef]
16. Highways Agency. *Corrections within Design Manual for Roads and Bridges, Use of Compressive Membrane Action in Bridge Decks*; Highways Agency: London, UK, 2007; pp. 35–51.
17. Kirkpatrick, J.; Rankin, G.I.B.; Long, A.E. The influence of compressive membrane action on the serviceability of beam and slab bridge decks. *Struct. Eng.* **1986**, *64B*, 6–9.
18. Kuang, J.S.; Morley, C.T. A plasticity model for punching shear of laterally restrained slabs with compressive membrane action. *Int. J. Mech. Sci.* **1993**, *35*, 371–385. [CrossRef]
19. Tong, P.Y.; Batchelor, B.V. Compressive membrane enhancement in two-way bridge slabs. *SP 30-12* **1972**, *30*, 271–286.
20. Hewitt, B.E.; de Batchelor, B.V. Punching shear strenght of restrained slabs. *J. Struct. Div.* **1975**, *101*, 1837–1853.
21. Lantsoght, E.O.L.; Van der Veen, C.; Koekkoek, R.T.; Sliedrecht, H. Capacity of prestressed concrete bridge decks under fatigue loading. In Proceedings of the FIB Symposium, Cracow, Poland, 27–29 May 2019.
22. Bennett, E.W.; Muir, S.E.S.J. Some fatigue tests of high-strength concrete in axial compression. *Mag. Concr. Res.* **1967**, *19*, 113–117. [CrossRef]
23. Lantsoght, E.O.L.; van der Veen, C.; de Boer, A. Proposal for the fatigue strength of concrete under cycles of compression. *Constr. Build. Mater.* **2016**, *107*, 138–156. [CrossRef]

24. Isojeh, B.; El-Zeghayar, M.; Vecchio, F.J. Fatigue resistance of steel fiber-reinforced concrete deep beams. *ACI Struct. J.* **2017**, *114*, 1215–1226. [CrossRef]
25. Teng, S.; Ma, W.; Tan, K.H.; Kong, F.K. Fatigue tests of reinforced concrete deep beams. *Struct. Eng.* **1998**, *76*, 347–352.
26. Teworte, F.; Hegger, J. Shear fatigue of prestressed concrete beams. In Proceedings of the IABSE 2011, London, UK, 23 September 2011; p. 8.
27. Muller, J.F.; Dux, P.F. Fatigue of prestressed concrete beams with inclined strands. *J. Struct. Eng.* **1994**, *120*, 1122–1139. [CrossRef]
28. Yuan, M.; Yan, D.; Zhong, H.; Liu, Y. Experimental investigation of high-cycle fatigue behavior for prestressed concrete box-girders. *Constr. Build. Mater.* **2017**, *157*, 424–437. [CrossRef]
29. Fujiyama, C.; Gebreyouhannes, E.; Maekawa, K. Present achievement and future possibility of fatigue life simulation technology for rc bridge deck slabs. *Soc. Soc. Manag. Syst. Internet J.* **2008**, *4*.
30. Harajli, M.H.; Naaman, A.E. Static and fatigue tests on partially prestressed beams. *J. Struct. Eng.* **1985**, *111*, 1602–1618. [CrossRef]
31. Xin, Q.; Dou, Y.; Chen, W. Load spectrum and fatigue life computer analysis of prestressed concrete bridges. *Int. J. Secur. Appl.* **2015**, *9*, 247–266. [CrossRef]
32. Rijkswaterstaat. *Guidelines Assessment Bridges—Assessment of Structural Safety of an Existing Bridge at Reconstruction, Usage and Disapproval*; RTD 1006:2013 1.1; Rijkswaterstaat: Utrecht, The Netherlands, 2013; p. 117. (In Dutch)
33. Van der Veen, C.; Bosman, A. *Vermoeiingssterkte Voorgespannen Tussenstort*; Stevin Report nr. 25.5-14-06; Delft University of Technology: Delft, The Netherlands, 2014; p. 65.
34. Lantsoght, E.O.L.; Van der Veen, C.; Koekkoek, R.T.; Sliedrecht, H. Fatigue testing of transversely prestressed concrete decks. *ACI Struct. J.* **2019**, in press. [CrossRef]
35. Koekkoek, R.T.; van der Veen, C.; de Boer, A. *Fatigue Tests on Post-Tensioned Bridge Decks*; Springer International Publishing: Cham, Switzerland, 2018; pp. 912–920.
36. Koekkoek, R.T.; van der Veen, C. *Measurement Report Fatigue Tests on Slabs Cast in-between Prestressed Concrete Beams*; Stevin Report 25.5-17-14; Delft University of Technology: Delft, the Netherlands, 2017; p. 196.
37. Lantsoght, E.O.L.; Van der Veen, C.; Koekkoek, R.T.; Sliedrecht, H. Punching capacity of prestressed concrete bridge decks under fatigue. *ACI Struct. J.* **2019**, in press. [CrossRef]
38. Castillo, E.; Fernandez-Canteli, A. *A Unified Statistical Methodology for Modeling Fatigue Damage*; Springer: Dordrecht, The Netherlands, 2009; p. 189.
39. Hanif, A.; Usman, M.; Lu, Z.; Cheng, Y.; Li, Z. Flexural fatigue behavior of thin laminated cementitious composites incorporating cenosphere fillers. *Mater. Des.* **2018**, *140*, 267–277. [CrossRef]
40. Hanif, A.; Kim, Y.; Park, C. Numerical validation of two-parameter weibull model for assessing failure fatigue lives of laminated cementitious composites—Comparative assessment of modeling approaches. *Materials* **2018**, *12*, 110. [CrossRef] [PubMed]
41. Lu, C.; Dong, B.; Pan, J.; Shan, Q.; Hanif, A.; Yin, W. An investigation on the behavior of a new connection for precast structures under reverse cyclic loading. *Eng. Struct.* **2018**, *169*, 131–140. [CrossRef]
42. Code Committee 351001. *Assessement of Structural Safety of an Existing Structure at Repair or Unfit for Use—Basic Requirements*; NEN 8700:2011; Civil Center for the Execution of Research and Standard, Dutch Normalisation Institute: Delft, The Netherlands, 2011; p. 56. (In Dutch)
43. CEN. *Eurocode—Basis of Structural Design*; NEN-EN 1990:2002; Comité Européen de Normalisation: Brussels, Belgium, 2002; p. 103.
44. Walraven, J.C. *Background Document for EC-2, Chapter 6.4 Punching Shear*; Delft University of Technology: Delft, The Netherlands, 2002; pp. 1–16.

Article

A Damage Model for Concrete under Fatigue Loading

Zhi Shan, Zhiwu Yu, Xiao Li, Xiaoyong Lv * and Zhenyu Liao

School of Civil Engineering & National Engineering Laboratory for High Speed Railway Construction & Engineering Technology Research Center for Prefabricated Construction Industrialization of Hunan Province, Central South University, 68 South Shaoshan Road, Changsha 410075, China
* Correspondence: lvxiaoyong@csu.edu.cn; Tel.: +86-0731-82656611

Received: 3 June 2019; Accepted: 25 June 2019; Published: 9 July 2019

Abstract: For concrete, fatigue is an essential mechanical behavior. Concrete structures subjected to fatigue loads usually experience a progressive degradation/damage process and even an abrupt failure. However, in the literature, certain essential damage behaviors are not well considered in the study of the mechanism for fatigue behaviors such as the development of irreversible/residual strains. In this work, a damage model with the concept of mode-II microcracks on the crack face and nearby areas contributing to the development of irreversible strains was proposed. By using the micromechanics method, a micro-cell-based damage model under multi-axial loading was introduced to understand the damage behaviors for concrete. By a thermodynamic interpretation of the damage behaviors, a novel fatigue damage variable (irreversible deformation fatigue damage variable) was defined. This variable is able to describe irreversible strains generated by both mode-II microcracks and irreversible frictional sliding. The proposed model considered both elastic and irreversible deformation fatigue damages. It is found that the prediction by the proposed model of cyclic creep, stiffness degradation and post-fatigue stress-strain relationship of concrete agrees well with experimental results.

Keywords: concrete; fatigue; damage model; mode-II microcracks; thermodynamics

1. Introduction

Fatigue is an essential mechanical behavior of concrete. In real life, a large number of concrete structures are subjected to fatigue loads, e.g., off-shore structures and bridges. Although the subjected fatigue loads are lower than the relevant materials' original strength, these structures experience a progressive degradation and subsequently an abrupt failure. In order to investigate these fatigue behaviors, several methods (e.g., fatigue life concepts [1–4] and phenomenological models [5–8]) were developed by researchers and were widely applied in structural engineering. However, during the designing and analysis of structures, these methods [1–8] are only limited to describing the fatigue behaviors at phenomenological and empirical levels without a comprehensive understanding and explanation of the internal mechanism for damage behaviors of concrete under fatigue loading.

The complex constitution of concrete results in a sophisticated damage evolution process during material under loading. Specifically, in the material, the arbitrary distribution of initial defects causes the localization of stresses, which further produce the complex evolution process of damage. Experimental studies [9–14] have been conducted to understand the damage mechanism referring to concrete under fatigue loading. In detail, some experimental results showed that local stresses near the defect cause the heterogeneous crack openings perpendicular to tensile loading, i.e., mode-I cracks. The mode-I cracks were well studied in a number of research papers [9–12]. In addition, [13] applied X-ray techniques to study the microcrack mechanism of concrete, and it was found that microcracks parallel to tensile loading (mode-II cracks) can occur even under pure global axial loading. Reference [14] found the mode-II cracks are able to create irreversible strains due to local stresses.

Furthermore, a series of relevant comprehensive works have been conducted by researchers in the solid mechanical field for more than a decade [15–23].

Moreover, the mechanism of the development for irreversible/residual strains in concrete under fatigue loading have been studied throughout several methods [6,12,24–49]. Concretely, on one hand, based on the micro-mechanics method [6,12,24–40], it is concluded that irreversible strains are produced by a series of types of cracking, which are distinguished as follows: (I) for a compressive case, the transversely propagating crushing band [24,25], the axial wedge-splitting cracks at hard inclusions in hardened cement paste [26], the interface cracks at inclusions [6], the pore-opening axial cracks [27,28], and the inclined wing-tipped frictional cracks (i.e., wing cracks) [29–33]; (II) for a tensile case, the irreversible opening of mode-I cracks due to the locking mechanisms of crack faces [34], the irreversible sliding-like openings of mode-II crack due to the toughness of crack faces [35,36], the irreversible frictional sliding over the crack surface [37,38], the irreversible cracking of the fracture process zone [12,39], and other cracking mechanisms [40]. On the other hand, based on the macro-mechanics method [41–49], researchers have rarely considered the comprehensive mechanism of concrete damage, since they are usually focused on the accurate characterization of macroscopic mechanical behaviors.

However, among the above-mentioned literature, certain essential damage behaviors are not well considered in the study of the mechanism for the development of irreversible/residual strains in concrete. Specifically, one type of those damage behaviors is mode-II microcracks, which has attracted the attention of researchers in the field of solids mechanics for decades [15–23].

Therefore, it is necessary to develop a continuum damage model for concrete under fatigue loading with the consideration of this damage behavior. In detail, this damage model is able to be established based on the micro-mechanics and continuum damage mechanics. Micro-mechanics enables us to understand damage behavior under multi-axial loading, and the continuum damage mechanics (CDM) method (i.e., a macro-mechanics method) offers us a convenient way to characterize the macro behavior.

This work develops the above-mentioned contributions [6,12,24–49] in two aspects, the description of the micro-mechanism for mode-II microcracks in multi-axial conditions and the thermodynamics-based modeling of damage behaviors in concrete under fatigue loading.

2. Microcrack Mechanism in Concrete under Multi-Axial Loading

In this section, we briefly recall here the main steps of the methodology followed by the literature [22] for the micro-mechanical description of mode-II microcracks. In addition, the random distribution of initial defects in concrete under multi-axial loading was considered in this work.

2.1. The Definition of the Mode-II Microcracks

Mode-II microcracks are the local shear stress-caused by microcracks on the crack face and in the nearby area of the micro-defects and the mode-I crack. This type of crack is different from the mode-I crack and the mode-II crack. The differences can be concluded as follows, the mode-II microcracks are the result of local shear stresses, which is distinguished from tensile stress-caused by the mode-I crack. Additionally, unlike the mode-II crack, mode-II microcracks usually appear on the face and nearby area of the micro-defects and mode-I crack.

2.2. The Causes for the Mode-II Microcracks under Multi-Axial Loading

When the concrete is subjected to a biaxial tensile load, a micro-cell within a representative volume element (RVE) was introduced and is shown in Figure 1. In detail, the stress flow curve becomes concentrated when it approaches the crack tip, and the plane stress on the plane horizontal and vertical to the direction of the crack propagation is able to be described by the normal and shear stress as follows, σ_h and τ_h, σ_v and τ_v, respectively (Figure 1c). Due to sufficient normal stress σ_v or stress intensity factor (SIF) K_I at the crack tip, the crack will initiate and grow through the direction where the maximum SIF exists (i.e., transverse to the direction of maximum principal tension, Figure 1d–e). Therefore, the crack type, namely the mode-I crack, decreases the effective load area of the micro-cell in the maximum principal tension direction, resulting in the stiffness degradation of the specimen [33–36,50].

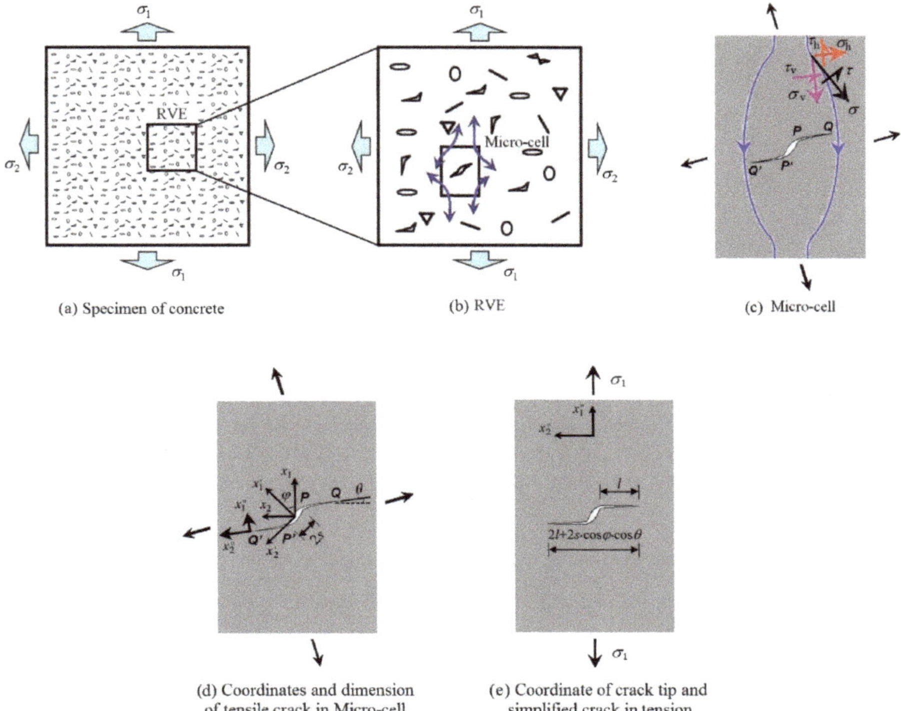

Figure 1. Sketch of stress flow curve around the random selected micro-cell of the representative volume element (RVE) in the specimen under multi-axial tension, and the related coordinates and dimension, where σ_1 denotes the maximum principal tensile stress.

However, mode-II microcracks have not been well considered in modeling concrete under multi-axial stresses in the literature [6,12,22–40]. Through different approaches, including experimental observations, mechanical analysis and atomic simulations [15–21], it is validated that the real crack (excepting some pre-existing cracks) in the material is blunt, caused by the appearance of mode-II microcracks.

Specifically, mode-II microcracks are produced by the local shear stresses (i.e., the shear stress τ_h in Figures 1c and 2a) on the face and nearby area of relevant cracks. In a biaxial tensile load case, the directions of local shear stresses are arbitrary due to the random location of initial defects. It is distinguished from that in a uniaxial tensile case [22]. Moreover, several researchers [34–40] observed that the mode-II microcracks, rather than the dislocation-induced plastic flow, appear in complex composite materials such as concrete. Further description and explanation can be found in the literature [6,51].

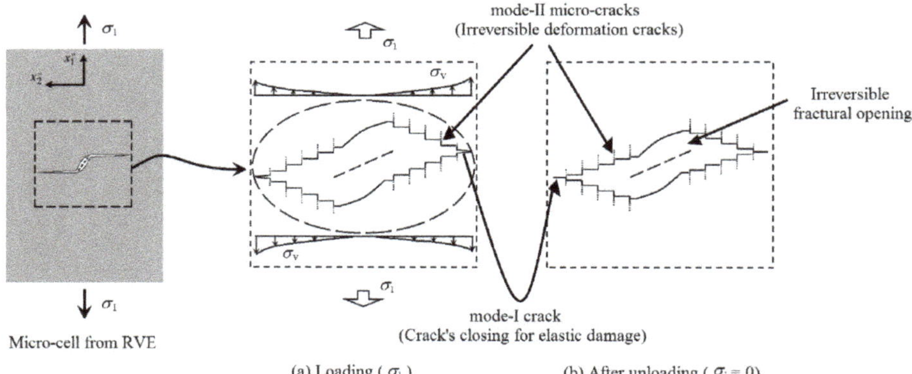

Figure 2. Sketch of microcracks under multi-axial tension, leading to both the stiffness degrading and the irreversible strain developing, where σ_1 denotes the maximum principal tensile stress.

Note that it is speculated that the local constraint condition around the micro defect is stable during the crack initiation and/or propagation under multi-axial tension; otherwise, the mode-II dominant failure will appear. The development of the mode-II microcrack leads to certain energy dissipation, and it may also release the tip stress concentration of the mode-I cracks, since it causes a relatively blunter crack tip.

2.3. Influence of the Mode-II Microcracks on the Irreversible Strains in Concrete

It is revealed that the mode-II microcracks are attributed to crack blunting and the irreversible deformation (Figure 2) of material even in brittle material such as glass [21]. In this section, we briefly recall the methodology followed by [22] for the micro-mechanical description and further develop it with consideration of stochastic properties in a multi-axial tension case, as follows.

For simplicity, we introduce a micro-cell damage model (Figure 3) considering the mode-II microcracks to describe the damage behaviors in concrete under biaxial tension. In Figure 3, the region near a certain defect is firstly highlighted and further discretized by amounts of micro-cells (micro-cell i, micro-cell $i + 1$, etc.). The behavior of each micro-cell is modeled by two sets of springs (spring type A and B). The spring type A can be stretched vertically along the direction of the maximum principal loading, and spring type B is attached to the middle of spring type A. Unlike spring type A, spring type B cannot be stretched but it can slip between two parallel sets of micro-cells. In such a micro-cell damage model, the elastic behavior and elastic deformation damage are described by spring type A, and the irreversible deformation damage is modeled by spring type B. After unloading, there is a micro deformation b and an irreversible strain $\varepsilon_{I,f}$ left in the material. The micro irreversible fractural opening is caused by mode-II microcracks illustrated in the micro-cell damage model.

In summary, the elastic deformation damage in the micro-cell damage model corresponds to the stiffness degradation, and the irreversible deformation damage is responsible for a certain part of the irreversible strain. Specifically, the local shear stresses produce mode-II microcracks on the crack face and nearby areas, which generate the micro deformation b and an irreversible strain $\varepsilon_{I,f}$ in the material (Figure 3).

Macro behaviors

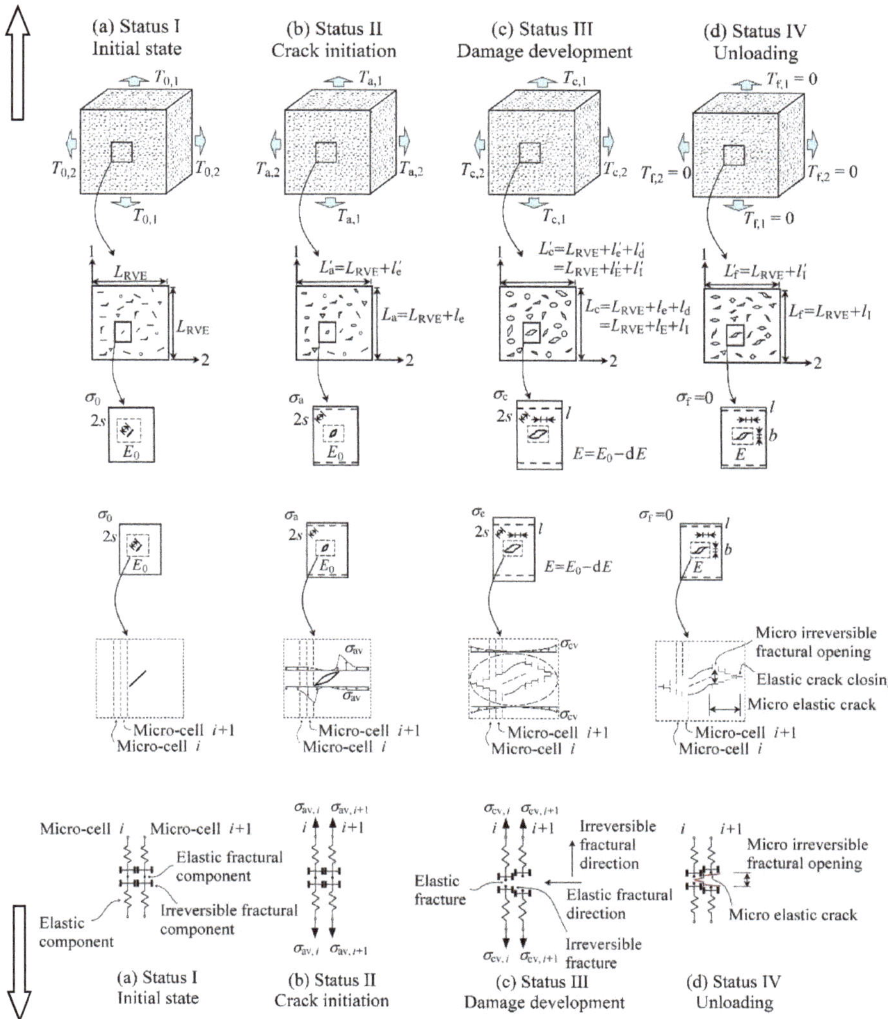

Micro mechanism

Figure 3. Macro behaviors and micro mechanism of a concrete cube under bi-axial tension from various loading statuses. This figure was developed based on [22] (Reproduced with permission from [John Wiley & Sons Ltd], 2016), however, the random distribution of initial defects was considered in this work.

2.4. Irreversible Strains in Concrete under Multi-Axial Loading

For the irreversible strains that are not induced by mode-II microcracks, a simplified frictional sliding model is developed in this work (Figure 4) for revealing the development of the irreversible strains in concrete under multi-axial tension based on the literature [37,38]. In detail, as illustrated in Figure 4, according to this model, the behavior of frictional sliding generally produces a new portion of irreversible deformation b' in the micro-cell of RVE (Figure 4).

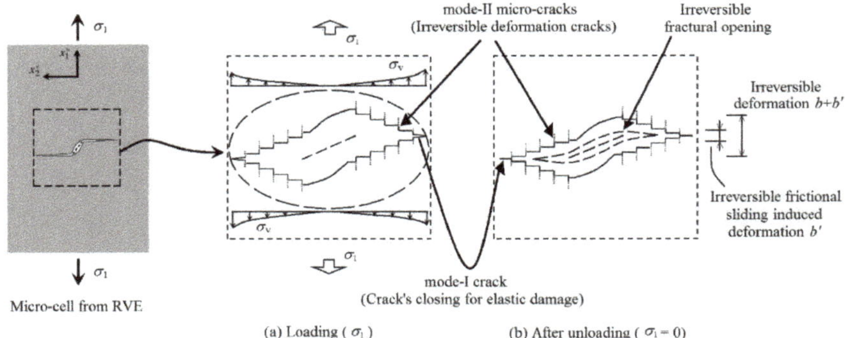

Figure 4. Sketch of tensile irreversible deformations due to both mode-II microcracks and irreversible frictional sliding, where σ_1 denotes the maximum principal tensile stress in multi-axial tension. This figure was developed based on [22] (Reproduced with permission from [John Wiley & Sons Ltd], 2016), however, the random distribution of initial defects was considered in this work.

Thus, in this work, the irreversible strains in concrete under multi-axial tension are produced by two mechanisms: the mode-II microcracks and irreversible frictional sliding. It is noted that the other mechanisms [6,12,24–36,39,40] are not employed in the work for the sake of simplicity. In addition, the irreversible deformation damages are assumed to consist of both mode-II microcracks and irreversible frictional sliding (see Figure 4).

For simplicity, the multi-axial stresses in the material are assumed to be classified into two stress spaces: the tension- and compression-dominant stress spaces (Figure 5). Precisely, the stress spaces are distinguished by the plane vertical to the stress line, which indicates the stresses on triaxis are equal to each other (see Figure 5). Figure 5 illustrates that the tension-dominant stress space consists of both the multi-axial tension space and a certain part of tension-compression space. The compression-dominant stress space represents the rest of the stress space. It is worth mentioning that the micro damage mechanisms are different when related to the above two dominant stresses. Concretely, for simplicity, the micro damage mechanism of concrete under tension-dominant stress is assumed to be similar to that under multi-axial tension developed in this work, and the micro damage mechanism of concrete under compression-dominant stress is assumed to be similar to that under multi-axial compression in [23].

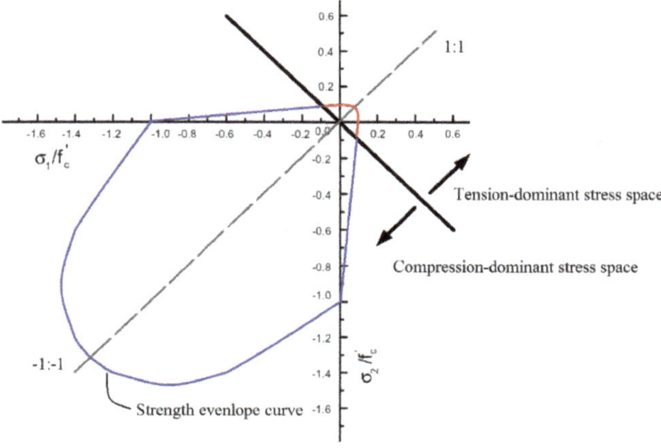

Figure 5. Sketch of the tension- and compression-dominant stress space in two-dimensional.

The effect of the roughness and friction of crack faces on the progressive damage and irreversible strains under fatigue compression is able to be concluded as follows, the roughness and friction of the crack faces for the initial inclined frictional crack in wing cracks [23,29–33] and mode-II microcracks will influence the irreversible behaviors: higher roughness and friction results in the later initiation of the crack and further leads to a lower amount of the irreversible strains.

It is worth mentioning that the new development of micro-mechanical descriptions related to mode-II microcracks in this work has been obtained in the following ways. Firstly, this work extended the description of the micro damage behaviors in concrete under multi-axial tension with the consideration of both the stochastic properties of initial cracks and the influences of mode-II microcracks. It is distinguished from the work in [22], which focused on an idealized model of the initial uniformly and horizontally distributed cracks under uniaxial tension, and from that in [23], which involved a model of wing crack under multi-axial compression. Secondly, this work introduced a simplified description of damage behaviors in concrete under tension-compression, which was not considered in the literature [22,23].

3. Thermodynamics Based Continuum Damage Mechanics Model

Physically, the damage propagation including both the expanded crack length $2l$ and the developed crack opening $b + b'$ in micro-scale in Figure 6 (discretely modeled by the micro damage model in Figure 6) is an irreversible thermodynamic process characterized in Figure 6. Both microscale behaviors are able to be idealized/unified and thermodynamics modeled by considering the stiffness degradation E_d and the irreversible strains development ε^I (Figure 6), respectively.

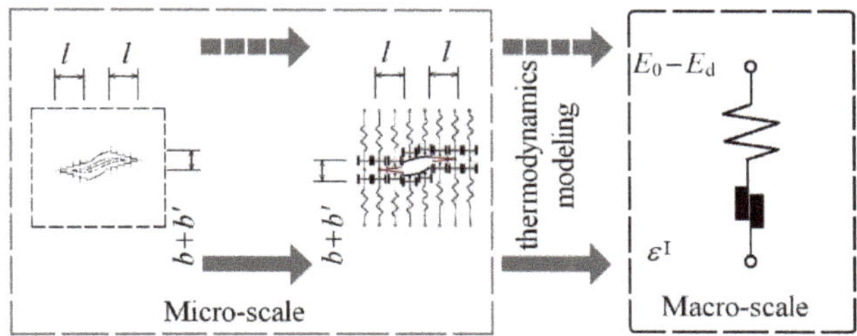

Figure 6. Thermodynamics interpretation of micro-scale damage behaviors. Specifically, the elastic deformation damage (mode-I cracks) produces the stiffness reduction E_d, and the irreversible deformation damage (both mode-II microcracks and irreversible friction sliding) causes the development of irreversible strains ε^I.

Thus, the complex microscale crack behaviors (Figure 6) are thermodynamically interpreted into a simple macroscale damage mechanics model (Figure 6), which obtained a thermodynamics based CDM model. In the following section, the definition of a new type of damage variable—the irreversible deformation fatigue damage variable—is firstly introduced and then the details for the model formulation are given.

3.1. Thermodynamics Interpretation

In this section, we briefly recall here the main steps of the methodology followed by [52] for the thermodynamics interpretation of the damage variable. This work developed the method from [52] for interpreting the damage variable under fatigue loading.

The infinite deformation behavior of concrete material with damage can be viewed within the framework of thermodynamics with internal state variables. The Helmholtz free energy per unit mass, in an isothermal deformation process at the current state of the deformation and material damage, is assumed as follows:

$$\Psi_n = \psi_n + \gamma_n \tag{1}$$

where the subscript n denotes the cyclic number of the fatigue loading ($n = 1, 2, 3, ..., N$), ψ denotes the strain energy or a purely reversible stored energy, while γ represents the irreversible energy associated with specific micro structural changes produced by damage (i.e., elastic deformation damage induced by mode-I fractures, and irreversible deformation damage due to both mode-II micro-cracks and irreversible fictional sliding, see Figures 4 and 6). An explicit presentation of the irreversible energy and its rate is generally limited by the complexities of the internal micro structural changes discussed in the recent section; however, only one internal variable damage (contains two components) is considered in this work. The damage contains two components, that is, elastic deformation damage induced by mode-I fractures, and irreversible deformation damage due to both mode-II micro-cracks and irreversible fictional sliding, contribute to stiffness degradation and irreversible strains development, respectively (see Figure 6). For the purpose of developing a schematic description of the concepts based on the proposed micro damage model, the uniaxial stress-strain curves are used in Figure 7. In Figure 7a, E_0 denotes the initial undamaged stiffness (relates to loading line OA_0). The strain and the stress at point A_0 are denoted by $\varepsilon_{0,1}$ and σ_{max}, respectively.

3.1.1. Interpretation in First and Second Loading Cycle

Firstly, considering the stress-strain response during the first loading cycle, the unloading curve $A_1 B_1$ is simplified by the line $A_1 B_1$ in Figure 7a,b in this work. At point A_1, the strain ε_1 and irreversible damage strain $\varepsilon_1{}^{di}$ exist in the specimen. The initial stiffness changes from E_0 to E_1. Even though these notations are for the uniaxial case, they are able to be used in indicial tensor notation in the equations without loss of generality. The total strain (described by line $OB_1 G_1 H_1$ in Figure 7a) is given as follows:

$$\varepsilon_1 = \varepsilon_1^{E} + \varepsilon_1^{I} = \left(\varepsilon_{0,1} + \varepsilon_1^{de}\right) + \varepsilon_1^{di} \tag{2}$$

where the subscript 1 denotes the cyclic number of the first fatigue loading.

The strain energy is expressed as follows (see the area described by points $B_1 A_1 H_1$ in Figure 7a)

$$\psi_1 = \frac{1}{2v}E_1 \cdot \left(\varepsilon_1^{E}\right)^2 = \frac{1}{2v}E_0 \varepsilon_{0,1} \varepsilon_1^{E} \tag{3}$$

$$\psi_1 = \psi_0^{e} + \psi_1^{de} \tag{4}$$

where $\psi_0{}^{e}$ denotes the initial strain energy (see the area $G_1 A_1 H_1$ in Figure 7a), and $\psi_1{}^{de}$ denotes the elastic deformation damage strain energy during the first cycle (see the area $B_1 A_1 G_1$ in Figure 7a), that is,

$$\psi_0^{e} = \frac{1}{2v}E_0 \cdot \left(\varepsilon_{0,1}\right)^2 \tag{5}$$

$$\psi_1^{de} = \frac{1}{2v}E_0 \varepsilon_{0,1} \varepsilon_1^{de} = \psi_1 - \psi_0^{e} = \frac{1}{2v}E_0 \varepsilon_{0,1}\left(\varepsilon_1^{E} - \varepsilon_{0,1}\right) \tag{6}$$

And the irreversible energy is expressed as follows (see the area $OA_0 A_1 B_1$ in Figure 7a)

$$\gamma_1 = \frac{1}{v}\left(\sigma_{max}\varepsilon_1^{di} + \frac{1}{2}\sigma_{max}\varepsilon_1^{de}\right) = \frac{1}{v}E_0 \varepsilon_{0,1}\left(\varepsilon_1^{d} - \frac{1}{2}\varepsilon_1^{de}\right) \tag{7}$$

$$\gamma_1 = \gamma_1^{di} + \gamma_1^{de} \tag{8}$$

where $\gamma_1{}^{di}$ denotes the irreversible-damage irreversible energy (see the area $OA_0 I_1 B_1$ in Figure 7a), and $\gamma_1{}^{de}$ denotes the elastic deformation damage irreversible energy (see the area $B_1 I_1 A_1$ in Figure 7a), that is,

$$\gamma_1^{\mathrm{di}} = \frac{1}{v}\sigma_{\max}\varepsilon_1^{\mathrm{di}} = \frac{1}{v}E_0\varepsilon_{0,1}\varepsilon_1^{\mathrm{di}} \tag{9}$$

$$\gamma_1^{\mathrm{de}} = \frac{1}{2v}\sigma_{\max}\varepsilon_1^{\mathrm{de}} = \frac{1}{2v}E_0\varepsilon_{0,1}\varepsilon_1^{\mathrm{de}} \tag{10}$$

In regard to stored energy λ (contains both the purely reversible stored energy ψ and the irreversible energy γ), one is able to obtain the formula as follows

$$\lambda_n = \psi_n + \gamma_n \tag{11}$$

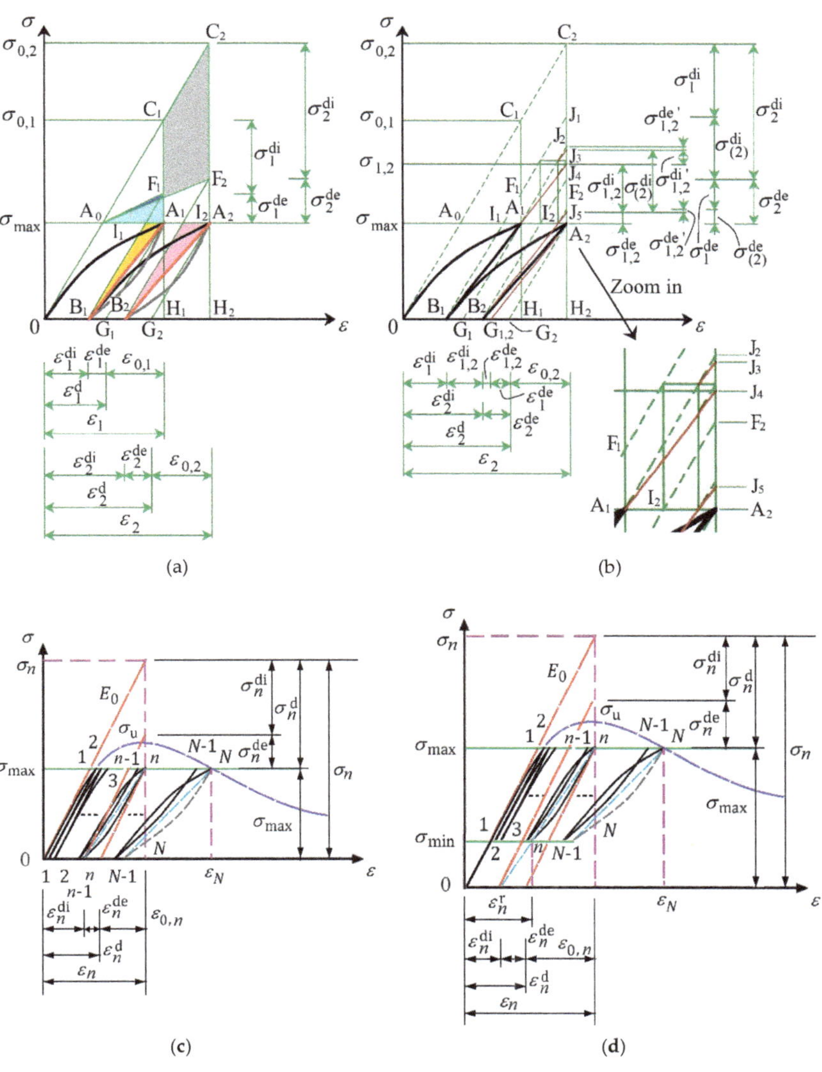

(a)

(b)

(c)

(d)

Figure 7. Sketch of the mechanical parameter definition for concrete under fatigue loading. (a) Energy dissipation; (b) Energy dissipation; (c) Energy dissipation ($\sigma_{\min} = 0$); (d) Energy dissipation ($\sigma_{\min} \neq 0$);

When the material is assumed to be a perfect elastic material, it undergoes a strain ε_1 and obtains a stored energy $\lambda_1{}^0 = 1/2(E_0\varepsilon_1{}^2) = \psi_1{}^0$ (i.e., the perfect material's purely reversible stored energy or strain energy) due to external loads. However, the material focused on in this work is a quasi-brittle material assumed to undergo both elastic deformation damage and irreversible deformation damage. It reduces a certain part of stored energy (denoted by the area $A_0F_1A_1$ in Figure 7a) caused by the elastic deformation damage (i.e., mode-I fracture in the proposed micro-cell damage model, see Figures 4 and 6). Additionally, another part of the stored energy (described by the area $A_0C_1F_1$ in Figure 7a) is also decreased, as a result of the irreversible damage (due to both mode-II micro-cracks and irreversible fictional sliding, see Figures 4 and 6). Thus, the damaged material's stored energy is derived as follows

$$\lambda_1 = \psi_1 + \gamma_1 = \lambda_1^0 - \lambda_1^{\mathrm{d}} \tag{12}$$

where

$$\lambda_1^{\mathrm{d}} = \lambda_1^{\mathrm{de}} + \lambda_1^{\mathrm{di}} \tag{13}$$

$$\lambda_1^{\mathrm{de}} = \frac{1}{2v}\sigma_1^{\mathrm{de}}\varepsilon_1^{\mathrm{d}} \tag{14}$$

$$\lambda_1^{\mathrm{di}} = \frac{1}{2v}\sigma_1^{\mathrm{di}}\varepsilon_1^{\mathrm{d}} \tag{15}$$

where $\lambda_1{}^{\mathrm{d}}$ denotes the total damage caused reduction of stored energy, $\lambda_1{}^{\mathrm{de}}$ denotes the elastic deformation damage (i.e., mode-I fracture) caused reduction of stored energy, $\lambda_1{}^{\mathrm{di}}$ denotes the irreversible damage (due to both mode-II micro-cracks and irreversible fictional sliding) caused reduction of stored energy.

Secondly, considering the stress-strain response during the second loading cycle, the unloading curve A_2B_2 is also simplified by the line A_2B_2 in Figure 7a,b. At point A_2, the strain ε_2 and irreversible damage strain $\varepsilon_2{}^{\mathrm{di}}$ exist in the specimen. The stiffness is changed from E_1 to E_2. Even though these notations are for the uniaxial case, they are able to be used in indicial tensor notation in the equations without loss of generality. The total strain (described by line $OB_2G_2H_2$ and $OB_1B_2G_{1,2}H_2$ in Figure 7b) is given as follows:

$$\varepsilon_2 = \varepsilon_2^{\mathrm{E}} + \varepsilon_2^{\mathrm{I}} = \left(\varepsilon_{0,2} + \varepsilon_2^{\mathrm{de}}\right) + \varepsilon_2^{\mathrm{di}} = \varepsilon_1^{\mathrm{di}} + \varepsilon_{1,2}^{\mathrm{di}} + \varepsilon_{1,2}^{\mathrm{de}} + \left(\varepsilon_{0,2} + \varepsilon_1^{\mathrm{de}}\right) \tag{16}$$

where $\varepsilon_{0,2} = \varepsilon_{0,1}$ (see Figure 7a).

The strain energy is expressed as follows

$$\psi_2 = \psi_1 + \psi_{1,2}^{\mathrm{de}} = \psi_0^{\mathrm{e}} + \psi_1^{\mathrm{de}} + \psi_{1,2}^{\mathrm{de}} \tag{17}$$

where the subscript 2 denotes the cyclic number of the second fatigue loading, $\psi_{1,2}{}^{\mathrm{de}}$ denotes the elastic deformation damage strain energy due to the additional elastic deformation damage during the second cycle (see the area $B_2A_2G_{1,2}$ in Figure 7b), that is,

$$\psi_{1,2}^{\mathrm{de}} = \frac{1}{2v}E_0\varepsilon_{0,1}\varepsilon_{1,2}^{\mathrm{de}} \tag{18}$$

Thus, the strain energy is expressed as follows (see the area described by points $B_2A_2H_2$ in Figure 7a)

$$\psi_2 = \frac{1}{2v}E_0\varepsilon_{0,1}\left(\varepsilon_1^{\mathrm{E}} + \varepsilon_{1,2}^{\mathrm{de}}\right) = \frac{1}{2v}E_0\varepsilon_{0,1}\varepsilon_2^{\mathrm{E}} \tag{19}$$

The irreversible energy γ_2 is expressed as follows

$$\gamma_2 = \gamma_1 + \gamma_{1,2} = \left(\gamma_1^{\mathrm{di}} + \gamma_1^{\mathrm{de}}\right) + \left(\gamma_{1,2}^{\mathrm{di}} + \gamma_{1,2}^{\mathrm{de}}\right) \tag{20}$$

where $\gamma_{1,2}{}^{di}$ denotes the irreversible deformation damage irreversible energy due to the additional irreversible deformation damage during the second cycle (see the area $B_1A_1I_2B_2$ in Figure 7b), and $\gamma_{1,2}{}^{de}$ denotes the elastic deformation damage irreversible energy due to the additional elastic deformation damage during the second cycle (see the area $B_2A_2G_{1,2}$ in Figure 7b), that is,

$$\gamma_{1,2}^{di} = \frac{1}{v}\sigma_{max}\varepsilon_{1,2}^{di} = \frac{1}{v}E_0\varepsilon_{0,1}\varepsilon_{1,2}^{di} \tag{21}$$

$$\gamma_{1,2}^{de} = \frac{1}{2v}\sigma_{max}\varepsilon_{1,2}^{de} = \frac{1}{2v}E_0\varepsilon_{0,1}\varepsilon_{1,2}^{de} \tag{22}$$

Thus, the irreversible energy γ_2 is derived as follows (see the area $OA_0A_2B_2$ in Figure 7b)

$$\begin{aligned}\gamma_2 &= \tfrac{1}{v}E_0\varepsilon_{0,1}\left(\varepsilon_1^{di} + \tfrac{1}{2}\varepsilon_1^{de} + \varepsilon_{1,2}^{di} + \tfrac{1}{2}\varepsilon_{1,2}^{de}\right) = \tfrac{1}{v}E_0\varepsilon_{0,1}\left[\left(\varepsilon_1^{di} + \varepsilon_{1,2}^{di}\right) + \tfrac{1}{2}\left(\varepsilon_1^{de} + \varepsilon_{1,2}^{de}\right)\right]\\ &= \tfrac{1}{v}E_0\varepsilon_{0,1}\left(\varepsilon_2^{di} + \tfrac{1}{2}\varepsilon_2^{de}\right)\end{aligned} \tag{23}$$

The stored energy λ_2 is derived as follows

$$\lambda_2 = \psi_2 + \gamma_2 = \lambda_2^0 - \lambda_2^d \tag{24}$$

where $\lambda_2{}^d$ denotes the total damage caused reduction of stored energy, that is,

$$\lambda_2^d = \lambda_1^d + \lambda_{1,2}^d = \left(\lambda_1^{di} + \lambda_1^{de}\right) + \left(\lambda_{1,2}^{di} + \lambda_{1,2}^{de}\right) \tag{25}$$

where $\lambda_{1,2}{}^{di}$ denotes the irreversible deformation damage caused by the reduction of stored energy due to the additional irreversible deformation damage during the second cycle (see the composite areas $C_1C_2J_1F_1$, $A_1J_3J_4$ and $A_1J_5A_2$ in Figure 7b), and $\lambda_{1,2}{}^{de}$ denotes the elastic deformation damage caused reduction of stored energy due to the additional elastic deformation damage during the second cycle (see the composite areas $F_1J_1J_2A_1$, $A_1J_2J_3$ and $A_1J_4J_5$ in Figure 7b), that is,

$$\begin{aligned}\lambda_{1,2}^{di} &= \tfrac{1}{v}\sigma_1^{di}\left(\varepsilon_2 - \varepsilon_1\right) + \tfrac{1}{2v}\sigma_{1,2}^{di}{}'\left(\varepsilon_2 - \varepsilon_1\right) + \tfrac{1}{2v}\sigma_{1,2}^{di}\left(\varepsilon_2 - \varepsilon_1\right)\\ &= \tfrac{1}{v}\left(\varepsilon_2 - \varepsilon_1\right)\left(\sigma_1^{di} + \tfrac{1}{2}\sigma_{1,2}^{di}{}' + \tfrac{1}{2}\sigma_{1,2}^{di}\right)\end{aligned} \tag{26}$$

$$\begin{aligned}\lambda_{1,2}^{de} &= \tfrac{1}{v}\sigma_1^{de}\left(\varepsilon_2 - \varepsilon_1\right) + \tfrac{1}{2v}\sigma_{1,2}^{de}{}'\left(\varepsilon_2 - \varepsilon_1\right) + \tfrac{1}{2v}\sigma_{1,2}^{de}\left(\varepsilon_2 - \varepsilon_1\right)\\ &= \tfrac{1}{v}\left(\varepsilon_2 - \varepsilon_1\right)\left(\sigma_1^{de} + \tfrac{1}{2}\sigma_{1,2}^{de}{}' + \tfrac{1}{2}\sigma_{1,2}^{de}\right)\end{aligned} \tag{27}$$

Thus, the total damage caused reduction of stored energy $\lambda_2{}^d$ is derived as follows (see the area $A_0C_2A_2$ in Figure 7b)

$$\lambda_2^d = \left(\lambda_1^{di} + \lambda_1^{de}\right) + \left(\lambda_{1,2}^{di} + \lambda_{1,2}^{de}\right) = \lambda_2^{di} + \lambda_2^{de} \tag{28}$$

$$\lambda_2^{di} = \frac{1}{2v}\sigma_2^{di}\varepsilon_2^d \tag{29}$$

$$\lambda_2^{de} = \frac{1}{2v}\sigma_2^{de}\varepsilon_2^d \tag{30}$$

where $\lambda_2{}^{di}$ denotes the irreversible deformation damage (due to both mode-II micro-cracks and irreversible fictional sliding) caused by the reduction of stored energy, $\lambda_2{}^{de}$ denotes the elastic deformation damage (i.e., mode-I fracture) caused by the reduction of stored energy (see Figure 7a).

3.1.2. Interpretation in nth Loading Cycle

Based on the recent thermodynamics interpretation in this work (see Equations (3), (7), (12–15), (19), (23), (24), (28)–(30)), by comparing Equation (19) with Equation (3) and replacing the cycle number 2 by n in Equation (19), it is possible to derive the strain energy after nth loading cycle, as follows:

$$\psi_n = \frac{1}{2v} E_0 \varepsilon_0 \varepsilon_n^E = \frac{1}{2v} E_0 \varepsilon_0 \left(\varepsilon_{n-1}^E + \varepsilon_{n-1,n}^{de} \right) \tag{31}$$

By comparing Equation (23) with Equation (7) and replacing the cycle number 2 by n in Equation (23), it is possible to derive the irreversible energy after nth loading cycle, as follows:

$$\gamma_n = \frac{1}{v} E_0 \varepsilon_0 \left(\varepsilon_n^{di} + \frac{1}{2} \varepsilon_n^{de} \right) = \frac{1}{v} E_0 \varepsilon_0 \left[\left(\varepsilon_{n-1}^{di} + \varepsilon_{n-1,n}^{di} \right) + \frac{1}{2} \left(\varepsilon_{n-1}^{de} + \varepsilon_{n-1,n}^{de} \right) \right] \tag{32}$$

By comparing Equations (24), (28–30) with Equations (12–15) and replacing the cycle number 2 by n in Equations (24), (28–30), it is possible to derive the damage caused by the reduction of stored energy after nth loading cycle, as follows:

$$\lambda_n = \lambda_n^0 - \lambda_n^d = \lambda_n^0 - \left(\lambda_n^{di} + \lambda_n^{de} \right) \tag{33}$$

$$\lambda_n^{di} = \frac{1}{2v} \sigma_n^{di} \varepsilon_n^d \tag{34}$$

$$\lambda_n^{de} = \frac{1}{2v} \sigma_n^{de} \varepsilon_n^d \tag{35}$$

Note that, considering both the micro structural changes (based on the proposed micro damage model, see Figures 3, 4 and 6) and the macro irreversible energy (see Equation (1)), this work is ruled by second thermodynamics law, that is,

$$\sigma \dot{\varepsilon} - v \dot{\psi} \geq 0 \tag{36}$$

3.1.3. Damage Variable Definition and Its Thermodynamics Interpretation

Based on the above thermodynamics interpretation and the damage variable definition method in Appendix A, the elastic deformation fatigue damage variable in the elastic strain space and total strain space (D_n^E and D_n^e), and the irreversible deformation fatigue damage variable (D_n^i) are defined, respectively, as follows (see Figure 7):

$$D_n^E = \frac{\Psi_n^{de}}{\Psi_n^e + \chi_n} = \frac{\psi_n^{de} + \gamma_n^{de}}{\left(\psi_{0,n}^e + \psi_n^{de} + \gamma_n^{de} \right) + \chi_n} = \frac{\varepsilon_n^{de}}{\varepsilon_n^E} \tag{37}$$

$$D_n^e = \frac{\Psi_n^{de}}{\Psi_n + \chi_n} = \frac{\psi_n^{de} + \gamma_n^{de}}{\left(\psi_{0,n}^e + \psi_n^{de} + \gamma_n^{de} + \gamma_n^{di} \right) + \chi_n} = \frac{\varepsilon_n^{de}}{\varepsilon_n} \tag{38}$$

$$D_n^i = \frac{\Psi_n^{di}}{\Psi_n + \chi_n} = \frac{\gamma_n^{di}}{\left(\psi_{0,n}^e + \psi_n^{de} + \gamma_n^{de} + \gamma_n^{di} \right) + \chi_n} = \frac{\varepsilon_n^{di}}{\varepsilon_n} \tag{39}$$

$$D_n = \frac{\Psi_n^d}{\Psi_n + \chi_n} = D_n^e + D_n^i = \frac{\varepsilon_n^{de} + \varepsilon_n^{di}}{\varepsilon_n} = \frac{\varepsilon_n^d}{\varepsilon_n} = \frac{\varepsilon_n - \varepsilon_{0,n}}{\varepsilon_n} = \frac{\varepsilon_n - \sigma_{max}/E_0}{\varepsilon_n} \tag{40}$$

where ε_n^d, ε_n^{de}, and ε_n^{di} denote the strain development caused by the total damage, the elastic deformation damage and the irreversible deformation damage, respectively (related to the total cracking, the mode-I cracking and the irreversible deformation cracking discussed in micro-mechanical description in this work, respectively), when the material is subjected to the nth cyclic loading (Figure 7); ε_n^E denotes the elastic strain, $\varepsilon_n^E = \varepsilon_{0,n} + \varepsilon_n^{de}$ ($\varepsilon_{0,n} = \varepsilon_0$); ε_n denotes the total strain, $\varepsilon_n = \varepsilon_{0,n} + \varepsilon_n^{de} +$

ε_n^{di}; and χ_n denotes the energy dissipation exists as a typical characteristics of the elastic behaviors, it accompanies and equals the initial strain energy $\psi_{0,n}^e$ (see Figures 7 and A1), that is,

$$\chi_n = \frac{1}{2v}\sigma_{max}\varepsilon_{0,n} = \frac{1}{2v}\sigma_{max}\varepsilon_0 = \psi_{0,n}^e = \psi_0^e \tag{41}$$

Note that Equation (40) illustrates that the damage evolution D_n depending on the total strain ε_n will be varied, when σ_{max} is changed during different fatigue loading processes.

The nonlinear stress-strain relation is described as follows:

$$\begin{aligned}\sigma &= \left(1 - D_n^E\right)E_0\left(\varepsilon_n - \varepsilon_n^{di}\right) = \left(1 - D_n^E\right)\left(1 - D_n^i\right)E_0\varepsilon_n \\ &= \left[1 - \left(D_n^e + D_n^i\right)\right]E_0\varepsilon_n = (1 - D_n)E_0\varepsilon_n\end{aligned} \tag{42}$$

The damage variables D_n^E (or D_n^e) and D_n^i are able to be used to characterize the stiffness degradation and irreversible strain development of concrete under fatigue loading, respectively, as follows (see Figure 7a–c),

$$E_n = \left(1 - D_n^E\right)E_0 = \left(1 - \frac{D_n^e}{1 - D_n^i}\right)E_0 = \frac{1 - D_n}{1 - D_n^i}E_0 \tag{43}$$

$$\varepsilon_n^{di} = D_n^i\varepsilon_n \tag{44}$$

Given that in the case of general engineering the value of the minimum stress σ_{min} is not equal to zero in concrete (Figure 7d), the residual strain ε_n^r related to the fatigue behaviors is distinguished from the irreversible strain ε_n^{di} by the following definition:

$$\varepsilon_n^r = \varepsilon_n - \frac{\sigma_{max} - \sigma_{min}}{E_n} \tag{45}$$

$$\varepsilon_n^{di} = \varepsilon_n - \frac{\sigma_{max}}{E_n} \tag{46}$$

which are obtained by the equations (Figure 7), respectively, as follows,

$$E_n = \frac{\sigma_{max} - \sigma_{min}}{\varepsilon_n - \varepsilon_n^r} \tag{47}$$

$$E_n = \frac{\sigma_{max}}{\varepsilon_n - \varepsilon_n^{di}} \tag{48}$$

Note that little research [6,10–12,41–49] has considered the difference between the residual strains and the irreversible strains, however, this difference is essential for characterizing the fatigue behaviors of concrete. Specifically, Equations (45) and (46) and Figure 7 illustrate that the value of the residual strains is usually higher than that of the irreversible strains.

Additionally, with Equations (9), (10), (14), (15), (17)–(19), (21), (22), (29), (30), (38) and (39) and Figure 7, it is able to correlate the mechanical parameters and damage variables to the thermodynamics parameters as follows:

$$\frac{\varepsilon_n^{de}}{\varepsilon_n^{di}} = \frac{\sigma_n^{de}}{\sigma_n^{di}} = \frac{D_n^e}{D_n^i} = \frac{\gamma_n^{de} + \psi_n^{de}}{\gamma_n^{di}} = \frac{\lambda_n^{de}}{\lambda_n^{di}} \tag{49}$$

Equation (49) shows that the two components of fatigue damage variable (D_n^e and D_n^i) are able to be correlated to the strain and stress decomposition and the energy dissipation, i.e., the energy ($\gamma_n^{de} + \psi_n^{de}$) and γ_n^{de}, λ_n^{de} and λ_n^{di}, respectively. Therefore, Equation (49) illustrates the thermodynamics interpretation of the newly defined damage variables.

3.2. Continuum Damage Mechanics Model

The damage evolution is the variation process of the micro structure in material under external load, i.e., the process of the crack initiation, development and converge, thus, it reveals the physical nature of certain material and it does not impact the stress status [53]. Hence, the fatigue damage evolution of concrete in uniaxial case is presumed to be applied on multi-axial case without loss of generality. Therefore, the constitutive model in scalar form (Equation (42)) is able to be extended into the tensor form, as follows:

$$\sigma_n = [\mathbf{I} - \mathbf{D}_n] : \mathbf{E}_0 : \varepsilon_n \tag{50}$$

where $\mathbf{D}_n$ denotes the tensor fatigue damage variable, which is used to model the nonlinearity of stress-strain response and can be expressed as follows:

$$\mathbf{D}_n = D_{i,n}^{\pm}\mathbf{P}_i^{\pm} = D_{1,n}^{+}\mathbf{P}_1^{+} + D_{1,n}^{-}\mathbf{P}_1^{-} + D_{2,n}^{+}\mathbf{P}_2^{+} + D_{2,n}^{-}\mathbf{P}_2^{-} + D_{3,n}^{+}\mathbf{P}_3^{+} + D_{3,n}^{-}\mathbf{P}_3^{-} \tag{51}$$

$$\begin{cases} \mathbf{P}_i^{+} = \mathrm{H}[\sigma_i(n)]m^{(i)} \otimes m^{(i)} \otimes m^{(i)} \otimes m^{(i)} \\ \mathbf{P}_i^{-} = \{1 - \mathrm{H}[\sigma_i(n)]\}m^{(i)} \otimes m^{(i)} \otimes m^{(i)} \otimes m^{(i)} \end{cases} \tag{52}$$

where i denotes the number of the principal stress direction, $i = 1, 2, 3$, for simplicity, in the uniaxial case, it can be omitted; H(x) denotes the Heaviside function, if $x > 0$, its value is 1, otherwise 0; + and − denote the tensile and compressive loading condition, respectively.

In the biaxial stress condition, the fatigue damage constitutive model in the principal stress direction is able to be described as follows:

$$\begin{Bmatrix} \sigma_{1,n} \\ \sigma_{2,n} \\ \tau_{12,n} \end{Bmatrix} = \frac{1}{1 - \mu^2} \begin{bmatrix} \alpha_1\left(1 - D_{1,n}^{\pm}\right) & & \\ & \alpha_2\left(1 - D_{2,n}^{\pm}\right) & \\ & & 1 \end{bmatrix} \begin{bmatrix} E_0 & \mu E_0 & \mu E_0 \\ \mu E_0 & E_0 & \mu E_0 \\ \mu E_0 & \mu E_0 & E_0 \end{bmatrix} \begin{Bmatrix} \varepsilon_{1,n} \\ \varepsilon_{2,n} \\ \varsigma_{12,n} \end{Bmatrix} \tag{53}$$

where α denotes the parameter considering the bia-compressive effects [22,53], its value can be obtained from [22]; μ denotes the Poisson ratio; τ denotes the shear stress; ς denotes the shear strain.

The cyclic creep and stiffness degradation in a three-stage process are characterized by the tensor fatigue irreversible deformation damage variable $\mathbf{D}_n^{\mathrm{i}}$ and the tensor fatigue elastic deformation damage variable $\mathbf{D}_n^{\mathrm{e}}$, respectively, and the post-fatigue stress-strain response is also described by the recently defined damage variables. As a result, the current stiffness and the irreversible/residual strain are able to be described as follows, respectively:

$$E_n = \frac{1 - D_n^{\pm}}{1 - D_n^{\mathrm{i}\pm}} E_0 \tag{54}$$

$$\varepsilon_n^{\mathrm{di}} = D_n^{\mathrm{i}\pm} \varepsilon_n \tag{55}$$

$$\varepsilon_n^{\mathrm{r}} = \varepsilon_n - \frac{\sigma_{\max} - \sigma_{\min}}{E_n} \tag{56}$$

The post-fatigue stress-strain response of concrete under monotonic uniaxial loading is assumed to be expressed as follows:

$$\sigma_n^{\mathrm{p}\pm} = \begin{cases} E_n \cdot \left(\varepsilon - \varepsilon_n^{\mathrm{di}}\right), & \varepsilon_n^{\mathrm{di}} \leq \varepsilon \leq \sigma_{\max}/E_n \\ \left[1 - k_N^{\pm} \cdot D_n^{f\mathrm{cu}\pm}\right] \cdot E_0 \cdot \left(\varepsilon - \varepsilon_n^{\mathrm{di}}\right), & \sigma_{\max}/E_n < \varepsilon \leq \varepsilon_N \\ \left(1 - D^{\mathrm{s}\pm}\right) \cdot E_0 \cdot \varepsilon, & \varepsilon > \varepsilon_N \end{cases} \tag{57}$$

where $D_n^{f\mathrm{cu}}$ denotes a simplified parameter related to strength reduction, $D_n^{f\mathrm{cu}\pm} = (1 - S^{\pm}) \cdot D_n^{0\pm}$, D_n^{0} denotes a newly introduced damage variable, which will be defined in Section 4.1; k_N denotes a modifying parameter considering the bound condition of fatigue failure surface [4], if $\varepsilon_N \leq \varepsilon \leq \varepsilon_{f\mathrm{cu}}$,

$k_N = (\varepsilon_N - \varepsilon)/(\varepsilon_N - \varepsilon_{fcu})$, otherwise, $k_N = 1$, and ε_{fcu} denotes the strain corresponding to the peak stress f_{cu} under monotonic uniaxial loading. Furthermore, the residual strength is described as follows:

$$f_n^{r\pm} = f_{cu}^{\pm} \cdot \left[1 - D_n^{fcu\pm} \right] \tag{58}$$

4. Verification and Discussions

4.1. Solution Procedure of CDM Model for Concrete under Fatigue Loading

In order to characterize the three-stage behaviors of concrete under fatigue loading considering different stress level S, based on the concept of the fatigue failure surface [4], a normalized fatigue damage variable $D_n{}^0$ is defined in this work as follows:

$$D_n^{0\pm} = \frac{D_n^{\pm} - D_1^{\pm}}{D_N^{\pm} - D_1^{\pm}} \tag{59}$$

where D_1 and D_N denote the damage variable of the joint points of the static stress-strain curve and the level line $\sigma = \sigma_{max}$, respectively, and they are able to be calculated by the model in [22] as follows:

$$\sigma^{\pm} = (1 - D^{s\pm})E_0\varepsilon^{\pm} \tag{60}$$

where $\sigma = \sigma_{max}$, $\varepsilon = \varepsilon_1$ or $\varepsilon = \varepsilon_N$, ε_1 and ε_N denote the strain of the joint points of the static stress-strain curve and the level line $\sigma = \sigma_{max}$, respectively, D^s denotes the damage variable of concrete under monotonic uniaxial loading, and it is able to be predicted by the model in [22] as follows:

$$D^{s\pm} = \frac{A_1^{\pm} - A_2^{\pm}}{1 + \left(\varepsilon - \varepsilon_0^{\pm}\right)^{p\pm}} + A_2^{\pm} \tag{61}$$

where A_1, A_2, ε_0, and p denote the parameters related to the damage evolution in concrete under monotonic uniaxial loading, and they are able to be calibrated by experimental results [22].

Thus, the fatigue damage variable of concrete is derived as follows:

$$D_n^{\pm} = D_1^{\pm} + D_n^{0\pm} \cdot \left(D_N^{\pm} - D_1^{\pm} \right) \tag{62}$$

By a set of trial and error procedures, the normalized fatigue damage variable is assumed to be modeled by the equation:

$$D_n^{0\pm} = A_3^{\pm} \cdot \left(-\frac{n/N}{n/N - A_4^{\pm}} \right)^{\frac{1}{A_5^{\pm}}} \tag{63}$$

where A_3, A_4, and A_5 denote the parameters related to the damage in concrete during fatigue loading, and they are able to be calibrated by the experimental results.

The literature [54,55] presumed that the reciprocal of irreversible deformation damage variable $1/D^i$ is linearly dependent on the total damage variable D. However, by applying the above linearly analysis, the resulted value of irreversible deformation damage variable D^i is overestimated in certain cases. For instance, it is found that the calculated irreversible deformation damage D^i is greater than the total damage D when the total damage D is approximately lower than 0.2, which is unreasonable (see Equation (40) and Figure 7). Hence, the relation between $1/D^i$ and D is assumed to be corrected as follows,

$$D_n^i = 1/\left(B_1 + B_2 \cdot D_n + B_3 \cdot D_n^2 \right) \tag{64}$$

where B_1, B_2 and B_3 denote the parameters related to the coupling of irreversible and elastic deformation damage in concrete under loading, and they are able to be calibrated by the experimental results of concrete under cyclic loading [54,55].

The fatigue life is estimated by the convenient method [56], such that,

$$\log N = 14.7 - 13.5 \frac{\sigma_{\max} - \sigma_{\min}}{f_{cu}^{\pm} - \sigma_{\min}}.$$ (65)

Therefore, this CDM model is able to characterize the mechanical behaviors of concrete under fatigue loading including following characteristics: the progressive stiffness degradation, development of cyclic creep and residual strain in a three-stage process, and the post-fatigue stress-strain response under monotonic loading.

4.2. Behaviors of Concrete under Fatigue Compression with Constant Amplitude

In order to verify the effectiveness of the proposed model, the predictions are obtained using a calibration method similar to that in the literature [45] and compared with the experimental results. A series of experiments of concrete under fatigue compression with a constant amplitude are conducted and the results are reported in [57]. In this section, a typical result is selected for model verification. By using the definition in Equations (38–40), the calibrated parameters are obtained and listed as follows, $E_0 = 32.3$ GPa, $f_{cu}^{-} = 49.3$ MPa, $A_1^{-} = 0$, $A_2^{-} = 1$, $\varepsilon_0^{-} = 1700.7$ με, $p^{-} = 2.138$, $A_3^{-} = 1.864$, $A_4^{-} = 6.961$, $A_5^{-} = 2.702$, $B_1 = 38.19$, $B_2 = -80.77$, and $B_3 = 44.66$. Figure 8 illustrates the agreement of the predicted and the experimental results. In detail, initially, the three stages evolution of cyclic creep (i.e., the total fatigue strains and residual strains) of both the predicted and experimental results are coincidental (Figure 8a). Additionally, the proposed model is able to reproduce the stiffness degradation during fatigue life (Figure 8b).

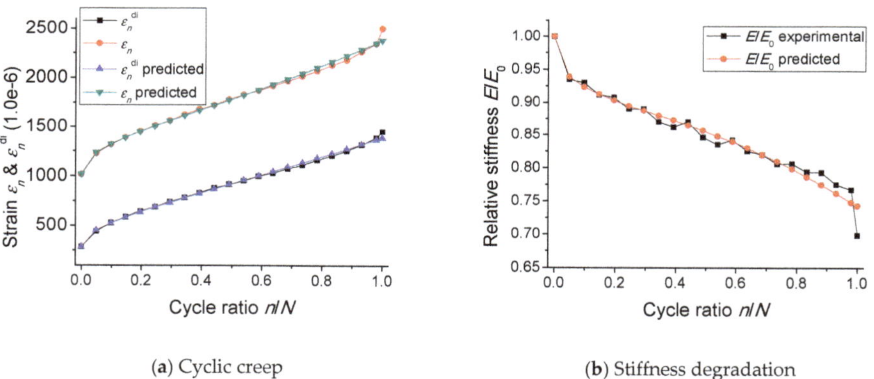

(a) Cyclic creep

(b) Stiffness degradation

Figure 8. Calibration of parameters in the model, and comparison between experimental [57] (The graphs are completely redrawn by authors).and predicted results.

4.3. Behaviors of Concrete under Fatigue Compression with Various Stress Levels

In order to model the behaviors of concrete under fatigue compression with various stress levels of constant amplitudes, a typical result [4] is used for analysis in this section. The parameters are calibrated as follows, $E_0 = 21.8$ GPa, $f_{cu}^{-} = 26.0$ MPa, $A_1^{-} = 0$, $A_2^{-} = 1$, $\varepsilon_0^{-} = 2306.5$ με, $p^{-} = 2.464$, $A_3^{-} = 0.713$, $A_4^{-} = 1.160$, $A_5^{-} = 5.439$. The cyclic creep and the fatigue strain [4] (defined as the fatigue strain = the total strain ε_n—the initial strain (i.e., the total strain in the first cycle ε_1)) are predicted and compared with experimental results in Figure 9.

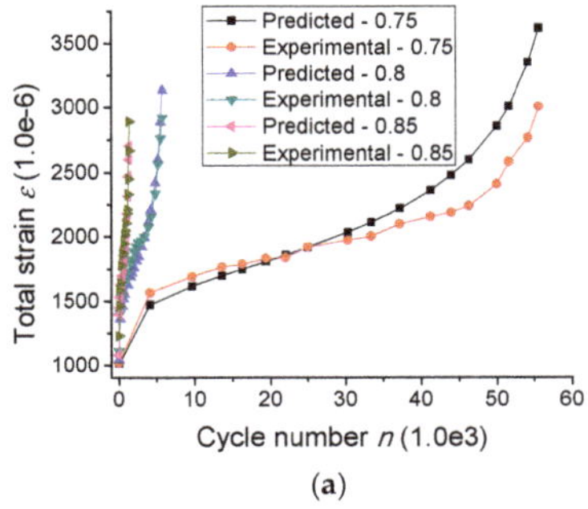
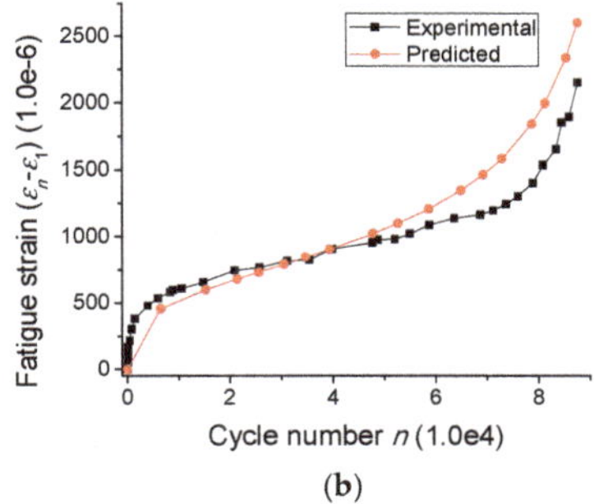

Figure 9. Comparison of total fatigue strains (**a**) and fatigue strains (**b**) between predicted and experimental [4] (The graphs are completely redrawn by authors). results, where the fatigue strain = the total strain ε_n—the initial strain (i.e., the total strain in the first cycle ε_1).

Figure 9 illustrates that the predictions agree with the experimental results. In detail, initially, both the total fatigue strains and the fatigue strains in predictions agree with the experimental results. Additionally, the proposed model is able to reproduce the three stages evolution of total fatigue strains and fatigue strains during fatigue life. It is noted that the tail results in predictions are slightly higher than the experimental results, since the strain due to static stress-strain response in concrete is developing faster than the total fatigue strains in experiments [4], which leads to the higher strains predictions (Figure 9) by using Equations (40) and (60) based on fatigue failure surface concept [4]. Therefore, the proposed model is able to characterize the behaviors of concrete under fatigue compression with various stress levels with constant amplitudes.

4.4. Behaviors of Concrete under Biaxial Fatigue Compression

In order to model the behaviors of concrete under biaxial fatigue compression with constant amplitude, a typical result [58] (Figure 10) is applied for analysis in this section. The parameters are calibrated as follows, $E_0 = 26.3$ GPa, $f_{cu}^- = 20.47$ MPa, $A_1^- = 0$, $A_2^- = 1$, $\varepsilon_0^- = 1568.1$ µε, $p^- = 2.218$, $A_3^- = 0.713$, $A_4^- = 1.160$, $A_5^- = 5.439$ and $\alpha_1 = 1.254$. Figure 10 illustrates that the predicted results of the proposed model agree well with the experimental results. Therefore, it can be concluded that the proposed model is applicable in the analysis of concrete under biaxial fatigue loading.

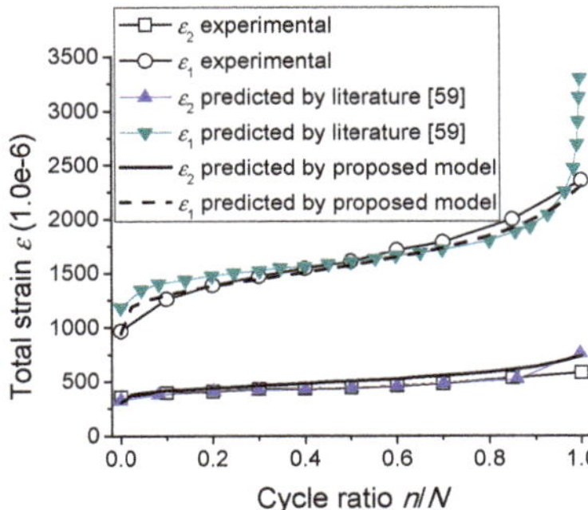

Figure 10. Comparison of cyclic creep under biaxial fatigue compression among predicted and experimental results [58] (The graphs are completely redrawn by authors).

4.5. Post-Fatigue Constitutive Behaviors of Concrete under Monotonic Loading

To verify the effectiveness of the proposed model in predicting the post-fatigue stress–strain response of concrete, the literature [59] is applied to calibrate the parameters, as follows, $E_0^- = 36.0$ GPa, $f_{cu}^- = 41.4$ MPa, $A_1^- = 0$, $A_2^- = 1$, $\varepsilon_0^- = 2153.8$ με, $p^- = 3.186$, $A_3^- = 0.713$, $A_4^- = 1.160$, $A_5^- = 5.439$. The predicted post-fatigue stress-strain responses are obtained in Figure 11a. Figure 11a illustrates that the post-fatigue stress–strain responses predicted by the model varies in a typical three stages way depending on the increasing of cycle ratio. In detail, the development of the initial strains (i.e., the residual strains) of the responses grows in a three stages way, and the variation of the initial (post-fatigue) stiffness and the residual strength experience in a similar way. Therefore, the proposed model is able to reproduce the post-fatigue constitutive behaviors.

4.6. Comparison among Proposed Model and Other Models

The proposed model is compared with the typical damage evolution models [60,61]. The parameters are calibrated by using the experimental results in the literature [61], such that, $E_0^- = 54.5$ GPa, $A_3^- = 0.6$, $A_4^- = 1.01$, $A_5^- = 9$, $B_1 = 2.29$, $B_2 = -1.23$, and $B_3 = 0$. Thus, the predictions are obtained in Figure 11b–d. Figures 10 and 11b illustrate that the predictions of the proposed model are more accurate than those of other damage evolution models [60,61]. In addition, the proposed model is able to reproduce the other behavior variations under fatigue loading (e.g., the development of cyclic creep (Figures 10 and 11c), stiffness degradation (Figure 11d) and post-fatigue constitutive behavior (Figure 11a), which are not well considered in the other models [60,61].

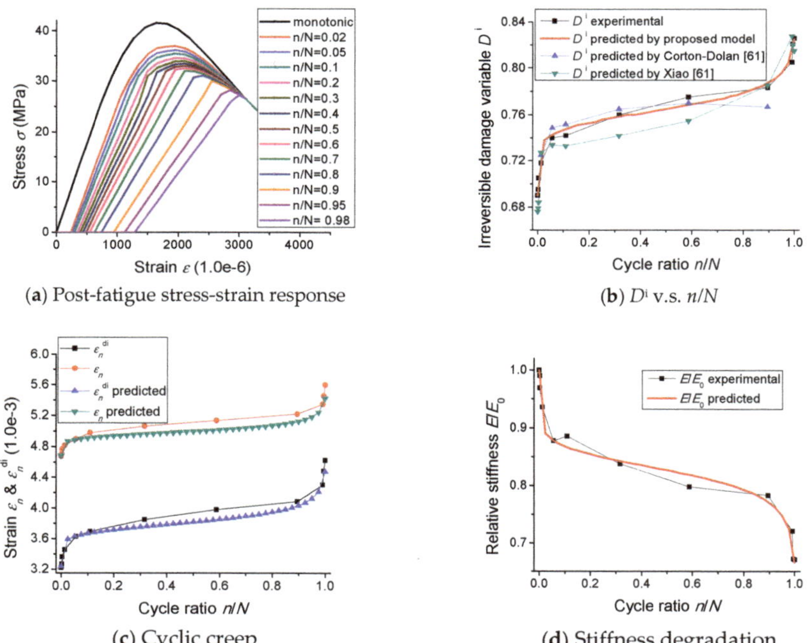

(**a**) Post-fatigue stress-strain response

(**b**) D^i v.s. n/N

(**c**) Cyclic creep

(**d**) Stiffness degradation

Figure 11. Post-fatigue stress–strain response predicted by the proposed model (n/N = 0.02~0.98, S = 0.66), and comparison of D^i, cyclic creep and stiffness degradation among predicted and experimental results [61] (The graphs are redrawn by authors).

Furthermore, the proposed models are capable of predicting the variations of cyclic creep, stiffness degradation, residual strength, and the stress–strain relationship under both fatigue loading and post-fatigue loading. However, few damage models [6,41–49] took all the characteristics above into account for relevant characterizing.

Additionally, the proposed model obtains a clear physical consideration based on the micro mechanical description of damage behaviors in concrete under multi-axial loading, and proposed a behavior characterizing method based on both the above-mentioned description and a thermodynamics-based CDM method. However, in the classical damage models [44,45], the yield concept cannot coexist with the loading/unloading irreversible strain concept [62] introduced in their framework. In detail, in the yield concept [44,45], there is only one yield surface for determining the plastic strains in the material. However, in the loading/unloading irreversible strain concept [62], each loading/unloading process (i.e., a loading cycle) obtains a corresponding irreversible strain surface for the development of irreversible strains.

5. Conclusions

In this work, a damage model with the concept of mode-II microcracks using thermodynamic interpretation of damage behaviors for concrete under fatigue loading was developed.

In detail, by applying the micromechanics method, a micro-cell-based damage model was introduced to understand the damage behavior. The mode-II microcracks were further introduced as a contributing part of irreversible/residual strains.

Additionally, by introducing the physical interpretation of the damage variable based on the thermodynamic method, a novel fatigue damage variable (irreversible deformation fatigue damage variable) was proposed to describe the irreversible strains. With this methodology, a continuum damage mechanics model considered both the elastic and irreversible deformation fatigue damages was developed.

It is found that the predictions of this model highly agreed with experimental results. This model is able to characterize the variations of cyclic creep, stiffness degradation, residual strength, and the post-fatigue stress-strain relationship of concrete. The model can also be used to analyze the behaviors of concrete under complex fatigue loads such as a multi-axial case.

Author Contributions: Conceptualization, Z.S., Z.Y., X.L. (Xiao Li), X.L. (Xiaoyong Lv) and Z.L.; methodology, Z.S., Z.Y., X.L. (Xiao Li), X.L. (Xiaoyong Lv) and Z.L.; software, Z.S., Z.Y., X.L. (Xiao Li), X.L. (Xiaoyong Lv) and Z.L.; validation, Z.S., Z.Y., X.L. (Xiao Li), X.L. (Xiaoyong Lv) and Z.L.; formal analysis, Z.S., Z.Y., X.L. (Xiao Li), X.L. (Xiaoyong Lv) and Z.L.; investigation, Z.S., Z.Y., X.L. (Xiao Li), X.L. (Xiaoyong Lv) and Z.L.; resources, Z.S., Z.Y., X.L. (Xiao Li), X.L. (Xiaoyong Lv) and Z.L.; data curation, Z.S., Z.Y., X.L. (Xiao Li), X.L. (Xiaoyong Lv) and Z.L.; writing—original draft preparation, Z.S., Z.Y., X.L. (Xiao Li), X.L. (Xiaoyong Lv) and Z.L.; writing—review and editing, Z.S., Z.Y., X.L. (Xiao Li), X.L. (Xiaoyong Lv) and Z.L.; visualization, Z.S., Z.Y., X.L. (Xiao Li), X.L. (Xiaoyong Lv) and Z.L.; supervision, Z.S. and Z.Y.; project administration, Z.S. and Z.Y.; funding acquisition, Z.S. and Z.Y.

Funding: This research was supported by the National Key R&D Program of China, grant number 2018YFD1100401; the National Natural Science Foundation of China, grant numbers 51808558, 51820105014, U1434204, 51378506, 51478478; the Natural Science Foundation of Hunan Province, China, grant number 2019JJ50800; the China Energy Investment Corporation, grant number SHGF-18-50.

Conflicts of Interest: The authors declare no conflict of interest.

Appendix A. Damage Variable Definition Based on Thermodynamics

(1)   Perfect elastic materials

The Helmholtz free energy of a perfect elastic material per unit mass is obtained as follows (see Figure A1a):

$$\Psi = \psi + \gamma = \psi = \psi_0^e \tag{A1}$$

And the energy dissipation χ is introduced in this work (see Figure A1), that is,

$$\chi = \frac{1}{2v}\sigma\varepsilon_0 = \frac{1}{2v}\sigma\varepsilon = \psi_0^{\mathrm{e}} \tag{A2}$$

where χ denotes that energy dissipation exists as a typical property of the elastic behaviors, it accompanies the initial strain energy $\psi_0{}^{\mathrm{e}}$, and both are equal to each other like the twins, see Figure A1.

There is no damage appearing in the material, thus, it is not necessary to define a damage variable. And the stress–strain relation is described as follows:

$$\sigma = E_0\varepsilon \tag{A3}$$

(2)   Elastic deformation damage materials

The Helmholtz free energy of an elastic deformation damage material is obtained as follows (see Figure A1b):

$$\Psi = \psi + \gamma = \left(\psi_0^{\mathrm{e}} + \psi^{\mathrm{de}}\right) + \gamma^{\mathrm{de}} \tag{A4}$$

where

$$\psi^{\mathrm{de}} = \gamma^{\mathrm{de}} = \frac{1}{2v}\sigma\varepsilon^{\mathrm{de}} \tag{A5}$$

And the energy dissipation χ is derived in this work (see Figure A1), that is,

$$\chi = \frac{1}{2v}\sigma\varepsilon_0 = \psi_0^{\mathrm{e}} \tag{A6}$$

The elastic deformation damage variable in the elastic strain space is defined in this work by two methods considering two different energy dissipation aspects, respectively, as follows:

$$D^{\mathrm{E}} = \frac{\psi^{\mathrm{de}}}{\psi} = \frac{\psi^{\mathrm{de}}}{\psi_0^{\mathrm{e}} + \psi^{\mathrm{de}}} = \frac{\varepsilon^{\mathrm{de}}}{\varepsilon_0 + \varepsilon^{\mathrm{de}}} = \frac{\varepsilon^{\mathrm{de}}}{\varepsilon^{\mathrm{E}}} \tag{A7}$$

$$D^{\mathrm{E}} = \frac{\Psi^{\mathrm{de}}}{\Psi + \chi} = \frac{\psi^{\mathrm{de}} + \gamma^{\mathrm{de}}}{\left(\psi_0^{\mathrm{e}} + \psi^{\mathrm{de}} + \gamma^{\mathrm{de}}\right) + \chi} = \frac{\varepsilon^{\mathrm{de}}}{\varepsilon^{\mathrm{E}}} \tag{A8}$$

where ε^{E} denotes the elastic strains, $\varepsilon^{\mathrm{E}} = \varepsilon_0 + \varepsilon^{\mathrm{de}}$.

The nonlinear stress–strain relation is described as follows:

$$\sigma = \left(1 - D^{\mathrm{E}}\right)E_0\varepsilon \tag{A9}$$

The damage variable D^{E} is able to be used to describe the stiffness degradation, that is (Figure A1b),

$$E = \left(1 - \frac{\varepsilon^{\mathrm{de}}}{\varepsilon}\right)E_0 = \left(1 - D^{\mathrm{E}}\right)E_0 \tag{A10}$$

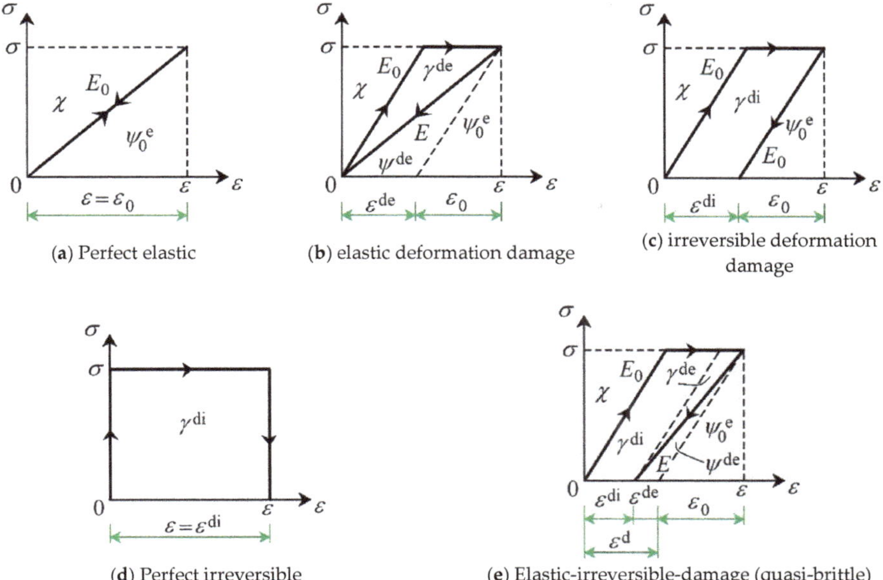

Figure A1. Sketch of the energy dissipation for various materials under uniaxial loading.

(3) Irreversible deformation damage materials

The Helmholtz free energy of an irreversible deformation damage material is obtained as follows (see Figure A1c):

$$\Psi = \psi + \gamma = \psi_0^e + \gamma^{di} \tag{A11}$$

And the energy dissipation χ is derived in this work (see Figure A1), that is,

$$\chi = \frac{1}{2\upsilon}\sigma\varepsilon_0 = \psi_0^e \tag{A12}$$

The irreversible deformation damage variable is defined in this work by the method considering energy dissipation, as follows:

$$D^i = \frac{\psi^{di}}{\Psi + \chi} = \frac{\gamma^{di}}{\left(\psi_0^e + \gamma^{di}\right) + \chi} = \frac{\varepsilon^{di}}{\varepsilon} \tag{A13}$$

where $\varepsilon = \varepsilon_0 + \varepsilon^{di}$.

The nonlinear stress–strain relation is described as follows:

$$\sigma = E_0\left(\varepsilon - \varepsilon^{di}\right) = \left(1 - D^i\right)E_0\varepsilon \tag{A14}$$

The damage variable D^i is able to be used to describe the irreversible strain development, that is (see Figure A1c),

$$\varepsilon^{di} = D^i\varepsilon \tag{A15}$$

Additionally, when the initial stiffness approaches an infinite value ∞, the material exhibits a prefect brittle-plastic behavior (see Figure A1d). Thus, the Helmholtz free energy is derived as follows,

$$\Psi = \psi + \gamma = \gamma = \gamma^{di} \tag{A16}$$

due to the energy dissipation $\chi = \psi_0^{\mathrm{e}} = 0$. The irreversible deformation damage variable is obtained, as follows:

$$D^{\mathrm{i}} = \frac{\psi^{\mathrm{di}}}{\Psi + \chi} = \frac{\gamma^{\mathrm{di}}}{\gamma^{\mathrm{di}}} = 1 \tag{A17}$$

And the irreversible strain is derived, as follows:

$$\varepsilon^{\mathrm{di}} = D^{\mathrm{i}}\varepsilon = \varepsilon \tag{A18}$$

Furthermore, the nonlinear stress-strain relation is described as follows:

$$\begin{cases} \sigma = \sigma \\ \varepsilon = \varepsilon^{\mathrm{di}} \end{cases} \tag{A19}$$

(4) Quasi-brittle materials (Elastic-irreversible deformation damage materials)

The Helmholtz free energy of a quasi-brittle material is obtained as follows (see Figure A1e):

$$\Psi = \psi + \gamma = \left(\psi_0^{\mathrm{e}} + \psi^{\mathrm{de}}\right) + \left(\gamma^{\mathrm{de}} + \gamma^{\mathrm{di}}\right) \tag{A20}$$

where

$$\psi^{\mathrm{de}} = \gamma^{\mathrm{de}} = \frac{1}{2v}\sigma\varepsilon^{\mathrm{de}} \tag{A21}$$

And the energy dissipation χ is derived in this work (see Figure A1), that is,

$$\chi = \frac{1}{2v}\sigma\varepsilon_0 = \psi_0^{\mathrm{e}} \tag{A22}$$

The elastic deformation damage variable in the elastic strain space is defined in this work by two methods considering two different energy dissipation aspects, respectively, as follows:

$$D^{\mathrm{E}} = \frac{\psi^{\mathrm{de}}}{\psi} = \frac{\psi^{\mathrm{de}}}{\psi_0^{\mathrm{e}} + \psi^{\mathrm{de}}} = \frac{\varepsilon^{\mathrm{de}}}{\varepsilon_0 + \varepsilon^{\mathrm{de}}} = \frac{\varepsilon^{\mathrm{de}}}{\varepsilon^{\mathrm{E}}} \tag{A23}$$

$$D^{\mathrm{E}} = \frac{\psi^{\mathrm{de}}}{\Psi^{\mathrm{e}} + \chi} = \frac{\psi^{\mathrm{de}} + \gamma^{\mathrm{de}}}{\left(\psi_0^{\mathrm{e}} + \psi^{\mathrm{de}} + \gamma^{\mathrm{de}}\right) + \chi} = \frac{\varepsilon^{\mathrm{de}}}{\varepsilon^{\mathrm{E}}} \tag{A24}$$

The elastic deformation damage variable in the total strain space is also defined in this work by the method considering energy dissipation, as follows:

$$D^{\mathrm{e}} = \frac{\psi^{\mathrm{de}}}{\Psi + \chi} = \frac{\psi^{\mathrm{de}} + \gamma^{\mathrm{de}}}{\left(\psi_0^{\mathrm{e}} + \psi^{\mathrm{de}} + \gamma^{\mathrm{de}} + \gamma^{\mathrm{di}}\right) + \chi} = \frac{\varepsilon^{\mathrm{de}}}{\varepsilon} \tag{A25}$$

where the total strain $\varepsilon = \varepsilon_0 + \varepsilon^{\mathrm{de}} + \varepsilon^{\mathrm{di}}$, and the elastic strain $\varepsilon^{\mathrm{E}} = \varepsilon_0 + \varepsilon^{\mathrm{de}}$.

The irreversible deformation damage variable is defined in this work by the method considering energy dissipation, as follows:

$$D^{\mathrm{i}} = \frac{\psi^{\mathrm{di}}}{\Psi + \chi} = \frac{\gamma^{\mathrm{di}}}{\left(\psi_0^{\mathrm{e}} + \psi^{\mathrm{de}} + \gamma^{\mathrm{de}} + \gamma^{\mathrm{di}}\right) + \chi} = \frac{\varepsilon^{\mathrm{di}}}{\varepsilon} \tag{A26}$$

The nonlinear stress–strain relation is described as follows:

$$\begin{aligned} \sigma &= \left(1 - D^{\mathrm{E}}\right)E_0\left(\varepsilon - \varepsilon^{\mathrm{di}}\right) = \left(1 - D^{\mathrm{E}}\right)\left(1 - D^{\mathrm{i}}\right)E_0\varepsilon \\ &= \left[1 - \left(D^{\mathrm{e}} + D^{\mathrm{i}}\right)\right]E_0\varepsilon = (1 - D)E_0\varepsilon \end{aligned} \tag{A27}$$

The damage variable D^{E} (or D^{e}) and D^{i} is able to be employed to characterize the stiffness degradation and irreversible strain development, respectively, as follows (see Figure A1e),

$$E = \left(1 - D^{\mathrm{E}}\right)E_0 = \left(1 - \frac{D^{\mathrm{e}}}{1 - D^{\mathrm{i}}}\right)E_0 = \frac{1 - D}{1 - D^{\mathrm{i}}}E_0 \tag{A28}$$

$$\varepsilon^{\mathrm{di}} = D^{\mathrm{i}}\varepsilon \tag{A29}$$

Note that the energy dissipation χ presents a typical property of materials' elastic behaviors. Figure A1 and Equations (A2), (A6), (A12) and (A22) show that the energy dissipation χ is equal to the initial strain energy $\psi_0{}^{\mathrm{e}}$, i.e., $\chi = \psi_0{}^{\mathrm{e}}$, and in a limit case without elastic behaviors in Figure A1d, $\chi = \psi_0{}^{\mathrm{e}} = 0$. Additionally, the damage variable definition method considering the energy dissipation χ in Equations (A8) and (A24) is as effective as that in Equations (A7) and (A23), respectively. Therefore, it is reasonable to apply the damage variable definition method considering the energy dissipation χ in this work.

References

1. Susmel, L. A unifying methodology to design un-notched plain and short-fibre/particle reinforced concretes against fatigue. *Int. J. Fatigue* **2014**, *61*, 226–243. [CrossRef]
2. Chen, Y.; Ni, J.; Zheng, P.; Azzam, R.; Zhou, Y.; Shao, W. Experimental research on the behaviour of high frequency fatigue in concrete. *Eng. Fail. Anal.* **2011**, *18*, 1848–1857. [CrossRef]
3. Guo, L.P.; Carpinteri, A.; Spagnoli, A.; Wei, S. Effects of mechanical properties of concrete constituents including active mineral admixtures on fatigue behaviours of high performance concrete. *Fatigue Fract. Eng. Mater. Struct.* **2010**, *33*, 66–75.
4. Kim, J.K.; Kim, Y.Y. Experimental study of the fatigue behavior of high strength concrete. *Cem. Concr. Res.* **1996**, *26*, 1513–1523. [CrossRef]
5. Nor, N.M.; Ibrahim, A.; Bunnori, N.M.; Saman, H.M.; Saliah, S.N.M.; Shahidan, S. Diagnostic of fatigue damage severity on reinforced concrete beam using acoustic emission technique. *Eng. Fail. Anal.* **2014**, *41*, 1–9. [CrossRef]
6. Bazant, Z.P.; Hubler, M.H. Theory of cyclic creep of concrete based on Paris law for fatigue growth of subcritical microcracks. *J. Mech. Phys. Solids* **2014**, *63*, 187–200. [CrossRef]
7. Saucedo, L.; Yu, R.C.; Medeiros, A.; Zhang, X.X.; Ruiz, G. A probabilistic fatigue model based on the initial distribution to consider frequency effect in plain and fibre reinforced concrete. *Int. J. Fatigue* **2013**, *48*, 308–318. [CrossRef]
8. Bazant, Z.P.; Panula, L. Practical prediction of time-dependent deformations of concrete. Part VI: Cyclic creep, nonlinearity and statistical scatter. *Mater. Struct.* **1979**, *12*, 175–183.
9. Breitenbucher, R.; Ibuk, H. Experimentally based investigations on the degradation process of concrete under cyclic load. *Mater. Struct.* **2006**, *39*, 717–724. [CrossRef]
10. Neville, A.M.; Hirst, G.A. Mechanism of cyclic creep of concrete. In *Douglas McHenry International Symposium on Concrete and Concrete Structures*; ACISP-55; Special Publication: London, UK, 1978; pp. 83–101.
11. Garrett, G.G.; Jennings, H.M.; Tait, R.B. The fatigue hardening behavior of cement-based materials. *J. Mater. Sci.* **1979**, *14*, 296–306. [CrossRef]
12. Le, J.L.; Bazant, Z.P. Unified nano-mechanics based probabilistic theory of quasibrittle and brittle structures: II. Fatigue crack growth, lifetime and scaling. *J. Mech. Phys. Solids* **2011**, *59*, 1322–1337. [CrossRef]
13. Slate, F.O.; Olsefski, S. X-Rays for Study of Internal Structure and Microcracking of Concrete. *J. Am. Concr. Inst.* **1963**, *60*, 575–588.
14. Abu Al-Rub, R.K.; Voyiadjis, G.Z. On the coupling of anisotropic damage and plasticity models for ductile Materials. *Int. J. Solids Struct.* **2003**, *40*, 2611–2643. [CrossRef]
15. Paskin, A.; Massoumzadeh, B.; Shukla, K.; Sieradzki, K.; Dienes, G.J. Effect of atomic crack tip geometry on local stresses. *Acta Met.* **1985**, *33*, 1987–1996. [CrossRef]
16. Dienes, G.J.; Paskin, A. Molecular dynamic simulations of crack propagation. *J. Phys. Chem. Solids* **1987**, *48*, 1015–1033. [CrossRef]

17. Gumbsch, P. An atomistic study of brittle fracture: Toward explicit failure criteria from atomistic modeling. *J. Mater. Res.* **1995**, *10*, 2897–2907. [CrossRef]

18. Gumbsch, P.; Beltz, G.E. On the continuum versus atomistic descriptions of dislocation nucleation and cleavage in nickel. *Model. Simul. Mater. Sci. Eng.* **1995**, *3*, 597–613. [CrossRef]

19. Zhou, Z.L.; Gu, J.L.; Chen, N.P.; Li, D.C.; Liu, H.Q. Comparison of finite element calculation and experimental study of elastic-plastic deformation at crack tip. *Acta Mech. Sin.* **1995**, *27*, 51–57.

20. Fischer, L.L.; Beltz, G.E. The effect of crack blunting on the competition between dislocation nucleation and Cleavage. *J. Mech. Phys. Solids* **2001**, *49*, 635–654. [CrossRef]

21. Hajlaoui, K.; Yavari, A.R.; Doisneau, B.; LeMoulec, A.; Vaughan, G.; Greer, A.L.; Inoue, A.; Zhang, W.; Kvick, Å. Shear delocalization and crack blunting of a metallic glass containing nanoparticles: In situ deformation in TEM analysis. *Scr. Mater.* **2006**, *54*, 1829–1834. [CrossRef]

22. Yu, Z.; Shan, Z.; Ouyang, Z.; Guo, F. A simple damage model for concrete considering irreversible mode-II microcracks. *Fatigue Fract. Eng. Mater. Struct.* **2016**, *39*, 1419–1432. [CrossRef]

23. Yu, Z.W.; Tan, S.; Shan, Z.; Tian, X.Q. X-ray computed tomography quantification of damage in concrete under compression considering irreversible mode-II microcracks. *Fatigue Fract. Eng. Mater. Struct.* **2017**, *40*, 1960–1972. [CrossRef]

24. Suresh, S.; Tschegg, E.K.; Brockenbrough, J.R. Fatigue Crack growth in cementitious composites under cyclic compressive loads. *Cem. Concr. Res.* **1989**, *19*, 827–833. [CrossRef]

25. Eliáš, J.; Le, J.L. Modeling of mode-I fatigue crack growth in quasi brittle structures under cyclic compression. *Eng. Fract. Mech.* **2012**, *96*, 26–36.

26. Bazant, Z.P.; Xiang, Y. Size effect in compression fracture: Splitting crack band propagation. *J. Eng. Mech.* **1997**, *123*, 207–213. [CrossRef]

27. Fairhurst, C.; Comet, F. Rock fracture and fragmentation. In *Rock Mechanics: From Research to Application. Proceedings of the U.S. Symposium on Rock Mechanics*; Einstein, H.H., Ed.; MIT Press: Cambridge, MA, USA, 1981; pp. 21–46.

28. Sammis, C.G.; Ashby, M.F. The failure of brittle porous solids under compressive stress state. *Acta Metall.* **1986**, *34*, 511–526. [CrossRef]

29. Horii, H.; Nemat-Nasser, S. Compression induced nonplanar crack extension with application to splitting, exfoliation and rock burst. *J. Geophys. Res.* **1982**, *87*, 6805–6822.

30. Kachanov, M. A micro crack model of rock in elasticity—Part I. Frictional sliding on micro cracks. *Mech. Mater.* **1982**, *1*, 19–27. [CrossRef]

31. Nemat-Nasser, S.; Obata, M. A microcrack model of dilatancy in brittle materials. *J. Appl. Mech.* **1988**, *55*, 24–35. [CrossRef]

32. Budiansky, B.; O'Connell, R.J. Elastic moduli of a cracked solid. *Int. J. Solids Strcut.* **1976**, *12*, 81–97. [CrossRef]

33. Ning, J.; Ren, H.; Fang, M. Reseasrch on the process of micro-crack damage evolution and coalescence in brittle materials. *Eng. Fail. Anal.* **2014**, *41*, 65–72.

34. Hu, G.; Liu, J.; Graham-Brady, L.; Ramesh, K.T. A 3D mechanistic model for brittle materials containing evolving flaw distributions under dynamic multiaxial loading. *J. Mech. Phys. Solids* **2015**, *78*, 269–297. [CrossRef]

35. Burr, A.; Hild, F.; Leckie, F.A. Micro-mechanics and continuum damage mechanics. *Arch. Appl. Mech.* **1995**, *65*, 437–456. [CrossRef]

36. Mazars, J.; Pijaudier-Cabot, G. Continuum damage theory—Application to concrete. *J. Eng. Mech.* **1989**, *115*, 345–365. [CrossRef]

37. Halm, D.; Dragon, A. An anisotropic model of damage and frictional sliding for brittle materials. *Eur. J. Mech. A Solids* **1998**, *17*, 439–460. [CrossRef]

38. Dragon, A.; Halm, D.; Desoyer, T. Anisotropic damage in quasi-brittle solids: Modelling, computational issues and applications. *Comput. Methods Appl. Mech. Eng.* **2000**, *183*, 331–352. [CrossRef]

39. Le, J.L.; Bazant, Z.P.; Bazant, M.Z. Unified nano-mechanics based probabilistic theory of quasibrittle and brittle structures: I. Strength, static crack growth, lifetime and scaling. *J. Mech. Phys. Solids* **2011**, *59*, 1291–1321. [CrossRef]

40. Feng, X.Q.; Gross, D. Three-dimensional micromechanical model for quasi-brittle solids with residual strains under tension. *Int. J. Damage Mech.* **2000**, *9*, 79–110. [CrossRef]

41. Xiao, J.Z.; Li, H.; Yang, Z.J. Fatigue behavior of recycled aggregate concrete under compression and bending cyclic loadings. *Constr. Build. Mater.* **2013**, *38*, 681–688. [CrossRef]

42. Pandolfi, A.; Taliercio, A. Bounding surface models applied to fatigue of plain concrete. *J. Eng. Mech.* **1998**, *5*, 556–564. [CrossRef]

43. Papa, E.; Taliercio, A. Anisotropic damage model for the multiaxial static and fatigue behavior for plain concrete. *Eng. Fract. Mech.* **1996**, *55*, 163–179. [CrossRef]

44. Alliche, A. Damage model for fatigue loading of concrete. *Int. J. Fatigue* **2004**, *26*, 915–921. [CrossRef]

45. Mai, S.H.; Le-Corre, F.; Forêt, G.; Nedjar, B. A continuum damage modeling of quasi-static fatigue strength of plain concrete. *Int. J. Fatigue* **2012**, *37*, 79–85. [CrossRef]

46. Lemaitre, J.; Lippmann, H. *A Course on Damage Mechanics*; Springer: Berlin, Germany, 1996.

47. Richard, B.; Ragueneau, F.; Cremona, C.; Adelaide, L. Isotropic continuum damage mechanics for concrete under cyclic loading: Stiffness recovery, inelastic strains and frictional sliding. *Eng. Fract. Mech.* **2010**, *77*, 1203–1223. [CrossRef]

48. Lu, P.Y.; Li, Q.B.; Song, Y.P. Damage constitutive of concrete under uniaxial alternate tension- compression fatigue loading based on double bounding surfaces. *Int. J. Solids Struct.* **2004**, *41*, 3151–3166. [CrossRef]

49. Najar, J. Brittle residual strain and continuum damage at variable uniaxial loading. *Int. J. Damage Mech.* **1994**, *3*, 260–276. [CrossRef]

50. Shan, Z.; Yu, Z.W. A fiber bundle-plastic chain model for quasi-brittle materials under uniaxial loading. *J. Stat. Mech. Theory Exp.* **2015**, *2015*, P11010. [CrossRef]

51. Yazdani, S.; Schreyer, H.L. Combined plasticity and damage mechanics model for plain concrete. *J. Eng. Mech.* **1990**, *116*, 1435–1450. [CrossRef]

52. Voyiadjis, G.Z.; Park, T. The kinematics of damage for finite-strain elasto-plastic solids. *Int. J. Eng. Sci.* **1999**, *37*, 803–830. [CrossRef]

53. Li, J.; Wu, J.Y.; Chen, J.B. *Stochastic Damage Mechanics of Concrete Structure*; Science Press: Beijing, China, 2017. (In Chinese)

54. Yu, Z.W.; Song, L.; Xie, Y.; Shan, Z. Experimental study on performance of CRTS III slab ballastless track system under service condition. In *Third Report: Numerical Simulation on Fatigue Performance of CRTS III Slab Ballastless Track Structure Under High Speed Train Load*; China State Railway Group Co., Ltd.: Beijing, China, 2015. (In Chinese)

55. Sima, J.F.; Roca, P.; Molins, C. Cyclic constitutive model for concrete. *Eng. Struct.* **2008**, *30*, 695–706. [CrossRef]

56. Research Group on Concrete Fatigue Behaviors. *Reliability Evaluation Method of Concrete Bending Members under Fatigue Loading*; China Architecture & Building Press: Beijing, China, 1994. (In Chinese)

57. Thomas, C.; Setien, J.; Polanco, J.A.; Lompillo, I.; Cimentada, A. Fatigue limit of recycled aggregate concrete. *Constr. Build. Mater.* **2014**, *52*, 146–154. [CrossRef]

58. Lu, P.; Song, Y.; Li, Q. Behavior of concrete under compressive fatigue loading with constant lateral stress. *Eng. Mech.* **2004**, *21*, 173–177.

59. *The National Standard of the People's Republic of China: Code for Design of Concrete Structures (GB50011-2010)*; China Building Industry Press: Beijing, China, 2010. (In Chinese)

60. Liang, J.S.; Ren, X.D.; Li, J. A competitive mechanism driven damage-plasticity model for fatigue behavior of concrete. *Int. J. Damage Mech.* **2016**, *25*, 377–399. [CrossRef]

61. Xiao, J. Theoretical and Experimental Investigation on Fatigue Properties of Rock under Cyclic Loading. Master's Thesis, Central South University, Changsha, China, 2009. (In Chinese).

62. Marigo, J. Modelling of brittle and fatigue damage for elastic material by growth of microvoids. *Eng. Fract. Mech.* **1985**, *21*, 861–874. [CrossRef]

applied
sciences

Article

Plain and Fiber-Reinforced Concrete Subjected to Cyclic Compressive Loading: Study of the Mechanical Response and Correlations with Microstructure Using CT Scanning

Jesús Mínguez [1,*], Laura Gutiérrez [2], Dorys C. González [1] and Miguel A. Vicente [1]

[1] E. Politécnica Superior, Universidad de Burgos, Campus Milanera (Edif. D), c/Villadiego s/n, 09001 Burgos, Spain

[2] Institut d'Aran, carretera Betrén s/n, 25530 Lleida, Spain

* Correspondence: jminguez@ubu.es; Tel.: +34-947-259-423

Received: 30 May 2019; Accepted: 25 July 2019; Published: 27 July 2019

Abstract: The response ranges of three principal mechanical parameters were measured following cyclic compressive loading of three types of concrete specimen to a pre-defined number of cycles. Thus, compressive strength, compressive modulus of elasticity, and maximum compressive strain were studied in (i) plain, (ii) steel-fiber-reinforced, and (iii) polypropylene-fiber-reinforced high-performance concrete specimens. A specific procedure is presented for evaluating the residual values of the three mechanical parameters. The results revealed no significant variation in the mechanical properties of the concrete mixtures within the test range, and slight improvements in the mechanical responses were, in some cases, detected. In contrast, the scatter of the mechanical parameters significantly increased with the number of cycles. In addition, all the specimens were scanned by means of high resolution computed tomography, in order to visualize the microstructure and the internal damage (i.e., internal micro cracks). Consistent with the test results, the images revealed no observable internal damage caused by the cyclic loading.

Keywords: fatigue; high performance concrete; fibre-reinforced high performance concrete; compressive stress; compressive modulus of elasticity; maximum compressive strain

1. Introduction

Progressive improvement of the compressive properties of concrete have led to the development of more and more slender concrete structures worldwide, resulting in a progressive reduction of component weight. In consequence, cyclic loads due to variable loading (transient loads and wind, among others) have to be closely studied as tolerance margins are reduced. In some cases, fatigue loads are crucial and the design of the structural element, as in the case of concrete wind turbine towers, depends on fatigue load levels.

Fatigue in concrete can be defined as a process of mechanical degradation leading to failure. Due to cyclic loads, crack initiation and propagation within the specimen occur, until final failure. However, during the application of the variable loads, a modification of the mechanical parameters of the concrete occurs, progressively varying the structural responses of an element.

The classic way to address the problem of fatigue in concrete is to determine the number of cycles (N) that a concrete element can support under given loading conditions. "N" is also called "fatigue life." A concrete element under cyclic loads collapses after a sufficient number of cycles, even if the maximum applied stress is less than its compressive strength.

The value of N depends on both the maximum and the minimum stress applied. The most common way to represent this relationship is through S–N curves. Fatigue strength (S) is defined as the fraction of the static strength that can be supported repeatedly over a number of cycles. In general, there is a huge scatter in the results of apparently identical specimens.

Traditionally, fatigue analysis has focused on the study of fatigue life, i.e., the number of cycles that the specimen can withstand before collapse [1–10], and almost no attention has been paid to how the mechanical parameters vary with the number of cycles [11–14].

Nevertheless, those variations represent an interesting approach to fatigue analysis, as most structures are subjected to a combination of cyclic loads, resulting in mechanical degradation, and extreme static loading. A proper design should guarantee the safety of the structure, taking into account the possible reduction of the mechanical capacity of the structure due to cyclic loads.

The addition of fibers inside the concrete mass provides a better behavior of concrete under fatigue loads. Fibers bridge the cracks, resulting in a significant increase of the fatigue life. Much research has been carried out in this field [15–21], and in all cases, the results have shown a relevant improvement of the fatigue response of the concrete elements.

It is commonly accepted that a fatigue test consists of applying a cyclic load until failure, with the condition that the maximum load is always below the strength of the concrete element. However, cyclic loads applied to concrete elements will result in a progressive deterioration of the microstructure, with the initiation and propagation of microcracks. This fuzzy damage, distributed over the whole specimen, is expected to result in a progressive variation of the macroscopic response, i.e., the mechanical parameters of the concrete. At the point in time when the specimen collapses during the fatigue test, it may be understood that the residual strength of the concrete specimen is exactly the maximum load that is applied. In consequence, a progressive reduction of the strength of the concrete occurs during the cyclic test.

It must be taken into account that real concrete structures are not subjected only to cyclic loading until failure. In fact, they are usually subjected to cyclic loading, and occasionally to extreme loading episodes. Cyclic loading does not usually causes a concrete structure to collapse, but it causes a variation in its mechanical parameters, altering its behavior under extreme loading episodes.

Concrete towers for wind turbines are a good example of this fact, since they are subjected to a "permanent" cyclic loading and eventual extreme loading events. Considering this fact, Urban et al. [22] studied the variation in the compressive modulus of elasticity of concrete towers for wind turbines caused by cyclic loading, and its consequence in terms of variation in the structural response to static and dynamic loading.

This phenomenon could also affect post-tensioned concrete beams. Variations in compressive strength, compressive modulus of elasticity, and maximum compressive strain due to cyclic loading alter the structural response, especially under static loading (instantaneous elastic deflection, deferred deflection, etc.). The dynamic response of the structural element is also modified (natural frequency, maximum vertical acceleration, impact factor, etc.).

Damage caused by cyclic loads could leave a trace in the interior of the concrete specimen, such as a crack pattern, and could therefore be studied using computed tomography (CT) technology.

The CT scan is a non-destructive technique to visualize the microstructure of materials based on X-ray properties. This technology is able to define the density of each specimen voxel (volumetric pixel) by assigning a shade of gray according to voxel density. Light shades of grey correspond to high densities, whereas dark shades of grey correspond to low densities. In recent years, many authors have conducted research on concrete microstructures with this technique, and many have, in particular, focused on the fiber orientation and fiber distribution inside concrete matrices [15,21,22]. Some other research works have focused on the spatial distributions of concrete voids and their effects on macroscopic properties [23–28]. A complete, state-of-the-art work on the use of computed tomography to explore the microstructure of materials in civil and mechanical engineering is presented in Vicente et al. [29,30].

This paper focused on study of the variation of the mechanical parameters of three concrete mixtures with a number of cycles: plain, steel-fiber-reinforced, and polypropylene-fiber-reinforced high-performance concrete (HPC, SFHPC, and PFHPC, respectively). In this case, the mechanical parameters under study were compressive strength (f_c), compressive modulus of elasticity (E_c), and maximum compressive strain ($\varepsilon_{c,max}$). The specimens were first subjected to cyclic compression loads over five complete cycles: 0; 2000; 20,000; 200,000; and 2,000,000. The specimens were then subjected to a static compression test up to failure, and the abovementioned mechanical parameters were tested in each specimen.

This paper shows an experimental procedure by which to evaluate the variation of the main mechanical parameters due to cyclic loads, and also the results obtained in the three different mixtures with the number of cycles.

Finally, having completed the compressive test, the specimens were scanned and the internal failure mechanisms were studied, using the information provided by the CT scan to gain insight into the resistance mechanisms of the fibers and their influence on the macroscopic response of the concrete specimen.

The structure of this paper is as follows: The experimental program is presented in Section 2, the experimental results are described and discussed in Section 3, the resistance mechanisms are described in Section 4, and, finally, the conclusions are presented in Section 5.

2. Experimental Program

2.1. Materials

A total of three high-strength concrete mixtures were cast. They all shared the same concrete matrix, and the difference was that one of them had no fibers, while the other two had either steel or polypropylene fibers. The mixtures were identified as plain high-performance concrete (HPC), steel-fiber-reinforced high-performance concrete (SFHPC), and polypropylene-fiber-reinforced high-performance concrete (PPFHPC).

A total of 40 test specimens of each mixture were cast in the form of cylinders with a diameter of 100 mm and a height of 200 mm. Table 1 shows the mixtures that were used.

Table 1. Concrete mixture.

Dosage	HPC	SFHPC	PPFHPC
Cement (kg/m^3)	400.0	400.0	400.0
Water (kg/m^3)	125.0	125.0	125.0
Superplasticizer (kg/m^3)	14.0	14.0	14.0
Nanosilica (kg/m^3)	6	6	6
Fine aggregate (kg/m^3)	800.0	800.0	800.0
Coarse aggregate (kg/m^3)	1080.0	1080.0	1080.0
Fiber (% by volume)	–	1%	1%

HPC is high-performance concrete; SFHPC is steel-fiber-reinforced high-performance concrete and PPFHPC is polypropylene-fiber-reinforced high-performance concrete.

The SFHPC contained a volume of 78.5 kg/m^3 of Dramix 3D 45/50BL hooked-ended steel fibers (BEKAERT, Kortrijk, Belgium), each of 50 mm in length, with a diameter of 1.05 mm, giving an aspect ratio of 45, a fiber tensile strength of 1115 MPa, and a Young's modulus of 200 GPa. In the case of PPFHPC, an amount of 9.1 kg/m^3 of monofilament polypropylene fibers Masterfiber 249 (BASF, Ludwigshafen am Rhein, Germany) was used. These fibers were 48 mm in length with a diameter of 0.85 mm, resulting in an aspect ratio of 56.5. The tensile strength was 400 MPa and the Young's modulus was 4.7 GPa. Both fibers were quite similar in terms of their geometry and the number of fibers inside the concrete mixture. The main difference was the structural behavior and overall stiffness.

MasterRoc MS 685 (BASF, Ludwigshafen am Rhein, Germany) nanosilica and a Glenium 52 (BASF, Ludwigshafen am Rhein, Germany) superplasticizer were considered. Siliceous aggregate was used for both fine aggregate and coarse aggregate, with a nominal maximum aggregate size of 4 mm for fine aggregate and 12 mm for coarse aggregate.

Mixing was done in a rotary mixer and the fibers were gradually sprinkled into the drum by hand. The specimens were cured for 180 days in a curing room at a constant relative humidity of 100% and an ambient temperature of 20 °C. The specimens were then removed from the curing room and held under laboratory conditions until testing. All the specimens were at least 300 days old when the test campaign began. Thus, the possible strength increase during the fatigue test was minimized.

2.2. Testing Campaign

As explained before, a total of 40 cylinders of each mixture were cast and divided into two series of 20 cylinders each. The two series were labeled as L-Series and H-Series, as explained later.

Additionally, three cylinders, 150 mm in diameter and 300 mm in height, were cast to define the concrete quality. These cylinders were kept in the curing room with the rest of the specimens until day 28, when they were tested under static compression. The tests were performed according to the European Standards [31,32].

The average compressive strength, f_{cm}, was 75.95 MPa and the characteristic compressive strength, f_{ck}, was 73.0 MPa. According to Eurocode 2 [33], the strength class was therefore C70/85.

The abovementioned 40 cylinders, 100 mm in diameter and 200 mm in height, were tested at an age of 180 days, as follows. Four specimens of each series, i.e., eight cylinders, were tested under static compression until failure just before starting the cyclic testing campaign, in order to define the real loads to be applied during the cyclic tests to the rest of the specimens. Tables 2 and 3 show the average values of the main mechanical parameters of the concrete for each series. The value in brackets represents the standard deviation.

Table 2. Mechanical parameters of the concrete before starting the cyclic tests. L-Series.

Mechanical Parameter	HPC	SFHPC	PPFHPC
f_c (MPa)	91.8 (0.8)	94.8 (2.1)	88.9 (1.1)
E_c (MPa)	33,326.3 (1184.4)	36,880.3 (2368.3)	35,690.5 (2568.0)
$\varepsilon_{c,max}$	0.0037 (0.0003)	0.0028 (0.0003)	0.0032 (0.0002)

Table 3. Mechanical parameters of the concrete before starting the cyclic test. H-Series.

Mechanical Parameter	HPC	SFHPC	PPFHPC
f_c (MPa)	94.6 (0.5)	102.7 (2.6)	92.4 (1.9)
E_c (MPa)	38,812.8 (1748.6)	43,093.2 (1410.7)	38,711.7 (2014.7)
$\varepsilon_{c,max}$	0.0029 (0.0002)	0.0032 (0.0001)	0.0026 (0.0002)

The compressive modulus of elasticity was obtained following the method described in standard EN 12390-13 [31]. The static compression test was then performed, as per standard EN 12390-3 [32], and the compressive strength and maximum compressive strain values were extracted.

The rest of the cylinders, that is, 16 specimens of each series, were tested under cyclic load up to a pre-defined number of cycles. The main parameters of the cyclic tests were:

- Stress level. In this case, two stress levels were considered, one per series. The compression load for the first series of cylinders ranged from 35% to 50% of the characteristic compressive strength, defined through the static tests described above. This first stress level was labeled "high level series" or "H-Series." For the second series, compression loading ranged from 25% to 40% of the characteristic compressive strength, defined again through the static tests described above. This second stress level was labeled "low level series" or "L-Series."

- Number of cycles. Four specimens of each series were subjected to cyclic loads of up to 2000 cycles. Another set of four specimens from each series was subjected to cyclic loads of up to 20,000 cycles.

A third set of four specimens from each series was tested up to 200,000 cycles, and the last set of four specimens, up to 2,000,000 cycles.

- Test frequency. In all cases, the test frequency was 6 Hz.

According to Model Code 2010 [34], the expected fatigue life for the H-Series was 1.5×10^3 cycles, while the expected fatigue life for the L-Series was 1.9×10^8 cycles. These values were obtained considering the characteristic compressive strength at 28 days, i.e., 70 MPa, which was the reference value for concrete design. On the contrary, considering the characteristic compressive strength at the age when the cyclic testing was carried out, the expected fatigue life for the H-Series was 1.8×10^8 cycles, while the expected fatigue life for the L-Series was 1.2×10^{11} cycles. From 28 days to the date when the cyclic testing began, concrete compression had increased around 30%, but the theoretical fatigue life had increased around 100,000 times for the case of the H-Series and around 1000 times for the case of the L-Series.

None of the specimens broke during the cyclic testing. Once the cylinders reached the pre-defined number of cycles, they were tested under static loading, first to obtain the compressive modulus of elasticity, and then the compressive strength and the maximum compressive strain. Figure 1 shows a scheme of the test procedure. The longitudinal axis represents the time.

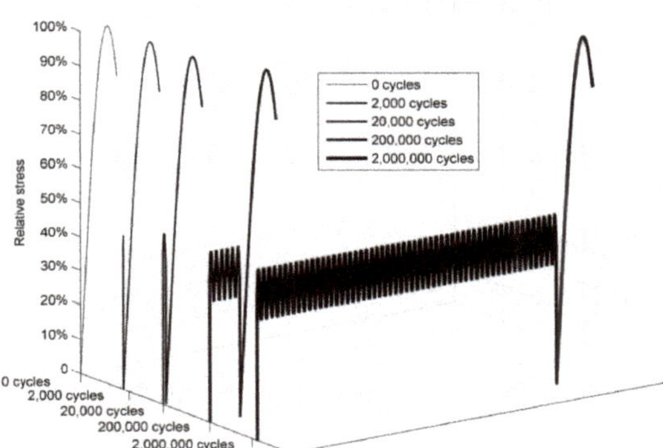

Figure 1. Testing procedure.

All the specimens were wrapped in polyvinyl chloride (PVC) cling film before the static post-cyclic tests, to prevent defragmentation once failure occurred. Moreover, an end-of-test condition was programmed into the test machine: the test ended when the vertical load dropped by more than 20% of the maximum load. In this case, the end-of-test conditions, in combination with the PVC cling film, meant that all of the specimen fragments were easily locatable.

2.3. CT Scanning

Having completed the mechanical tests, all the specimens were scanned in order to visualize the microstructure of the concrete. CT scan technology generated an image for close up examination of the fiber locations and orientations in each specimen and the internal microcracks, so that the resistance mechanisms, i.e., the mechanisms used by the specimen to withstand the compressive load, could be observed.

In this research, the cubic specimens were CT scanned using a GE Phoenix v|tome|x device (General Electric, Boston, MA, USA) equipped with a 300 kV/500 W tube. The intensity-controlled X-ray source emits a cone ray that is received by an array of detectors, which measures the loss of X-ray intensity, depending on the density of the matter along its path. Using a post-processing software package, a total of 1334 2D slices with pixel sizes of 2048 × 2048 (Figure 2) were obtained throughout the height of the specimen from the CT scan data. In this case, the horizontal resolution was approximately 75×75 μm^2 and the vertical distance between the slices was around 75 μm. A total of 2667 images were obtained from each cylinder. After that, a 3D image of the whole specimen was generated using all the above mentioned 2D images. The post-processing software assigns a grey value to each voxel, from 0 (which belongs to the least dense voxel, i.e., voids and cracks) to 255 (which belongs to the densest voxel, i.e., steel). The scanning process results in a file that includes the X, Y, and Z Cartesian coordinates of the voxel center of gravity and an integer number, from 0 to 255, with regards to its density. The total number of voxels in a specimen was around 4.8×10^9.

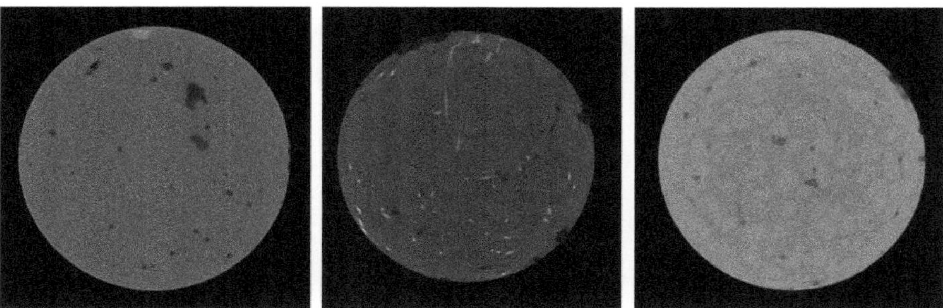

Figure 2. Slices belonging to different mixtures. From left to right, HPC, SFHPC, and PPFHPC.

3. Experimental Results

The three main mechanical parameters obtained from the static post-cyclic tests were the compressive strength (f_c), the compressive modulus of elasticity (E_c), and the maximum compressive strain ($\varepsilon_{c,max}$) (Figure 3).

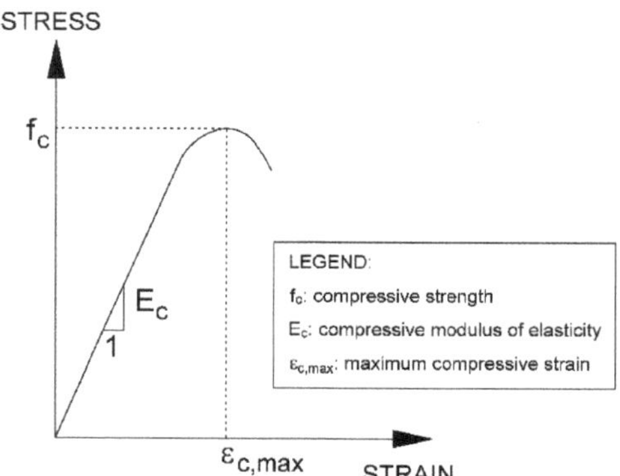

Figure 3. Definition of the three main mechanical parameters.

It was assumed that the data distribution pattern of the same set of four cylinders (sharing the same stress level, the same concrete mixture, and the same number of cycles) would follow a normal distribution.

Using those hypotheses, the following three main values were obtained: the average value, showing a probability of not being exceeded by 50% ($p = 0.50$); the minimum value, showing a probability of not being exceeded by 5% ($p = 0.05$); and the maximum value, showing a probability of not being exceeded of 95% ($p = 0.95$).

The relative value was defined as the quotient between the value and the average value at 0 cycles, for better visualization of their evolution with the number of cycles. Moreover, three tendency lines were fitted (using the least square method), for the average, maximum, and minimum relative values, respectively.

These data measurements, compressive strength, compressive modulus of elasticity, and maximum compressive strain are discussed below.

3.1. Compressive Strength

Figures 4–6 show the range of compressive strengths by number of cycles for the three mixtures and for both series.

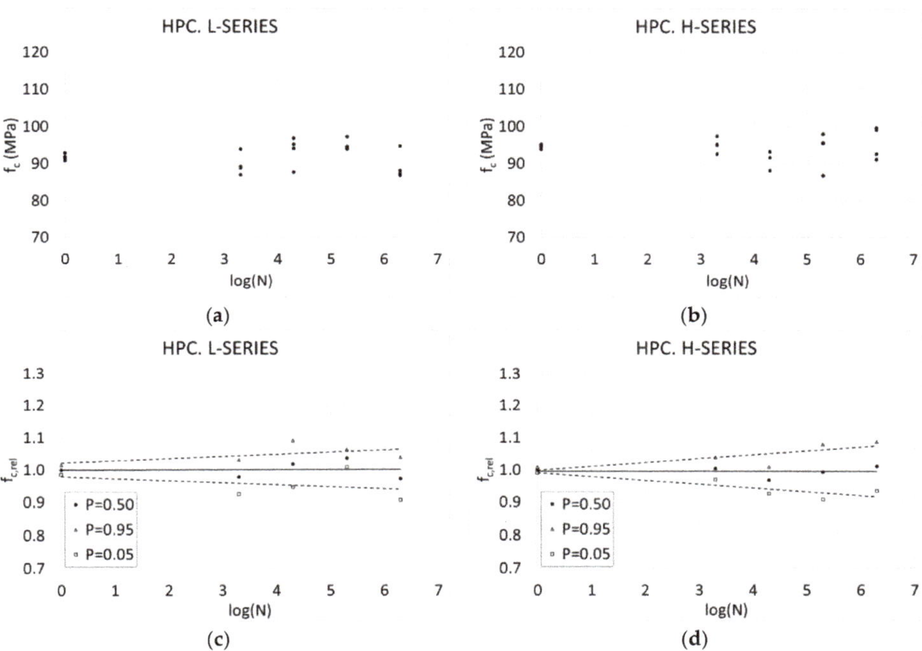

Figure 4. Variation of compressive strength by number of cycles. HPC (**a**) L-Series. Raw values. (**b**) H-Series. Raw values. (**c**) L-Series. Maximum, minimum, and average relative values, and tendency lines. (**d**) H-Series. Maximum, minimum, and average relative values, and tendency lines.

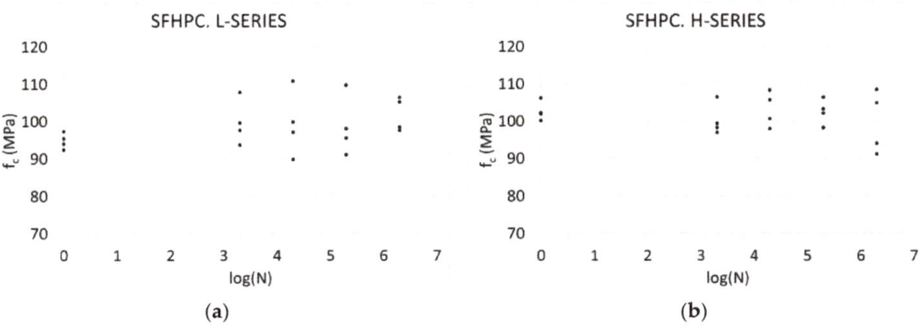

Figure 5. *Cont.*

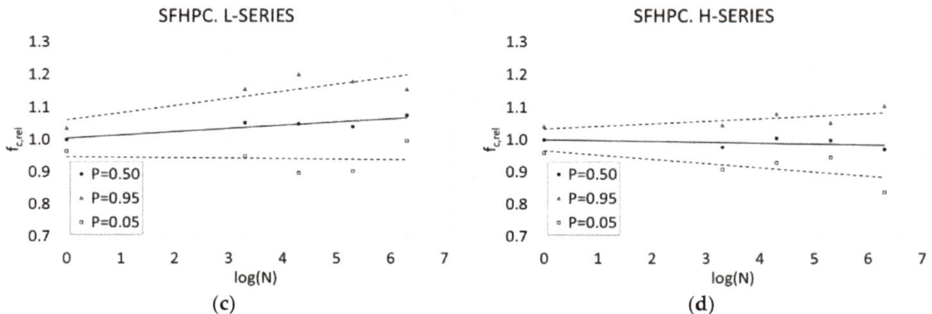

Figure 5. Variation of compressive strength by number of cycles. SFHPC (**a**) L-Series. Raw values. (**b**) H-Series. Raw values. (**c**) L-Series. Maximum, minimum, and average relative values, and tendency lines. (**d**) H-Series. Maximum, minimum, and average relative values, and tendency lines.

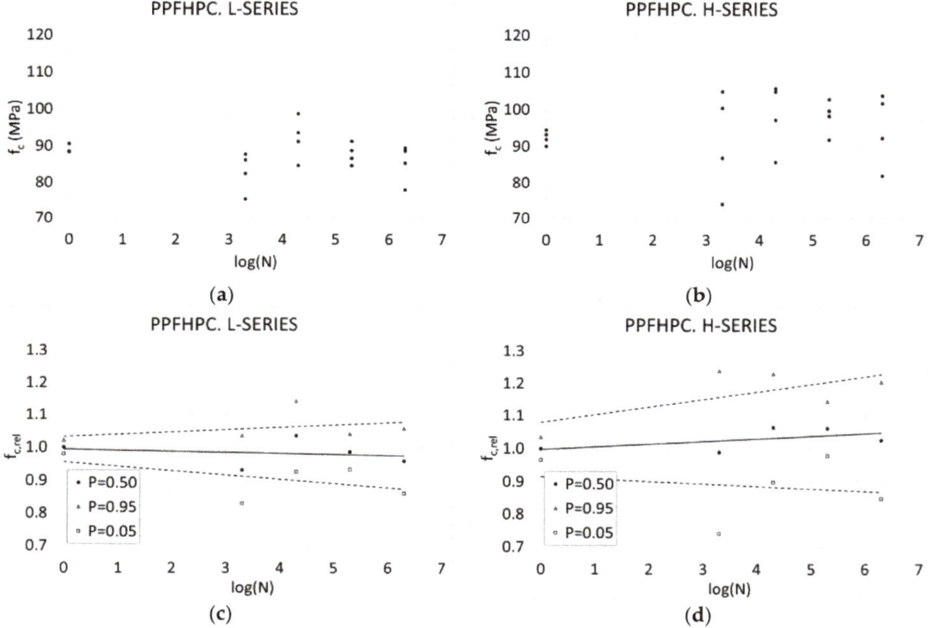

Figure 6. Variation of compressive strength by number of cycles. PPFHPC (**a**) L-Series. Raw values. (**b**) H-Series. Raw values. (**c**) L-Series. Maximum, minimum, and average relative values, and tendency lines. (**d**) H-Series. Maximum, minimum, and average relative values, and tendency lines.

Figures 4–6 show some interesting results. First, it should be highlighted that although the expected fatigue life was 1.9×10^8 cycles for the L-Series and 1.5×10^3 cycles for the H-Series, no collapse occurred during the cyclic tests. It was especially noticeable for the H-Series, where the number of cycles was, in all cases, greater than the expected fatigue life. When the expected fatigue life corresponding to the real concrete quality at the beginning of the cyclic tests is considered, the results are more consistent.

The results also showed no significant variation in the average compressive strength of the concrete by the number of cycles. In all cases, the variation was less than 7%. This behavior was observed for all the mixtures and for both series. It means that, in this case, any damage caused by cyclic loading was very small, and no appreciable mechanical consequences were observed.

On the contrary, a relevant increase of the scatter of the results was observed. Specimens not previously subjected to cyclic loads showed very low scatter, while the rest of the specimens, in general, showed a higher scatter. In general, the scatter increased more than two-fold, and this variation is quite similar in all cases.

This phenomenon can be explained as follows. Concrete specimens have internal stress caused by the curing process, shrinkage, etc. Cyclic loads release the internal stresses, resulting in microcracking. This phenomenon has been observed in previous research [35–37]. Depending on the real internal stress field, the releasing process is more or less intense and more or less microcracking occurs, resulting in different macroscopic responses.

In some cases, this internal stress relaxation is not harmful and a greater compressive strength is observed.

This scatter was higher in the case of the mixtures including fiber, which reveals that the fibers introduced additional restrictions to the concrete mixture during the curing process, resulting in more internal stress. Cyclic loads released this stress, and the macroscopic result was a higher scatter.

The mixture including steel fibers showed, in general, higher compressive strength, which denotes that steel fibers help to increase the mechanical capacity of concrete. In contrast, the mixtures without fibers and with polypropylene fibers showed similar results.

3.2. Compressive Modulus of Elasticity

Figures 7–9 show the results of the variation of the compressive modulus of elasticity with the number of cycles, for the three mixtures and for both series.

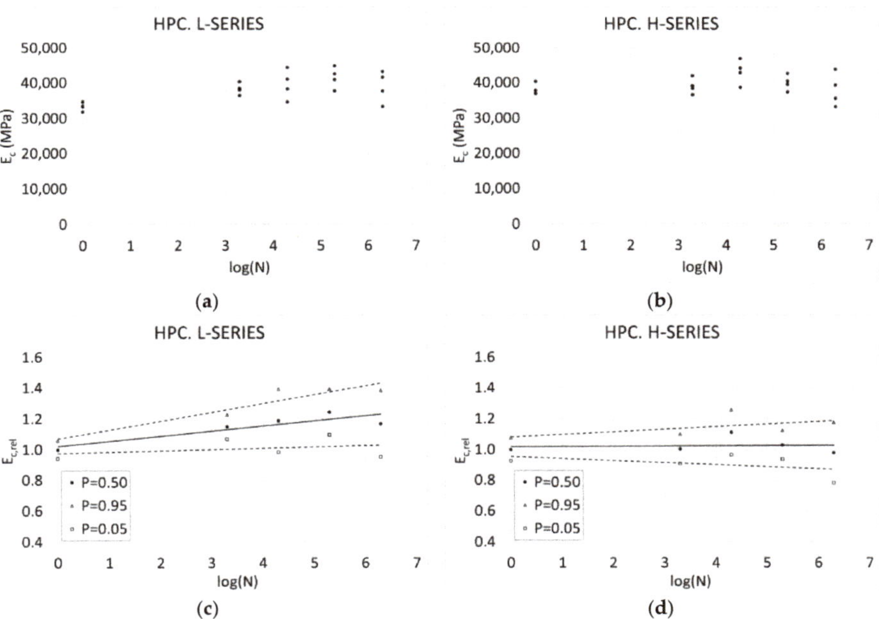

Figure 7. Variation of the compressive modulus of elasticity by number of cycles. HPC (**a**) L-Series. Raw values. (**b**) H-Series. Raw values. (**c**) L-Series. Maximum, minimum, and average relative values, and tendency lines. (**d**) H-Series. Maximum, minimum, and average relative values, and tendency lines.

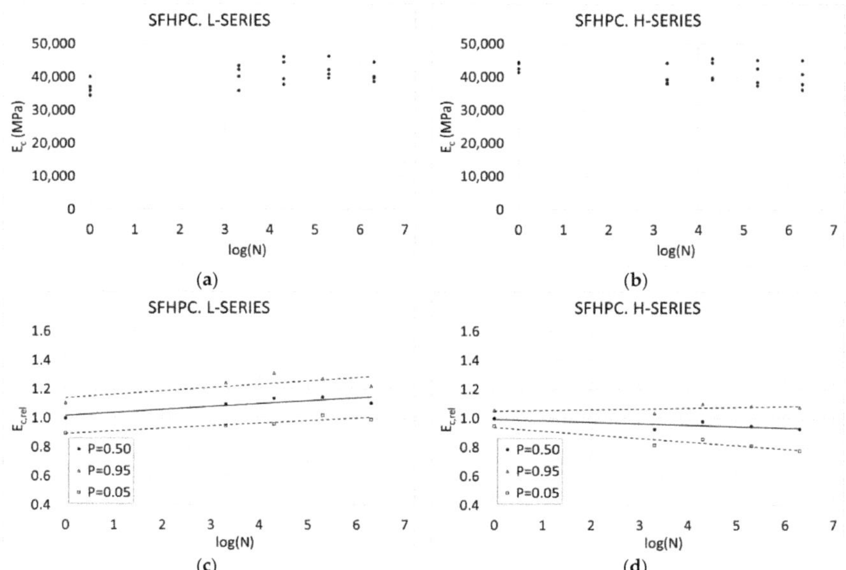

Figure 8. Variation of the compressive modulus of elasticity with the number of cycles. SFHPC (**a**) L-Series. Raw values. (**b**) H-Series. Raw values. (**c**) L-Series. Maximum, minimum, and average relative values, and tendency lines. (**d**) H-Series. Maximum, minimum, and average relative values, and tendency lines.

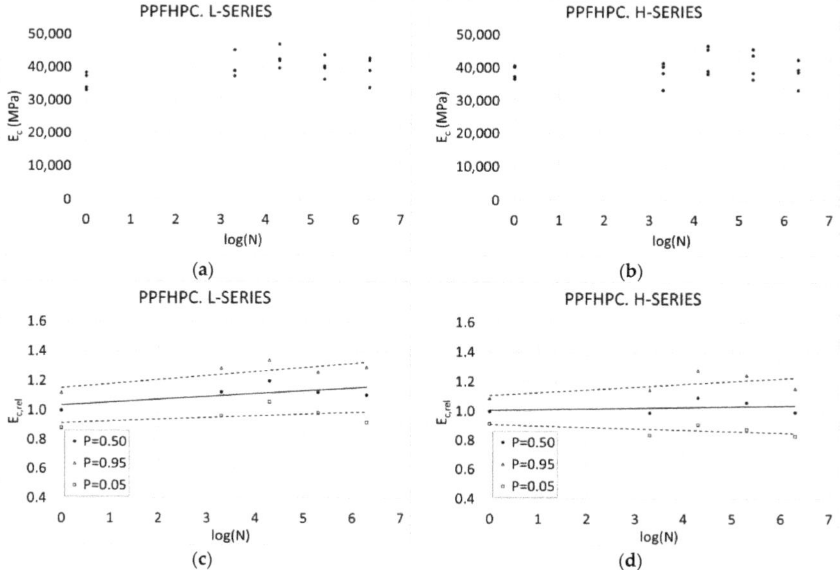

Figure 9. Variation of the compressive modulus of elasticity with the number of cycles. PPFHPC (**a**) L-Series. Raw values. (**b**) H-Series. Raw values. (**c**) L-Series. Maximum, minimum, and average relative values, and tendency lines. (**d**) H-Series. Maximum, minimum, and average relative values, and tendency lines.

In contrast to the results depicted in Figures 4–6, Figures 7–9 show that there was no significant variation in the average compressive modulus of elasticity of concrete with the number of cycles. The L-Series showed a slight increase of the compressive modulus of elasticity, up to 15–20%. On the other hand, the H-Series showed a flat behavior, with a variation below 5%.

As with compressive strength, the compressive modulus of elasticity was not adversely affected by cyclic loading, and no relevant mechanical consequences were observed.

In a deeper analysis, it may be observed that the variation of the compressive modulus of elasticity did not follow a monotonic tendency. First, an initial slight increase of the compressive modulus of elasticity up to a maximum value was observed, which was located between 20,000 and 200,000 cycles, depending on the mixture and on the series. It then started to decrease.

In this case, an increase of the scatter of the results was observed, although it is not very relevant.

As happened with compressive strength, the internal stress released by the cyclic loads during the first cycles resulted in microcracking. In this case, this stress relaxation resulted in an increase of the compressive modulus of elasticity. Under low and moderate stress levels, which are the ones used to measure the compressive modulus of elasticity, the results show that the presence of fibers had no influence, since no different tendency was observed between the mixtures with and without fibers.

The scatter was also similar in all cases, which denotes that fibers had no relevant influence on the compressive modulus of elasticity, i.e., the evolution of the microcracking and its impact on the compressive modulus of elasticity depends on concrete matrix only. The scatter depends on concrete matrix only, resulting in a similar variation in all cases, regardless of the presence or absence of fibers, and the type of fibers.

Moreover, all the mixtures and both series showed very similar values for the compressive modulus of elasticity (in the surrounding of 40,000 MPa), implying that this parameter was not sensitive to the presence of fibers.

3.3. Maximum Compressive Strain

Figures 10–12 show the results of the variation of the maximum compressive strain with the number of cycles, for the three mixtures and for the two series.

Figures 10–12 show some interesting results, complementary to those extracted from Figures 4–9. First, the results show that, on one side, there was a slight decrease of the average maximum compressive strain of concrete by number of cycles in the case of the L-Series, with a maximum variation of around 18%. On the other side, there was a slight increase of this parameter in the case of the H-Series, with a maximum variation of around 20%. However, this variation was not very relevant. As with compressive strength and the compressive modulus of elasticity, any damage caused by the cyclic loading was very slight, with no observable mechanical distress.

In a deeper analysis, it may be observed that there was an initial slight decrease of the maximum compressive strain up to a minimum value, located between 20,000 and 200,000 cycles depending on the mixture and on the series, after which it started to increase. This expected tendency was quite the opposite to the compressive modulus of elasticity.

The mixtures, in general, showed a very linear behavior up to their compressive strength. After this point, the plain concrete sample shattered and the fiber-reinforced concrete underwent dramatic softening. In this situation, and assuming that the compressive strength remained almost constant, the higher the compressive modulus of elasticity, the weaker the maximum compressive strain (see Figure 3).

In this case, increased, although not very relevant, scatter was observed.

As occurred with the previous parameters, the relaxation of the internal stress was caused by microcracking due to the cyclic loading. In this case, this stress relaxation resulted in a decrease of the maximum compressive strain.

The scatter was similar in all mixtures, meaning that the fibers had no relevant influence on the maximum compressive strain values. On the contrary, a few added differences between the maximum compressive strains were observed; the average values of that parameter varied between 0.0025 and 0.0035, showing no clear observable tendency.

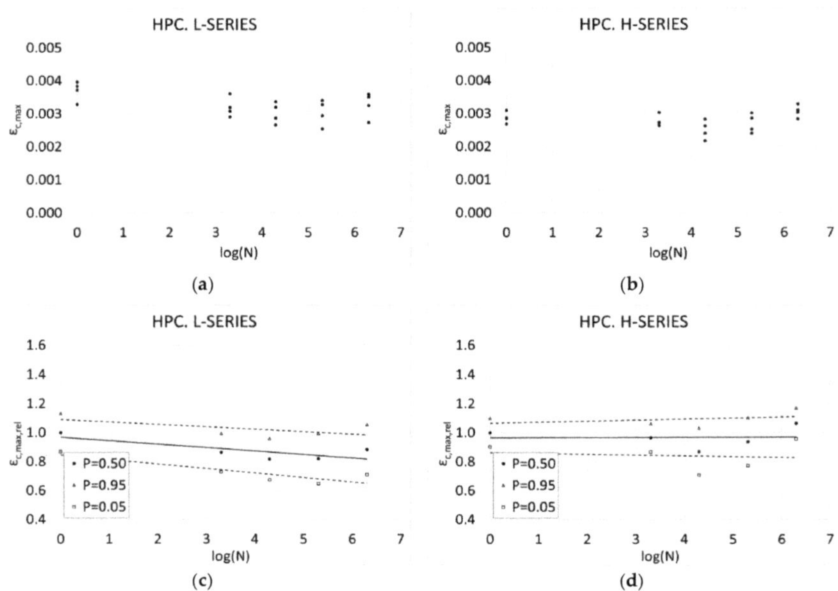

Figure 10. Range of maximum compressive strain by number of cycles. HPC (**a**) L-Series. Raw values. (**b**) H-Series. Raw values. (**c**) L-Series. Maximum, minimum, and average relative values, and tendency lines. (**d**) H-Series. Maximum, minimum, and average relative values, and tendency lines.

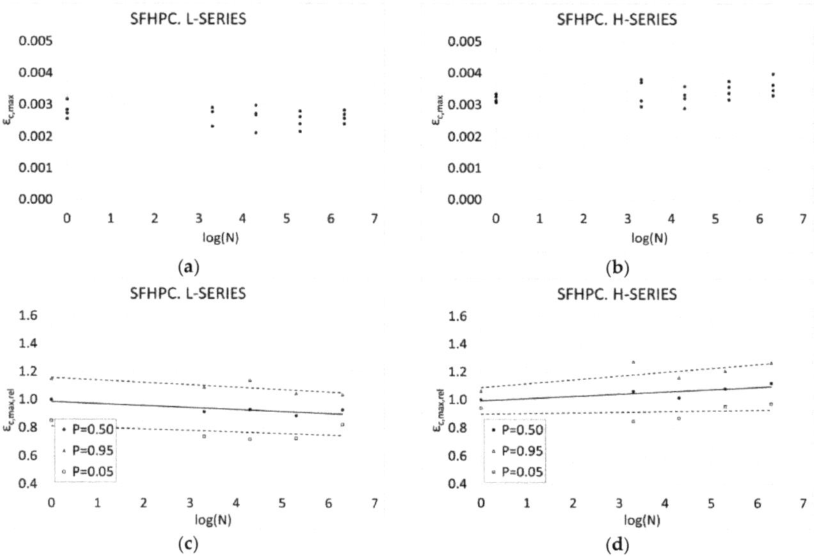

Figure 11. Range of maximum compressive strain by number of cycles. SFHPC (**a**) L-Series. Raw values. (**b**) H-Series. Raw values. (**c**) L-Series. Maximum, minimum, and average relative values, and tendency lines. (**d**) H-Series. Maximum, minimum, and average relative values, and tendency lines.

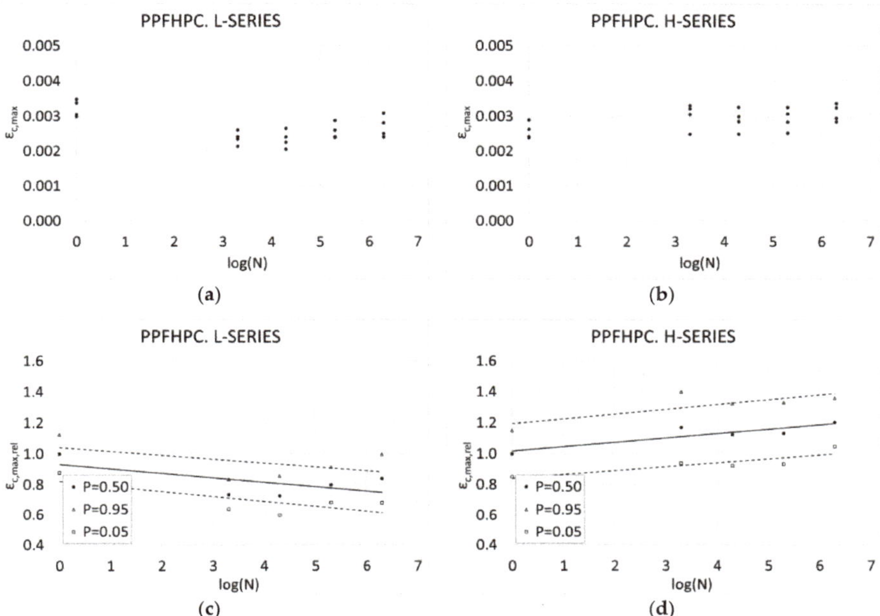

Figure 12. Range of the maximum compressive strain by number of cycles. PPFHPC (**a**) L-Series. Raw values. (**b**) H-Series. Raw values. (**c**) L-Series. Maximum, minimum, and average relative values, and tendency lines. (**d**) H-Series. Maximum, minimum, and average relative values, and tendency lines.

4. Resistance Mechanisms

The results described above revealed that the fibers had an influence on the fatigue behavior of the specimens, which seemed to be related to the microstructure of the mixture. In this section, the resistance mechanisms of the specimens, i.e., the mechanisms used by the specimen to withstand the compressive load, were studied.

Using CT scan technology, it was possible to observe the microstructure and the failure mechanisms within the concrete, in order to extract relevant information for a better understanding of the macroscopic response of the specimens. As explained before, the grey scale of the slides obtained through the scanning process ranged from 0 (which belongs to the least dense voxel, i.e., voids and cracks) to 255 (which belongs to the densest voxel, i.e., steel).

The static test was carried out under displacement control. During testing, the failure criterion was fixed at a 5% reduction of the load, in order to prevent shattering of the cylinders. In addition, the specimens were wrapped in plastic film, to prevent defragmentation. Consequently, only small cracks were observed in the broken cylinders.

The cracks in the HPF were mainly close to the lateral border of the cylinder. They tended to be long but few in number, and they passed through the aggregates. Only a few cracks were found in the internal cores of the specimen. This phenomenon was observed in all the specimens, regardless of the number of cycles previously applied (i.e., the damage level) and the stress level during the cyclic tests (Figures 13 and 14).

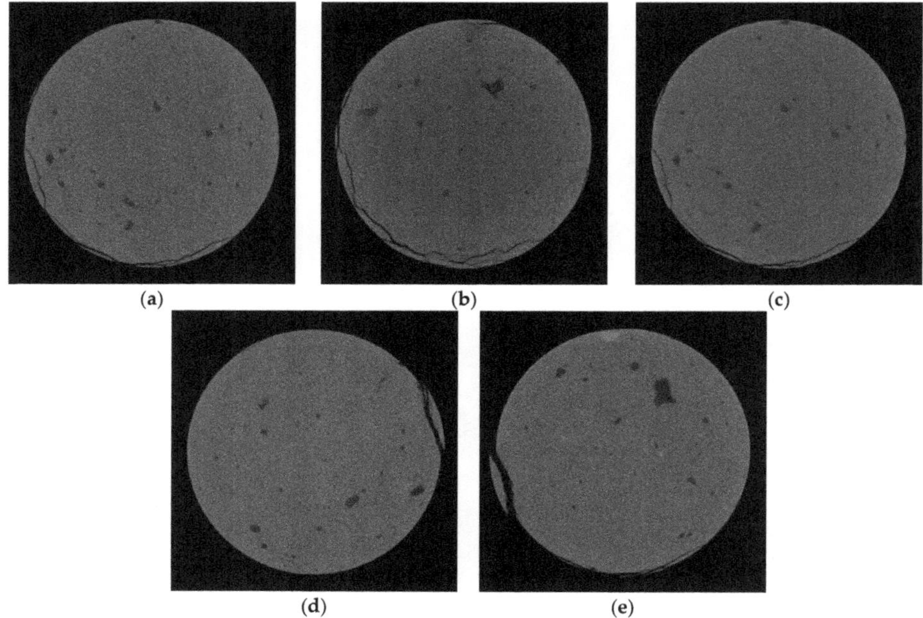

Figure 13. Examples of cracking in HPC specimens. L-Series under the five different sets of cycles: (**a**) 0 cycles; (**b**) 2000 cycles; (**c**) 20,000 cycles; (**d**) 200,000 cycles; (**e**) 2,000,000 cycles.

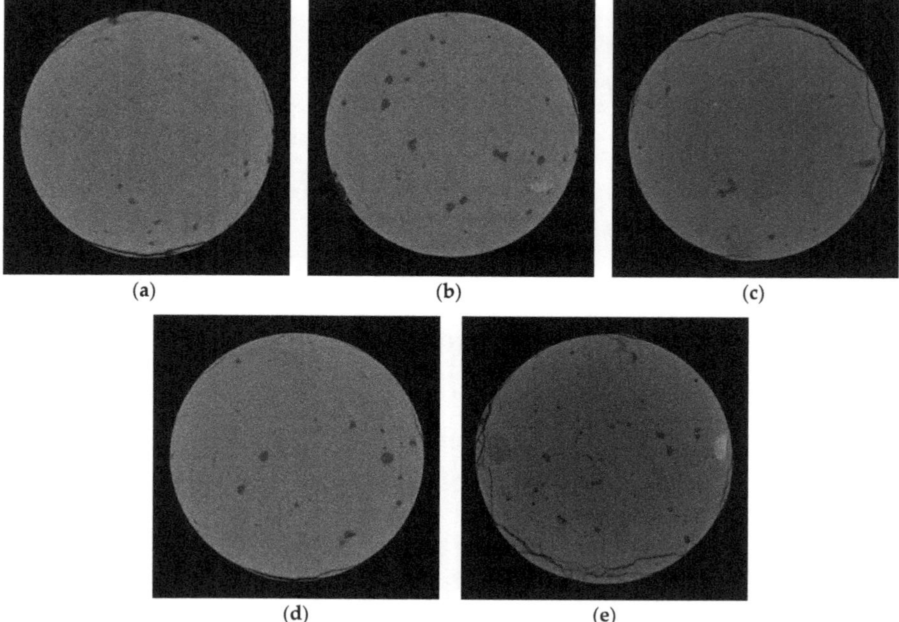

Figure 14. Examples of cracking in HPC specimens. H-Series the five different sets of cycles: (**a**) 0 cycles: (**b**) 2000 cycles; (**c**) 20,000 cycles; (**d**) 200,000 cycles; (**e**) 2,000,000 cycles.

Figures 13 and 14 reveal that, in general, the cyclic tests resulted in low damage levels. The crack surfaces tended to be parallel to the compressive load, showing no significant influence on the behavior of the concrete under static loads.

This finding agrees with the macroscopic behavior of the specimens. The three main mechanical parameters analyzed in this paper showed a slight or null variation with the number of cycles, and no relevant internal damage was observed in the specimens.

In the case of the SFHPF, the cracks showed the same pattern as in the previous case. Similarly to the previous cases, the cracks, which tended to be long but few, were mainly close to the lateral borders of the cylinder. In this case, no fibers were detected in the internal core of the specimen. Again, no differences in the crack patterns of the specimens related to the number of cycles or the stress level during the cyclic tests were observed (Figures 15 and 16).

In this case, as steel fibers are very stiff, the damage due to cyclic loading was mainly concentrated in the lateral border region, where the fibers could not bridge the cracks, rather than within the interior. Again, the crack surfaces tended to be parallel to the compressive load and had no influence on the behavior of the concrete under static loads.

Again, an agreement between the analysis of the CT scan images and the macroscopic behavior of the specimens can be observed. On one hand, the three main mechanical parameters analyzed showed a slight or null variation with the number of cycles, and, on the other hand, no relevant internal damage was observed in the specimens.

In the case of PPFHPC, the cracks showed a similar pattern to the two previous cases, although the cracks were not only concentrated close to the lateral border, but some of them penetrated towards the internal core of the specimens. Again, the cracks tended to be lengthy, and more numerous cracks could be observed. There were no significant differences between the specimens, and neither the number of cycles nor the stress level appeared to have influenced the crack pattern (Figures 17 and 18).

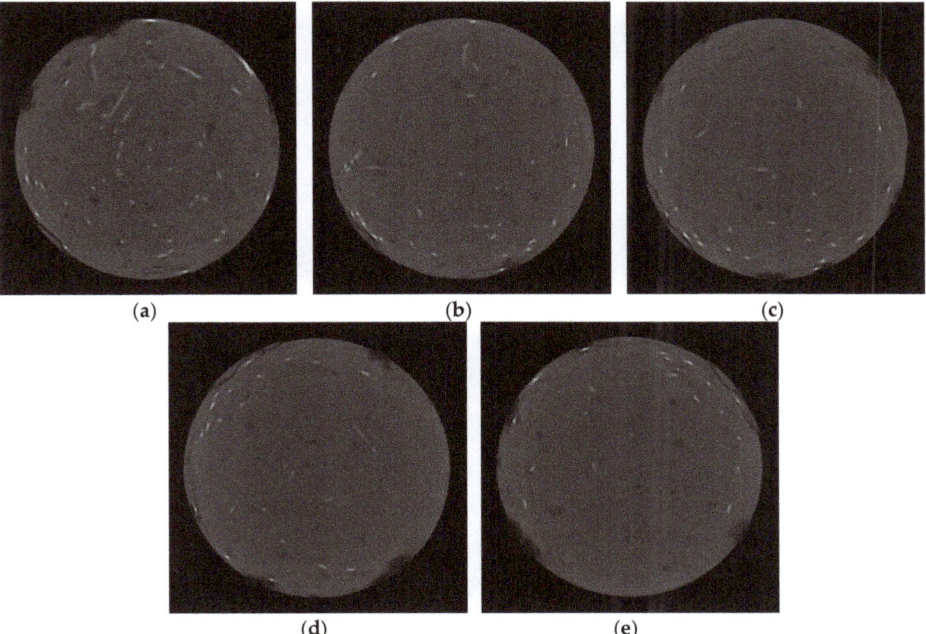

Figure 15. Examples of cracking in specimens of SFHPC. L-Series under the five different sets of cycles: (**a**) 0 cycles; (**b**) 2000 cycles; (**c**) 20,000 cycles; (**d**) 200,000 cycles; (**e**) 2,000,000 cycles.

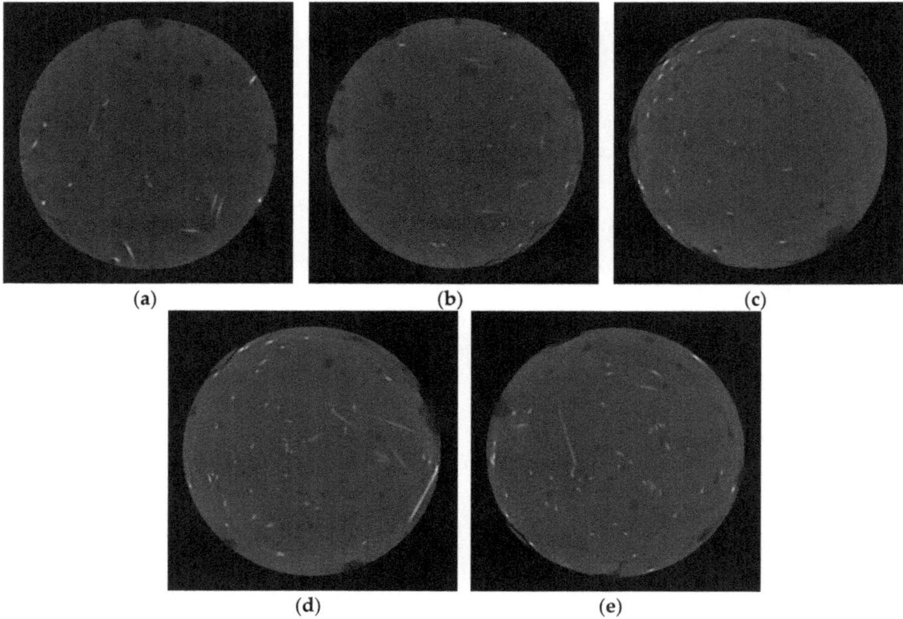

Figure 16. Examples of cracking in specimens of SFHPC. H-Series under the five different sets of cycles: (**a**) 0 cycles; (**b**) 2000 cycles; (**c**) 20,000 cycles; (**d**) 200,000 cycles; (**e**) 2,000,000 cycles.

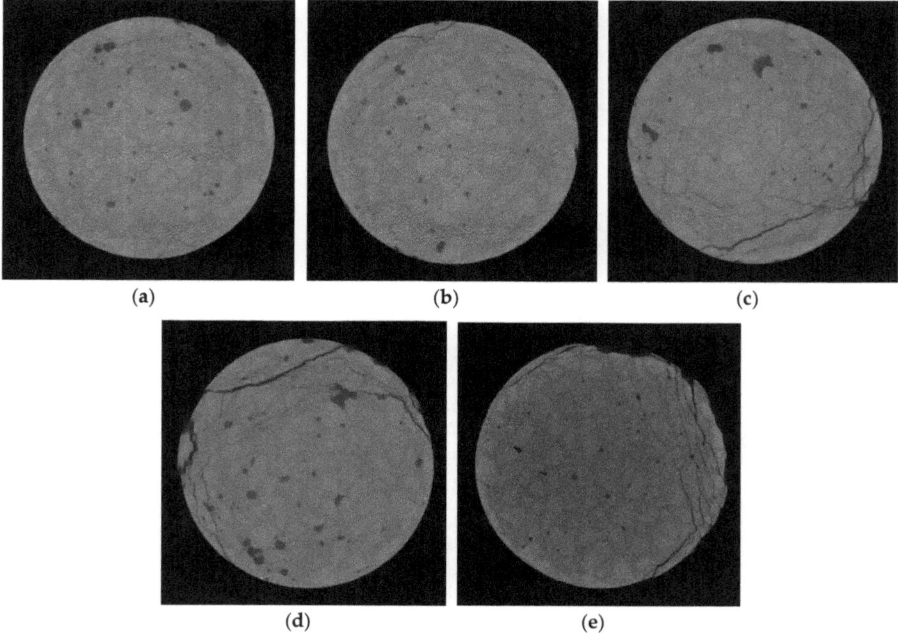

Figure 17. Examples of cracking in specimens of PPFHPC. L-Series under the five different sets of cycles: (**a**) 0 cycles; (**b**) 2000 cycles; (**c**) 20,000 cycles; (**d**) 200,000 cycles; (**e**) 2,000,000 cycles.

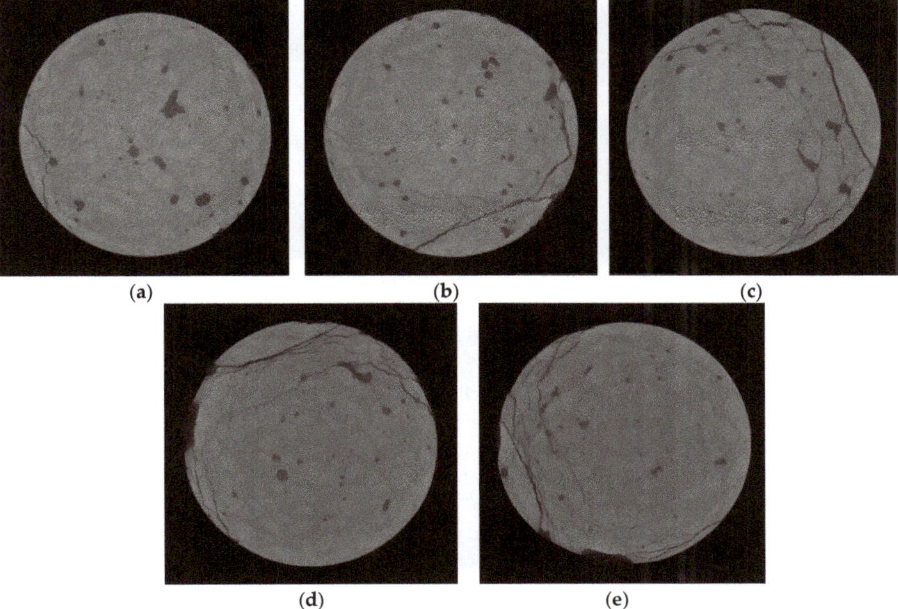

Figure 18. Examples of cracking in specimens of PPFHPC. H-Series under the five different sets of cycles: (**a**) 0 cycles; (**b**) 2000 cycles; (**c**) 20,000 cycles; (**d**) 200,000 cycles; (**e**) 2,000,000 cycles.

In this case, the polypropylene fibers had a very low stiffness, and the cyclic loads resulted in internal damage. Consequently, some cracks were observed within the specimens. As in the previous cases, the crack surfaces were mainly vertical, i.e., parallel to the compressive loads, and they had no influence on macroscopic behavior under compressive loads.

It can be concluded that in all mixtures, cracks were mostly placed at the surface. This fact seems to have been due to the concrete matrix. The scale effect on concrete specimens is well known, and many examples of this phenomenon can be observed in concrete elements [38–40]. In this case, the scale effect conducted to a lower amount of coarse aggregates and fibers at the lateral border [41]. This would be the reason why the cracks mainly appeared at the surface.

As in the previous cases, the three main mechanical parameters did not show relevant variation with the number of cycles, which agrees with the internal damage observed.

For the three mechanical parameters studied in this paper, the results revealed that the scatter is a parameter clearly dependent on the number of cycles. This fact can be observed in all cases, i.e., in all the mixtures and in both series.

5. Conclusions

In this paper, the range of values of three mechanical parameters of concrete, following the application of five series of cyclic compressive loading, was studied. Three concrete mixtures—plain, steel-fiber-reinforced, and polypropylene-fiber-reinforced high-performance concrete (HPC, SFHPC, and PPFHPC)—were studied. The specimens were subjected to either high stress or low stress levels, and to five different series of cycles: 0, 2000; 20,000; 200,000, and 2,000,000 cycles. All the specimens survived the cyclic test. They were then subjected to a static compression test up to failure, and the compression strength, the compressive modulus of elasticity, and the maximum compressive strain were all measured.

According to Model Code 2010 [34], the specimens were subjected to relevant fatigue damage, especially in the case of high stress level series.

In addition, before starting the cyclic tests and after the static tests, all the specimens were scanned, in order to visualize the internal damage caused by the cyclic loads. A high-resolution CT scan device was used to do so.

The mechanical results showed no relevant damage caused by the cyclic loading. Moreover, the mechanical parameters of the concrete mixtures showed no significant reduction, and even a slight improvement in some cases.

Regarding the compressive strength, the average value of this mechanical parameter did not significantly vary with the number of cycles, and, in some cases, a slight increase was observed. In addition, the scatter showed a relevant increase with the damage levels, equitable with the number of cycles.

The compressive modulus of elasticity showed a similar behavior. The L-Series showed a slight increase of the compressive modulus of elasticity, while the H-Series showed a flat behavior. Again, the scatter of the compressive modulus of elasticity significantly increased with the number of cycles.

Regarding the maximum compressive strain, the average value showed no clear tendency. In case of the L-Series, an overall decrease with the number of cycles was observed, while in the case of the H-Series, an overall increase with the number of cycles occurred. Again, the scatter showed a progressive increase with the number of cycles.

It can be noticed that for the three mechanical parameters studied in this paper, the results revealed that the scatter is a parameter clearly dependent on the number of cycles. This fact was observed in all cases, i.e., in all the mixtures and in both series.

Moreover, the variation of the mechanical parameters with the damage hardly appeared to be related to the stress level and to the concrete mixture.

The CT scan images revealed no observable internal damage (i.e., internal microcracking) due to cyclic loading, only an observably small number of long cracks. They mainly propagated close to the lateral area of the cylinders and appeared to have been caused by the static testing and not caused by the previous cyclic test, because no differences were observed between the specimens with and without cyclic damage (i.e., the ones subjected to 0 cycles).

Internal cracks were only observed for the PPFHPC specimens, which can be explained by the lower stiffness of the polypropylene fibers.

It was concluded that, contrary to the predictions of the Model Code [34], the cyclic loads tested in this research caused little or no internal damage (i.e., internal micro-cracking), and no degradation of the microstructure was observed. Both the L-Series and H-Series were less aggressive to the concrete specimens, and even at 2,000,000 cycles no relevant internal damage was observed and the variation of the mechanical properties of the concrete with the number of cycles was small. It can be highlighted that, in the case of the H-Series, the Model Code [34] predicted an expected fatigue life of 1.5×10^3 cycles, which means that all the numbers of cycles tested, i.e., 2000, 20,000, 200,000, and 2,000,000 cycles, were beyond their fatigue life. The truth is that none of the specimens collapsed during the cyclic test and their residual mechanical properties did not vary significantly with respect to the ones showed by the specimens not previously subjected to cyclic loads.

Author Contributions: J.M. and L.G. developed and carried out the laboratory testing, J.M. also worked on the analysis of the CT scan images, D.C.G. and M.A.V. analyzed the data and wrote the paper.

Funding: This research was funded by the Ministerio de Ciencia e Innovación (Spain), grant number BIA2004-08033.

Conflicts of Interest: The authors declare no conflict of interest.

References

1.   Hsu, T.T.C. Fatigue of plain concrete. *ACI J.* **1981**, *78*, 292–305.

2. Naaman, A.E.; Hammoud, H. Fatigue Characteristics of High Performance Fiber-reinforced Concrete. *Cem. Con. Comp.* **1998**, *20*, 353–363. [CrossRef]

3. Kim, J.-K.; Kim, Y.Y. Experimental study of the fatigue behaviour of high strength concrete. *Cem. Con. Res.* **1996**, *10*, 1513–1523. [CrossRef]

4. Cachim, P.B.; Figueiras, J.A.; Pereira, P.A.A. Fatigue behavior of fiber-reinforced concrete in compression. *Cem. Con. Comp.* **2002**, *24*, 211–217. [CrossRef]

5. Singh, S.P.; Singh, B.; Kaushik, S.K. Probability of fatigue failure of steel fibrous concrete. *Mag. Con. Res.* **2005**, *57*, 65–72. [CrossRef]

6. Singh, S.P.; Mohammadi, S.G.; Kaushik, S.K. Prediction of mean and design fatigue lives of steel fibrous concrete beams in flexure. *Adv. Struc. Eng.* **2007**, *10*, 25–36. [CrossRef]

7. Medeiros, A.; Zhang, X.; Ruiz, G.; Yu, R.C.; de Souza, M. Effect of the loading frequency on the compressive fatigue behavior of plain and fiber reinforced concrete. *Int. J. Fatigue* **2015**, *70*, 342–350. [CrossRef]

8. Poveda, E.; Ruiz, G.; Cifuentes, H.; Yu, R.C.; Zhang, X. Influence of the fiber content on the compressive low-cycle fatigue behavior of self-compacting SFRC. *Int. J. Fatigue* **2017**, *101*, 9–17. [CrossRef]

9. Vicente, M.A.; González, D.C.; Mínguez, J.; Tarifa, M.A.; Ruiz, G.; Hindi, R. Influence of the pore morphology of high strength concrete on its fatigue life. *Int. J. Fatigue* **2018**, *112*, 106–116. [CrossRef]

10. Ruiz, G.; de la Rosa, A.; Wolf, S.; Poveda, E.I. Model for the compressive stress-strain relationship of steel fiber-reinforced concrete for non-linear structural analysis. *HyA* **2018**, *69*, 75–80. [CrossRef]

11. Vicente, M.A.; González, D.C.; Mínguez, J.; Martínez, J.A. Residual modulus of elasticity and maximum compressive strain in HSC and FRHSC after high-stress-level cyclic loading. *Struct. Conc.* **2014**, *15*, 210–218. [CrossRef]

12. Liu, F.; Zhou, J. Research on fatigue strain and fatigue modulus of concrete. *Adv. Struct. Eng.* **2017**, *2017*, 6272906. [CrossRef]

13. Isojeh, B.; El-Zeghayar, M.; Vecchio, F.J. Concrete Damage under Fatigue Loading in Uniaxial Compression. *ACI Mater. J.* **2017**, *114*, 225–235. [CrossRef]

14. Zanuy, C.; Albajar, L.; de la Fuente, P. The fatigue process of concrete and its structural influence. *Mater. Const.* **2011**, *303*, 385–399. [CrossRef]

15. Vicente, M.A.; Ruiz, G.; González, D.C.; Mínguez, J.; Tarifa, M.; Zhang, X. CT-Scan study of crack patterns of fiber-reinforced concrete loaded monotonically and under low-cycle fatigue. *Int. J. Fatigue* **2018**, *114*, 138–147. [CrossRef]

16. González, D.C.; Moradillo, R.; Mínguez, J.; Martínez, J.A.; Vicente, M.A. Postcracking residual strengths of fiber-reinforced high-performance concrete after cyclic loading. *Struct. Concr.* **2018**, *19*, 340–351. [CrossRef]

17. Parvez, A.; Foster, S.J. Fatigue behavior of steel-fiber-reinforced concrete beams. *J. Struct. Eng.* **2015**, *141*, 04014117. [CrossRef]

18. González, D.C.; Vicente, M.A.; Ahmad, S. Effect of cyclic loading on the residual tensile strength of steel fiber–reinforced high-strength concrete. *J. Mater. Civ. Eng.* **2015**, *27*, 04014241. [CrossRef]

19. Minguez, J.; González, D.C.; Vicente, M.A. Fiber geometrical parameters of fiber reinforced high strength concrete and their influence on the residual postpeak flexural tensile strength. *Constr. Build. Mater.* **2018**, *168*, 906–922. [CrossRef]

20. Ribolla, E.L.M.; Hajidehi, M.R.; Rizzo, P.; Scimeni, G.F.; Spada, A.; Giambanco, G. Ultrasonic inspection for the detection of debonding in CFRP-reinforced concrete. *Struct. Inf. Eng.* **2018**, *14*, 807–816. [CrossRef]

21. Herrmann, H.; Pastorelli, E.; Kallonen, A.; Suuronen, J.P. Methods for fibre orientation analysis of X-ray tomography images of steel fibre reinforced concrete (SFRC). *J. Mater. Sci.* **2016**, *51*, 3772–3783. [CrossRef]

22. Vicente, M.A.; González, D.C.; Mínguez, J. Determination of dominant fibre orientations in fibre-reinforced high strength concrete elements based on computed tomography scans. *Nondestr. Test. Eval.* **2014**, *29*, 164–182. [CrossRef]

23. Mínguez, J.; Vicente, M.A.; González, D.C. Pore morphology variation under ambient curing of plain and fiber-reinforced high performance mortar at an early age. *Const. Build. Mater.* **2019**, *198*, 718–731. [CrossRef]

24. Vicente, M.A.; Mínguez, J.; González, D.C. Variation of the pore morphology during the early age in plain and fiber-reinforced high-performance concrete under moisture-saturated curing. *Materials* **2018**, *12*, 975. [CrossRef] [PubMed]

25. Chandrappa, A.K.; Biligiri, K.P. Pore structure characterization of pervious concrete using X-ray microcomputed tomography. *J. Mater. Civ. Eng.* **2018**, *30*, 04018108. [CrossRef]

26. Lu, H.; Peterson, K.; Chernoloz, O. Measurement of entrained air-void parameters in Portland cement concrete using micro X-ray computed tomography. *Int. J. Pavement Eng.* **2018**, *19*, 109–121. [CrossRef]

27. Wang, Y.-S.; Dai, J.-G. X-ray computed tomography for pore-related characterization and simulation of cement mortar matrix. *NDT E Int.* **2017**, *86*, 28–35. [CrossRef]

28. Moradian, M.; Hu, Q.; Aboustait, M.; Ley, M.T.; Hanan, J.C.; Xiao, X.; Scherer, G.W.; Zhang, Z. Direct observation of void evolution during cement hydration. *Mater. Des.* **2017**, *136*, 137–149. [CrossRef]

29. Vicente, M.A.; Mínguez, J.; González, D.C. The use of computed tomography to explore the microstructure of materials in civil engineering: From rocks to concrete. In *Computed Tomography—Advanced Applications*; Halefoglu Ahmet Mesrur, InTech: London, UK, 2017.

30. Vicente, M.A.; González, D.C.; Mínguez, J. Recent advances in the use of computed tomography in concrete technology and other engineering fields. *Micron* **2019**, *118*, 22–34. [CrossRef]

31. European Committee for Standardization (CEN). *Testing Hardened Concrete—Part 13: Determination of Secant Modulus of Elasticity in Compression*; EN 12390-13; CEN: Brussels, Belgium, 2013.

32. European Committee for Standardization. *Testing Hardened Concrete—Part 3: Compressive Strength of Test Specimens*; EN 12390-3; CEN: Brussels, Belgium, 2009.

33. European Committee for Standardization. *Eurocode 2, Design of Concrete Structures*; CEN: Brussels, Belgium, 2004.

34. International Federation for Structural Concrete. Model Code for Concrete Structures. In *FIB Bulletin 65*; Ernst Sohn: Lausanne, Switzerland, 2010.

35. Bazant, Z.P.; Hubler, M.H. Theory of cyclic creep of concrete based on Paris law for fatigue growth of subcritical microcracks. *J. Mech. Phys. Sol.* **2014**, *63*, 187–200. [CrossRef]

36. Soroushian, P.; Elzafraney, M. Damage effects on concrete performance and microstructure. *Cem. Con. Comp.* **2004**, *26*, 853–859. [CrossRef]

37. Suresh, S. Mechanics and micromechanisms of fatigue crack growth in brittle solids. *Int. J. Fract.* **1990**, *42*, 41–56. [CrossRef]

38. Darlington, W.J.; Ranjith, P.G.; Choi, S.K. The Effect of Specimen Size on Strength and Other Properties in Laboratory Testing of Rock and Rock-Like Cementitious Brittle Materials. *Rock Mech. Rock Eng.* **2011**, *44*, 513. [CrossRef]

39. Ferro, G.; Carpinteri, A. Effect of Specimen Size on the Dissipated Energy Density in Compression. *J. Appl. Mech.* **2008**, *75*, 041003. [CrossRef]

40. Carpinteri, A.; Ferro, G.; Monetto, I. Scale effects in uniaxially compressed concrete specimens. *Mag. Con. Res.* **1999**, *51*, 217–225. [CrossRef]

41. Zheng, J.J.; Li, C.Q.; Jones, M.R. Aggregate distribution in concrete with wall effect. *Mag. Con. Res.* **2003**, *55*, 257–265. [CrossRef]

Article

Fatigue in Concrete under Low-Cycle Tensile Loading Using a Pressure-Tension Apparatus

Sayed M. Soleimani [1,*], Andrew J. Boyd [2], Andrew J.K. Komar [3] and Sajjad S. Roudsari [4]

[1] Department of Civil Engineering, Australian College of Kuwait, P.O. Box 1411, Safat 13015, Kuwait
[2] Department of Civil Engineering and Applied Mechanics, McGill University, 817 Sherbrooke Street West, Montreal, QC H3A 0C3, Canada
[3] Project Coordinator, Aecon, 20 Carlson Ct, Etobicoke, ON M9W 7K6, Canada
[4] Department of Computational Science and Engineering, North Carolina A&T State University, 1601 E, Market Street, Greensboro, NC 27411, USA
* Correspondence: s.soleimani@ack.edu.kw; Tel.: +965-2537-6111

Received: 15 July 2019; Accepted: 6 August 2019; Published: 7 August 2019

Abstract: Fatigue due to low-cycle tensile loading in plain concrete was examined under different conditions using the pressure-tension apparatus. A total of 22 wet or dry standard concrete cylinders (100 mm × 200 mm) were tested. By definition, low-cycle loading refers to the concept of multiple load cycles applied at high stress levels (i.e., a concrete structure subjected to seismic loading). Results suggest that concrete samples subjected to low-cycle tensile loading will fail after a relatively low number of cycles of loading and at a lower magnitude of stress compared to the maximum value applied during cyclic loading. Furthermore, non-destructive testing was employed in order to ascertain the extent of progressive damage inflicted by tensile loading in concrete specimens. It was found that ultrasonic pulse velocity is a viable technique for evaluating the damage consequential of loads applied to concrete, including that resultant from low levels of tensile stress (i.e., as low as 10% of its maximum tensile capacity). Additionally, finite element analysis was performed on a modeled version of the pressure-tension apparatus with a sample of concrete, which has yielded similar results to the experimental work.

Keywords: concrete; fatigue; tension; pressure-tension apparatus; nondestructive testing; ultrasonic pulse velocity; ABAQUS FEA

---

1. Introduction

The pressure tension testing method was originally developed by the Building Research Council as a new method of examining anisotropic loading conditions on materials [1]. The pressure tension apparatus uses standard 100 mm × 200 mm concrete cylinders as samples. Concrete core samples can also be used. The size of the sample allows this test method to be adapted to common concrete testing procedures used in industry. The test uses a pressurized gas equally applied along the curved surface of the cylinder, with rubber O-rings at either end of the specimen to hold the gas between the testing chamber and the concrete cylindrical sample surface, as shown in Figure 1. The two flat ends of the cylinder are left open to the atmosphere, which causes a net induced tension field to arise within the concrete sample parallel to the longitudinal axis of the cylinder. Komar [2] devised and implemented a novel instrumentation system that allowed for fully automated control of the loading rate and variety of loading conditions, including ramped loading at different rates, cyclic loading, constant loading rate (such as creep), and controlled unloading. This study uses the method of cyclic loading.

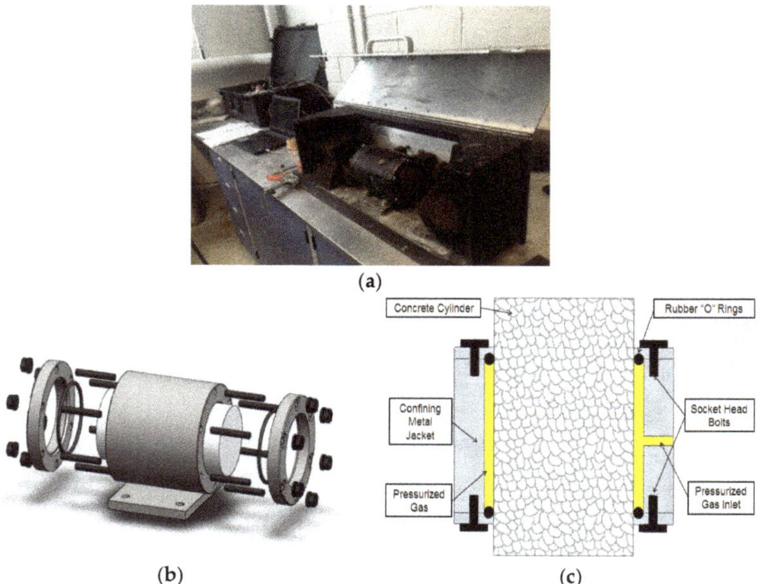

(a)

(b)   (c)

Figure 1. (**a**) Photograph of pressure-tension testing machine; (**b**) Exploded view of the pressure-tension apparatus with concrete cylinder in machine; (**c**) Cross section of pressure-tension apparatus and concrete cylinder [2].

In this test method, a tension field will be produced from a compressive load. It can be understood as a net effective stress which is consistent with principles first developed for soil mechanics [3]. Concrete is a two-phase material with a solid phase consisting of the hydrated cementitious matrix and the aggregates, and a liquid phase consisting of the pore water. The different reactions of the two phases to a biaxial stress gives rise to the pressure tension effect: the liquid phase reacts in a hydrostatic manner, in contrast to the solid phase which reacts in the directions of the applied load. All the stresses in the plane of gas loading cancel each other out, with a net tension field arising along the axis of the cylinder [4], which is shown in Figure 2. The pressure tension apparatus uniformly increases the gas pressure applied to the concrete cylinder until the point where the solid phase of the specimen can no longer remain together, at which point the concrete cylinder breaks at the weakest plane along its length. The gas pressure at the moment of failure is taken as the ultimate tensile strength of the concrete.

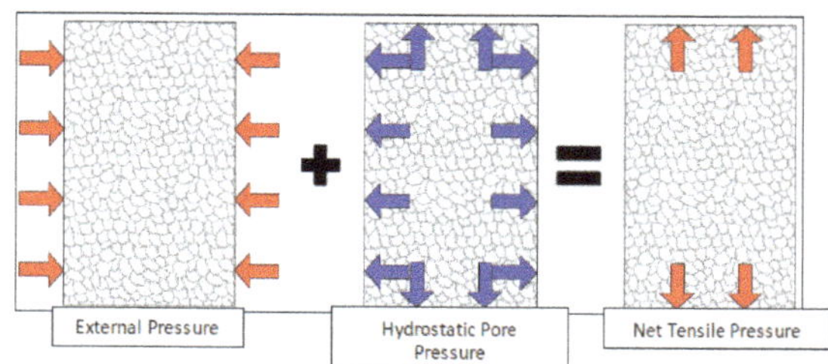

Figure 2. Pressure tension effect [2].

Rashidi et al. [5] conducted an experimental study to evaluate the strength of concrete subjected to Indirect Diametrical Tensile (IDT), Unconfined Compressive Strength (UCS) and submaximal modulus testing methods. Their experimental design involved the application of a specific strain rate of 1 mm/min on concrete specimens. IDT employs a controlled loading condition to split the cylinder into two half-cylinders. In contrast, the present experimental design uses gas pressure to achieve the same mode of failure.

Ultrasonic Pulse Velocity (UPV) provides information about the evolution of the pore space of concrete, for both hydration and changes due to durability issues. Komar [2] used UPV to track the hydration process during curing; they found UPV could distinguish between different pore structures at early ages and could track the evolution of the pore structures over the curing period. UPV can be divided into three categories: Direct (sensors opposite to each other), indirect (adjacent sensors), and semi-direct (sensors at right angles) transmission. Ultrasonic waves have been used to predict and evaluate concrete strength and its properties in several studies [6]. However, this method can also be used to detect the internal deficiencies of concrete such as cracks. In this experimental work, the UPV method is used to determine the presence of cracks in concrete after a low magnitude of tensile stress is applied. The UPV test is a measurement of the continuity of the solid part of the concrete through which the stress waves propagate, and it is well-suited for evaluating any changes to this solid structure due to ongoing deterioration. The test apparatus produces and applies a pulse into the concrete by means of a pulse transducer and receiver and encompasses the ability to accurately measure the amount of time the pulse takes to pass through the concrete.

2. Materials and Methods

A total of 22 standard concrete cylinders (100 mm × 200 mm) were cast and tested. The concrete used in the cylinders was made using the mix design shown in Table 1. The specimens were molded and compacted based on ASTM C31 [7]. After 24 h of curing in their molds, the cylinders were removed and stored in ambient laboratory conditions for over 2 years.

Table 1. Mix design of the concrete specimens for this study (water-to-cement ratio = 0.45).

Component	Amount	Unit
Water	15.866	kg
Cement	29.282	kg
Coarse aggregates	72.872	kg
Fine aggregates	79.093	kg
Superplasticizer	0.146	Liter

In a pressure-tension test, a destructive test method that specifically evaluates the tensile capacity of concrete, a cylindrical steel jacket is used to hold a concrete cylinder serving as the specimen. A rubber O-ring is used at each end to keep an airtight seal and prevent possible gas leakage. With the ends then left free, gas pressure is incrementally applied to the curved surface of the concrete cylinder until the specimen ultimately fractures across a single plane perpendicular to the axis of the concrete specimen. The value of gas pressure at fracture thus corresponds to the maximum tensile capacity of the concrete. Compressed air is used as the loading medium for all specimens in this experimental work. The air was pressurized using a portable air compressor with a maximum available pressure of about 17 MPa, which was stored in a pressurized gas vessel attached to the compressor and the testing machine. A variable check valve was used to keep the pressure head constant in order to minimize inconsistency in the test caused by pressure variance in the storage tank. A needle valve was used as the controller for the flow rate of the compressed gas into the testing chamber. The valve was operated by a mechanical actuator which was controlled using custom software and monitored using a digital acquisition device (DAQ). Measurement of pressure within the test chamber was obtained using a digital pressure transducer directly adjacent to the chamber, which was used as a control input in the software. The pressure transducer had a measured error of ±0.05 MPa. The DAQ had a sensing resolution of approximately 0.1 s [2,6]. All pressure-time plots were recorded and stored, and the data point immediately preceding failure was used as the failure load.

The implemented control algorithm is a modified proportional integral derivative function which automatically varies the valve position based on the detected error between the measured pressure and the set point. The set point is an arbitrary time-dependent function, and all tests carried out in this study used a uniformly increasing pressure-time signal with a rate of about 0.02 to 0.03 MPa/second, comparable to loading rates in both ASTM C 496 and ASTM C 33 [6]. A higher rate of about 0.10 to 0.15 MPa/second was used during the cyclic loading of the specimens to expedite the process.

Rubber O-rings with a diameter of 10 mm were used to seal the interface between the test chamber and the concrete sample (Figure 1). One or two layers of polyvinylchloride tape was applied on both ends of the cylinder to provide a better seal between the O-rings and the cylinder, which occasionally had voids or deterioration defects that could cause leakage around the O-rings. The O-rings were torqued with a series of six lug nuts bearing on a steel plate in a manner that applied equal pressure on them and ensured a tight seal for the loading gas.

Pressure tension testing is affected by the moisture content of the specimens; samples with a higher water content fail at a higher tensile resistance as compared to dry specimens [2]. For dry samples, the failure mode in pressure tension testing changes from a single crack to a volumetric failure at a lower applied tensile stress [2,8].

Fatigue loading can be categorized into low-cycle and high-cycle loading. Low-cycle loading encompasses the introduction of a few high-stress load cycles such as the application of the earthquake loads on a structure. Conversely high-cyclic loading is categorized by the introduction of a large number of low-stress load cycles, such as the live load applications on the bridges and airport pavements.

The flexural test method is the most commonly used method for fatigue testing. Comparatively, compressive fatigue tests have been studied to a lesser extent. After the emergence of nonlinear fracture mechanics toward evaluating the performance of concrete, some researchers have shown interest in the fatigue characteristics of concrete in tension [9–11]. Nevertheless, there are very limited studies available on the behavior of plain concrete under direct tension fatigue loading [12–15]. The challenges associated with direct tensile testing on concrete is also a primary reason for limited and often conflicting data availability.

Furthermore, non-destructive testing (NDT) was employed to evaluate the progressive damage inflicted by tensile loading on concrete cylinders using UPV apparatus. A Proceq TICO model of automatic UPV was used for this experimental study. Three measurements of transit times along the axis of the cylinder were recorded, and the average of these three values was noted. Length measurements from the cylinders (i.e., height of the concrete cylinder specimens) were used to convert the travel times into the UPV values reported. A commercially available water-based lubricant was used to ensure a good connection between the transmitter/receiver and the greatly variable surface of the concrete samples.

In this study, two piezoelectric sensors are placed at opposite ends of the test specimen (i.e., direct method). Electronic pulses are generated by one sensor, while the other sensor measures the time taken by the pulse to transmit through the concrete. Knowing the distance traveled (i.e., the height of the concrete cylinder specimen in this case), transmission velocity will be calculated based on the resultant value and the condition of the concrete may be determined.

3. Experimental Results and Discussion

3.1. Tensile Strengths (No Cyclic Loads)

The first trial of samples consisted of six standard concrete cylinders loaded up to failure (Figure 3). The first two were tested in dry conditions, while the other four were tested in wet conditions (they have been submerged in water for few days before the test). The results of each individual test are presented in Table 2. The statistical analysis exhibits that the average tensile strength of the dry specimens was lower than that of the wet specimens. In contrast, the coefficient of variation for the wet specimens is much bigger than that of the dry specimens. The recommended coefficient of variation in the ASTM C 496 standard is 14%. Given the limited number of tests conducted in this study, it seems that the pressure tension test on dry specimens will give a conservative and reliable result as compared to the wet specimens. More test results are required to attest this statement.

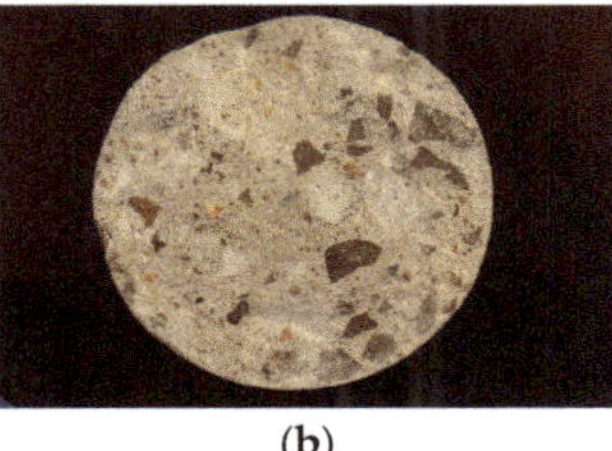

(a)   (b)

Figure 3. Broken specimen under tension using pressure-tension apparatus: (**a**) Elevation view; (**b**) Cross-sectional view.

Table 2. Tensile strength of concrete exposed to ambient conditions for two years.

Specimen No.	Condition	Ultimate Tensile Strength (MPa)	Average (MPa)	Standard Deviation (MPa)	Coefficient of Variation (%)
1	Dry	2.90	3.02	0.16	5.39
2	Dry	3.13			
3	Wet	3.83	4.45	0.76	17.15
4	Wet	4.33			
5	Wet	4.08			
6	Wet	5.55			

3.2. Tensile Strengths (Cyclic Loads)

Twelve wet and four dry concrete cylinders (a total of sixteen specimens) have been tested under a cyclic loading condition. Since the tensile strength of the specimen can be varied substantially for the wet specimens as shown in Table 2, finding the fatigue threshold associated with a low-cycle loading protocol can be quite challenging. The results are shown in tabular format for these specimens.

Table 3 shows the cyclic test results for two wet specimens (No. 7 and No. 8). Each one failed during the cyclic loading and under a tensile stress that was below the maximum applied tensile stress in previous cycle(s). Both specimens have not been loaded prior to the cyclic loads. It is worth mentioning that the stress at failure is also less than the average tensile strength of 4.45 MPa (as indicated in Table 2).

Table 3. Cyclic load results for specimens 7 and 8.

Specimen No.	Condition	Failed during the Load Cycle No.	Max. Tensile Stress Applied in Previous Cycles (MPa)	Stress at Failure (MPa)
7	Wet	2	3.43	3.23
8	Wet	12	3.13	3.03

Specimens No. 9 to No. 14 have been subjected to 100 cycles of tensile loads with a maximum tensile stress of 2.23 MPa to 3.01 MPa. Then, they have been loaded up to the failure point and the tensile strengths have been recorded to see the effect of the cyclic loading on the final tensile strength. The results are tabulated in Table 4. It seems that the tensile strength of concrete will not be affected if the maximum tensile stress in the cyclic loading is below a threshold (i.e., low-cycle loading). This threshold can be more than 62% of the tensile strength (based on the limited number of test results presented in Table 4) but more test results are required for verification. The difference between these specimens and those reported in Table 3 is that they have been subjected to a cyclic loading for 100 cycles and then tested up to the failure point. The specimens in Table 3 failed during the cyclic loading.

Table 4. Cyclic load results for specimens 9 to 14.

Specimen No.	Condition	Max. Tensile Stress Applied in a Cyclic Loading of 100 Cycles (MPa) [A]	Stress at Failure (MPa) [B]	[A]/[B]
9	Wet	2.23	5.37	0.42
10	Wet	2.67	4.34	0.62
11	Wet	2.72	4.89	0.56
12	Wet	2.83	4.74	0.60
13	Wet	2.88	5.62	0.51
14	Wet	3.01	6.05	0.50

Specimens No. 15 to No. 18 have been subjected to a cyclic loading with a maximum of 20 cycles. The cyclic loading started at a low level (about 50% of the average tensile strength of the wet specimens as indicated in Table 2) and the maximum load in the next 20 cycles has been increased gradually (by about 5% of the average tensile strength of the wet specimens as indicated in Table 2). The purpose of this test method was to prove that there is a threshold for the cyclic loading at which the fatigue failure will happen at a low-cycle loading. Test results for these specimens are shown in Table 5.

Table 5. Cyclic load results for specimens 15 to 18.

Specimen No.	Condition	Failed during the Load Cycle No.	Max. Tensile Stress Applied in Previous Cycles (MPa)	Stress at Failure (MPa)
15	Wet	2	4.20	4.15
16	Wet	2	4.42	4.40
17	Wet	14	4.59	4.49
18	Wet	3	4.87	4.10

Table 6 shows the test results for specimens No. 19 to No. 22. They have been tested in a dry condition. The same procedure explained for testing specimens No. 15 to No. 18 has been followed for these tests. It is clear that the condition of the concrete cylinders (i.e., wet versus dry) plays an important role in the tensile capacity of the concrete in cyclic loading (Tables 3–6) and non-cyclic loading (Table 2). As an example, the cyclic loading on specimen No. 19 is shown in Figure 4, in which the specimen failed during the sixth cycle.

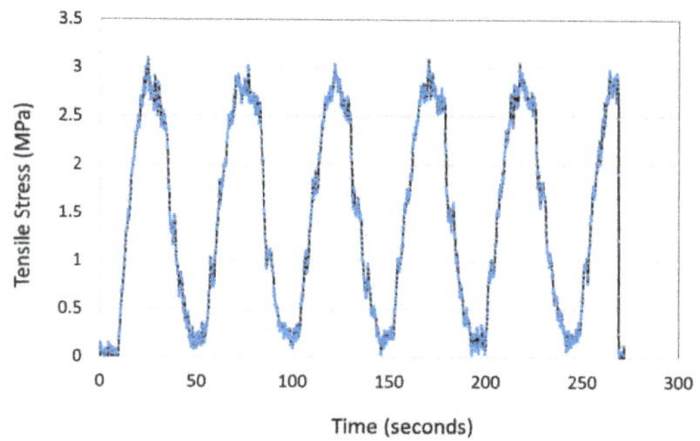

Figure 4. Cyclic loading on specimen No. 19 (failed during the 6th cycle).

Table 6. Cyclic load results for specimens 19 to 22.

Specimen No.	Condition	Failed during the Load Cycle No.	Max. Tensile Stress Applied in Previous Cycles (MPa)	Stress at Failure (MPa)
19	Dry	6	3.11	2.90
20	Dry	3	2.39	2.23
21	Dry	5	2.96	2.80
22	Dry	4	2.85	2.72

3.3. UPV Test Results and Discussions

The average ultrasonic pulse velocity measurements (the mean value of three measurements) versus the applied tensile stress can be found in Figures 5–7. The UPV measurements took place after the tensile force was removed. For all cases, a general downward trend is apparent from the data, indicating decreased pulse velocities associated with an increasing of the applied tensile stresses. Higher UPV measurements are indicative of an internal structure which is relatively denser (thus allowing for faster transmission of the pulse). A downward trend as seen in Figures 4–6 shows that the application of a tensile stress on a concrete cylinder is affecting the internal structure so as to cause an increase in travel times and a lower UPV measurement.

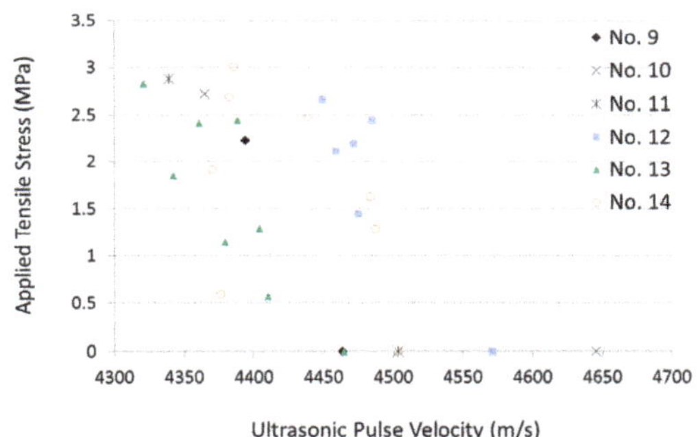

Figure 5. Ultrasonic Pulse Velocity (UPV) (m/s) vs. applied tensile stress (MPa) for Specimens No. 9 to No. 14.

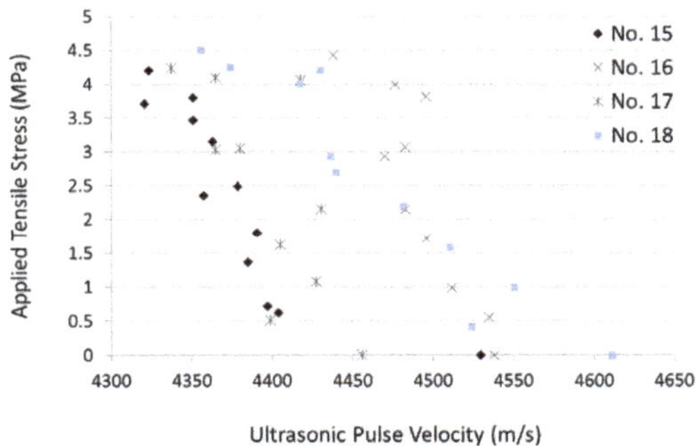

Figure 6. UPV (m/s) vs. applied tensile stress (MPa) for Specimens No. 15 to No. 18.

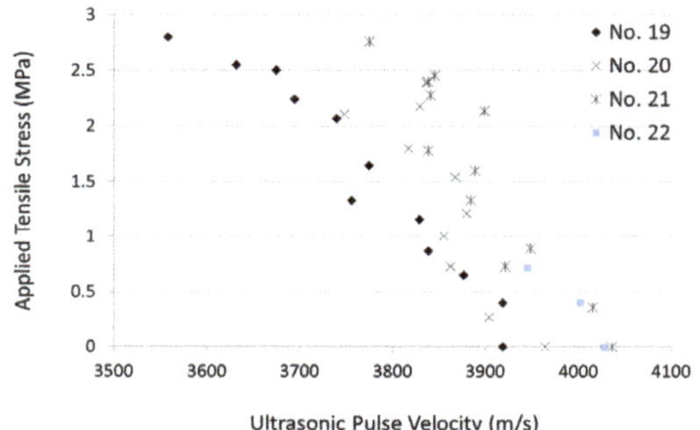

Figure 7. UPV (m/s) vs. applied tensile stress (MPa) for Specimens No. 19 to No. 22.

4. Finite Element Modeling

ABAQUS FEA software, a commercially available finite-element analysis program, was used to model the concrete cylinder. The Concrete Damage Plasticity Model (CDPM) was used to define the parameters of the concrete—these parameters are computed using MATLAB toolbox [16].

For concrete modeling, 3D 8-node Linear Isoparametric Elements with reduced integration (C3D8R) have been utilized. In order to model the gas pressure, two parts are defined; first a solid element with C3D8R features; and second the Eulerian type with EC3D8R (an 8-node linear Eulerian brick, reduced integration, hourglass control). It should be noted that the solid part covers both Eulerian and concrete cylinder. Figure 8 displays the geometry of Finite Element Modeling (FEM).

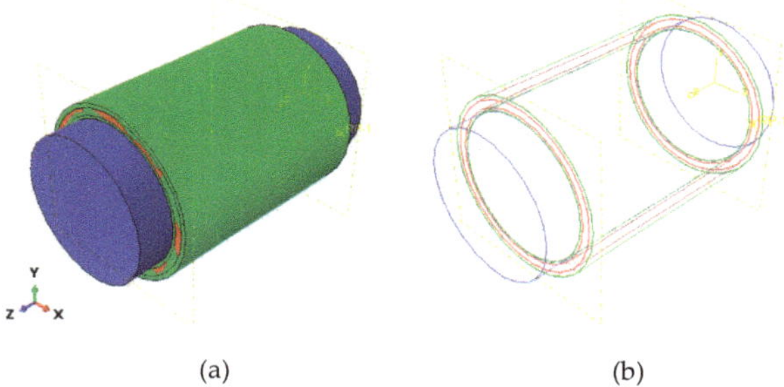

(a)   (b)

Figure 8. Geometrical feature of the Finite Element Modeling (FEM): (**a**) Solid volume representation; (**b**) Outline representation.

The material properties for gas pressure (Eulerian Type) is considered to be Equation of State (EOS) with the type of Ideal Gas. The gas constant and ambient pressure are 278×10^8 mJ/(tonne K) and 0.101325 MPa, respectively. Also, Non-Linear Dynamic Explicit analysis is used for simulation. The contact behavior has both Tangential (frictionless) and Normal (hard contact) behaviors. The Discrete Field was converted to Volume Fraction Tools to make Eulerian part as pressure. The initial gas pressure is applied as Predefined Stress and gravity load. Also, the cyclic pressure is assigned based on the amplitude. The amplitude is displayed in Figure 9 and the stress distribution in concrete cylinder is shown in Figure 10.

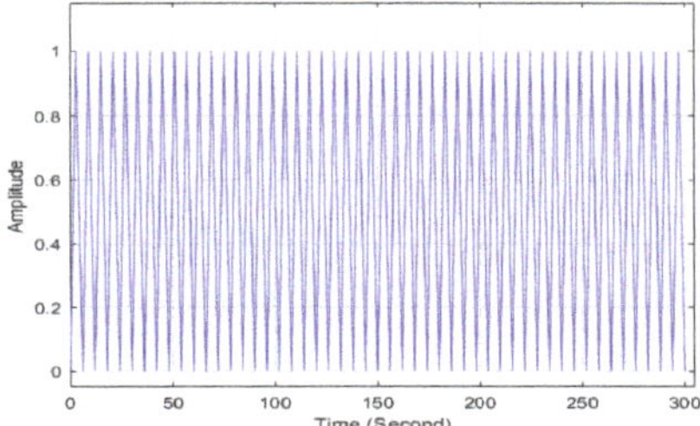

Figure 9. Time vs. amplitude graph.

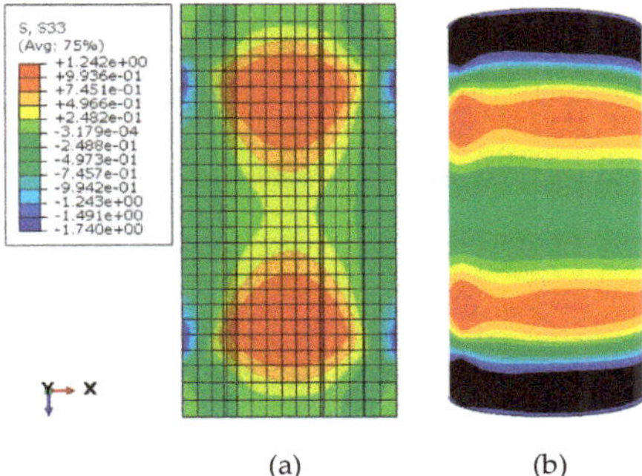

(a)   (b)

Figure 10. Stress distribution in the FEM: (**a**) Longitudinal section; (**b**) Volumetric view.

When cyclic loading was applied at low pressure, the concrete sample did not fail in tension. Then, the maximum pressure in the cyclic loading was increased step-by-step in the FEM sample until the observation of fatigue failure after few cycles to simulate the low-cycle fatigue failure. Interestingly, the FEM sample failed after 13 cycles that is very similar to the experimental result of sample 17 (Table 5). Also, the sample in FEM failed under a pressure that was lower than the maximum applied pressure in previous cycles. FEM and experimental results of specimen No. 17 are compared in Figure 11.

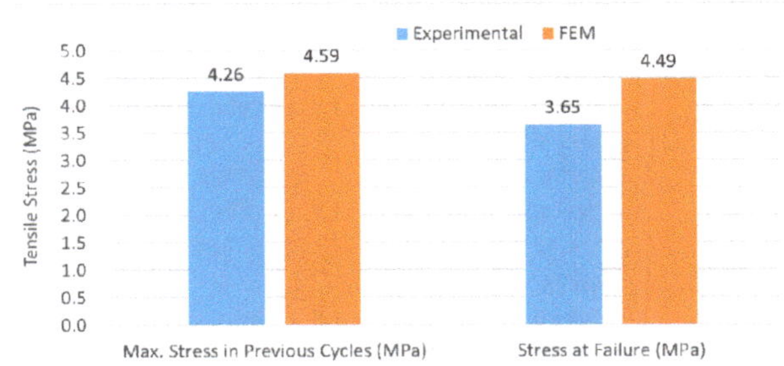

Figure 11. Comparison between FEM and experimental results for Specimen No. 17.

5. Conclusions

A total of 22 standard concrete cylinders were tested under tension using a pressure-tension apparatus. Based on the limited number of test results reported in this paper, one can conclude that the tensile strength of the dry specimens was lower than that of the wet specimens with a lower coefficient of variation in the test results. Therefore, it will be safer and more reliable to test the specimens in dry conditions.

It was shown that failure in tension due to fatigue can happen as early as the second cycle of loading if the concrete specimen has been loaded up to a high percentage of its tensile capacity (most probably more than 62% of its tensile capacity).

ABAQUS FEA software was used to model the pressure-tension test. It was shown that the failure under low-cycle tensile force can be predicted using appropriate parameters in the software. It was also found that concrete specimens subjected to a low-cycle loading protocol always fail at a lower stress level compared to the maximum value applied in the previous cycles. This was the case for both the experimental results as well as the FEM. With respect to the quantification of the effects of fatigue on the performance of concrete, comparison of the experimental and FEM results illuminates the maximum stress before failure to be 7.7% higher in FEM. Additionally, the stress at failure was 23.0% higher in FEM compared to the experimental result.

NDT was also employed to evaluate the progressive damage inflicted by tensile load in concrete using the UPV apparatus. It was shown that the application of a tensile stress on a concrete affects its internal structure and causes an increase in travel times and a lower UPV measurement. The experimental results presented in this paper prove that the UPV method can be used effectively for the detection of cracks in hardened concrete subject to tensile stresses—as low as about 0.4 MPa.

Author Contributions: S.M.S. and A.J.K.K. carried out the tests. S.M.S. and A.J.B. supervised the project. S.M.S. prepared the manuscript and analyzed the data. S.S.R. conducted the finite element analysis. All authors read and approved the final manuscript.

Funding: This research received no external funding.

References

1. Clayton, N.; Grimer, F. The diphase concept, with particular reference to concrete. In *Developments in Concrete Technology*; Lydon, F.D., Ed.; Applied Science Publisher: Basel, Switzerland, 1979; pp. 283–317.
2. Komar, A.J.K. Development of Evaluation Procedures for Evaporative Transport Based Deterioration. Ph.D. Thesis, McGill University, Montreal, QC, Canada, 2017.
3. Terzaghi, K.; Peck, R.B. *Soil Mechanics in Engineering Practice*; Wiley: New York, NY, USA, 1967.
4. Uno, T.; Fujikake, K.; Mindess, S.; Xu, H. The nitrogen gas tension test of concrete, Part 1: Effect of boundary conditions and axial strain response. *Mater. Struct.* **2010**, *44*, 857–864. [CrossRef]
5. Rashidi, M.; Ashtiani, R.S.; Si, J.; Izzo, R.P.; McDaniel, M. A Practical Approach for the Estimation of Strength and Resilient Properties of Cementitious Materials. *Transp. Res. Rec. J. Transp. Res. Board* **2018**, *2672*, 152–163. [CrossRef]
6. Komar, A.J.K.; Hartell, J.; Boyd, A.J. Pressure Tension Test: Reliability for Assessing Concrete Deterioration. In Proceedings of the 7th International Conference on Concrete under Severe Conditions—Environment and Loading, RILEM, Nanjing, China, 23–25 September 2013.
7. ASTM International. *ASTM C31/C31M-19, Standard Practice for Making and Curing Concrete Test Specimens in the Field*; ASTM International: West Conshohocken, PA, USA, 2019.
8. Xu, G.W.; Komar, A.J.K.; Boyd, A.J. Tensile Strength of Plain Concrete under Sustained Loading by PT Machine. *Constr. Build. Mater.* **2019**, *209*, 260–269. [CrossRef]
9. Saito, M. Characteristics of microcracking in concrete under static and repeated tensile loading. *Cem. Concr. Res.* **1987**, *17*, 211–218. [CrossRef]
10. Cornelissen, H.A.W. Fatigue failure of concrete in tension. *Heron* **1984**, *29*, 1–68.
11. Zhang, J.; Stang, H.; Li, V.C. Experimental study on crack bridging in FRC under uniaxial fatigue tension. *J. Mater. Civ. Eng.* **2000**, *12*, 66–73. [CrossRef]
12. Chen, X.; Shi, D.; Li, S.; Fan, X.; Lu, J. Influence of Loading Sequence on Low Cycle Fatigue Behavior of Normal Weight Concrete under Direct Tension. *J. Test. Eval.* **2019**, *47*, 19. [CrossRef]
13. Chen, X.; Bu, J.; Fan, X.; Lu, J.; Xu, L. Effect of Loading Frequency and Stress Level on Low Cycle Fatigue Behavior of Plain Concrete in Direct Tension. *Constr. Build. Mater.* **2017**, *133*, 367–375. [CrossRef]
14. Isojeh, B.; El-Zeghayar, M.; Vecchio, F.J. Fatigue Behavior of Steel Fiber Concrete in Direct Tension. *J. Mater. Civ. Eng.* **2017**, *29*, 4017130. [CrossRef]

15. Zhao, D.; Gao, H.; Liu, H.; Jia, P.; Yang, J. Fatigue Properties of Plain Concrete under Triaxial Tension-Compression-Compression Cyclic Loading. *Shock Vib.* **2017**, *2017*, 1–10. [CrossRef]
16. Sayyar Roudsari, S.; Hamoush, S.A.; Soleimani, S.M.; Madandoust, R. Evaluation of large-size reinforced concrete columns strengthened for axial load using fiber reinforced polymers. *Eng. Struct.* **2019**, *178*, 680–693. [CrossRef]

Article

Comprehensive Evaluation of Fatigue Performance of Modified Asphalt Mixtures in Different Fatigue Tests

Kun Li [1], Ming Huang [2,*], Haobai Zhong [3] and Benliang Li [3]

[1] School of Transportation Science and Engineering, Beihang University, Beijing 100191, China; likunshd@163.com
[2] Shanghai Municipal Engineering Design Institute (Group) Co., Ltd., Shanghai 200092, China
[3] Key Laboratory of Road and Traffic Engineering of the Ministry of Education, Tongji University, Shanghai 201804, China; 1733235@tongji.edu.cn (H.Z.); 13719145727@163.com (B.L.)
*   Correspondence: huangming@tongji.edu.cn

Received: 3 April 2019; Accepted: 28 April 2019; Published: 6 May 2019

Featured Application: Three commonly used fatigue test methods for asphalt mixtures, namely the four-point bending beam fatigue test, the two-point bending trapezoidal beam fatigue test, and an overlay tester, were selected for a fatigue test of six kinds of mixtures, including newly developed modified asphalt. The research is supposed to bring about a comprehensive evaluation of the fatigue performance of a modified asphalt mixture to provide data support for the design of fatigue in an asphalt mixture in the near future.

Abstract: The four-point bending beam fatigue test (4PB), two-point bending trapezoidal beam fatigue test (Trapezoidal Beam), and Overlay Tester (OT) are used to evaluate the fatigue performance of six kinds of asphalt mixtures that are widely used in engineering, and newly developed ones. The result shows that, in all three kinds of fatigue tests, the fatigue performances of the 6% SBS (styrene-butadiene-styrene block copolymer) modified asphalt mixture is the best, and those of the 10% WPE (waxed polyethylene) + 3% SBS, 4% SBS + 0.4% PA610, and 4% SBS modified asphalt mixture are good. The fatigue performances of the warm modified mixing agent and the base asphalt mixture are the worst. An increase in SBS content can effectively improve the fatigue performance of the asphalt mixture. The fatigue performance of the SBS-modified asphalt mixture can be improved by the addition of WPE and PA610. In different tests, the ranking of fatigue performance of the asphalt mixture is similar, and the specific ranking is slightly different. The three different fatigue tests can be used simultaneously to obtain a more comprehensive and objective evaluation in the R&D process for a new modified asphalt. The three fatigue tests process shows that more precise forming and cutting technology is needed, as the strain range used in the 4PB test is very wide, and the number of samples used in each group is small. The preparation of the Trapezoidal Beam test samples is complex; the amount of test data is huge and has high precision, which is suitable for scientific research instead of a field laboratory, and the strain range of the test is moderate in the three methods. The strain range of the OT test is the narrowest; the test specimen is relatively simple to prepare, and the fatigue performance of a specific modified asphalt mixture can be obtained quickly in a simple laboratory.

Keywords: fatigue life; modified asphalt mixture; four-point bending beam fatigue test; two-point trapezoidal beam fatigue test; overlay tester

---

1. Introduction

Fatigue is a well-known behavior due to the iterative loading and unloading at stress levels that are below the ultimate strength of the asphalt pavement. Thus, the fatigue performance of an asphalt mixture is an important criterion for evaluating the asphalt mixture. Some studies [1–4] have shown that the fatigue performance of a modified asphalt mixture is obviously better than that of a base asphalt mixture due to the addition of an SBS modifier. Some studies [2–7] show that the addition of PE (polyethylene), PA (polyamide), and warm mixing agents will change the fatigue performance of the asphalt mixture. Many scholars [4–12] have chosen different fatigue setups to carry out fatigue tests. However, fatigue test methods have different accuracy and pertinence in different test environments and conditions. It is necessary to compare the fatigue performance of different modified asphalt mixtures after analyzing the principles of different fatigue tests to make up for defects of understanding in design and blindness in the choice of asphalt.

In this study, the following three representative fatigue test methods are selected to test the fatigue life of six kinds of modified asphalt mixtures and base asphalt mixtures: the four-point bending beam fatigue test (4PB), which is an ASTM standard test and widely accepted in China [13]; the two-point trapezoidal beam fatigue test (Trapezoidal Beam), which is widely used in Europe [14]; and Overlay Tester (OT) tension, which is mainly used in Texas, USA [15]. Different test methods can provide a more comprehensive reference for fatigue performance evaluation of different kinds of asphalt mixtures from multiple dimensions, avoiding the evaluation deviation caused by defects in a single fatigue test. At the same time, it is easier to find a suitable fatigue method to evaluate the effects of various modifiers, especially for the newly developed ones, on the fatigue performance of asphalt mixtures, and the advantages and disadvantages of the same asphalt mixtures in different test methods can be analyzed.

2. Materials and Experiment Design

2.1. Asphalt and Its Preparation Technology

Six types of asphalts were selected for the experiment, including base asphalt and modified ones, which cover commonly used types in engineering and newly developed ones. The base asphalt is SK70. The modified asphalt used in the test contains two types of SBS-modified asphalt (SBS 4 and SBS 6), warm mixing agent modified asphalt (WA), waxed PE (waxed polyethylene) and SBS composite modified asphalt (WPES), and SBS + nylon fiber 610 (PA610) modified asphalt (4S4P). The modification scheme of modified asphalt is shown in Table 1, and the asphalt used in the test and its index are shown in Table 2.

Table 1. Modification scheme of modified asphalt.

Modified Asphalt	Abbreviation	Formula	Preparation Process
SBS Modified asphalt	SBS4 SBS6	SK70$^{\#}$ + 4% SBS SK70$^{\#}$ + 6% SBS	The asphalt was heated to 175 °C, and SBS was added with stirring for 1 h. After adding stabilizer, the mixture was stirred for 0.5 h.
Modified bitumen with warm mixing agent	WA	SK70$^{\#}$ + 14% Warm agents	The asphalt was heated to 145 °C, and warm mixing agent was added with stirring for 1 h.
WPES composite modified asphalt	WPES	10% WPE + 3% SBS	The WPE asphalt was heated to 175 °C, SBS was added with stirring for 1 h. After adding stabilizer, the mixture was stirred for 0.5 h.
PA610 composite modified asphalt	4S4P	4% SBS + 0.4% PA610	The SBS modified asphalt was heated to 175 °C, and PA610 was added with stirring for 1 h.

Table 2. Asphalts of the test and their properties.

Code of Asphalt	Penetration (25 °C, 100 g, 5 s)/0.1 mm	Softening Point ($T_{R\&B}$)/°C	Ductility (5 °C, 5 cm/s)/cm	Viscosity (135 °C)/Pa·s
SK70	64	50.9	0	0.428
SBS4	45	73.9	73.9	2.378
SBS6	36	90	90	2.94
WA	52	56	0	0.402
WPES	68	56.2	53	2.021
4S4P	43.1	82.6	33.7	3.056

2.2. Aggregate and Gradation

Aggregate, gradation, and mixture designs are as follows: coarse aggregate (particle size >2.36 mm) is made of basalt gravel, fine aggregate (particle size <2.36 mm) is made of manufactured sand, and filler is made of mineral powder. The vast majority of mineral fillers are made of limestone mainly because the limestone powder combines with asphalt strongly [16]. The aggregate indexes meet the requirements of JTG F40-2004 "Technical Specification for Asphalt Pavement Construction of Highway" [17]. The asphalt content of the asphalt mixture is 4.8%, and the gradation is AC-13. The design gradation is shown in Table 3, which contains a group of gravity data of the aggregates in different particle sizes.

Table 3. Gradation of the asphalt mixture and gravity of aggregates of different particle sizes.

Sieve Size/mm	Mineral Powder	16	13.2	9.5	4.75	2.36	1.18	0.6	0.3	0.15	0.075
Passing rate (by mass)	-	100	92	78	52	36	24	17	11	8	7
Gravity	2.786		2.865			2.856			2.845		

2.3. Mix Design

The optimum asphalt content for the different modified asphalt mixtures has been determined by the Marshall mix design. For consistency with the applications and studies in China [1,4,7,9,18], we chose the Marshall compaction method, T0702-2011, of Chinese specification [13]. The design indexes are shown in Table 4.

Table 4. Asphalt design indexes.

Code of Asphalt	Optimum Asphalt Content/%	Average of Air Void/%
SK70	4.2	4.0
SBS4	4.5	4.1
SBS6	4.6	4.0
WA	4.5	4.0
WPES	4.8	4.1
4S4P	5.1	3.9

2.4. Three Fatigue Test Methods

The 4PB fatigue test derives from the ASTM D7460 [19] standard of the United States. The mixture specimens were prepared in a trabecular (380 mm (long) × 63.5 mm (wide) × 50 mm (height)). The loading waveform is a half sinusoidal wave, the standard loading frequency is 10 Hz, the equivalent fatigue temperature of pavement in China is 15 °C, and the 50th loading was used as the initial stiffness modulus. When the normalized stiffness ratio N_M reaches the maximum value, the loading times M, which cause fatigue failure of the specimens, is considered as the fatigue life of the asphalt mixture, in which N_M is calculated according to the following equation:

$$N_M = \frac{S_i \times N_i}{S_0 \times N_0},\tag{1}$$

where S_i is the modulus of stiffness under the x-th loading, N_i are loading times, S_0 is the initial loading stiffness modulus (the 50th stiffness modulus is defined as the initial loading modulus), and N_0 are the initial loading times (taken to be 50).

This method was first proposed by Rowe et al. [11] to form the standard ASTM D7460. Compared with the standard method in which the stiffness modulus changes to 50% of the initial stiffness modulus as the end of the test, the fatigue life variation coefficient is smaller, which can reflect the fatigue life of a modified asphalt mixture better [20]. The 10 beam shape samples were tested in each of the asphalt mixes, and the variance of 10% are the valid data.

The fatigue test of Trapezoidal Beam was adopted from the European EN 12697-24 [21] standard. The mixture specimens were prepared to be 25 mm × 25 mm at the top, 25 mm × 56 mm at the bottom, and 250 mm high. The loading waveform is a sinusoidal wave, the standard loading frequency is 25 Hz, and the equivalent fatigue temperature of pavement in China is 15 °C. The initial stiffness modulus is the 100th loading, and when the modulus of stiffness decreases to 50% of the initial value, the number of loading times is taken as the fatigue life. The three beam samples were tested in each of the asphalt mixes, and the variance of 20% are the valid data.

The Overlay Tester fatigue test adopts the TEX-248-F [15] standard of Texas, USA. The mixture specimen is a cylinder, which is 150 mm in diameter, 38 mm in height, and 76 mm in width. The loading waveform is a triangular wave, the standard loading period is 0.1 Hz, and the test temperature is 25 °C. The judgment for the end of the experiment is that the initial force decreases to 93% of the initial one or the loading times reach 1000. So, the number of loading times taken to reach 93% of the initial force or loading times of 1000 will be taken as the fatigue life of the specimen. The three samples were tested in each of the asphalt mixes, and the final results were averaged and are included in the chart.

Figure 1 shows the three fatigue testing apparatuses and Figure 2 shows the principles of the three fatigue test methods.

Figure 1. *Cont.*

(c)

Figure 1. Three fatigue testing apparatuses: (**a**) four-point beam (4PB) fatigue test, (**b**) two-point trapezoidal beam test (Trapezoidal Beam), and (**c**) Overlay Tester.

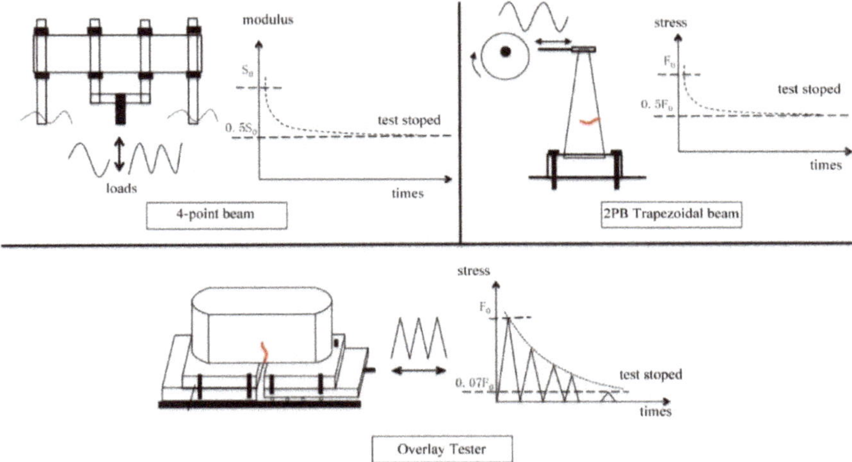

Figure 2. The loading mode and model diagram of three loading modes of the fatigue test.

3. Results and Discussion

3.1. Fatigue Test Results

The fatigue tests of 4PB and Trapezoidal Beam are under a different strain level, while the OT tests keep the same strain at 300×10^{-6}. All tests should be repeated at least four times, and the average values of the three results with the smallest coefficient of variation should be taken as the test results. The results of all the three fatigue tests are shown in Figures 3–5.

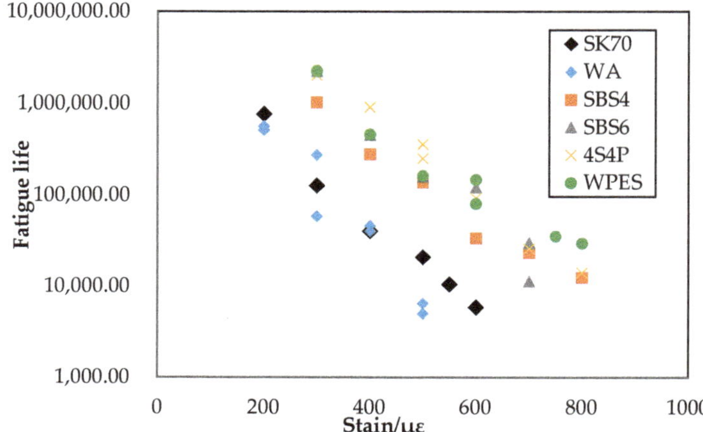

Figure 3. The fatigue life of different asphalt mixtures of 4PB at varied strain levels.

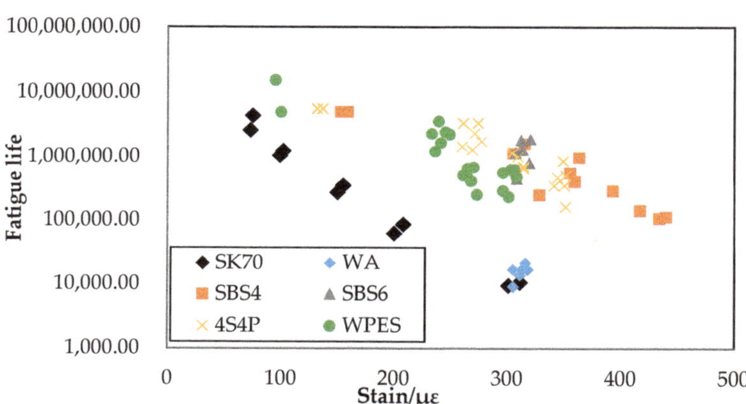

Figure 4. The fatigue life of different asphalt mixtures of Trapezoidal Beam at varied strain levels.

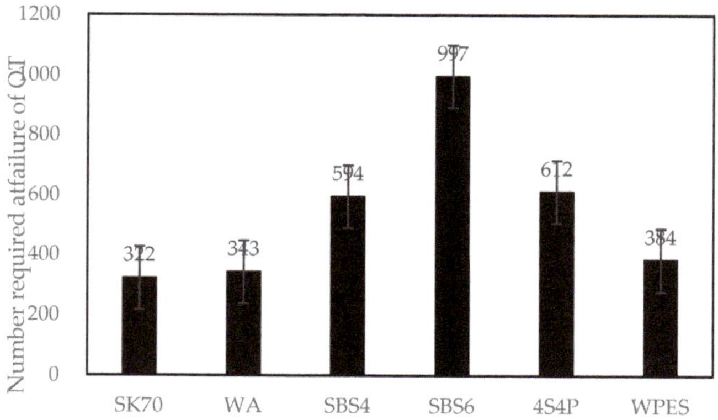

Figure 5. The fatigue life of different asphalt mixtures of Overlay Tester (OT) at a strain of 300×10^{-6}.

3.2. Separate Analysis of the Results

4PB: The strain range used in the 4PB test is very wide, and number of samples used in each group is small, which requires high precision. Therefore, more precise forming and cutting technology is needed. The fatigue life of the six asphalt mixtures decreases gradually with the increase of strain. The fatigue properties of SK70 and WA are obviously poor, SBS4 is in the middle, and 4S4P, WPES, and SBS6 are the best ones. The specific performance order is slightly different with the change of strain, and WPES performs best at high strain. With the increase of SBS content, the fatigue performance of the asphalt mixture increases continuously. The addition of WPE and PA610 improves the fatigue performance of the modified asphalt mixture in varying degrees.

Trapezoidal Beam: The preparation of test samples is complex and the amount of test data is huge. It has high precision and is suitable for scientific research instead of a field laboratory. The strain range of the test is moderate in the three methods. The fatigue life of the six asphalt mixtures decreases gradually with the increase of strain, and the fatigue performance of SK70 and WA are the worst, and 4S4P, SBS4, and SBS6 are the best. The specific performance ranking varies slightly with strain, which is similar to the conclusion for 4PB. The values obtained from the Trapezoidal Beam tests are more concentrated in regions with a smaller strain than 4PB.

Overlay Tester: The strain range of the OT test is the narrowest. The test specimen is relatively simple to prepare and the test process takes less time. It is easier to popularize than the above two test methods. Similarly, the OT test again verified that the fatigue performance of SK70 and WA are poor. The fatigue performance of SBS4, 4S4P, and WPES are in the middle, and SBS6 showed the best fatigue performance. With the increase of SBS content, the fatigue life of the asphalt mixture increased. The fatigue life of 4S4P was better than that of SBS4. This result confirmed the results of the Trapezoidal Beam tests but differed from that of 4PB.

3.3. Comparative Analysis of Three Kinds of Fatigue Tests

In order to analyze the comprehensive evaluation results of different asphalt mixtures in the three fatigue tests, the fatigue tests at or nearby 300×10^{-6} strain were unified and chosen to compare the fatigue life of different asphalt mixtures under the three different tests. According to the order of fatigue life of different mixtures, the mixture with the best fatigue performance scores 6 points and the worst scores 1 point. The total scores of the three test methods for various asphalt mixtures are given by analogy. The results are shown in Table 5.

Table 5. The fatigue life and score of different asphalt mixtures in different fatigue tests.

	Asphalt	SK70	WA	SBS4	4S4P	WPES	SBS6
	Fatigue life	118,340	85,870	1,123,940	2,055,560	2,265,690	2,081,300
4PB	Ratio of fatigue life to SK70 mixture	1	0.73	9.5	17.37	19.15	17.59
	Score	2	1	3	4	6	5
	Fatigue life	10,441	14,779	890,451	561,510	412,352	1,331,028
Trapezoidal Beam	Ratio of fatigue life to SK70 mixture	1	1.42	85.28	53.78	39.49	127.48
	Score	1	2	5	4	3	6
	Fatigue life	322	343	594	612	384	997
OT	Ratio of fatigue life to SK70 mixture	1	1.07	1.84	1.9	1.19	3.1
	Score	1	2	4	5	3	6
	Total score	3	5	12	13	12	17

1. In the 4PB fatigue test, WPES has the best fatigue life, which is 19.03 times of the fatigue life of the base asphalt mixture. In the Trapezoidal Beam fatigue test, SBS6 has the best fatigue life, which is 127.48 times of the fatigue life of the base asphalt mixture. In the OT fatigue test, SBS6 has the best fatigue life, which is 3.1 times of the fatigue life of base asphalt mixture.

2. According to the total score, the fatigue performance of the asphalt mixture is ranked: SBS6 > 4S4P > WPES ≈ SBS4 > WA > SK70.

3. In the three fatigue tests, the fatigue lives of SK70 and WA are significantly lower than that of the other four asphalt mixtures, which is also identical to the experience in practical engineering [22–24], that is, the fatigue life of the base asphalt mixtures is not significantly improved by a warm mixing agent [25,26]. The fatigue performance of the asphalt mixture is improved by adding SBS, and the fatigue performance of the asphalt mixture is improved more obviously with the increase of SBS content. The addition of WPE and PA610 also improves the fatigue performance of the asphalt mixture in varying degrees. Considering fatigue performance and cost saving, these two new modification methods can replace part of the SBS-modified asphalt.

4. In the three fatigue tests, the fatigue performance of the mixture can be roughly distinguished, but the specific rankings are different. The test samples, waveforms, and frequencies are different, so the evaluation results of fatigue life values cannot be directly transformed. However, in the research and development of new materials, the three different fatigue tests can be used simultaneously to obtain a more comprehensive and objective evaluation.

4. Conclusions

In this research program, test schemes are designed for the fatigue performance of several modified asphalt mixtures in three fatigue tests. According to the experimental test results and the statistical analysis findings, the following conclusions can be drawn:

1. In the 4PB fatigue test, the fatigue performance ranking is WPES > SBS6 > 4S4P > SBS4 > SK70 > WA; in the Trapezoidal Beam fatigue test, the fatigue performance ranking is SBS6 > SBS4 > 4S4P > WPE + SBS > WA > SK70; in the OT fatigue test, the fatigue performance ranking is SBS6 > SBS4 > WPE + SBS > 4S4P > SK70.

2. According to the total score, the fatigue performance of the asphalt mixture is ranked: SBS6 > 4S4P > WPES ≈ SBS4 > WA > SK70.

3. In all three fatigue tests, the fatigue lives of SK70 and WA are significantly lower than that of the other four asphalt mixtures, which means that the fatigue life of the base asphalt mixtures is not significantly improved by the warm mixing agent. The fatigue performance of the asphalt mixture is improved by adding SBS, and the fatigue performance of asphalt mixture is improved more obviously with the increase of SBS content. The addition of WPE and PA610 also improves the fatigue performance of the asphalt mixture in varying degrees. Considering fatigue performance and cost saving, these two new modification methods can replace part of SBS-modified asphalt.

4. In all three fatigue tests, the fatigue performance of the mixture can be distinguished, but it differs in specific rankings. Although the evaluation results of the three fatigue life values cannot be directly transformed, the three different fatigue tests can be used simultaneously to obtain a more comprehensive and objective evaluation in the R&D process of a new modified asphalt.

5. More precise forming and cutting technology is needed because the strain range used in the 4PB test is very wide, and number of samples used in each group is small. The preparation of the Trapezoidal Beam test samples is complex; the amount of test data is huge and has high precision, which is suitable for scientific research instead of the field laboratory, and the strain range of the test is moderate in the three methods. The strain range of the OT test is the narrowest; the test specimen is relatively simple to prepare, and fatigue performance of a specific modified asphalt mixture can be obtained quickly in a simple laboratory. It is easier to popularize than the above two test methods.

Author Contributions: Conceptualization, K.L.; methodology, M.H., H.Z. and B.L.; validation, K.L. and B.L.; formal analysis, M.H. and H.Z.; investigation, H.Z. and B.L.; resources, H.Z. and B.L.; data curation, M.H. and B.L.; writing—original draft preparation, H.Z. and B.L.; writing—review and editing, K.L. and M.H.; visualization, B.L.; supervision, M.H..; project administration, K.L. and M.H.; funding acquisition, K.L.

Funding: This research received no external funding.

Acknowledgments: The authors would like to thank Huang Xin from Beihang University for helpful and constructive prophase studies.

Conflicts of Interest: The authors declare no conflict of interest.

References

1. Yu, J. *Research on Fatigue Performance of Asphalt Mixture*; South China University of Technology: Guangzhou, China, 2005.
2. Fontes, L.P.T.L.; Trichês, G.; Pais, J.C.; Pereira, P.A.A. Evaluating permanent deformation in asphalt rubber mixtures. *Constr. Build. Mater.* **2010**, *24*, 1193–1200. [CrossRef]
3. Kim, T.W.; Baek, J.; Lee, H.J.; Choi, J.Y. Fatigue performance evaluation of SBS modified mastic asphalt mixtures. *Constr. Build. Mater.* **2013**, *48*, 908–916. [CrossRef]
4. Li, X.; Clyne, T.; Reinke, G.; Johnson, E.N.; Gibson, N.H.; Kutay, M.E. Laboratory Evaluation of Asphalt Binders and Mixtures Containing Polyphosphoric acid. *Transp. Res. Rec. J. Transp. Res. Board.* **2011**, *2210*, 47–56. [CrossRef]
5. Kristjansdottir, O. *Warm Mix Asphalt*; University of Washington: Seattle, WA, USA, 2006.
6. Modarres, A.; Hamedi, H. Developing laboratory fatigue and resilient modulus models for modified asphalt mixes with waste plastic bottles (PET). *Constr. Build. Mater.* **2014**, *68*, 259–267. [CrossRef]
7. Huang, W.; Li, B.; Huang, M. Evaluation of Self-healing of Asphalt Mixture through Four-Point Bending Fatigue Test. *J. Build. Mater.* **2015**, *18*, 572–577.
8. Vaitkus, A.; Čygas, D.; Laurinavičius, A.; Vorobjovas, V.; Perveneckas, Z. Influence of warm mix asphalt technology on asphalt physical and mechanical properties. *Constr. Build. Mater.* **2016**, *112*, 800–806. [CrossRef]
9. Huang, S.C.; Turner, T.F.; Miknis, F.P.; Thomas, K.P. Long-Term Aging Characteristics of Polyphosphoric acid-Modified Asphalts. *Transp. Res. Rec. J. Transp. Res. Board.* **2008**, *2051*, 1–7. [CrossRef]
10. Al-Khateeb, G.; Shenoy, A. A distinctive fatigue failure criterion. *J. Assoc. Asphalt Paving Technol.* **2004**, *73*, 585–622.
11. Rowe, G.M.; Bouldin, M.G. Improved techniques to evaluate the fatigue resistance of asphaltic mixtures. In Proceedings of the 2nd Euro Asphalt and Euro Bitume Congress, Barcelona, Spain, 20–22 September 2000.
12. Huang, M.; Wen, X.; Huang, W. Fatigue Performance of Asphalt Mixtures with Different Design Goals. *J. Tongji Univ. Nat. Sci.* **2016**, *44*, 572–579.
13. *JTG E20-2011—Standard Test Method of Bitumen and Bituminous Mixtures for Highway Engineering*; China Communications Press: Beijing, China, 2011.
14. British Standards Institution. *Bituminous Mixtures—Test Methods for Hot Mix Asphalt—Part 24: Resistance to Fatigue*; BSI: London, UK, 2012.
15. *TEX-248-F—Test Procedure for the Overlay Test. TEX-248-F*; Texas Department of Transportation: Austin, TX, USA, 2013; pp. 15–37.
16. British Standards Institute. *Bituminous Mixtures—Test Method for Hot Mix Asphalt—Part 2: Determination of Particle Size Distribution*; British Standards Institute: London, UK, 2002.
17. *JTG F40-2004—Technical Specification for Asphalt Pavement Construction of Highway*; China Communications Press: Beijing, China, 2004.
18. Qi, X.; Shenoy, A.; Al-Khateeb, G.; Arnold, T.; Gibson, N.; Youtcheff, J.; Harman, T. Laboratory Characterization and Full-Scale Accelerated Performance Testing of Crumb Rubber Asphalts and Other Modified Asphalt Systems. In Proceedings of the Asphalt Rubber 2006 Conference, Palm Springs, CA, USA, 25–27 October 2006.
19. *ASTM D7460-10—Standard Test Method for Determining Fatigue Failure of Compacted Asphalt Concrete Subjected to Repeated Flexural Bending*; ASTM International: West Conshohocken, PA, USA, 2010.
20. Huang, M.; Huang, W. Laboratory Investigation on Fatigue Performance of Modified Asphalt Concretes Considering Healing. *Constr. Build. Mater.* **2016**, *113*, 68–76. [CrossRef]
21. Pérez-Jiménez, F.; Valdés, G.A.; Botella, R.; Miró, R.; Martínez, A. Approach to fatigue performance using Fénix test for asphalt mixtures. *Constr. Build. Mater.* **2012**, *26*, 372–380. [CrossRef]
22. Tsai, B.W.; Jones, D.; Harvey, J.T.; Monismith, C.L. *Reflective Cracking Study: First-Level Report on Laboratory Fatigue Testing*; Research Report UCPRC-RR-2006-08; Institute of Transportation Studies, University of California: Davis, CA, USA, 2008.
23. El-Basyouny, M.M.; Witczak, M. *Calibration of Alligator Fatigue Cracking Model for 2002 Design Guide*; Transportation Research Record No. 1919; Transportation Research Board, National Academies: Washington, DC, USA, 2005; pp. 77–86.

24. Jones, D.; Harvey, J.; Monismith, C. *Reflective Cracking Study: Summary Report*; Report UCPRC-SR-2007-01; University of California Pavement Research Center: Davis, CA, USA, 2007.
25. Safaei, F.; Lee, J.-S.; do Nascimento, L.A.H.; Hintz, C. Implications of warm-mix asphalt on long-term oxidative ageing and fatigue performance of asphalt binders and mixtures. *Road Mater. Pavement Des.* **2014**, *15*, 45–61. [CrossRef]
26. Goh, S.W.; You, Z. *Evaluation of Warm Mix Asphalt Produced at Various Temperatures through Dynamic Modulus Testing and Four Point Beam Fatigue Testing*; Geotechnical Special Publication No. 212; ASCE: Reston, VA, USA, 2011; pp. 123–129.

Article

Indices to Determine the Reliability of Rocks under Fatigue Load Based on Strain Energy Method

Huanran Fu [1,2,3], Sijing Wang [1,2,*], Xiangjun Pei [4] and Weichang Chen [1,2,3]

[1] Key Laboratory of Shale Gas and Geoengineering, Institute of Geology and Geophysics, Chinese Academy of Sciences, Beijing 100029, China; fhrforest@126.com (H.F.); chenweichang1989@163.com (W.C.)
[2] Institutions of Earth Science, Chinese Academy of Sciences, Beijing 100029, China
[3] College of Earth Science, University of Chinese Academy of Sciences, Beijing 100049, China
[4] State Key Laboratory of Geohazard Prevention and Geoenvironment Protection, Chengdu University of Technology, Chengdu 610059, China; peixj0119@tom.com
* Correspondence: wangsijing@126.com; Tel.: +86-10-8299-8405

Received: 14 November 2018; Accepted: 15 January 2019; Published: 22 January 2019

Abstract: Rock is a complicated material which includes randomly distributed grains and cracks. The reliability of rocks under fatigue load is very important during the construction and operation of rock engineering. In this paper, we studied the deformation and failure process of red sandstone under fatigue load in a laboratory based on a new division method of strain energy types. The traditional elastic strain energy density is divided into two categories: grain strain energy density and crack strain energy density. We find that the proportion of the grain strain energy density to total strain energy density can be used as an indicator of rock yield and the proportion of the crack strain energy density to total strain energy density can be used as an indicator of rock failure. Subsequently, through extensive literature research, we found that such a phenomenon is widespread. We also find the proportion of grain strain energy density to total strain energy density when yielding is affected by rock types and elastic modulus. The proportion of crack strain energy density to total strain energy density in the pre-peak stage is stable and not affected by rock types and elastic modulus, which is about 0.04~0.13. These findings should be very helpful for rock stable state judging in rock engineering.

Keywords: reliability of rocks; fatigue load; strain energy; red sandstone; distribution of strain energy; indices

---

1. Introduction

Rock is a kind of complicated material composed of mineral grains and cracks [1–4]. The deformation and failure process of rock is mainly controlled and affected by cracks [5–11]. During earthquakes, mining, and engineering construction, rock suffers deformation, producing fractures or failure as a result of fatigue load, which can impacts human lives and the environment [12–14]. Studies on the mechanical properties of rocks under fatigue loads are usually conducted using stress-based, strain-based, or strain energy-based approaches [15–18]. Due to the essence of rock material, deformation and failure is an energy-driven process [19] and energy evolution runs through the whole process of rock deformation and failure. The analyses based on the strain energy based-approach proved to be an effective approach [20–23]. In rocks, yield point and peak stress point are very important in evaluating the mechanical properties of rocks [9,24–30]. During the yield stage, rock deformation is not easily controlled by stress, and the rock structure becomes unstable. When the deformation of rock exceeds the peak stress point, the whole rock is destroyed. Therefore, it is more significant to determine whether the rock enters the yield state than whether it breaks in engineering practice. From the point of view of elastic–plastic mechanics, the mechanical energy input from outside is divided into two parts in rock without considering heat exchange. One part is elastic strain energy, which accumulates in the rock. The other part is plastic strain energy, which leads to the cracks growing, structural changes, and unrecoverable deformations. When some kind of strain energy reaches a certain state, the rock also enters a corresponding deformation or failure stage [31]. Therefore, the study of the effects of elastic and plastic strain energy on the rock deformation process can deepen our understanding of the mechanical behavior of rocks.

Presently there are no studies in literature that described or proposed a method to determine whether rock has entered the yield stage or failure stage from the point of view of strain energy. Some studies have focused on the evolution of elastic and plastic strain energy in rock and the associated damage to rock structures [32–35]. Some studies also analyzed the external factors that affect the evolution of elastic energy and plastic strain energy [36,37]. In addition, some studies analyzed the deformation and failure behavior of rocks under different fatigue loads [13,38–40]. These studies provide a good baseline for understanding the relationship between rock and strain energy and the mechanical behavior of rocks under fatigue loads. However, most of these studies are based on the traditional elastic and plastic strain energy division method, which cannot reflect the strain energy effect of rock grains and cracks in the process of rock deformation; hence, it is difficult to make a break through. At the same time, almost all of these studies only explain the phenomena, and do not give indices of strain energy when the rock enters the yielding stage or failure. Therefore, with the further research, the new strain energy division method and the index to determine the rock from stable to unstable stage should to be discussed.

Focusing on the abovementioned questions, this study tries to give indices to determine whether rocks have entered the yielding stage or failure. Through the new classification of strain energy density types, we found that the ratio of some kinds of strain energy densities to corresponding total strain energy densities obtained by the new method could be used as the index to determine whether the rock has entered the yield stage or failed. We subsequently collected experimental data from extensive literature to confirm this finding. We believe that this discovery is of great significance in rock engineering and earthquake prediction for judging or predicting rock stability.

2. Classification and Calculation Methods of Rock Strain Energy by Tests Results

2.1. Classifications and Methods of CalculatingStrain Energy

It is of great importance to divide suitable strain energy types to explain the elastic–plastic transformation and the mechanical properties of rocks. Although many methods in rock researches are derived from metal, the unique properties of properties of rocks clearly makes the two materials to be treated differently in some instances. In this study, we use the strain energy density u_i to represent the strain energy in rocks.

In conventional rock cyclic loading and unloading studies, the researchers only divided the strain energy density into elastic strain energy density u_i^e and plastic strain energy density u_i^p, which as shown in Figure 1a. The deformation of rock consists of two parts: deformation of rock grains and deformation of cracks. Cai et al. [3] pointed out that during the unloading process, due to the previously compacted cracks and the opening of newly formed crack as a result of decrease in load, the slope of the unloading curve would gradually slow down. The deformation rebound of rock, at the beginning of unloading, was mainly due to the elastic deformation of grains because the crack in the rock cannot immediately change from compression to opening. Therefore, in the pre-peak stage, in order to obtain the elastic strain energy density of rock grains when unloading. We draw a line section with the slope of unloading modulus obtained at the initial stage of unloading from the unloading point A, and then cross the strain axis at point C, as the line AC shown in Figure 1b. So in this paper, the elastic strain energy density u_i^e is divided into two parts. One part is the elastic strain energy density of rock grains u_i^g, the second part is the crack strain energy density u_i^c. The u_i^c represents the strain energy density generated by the recoverable deformation of cracks. The method of dividing strain energy in post-peak stage is the same as that in pre-peak stage which are shown in Figure 1c,d.

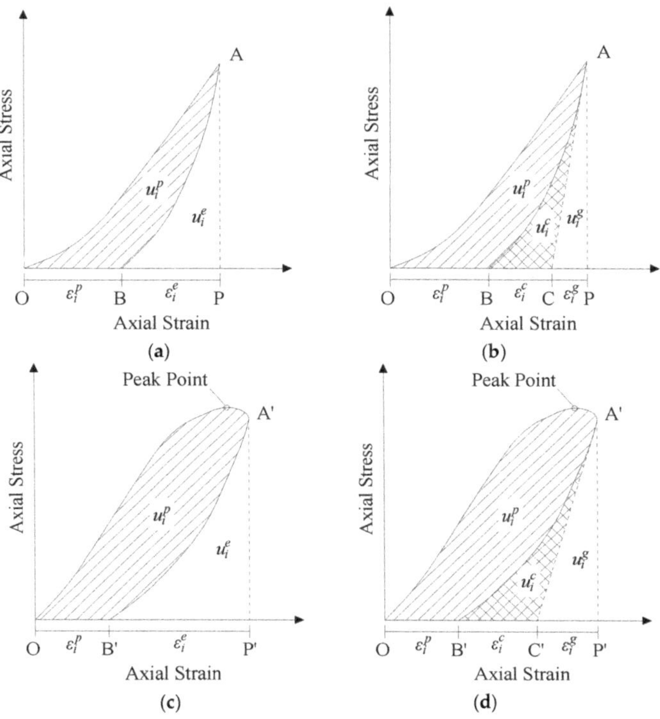

Figure 1. Comparison of two methods for dividing strain energy in pre-peak stage and post-peak stage: (**a**) The pattern of strain energy division in pre-peak stage which has been used in the former studies; (**b**) the new pattern of strain energy division in pre-peak stage which has been used in this study; (**c**) the pattern of strain energy division in post-peak stage which has been used in the former studies; (**d**) the new pattern of strain energy division in post-peak stage which has been used in this study.

Thus, the relationship between these kinds of strain energy density can be written as

$$u_i = u_i^p + u_i^e = u_i^p + \left(u_i^g + u_i^c \right), \tag{1}$$

According to Figure 1b,d, the calculation method of strain energy density in a loading and unloading process can be written as follows:

$$u_i = \int_0^{\varepsilon_{i1}} \sigma_{i1} d\varepsilon_{i1} + \int_0^{\varepsilon_{i2}} \sigma_{i2} d\varepsilon_{i2} + \int_0^{\varepsilon_{i3}} \sigma_{i3} d\varepsilon_{i3}, \tag{2}$$

$$u_i^e = \int_a^b \sigma_{i1}' d\varepsilon_{i1}^e + \int_a^b \sigma_{i2}' d\varepsilon_{i2}^e + \int_a^b \sigma_{i3}' d\varepsilon_{i3}^e, \tag{3}$$

$$u_i^p = u_i - u_i^e, \tag{4}$$

$$u_i^g = u_{i1}^g + u_{i2}^g + u_{i3}^g = \frac{1}{2}\sigma_{i1}\varepsilon_{i1}^g + \frac{1}{2}\sigma_{i2}\varepsilon_{i2}^g + \frac{1}{2}\sigma_{i3}\varepsilon_{i3}^g, \tag{5}$$

$$u_i^c = u_i^e - u_i^g, \tag{6}$$

In these equations, u represents strain energy density, σ and σ' represent the loading stress and unloading stress respectively; ε represents the strain; subscript i represents a rock unit; numbers 1, 2, and 3 represent three principal stress directions; and superscript p, e, g, and c represent plastic strain, elastic strain, elastic strain of rock grains, and elastic strain of cracks, respectively. The letters a and b represent the strain values corresponding to the stress unloading to the lowest point and the unloading start point, respectively.

2.2. Testing Schemes and Experimental Data

The test schemes adopted cyclic loading and unloading tests under different confining pressures and conducted with the MTS-815 servo-controlled testing machine in the State Key Laboratory of Geo-hazard Prevention and Geo-environment Protection (SKLGP), Chengdu, Sichuan, China. The fresh red sandstone blocks, which were taken from a quarry in Leshan, Sichuan, China were selected for this test. The blocks were cut into several cylindrical samples with 50mm in diameter and 100mm in height. The average density of dried red sandstone specimens is about 2.06 g/cm3 and the main mineral constituents are calcite (78.2%) and quartz (0.5%). The X-ray diffraction pattern is shown in Figure 2.

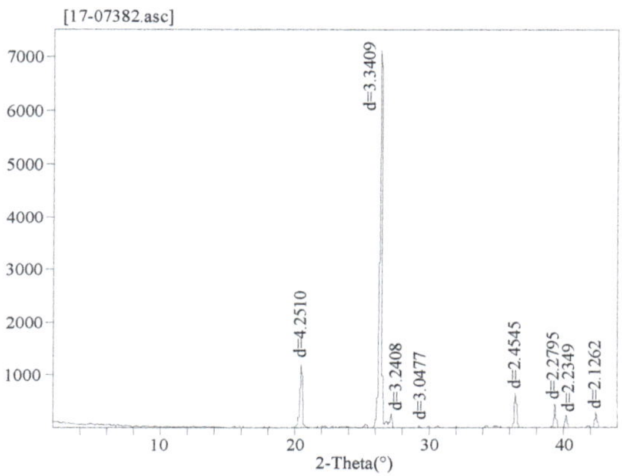

Figure 2. X-ray diffraction pattern of sandstone samples in this study.

In this study, cyclic loading and unloading tests under confining pressures of 0MPA, 5 MPa, 10 MPa, and 20 MPa were designed. By comparing the uniaxial compression deformation stages of rocks classified by some researchers [2,41,42] and the deformation stages of cyclic loading and unloading test, stress–strain curves can be divided into 5 stages: (I) crack closure, (II) elastic deformation, (III) crack initiation and stable crack growth,(IV) crack damage and unstable crack growth, and(V) failure and post-peak behavior. In general, the point between stage III and IV is called the crack damage point and be treated as the yield point and the point between stage IV and V is called the peak point.

During the cyclic loading and unloading test, the rock samples were subjected to axial force with the loading rate of 1kN/s. When the axial force reaches 10 kN, we stopped the loading and start the unloading process. The unloading force was also set at 1kN/s. In order to avoid the impact of the testing machine on the sample during the reloading process, the sample is loaded when the axial force is 1 kN in the successive unloading process. For subsequent unloading cycles, we increased the unloading force by 10 kN, i.e., during the first cycle the unloading force was set at 10 kN and was 20 kN during the second cycle. In order to obtain more cycles when the rock deformation enters the yield stage, we decreased the unloading force of subsequent cycle to 5 kN. We adopted this loading method until the rock deformation reaches the peak point and failure occurred. During post peak stage, we controlled the unloading points manually. This makes the unloading force different for each sample. Despite this approach, the difference of strain energy density evolution in the post-peak and pre-peak stage is still obvious. A representative cyclic loading and unloading stress–strain curve is shown in Figure 3.

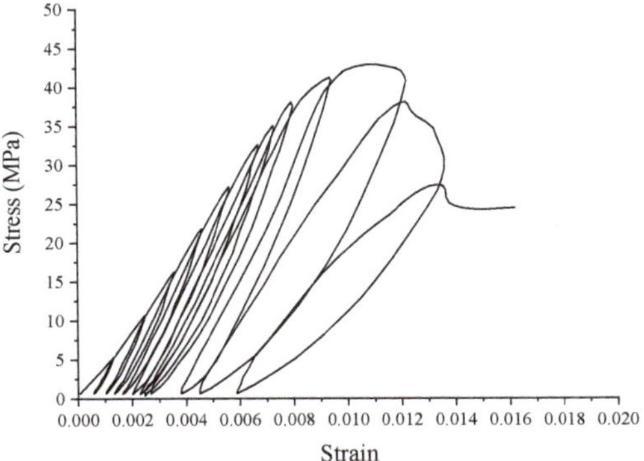

Figure 3. A representative cyclic loading and unloading stress-strain curve under 5 MPa confining pressure.

The mechanical properties of red sandstone samples under different confining pressures are shown in Table 1. In Table 1 σ_3 represents confining stress, E represents the average elastic modulus, the σ_{cd} and ε_{cd} represents the average crack damage stress and strain respectively, the σ_p and ε_p represents the average peak strength and strain respectively.

Table 1. The stress and strain at each stage of red sandstone sample's deformation under different confining pressures.

σ_3 (MPa)	E (GPa)	σ_{cd} (MPa)	ε_{cd}	σ_p (MPa)	ε_p
0	2.79	18.69	0.0064	24.92	0.0093
5	4.42	30.95	0.0068	38.15	0.0109
10	5.27	38.23	0.0071	49.75	0.0118
20	8.75	51.84	0.0083	67.87	0.0161

3. Test Results

3.1. The Evolution of Strain Energy

Based on the strain energy types mentioned above and the laboratory test results, we obtained the strain energy evolution of red sandstone samples under different confining pressures. It should be noted that the data of these kinds of strain energies in the post-peak stages cannot be obtained in uniaxial compression tests, and two sets of post-peak strain energy data are obtained in the confining pressure tests of 5 MPa. A set of post-peak strain energy data is obtained in the confining pressure tests of 10 MPa and 20 MPa, respectively. According to the results of evolution of strain energy shown in Figures 4 and 5, it can be found that the evolution of strain energies has some characteristics as follows:

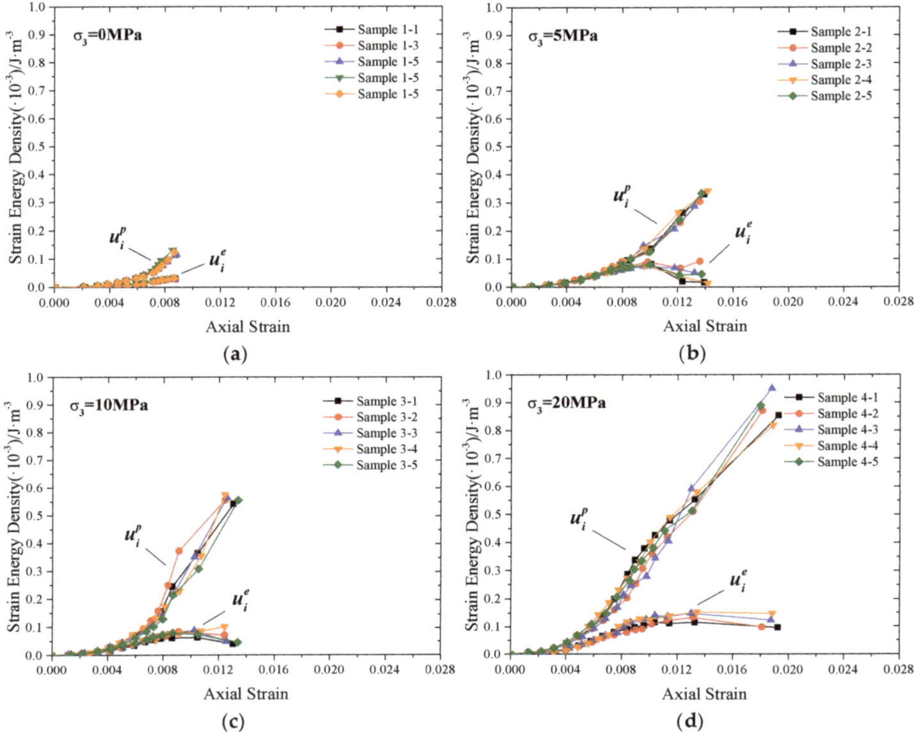

Figure 4. The evolution of u_i^p and u_i^e of red sandstone samples under different confining pressures: (**a**) Confining pressure is 0 MPa; (**b**) Confining pressure is 5 MPa; (**c**) Confining pressure is 10 MPa; (**d**) Confining pressure is 20 MPa.

(1) The characteristics of u_i^p.

The u_i^p grows in quadratic form during the whole deformation process. With the increase in strain, the numerical growth rate of u_i^p is accelerated especially as it enters the yield stage. The value of u_i^p is affected by the confining pressure, the higher the confining pressure the higher the value of u_i^p.

(2) The characteristics of u_i^e.

In the crack closure stage, the growth rate of u_i^e is accelerated, and the growth rate in elastic stage is almost constant. After entering the yield stage, the growth rate of u_i^e decreases, but the value of u_i^e still increases. In the post-peak stage, the growth rate of u_i^e becomes negative. With the increase of confining pressure, the values of u_i^e increase, and the attenuation effect in the post-peak stage is also inhibited.

(3) The characteristics of u_i^g.

The value of u_i^g increases in S-shape during the deformation process and decreases after peak stress, which is similar to the stress–strain curve of rocks. The growth rate of u_i^g decreases once the sample enters the yield stage. For some samples, the value of u_i^g will still increase slightly due to confining pressure in the early post-peak stage. The value of u_i^g will also increase with the increase of confining pressure.

(4) The characteristics of u_i^c.

The value of u_i^c increases slightly with the increase of strain before the peak point. After entering the post-peak stage, the u_i^c of each sample is highly discrete. For most samples, the u_i^c decrease in the post-peak stage and even becomes negative, only a small number of them increase. This may be as a result of the effects of confining pressures.

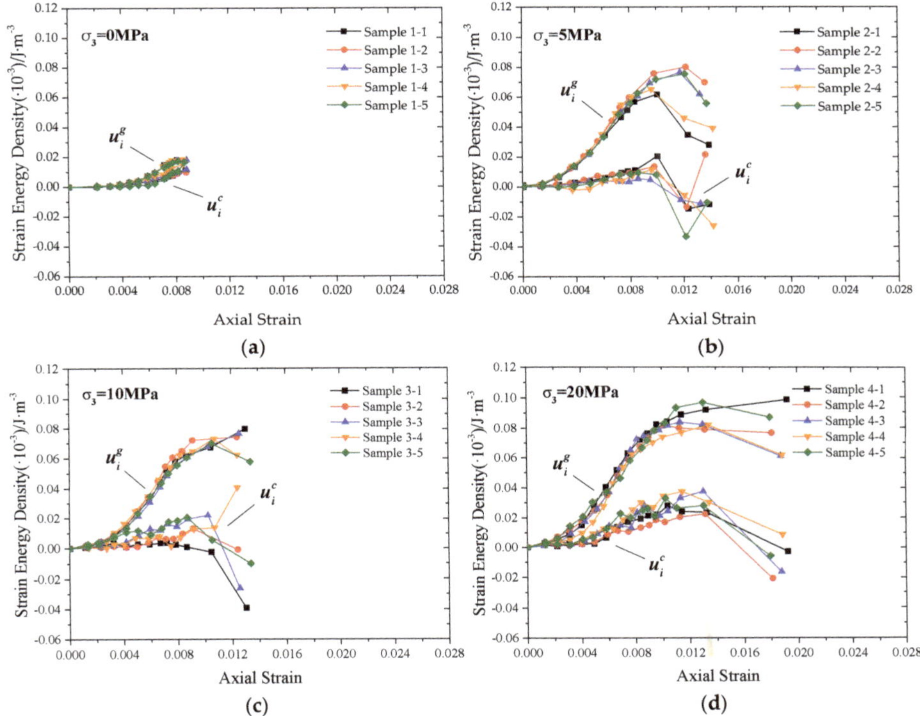

Figure 5. The evolution of u_i^g and u_i^c of red sandstone samples under different confining pressures: (**a**) Confining pressure is 0 MPa; (**b**) Confining pressure is 5 MPa; (**c**) Confining pressure is 10 MPa; (**d**) Confining pressure is 20 MPa.

The results of Figures 4 and 5 show that under the same test conditions, the same kind of strain energy of different samples shows a highly concentrated behavior, and the rock yielding or failure is because the strain energy reaching a certain value. However, due to different confining pressures, different strain energy is required for rock entering different deformation stages, which is relatively limited in application.

3.2. The Distribution of Strain Energy

Inspired by the above results, we check whether there is a relationship between the total strain energy absorbed by rock and other forms of strain energy in different deformation stages. Therefore, we divided u_i^p, u_i^e, u_i^g, and u_i^c at different deformation moments by u_i at the corresponding moment. We then obtained the distribution law of the proportion of different strain energy to the total strain energy in the entire deformation process, which is shown in Figure 6. It is worth pointing out that the first two points of each curve are in the crack closure stage, which cannot reflect the properties of materials from the point of view of elastic-plastic mechanics, so they can be neglected in elastic–plastic analysis.

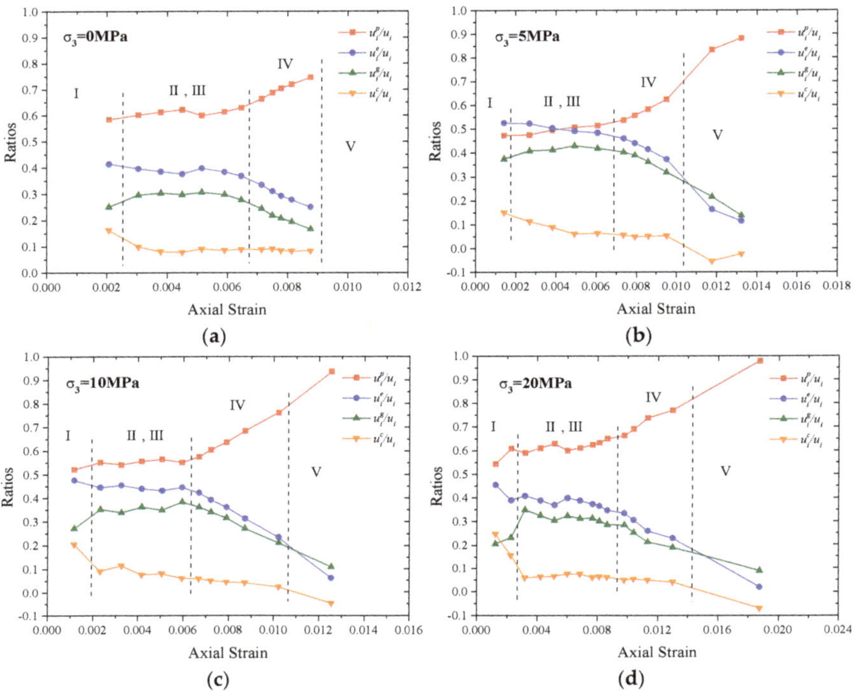

Figure 6. The distribution of strain energies under different confining pressures obtained after normalization: (**a**) Confining pressure is 0 MPa; (**b**) Confining pressure is 5 MPa; (**c**) Confining pressure is 10 MPa; (**d**) Confining pressure is 20 MPa.

Figure 6 and Table 1, reveal the following characteristics about the distribution of strain energies:

(1) Throughout the deformation process, the proportion of the same type of strain energy to the total strain energy under different confining pressures is concentrated in a certain range and shows the same growth trend.

(2) Before the yield stage, the ratio of u_i^e and u_i^p to the u_i is balanced. After entering the yield stage, the proportion of u_i^p and u_i^e increases and decreases rapidly, respectively.

(3) The ratio of u_i^g to u_i rises rapidly to a certain level at the beginning of deformation, and remains relatively stable until it enters the yield stage. After entering the yield stage, the ratio of u_i^g to u_i decreases rapidly.

(4) Except the first two points, the proportion of u_i^c in the pre-peak stages almost remains a constant value, and decreases rapidly in the post-peak stage.

3.3. Discussions

Based on the above results, it can be found that the evolution of strain energy is related to the deformation and failure process of red sandstone. For a sample under a confining pressure, the plastic strain energy or elastic strain energy required to reach the yield stage is almost certain. The yield conditions can be constructed by testing the values of strain energy satisfied by different rocks under different test conditions. More importantly, when the proportion of u_i^p, u_i^e, u_i^g, and u_i^c in u_i is plotted, it is found that the distribution of strain energy will appear as obvious points before or after the yield stage and pre-peak or post-peak stage.

As mentioned at the beginning in this paper, rock is a natural material composed of rock grains and cracks. Under the action of external force, the action of rock grains is to resist external force, and then weaken the damage as a result of external force to the structure. The crack in rock is a kind of existence that weakens the ability of rock to resist external force. We can assume that even though there are two opposing functions in rock, as long as their relationship is balanced, the strength of rock will increase with the increase of external forces within the energy storage range of rock grains. If this assumption is extended to the strain energy, that is, the ratio of strain energy density of different forms to total strain energy density are relatively stable before the big changes in rock structure. If the proportion of one type of strain energy density to total strain energy density changes significantly, the rock will enter the corresponding deformation stages, which is also consistent with the test results.

According to the definition of u_i^g and u_i^c, it can be seen that u_i^g represents the elastic strain energy stored in rock grains in a rock unit, which can be released with unloading or rock structure damage. The u_i^c represents the strain energy released by the recoverable deformation part of cracks. Therefore, we presume that the growth rate of the proportion of u_i^g to u_i becomes negative, which can be used as an indicator to judge the rock entering the yield stage. We also noticed that even under different confining pressures, the ratio of u_i^c is almost the same in the pre-peak stage, which was a constant, and its ratio is about $0.06(\pm0.02)$. Therefore, we consider that whether the large-scale abrupt change of this proportion can be used as the basis for judging rock failure.

4. Verification and Extension of Strain Energy Index

Due to the limitation of the type and quantity of the test samples, we cannot draw general conclusions based on the experimental results only. Therefore, we first discuss the relationship between strain energy and rock deformation mechanism, and then obtained relevant data from published literature to verify the conclusions and assumptions in previous section.

4.1. The Relationship between Strain Energy and the Deformationmechanism of Rocks

Rock deformation process is related to the process of crack initiation and expansion. When the damage is within the range of rock resistance, the rock is stable, otherwise it becomes unstable. From the point of strain energy, the plastic strain energy reflects the energy consumed by the irreversible deformation of rock, including the irreversible deformation of most cracks and the displacement of a small part of rock grains. The elastic strain energy is mainly reflected by the elastic deformation of rock grains and a small number of recoverable cracks.

During the deformation of rock, the material properties of rock are activated with the closure of cracks. In the elastic stage, the elastic strain energy accumulated by rock particles is linear with the total strain energy, and the proportion of the total strain energy is relatively constant. Only after the yield stage, more and more energy is consumed by plastic deformation due to the unsteady propagation of cracks, so the proportion of elastic energy of rock grains decreases. For crack strain energy, the proportion of rock in the whole strain energy is relatively constant before rock failure because rock still has a relatively complete shape. When the rock has failed, the rock no longer has a complete shape, and the proportion of crack strain energy changes rapidly.

4.2. Collection of Experimental Results from PreviousStudies

We selected 94 rock cyclic loading and unloading tests of five types of rocks from published data for analysis. The basic information of the samples and test conditions are presented in Table 2. Although some of the data did not perform cyclic loading and unloading at the post-peak stage, it was impossible to analyze the post-peak condition, but the analysis in the pre-peak stage was still effective.

Table 2. Basic information of the samples and test conditions from literature.

No.	Rock Type	Number of Data	σ_3 (MPa)	References
1	Marble	6	0	[43]
2	Marble	4	40	[44]
3	Granite	4	10	[32]
4	Coal	1	0	[44]
5	Coal	1	10	[21]
6	Sandstone	1	10	[21]
7	Granite	1	10	[21]
8	Marble	1	5	[37]
9	Sandstone	1	0	[45]
10	Sandstone	1	0	[17]
11	Sandstone	2	0	[46]
12	Sandstone	6	0	[47]
13	Sandstone	1	0	[48]
14	Basalt	5	0	[49]
15	Granite	1	0	[2]
16	Sandstone	37	0	[23]
17	Coal	2	3	[22]
18	Marble	1	20	[50]
19	Marble	1	35	[50]
20	Marble	1	50	[50]
21	Marble	2	5	[51]
22	Marble	1	20	[51]
23	Marble	1	40	[51]
24	Granite	3	2	[52]
25	Granite	1	5	[52]
26	Granite	1	7	[52]
27	Granite	1	40	[53]
28	Granite	1	1	[54]
29	Granite	1	10	[54]
30	Granite	1	20	[54]
31	Granite	1	30	[54]

After analyzing the obtained data, it is found that the strain energy evolution of various samples is consistent with the strain energy evolution of red sandstone obtained in Section 3. For almost all samples, the proportion of grain strain energy will decrease after entering the yield stage. In this study, we choose the elastic modulus as abscissa axis because elastic modulus is affected by confining pressure, porosity, compactness of rock grains arrangement, and other comprehensive factors, which is the general embodiment of rock elastic properties. By analyzing the collected data, we observed that the ratio of grain strain energy to total strain energy increases with the increase in elastic modulus, different rocks show the distinct classification. The proportion of u_i^g to u_i of the same rock typeis linearly increasing with the elastic modulus. The relations between elastic modulus and proportion of u_i^g to u_i of different rocks while entering the yield stage are shown in Figure 7.

Let y denote the ratio of u_i^g to u_i when yielding and x denote the elastic modulus. Linear fitting of the data of different rocks in Figure 7 are given in the following equations:

The coal is

$$y = 0.08278x + 0.38472, R^2 = 0.99638 \tag{7}$$

The sandstone is

$$y = 0.03025x + 0.04219, R^2 = 0.53828 \tag{8}$$

The basalt is

$$y = 0.01229x + 0.41971, R^2 = 0.74461 \tag{9}$$

The marble is

$$y = 0.00753x + 0.12095, R^2 = 0.7071 \tag{10}$$

Finally, the granite is

$$y = 0.00603x + 0.31933, R^2 = 0.88499 \tag{11}$$

From the results of linear fitting, it can be seen that the slope of linear equation from Equation (7) to Equation (11) decreases successively, while the uniaxial compressive strength of rock increases successively. It should be noted that the uniaxial compressive strength referred to here is not the uniaxial compressive strength specific to a particular rock, but the uniaxial compressive strength commonly used to classify rocks. This is because the uniaxial compressive strength of the sandstone, in some cases, is higher than that of the marble. The classification of rock uniaxial strength is stated in China's national standards [55,56].

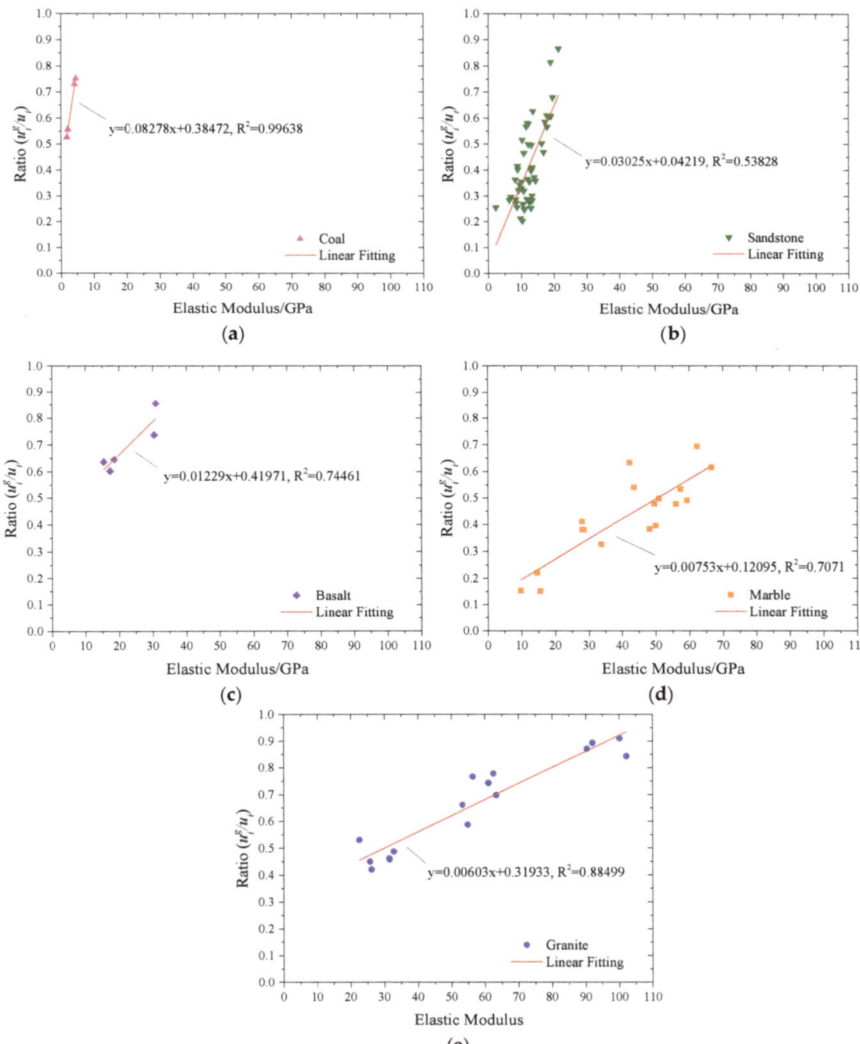

Figure 7. The relation between the proportion of grain strain energy density to total strain energy density and elastic modulus of different rock while entering the yield stage: (**a**) Coal; (**b**) Sandstone; (**c**) Basalt; (**d**) Marble; (**e**) Granite.

Through the analysis of the ratio of the u_i^c to u_i in these rock samples, it is found that this ratio is nearly stable in a range during the pre-peak period, and will not change until failure (Figure 8). In Figure 8, the data are concentrated in the range of 0.04~0.13 and are not affected by rock type and elastic modulus. The maximum value is 0.1534 and the minimum value is 0.0366.

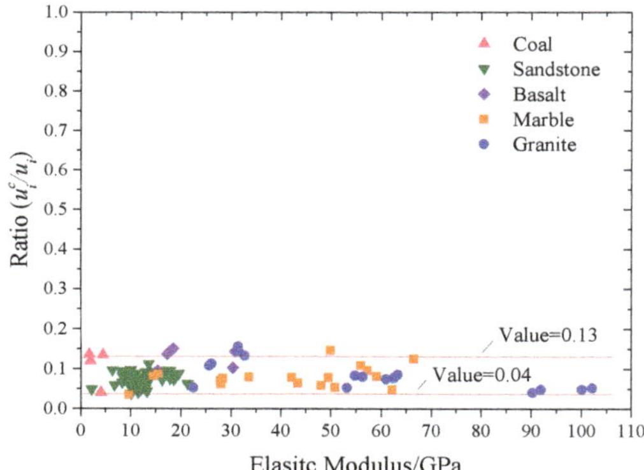

Figure 8. The proportion of crack strain energy density to total strain energy density of different rock samples in the pre-peak stage.

However, due to the limitation of sample data, this finding needs more sample data and more rock types to confirm. We believe that by analyzing enough experimental data of different rock types, we can get more accurate and comprehensive evolution and proportion of grain strain energy density and crack strain energy density to total strain energy density, and can use these two indices to predict or analyze the stability or failure of rocks.

5. Conclusions

According to a series of analysis of test results and literature data, this work studied the evolution of strain energy in the process of rock deformation and its relationship with deformation mechanism classified by new strain energy types. It is observed that the proportion of grain strain energy density and crack strain energy density to total strain energy density can be used as the indices for determine rock stability or failure. The main findings of this study are as follows:

(1) The elastic strain energy density of rock is composed of grain strain energy density and crack strain energy density according to the new division method. Both grain strain energy density and crack strain energy density increase before rock failure, but the growth rate of grain strain energy density decreases after it enters the yield stage.

(2) From the point of view of energy distribution, the distribution of grain strain energy density is stable before the yield stage and the proportion of grain strain energy density decreases after entering the yield stage. This indicates that more strain energy density is allocated to the plastic strain energy density, which results in the decrease of grain strain energy density. This phenomenon is consistent with the deformation mechanism of cracks expansion after rock enters the yield stage. However, the distribution of crack strain energy is stable during the pre-peak deformation stage, and it decreases only after the peak stage. We believe that this phenomenon is related to the integrity of the rock because no matter how many cracks grow in the rock, as long as the rock has not failed, the restorable part of the crack is coordinated with the total strain energy density.

(3) Through the analysis of the data obtained from other literatures, it is found that the proportion of grain strain energy density to total strain energy density can be used as the index to determine

the stability of rock. The proportion of crack strain energy density to total strain energy density can be used as the index to determine rock failure.

(4) The proportion of grain strain energy density to total strain energy density is concentrated in the same rock type, which has linear relationship with elastic modulus. The slope of the linear equation between different rocks and elastic modulus will decrease with the increase of uniaxial compression strength of rock under normal conditions.

(5) The proportion of crack strain energy density to total strain energy density is stable in the range of 0.04~0.13, which is independent of rock type and elastic modulus.

It should be noted that these findings are based on limited data, but we believe these data have already opened the door to a new research area. Later, we will enrich these research findings by obtaining more data and popularizing it for engineering applications.

Author Contributions: Conceptualization, H.F. and S.W.; Methodology, H.F.; Validation, H.F., S.W. and X.P.; Formal analysis, H.F. and W.C.; Investigation, S.W. and H.F.; Resources, X.P.; Writing—original draft preparation, H.F.; Writing—review and editing, H.F. and S.W.; Visualization, H.F. and W.C.; Supervision, S.W. and X.P.; Project administration, H.F.; Funding acquisition, S.W.and X.P.

Funding: This research was funded by the Opening Fund of State Key Laboratory of Geohazard Prevention and Geoenvironment Protection (Chengdu University of Technology) No. SKLGP2016K001; National Key R&D Program of China No. 2017YFC1501002; the National Natural Science Foundation of China No. 41572302.

Acknowledgments: We would like to thank Cheng Cheng's suggestions and Afolagboye lekan olatayo's help in improving the English of manuscript. We also would like to thank the anonymous reviewers for their helpful comments.

References

1. Gu, D. *Fundamentals of Rock Engineering Geomechanics*; Science Press: Beijing, China, 1979.
2. MARTIN; CHANDLER. The progressive fracture of Lac du Bonnet granite. *Int. J. Rock Mech. Min. Sci. Geomech. Abstr.* **1994**, *31*, 643–659. [CrossRef]
3. Meifeng, C.; Manchao, H.; Dongyan, L. *Rock Mechanics and Engineering*; Science Press: Beijing, China, 2002; pp. 52–61.
4. Wang, S. Geological nature of rock and its deduction for rock mechanics. *Chin. J. Rock Mech. Eng.* **2009**, *28*, 433–450.
5. Lockner, D. The role of acoustic emission in the study of rock fracture. *Int. J. Rock Mech. Min. Sci. Geomech. Abstr.* **1993**, *30*, 883–899. [CrossRef]
6. Yamada, I.; Masuda, K.; Mizutani, H. Electromagnetic and acoustic emission associated with rock fracture. *Phys. Earth Planet. Inter.* **1989**, *57*, 157–168. [CrossRef]
7. Rudajev, V.; Vilhelm, J.; LokajíČek, T. Laboratory studies of acoustic emission prior to uniaxial compressive rock failure. *Int. J. Rock Mech. Min. Sci. Geomech. Abstr.* **2000**, *37*, 699–704. [CrossRef]
8. Shah, K.R.; Labuz, J.F. Damage mechanisms in stressed rock from acoustic emission. *J. Geophys. Res. Solid Earth* **1995**, *100*, 15527–15539. [CrossRef]
9. Wawersik, W.R.; Fairhurst, C. A study of brittle rock fracture in laboratory compression experiments. *Int. J. Rock Mech. Min. Sci. Geomech. Abstr.* **1970**, *7*, 561–575. [CrossRef]
10. Carpinteri, A.; Lacidogna, G.; Corrado, M.; Battista, E.D. Cracking and crackling in concrete-like materials: A dynamic energy balance. *Eng. Fract. Mech.* **2016**, *155*, 130–144. [CrossRef]
11. Iturrioz, I.; Lacidogna, G.; Carpinteri, A. Acoustic emission detection in concrete specimens: Experimental analysis and simulations by a lattice model. *Int. J. Damage Mech.* **2013**, *23*, 327–358. [CrossRef]
12. Zhu, W.; Feng, C. Constitutive model of energy dissipation and its application to stability analysis of ship lock slope in three gorges project. *Chin. J. Rock Mech. Eng.* **2000**, *19*, 261–264.
13. Wang, J.A.; Park, H.D. Comprehensive prediction of rockburst based on analysis of strain energy in rocks. *Tunn. Undergr. Space Technol. Inc. Trenchless Technol. Res.* **2001**, *16*, 49–57. [CrossRef]
14. Xiao, J.Q.; Ding, D.X.; Jiang, F.L.; Xu, G. Fatigue damage variable and evolution of rock subjected to cyclic loading. *Int. J. Rock Mech. Min. Sci.* **2010**, *47*, 461–468. [CrossRef]

15. Hua, A. Energy analysis of surrounding rocks in underground engineering. *Chin. J. Rock Mech. Eng.* **2003**, *22*, 1054–1059.

16. Yu, M.H. Advances in strength theories for materials under complex stress state in the 20th Century. *Adv. Mech.* **2004**, *34*, 529–560. [CrossRef]

17. Heping, X.; Yang, J.; Liyun, L.; Ruidong, P. Energy mechanism of deformation and failure of rock masses. *Chin. J. Rock Mech. Eng.* **2008**, *27*, 1729–1740.

18. Carpinteri, A.; Corrado, M.; Lacidogna, G. Three different approaches for damage domain characterization in disordered materials: Fractal energy density, b-value statistics, renormalization group theory. *Mech. Mater.* **2012**, *53*, 15–28. [CrossRef]

19. Xie, H.P.; Yang, J.U.; Li-Yun, L.I. Criteria for strength and structural failure of rocks based on energy dissipation and energy release principles. *Chin. J. Rock Mech. Eng.* **2005**, *24*, 3003–3010.

20. Cai, M.; Dong, J.I.; Guo, Q. Study of rockburst prediction based on in-situ stress measurement and theory of energy accumulation caused by mining disturbance. *Chin. J. Rock Mech. Eng.* **2013**, *32*, 1973–1980.

21. Zhang, Z.; Feng, G. Experimental investigations on energy evolution characteristics of coal,sandstone and granite during loading process. *J. China Univ. Min. Technol.* **2015**, *44*, 416–422.

22. Jiang, C.; Duan, M.; Yin, G.; Wang, J.G.; Lu, T.; Xu, J.; Zhang, D.; Huang, G. Experimental study on seepage properties, AE characteristics and energy dissipation of coal under tiered cyclic loading. *Eng. Geol.* **2017**, *221*, 114–123. [CrossRef]

23. Meng, Q.; Zhang, M.; Han, L.; Pu, H.; Chen, Y. Acoustic Emission Characteristics of Red Sandstone Specimens Under Uniaxial Cyclic Loading and Unloading Compression. *Rock Mech. Rock Eng.* **2018**, *51*, 1–20. [CrossRef]

24. Cook, N.G.W. The failure of rock. *Int. J. Rock Mech. Min. Sci. Geomech. Abstr.* **1965**, *2*, 389–403. [CrossRef]

25. Wawersik, W.R.; Brace, W.F. Post-failure behavior of a granite and diabase. *Rock Mech.* **1971**, *3*, 61–85. [CrossRef]

26. Bukowska, M. Post-peak failure modulus in problems of mining geo-mechanics. *J. Min. Sci.* **2013**, *49*, 731–740. [CrossRef]

27. Bogusz, A.; Bukowska, M. Stress-strain characteristics as a source of information on the destruction of rocks under the influence of load. *J. Sustain. Min.* **2015**, *14*, 46–54. [CrossRef]

28. Pasternak, E.; Dyskin, A.V.; Sevel, G. Chains of oscillators with negative stiffness elements. *J. Sound Vib.* **2014**, *333*, 6676–6687. [CrossRef]

29. Pasternak, E.; Dyskin, A. Dynamic Instability in Geomaterials Associated with the Presence of Negative Stiffness Elements. In *Bifurcation and Degradation of Geomaterials in the New Millennium*; Springer: Cham, Switzerland, 2015.

30. Zhang, X.; Tahmasebi, P. Micromechanical evaluation of rock and fluid interactions. *Int. J. Greenh. Gas Control* **2018**, *76*, 266–277. [CrossRef]

31. Bingye, X.; Xinsheng, L. *Application Elastic- Plastic Mechanics*; Tsinghua University Press: Beijing, China, 1995.

32. Xingguang, Z.; Pengfei, L.; Like, M.; Rui, S.; Ju, W. Damage and dilation characteristics of deep granite at beishan under cyclic loading-unloading conditions. *Chin. J. Rock Mech. Eng.* **2014**, *33*, 1740–1748.

33. Guo, S.; Qi, S.; Zhan, Z.; Zheng, B. Plastic-strain-dependent strength model to simulate the cracking process of brittle rocks with an existing non-persistent joint. *Eng. Geol.* **2017**, *231*, 114–125. [CrossRef]

34. Zhang, C.; Tu, S.; Zhang, L. Analysis of Broken Coal Permeability Evolution Under Cyclic Loading and Unloading Conditions by the Model Based on the Hertz Contact Deformation Principle. *Transp. Porous Med.* **2017**, *119*, 1–16. [CrossRef]

35. Fengqiang, G.; Jingyi, Y.; Xibing, L. A new criterion of rock burst proneness based on the linear energy storage law and the residual elastic energy index. *Chin. J. Rock Mech. Eng.* **2018**, *37*, 1993–2014.

36. Bagde, M.N.; Petroš, V. Fatigue and dynamic energy behaviour of rock subjected to cyclical loading. *Int. J. Rock Mech. Min. Sci.* **2009**, *46*, 200–209. [CrossRef]

37. Hui, Z.; Fanjie, Y.; Chuanqing, Z.; Rongchao, X.; Kai, Z. An elastoplastic coupling mechanical model for marble considering confining. *Chin. J. Rock Mech. Eng.* **2012**, *31*, 2389–2399.

38. Ma, L.J.; Liu, X.Y.; Wang, M.Y.; Xu, H.F.; Hua, R.P.; Fan, P.X.; Jiang, S.R.; Wang, G.A.; Yi, Q.K. Experimental investigation of the mechanical properties of rock salt under triaxial cyclic loading. *Int. J. Rock Mech. Min. Sci.* **2013**, *62*, 34–41. [CrossRef]

39. Zhou, Z.L.; Zhi-Bo, W.U.; Xi-Bing, L.I.; Xiang, L.I.; Chun-De, M.A. Mechanical behavior of red sandstone under cyclic point loading. *Trans. Nonferrous Met. Soc. China* **2015**, *25*, 2708–2717. [CrossRef]

40. Li, D.; Sun, Z.; Xie, T.; Li, X.; Ranjith, P.G. Energy evolution characteristics of hard rock during triaxial failure with different loading and unloading paths. *Eng. Geol.* **2017**, *228*, 270–281. [CrossRef]

41. Brace, W.F.; Paulding, B.W.; Scholz, C. Dilatancy in the fracture of crystalline rocks. *J. Geophys. Res.* **1966**, *71*, 3939–3953. [CrossRef]

42. Bieniawski, Z.T. Mechanism of brittle fracture of rock: Part I—Theory of the fracture process. *Int. J. Rock Mech. Min. Sci. Geomech. Abstr.* **1967**, *4*, 395–406. [CrossRef]

43. You, M. Experimental study on strengthening of marble specimen in cyclic loading of uniaxial or pseudo-triaxial compression. *Chin. J. Solid Mech.* **2008**, *29*, 66–72.

44. Fukun, X.; Zhiliang, S.; Gang, L.; Ze, Z.; Fengrui, Z. Relationship between hysteresis loop and elastoplastic strain energy during cyclic loading and unloading. *Chin. J. Rock Mech. Eng.* **2014**, *33*, 1791–1797.

45. Zhen, L.I.; Zhou, H.; Yang, F.J.; Zhao, H.B.; Zhong-Liang, R.U. Elastoplastic coupling strain definition and constitutive function. *Rock Soil Mech.* **2018**, *39*, 917–925.

46. Zhizhen, Z.; Feng, G. Research on nonlinear characteristics of rock energy evolution under uniaxial compression. *Chin. J. Rock Mech. Eng.* **2012**, *31*, 1198–1207.

47. Meng, Q.; Zhang, M.; Han, L.; Pu, H.; Nie, T. Effects of Acoustic Emission and Energy Evolution of Rock Specimens Under the Uniaxial Cyclic Loading and Unloading Compression. *Rock Mech. Rock Eng.* **2016**, *49*, 1–14. [CrossRef]

48. Zhu, X.; Li, Y.; Wang, C.; Sun, X.; Liu, Z. Deformation Failure Characteristics and Loading Rate Effect of Sandstone Under Uniaxial Cyclic Loading and Unloading. *Geotech. Geol. Eng.* **2018**, 1–8. [CrossRef]

49. Heap, M.J.; Vinciguerra, S.; Meredith, P.G. The evolution of elastic moduli with increasing crack damage during cyclic stressing of a basalt from Mt. Etna volcano. *Tectonophysics* **2009**, *471*, 153–160. [CrossRef]

50. Yang, S.Q.; Xu, P.; Ranjith, P.G.; Chen, G.F.; Jing, H.W. Evaluation of creep mechanical behavior of deep-buried marble under triaxial cyclic loading. *Arabian J. Geosci.* **2015**, *8*, 6567–6582. [CrossRef]

51. Zhou, H.; Zhang, K.; Feng, X.T. Experimental study on progressive yielding of marble. *Mater. Res. Innov.* **2015**, *15*, s143–s146. [CrossRef]

52. Jun, L.; Da-wei, H.; Hui, Z.; Jing-jing, L.; Tao, L.; Lin-ken, S. Gas permeability of granite in triaxial cyclic compression tests. *Rock Soil Mech.* **2018**, *40*, 1–8.

53. Zhang, Z.; Zhu, J.; Wang, B.; Feng, Z.; Bo, L.U.; Zhang, L.; Jiang, Y. The damage and shear dilation property evolution based on energy dissipation mechanism of gneissic granite. *Chin. J. Rock Mech. Eng.* **2018**, *37*, 3441–3448.

54. Peng, Z.; Hongguang, J.I.; Sun, L.; Zhang, Z.; Yu, G.; Hua, J.; Chengjiang, L.I. Experimental study of characteristics of irreversibility and fracture precursors of acoustic emission in rock under different confining pressures. *Chin. J. Rock Mech. Eng.* **2016**, *35*, 1333–1340.

55. China, MOHURD. Code for Hydropower Engineering Geological Investigation. Available online: http://www.mohurd.gov.cn/wjfb/201702/t20170224_230737.html (accessed on 18 August 2016).

56. China, MOHURD. *Standard for Engineering Classification of Rock Mass*; China Planning Press: Beijing, China, 2015. Available online: http://www.mohurd.gov.cn/wjfb/201508/t20150829_224347.html (accessed on 27 August 2014).

 *applied sciences*

Article

Comparative Fatigue Life Assessment of Wind Turbine Blades Operating with Different Regulation Schemes

Brian Loza [1], Josué Pacheco-Chérrez [1], Diego Cárdenas [1], Luis I. Minchala [2] and Oliver Probst [1,*

[1]   School of Engineering and Science, Tecnologico de Monterrey, Ave. Eugenio Garza Sada 2501, Monterrey 64849, N.L., Mexico; a00824137@itesm.mx (B.L.); a00824133@itesm.mx (J.P.-C.); diego.cardenas@tec.mx (D.C.)

[2]   Department of Electric, Electronics and Telecommunications, Universidad de Cuenca, Ave. 12 de Abril y Agustin Cueva, Cuenca 010203, Ecuador; ismael.minchala@ucuenca.edu.ec

*   Correspondence: oprobst@tec.mx

Received: 18 July 2019; Accepted: 11 October 2019; Published: 31 October 2019

Abstract: A comparative evaluation of the fatigue damage occurring in the blades of small wind turbines, with different power regulation schemes, has been conducted for the first time. Three representative test cases were built, one based on stall regulation and two using pitch regulation. The power curves were tuned to be identical in all cases, in order to allow for a direct comparison of fatigue damage. A methodology combining a dynamic simulation of a wind turbine forced by stochastic wind speed time series, with the application of the IEC 61400-2 standard, was designed and applied for two levels of turbulence intensity. The effect of the wind regime was studied by considering Weibull-distributed wind speeds with a variety of parameter sets. Not unexpectedly, in typical wind regimes, stall regulation led to a generally higher fatigue damage than pitch regulation, for similar structural blade design, but the practical implications were smaller than thought previously. Given the need for cost-effective designs for small wind turbines, stall regulation may be a viable alternative for off-grid applications.

Keywords: fatigue life; small wind turbine; stall regulation; pitch regulation; aeroelastic simulation

---

1. Introduction

In the past few decades, the use of renewable energy sources has increased significantly, due a combination of a gradual paradigm shift (driven by environmental degradation and the depletion of conventional fossil fuel reserves) and more competitive pricing in key technologies, in particular, solar photovoltaic and utility-scale wind. Regarding the latter, it is worth mentioning that the global installed wind power capacity in 2017 was 500 GW [1], and it is expected to reach 1013 GW by 2025 [2].

Small-scale wind energy generation, on the other hand, has remained a niche technology, in spite of its significant potential for supplying small and isolated loads [3], households, or off-grid communities with no access to the electricity distribution network, and it contributes to clean energy goals in residential, urban [4], and industrial settings [5]. Small-scale wind energy generation is often recognized to be a major component in overcoming energy poverty [6], both on a stand-alone basis and in conjunction with other generators, such as solar photovoltaic, diesel generators, and storage [7]. Small-scale wind energy generation is ideal for balancing solar generation and reducing the size of backup generators (if any) and storage facilities in hybrid systems. However, given the dramatic reduction observed over the last decade in the cost of solar photovoltaic panels and systems [8], the pressure for cost reduction in small wind systems has risen considerably. Though small wind turbines do not necessarily have to compete directly with solar photovoltaic panels and systems, but rather with the storage units and systems that they are likely to replace or downsize, the need for building cost-effective small wind turbines, capable of reliably functioning over 20 years or more, with minimal user intervention for maintenance and repair, continues to be a hot topic.

Unlike their utility-scale counterparts of the multiple-MW class, small wind systems generally operate unsupervised. Moreover, in autonomous applications, the energy security of the user community critically depends on the availability of the wind turbine, so reliability, rather than energy production, becomes the individual most important design consideration, apart from the need for building turbines and systems with a low initial investment cost. Given that the regulation of power and rotor speed is a major design decision for wind turbines, affecting reliability and cost alike, it is worth considering some alternatives. Early commercial turbine designs, such as the ones deployed in California during the 1980s, generally relied on stall regulation, a scheme where the turbine rotor aerodynamics is designed in such a way that flow detachment is obtained at a critical wind speed, leading to a drastic reduction in rotor torque and, thereby, limiting both power and rotor speed. This scheme was largely improved in the designs of the then emergent Danish manufacturers, leading to a scheme generally called the Danish concept [9]. This design choice survived in commercial designs well into the early 2000s, when large commercial wind turbines started to rely exclusively on pitch regulation, a scheme where the turbine blades are rotated around their long axes, either passively or actively, collectively or individually, in order to adjust the blade lift forces and thereby the rotor torque in a continuous way. Interestingly, there has recently been a renewed interest in stall regulation for multi-megawatt wind turbines, specifically designed for fatigue load mitigation; see [10] for a recent approach.

Pitch regulation is now practically an industry standard in large-scale wind turbines, but is still a rather incipient technology in the small wind world, although passively operated pitch schemes, used, e.g., in the Jacobs wind turbines, have been around for about half a century. Stall regulation, on the other hand, though largely extinct at the utility scale, has made a relatively recent re-appearance in small wind turbines. While the Danish concept relied on a fixed rotor speed (dictated by the fixed electricity frequency of the utility grid to which the turbine was interconnected), modern small wind systems are not limited by this restriction, given the availability of low-cost power electronics devices capable of adjusting the generator torque in a continuous way, thereby allowing for a variety of design choices, including the possibility of obtaining a flat nominal power curve, such as in pitch-controlled systems.

Currently, many commercial small wind turbines operate with yet a different power/rotor speed regulation system called furling [11], a robust scheme based on the interplay of aerodynamics forces and moments acting on the rotor, mounted somewhat eccentrically from the vertical rotation axis, and a tail vane. Furling has been demonstrated to allow for reliable designs withstanding two decades or more of operation; however, the trade-offs are considerable, including a sharp drop-off of the output power after the onset of furling, hysteresis between the operational curves for increasing and decreasing wind speeds, respectively, and a quite complex aerodynamic state during furling, particularly in a highly turbulent environment, typically encountered by small wind turbines. Furling will not be discussed in the current paper; the interested reader is referred to [11,12] for an introduction.

Independent of the regulation scheme selected, fatigue loading is always a major factor for wind turbine life, particularly the rotor blades. Fatigue loading arises as a combination of the stochastic nature of the wind field impinging on the rotor and rotational sampling [13] caused by the finite spatial coherence of the wind field. Excessive fatigue loads lead to a reduction in the blade life and increase maintenance costs and financial losses [14], making a careful design for fatigue an important part of the (small) wind turbine design process. Given that fatigue loading is a major concern in wind turbines, a large body of literature exists on the subject, with an emphasis on utility-scale turbines. While much of the knowledge on fatigue damage in wind turbine blades has been generated by testing, both of the composite materials commonly used [15,16] and of components and systems, design procedures based on a combination of simulations and standards are becoming more common.

This is particularly important in the field of small wind turbines where extensive testing is not cost effective for manufacturers.

Ardila et al. [17] used load cases for power production and parked turbines recommended in the IEC 61400-1 standard [18] for the fatigue life assessment of a blade designed for a 5-MW wind turbine; their approach considers a combination of aeroelastic simulation tools such as FAST and HAWC2 for the generation of fatigue loading and fatigue assessment tools developed by the authors. As in most works, the Palmgren–Miner (PM) rule is used for cumulative damage calculation and the Goodman method for computing the allowed load cycle number, and the rainflow method is applied to determine the number of events with a given combination of mean stress and stress range. A similar approach was used by Hayata et al. [19], who also studied the fatigue life of the blade of a 5-MW wind turbine. Kong et al. [20], as part of an effort to design a 750-kW wind turbine in South Korea, developed a simplified fatigue design procedure, using empirical formulae for fatigue loading in combination with results from materials testing; Goodman diagrams obtained from laboratory testing and the PM rule were then used to calculate the expected fatigue life. The work in [16] presented common methods for calculating damage on the blades. The most common method is the Palmgren–Miner (PM) rule [21], in which, if the damage parameter, D, reaches the value of one, the existence of damage is concluded. Currently, there are several other methods, which are variations of the PM rule, such as the Marco–Starkey rule [22]. Currently, standardized methods for estimating the fatigue life of wind turbine rotors are available [23], providing the designer with a simplified conceptual framework useful for practical design decisions. As discussed by Evans et al. [24], the small wind turbine standard IEC-61400-2 [23] contemplates three possible methods for fatigue life assessment, one based on operational measurements, another using aeroelastic simulations, and, finally, a simplified procedure to be used by designers and small wind turbine manufacturers without access to sophisticated aeroelastic modeling tools. Evans et al. found that the simplified procedure in their study case produced extremely conservative results, leading to blade designs that are likely not cost-effective. Given the need for small wind turbine systems to reduce their costs while maintaining a level of reliability, other approaches are needed.

Derived from the observation above, the following two-fold objective of the present work was formulated: first, a practical simulation-based methodology for assessing fatigue life in small wind turbines, consistent with the recommendations of the standard IEC 61400-2, is proposed and demonstrated. This approach is believed to simplify the design process of small wind turbines greatly, allowing for a greater accuracy (as compared to the option of the simplified design method proposed in the standard) or a faster design process (as compared to conventional processes). Having introduced the methodology, a comparative study of the fatigue behavior of stall- and pitch-regulated small-scale wind turbines is carried out. The stall-regulated rotor was designed as part of the project P09 of the Mexican Center for Innovation in Wind Energy (CEMIE Eólico); the study of different power/rotor speed control schemes was part of the project P19 of the CEMIE Eólico. The design of the stall-regulated rotor resulted from an co-optimization of three objectives: (a) a high power coefficient at partial load (i.e., below nominal power), (b) overspeed control to guarantee structural stability up to 20 m/s without the need for additional braking systems, and (c) a high starting torque to overcome the generator cogging torque at low wind speeds. The third requisite somewhat restricted the power production at medium wind speeds, but enabled the rotor to work with a variety of generators. The detailed design considerations will be described elsewhere and are not part of the present work.

The comparative assessment of stall- and pitch-regulated small-scale wind turbines was partly motivated by the fact that the CEMIE Eólico considered both technologies. However, while investigating the subject, it became apparent that comparative assessments of the impact of power/rotor speed regulation on fatigue damage are very scarce. To the best knowledge of the authors, only Kennedy et al. [25] addressed this issue, albeit in the context of tidal, not wind turbines. As opposed to [25], in this work, all wind turbine designs studied have (almost) the same power curve, leading to an identical energy production for any given wind regime (at least under steady-state conditions) and allowing for a direct comparison of the analyzed schemes.

This paper is organized as follows: In Section 2, the basic relations used to calculate the fatigue life of small wind turbine blades according to the simplified methodology in standard IEC 61400-2 are briefly reviewed, followed by an overview of the complete methodology, including the different design and simulation tools available in the public-domain suite QBlade [26]. A general description of the blade geometry and materials, as well as the design curves for power, rotor speed, and pitch angle is also provided for the case of the stall-regulated wind turbine and two pitch-regulated alternatives. The full information about the outer geometry of both blades is provided in the Appendix A. A brief description about the setup of the wind field generator and the dynamic simulation, using FAST, is also included. The results obtained with the methodology proposed in this work are presented and discussed in some detail in Section 3. The summary and conclusions are presented in Section 4.

2. Methodology

2.1. Blade Damage Model

Fatigue damage occurs when the residual strength of a piece of material subjected to periodic loading equals the maximum stress within a cycle period [27]; the residual strength is a function of the degradation of the properties of a material, occurring in response to static or dynamic loads. In the case of fatigue-only damage, there is a distinct number of load cycles (N), after which the material fails if excited with a given stress amplitude and mean. In a homogeneous sample, often, a characteristic function governing the decay of the residual strength with the number of cycles can be determined, either experimentally or through micro-mechanical simulation, although the details of fracture mechanics, particularly in composite materials, turn even this well-defined problem into an undertaking with considerable complexity. In the case of more realistic geometries, such as composite beams, the complexities of a non-homogeneous stress field can be addressed by Progressive Failure Analysis (PFA) (see, e.g., [28,29]), while still assuming that the results for fatigue degradation obtained from a (small) homogeneous sample remain approximately valid, an approach (as most approximate engineering procedures) which is not without criticism. While PFA may or may not be completely adequate, engineering design approaches for wind turbine fatigue design, as already discussed in the literature review, are still considerably more simplified and simply focus on one critical location within the structure (generally located at the blade root), limiting themselves to calculating the number of cycles at which the blade root material (as determined from a small sample) fails. Though, in practice, the order in which different stress amplitudes and mean values occur might be of relevant, the standard procedures, such as the ones recommended in the IEC61400-2 norm, neglect such effects. Fatigue damage is measured by a metric such as the cumulative damage D in the Palmgren–Miner (PM) rule [16]:

$$D = \sum_i \frac{n_i}{N_i} \qquad (1)$$

where,

D :   the cumulative damage ($D = 1$ when a fault occurs)
n_i :   the number of load cycles for a given stress amplitude/mean combination
N_i :   the number of permissible load cycles for a given stress amplitude/mean combination

The fatigue life estimation then requires determining the number of load cycles n_i [30]. As in most fatigue research, the rainflow method was used to extract the number of events for a given stress amplitude/mean combination from the relevant input time series. As mentioned above, the stress time series obtained from an aeroelastic simulation of the blade under normal operation conditions at different average wind speeds, as extracted at the blade root, was used as input to the rainflow counting algorithm in all occasions.

In order to calculate the permissible number of cycles for a given stress state, based on the Goodman method, the following formula was used [23]:

$$N = \left[\frac{X_t + |X_c| - |2\gamma_{Ma}\sigma_{1,m} - X_t + |X_c||}{2\left(\gamma_{Mb}/C_{1b}\right)\sigma_{1,a}} \right]^m \tag{2}$$

where,

X_t :	maximum tensile strength (MPa)
X_c :	maximum compression strength (MPa)
$\sigma_{1,m}$:	average atress of a cycle calculated by the rainflow method (MPa)
$\sigma_{1,a}$:	stress amplitude of a cycle calculated by the rainflow method (MPa)
m :	slope of the $S - N$ diagram
γ_{Ma,M_b} :	safety factors

The definition of the safety factors, as well as m, was given in [31]. The value assigned to m depends on the material of the blade. Table 1 shows some values assigned to m for three different materials. The safety factors are calculated according to:

$$\gamma_{Mx} = \gamma_{M0} \prod_i C_{ix} \tag{3}$$

where $\gamma_{M0} = 1.35$, and the values assigned to C_{ix} are selected according to the material used in the blade manufacturing. Table 2 shows some values of C_{ix} used for materials reinforced with polymeric fiber.

Table 1. Values of the parameter m [31].

Material	m
Films with polyester resin	9
Films with epoxy resin	10
Polymeric films reinforced with carbon	14

Table 2. Values of safety factors [31].

	Value	Description
C_{1a}	1.35	Influence of aging
C_{2a}	1.1	Temperature effect
C_{3a}	1.1	Films produced by resin infusion
	1.2	Films produced by pressure
C_{4a}	1.0	Post-cured film
	1.1	Non-post-cured film
C_{2b}	1.1	Temperature effect
	1.0	Unidirectional reinforcement
C_{3b}	1.1	No tissues
	1.2	Tissues and mat
C_{4b}	1.0	Post-cured film
	1.1	Non-post-cured film
C_{5b}	1.0 to 1.2	

Equivalent Stress

The section experiencing the greatest stress along the rotor generally corresponds to the blade root, given that the blade bending moments are the largest here [32]. In the case of a circular root section of the blade, the equivalent stress can be written as follows [23,33]:

$$\sigma_{eqB} = \frac{F_{zB}}{A_B} + \frac{\sqrt{M_{xB}^2 + M_{yB}^2}}{W_B} \tag{4}$$

where,

F_{zB} :   axial force (rotor thrust) (N)
M_{xB} :   edgewise moment (N·m)
M_{yB} :   flapwise moment (N·m)
A_B :   root cross section (m^2)
W_B :   modulus for the root section of the blade (m^3)

The modulus for a circular area, with $A_B = 0.0123$ m^2, is calculated as follows:

$$W_B = \frac{\pi}{4} \left(\frac{r_{ext}^4 - r_{int}^4}{r_{ext}} \right) = 6.54 \times 10^{-5} \text{ m}^3 \tag{5}$$

where r_{int} and r_{ext} are the interior and exterior radii of the blade root, respectively.

2.2. Methodology for Calculating Fatigue Damage

Figure 1 shows a general scheme of the proposed methodology used in this work. The primary output is the predicted fatigue damage $D(v_i, \text{TI}, \text{reg})$ of the blade for a given average wind speed v_i, a given level of Turbulence Intensity (TI) (high or low), and the turbine configuration, where "reg" stands for the turbine regulation scheme used,

$$T[\text{years}] = \frac{\tau[\text{min}]}{24 \times 60 \times 365 \times D\left(v_i, \text{TI}, \text{reg}\right)} \tag{6}$$

where τ is the simulation time in minutes. In this work, $\tau = 10$ min in all cases. An average fatigue life can then be calculated by convoluting the fatigue damage for a given wind speed class v_i with a Weibull distribution function $f_{\text{Weibull}}(v; v_s, k)$, where v_s is the Weibull scale factor in m/s and the dimensionless quantity k is the Weibull shape factor. Note that the average wind speed $\bar{v}$ is related to the shape factor by $\bar{v} = v_s \Gamma (1 + 1/k)$, where Γ is the gamma function.

The proposed methodology combines two software packages, the public-domain suite QBlade and a set of codes implemented in MATLAB, shown on the left- and right-hand side of Figure 1, respectively. QBlade can be used for (1) simulating and designing airfoils, (2) simulating rotor aerodynamics using Blade Element Momentum (BEM) theory, (3) simulating steady-state operation based on prescribed curves for rotor speed vs. wind speed and blade pitch angle vs. wind speed, (4) simulating stochastic wind fields, and (5) conducting aeroelastic simulations using the rotor setup and operational characteristics defined in Steps (1)–(3). Step (5) is carried out with the tool FAST [34], developed by NREL and embedded in QBlade, whereas Step (1) is based on another public-domain tool, XFoil. In addition to Tools (1)–(5) QBlade allows for finite-element simulation of the mechanical state of the blade for given stationary load conditions, which can be used to conduct a preliminary structural design of the blade.

In all three turbine configurations studied in this work, the QBlade tools (1)–(3) were used iteratively to reach the given design objectives. In the case of the stall design, as mentioned before, additional evaluations were carried out outside of the QBlade framework for an optimization of the start-up behavior, which is not part of the present work. For all three turbine configurations, an electric nominal output power of 10 kW was specified, to be reached at a rotor speed of about 130 rpm. In the case of a pitch-regulated wind turbine, the requirement of a constant nominal rotor speed can be met always, at least theoretically, by adjusting the pitch angle correspondingly, although an optimal design is far from trivial. In a stall-regulated turbine, however, this additional degree of freedom is not available, so all regulation has to be conducted through the prescription of the rotor speed, which is achieved in practical terms by the control of the generator torque (for a modern application of small wind turbine control through generator torque control, see, e.g., [35]).

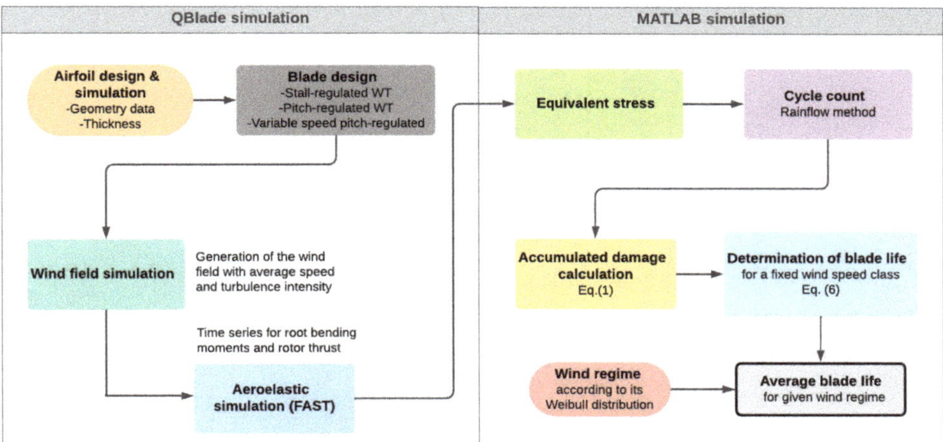

Figure 1. Overall methodology used in this work.

2.2.1. Blade Characteristics and Geometry

For the stall-regulated turbine simulated in this work, airfoils from the series DU-210_xx were considered. The selection of this series was in part motivated by its favorable starting performance. On the other hand, both pitch-regulated turbines were based on a design using one single airfoil, namely the NACA 4412. Figure 2 shows representative airfoils used for stall and pitch design, respectively. Figure 3 shows a 3D rendering of both blades. The chord and twist angle functions are shown in Figure 4. The blade designed for pitch control had a roughly optimal chord function, with a transition to a circular root section to allow for convenient flanging to the rotor hub (as usual in wind turbine engineering, an "optimal" blade is one with an optimal power extraction at a constant tip-speed ratio, considering negligible airfoil drag). The twist function was obtained from a performance optimization using QBlade, taking the twist function proposed in [36] as a starting point, and a smoothing process to allow for a manufacturable blade. The stall design, on the other hand, was based on the requirement that at a given wind speed, all blade sections simultaneously enter the stall regime, thereby leading to a sharp drop in the power coefficient of the rotor. This was achieved using the optimization feature provided by the QBlade suite. As discussed above, the DU210xx airfoil series was selected in part because of the relative sharp drop in the lift coefficient for angles of attack larger than a critical angle (the stall angle), which greatly facilitates a good rotor stall behavior. The other reason for choosing the DU201xx airfoil series lies with its favorable starting performance, as mentioned before.

The geometry of the stall blade considered previous designs from the authors. Table A1 shows the densities, Young's module, the maximum tension strength (X_t), and the maximum compression strength (X_c). Table A2 shows information about the geometry of the stall blade, as well as the aerodynamic profiles for each section and the Reynolds numbers. The blades have a length of 3.58 m, divided into 30 sections or blade stations. The blades were manufactured with two distinct kinds of Fiber Glass-Reinforced Polymeric (FGRP) and epoxy resin. The root and shell of the blade were built from tri-axial material, while the crossbar used a bidirectional material.

The design of the pitching blade was inspired by a blade design published by Hansen et al. [36]. To adapt this model to the present work, the original blade, with a total length of 40 m and divided into 14 sections, was scaled to match the length of the stall-regulated blade. The geometry was interpolated to 30 blade stations. To reach the final blade design, the scaled and interpolated chord function was used directly, whereas the twist distribution was used as a starting point for a final twist optimization using QBlade [26]. The same characteristics of the materials used in the stall blade were considered at a given blade section. The data of the geometry of the pitching blade are shown in Table A3. Finally, for both blade designs, the same Reynolds numbers were considered. The masses of the blades were 19.835 kg for the stall blade and 21.642 kg for the pitching blade. Figure 3 illustrates the blade geometry for each case.

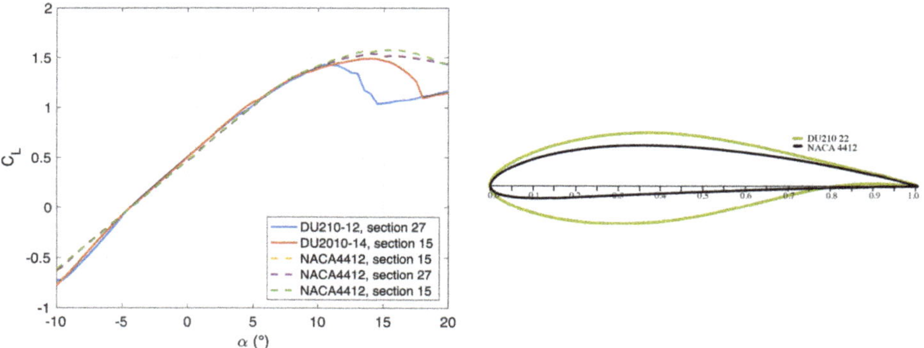

Figure 2. Representative airfoils for turbines using stall (DU-210_xx airfoils) and pitch regulation (NACA 4412 airfoil), respectively.

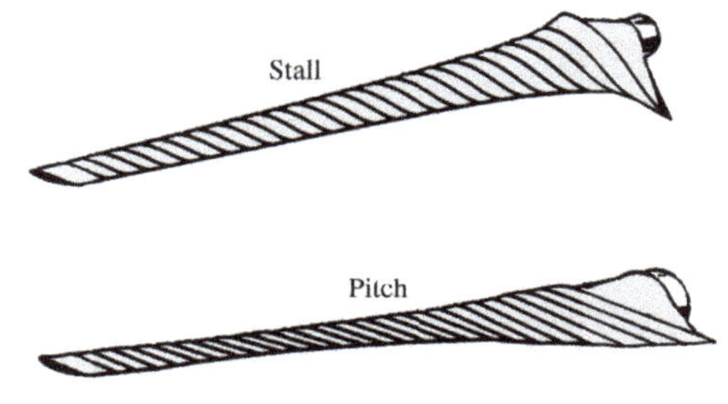

Figure 3. Blade geometry.

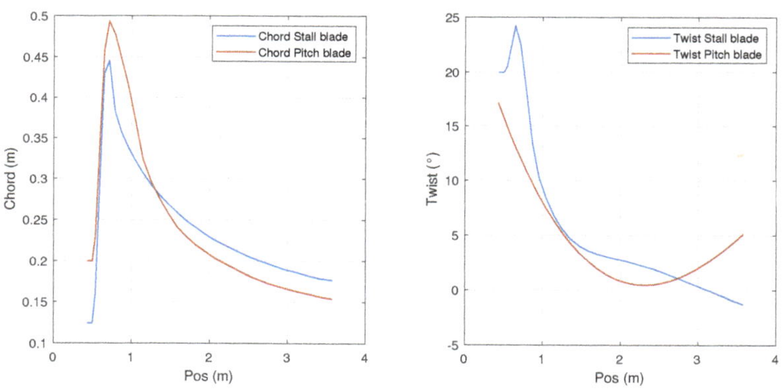

Figure 4. Chord and twist functions for the stall and pitching blades.

2.2.2. Steady-State Aerodynamic Simulation and Operational Characteristics

Blade Element Momentum (BEM) theory, as implemented in QBlade [26], was used to determine the aerodynamic behavior of the blades. BEM is based on the balances of the linear and angular momentum between the air mass flowing through annular rotor disk sections and the aerodynamic forces and moments acting on the blade sections located at a given radial position, determined from empirical or simulated airfoil data, plus a number of empirical corrections, such as for rotor thrust under heavy loading conditions and tip and rotor losses. The local differences in Reynolds number along the blade radius can be taken into account by simulating each required airfoil with a different Reynolds number, depending on the blade station where this airfoil is going to be placed. While in utility-scale wind turbines, the effect of making this distinction is marginal, given the large Reynolds numbers prevailing at most parts of the blades, in small-scale wind turbines, this differentiation is critical, particularly near the root where the Reynolds number may be as low as several tens of thousands. The solution of the local momentum equations allows obtaining the rotor torque, the extracted mechanical power, and the rotor thrust. The main parameters of the BEM simulations were: (1) air density: $\rho = 1.225$ kg/m^3, (2) air viscosity: $1.456 \times 10^{-4} \frac{\text{kg}}{\text{m·s}}$, and (3) elements and iterations: 40 and 100, respectively.

The output of the BEM calculations was then combined with prescribed functions for the rotor speed $n(v)$ and the blade pitch angle $\beta(v)$ as a function of the free-stream wind speed v to define the steady-state operational characteristics of each wind turbine, assuming a constant efficiency of all other components such as the electric generator. The determination of these functions was conducted outside QBlade and was based on the considerations described at the beginning of Section 2.2. The following results were obtained (see Figure 5 for an illustration and Table 3 for a summary of the turbine characteristics): in all cases, the regulation scheme was designed to obtain optimal operation at partial load, from just after cut-in up to a free-stream wind speed of 8 m/s; in this regime, all turbines operated at an optimal wind speed ratio λ_{opt} of about six, where $\lambda = \Omega R/v$, Ω being the rotor speed and R the rotor radius. From 8 m/s–10 m/s the regulation of all three turbines continued to be identical, with the shaft frequency now being set to its nominal value of 130 rpm and the pitch angle remaining at zero degrees in all cases. The effect of reducing the rotor speed was the same for all, a reduction in the power coefficient (the proxy for turbine efficiency), which made up for part of the cubic increase in power density with wind speed, resulting in a practically linear power curve in this section (Figure 5a). For wind speeds higher than 10 m/s, the three regulation schemes now started to differentiate themselves, as described below:

- Stall-regulated wind turbine: The pitch angle cannot be varied in this design, so it remained at zero degrees for all wind speeds. By operating the wind turbine at its maximum power point up to a wind speed 8 m/s and then setting the rotational speed to a constant value of 130 rpm, the turbine slope of the P(v) curve was limited, and a smooth transition to the nominal regime starting around 11–12 m/s was guaranteed. This design ensured a reliable operation of the turbine, avoiding large excursions in the opposing torque and the possibility of overheating. For higher wind speeds, the rotor speed had to be increased slightly (up to about 145 rpm for $v = 14.5$ m/s) to compensate for a slight drop in the power coefficient, resulting from the stall design. In practice, this is achieved by a slight reduction in generator torque. This small increase in rotor speed (from 130 rpm–145 rpm) is hardly a concern from a structural load perspective. Note that a classical stall-regulated wind turbine, based on the Danish concept, would have a drop in output power, since the rotor speed cannot be changed under that scheme. For higher wind speeds, the rotor speed has to be lowered again, with the lowest value occurring at $v = 20$ m/s, where n is reduced to about 120 rpm. For still higher wind speeds, practical limits are encountered with heat dissipation at the electric generator, so, typically, an additional braking system is recommended for $v > 20$ m/s. In the current case, a passive emergency pitching system was designed, the description of which is, however, outside the scope of the present study.
- Pitch-regulated wind turbine with constant rpm at nominal power: This scheme is operated in close analogy with a utility-sized wind turbine, where the power curve is roughly determined by two regimes: one with zero pitch angle and linearly increasing rotor speed (the partial load or optimal regime) and the nominal load regime where the rotor speed is held constant (by rotor

torque control), and all remaining regulation is done by pitching. In the current case, the two mentioned regimes exist, but a third —hybrid— regime was introduced, where a constant rotor speed coexists with a zero pitch angle. This "stall-regulation"-type behavior was achieved by deviating from optimal operation at a relatively low wind speed (8 m/s), well below the turbine design wind speed of 11 m/s, where the nominal power was attained. It should be noted that the pitching curve $\beta(v)$ showed a gentle almost monotonic increase from zero to about six degrees at 20 m/s.

Table 3. Operational characteristics of the three wind turbine designs studied in this work.

	Stall Regulation	Pitch Regulation	Variable Speed Pitch Regulation
Nominal power	10 kW	10 kW	10 kW
$V_{\text{cut-in}}$, V_d, $V_{\text{cut-out}}$	3 m/s–11 m/s–20 m/s	3 m/s–11 m/s–20 m/s	3 m/s–11 m/s–20 m/s
Rotational speed	Variable	Nominal: 130 rpm	Variable
Pitch angle (β)	0	Variable	Variable
Variable losses	13%	13%	13%
λ_{opt}	6.1	6.21	6.21

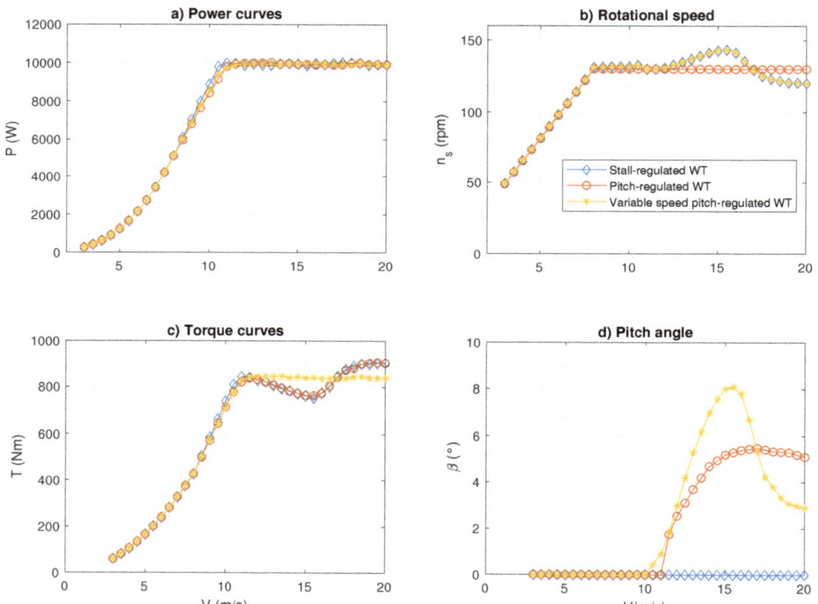

Figure 5. Operation curves of the WTs' designs: power curve, rotational speed, torque curves, and pitch angle.

- Pitch-regulated wind turbine with variable rpm at nominal power: This scheme was introduced for comparative purposes and represented a hybrid version between the former two approaches. As shown in Figure 5b, in this scheme, the rotor speed was varied in a fashion identical to the one used for stall regulation and a corresponding adjustment in the pitch angle curve (Figure 5d). Though there did not seem to be an objective, practical reason for using this strategy, using the same rotor speed curve is a useful tool for assessing the effect of the rotor thrust on the rotor dynamics, as simulated by FAST, and, consequently, on the fatigue loading and damage.

2.2.3. Wind Field Simulation

An important step in the creation of a dynamic aeroelastic simulation of a wind turbine rotor is the generation of a time-varying wind field, large enough to cover the whole rotor. Using stochastic wind fields, as opposed to a homogeneous wind speed over the swept rotor area, makes a critical difference for the purposes of fatigue assessments. As shown below, because of the finite (non-zero and not one) spatial coherence of real-life wind fields, the periodic rotation of the wind turbine translates into a time-varying stochastic wind speed signal with periodic components at any given location along the rotor blade. These periodic components correspond to the fundamental rotor frequency and its harmonics. Evidently, such periodic loading has a significant impact on fatigue loading.

QBlade has a computational tool for the creation of such wind fields based on stochastic temporospatial simulation. The following parameters were used in the present work:

- Simulation time (s): All simulations were performed for 600 s (10 min)
- Rotor radius (m): This parameter must be slightly larger than the rotor radius, to avoid convergence problems in the FAST simulation; therefore, R was set to 4 m
- Hub height of the turbine (m): 15 m
- Average speed (m/s): This parameter was varied from 3 m/s–20 m/s in order to assess the fatigue damage of each of the turbines for a large range of wind speed classes. The final set of simulations used for this work was limited to the range between 7 m/s and 20 m/s since the damage for lower wind speeds was found to be practically zero
- Turbulence intensity (%): Two sets of Turbulence Intensities (TI) were assumed, one corresponding to "high" and another to "low" TI. Given that TI depends on the average speed, the following formula, taken from the standard IEC61400-2, was used to adjust TI for each run:

$$\sigma_v[\text{m/s}] = \frac{15 + aV_{\text{hub}}}{1 + a} \times I_{15},\tag{7}$$

where TI $= \sigma_v/\bar{v}$, $a = 2$, and $I_{15} = 0.16$ for low turbulence intensity, as well as $a = 3$ and $I_{15} = 0.18$ for high TI.

Figure 6 shows a spatial wind field map for one instant taken from the simulation of 600 seconds with an average speed of 10 m/s and a high level of turbulence intensity (corresponding to 21% according to Equation (7).

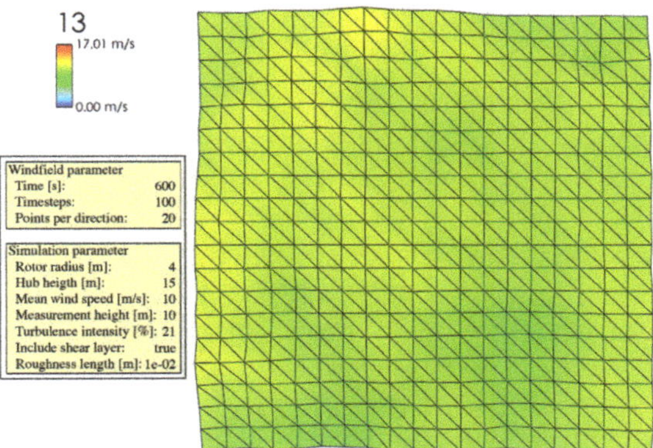

Figure 6. Field wind simulation $\left(10\frac{m}{s}\right)$ with high turbulence intensity.

2.2.4. Aeroelastic Simulation

The last simulation step in QBlade corresponds to the aeroelastic simulation of each of the three wind turbines described above using FAST [34]. FAST is a simulator developed and maintained by the National Renewable Energy Laboratory (NREL) for the simulation of extreme and fatigue loads and is often considered a standard for aeroelastic wind turbine simulations. FAST is embedded in QBlade, which offers the user the advantage of being able to import directly the turbine configuration obtained from the previous design into FAST, and it conducts the simulation with a relatively minor configuration effort. However, a basic understanding of the parameters required for the setup of the simulations is vital. A selection of some of the relevant parameters is shown below:

- StallMode [STEADY/BEDDOES]: This parameter is a flag indicating whether the Leishman–Beddoes dynamic stall model is to be used [34]. In the present work, StallMode was set to BEDDOES for the simulations of the stall-regulated turbine, whereas STEADY was used for the pitch simulations
- InfModel [EQUIL/DYNIN]: This parameter decides whether a generalized Dynamic-wake model (DYNIN) or an Equilibrium-inflow model (EQUIL) should be used [34]. In this work, the EQUIL model was selected throughout.
- HLModel [PRANDTL/NONE]: When using the equilibrium inflow model, this parameter allows including or disabling hub-loss calculations [34]. The options NONE was used throughout this work.

Most of the FAST parameters were found to have a negligible effect on the obtained results, with the notable exception of the StallMode parameter, where the wrong setting led to instabilities. Once a FAST simulation was correctly configured, the following output time series were obtained: wind speed at the hub height, output power and torque, axial force (rotor thrust), and moments for flapwise and edgewise bending. Figure 7 shows the time series results of the bending moments, axial force, and wind speed. By using the time series from Figure 7 and applying Equation (4), the equivalent stress over the blade root can be obtained. Figure 7 shows the corresponding time series.

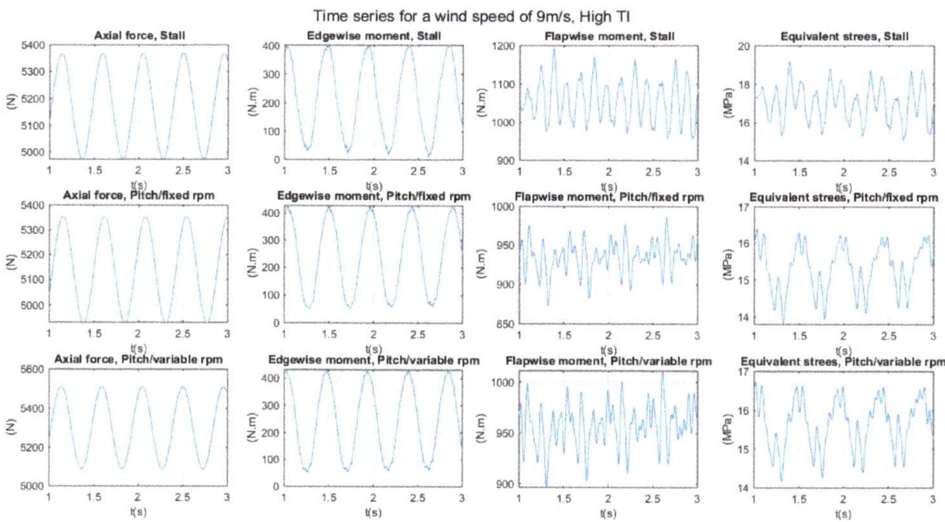

Figure 7. Results of the aeroelastic simulation for an average wind speed of 9 m/s and high turbulence intensity: time series.

2.2.5. Cycle Counting and Damage Calculation

The last four stages of the proposed methodology were implemented in a MATLAB script. The input data corresponded to the equivalent stress time series (Figure 7). The first step was cycle counting, which was done with the rainflow method; the MATLAB implementation was used for this purpose. The output matrix contained the number of events for a given stress amplitude/average stress value combination.

The second calculation processed the output of the rainflow analysis by means of the Goodman method to obtain the number of permissible cycles for each combination of stress amplitude and average value (Equation (2)). The maximum values for the tension and compression resistance are presented in Table A1 for the root section. The value of the m parameter in Equation (2) was set to 10, corresponding to epoxy resin composites, according to [31] (see Table 1). The safety factors were taken from Table 2. Additionally, $\gamma_{M0} = 1.35$. Table 4 presents the calculation of the safety factors.

Table 4. Calculation of the safety factors.

C_{1a}	1.35	
C_{2a}	1.1	
C_{3a}	1.2	$\gamma_{Ma} = \gamma_{M0} \prod_{i=1}^{4} C_{ia} = 2.64627$
C_{4a}	1.1	
C_{2b}	1.1	
C_{3b}	1.0	
C_{4b}	1.1	$\frac{\gamma_{Mb}}{C_{1b}} = \gamma_{M0} \prod_{i=2}^{5} C_{ib} = 1.9602$
C_{5b}	1.2	

2.2.6. Fatigue Life Estimation

In order to build a consistent fatigue life vs. wind speed curve, some post-processing is necessary. The reason for this is two-fold: (i) Given the relatively short simulation time for each wind speed class, each 10-min time series is not necessarily fully representative of the underlying stochastic process generating the wind field. This problem can be addressed by repeatedly running the wind field generator and FAST with the same parameters and a different seed and then averaging the results. However, since the current version of QBlade does not allow for batch mode processing, there are practical limits to this approach; (ii) for the given typical values of average wind speed and turbulence intensity, there is a significant overlap of the corresponding time series with neighboring classes; therefore, a mobile weighted average is an appropriate solution. In order to obtain the results presented in this work, wind fields were generated for wind speed averages in increments of 0.5 m/s. Then, the probability density function, for each hub height wind speed time series, was determined and used as the weighting function for determining the weighted average of the fatigue damage at the center wind speed class. This approach effectively increased the sampling horizon for each wind speed class and simultaneously accounted for the finite distribution of each time series.

3. Results and Discussion

We will start the Results Section by first inspecting some of the time series outputs produced directly by the aeroelastic simulation. Figure 7 shows a short stretch of the time series results obtained for the case of an average wind speed of 9 m/s and a high turbulence intensity. It can be seen that the axial force (rotor thrust) and the in-plane ("edgewise") bending moment showed an almost perfectly sinusoidal signal with a period of a little under half a second. By comparison to the power curve, it can be seen that this frequency was consistent with the rotor speed of 130 rpm, expected by design for a wind speed of 9 m/s. The amplitude of the rotor thrust oscillation was only about 5% of the average thrust, which is consistent with the fact that the rotor thrust always created a force in the downward direction, with a minor periodic variation, arising from the fact that the wind speed was higher at the top than at the bottom position. The amplitude of the edgewise oscillations, on the other hand, was almost 100% of the average value. This fact can be qualitatively understood by the observation that at the upper position, the blade sees the highest wind speed (and, consequently, also the highest in-plane force component), at the blade tip where the leverage is the highest, and conversely, the lowest wind speed (and therefore the smallest force) near the root where the leverage is the smallest. At the lower position, on the other hand, the point with the lowest wind speed has the highest leverage whereas the point with the smallest leverage has the highest wind speed. Evidently, this leads to a small in-plane bending moment at the bottom position of the blade and a high bending moment at the top. In any case, no significant difference was observed among the three regulation schemes. Note that the terrain roughness length used in the generation of the stochastic wind field was $z_0 = 0.01$ m, corresponding to a power law exponent of 0.136.

The situation at the out-of-plane or flapwise bending moment, as well as the resulting effective stress, is quite different. Not only were the average values higher for the stall-regulated wind turbine, but also could the presence of strong harmonics be detected, with a much stronger presence of harmonics in the case of stall regulation as compared to both pitch-regulation schemes, which had a nearly identical behavior between them, not unexpectedly, given the very similar regulation characteristics at 9 m/s.

After receiving some initial insights from inspecting the time series, it is interesting to study the power spectra. In Figure 8, the primary outputs of the FAST simulation (thrust and bending moments) are shown for all three regulation schemes, again for the case of a 9-m/s wind speed. In the case of the axial force signal, one prominent peak at the primary rotor frequency can be observed, consistent with the almost perfectly sinusoidal signal shown in Figure 7. The situation at the bending moments, however, was quite different, as evidenced by the rich harmonic sequence in both sets of spectra. Whereas the periodicity of the rotor thrust signal can be plausibly interpreted as a result of the different wind speeds at the top and bottom positions of the blade, the periodicity of the bending moments can likely be attributed, at least in part, to the rotational sampling effect described above, which is expected to display a comb of harmonics. Regarding a possible distinction between the three regulation schemes, it can be noted that not only the stall regulation showed again the highest peak at the fundamental rotor frequency, but also was the first harmonics was stronger than those in the signals obtained from pitch regulation. Interestingly, the fourth harmonics seemed to be stronger in the pitch-regulated signal, but its overall contribution was small.

Having collected these initial insights, some of the results of the rainflow analysis will now be inspected. In Figure 9, the results obtained with the rainflow counting analysis of the stress time series obtained with each of the regulation schemes are shown (upper row), as well as short segments of the corresponding time series. Additionally, the inflection points as identified by the rainflow algorithm are marked for clarity. Prior to generating these diagrams, the stress time series were subjected to some moving-average filtering in order to eliminate the large number of very small-amplitude cycles, which do not contribute to fatigue damage, but obscure the direct comparison between the bivariate histograms.

As shown in Figure 9, the higher load level in the stall-regulation scheme, with vanishing differences between the two pitch-regulated schemes, become apparent. Not only is the average stress level higher (as evidenced by both the bivariate histogram and the time series segment), but also are the stress cycles deeper. The situation changed quite drastically when a higher average wind speed, specifically 18 m/s, was considered. As shown in Figure 10, at this wind speed, the stall-regulated signal now showed a significantly smaller average stress and also a smaller depth of the stress cycles. As opposed to the 9-m/s case, the two pitch regulation schemes now showed some differentiation, with the variable rpm pitch variant exhibiting a lower average stress, very similar to the one experienced by the stall regulated turbine, albeit with a higher alternating component.

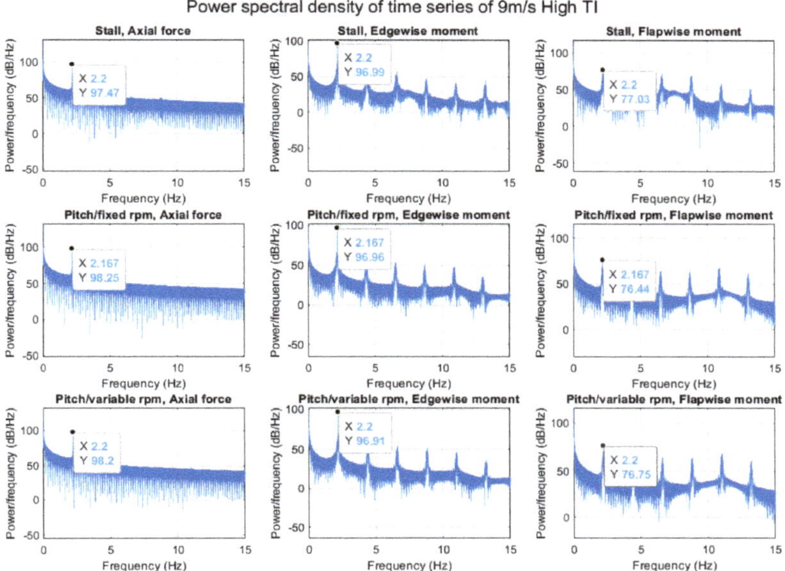

Figure 8. Results of the aeroelastic simulation for an average wind speed of 9 m/s and high turbulence intensity: power spectra.

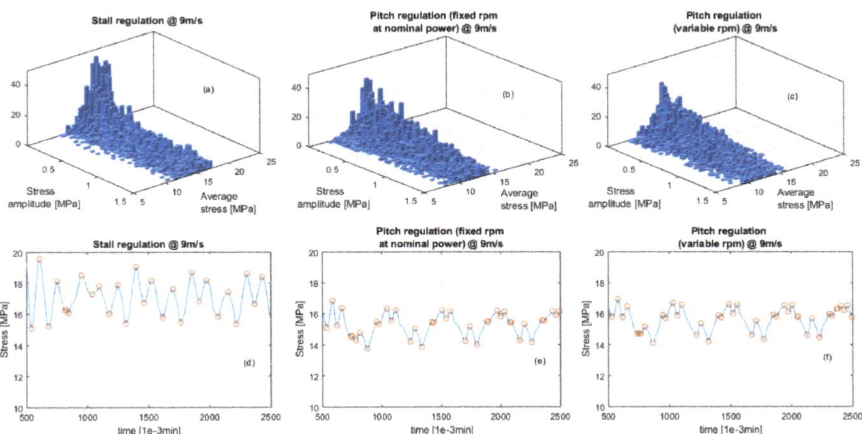

Figure 9. Results of the rainflow analysis for 9 m/s. Upper row (**a–c**): results of the rainflow analysis for the three regulation schemes. Lower row (**d–f**): corresponding time series with inflection points marked.

After these preparations, a fatigue life diagram can now be constructed as a function of the wind speed class. Figure 11 shows such a diagram for low (Figure 11a) and high levels of the turbulence intensity (Figure 11b). A number of things can be observed: first of all, and not unexpectedly, fatigue life decayed rapidly with wind speed; note the logarithmic scale of the figure. For each increase in wind speed by about 2 m/s, a reduction in fatigue life by one order of magnitude was observed, so evidently higher wind speeds put a higher strain on the rotor. Secondly, higher turbulence intensity generally led to higher fatigue damage, also an expected finding. This effect was also substantial: the higher turbulence cases showed a fatigue life about half an order of magnitude smaller than their low-turbulence counterparts, except for very high wind speeds (16 m/s and higher). This is clear evidence of the fact that the stress cycle amplitude (rather than the average stress) is the main driver behind fatigue degradation, at least for most of the wind speeds encountered in a normal wind climate.

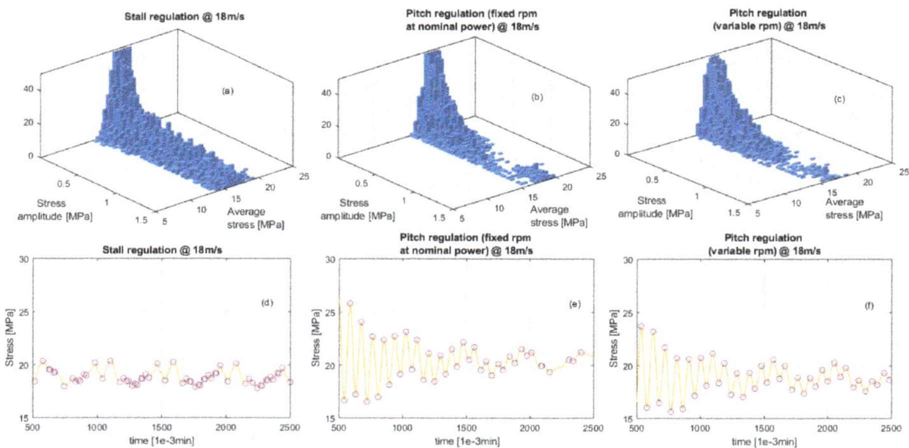

Figure 10. Results of the rainflow analysis for 18 m/s. Upper row (**a–c**): results of the rainflow analysis for the three regulation schemes. Lower row (**d–f**): corresponding time series with inflection points marked.

Thirdly, and most importantly in the context of the present study, the type of power/rotor speed regulation makes a difference, and a great one. For medium wind speeds (7 m/s–10 m/s), the stall-regulated turbine experiences a fatigue damage about an order of magnitude larger than the pitch-regulated wind turbines. This differentiation was, again, expected since stall-regulated wind turbines experience a far more turbulent environment in their regulation regime, as compared to pitch-controlled turbines where flow remains attached most of the time. In the present case, the difference between the design operating conditions in the wind speed range between 8 m/s and 10 m/s was actually relatively small between the stall- and pitch-regulated turbines, since the pitch designs used in this work shared a "soft stall" with the stall turbine in that wind speed range, so the difference in fatigue life observed in Figure 11 was larger than one would expect. A plausible explanation for this is the use of the stall model used in the aeroelastic simulations; as mentioned above, reliable simulations without instabilities require the use of the Leishman–Beddoes dynamic stall model in the FAST program, whereas the pitch-regulated wind turbines were always simulated assuming steady flow conditions, even in the regime between 8 m/s and 10 m/s, where "soft stall" conditions could arise.

The difference in fatigue life between the stall- and the pitch-regulated turbines became gradually smaller with wind speed and even inverted its sign at very high wind speeds. This cross-over occurred at 17 m/s for the low turbulence intensity cases and at 16 m/s for high TI. Similarly, the fatigue life curve for the pitch-regulated turbine with variable rotor speeds started to separate from the fixed-rpm curve, after being almost indistinguishable for all lower wind speeds. This joint behavior of both the stall and the variable-rpm pitch turbine clearly pointed to a common rotor speed-related root cause. The fact that this departure from the general trend of the fatigue life curves occurred for smaller wind speed classes in the case of high turbulence is readily attributed to the fact that the filter mask for the moving-average calculation used to get to Figure 11 had a larger spread to neighboring wind speed classes in this case, leading to an effectively larger averaging window. Therefore, the effect of the rpm reduction occurring in the stall- and variable-rpm pitch turbines manifested itself at lower wind speeds. Other than that, the underlying effect appears to be the same. The most likely candidate for an explanation was the axial force or rotor thrust. At high wind speeds, the average thrust diminishes again in the case of the stall and variable-rpm pitch regulation, below the average thrust value of the fixed-rpm pitch-regulated wind turbine (not shown for brevity). This mechanism is likely to explain the observed behavior at very high wind speeds.

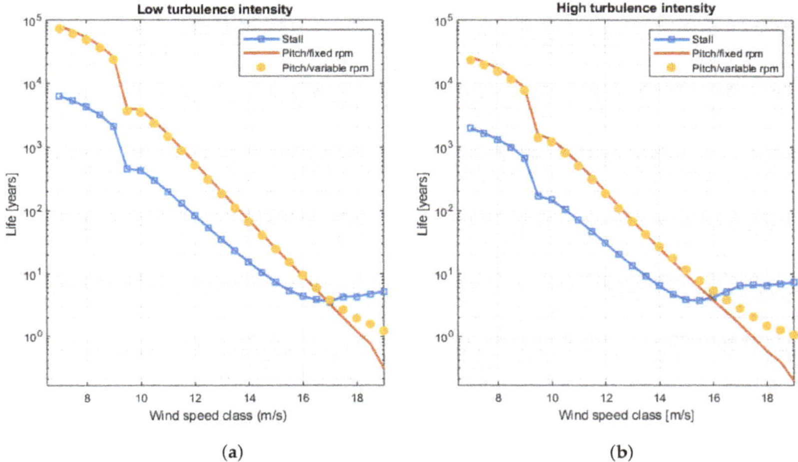

Figure 11. Fatigue life as a function of the wind speed class for all three power regulation schemes. (a) Low turbulence intensity. (b) High turbulence intensity.

Now, after the differences in fatigue life between the three regulation schemes and their underlying root causes have been discussed with some detail, it is convenient to close the discussion with some considerations regarding the practical implications of the regulation schemes introduced in this work. A simple way of doing this is by convoluting the damage function $D(v, \mathrm{TI}, \mathrm{reg})$ with a Weibull distribution function characteristic of a given wind regime, as described above, and then applying Equation (4) in order to obtain the corresponding (average) fatigue life. As mentioned before, the primary damage function was assumed to be zero for wind speed averages lower than 7 m/s, meaning ($D(v < 7\,\mathrm{m/s}, \mathrm{TI}, \mathrm{reg}) = 0$; additionally, only wind speed classes up to 20 m/s were considered; for higher values, the turbines were assumed to be parked, and fatigue damage accumulated during parked conditions was assumed to be negligible; therefore, ($D(v > 20\,\mathrm{m/s}, \mathrm{TI}, \mathrm{reg}) = 0$. Damage resulting from extreme events cannot, of course, be ruled out, but this is outside the scope of the current study.

In Figure 12, the results for the average fatigue life are shown for the cases of low TI levels. The upper rows show the fatigue as a function of the Weibull scale factor v_s, whereas the lower row has the corresponding Weibull probability density functions for selected v_s values (3 m/s, 5 m/s, and 7 m/s). The columns are organized according to their Weibull shape factor ($k = 2, 1.5, 3$). In the standard case of $k = 2$ (also referred to as a Rayleigh distribution), the normal ordering was largely conserved, with the stall-regulated turbine showing the lowest average fatigue life up to Weibull scale factors of about 7 m/s. Given that most small wind turbines operated at sites within this wind speed range, it is safe to say that normally stall regulation will lead to lower fatigue life. The differentiation was somewhat exacerbated in the case of $k = 3$, where the average fatigue life of stall turbines remained consistently about an order of magnitude below that of their pitch-regulated counterparts. This stronger differentiation can be attributed to the narrower distributions obtained in the case of $k = 3$. Conversely, sites with a low shape factor, as illustrated by the $k = 1.5$ case, showed wider distributions, leading to a cross-over between the stall and pitch fatigue life curves, at Weibull shape factors as low as about 4.5 m/s. The situation at high TI levels was analogous to the one at low TI; therefore, the corresponding graphs were omitted for brevity.

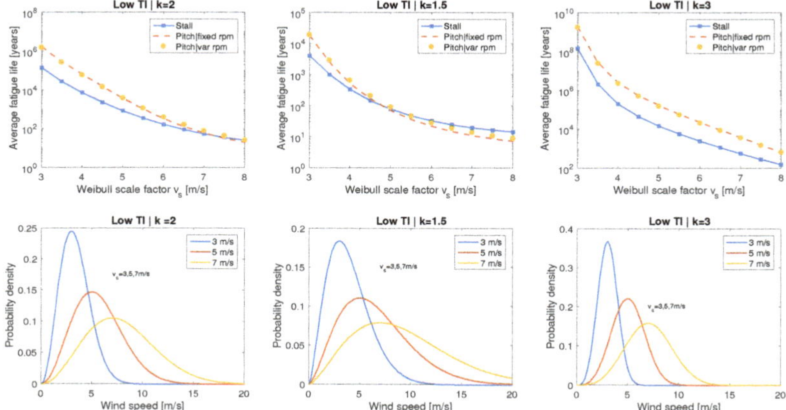

Figure 12. Average fatigue life as a function of the Weibull scale factor v_s for all three power regulation schemes, different Weibull shape factors, and low turbulence intensity. Upper row: average fatigue life. Lower row: corresponding Weibull distributions for selected v_s values.

4. Summary and Conclusions

A comparative fatigue study was conducted for three small wind turbine power/rotor speed control schemes, one based on stall regulation and two others on pitch regulation. The rotor of the stall-regulated turbine was designed previously to co-optimize the objectives of high power coefficient at partial load, a flat power curve under nominal conditions with small excursions of the rotor speed, and favorable starting performance. The first pitch-controlled turbine design was based on a published large-scale blade, which was downscaled to the blade length of the stall design and afterwards fine-tuned. A second pitch-regulated wind turbine was designed by emulating the rotor speed function of the stall-regulated turbine; the objective was to separate any rpm-related effects in the fatigue damage from the intrinsic behavior of the stall and pitch regulation, respectively.

A methodology for fatigue damage assessment based on a combination of an aeroelastic simulation and the simplified model proposed in the IEC-61400-2 standard was designed and implemented. Fatigue life curves for two levels of turbulence intensity (adjusted to their respective average wind speeds) were calculated for all three regulations schemes as a function of the wind speed class. An exponential decay of the fatigue life was observed over a large range of average wind speeds, from 7 m/s to about 16 m/s, for all schemes. While the fixed-rpm pitch-regulated schemes continued to decay with the same law for even higher wind speeds, the fatigue life curves for the stall-regulated turbine were found to level off afterwards; the reduced rotor thrust coefficient, resulting from a reduced rotor speed at high wind speeds, was identified as the main underlying cause. Consistently, a similar behavior was observed for the pitch design with variable rotor speed.

Average fatigue life curves as a function of the Weibull scale factor were calculated for all cases and for different Weibull shape factors. Under typical wind regime conditions and small to moderate average wind speeds, the stall turbine exhibited a larger average fatigue damage, as expected. This difference was less pronounced for wind regimes with smaller shape factors, given the broader wind speed distributions. In all cases, both pitch- and stall-regulated wind turbines showed an average fatigue life in excess of several decades for wind speeds typically encountered in small wind turbine locations.

Author Contributions: Conceptualization, O.P.; data curation, B.L. and J.P.-C.; formal analysis, O.P.; funding acquisition, O.P.; investigation, B.L., J.P.-C., and L.I.M.; methodology, D.C. and O.P.; project administration, O.P.; resources, D.C. and L.I.M.; software, B.L.; supervision, O.P.; visualization, B.L. and O.P.; writing, original draft, L.I.M. and O.P.; writing, review and editing, O.P.

Funding: Three of the authors (D.C., L.I.M., and O.P.) acknowledge funding from FSE/CONACYT through the project P19 ("Control strategies for reliability in small wind turbines") of the Mexican Center for Innovation in Wind Energy (CEMIE Eólico). B.L. and J.P.-C. acknowledge a tuition scholarship for their M.Sc. studies at Tecnológico de Monterrey and a stipend from CONACYT/PNPC under Program Number 5602.

Acknowledgments: D.C., L.I.M, and O.P appreciate the opportunity to participate in project P09 ("Design of wind turbine blades") of the CEMIE Eólico. D.C. and O.P. would also like to thank the Energy and Climate Change group at Tecnológico de Monterrey for support. L.I.M. appreciates support from Tecnológico de Monterrey and Universidad de Cuenca.

Conflicts of Interest: The authors declare no conflict of interest.

Appendix A

Table A1. Blade materials.

	Tri-Axial Material	Bidirectional Material
Density (kg/m^3)	1900	1810
Young module (GPa)	26.9	11.8
X_t (MPa)	131	128
X_c (MPa)	599	131

Table A2. Stall wind turbine blade geometry.

Pos (m)	Chord (m)	Twist (°)	Airfoil	Reynolds Number
0	0.125	20	Circular	100,000
0.1	0.125	20	Circular	100,000
2.43	0.484	25	DU210_22	794,126
0.368	0.376	18	DU210_22	668,736
0.473	0.35	11.53	DU210_21	660,289
0.583	0.33	9.09	DU210_20	657,824
0.683	0.312	7.25	DU210_19	650,773
0.793	0.297	5.91	DU210_19	648,334
0.893	0.284	4.93	DU210_18	644,009
1.003	0.273	4.25	DU210_17	643,540
1.103	0.263	3.77	DU210_17	640,649
1.213	0.254	3.43	DU210_16	639,978
1.313	0.246	3.19	DU210_16	637,957
1.423	0.239	3	DU210_14	638,624
1.523	0.232	2.83	DU210_14	636,073
1.633	0.226	2.66	DU210_14	636,483
1.733	0.221	2.48	DU210_14	637,019
1.843	0.216	2.28	DU210_14	637,953
1.943	0.211	2.06	DU210_14	636,508
2.053	0.207	1.81	DU210_14	638,509
2.153	0.203	1.54	DU210_12	638,453
2.263	0.199	1.26	DU210_12	638,856
2.363	0.196	0.97	DU210_12	640,628
2.473	0.192	0.68	DU210_12	639,616
2.573	0.189	0.387	DU210_12	640,226
2.683	0.187	0.097	DU210_12	644,794
2.783	0.184	−0.197	DU210_12	644,428
2.893	0.181	−0.505	DU210_12	644,546
2.993	0.179	−0.85	DU210_12	646,829
3.153	0.177	−1.241	DU210_12	654,205

Table A3. Pitch wind turbine blade geometry

Pos (m)	Chord (m)	Twist (°)	Foil	Reynolds Number
0	0.2	20	Circular	100,000
0.1	0.2	20	Circular	100,000
0.243	0.503217	17.86	NACA 4412	794,126
0.368	0.475326	16.49	NACA 4412	668,736
0.473	0.435994	14.4	NACA 4412	660,289
0.583	0.396337	12.78	NACA 4412	657,824
0.683	0.333628	11.59	NACA 4412	650,773
0.793	0.304067	10.28	NACA 4412	648,334
0.893	0.282357	9.38	NACA 4412	644,009
1.003	0.264434	8.48	NACA 4412	643,540
1.103	0.246997	7.77	NACA 4412	640,649
1.213	0.234625	6.94	NACA 4412	639,978
1.313	0.226237	6.35	NACA 4412	637,957
1.423	0.21686	5.72	NACA 4412	638,624
1.523	0.210705	5.13	NACA 4412	636,073
1.633	0.203875	4.59	NACA 4412	636,483
1.733	0.198556	4.09	NACA 4412	637,019
1.843	0.193004	3.62	NACA 4412	637,953
1.943	0.188138	3.26	NACA 4412	636,508
2.053	0.182872	2.89	NACA 4412	638,509
2.153	0.178818	2.59	NACA 4412	638,453
2.263	0.174903	2.35	NACA 4412	638,856
2.363	0.171661	2.19	NACA 4412	640,628
2.473	0.168592	2.12	NACA 4412	639,616
2.573	0.165971	2.125	NACA 4412	640,226
2.683	0.163438	2.23	NACA 4412	644,794
2.783	0.161215	2.4	NACA 4412	644,428
2.893	0.158969	2.68	NACA 4412	644,546
2.993	0.156959	3.06	NACA 4412	646,829
3.153	0.153977	3.77	NACA 4412	654,205

References

1. *Global Wind Statistics 2017*; Technical Report; Global Wind Energy Council: Brussels, Belgium, 2018.
2. Gupta, A.; McIntyre, A. *Hype Cycle for Sustainability Technology*; Technical Report; Gartner: Stamford, CT, USA, 2017.
3. Fleck, B.; Huot, M. Comparative life-cycle assessment of a small wind turbine for residential off-grid use. *Renew. Energy* **2009**. [CrossRef]
4. Battisti, L.; Benini, E.; Brighenti, A.; Dell'Anna, S.; Castelli, M.R. Small wind turbine effectiveness in the urban environment. *Renew. Energy* **2018**, *129*, 102–113. [CrossRef]
5. Pinheiro, E.; Bandeiras, F.; Gomes, M.; Coelho, P.; Fernandes, J. Performance analysis of wind generators and PV systems in industrial small-scale applications. *Renew. Sustain. Energy Rev.* **2019**, *110*, 392–401. [CrossRef]
6. Dimitriou, A.; Kotsampopoulos, P.; Hatziargyriou, N. Best practices of rural electrification in developing countries: Technologies and case studies. *MedPower* **2014**, *2014*, 1–5. [CrossRef]
7. Parida, A.; Choudhury, S.; Chatterjee, D. Microgrid Based Hybrid Energy Co-Operative for Grid-Isolated Remote Rural Village Power Supply for East Coast Zone of India. *IEEE Trans. Sustain. Energy* **2018**, *9*, 1375–1383. [CrossRef]
8. Kavlak, G.; McNerney, J.; Trancik, J.E. Evaluating the causes of cost reduction in photovoltaic modules. *Energy Policy* **2018**, *123*, 700–710. [CrossRef]
9. Soter, S.; Wegener, R. Development of Induction Machines in Wind Power Technology. In Proceedings of the 2007 IEEE International Electric Machines Drives Conference, Antalya, Turkey , 3–5 May 2007; Volume 2, pp. 1490–1495. [CrossRef]
10. Macquart, T.; Maheri, A. A stall-regulated wind turbine design to reduce fatigue. *Renew. Energy* **2019**, *133*, 964–970. [CrossRef]

11. Audierne, E.; Elizondo, J.; Bergami, L.; Ibarra, H.; Probst, O. Analysis of the furling behavior of small wind turbines. *Appl. Energy* **2010**, *87*, 2278–2292. [CrossRef]

12. Arifujjaman, M.; Iqbal, M.T.; Quaicoe, J.E. Energy capture by a small wind-energy conversion system. *Appl. Energy* **2008**, *85*, 41–51. [CrossRef]

13. Burlibaşa, A.; Ceangă, E. Rotationally sampled spectrum approach for simulation of wind speed turbulence in large wind turbines. *Appl. Energy* **2013**, *111*, 624–635. [CrossRef]

14. Yang, W.; Peng, Z.; Wei, K.; Tian, W. Structural health monitoring of composite wind turbine blades: Challenges, issues and potential solutions. *IET Renew. Power Gener.* **2017**, *11*, 411–416. [CrossRef]

15. Rubiella, C.; Hessabi, C.A.; Fallah, A.S. State of the art in fatigue modelling of composite wind turbine blades. *Int. J. Fatigue* **2018**, *117*, 230–245. [CrossRef]

16. Post, N.L.; Case, S.W.; Lesko, J.J. Modeling the variable amplitude fatigue of composite materials: A review and evaluation of the state of the art for spectrum loading. *Int. J. Fatigue* **2008**. [CrossRef]

17. Ardila, O.G.C.; Lennie, M.; Branner, K.; Pechlivanoglou, G.; Nayeri, C.; Paschereit, C.O. Comparing Fatigue Life Estimations of Composite Wind Turbine Blades using different Fatigue Analysis Tools. In Proceedings of the 20th International Conference on Composite Materials, Copenhagen, Denmark, 19–24 July 2015; pp. 1–14.

18. IEC_61400-1. *Wind Energy Generation Systems—Part 1: Design Requirements*; International Electrotechnical Commission (IEC): Geneva, Switzerland, 2019.

19. Hayat, K.; Asif, M.; Ali, H.T.; Ijaz, H.; Mustafa, G. Fatigue life estimation of large-scale composite wind turbine blades. In Proceedings of the 2015 12th International Bhurban Conference on Applied Sciences and Technology (IBCAST), Islamabad, Pakistan, 13–17 January 2015, pp. 60–66. [CrossRef]

20. Kong, C.; Kim, T.; Han, D.; Sugiyama, Y. Investigation of fatigue life for a medium scale composite wind turbine blade. *Int. J. Fatigue* **2006**. [CrossRef]

21. Wilkins, E.W.C. Cumulative damage in fatigue. In Proceedings of the Colloquium on Fatigue/Colloque de Fatigue/Kolloquium über Ermüdungsfestigkeit, Stockholm, Sweden, 25–27 May 1956. [CrossRef]

22. Vassilopoulos, A.P.; Keller, T. *Fatigue of Fiber-Reinforced Composites*; Engineering Materials and Processes; Springer: London, UK, 2011. [CrossRef]

23. IEC_61400-2. *Wind Turbines—Part2: Design and Requirements for Small Wind Turbines*; International Electrotechnical Commission (IEC): Geneva, Switzerland, 2006.

24. Evans, S.; Bradney, D.; Clausen, P. Assessing the IEC simplified fatigue load equations for small wind turbine blades: How simple is too simple? *Renew. Energy* **2018**, *127*, 24–31. [CrossRef]

25. Kennedy CR, Jaksic V, L.S.B.C. Fatigue life of pitch- and stall-regulated composite tidal turbine blades. *Renew. Energy* **2018**, *121*, 688–699. [CrossRef]

26. Marten, D.; Wendler, J. *QBlade Guidelines v0.6*; Technical Report; Berlin, Germany, 2013. Available online: https://pdfs.semanticscholar.org/1885/8285e209c482c74c368af688e23bed080971.pdf (accessed on 28 December 2018).

27. Nijssen, R.P.L. Fatigue Life Prediction and Strength Degradation of Wind Turbine Rotor Blade Composites. Ph.D. Thesis, Delft University of Technology, Delft, The Netherlands, 2007.

28. Rivera, J.A.; Aguilar, E.; Cárdenas, D.; Elizalde, H.; Probst, O. Progressive failure analysis for thin-walled composite beams under fatigue loads. *Compos. Struct.* **2016**, *154*, 79–91. [CrossRef]

29. Cardenas, D.; Elizalde, H.; Probst, O. Progressive failure analysis of wind turbine blades under stochastic fatigue loads. *Ingeniería Mecánica Tecnología y Desarrollo* **2017**, *5*, 459–473.

30. Mikel Mendia. Simulación de Ensayos a Fatiga en Palas de Aerogeneradores. Ph.D. Thesis, Universidad Pública de Navarra, Navarra, Spain, 2010.

31. *Guideline for the Certification of Wind Turbines*; Technical Report; Germanischer Lloyd: Hamburg, Germany, 2010.

32. Lee, H.G.; Kang, M.G.; Park, J. Fatigue failure of a composite wind turbine blade at its root end. *Compos. Struct.* **2015**, *133*, 878–885. [CrossRef]

33. Wu, J.; Lai, F. Fatigue Life Analysis of Small Composite Sandwich Wind Tuebine Blades. *Procedia Eng.* **2011**, *14*, 2014–2020. [CrossRef]

34. Jonkman, B.J.; Jonkman, J.M. FAST v8.16.00a-bjj User's Guide. Nrel. 2016; 58p. Available online: https://wind.nrel.gov/nwtc/docs/README_FAST8.pdf (accessed on 28 December 2018).

35. Garcia, M.; San-Martin, J.C.; Favela-Contreras, A.; Minchala, L.I.; Cardenas-Fuentes, D.; Probst, O. A simple approach to predictive control for small wind turbines with an application to stress alleviation. *Control. Eng. Appl. Inform.* **2018**, *20*, 69–77.

36. Hansen, M.H.; Hansen, A.; Larsen, T.J.; Øye, S.; Sørensen, P.; Fuglsang, P. *Control Design for a Pitch-Regulated, Variable Speed Wind Turbine*; Technical Report; Risø National Laboratory: Roskilde, Denmark, 2005.

MDPI

St. Alban-Anlage 66

4052 Basel

Switzerland

Tel. +41 61 683 77 34

Fax +41 61 302 89 18

www.mdpi.com

Applied Sciences Editorial Office

E-mail: applsci@mdpi.com

www.mdpi.com/journal/applsci